Alkaloids

Biochemistry, Ecology, and Medicinal Applications

Alkaloids

Biochemistry, Ecology, and Medicinal Applications

Edited by

Margaret F. Roberts
The University of London
London, England

and

Michael Wink
University of Heidelberg
Heidelberg, Germany

PLENUM PRESS • NEW YORK AND LONDON

Library of Congress Cataloging-in-Publication Data

Alkaloids : biochemistry, ecology, and medicinal applications / edited by Margaret F. Roberts and Michael Wink.
p. cm.
Includes bibliographical references and index.
ISBN 0-306-45465-3
1. Alkaloids. I. Roberts, M. F. (Margaret F.) II. Wink, Michael.
QP801.A34A44 1998
572'.549--dc21 98-17602
CIP

ISBN 0-306-45465-3

A Division of Plenum Publishing Corporation
233 Spring Street, New York, N.Y. 10013

http://www.plenum.com

10 9 8 7 6 5 4 3 2 1

Printed in the United States of America

Contributors

J. C. Braekman • Laboratory of Bio-Organic Chemistry, Free University of Brussels, 50-1050 Brussels, Belgium

D. Daloze • Laboratory of Bio-Organic Chemistry, Free University of Brussels, 50-1050 Brussels, Belgium

Rainer Ebel • Julius von Sachs Institute for Biological Sciences, Department of Pharmaceutical Biology, University of Würzburg, D-97082 Würzburg, Germany

U. Eilert • Institute for Pharmaceutical Biology, Technical University of Braunschweig, D-38106 Braunschweig, Germany

A. H. C. Hoult • Department of Agronomy and Soil Science, University of New England, Armidale, NSW 2351, Australia

J. V. Lovett • Grains Research and Development Corporation, Kingston, ACT 2604, Australia

Isamu Murakoshi • Faculty of Pharmaceutical Sciences, Laboratory of Molecular Biology and Biotechnology, Research Center of Medicinal Resources, Chiba University, Inage-ku, Chiba 263, Japan

H. D. Neuwinger • Department of Chemistry, University of Heidelberg, D-69120 Heidelberg, Germany (Retired). *Present address:* Hauptstrasse 190, D-68789 St. Leon-Rot, Germany

J. M. Pasteels • Laboratory of Animal and Cellular Biology, Free University of Brussels, 50-1050 Brussels, Belgium

Peter Proksch • Julius von Sachs Institute for Biological Sciences, Department of Pharmaceutical Biology, University of Würzburg, D-97082 Würzburg, Germany

Margaret F. Roberts • The Centre for Pharmacognosy, School of Pharmacy, University of London, London WC1N 1AX, England

Richard J. Robins • Laboratoire d'Analyse Isotopique et Électrochimique de Metabolismes, CNRS UPRES-A 6006, University of Nantes, BP 92208, F-44322 Nantes Cedex 03, France

Kazuki Saito • Faculty of Pharmaceutical Sciences, Laboratory of Molecular Biology and Biotechnology, Research Center of Medicinal Resources, Chiba University, Inage-ku, Chiba 263, Japan

T. Schmeller • Institute for Pharmaceutical Biology, University of Heidelberg, D-69120 Heidelberg, Germany

F. R. Stermitz • Department of Chemistry, Colorado State University, Fort Collins, Colorado 80523

R. Verpoorte • Division of Pharmacognosy, Leiden/Amsterdam Center for Drug Research, University of Leiden, 2300 RA Leiden, The Netherlands

Peter G. Waterman • Phytochemistry Research Laboratories, University of Strathclyde, Glasgow G1 1XW, Scotland

Michael Wink • Institute for Pharmaceutical Biology, University of Heidelberg, D-69120 Heidelerg, Germany

Preface

We have tried, in this book, to provide a survey of recent scientific thinking on the biology of that heterogenous group of secondary metabolites, the alkaloids. To this end, we have prevailed on scientists who are specialists in their particular areas to define the current thinking in their areas of expertise. G. R. Waller and E. K. Nowacki edited *Alkaloid Biology and Metabolism in Plants* in 1978. The present volume can be regarded as a sequel.

The book is designed for use by advanced students and professional workers in the agricultural, biological, and pharmaceutical sciences, natural products chemistry, biochemistry, and botany. Each chapter is referenced with its own set of review and specialist references. Chapters are arranged in four parts: Part I is an introduction covering some of the historical uses of alkaloids; Part II considers the biochemistry of alkaloid production in plants; Part III discusses ecology and includes marine invertebrates, animals, and plant parasites; and Part IV studies alkaloids as antimicrobials and reviews the current medicinal use. The volume is an attempt to put into one book current thoughts in these areas and as such cannot be exhaustive on such a large and diverse group of constituents. For other details the reader should consult such reviews as *The Alkaloids,* first edited by R. H. Manske, later by A. Brossi and G. A. Cordell, 1952–1996, Academic Press, New York; *Alkaloids: Chemical and Biological Perspectives,* edited by S. W. Pelletier, 1983–1996, Pergamon Press, Oxford; and The Chemical Society Specialist Reports on Natural Products.

This book has brought together contributions from friends and colleagues from many parts of the world. As editors, we would like to thank all those who have taken part in the writing and preparation of this book.

M. F. Roberts
M. Wink

Contents

Chapter 3

Alkaloids in Arrow Poisons

H. D. Neuwinger

Part II. Biochemistry

Chapter 4

Chemical Taxonomy of Alkaloids

Peter G. Waterman

Chapter 5

Enzymology of Alkaloid Biosynthesis

Margaret F. Roberts

Chapter 12

Modes of Action of Alkaloids

Michael Wink

Chapter 13

Plant Parasites

F. R. Stermitz

Part IV. Alkaloids in Medicine

Chapter 17

Antimicrobially Active Alkaloids

R. Verpoorte

Chapter 18

Utilization of Alkaloids in Modern Medicine

T. Schmeller and Michael Wink

Alkaloids

Biochemistry, Ecology, and Medicinal Applications

Chapter 1

Introduction

Margaret F. Roberts and Michael Wink

1. INTRODUCTION

The alkaloids are one of the most diverse groups of secondary metabolites found in living organisms and have an array of structure types, biosynthetic pathways, and pharmacological activities. Although alkaloids have been traditionally isolated from plants, an increasing number are to be found in animals, insects, and marine invertebrates and microorganisms. Many alkaloids have been used for hundreds of years in medicine and some are still prominent drugs today. Hence, this group of compounds has had great prominence in many fields of scientific endeavor and continues to be of great interest today. Why do plants expend so much of their vital resources on the biosynthesis of alkaloids? How do they produce alkaloids and what are the mechanisms of regulation of biosynthesis and location within the plant? How does the plant store these substances which can occur at levels toxic to the producing cell itself? It has to be assumed that they have an important role in a plant's survival. We hope this book which is organized in four main parts will give some insight into these questions.

2. HISTORICAL IMPORTANCE OF ALKALOIDS

Human recogntion of alkaloids is as old as civilization, since these substances have been used as drugs, in potions, medicines, teas, poultices, and poisons for 4000 years (see Chapters 2 and 3). It is likely that, in the hunt for food and in dealings with enemies, particular use was made of plants containing alkaloids for arrow poisons and this use probably preceded their medicinal use. Even today these poisons are still in use in Africa and South America. Arrow poisons have provided men with ouabain and k-strophanthin

Margaret F. Roberts • The Centre for Pharmacognosy, School of Pharmacy, University of London, London WC1N 1AX, England. *Michael Wink* • Institute for Pharmaceutical Biology, University of Heidelberg, D-69120 Heidelberg, Germany.

Alkaloids: Biochemistry, Ecology, and Medicinal Applications, edited by Roberts and Wink. Plenum Press, New York, 1998.

for acute cardiac insufficiency, physostigmine for the treatment of glaucoma and myasthenia gravis, *d*-tubocurarine as a muscle relaxant in anesthesia, reserpine as an antihypertensive and psychotropic drug, and ajmaline in cases of cardiac rhythm disturbances. Sooner or later arrow poisons will disappear and there is a real need to continue to evaluate these poisons for active constituents.

As far as the use of alkaloids in medicine is concerned, it is only relatively recently (early 19th century) that therapeutically active substances have been isolated. The first crude drug to be investigated chemically was opium, the dried latex of the opium poppy, *Papaver somniferum*, a drug that had been used for centuries for both its analgesic and narcotic properties. In 1803 Derosne isolated a semipure alkaloid from opium and in 1805 Sertürner isolated and characterized this constituent as morphine and recognized its basic nature. It was another 12 years before this lead was pursued. Between the years 1817 and 1820 the laboratory of Pelletier and Caventou at the Faculty of Pharmacy in Paris isolated so many active principles from crude drugs that even today no other laboratory has isolated as many active principles of pharmaceutical importance. Among the alkaloids obtained in this brief period were strychnine, emetine, brucine, piperine, caffeine, quinine, cinchonine, and colchicine. These alkaloids are the cornerstone of all that has occurred in alkaloid chemistry to the present day (Cordell, 1983). In 1939 the number of alkaloids that had been isolated and structurally identified was on the order of 200 (Manskse, 1950). By 1989, the *Dictionary of Alkaloids* (Southon and Buckingham, 1989) listed details of 10,000 alkaloids and new structures continue to be found.

3. BIOCHEMISTRY

To appreciate this section of the book it is perhaps in order to define what is considered to be an alkaloid and to make a short comment on their basic chemistry and biogenetic origins.

3.1. Classification of Alkaloids

A simple general definition of an alkaloid has been suggested by Pelletier (1983): "An alkaloid is a cyclic compound containing nitrogen in a negative oxidation state which is of limited distribution in living organisms." This definition includes both alkaloids with nitrogen as part of a heterocyclic system as well as the many exceptions with extracyclic bound nitrogen such as colchicine or capsaicin. A basic character is no longer a prerequisite for an alkaloid and the chemistry of the nitrogen atom allows for at least four groups of nitrogenous compounds.

1. Secondary and tertiary amines which are more or less protonated and therefore hydrophilic at $pH < 7.0$ or the more general case where they are lipophilic and unprotonated at $pH > 8.0$. This is the classical alkaloid type.
2. Quaternary amino compounds which are very polar, charged at all pH values, and have to be isolated as salts, e.g., berberine and sanguinarine.
3. Neutral amino compounds, which include the amide-type alkaloids such as colchicine, capsaicin, and most lactams, e.g., ricinine.

4. *N*-oxides, which are generally highly water soluble, are frequently found in many alkaloid classes, the pyrrolizidine group of alkaloids being rich in this particular alkaloid type.

3.2. Biogenetic Grouping of Alkaloids

Because alkaloids are such a diverse grouping of chemical constituents, it is convenient to classify them according to their biogenetic origins and this is the method used throughout this book. For example, the indole alkaloids are tryptophan derived and may be further grouped within this category as either nonterpenoid or terpenoid indoles (iridoid). Within the latter group there are numerous subgroups which depend on the mode of cyclization after the removal of glucose from strictosidine. Eight subgroups are identifiable: coryantheans, aspidospermatans, ibogans, strychnans, plumerans, eburnans, vallesiachotamans and apparicines.

Not all alkaloids are strictly amino acid derived and therefore it is possible to recognize four groups:

1. Alkaloids derived from amino acids such as ornithine/arginine, lysine, histidine, phenylalanine/tyrosine, tryptophan, anthranilic acid, and nicotinic acid
2. Purine alkaloids, such as the xanthine caffeine
3. Aminated terpenes, e.g., the diterpene aconitine or the triterpene solanine
4. Polyketide alkaloids where nitrogen is introduced into a polyketide carbon skeleton as in coniine and the coccinellines

These last two alkaloid groups are increasing in size as more insect and marine organisms are further investigated. Also, in these latter two groups the character of the nonnitrogen analogues may influence or dominate the properties of the substance as may be observed with the saponinlike properties of the steroid alkaloids of the Solanaceae.

The production of alkaloids has been most studied in plants where experiments with radiolabeled precursors have laid the foundations for our understanding of the biosynthesis of most of the important groups of alkaloids. More recently the isolation and study of the enzymes involved in these pathways has served to further understanding of this area (see Chapters 5 and 8). The development of plant cell culture techniques has permitted bringing the plant into the laboratory without regard for season (see Chapter 7 and 9). This development, alone, gave great impetus to the isolation of enzymes and ultimately allowed the biochemistry of alkaloid production to be successfully studied at the gene level (see Chapter 6).

3.3. Occurence and Distribution

The major source of alkaloids in the past has been the flowering plants, the Angiospermae, where about 20% contain these constituents. In recent years an increasing number of examples of alkaloids have come from animals, insects, marine organisms, microorganisms, and lower plants. Alkaloids from terrestial animals such as muscopyridine from the musk deer and castoramine from the Canadian beaver have been recorded. However, in the latter case the alkaloids may have been acquired as a result of eating water lilies of the *Nuphar* species. The amphibians are of particular interest as a remarkable diversity of toxic or noxious alkaloids are found in the skin or in skin exudates, e.g., bufotalin from the common toad, *Bufo* (Daly and Spande, 1983). Arthropods, particularly insects, are another source of interesting alkaloids which act as attractants or pheromones and defensive agents, e.g., the trail pheromone, methyl-4-methylpyrrole-2-carboxylate, in the *Atta* species of leaf cutting ants (Jones and Blum, 1983). Marine organisms (algae and marine invertebrates) have yielded a great diversity of alkaloids; examples are saxitoxin, a neurotoxic constituent of the red tide *Gonyaulax catenella,* and the brominated isoxazoline alkaloids of the yellow sponge *Vergonia aerophobia* (Cordell, 1983). Microorganisms have also been found to contain alkaloids, some examples being the *Aspergillus* alkaloids (Yamamoto and Arai, 1986), pyocyanine from *Pseudomonas aeruginosa*, and chanoclavine-I from the ergot fungus, *Claviceps purpurea* (Cordell, 1983, see Chapter 18).

Alkaloids have restricted distribution in plants, microbial and animal species and where these constituents occur each organism has its own genetically defined alkaloid pattern which may in some instances be useful biogenetically as chemical characters in systematics (see Chapter 4).

4. ECOLOGY

It is now generally accepted that secondary metabolites have a role in the survival of the organism. In plants these compounds are involved as attractants to ensure pollination (flower color, scent, and nectar) and are found to play an important role in plant interactions with animals and higher and lower plants. Alkaloids are now generally considered to be part of an elaborate system of chemical defense in plants and indeed the same seems to be true in vertebrates, invertebrates, and microoganisms (see Chapters 11 and 12).

Most alkaloids are physiologically active compounds having a variety of toxic effects on animals. Toxic alkaloids are frequently found as part of conspicuous, often violent, insect defense systems and these may be synthesized by the insect or acquired as part of their diet. However, plants sequester alkaloids for use as a passive defense mechanism by acting as feeding deterrents for predating insects (Bernays, 1983).

The bitter taste of many alkaloids may be of importance as a feeding deterrent; some alkaloids, e.g., quinine, strychnine, and brucine, are extremely bitter. There are, however, little available data on the role of bitterness in plant/animal relationships—being for the most part based on human preferences—although it has been suggested that the bitterness of alkaloids is a universal feeding deterrent in plant foodstuffs (Bate-Smith, 1972). Our studies with geese did not support this view; these birds have a distaste for essential oils but tolerate bitter alkaloids (Wink *et al.*, 1993).

Other examples of alkaloids as defensive chemicals in animals are to be found in marine ecosystems. Marine sponges may produce ichthyotoxic and deterrent alkaloids such as latruculins (Groweiss *et al.*, 1983) and a novel type of bisquinolizidine alkaloid, the petrosins, which are assumed to be part of a chemical defense system (Van Soest *et al.*, 1994; see Chapter 16).

There are few reports of alkaloids involved in allelopathy although alkaloids were suggested as contributing to the weed status of *Datura stramonium* (Levitt and Lovett, 1984) and the lupin alkaloids in inhibition of seed germination (Wink, 1983; see Chapter 14).

In general, alkaloids are regarded as part of the plant's constitutive defense. This means that in certain plants alkaloids occur in a species-specific and genetically programmed accumulation pattern, which remains unaffected by herbivory, microbial attack, and mechanical damage, or stress. Alkaloids as phytoalexin are rare and in most cases evidence for this role is still somewhat controversial. In cell cultures derived from plants, elicitation of the production of alkaloids by various techniques can produce quite startling increases in yield (see Chapter 9).

Important to the role of alkaloids in chemical defense is the location of the alkaloids within the plant. For protection the alkaloids must be localized at the site to be protected and be at a sufficient concentration to be an adequate defense. Over 100 years ago botanists noticed that alkaloids, as are other secondary metabolites, are usually found where herbivore attack would most affect the plant's fitness, i.e., in inflorescences and young growing tips and the peripheral cell layers of stems and roots (Kerner von Maurilaun, 1890; Hartmann, 1991). These findings have been repeatedly confirmed in the intervening years. Alkaloids, although apparently present at low levels (0.1–2% dry weight), are mostly concentrated in specialist parts of the plant and the immediate concentration is much greater. A good example is the concentration of morphine which, in whole poppy capsules, is about 0.3–1.5% but in crude opium, the air dried poppy latex, is 25% or more. In the latex the alkaloids occur in specialized vesicles (Roberts *et al.*, 1983; Homeyer and Roberts, 1984). The level of monoterpene alkaloids in *Rauwolfia* root is about 0.8–2% but is more than 5% in the root bark, suggesting peripheral location in the root bark parenchyma. Quinolizidine and pyrrolizidine alkaloids are concentrated in the peripheral regions of stems of *Lupinus* and *Senecio* and are 10–20% higher than the mean values for the whole stem (Hartmann, 1991). Frequently, within an organ, the alkaloids are found within specialist cell layers like the beaker cells in *Conium* fruits (Fairbairn and Suwal, 1961).

Within the cell, the alkaloids are frequently found within the vacuole and complex mechanisms exist for their uptake into the vacuole and subsequent sequestration. These mechanisms allow for potentially toxic levels of alkaloids to be safely stored away from the cytoplasm (see Chapter 10). Frequently biosynthesis occurs remote from the site of storage and hence complex methods of translocation are required to allow for maximum plant protection, if this indeed is the *raison d'être* for their existence. This constitutive alkaloidal system with so many specific metabolic mechanisms suggests direct involvement of alkaloids as part of a chemical defense that has evolved as a result of selective pressure by predation. Although there is some evidence to support this theory from investigations of *Senecio* and *Lupinus* and their predators, there is a need to investigate other alkaloidal plants and their specific predators, particularly those that have learned to cope with the defense barrier (Hartmann 1991; Wink, 1993).

5. PHARMACOLOGICAL ACTIVITIES

Alkaloids generally exert pharmacological activity particularly in mammals such as humans. Even today many of our most commonly used drugs are alkaloids from natural sources and new alkaloidal drugs are still being developed for clinical use (e.g., taxol from *Taxus baccata*). Most alkaloids with biological activity in humans affect the nervous system, particularly the action of the chemical transmitters, e.g., acetylcholine, epinephrine, norepinephrine, γ-aminobutyric acid (GABA), dopamine, and serotonin (see Chapter 18).

Many alkaloids serve as models for the chemical synthesis of analogues with better properties. Important examples are hyoscyamine and scopolamine (*Atropa belladonna* and *Datura* species) as models for synthetic parasympatholytic agents; physostigmine (*Physostigma venenosum*) for synthetic parasympathomimetic agents; tubocurarine (*Chondodendron tomentosum*) for skeletal muscle relaxants; cocaine (*Erythroxylon coca*) for local anesthetics; morphine (*P. somniferum*) for analgesics; and codeine (*Papaver somniferum*) for antitussive agents. Alkaloids have many other pharmacological activities including antihypertensive effects (many indole alkaloids), antiarrhythmic effects (quinidine, ajmaline, sparteine), antimalarial activity (quinine), and anticancer actions (dimeric indoles, vincristine, vinblastine). These are just a few examples illustrating the great economic importance of this group of plant constituents (Cordell, 1983).

Antibiotic activities are common for alkaloids and some are even used as antiseptics in medicine, e.g., berberine in ophthalmics and sanguinarine in toothpastes; however, it is difficult to know the extent to which alkaloids give antimicrobial protection in the plant (Cordell, 1983).

For the past 40 centuries, mankind has lived with and made use of alkaloid bearing plants. For the past 200 years, an increasing number of scientists in an ever-widening number of disciplines have devoted themselves to elucidating the details of how and why alkaloids have the effects they do as well as how and why they provide survival benefit to the producers. We have tried, in this book, to provide a brief review of the results of those efforts.

REFERENCES

Bate-Smith, E. C., 1972, Attractants and repellants in higher animals, *Annu. Proc. Phytochem. Soc.*, **8**:45–56.

Cordell, G. A., 1983, *Introduction to Alkaloids: A Biogenic Approach,* Wiley, New York.

Daly, J. W., and Spande, T. F., 1986, Amphibian alkaloids: Pharmacology and biology, in: *Alkaloids: Chemical and Biological Perspectives,* Vol. 3 (S. W. Pelletier, ed.), Wiley, New York, pp. 1–274.

Fairbairn, J. W., and Suwal, P. N. 1961, The alkaloids of hemlock (*Conium maculatum*). I. Evidence for a rapid turnover of major alkaloids, *Phytochemistry* **1**:38–46.

Garson, M., 1994, The biosynthesis of sponge secondary metabolites: Why is it important?, in: *Sponges in Time and Space* (R. W. M. van Soest, Th. M. G. van Kempen, and J. C. Brackman, eds.), Balkema, Roterdam, pp. 427–440.

Groweiss, A., Shmueli, U., and Kashman, Y., 1983, Marine Toxins of *Latrunculia Magnicia, J. Org. Chem.* **48**:3512–3516.

Hartmann, T. 1991, Alkaloids, in: *Herbivours: Their Interaction with Secondary Metabolites, Vol. 1* (G. A. Rosenthal and M. R. Berenbaum, eds.), Academic Press, San Diego, pp. 79–116.

Homeyer, B. C., and Roberts, M. F., 1984 Alkaloid sequestration by *Papaver somniferum* L. latex, *Z. Naturforsch.* **39c**:876–881.

Jones, T. H., and Blum, M. S., 1983, Arthropod alkaloids: Distributions, functions and chemistry, in: *Alkaloids: Chemical and Biological Perspectives,* Vol. 1 (S. W. Pelletier, ed.), Wiley, New York, pp. 33–84.

Kerner von Marilaun, A., 1890, *Pflamzenleben* (2 volumes), Verlag des Bibliographischen Instituts, Leipzig, Wein.

Levitt, H., and Lovett, J. V., 1984, Activity of allelochemicals of *Datura stramonium* L., thornapple, in contrasting soil types, *Plant Soil* **79:**181–189.

Manskse, R. H. F., 1950, *The Alkaloids,* Vols. 1–5, Academic Press, San Diego.

Pelletier, S. W., 1983, The nature and definition of an alkaloid, in: *Alkaloids: Chemical and Biological Perspectives*, Vol. 1 (S. W. Pelletier, ed.), Wiley, New York, pp. 1–31.

Southon, I. W., and Buckingham, J., (eds.), 1989 *Dictionary of Alkaloids*, Chapman & Hall, London.

Van Soest, R. W. M., Van Kempen, T. M. G., and Braekman, J. C., (eds.), 1994, *Sponges in Time and Space*, Balkema, Rotterdam.

Wink, M., 1983, Inhibition of seed germination by quinolizidine alkaloids. Aspects of allelopathy in *Lupinus albus* L., *Planta* **158:**365–368.

Wink, M., Hofer, A., Bilfinger, M., Englert, E., Martin, M., and Schneider, D., 1993 Geese and plant dietary allelochemicals—Food palatability and geophagy, *Chemoecology* **4:**93–107.

Yamamoto, Y., and Arai, K., 1986, Alkaloidal substances in *Aspergillus* species, in: *Alkaloids,* vol. 29 (A. Brossi, ed.), Academic Press, San Diego, pp. 185–263.

Part I

Historical and Cultural Perspectives

Chapter 2

A Short History of Alkaloids

Michael Wink

1. INTRODUCTION

It may be assumed that plants developed a wide variety of secondary metabolites during hundreds of millions of years of evolution as a means of defending themselves against herbivores, microorganisms, viruses, and other plants (see Chapter 11). Among the more than 50,000 natural products that are known today, over 12,000 are alkaloids.

It is not pure chance that many alkaloids interact with receptors of neurotransmitters and thus with signal transduction from nerve cells to other neurons or muscles. During a process that we could term "evolutionary molecular modeling," compounds have been shaped and selected in such a way that they can effectively bind to receptors and either activate or inhibit them. A receptor and its natural ligand can be regarded as a lock and key that have to be closely tuned in order to function properly. A natural product that mimics a natural ligand would be a picklock, which could be used to manipulate, i.e., to either close or open the gate. The coevolutionary process could be to produce and optimize picklocks for several locks.

As receptors are important targets and their disturbance dangerous to an animal, the production of these compounds as protection against herbivores would be beneficial for a plant. Other compounds have been evolved in plants which interfere with vital functions of cells, organs, or the complete organism. Important molecular targets include DNA, RNA, and related enzymes, protein biosynthesis, membrane integrity and ion channels, electron chains, and signal transduction pathways. At the organ level, allelochemicals may affect the brain, muscles, heart and the circulatory system, lungs, kidneys, liver, bones, and related physiology such as digestion, diuresis, respiration, blood circulation, hormonal interactions, and reproduction, or the maintenance of homeostasis (Mann, 1992; Wink, 1993a; see Chapters 12 and 18).

Because humans are herbivores to some degree, we have had to cope with the toxic

Michael Wink • Institute for Pharmaceutical Biology, University of Heidelberg, D-69120 Heidelberg, Germany.

Alkaloids: Biochemistry, Ecology, and Medicinal Applications, edited by Roberts and Wink. Plenum Press, New York, 1998.

properties of plants. Using our intelligence, we learned to avoid or to use them as poisons (for murder, euthanasia, executions, and hunting) or for medicine. The latter aspect is only the other side of the coin: A natural product that functions as a protective poison for the plant at a high concentration, is often a useful medicine for *Homo sapiens* at lower doses.

Alkaloids and alkaloid-producing plants are especially well known for their toxic and sometimes psychomimetic, euphoric, and hallucinogenic properties. Consequently, several of them have been known to mankind for several thousand years. A good subtitle for this chapter is "Murder, magic and medicine," the title of a stimulating book by John Mann (1992).

2. ALKALOIDS AND ALKALOID-PRODUCING PLANTS IN ANTIQUITY

In developing civilizations, knowledge of the medicinal and toxic properties of plants was closely interwoven with mythology. The god of medicine and therapy was Asclepius (with a snake as insignia, which is still in use today). Knowledge of the intrinsic activities of alkaloids and alkaloid-producing plants was often restricted to a small elite: priests, medicine men, shamans, magicians, or witches. Because overdosing could be lethal for a patient or the shaman himself, the use of plant extracts demanded expertise in dose effects. At least some of the substances and plants that were used as poisons or medicines in antiquity have been successfully converted into clinically acceptable drugs of today (see Chapter 18).

2.1. Use of Alkaloid-Producing Plants in Early Medicine

The earliest evidence that humans used alkaloid-producing plants is derived from Assyrian clay tablets, written in cuneiform characters. On these 4000 year old plates, about 250 different plants are mentioned, including a number of alkaloid-containing plants, e.g., *Papaver somniferum, Atropa belladonna*, and *Mandragora officinarum.* As in Mesopotamia, empirical knowledge of medicinal plants developed in China; the Emperor Shen Nung described 365 drugs in about 3000 B.C. In India the traditional medicine was documented in the Ayurveda in about 900 B.C. Its main remedy was "soma," which had to be collected under moonlight and had to be extracted while praying. The effect of soma included slight intoxication and a feeling of strength, courage, and sexuality. The identity of the source plant has been disputed: Its main components could have been the hallucinogenic plants *Datura metel, Cannabis sativa, Sarcostemma acidum, S. brevistigma,* or more likely the mushroom *Amanita muscaria* (Schultes and Hoffman, 1987; Mann, 1992). The Ayurveda suggested *Datura* against mental illness, fever, tumors, eczema, infections, diarrhea and as an aphrodisiac. Besides *Datura,* other alkaloidal plants employed were *Aristolochia indica, Nymphaea nelumbo*, and *Aconitum napellus. Papaver somniferum* appears to have been unknown (Bellamy and Pfister, 1992).

From these older cultures, some of the knowledge eventually reached the Mediterranean countries through traders and migrations. In about 1550 B.C. medicinal plants were described in the Ebers papyrus [Professor Ebers a German Egyptologist, had obtained the

Figure 1. Dioscorides obtains mandrake from the goddess Euresis (goddess of explorations). After a colored plate from 512, Vienna; by permission of Hirmer-Verlag and the author.

papyri in 1872 in Luxor and had recognized its content and value (Joachim, 1890)]): In addition to recipes for purgatives (including castor oil, rhubarb, aloe, and senna), some alkaloidal plants were mentioned: henbane (*Hyoscyamus niger*), pomegranate (*Punica granatum*), and poppy (*Papaver somniferum*) (Mann, 1992; Schmidt, 1927; Gessner, 1974; Baumann, 1986).

The ancient Greeks could rely on Sumerian, Egyptian, and Indian traditions. Early "pharmacognosy" reached a summit with Hippocrates (460–377 B.C.), who critically reviewed more than 200 medicinal plants (overall, more than 1000 plant species were known to ancient Greeks and Romans). Hippocrates based his medicine on empirical data and freed medicine from the mythical past. He influenced Western medicine substantially and the "oath of Hippocrates" is still relevant today. Aristotle (384–322 B.C.) was the most important philosopher of his time. His friend and disciple Theophrastus of Eresos (372–287 B.C.) compiled their research on plants in a first *Historia Plantarum.* In this major text the available knowledge on more than 450 plant species was systematically summarized. During the next centuries Greek science concentrated on astronomy and mathematics and not on botany. But the Greek knowledge was passed to the Romans Polybius and Pliny (23–79 A.D.). Pedanius Dioscorides (ca. 40–90 A.D.) (see Fig. 1) produced the famous *De materia medica* in 78 A.D. which described more than 500 medicinal plants (roughly 8% of the plants that occur in the eastern Mediterranean) and their uses in detail, even mentioning the various synonyms. Dioscorides recognized plants for over 50 medical indications. This *materia medica* represents a rather modern approach, since the importance of effectivity and doses was already recognized. The work was used as the standard textbook until the Middle Ages and served as the base for various herbals. The Greek Galen (129–199 A.D.) combined the wisdom of Aristotle, Theophrastus, and Hippocrates with the anatomical and physiological knowledge developed in Pergamum, Izmir, Corinth, and Alexandria. Besides 304 medicinal plants, Galen employed complex mixtures, which were specially devised for each therapy; he thus became the founder of "galenics" (Krafft and Meyer-Abich, 1970; Mann, 1992; Schmidt, 1927; Lewin, 1984; Gessner, 1974; Baumann, 1986; Bellamy and Pfister, 1992).

In Arabia, Avicenna (980–1037 A.D.) based his medicine on Hippocrates, Dioscorides, and Galen. He was followed by Ibn al-Baitar (1197–1248) who recognized 1400 drugs and medicinal plants, combining the work of Dioscorides and that from the Far East. In Europe, several herbals appeared from the Middle Ages onwards, which eventually led to modern medicine and pharmacy, which often forget their roots (Bellamy and Pfister, 1992). Most influential was Paracelsus (Theophrastus Bombast von Hohenheim; 1493–1541): He realized that drugs contained active substances, the "Arkanum." Paracelsus demanded that the Arkanum be separated from nonactive ingredients of a drug and that it be administered using the correct dose. In addition, Paracelsus favored "Simplice," instead of using complex mixtures.

In modern medicine, several alkaloids (e.g., colchicine, serpentine, ajmaline, reserpine, ergobasine, ergotamine, quinine, cinchonine, sparteine, ephedrine, lobeline, caffeine, berberine, sanguinarine, tubocurarine, strychnine, papaverine, codeine, thebaine, noscapine, yohimbine, atropine, scopolamine, morphine, vinblastine, vincristine, taxol, camptothecin) and alkaloid-containing drugs are still in use to treat various diseases (see Chapters 12 and 17). This is a clear indication that our ancestors' empirical knowledge of medicinal plants was not rubbish—as is sometimes suggested—but that it was often based on active plant constituents.

2.2. Role of Alkaloids and Alkaloidal Plants for "Murder" and "Magic"

Alkaloidal plants were also used by hunters, priests, witches and magicians. Some alkaloids that interfere with the transmission of nerve impulses to the muscles and thus lead to immediate paralysis, such as aconitine (**9**) (which was still in use in medieval Europe), *Taxus* alkaloids, atropine (**14**), toxiferine (**1**), and tubocurarine (**2**), played a significant role as arrow poisons for hunting and warfare. The ancient Greek word *toxicon* originally meant "poison for arrows" (Mann, 1992). In Central and South America, frogs of the genera *Phyllobates* and *Dendrobates* secrete toxic alkaloids, such as batrachotoxin (**3**) or pumiliotoxins. Batrachotoxin is five times more potent than tetrodotoxin (**4**) and achieves its toxicity by activating Na^+ channels, thus disturbing neurotransmission. Natives used, and still use, these toxins as potent arrow poisons and the secretion of one frog provides poison for 50 arrows (Bisset, 1989). For more details consult Chapter 3.

Before and during the Middle Ages, several other alkaloids were used for murder and executions; infamous are coniine (**32**), aconitine, atropine, strychnine (**5**), colchicine (**30**), and many others (see below). Some animals also produce toxic alkaloids (see Chapter 15);

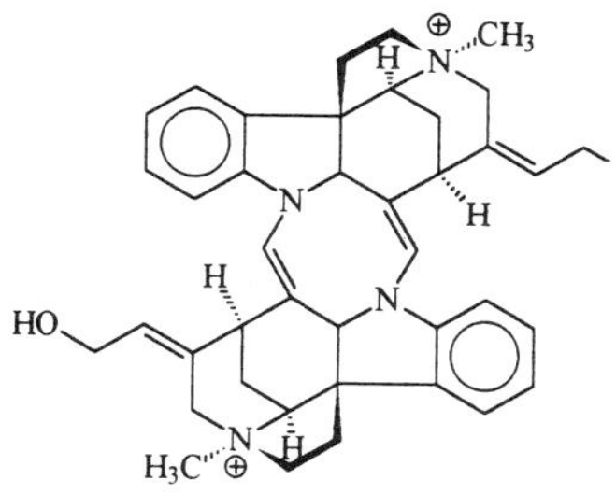

toxiferine I

1

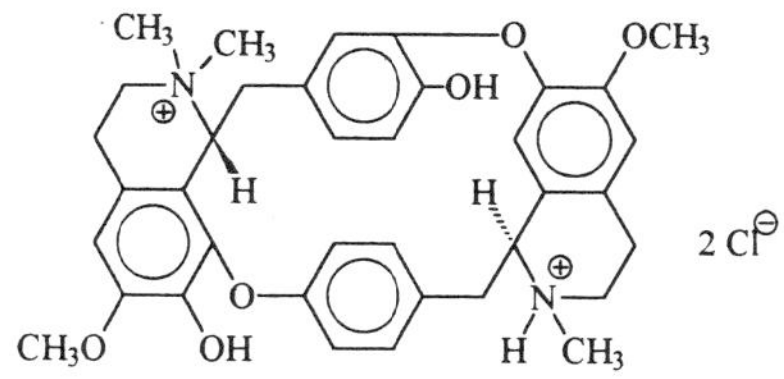

(+)-tubocurarine chloride

2

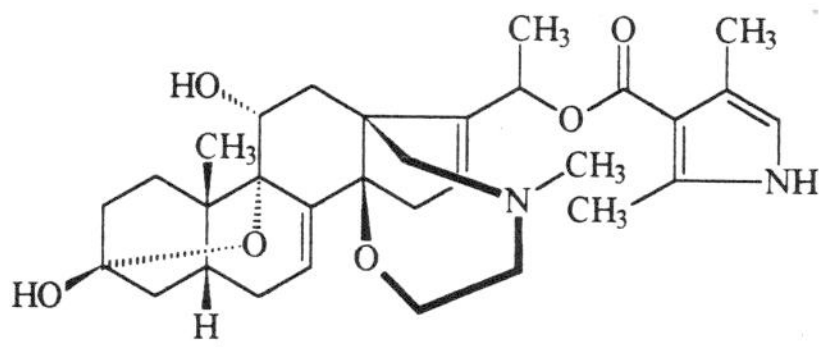

batrachotoxin

3

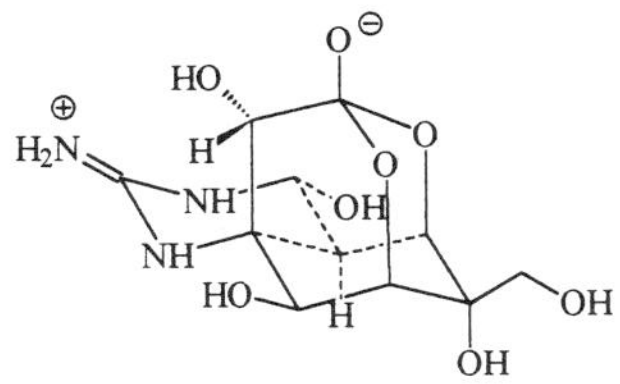

tetrodotoxin

4

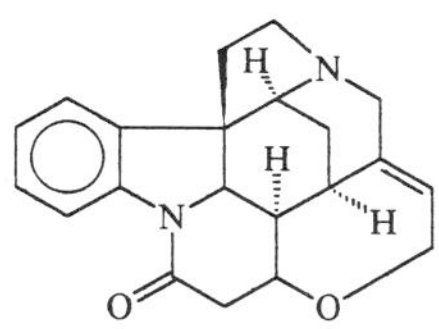

strychnine

5

a well-known example is the puffer fish and other members of the family Tetraodontidae. Known in ancient Egypt (2500 B.C.) and China (200 B.C.), the fugu or puffer was used for murder and suicide. The toxic ingredient is the alkaloid tetrodotoxin, which inhibits nerve transmission by blocking Na^+ channels.

Many other alkaloids and alkaloid-producing plants have been discovered to possess euphoric, psychomimetic and hallucinogenic properties (Schultes and Hofmann, 1980, 1987). Each continent (especially South America, Europe and Africa) has its own set of psychotropic plants, but common to all of them is that the alkaloids involved usually affect signal transduction in neurons of the brain: by modulating either the receptors of neurotransmitters or hormones [e.g., hyoscyamine, scopolamine, nicotine (**44**), muscarine (**10**), ephedrine (**37**), morphine (**45**), psilocin (**53**), psilocybin (**52**), mescaline (**39**), *N,N*-dimethyltryptamine (**6**), ergot alkaloids], the degradation of neurotransmitters [Monoamine oxidase (MAO), ACh esterase: e.g., β-carboline alkaloids, physostigmine, galanthamine (**8**)] or their reuptake into the presynaptic neuron and/or secretory vesicles [e.g., cocaine (**8**), reserpine (**7**)].

The hallucinogenic properties of plants must have puzzled early man and it is plausible that corresponding descriptions found their way into mythology. Female gods

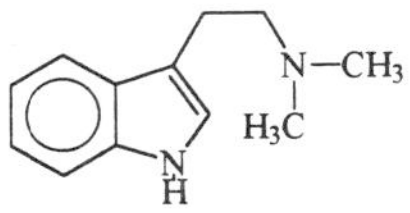

N, N-dimethyltryptamine

6

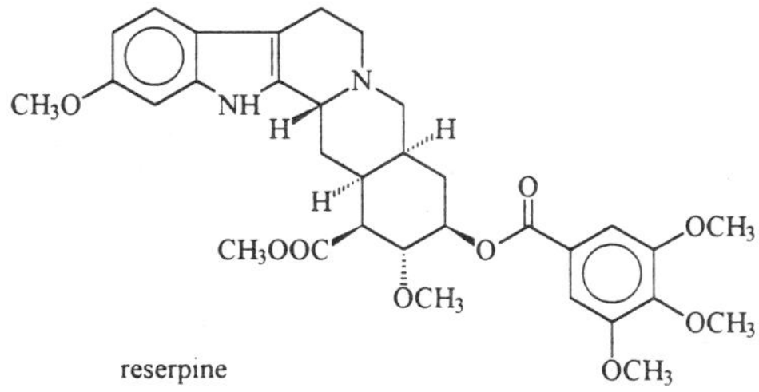

reserpine

7

have been described very early in this context. One of them is Kybele, the goddess of animals, mountains, and medicine. Originating in Phrygia the Kybele cult reached Greece around 1500 B.C. and Rome in 200 B.C., where it remained until 400 AD. It is likely but not documented that the Kybele cult employed hallucinogenic alkaloid drugs. Near the Black Sea, Hecate was worshiped as the goddess of witchcraft. Helpers of Hecate were the Pharmakides (our modern "pharmacists"?), witches, experienced in medicinal and toxic plants. It has been said that Hecate was the first to use *Aconitum* as a deadly poison testing it when having guests for a party. Hecate had a "botanical garden" on the island of Colchis where the following alkaloid plants were kept: Akoniton (*Aconitum napellus*), Diktamnon (*Dictamnus albus*), Mandragores (*Mandragora officinarum*), Mekon (*Papaver somniferum*), Melaina (*Claviceps purpurea*), Thryon (*Atropa belladonna*), and *Colchicum* spp. (Merk-Schäfer, 1997).

"Daughters" of Hecate were Circe and Medea; especially Circe has been portrayed in the *Odyssey* by Homer. She kept the tame lions as Kybele did and also used hallucinogenic drugs. The story of Ulysses and his comrades has often been retold. Arriving at the island of Circe, they were offered wine that was mixed with *Mandragora* (which has been appropriately named "Circaia") and other solanaceous plants containing tropane alkaloids (for a more detailed account, consult sections on *Atropa, Datura, Hyoscyamus,* and *Mandragora*). The effects were hallucinations, suggesting that they were pigs. But Ulysses was as clever as Circe and had used the plant Moly (which was given to him by Hermes) as an antidote; so Circe had no power over Ulysses. We can speculate that Ulysses used bulbs of some lilies (such as daffodils or snowdrops) that contain the alkaloid galanthamine (**8**). This alkaloid is an active inhibitor of acetylcholine esterase and

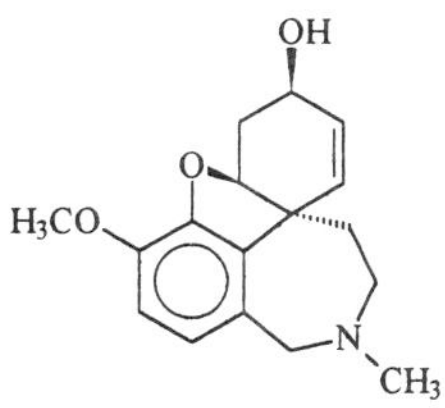

galanthamine

8

would counteract the activity of tropane alkaloids at the muscarinic acetylcholine receptor (AChR). At present, galanthamine is being used to treat patients with Alzheimer's. Galanthamine helps to maintain a higher level of acetylcholine, which is essential for signal transduction in the brain, despite the progressive destruction of neurons.

In Central and South America the use of hallucinogenic plants followed a different (although in essence similar) route: Sources indicate that medicinal and psychotropic plants and mushrooms have been used by Indians for at least 3000 years. Written evidence originates from the 16th century, e.g., from Hernandez (*Rerum Medicarum Novae Hispanieae Thesaurus*) and Bernardino de Sahagún (*Historia general de las cosas de Nueva España*). A rich knowledge had been accumulated by the Aztecs, who used toaloatzin (*Datura*), ololiuqui (*Turbina corymbosa*), teonanacatl (*Psilocybe mexicana*), peyotl (*Lophophora williamsii*), and several other drugs for divination and magico-religious purposes. We can speculate that psychotropic plants were first discovered by Indians of lower social status but that this knowledge was taken over and kept by the shamans where it has remained exclusively until today. Besides shamans, brujos (witch doctors), yerberos (herbalists), and curanderos (healers) still exist in modern Mexico. Whereas the former two are more concerned with magic and divination, the herbalists and healers, who are lower in social status, collect plants for medicinal applications (Mann, 1992). Although Central and South America were Christianized four centuries ago, some of the magico-religious rites of the Aztecs still exist today (albeit in Christian "costume," i.e., connected with the Virgin Mary, or Saints Peter and Paul) (Mann, 1992; Schultes and Hofmann, 1980, 1987).

An account of more than 90 plant species that have been used for stimulatory, euphoric, psychomimetic and hallucinogenic properties can be found in Schultes and Hofmann (1980, 1987).

Besides for murder, executions, and hallucinogens, a number of plants were used as abortifacients in antiquity and medieval ages, mostly plants rich in essential oils and other terpenes, but also in alkaloids; plants such as yew (*Taxus baccata*) which contains several diterpene alkaloids [including the anticancer alkaloid taxol (**59**)]. Also, aphrodisiacs were famous and sought after: Most of them contain extracts of *Mandragora* and of other plants with tropane alkaloids, such as *Hyoscyamus, Atropa,* and *Datura*. Aphrodite, the goddess of love, was called "mandragoritis" accordingly (Mann, 1992; Schmidt, 1927; Merk-Schäfer, 1997; Gessner, 1974; Baumann, 1986).

3. ALKALOIDAL PLANTS AND FUNGI PLAYING AN IMPORTANT ROLE IN THE HISTORY OF MANKIND

Some of the alkaloidal plants that have been used for murder, magic, and medicine are listed alphabetically and briefly characterized below (Mann, 1992; Schmidt, 1927; Merk-Schäfer, 1997; Gessner, 1974; Baumann, 1986; Kollmann-Hess, 1988; Schultes and Hofmann, 1980, 1987; Bellamy and Pfister, 1992):

3.1. *Aconitum napellus* (Family Ranunculaceae)

A. napellus, A. vulparia, and other aconites accumulate the diterpene alkaloid aconitine (**9**) and related compounds in all parts of the plant. Because aconitine is highly lipophilic, it can be resorbed even through the skin. Aconitine reduces the secretion of acetylcholine presynaptically and activates Na^+ channels and thus leads to rapid paralysis

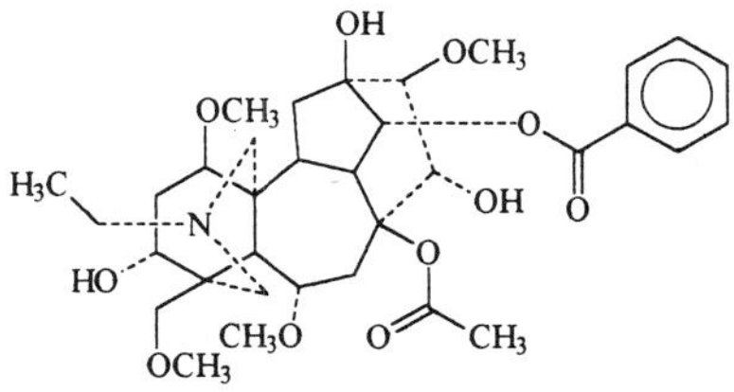

aconitine

9

and anesthesia. Because of these properties, extracts of aconite were employed as arrow poisons by early man in Europe, Alaska, and Asia. As aconitine is also highly toxic when given orally, it was used as a deadly poison to remove criminals and "disliked" contemporaries. In Rome, the cultivation of *Aconitum* was banned in the end (because it had been used for murder too deliberately) and the possession of the plant became a capital offense. On the Greek island of Chios, aconite was allowed for euthanasia of old and infirm men (Mann, 1992).

First effects of *Aconitum* intoxication are burning and paresthesia of mouth and throat, followed by insensitivity of fingers and toes, accompanied by cold sweat, general coldness, and severe diarrhea. Death sets in after 3 hours through respiratory failure. Aconite has been mentioned as a deadly poison by Juvenal, Ovid, Plutarch, and Theophrastus. The latter knew where the best aconite grew in Greece and that the plant protected itself against herbivores, such as sheep. As described in Chapter 11, we now believe that the protective effect is indeed the biological function of this and other alkaloids. Extracts of aconite were also used, with raw meat as a bait, to kill wolves, foxes, and other animals (which is mirrored in the name *A. vulparia* and the English name *wolfsbane*).

Aconitine has been used in folk medicine as an anesthetic, and as a remedy against pneumonia and heart disease.

3.2. *Amanita muscaria* (Family Amanitaceae)

The fly mushroom is common in Europe, Asia, and North America and accumulates active nitrogen-containing compounds that could be classified as alkaloids (*sensu lato*), such as muscarine (**10**), muscimol (**11**), and ibotenic acid (**12**). Whereas muscarine affects the "muscarinic" AChR, muscimol and ibotenic acid interfere with the activity of the neurotransmitter GABA (see Chapter 12).

The fly agaric looks spectacular and, because of its alkaloids, produces strong hallucinogenic effects. *Amanita muscaria* might be the oldest hallucinogen of mankind and was

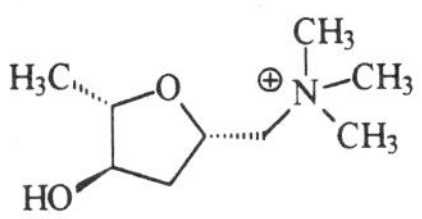

muscarine

10

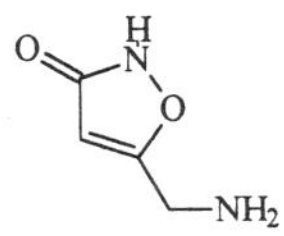

muscimol

11

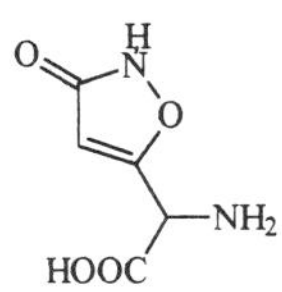

ibotenic acid

12

perhaps the active ingredient of soma in ancient India. Later it was also widely used as a hallucinogen in parts of Asia and by a few Indian tribes of North America (as a tradition deriving from early men who had migrated to America from Siberia via the Bering Strait?). It is therefore not surprising that the fly mushroom regularly occurs in mythology and even in children's fairy tales.

The utilization of *Amanita* as a hallucinogen in Siberia has been described by several European travelers. George Steller wrote in 1774 (after Wasson, 1968):

> The fly agarics are dried, then eaten in large pieces without chewing them, washing them down with cold water. After about half an hour the person becomes completely intoxicated and experiences extraordinary visions. Those who cannot afford the fairly high price of the mushrooms drink the urine of those who have eaten it, whereupon they become as intoxicated, if not more so. The urine seems to be more powerful than the mushroom, and its effect may last through the fourth and fifth man.

It is possible that ibotenic acid is converted to muscimol (which is a stronger hallucinogen) in the body, which would explain these observations.

While the fly agarics are still employed widely in Siberia (instead of alcohol), it has been speculated that it was also in use in northern and central Europe: On an "*Amanita*-trip" people experience the feeling of extreme power and strength, which reminds one of descriptions of the Norse berserkers. It might even be speculated that Miraculix or Getafix (a figure known to everybody who has indulged in Asterix and Obelix) used *Amanita* in his magic potion which made his people invincible.

3.3. *Areca catechu* (Family Arecaceae)

Areca is a palm that occurs in Southeast Asia. Its seeds are rich in alkaloids, such as arecoline, which displays parasympathomimetic properties; i.e., arecoline activates the muscarinic AChR (see Chapter 12).

In Southeast Asia and East Africa, *Areca* seeds are ingested as "betel," which is a combination of leaves of the vine *Piper betel*, slices of *Areca* seeds, and lime. For several hundred years, it has been a common habit of more than 200 million people to chew betel. The alkaloids, which are converted into the free base in the buccal cavity, are directly absorbed and quickly pass the blood–brain barrier. Betel produces a stimulatory and relaxing feeling similar to that of alcohol. Arecoline is used in veterinary medicine as a drug against intestinal worms of cattle and dogs.

3.4. *Aristolochia clematitis* (Family Aristolochiaceae)

The main alkaloid of *Aristolochia* is aristolochic acid (**13**), which stimulates phagocytes and wound healing. Higher doses lead to hypotension and convulsions; death is caused by respiratory failure. Because of the carcinogenic properties of aristolochic acid (see Chapter 12) extracts of *Aristolochia* are no longer allowed in pharmaceutical preparations. *Aristolochia* was known and used in ancient times. Dioscorides, who recognized several *Aristolochia* species, described the following applications: Extracts of *Aris-*

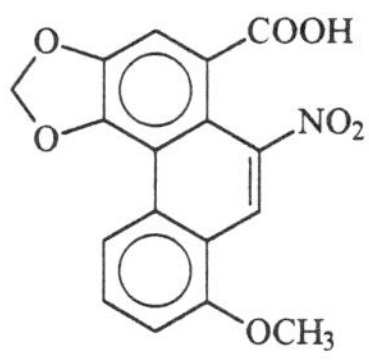

aristolochic acid

13

tolochia were used by women in childbirth, against snakebites and intoxication. In addition, *Aristolochia* was said to help against internal inflammations and in wound healing. Later the extracts were used as a folk-medicine abortifacient (Merk-Schäfer, 1997).

3.5. *Atropa belladonna* (Family Solanaceae)

Deadly nightshade is widely distributed in Europe. It contains tropane alkaloids, such as hyoscyamine as a main and scopolamine as a less abundant component in all parts of the plant including the attractive violet-colored fruits. Extracts of *Atropa* have been used for poisoning (five to ten berries will kill a person) or for its hallucinogenic/aphrodisiac properties.

These activities of atropine (**14**) (and scopolamine) can be explained at the molecular level: L-Hyoscyamine (atropine is a racemic mixture of D- and L-hyoscyamine) blocks the muscarinic AChR, which leads to an inhibition of sweat and saliva production and to relaxation of smooth muscles, especially those of stomach and intestine ("parasympatholytic"). Because hyoscyamine can pass the blood–brain barrier by diffusion, the central nervous system is also affected.

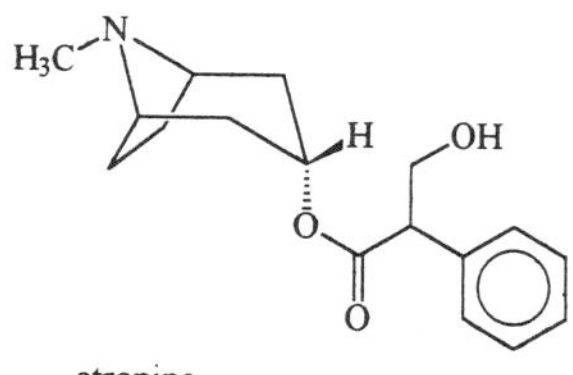

atropine

14

The toxicity of *Atropa* extracts must have been long known as early men used them as arrow poisons. Its use for murder was quite common in ancient Rome; Livia, the wife of Emperor Augustus, and Agrippina (wife of Claudius) used *Atropa* to murder their contemporaries (Mann, 1992). Dioscorides called the plant "Strychnos manikos" or "Thryon" (= plant from Colchic magic gardens). Dioscorides reported that 1 drachma (= 3.4 g) of roots extracted in wine would lead to hallucinations. After 2 drachmas hallucinations would continue for up to 4 days and 4 drachmas would kill a person. When applied to the eyes, dilation of the pupils takes place. Because wide pupils appeared to make women more attractive (reflected in the Latin name *A. belladonna*), *Atropa* was used as a cosmetic drug until the Renaissance.

Although it has been argued that the witch-hunt was an invention of the Church, used to encourage good Christian behavior, the connection of witchcraft with plants containing tropane alkaloids has recently become quite clear. During the Middle Ages, *Atropa* (as well as *Hyoscyamus, Datura,* and *Mandragora*) was one of the main components of potions and ointments prepared by drug addicts, better known as witches. The well-documented witch-hunt that persisted through a couple of centuries may thus have been, in part, a reaction to drug abuse, as people using *Atropa* and *Hyoscyamus* had a changed consciousness, behaved strangely, and lived in another world (remarkably reminiscent of today's cocaine or heroine users).

If people take tropane alkaloids, hallucinations with orgiastic adventures may result. Surprisingly, the feeling of being able to fly was produced when extracts from plants with tropane alkaloids combined with fats and oil were applied on skin (especially of armpits, external genitalia, but also into vagina and rectum). This is reflected in the well-known pictures of witches flying on a broomstick. For a vivid account of these phenomena see Mann (1992).

The dream of being an animal (remember the description in the *Odyssey* where the comrades of Ulysses were converted into pigs by Circe?) is caused when extracts with tropane alkaloids are taken orally. Higher doses, however, lead to intoxication, which starts with dryness in the cavity of the mouth, followed by a disturbance of vision, later by hallucinations, respiratory failure, and death. It should be noted that witches' potions also contained other compounds with hallucinogenic properties, such as skins from toads (*Bufo bufo*) with *N*-methyltryptamine and bufotenine (**15**) (besides bufotoxine, an esterified bufadienolide), and extracts from several plants containing tropane alkaloids, such as *Datura, Hyoscyamus, Scopolia,* and *Mandragora.*

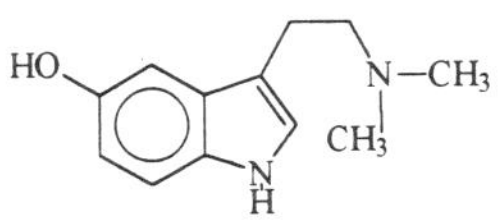

bufotenine

15

Extracts of *Atropa* were used by doctors to anesthetize their patients during surgery (sometimes combined with alkaloid extracts of *Papaver somniferum*; "theriac"). This application came to a stop during the Middle Ages when witch hunting started. As patients would tell their sometimes erotic dreams after surgery, the risk accused of being a sorcerer became too great for surgeons. For many years thereafter no anesthetic was available until the advent of modern compounds such as ether (1846), chloroform, and other products. Atropine is still in pharmaceutical use today as a mydriatic drug in ophthalmology and as an antispasmodic (see Chapter 18).

Atropine was first isolated as a pure compound in 1833 by Mein and by Geiger and Hesse. After research by Liebig in 1833, Lossen in 1864, Ladenburg in 1883, and Merling in 1891, it was Willstätter in 1889 who finally obtained the correct structure of atropine. L-Hyoscyamine and scopolamine were first isolated by Schmidt in 1890. The latter was named hyoscine (**16**) or scopoline by Ladenburg and Bender and atroscine and scopolamine by Hesse in 1896. The correct structure was elucidated by Gadamer and Hammer in 1921 and Hess and Wahl in 1922 (complete citations in Kollmann-Hess, 1988).

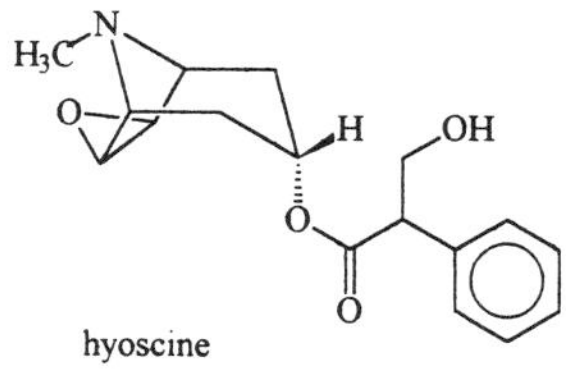

hyoscine

16

3.6. *Banisteriopsis caapa, B. inebrians* (Family Malpighiaceae)

Banisteriopsis is a woody climber of the Amazon and Orinoco River forests, known locally as "ayahuasca" or "caapi." The plants contain β-carboline alkaloids, such as harmine (**17**), harmaline (**18**), and tetrahydroharmine, which block serotonin receptors and monoamine oxidase (MAO) in the brain. Indians have discovered that a potion prepared from the alkaloid-rich bark of this liana has hallucinogenic properties. Starting with dizziness and nausea, a euphoric ecstasy follows, sometimes accompanied by aggressive behavior. Typical are intense color visions and dreams. Indians have called ayahuasca the "light of their souls" and use it as a general remedy against illness. In addition to euphoria, a pronounced aphrodisiac component is apparent, e.g., during initiation rites for adolescent boys (Schultes and Hofmann, 1987; Mann, 1992).

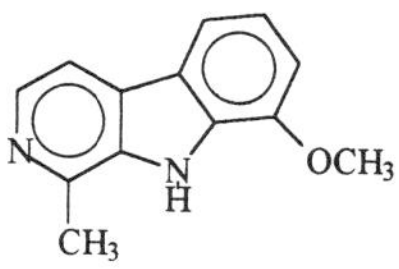

harmine

17

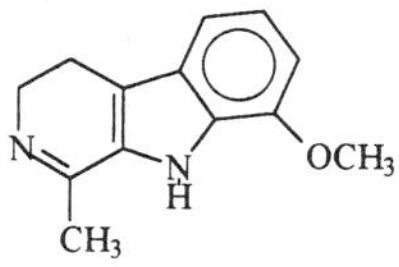

harmaline

18

3.7. *Buxus sempervirens* (Family Buxaceae)

The box tree originates from the Mediterranean area. It contains toxic alkaloids derive from the cholesterol skeleton, such as cyclobuxine (**19**) and buxanine (**20**). The box tree was devoted to Kybele and Hades in antiquity; several old illustrations show Circe holding a branch of *Buxus* in her hands when she welcomed Ulysses. As it is an evergreen plant, it was a symbol for immortality and even today, the box tree is a favored plant in European cemeteries. The box tree was known under the name "Lykion" in ancient Greece (perhaps this name was applied also to a few other plants) and extracts were employed (according to Dioscorides) to treat wounds, jaundice, and intoxications. During medieval times, extracts from *Buxus* were used against gout, rheumatism, and malaria. Intoxication in humans and animals are not uncommon: An initial excitement is followed by growing immobilization and finally death is caused by paralysis, i.e., respiratory failure.

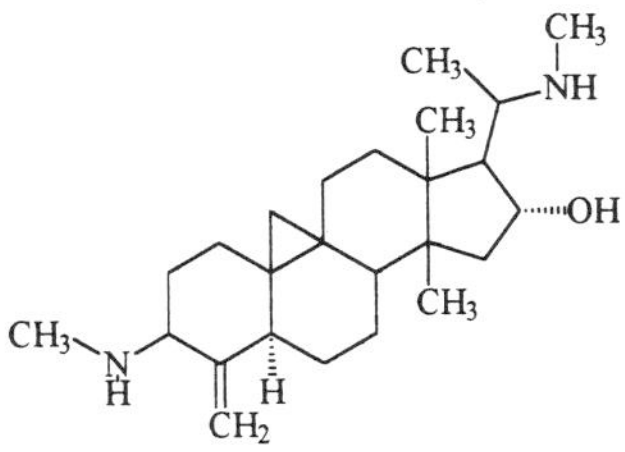

cyclobuxine

19

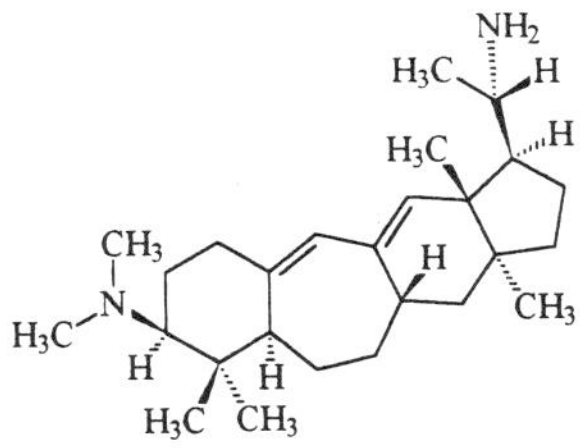

buxamine

20

3.8. *Camellia sinensis* (Family Theaceae)

Tea is a shrub of eastern Asia and produces methylxanthines, such as caffeine (**21**), theophylline (**22**) and theobromine (**23**) in the leaves. Caffeine acts as a central stimulant and activates circulation and respiration. Theophylline, especially, has bronchodilator activity (use in asthma patients). Molecular targets are phosphodiesterase (an enzyme that inactivates the second messenger cAMP) and adenosine receptors which are inhibited. A cup of tea contains ca. 40 mg of caffeine, enough to cause a mild central stimulation.

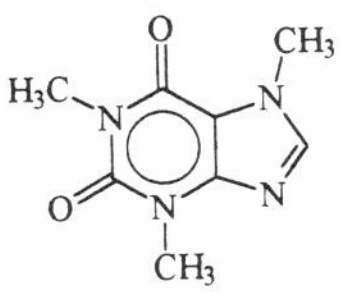

caffeine

21

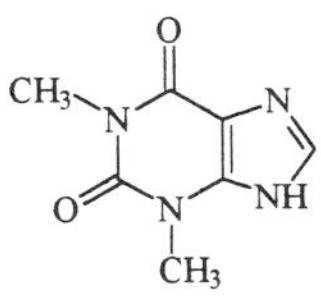

theophylline

22

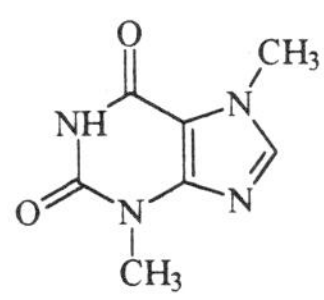

theobromine

23

Tea was first described in 350 B.C. by Kuo P'o as a precious commodity of ancient China. Buddhist monks drank tea to sustain them through long hours of meditation. From China tea came to Japan where it became an integral part of the Japanese way of life. Tea was unknown to Europeans until 1550 and tea import started in the early part of the 17th century through the English East India Company. Commercial plantations commenced in 1826 on Java by the Dutch and in 1836 in India by the English.

Another plant with caffeine is *Coffea arabica* (family Rubiaceae) (a native of Ethiopia) which was discovered much later and first plantations started in Yemen during the ninth century A.D. *Coffea* was introduced to England as "kahveh" in 1601 and became famous throughout Europe. Mass cultivations started in the 18th century, especially in South America. A cup of coffee contains ca. 80 mg of caffeine.

Theobroma cacao (family Sterculiaceae) originates from Central America and its seeds are rich in theobromine and caffeine. The Mayas and Aztecs recognized "chocolatl" as a food of the gods with aphrodisiac properties. When Cortes invaded America, he conquered the Aztec kingdom of Montezuma and got hold of cacao. Cacao was introduced in Europe as a refreshing and reputedly aphrodisiac drink and by 1700 there were 2000 chocolate houses in London alone (Mann, 1992).

Other plants with caffeine include *Ilex paraguariensis* (family Aquifoliaceae), *Cola nitida* (family Sterculiaceae), and *Paullinia cupana* (family Sapindaceae). All of these plants are exploited by man because of their alkaloids and the corresponding stimulatory effects. At present they are widely used by mankind, and our literature, music, arts, and thus our culture (think of the Viennese coffeehouses and their influence on artists and scientists) have certainly been positively influenced by caffeine and other purine alkaloids.

3.9. *Chelidonium majus* (Family Papaveraceae)

Chelidonium is a widespread Eurasian plant that is characterized by a yellow to orange-colored latex. The color is caused by protoberberine and benzophenanthridine alkaloids, such as protopine (**24**), chelidonine, chelerythrine (**25**), berberine (**26**) and sanguinarine (**27**) which are sequestered in latex vesicles (see chapter 10). On injury the coloured latex exudes from all parts of the plants. It is likely that such a spectacular phenomenon was known in antiquity. A legend (according to Pliny) says that swallows had blind young in their nest. When the parents brought a twig of *Chelidonium* and

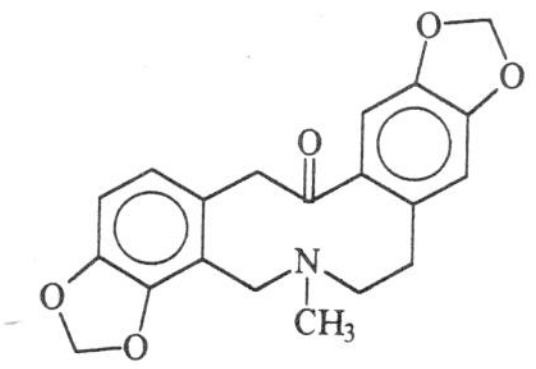

protopine

24

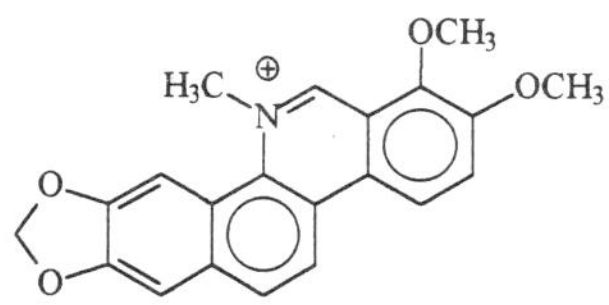

chelerythrine

25

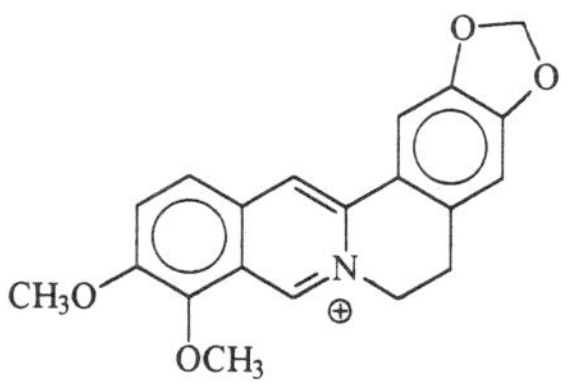

berberine

26

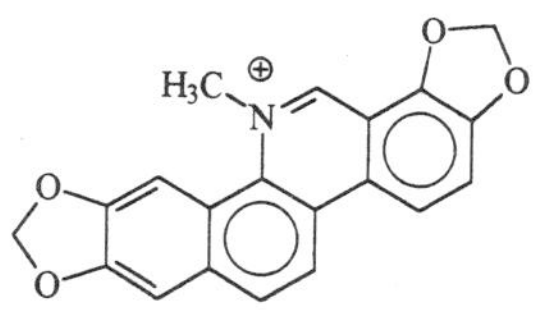

sanguinarine

27

applied the milky juice to the eyes of their young, the eyes recovered and the young could see. It has been suggested that this legend is reflected in the latin name *Chelidonium* as "chelidon" was the old name for swallow. Dioscorides recognized this plant under the names Paionia, Krataia, or Philomedeion (the Romans called it "Fabium," the Celts "Thona") and had recommended it for better eyesight, against jaundice and eczema (probably herpes simplex infections). In addition, the latex was used to treat warts of all sorts and was even used against tumors. In medieval times symbolism was used to find the application of plant drugs. As the extracts were yellow, it was thus plausible to use them against jaundice (even today extracts of *Chelidonium* are taken as a cholagogic spas-modicum in phytotherapy; see Chapter 18). Because it worked (as we know today, the alkaloids have antispasmodic properties, are antiseptic, and eliminated via the bile duct), *Chelidonium* was considered a proof of the power of symbolism. Can we also explain the widespread use of *Chelidonium* against ophthalmia, eczema, and warts? Berberine, chelidonine, and sanguinarine intercalate in DNA and inhibit a number of enzymes, including RNA and DNA polymerases. These properties would produce antiviral and antibacterial effects, and because viruses are the cause of warts, herpes lesions, or eye troubles, a therapeutic effect would be plausible. This example shows that the traditional utilization of plant drugs may sometimes (but not necessarily always) have a solid base that can be explained today in molecular terms.

Chelidonium alkaloids were first isolated by Schmidt in 1888. Chelidonine, sanguinarine, and chelerythrine were obtained in pure form by Schmidt and Selle in 1890 and Fischer in 1901 and their structures were elucidated mainly by Bruchhausen and coworkers in 1930–1931. Berberine, which occurs in *Chelidonium* but also in many members of the Berberidaceae, was first isolated in 1824 from *Geoffroya jamaicensis* by Hüttenschmidt. Elucidation of its structure was finally achieved by Perkin and Robinson in 1910 (citations in Kollmann-Hess, 1988).

3.10. *Claviceps purpurea* (Family Hypocreaceae)

Ergot is a fungus that grows on cereals, especially rye. The black-violet developmental stage is rich in indole alkaloids derived from tryptophan, such as ergotamine (**28**) and

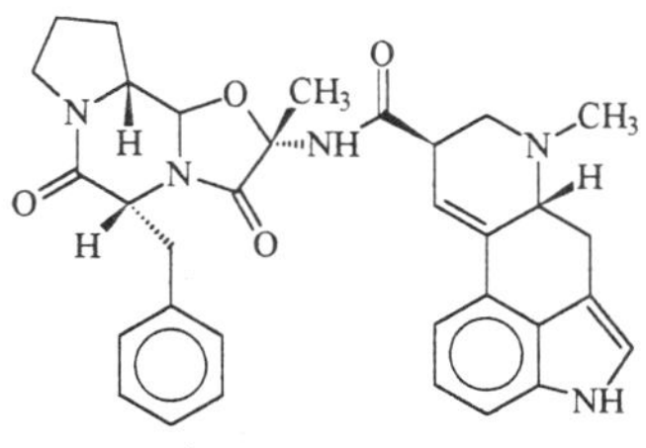

ergotamine

28

ergotoxine. Because these alkaloids can interact with several neurotransmitters, especially serotonin, but also noradrenaline and dopamine, a wide variety of pharmacological effects have been described (see Chapters 12 and 18). Besides central nervous disturbances (leading to convulsions and twitching of the body, i.e., epileptic seizures), chronic toxicity causes painful vasoconstriction. The results are that legs and arms go numb and eventually become gangrenous, i.e., they die off or have to be cut away.

The toxic effects of ergot ("ergotism") have long been known: It has been speculated that the 4000-year-old Eleusinian Mysteries of ancient Greece were connected with ergot-induced hallucinations (Hofmann *et al.,* 1978). Assyrians (about 600 B.C.) called it "a noxious pustule in the ear of grain." The Parsi (about 350 B.C.) wrote of "noxious grasses that cause pregnant women to drop the womb and die in childbed" (Mann, 1992). The Hebrews recognized ergot as "son" or "sonin" and knew its hallucinogenic and toxic properties. Alarming mass intoxications (in 994 A.D. about 40,000 people were reported to have died because of ergotism or the "holy fire" as it was also called) occurred during medieval times when rye was grown in larger quantities and many people ingested contaminated flour. Mass poisoning was observed even in the 20th century, e.g., in 1926–1927 11,319 cases of ergotism were observed in a population of 506,000 in the vicinity of Sarapol (near the Ural region). The dramatic and cruel effects of ergotism can be seen in many paintings of the old masters.

Hippocrates and his disciples recognized ergot as "Melanthion" (a name also used for *Nigella sativa*) and described its use to stop bleeding after childbirth. Even today, ergometrine (**29**) and derivatives are used in obstetrics and ergotamine (**28**) to treat migraine (see Chapter 18).

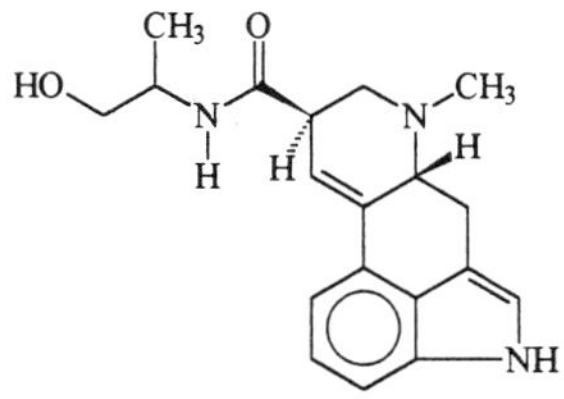

ergometrine

29

Why a fungus produces alkaloids that are obviously directed against higher animals has been an enigma for a long time. During the last 10 years a plausible answer has been put forward by ecologists: *Claviceps* and related fungi not only infect cereals but a wide set of other grasses (Poaceae). In a simple field experiment it could be shown that a meadow where grasses were highly infected was significantly less grazed by herbivores than a neighboring plot devoid of fungus infection. It was therefore suggested that the *Claviceps*–plant interaction is of symbiotic and not of parasitic origin: as the fitness of the host plant is only slightly diminished, the plants gain a fitness component through the fungal toxins (it should be remembered that grasses are usually not toxic) (see Chapter 11).

3.11. *Colchicum autumnale* (Family Liliaceae)

The meadow saffron is another Eurasian plant that did not escape the attention of our ancestors. Its biology is exceptional in that it flowers in autumn (thus the Latin name) and escapes winter frosts hidden 20–40 cm below ground. In spring, shoots appear above ground and with them the fruits that distribute many black seeds in early summer. All parts of the plants accumulate the alkaloid colchicine (**30**) which is accompanied by several minor alkaloids, such as demecolcine (**31**). *Colchicum* and its extracts are extremely toxic

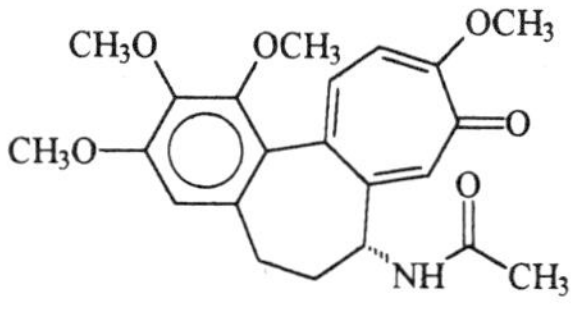

colchicine

30

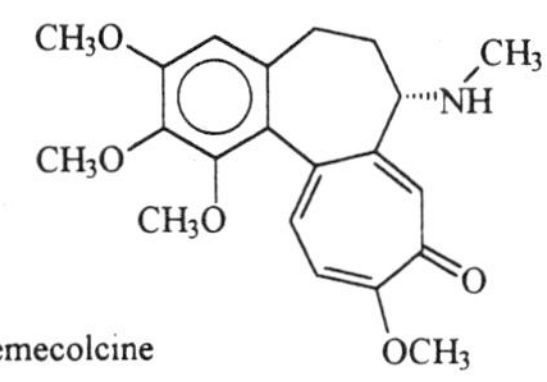

demecolcine

31

(LD_{50} in rats: 1.6 mg/kg, i.v.) and intoxications have been known from animals and humans. Dioscorides warned that bulbs of *Colchicum* are highly toxic and could be mistaken for bulbs of *Allium*; furthermore, they taste attractively sweet. Intoxications by colchicine and *Colchicum* start with nausea and vomiting, followed by watery and bloody diarrhea. An increasing paralysis is followed by death caused by respiratory failure. Today, the mechanisms of toxicity are well established: Colchicine binds to tubulin and prevents the formation of microtubules. As microtubules are important for cell division and cell movements, colchicine can block important targets in cells. The antimitotic activities could make colchicine a candidate for a cytostatic drug, but its side effects are too strong. However, colchicine is used in plant breeding to produce polyploidy in crop plants (which are usually bigger than the original diploid organism). In spite of its toxicity colchicine and extracts from *Colchicum* are used in medicine today to treat acute gout. Normally, macrophages would move to the site of gout inflammation, which is caused by a deposition of uric acid crystals. Because macrophages would release lactic acid there, the lowered hydrogen ion concentration would induce the precipitation of even more uric acid. This positive feedback mechanism is a vicious circle, as an increased inflammation will attract more macrophages and more macrophages will increase uric acid crystals, etc. Because colchicine blocks the movement of macrophages, it can disrupt this feedback mechanism.

3.12. *Conium maculatum* (Family Umbelliferae)

Hemlock occurs widely in Europe and it accumulates toxic piperidine alkaloids: coniine (**32**), coniceine, and *N*-methylconiine in all plant parts; with the fruits especially rich in alkaloids. Coniine affects the nicotinic acetylcholine receptor, as does nicotine. It acts as a strong toxin in animals. After nausea, vomiting, and diarrhea, the intoxication is followed by mental disturbance and paralysis which starts at the tip of arms and legs and ends with respiratory failure.

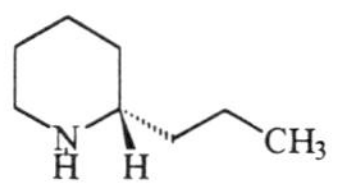

coniine

32

Extracts of *Conium* were frequently used for murder, suicide (which was not stipulated in antiquity), and execution (according to Dioscorides, Juvenal, Libanius, Theophrastus, and Pliny). The most famous victim was Socrates, who was forced to drink hemlock juice in 399 B.C, because—in the eyes of the people of Athens—he had taught about the gods in a contemptuous way and had spoiled the youth. According to Tacitus, *Conium* extracts were used by Nero to murder his stepbrother Britannicus, after the effectiveness of the poison had been tested on a billy goat.

Extracts of *Conium* ("Koneion") were also employed to treat ulcers and eczema and as an antaphrodisiac; according to Dioscorides, application on the testes of boys and the breasts of girls would inhibit the development of these organs.

3.13. *Datura metel* (Family Solanaceae)

The thornapple and its relatives have a long tradition as medicinal, but also hallucinogenic, plants which were used in Europe, Africa, Arabia, Asia, and America. All parts of the plant produce tropane alkaloids, with hyoscyamine and scopolamine as major alkaloids. The seeds are especially rich in alkaloids.

The pharmacology of tropane alkaloids has already been discussed under *Atropa* (see above), which would make the psychic (sensation of flying; transformation into animals), medicinal, and toxic effects of *Datura* alkaloids plausible.

Datura must have been known very early in human history; its possible use in "soma" in India and its appearance in the Ebers papyrus have already been mentioned. The old Greeks recognized "thryon" and "persicon," although it is not clear whether *Datura* or *Atropa* was meant. It has been speculated that *Datura* alkaloids were responsible for the loss of Mark Anthony's army in 36 B.C.: While campaigning against the Pashtun in Asia Minor, his troops ran out of food and had to eat unfamiliar plants. On one occasion the soldiers probably ate *Datura*, "which killed them after driving them mad." Medicinally, leaves of *Datura* were used in the 19th and 20th centuries as herbal cigarettes to treat patients with asthma and other respiratory conditions. A modern equivalent of this is the drug ipratropium (a synthetic structural analogue of hyoscyamine) which is inhaled by asthma patients (see Chapter 18).

In the New World we find *Datura stramonium* and other herbs in addition to the genus *Brugmansia* (shrub- and treelike plants, formerly also classified as *Datura*). Both *Datura* and *Brugmansia* were used by the Aztecs and modern Indians in the same way as tropane plants in the Old World. Ointments administered on the skin resulted in the "ability to fly." Extracts were also used for infanticide and murder (for a detailed description, see Mann, 1992). Angel's-trumpets have been misused recently by teenagers in the United States, because the content of tropane alkaloids (atropine, scopolamine, and meteloidine) induces exciting visions. But the wrong dosage often leads to severe intoxications: In Florida alone, in 1994, 112 teenagers were admitted to hospital suffering the side effects of *Brugmansia* intake.

The use of these plants for crime, seduction, and as narcotic drugs is easily revealed by some of the common names such as "herbe aux sorciers" or "herbe au diable."

3.14. *Delphinium consolida* (Family Ranunculaceae)

Several larkspur species occur in Eurasia. Most of them contain diterpene alkaloids such as delphinine (**33**), delcosine (**34**), or delsoline (**35**). They share a number of chemi-

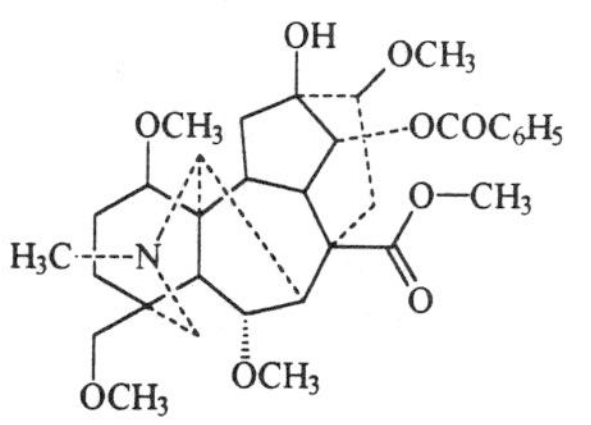

delphinine

33

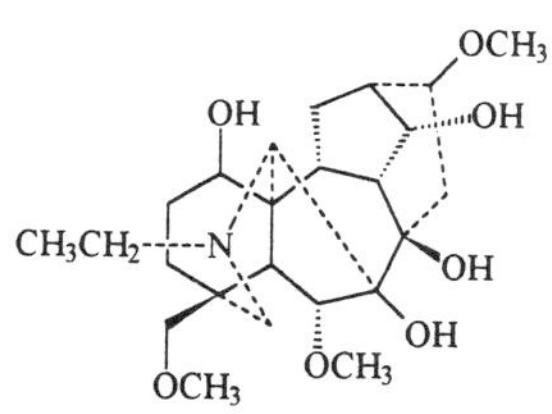

delcosine

34

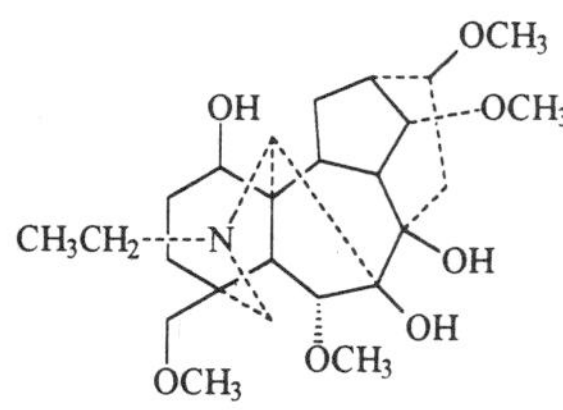

delsoline

35

cal features with aconitine and are also very toxic to animals. Symptoms are brachycardia, hypotonia and paralysis. The latter aspect must have been known in ancient Greece as Dioscorides mentions that "Delphinion" is also named "Paralysis." The same author describes the application of extracts from *Delphinium* against ectoparasites, such as lice and taken with wine as an antidote against scorpion stings.

3.15. *Dictamnus albus* (Family Rutaceae)

D. albus occurs in warm Eurasian habitats and is rich in furanoquinoline alkaloids [e.g., dictamnine (**36**)] and furanocoumarins (e.g., bergapten and psoralen). Extracts of

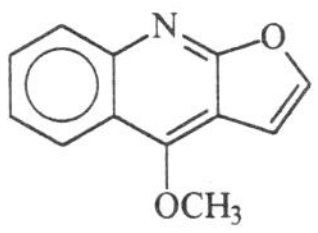

dictamnine

36

Dictamnus have been used for wound healing, especially wounds caused by spears or arrows. Dioscorides recognized the plant under the name "Beluakos" (remedy against arrow wounds) and "Artemidion" (after Artemis, the goddess of hunters). In addition, extracts from *Dictamnus* were used as an abortifacient.

3.16. *Ephedra fragilis* (Family Ephedraceae)

E. fragilis occurs in the eastern Mediterranean and produces ephedrine (**37**) and related alkaloids in all tissues. Dioscorides recommended its use to treat bronchial prob-

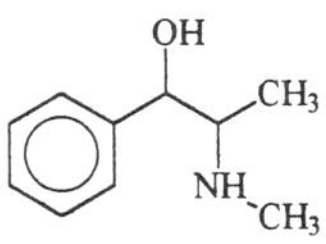

ephedrine

37

lems, as did the Chinese who employed the plant against asthma 5000 years ago (herbal of the Emperor Shen Nung). Because ephedrine binds to adrenergic receptors (see Chapter 12) and has spasmolytic properties, it is still used today to treat asthma and cough (see Chapter 18). Indians and settlers of North America prepared a stimulatory tea from *E. trifuera*, a plant rich in norpseudoephedrine.

A plant with related compounds is the African shrub, qat [*Catha edulis* (family Celastraceae)] which is found in Arabia (Yemen) and Ethiopia. Qat accumulates alkaloids such as norpseudoephedrine, which affect adrenergic receptors. A central nervous stimulatory activity and the alleviation of hunger and fatigue have been documented. Leaves of *C. edulis* are either chewed or the infusion made with hot water is imbibed. Long term usage leads to addiction. In Yemen, where qat is a regular drink, a high proportion of bachelors may be the consequence of an antaphrodisiac effect of the alkaloids present (Mann, 1992).

Ephedrine was first isolated in 1890 by Nagai; the structure of 1*R*,2*S* (−)-ephedrine and the other three stereoisomers were elucidated by Schmidt between 1907 and 1914 and Späth and Göring in 1920 (citations in Kollmann-Hess, 1988).

3.17. *Erythroxylum coca* (Family Erythroxylaceae)

Erythroxylum coca and related shrubs occur in South America (between 1500 and 6000 m above sea level) and accumulate cocaine (**38**) and related alkaloids in all parts of

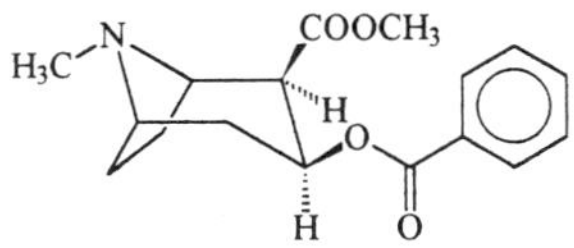

cocaine

38

the plant, especially the leaves. Cocaine inhibits the reuptake of the neurotransmitter dopamine into the synapses of the brain: Normally, dopamine is quickly released from dopamine receptors in the postsynaptic neuron and taken up in the presynaptic neuron by a special carrier mechanism. In the neuron it is either degraded by MAO or transported into secretory vesicles. When the reuptake is blocked by cocaine, dopamine concentration in the synaptic gap is increased leading to a prolonged activation of dopamine receptors. Because they are involved with the control of pleasure response, cocaine stimulates the so-called pleasure centers.

The stimulant properties of coca leaves must have been known for a long time; the mummy of a Peruvian king in the Nazca area, dated 500 A.D., was accompanied by several bags of coca leaves (Mann, 1992). During the Inca civilization (ca. tenth century A.D.) the utilization of coca leaves was well established but restricted to a few and used to silence hunger, to provide vigor, and to help to forget the miseries of life. Coca was an integral part of religious ceremonies and initiation rites and the shamans used it to induce trance-like states to communicate with the gods. After the Incas were conquered by Pizarro in 1533, coca use was no longer restricted to the shamans and Indians started coca-chewing everywhere to enjoy its stimulatory effects.

The conquistadors brought coca back to Europe and it was even celebrated as an "elixir of life." In the 1860s Angelo Mariani successfully invented the "Vin Mariani" which contained coca extracts and was said to have analgesic, anesthetic, and carminative properties. The wine was officially approved by the Vatican under Pope Leo XIII (Mann, 1992). In the United States, a "Peruvian wine of Coca" was sold from Sears, Roebuck and Co. with similar indications. Even more famous was Coca Cola, which originally contained true extracts of *E. coca* together with caffeine-containing *Cola nitida* and wine and was the equivalent to the Vin Mariani. In 1886, the wine was replaced by sugar syrup and the beverage was called Coca Cola. In 1904, coca extracts had to be removed from Coca Cola, but the trade name has been maintained until today (Mann, 1992).

Cocaine was isolated in the 1860s and became available as an extracted substance to a wider audience. In 1884, S. Freud experienced the stimulatory and euphoric properties of cocaine and in the same year his assistant C. Köller demonstrated its anesthetic properties and introduced cocaine as a local anesthetic for eye surgery. In 1915 cocaine was purified and used in large quantities for medicinal purposes, by the Merck company in Darmstadt, for example.

It was in the 20th century that cocaine became an infamous drug, which was first experienced by the "intelligentsia," such as Aldous Huxley. At present, cocaine is illegally produced in South America but exported for illicit purposes to North America, Europe, Australia, and Asia. Cocaine is applied by snorting (via the mucous membranes of the nose—facilitating an immediate passage to the brain), by smoking, or intravenously. While cocaine provides a powerful stimulatory effect on the pleasure centers of the brain, it also leads to addiction and the long term consequences of drug abuse. It has been estimated that more than 35 million Americans have taken cocaine at least once and cocaine abuse is a severe problem in many countries of the world at present.

More than 15 million Indians, the so-called "coquereos," regularly chew coca leaves (ca. 50 g fresh leaves per day, equivalent to 0.4 g cocaine) together with a little lime, which converts cocaine into its free base for better resorption. An increased endurance, a suppression of hunger, and an easing of life's difficulties are experienced. Cocaine ingested orally does not primarily affect the dopamine reuptake but interferes with the reuptake of noradrenaline and serotonin (Mann, 1992).

3.18. *Hyoscyamus niger, H. albus, H. muticus* (Family Solanaceae)

Henbane (*H. niger*) originally occurred around the Mediterranean basin, but as a ruderal plant it has been widely distributed in Eurasia. As are *Atropa* and *Datura*, henbane is rich in tropane alkaloids such as hyoscyamine and scopolamine; the latter alkaloid constitutes about 40% of total alkaloids. As a consequence *H. niger* is one of the infamous herbs used by witches and magicians; its hallucinogenic, analgesic, and toxic effects are similar to those described above, for *Atropa*.

The three *Hyoscyamus* species were well known to ancient Greeks and Romans: Dioscorides recommended *H. albus* for medicinal purposes, such as an analgesic, whereas *H. niger* and *H. muticus* would produce insanity and lethargy if eaten at a dose of greater than one tryblion (= 2 g). The oil from seeds was used to treat earache.

According to Pliny, henbane was known in Greece as "Herba Appolinaris" and taken by the priestesses of Appollon for producing their oracles. The oracle was named "Pytho" and the priestess "Pythia." The term "Pythonion" for henbane reflects these connections. The priestesses of the Delphic oracle were said to have inhaled smoke from smoldering henbane (Mann, 1992). In this context it is remarkable that scopolamine (which is abundant in henbane) was used in modern times for "brain-washing". Because of the higher scopolamine content in henbane, its hallucinogenic properties are more pronounced than those of *Atropa*. In antiquity, extracts were taken with wine and visions of transformations into pigs and wolves were frequently reported.

During medieval times, tropane alkaloids (especially scopolamine) were applied as ointments and resulted in the sensation of flying. Because of massive abuse of henbane and other drugs, many people became conspicuous and were either burnt as witches or put to death by an executioner. It has been reported that the condemned were given a drink with henbane to sedate and tranquilize them to make death less agonizing.

Another utilization of henbane extracts was found in beer brewing: as tropane alkaloids taste bitter, they were used as a substitute for hops and in addition would enforce the effects of ethanol. But modern laws in many countries no longer allow the addition of extracts from tropane alkaloid plants.

In modern medicine, scopolamine is used as a sedative before surgery, in ophthalmology, and to prevent seasickness (applied as a transdermal plaster behind the ear conch; see Chapter 18).

3.19. *Lophophora williamsii* (Family Cactaceae)

The peyote cactus occurs in Mexico and produces mescaline (**39**), besides several other simple isoquinoline alkaloids. Mescaline (first isolated in 1896) can be viewed as an

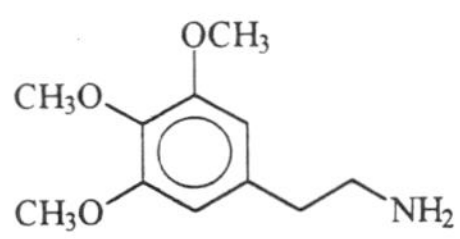

mescaline

39

analogue of the catecholamine-type neurotransmitters, although its activities resemble those of LSD (which mainly interacts with serotonin receptors). Peyotl, as the plant was known to the Aztecs, has substantial psychedelic properties and was widely used as a hallucinogen by Indians of Central America. Around 1880 the peyotl cult was introduced to Indians of North America. Even today, mescal buttons can be obtained legally via mail order for medico-religious purposes by North American Indians (Mann, 1992). Mescaline even gained literary celebrity through A. Huxley (1968) in his book *The Doors of Perception*. Mescaline, but especially the newer designer drugs such as ecstasy, have found their way into the narcotic drug scene of today. Mescaline is also a major compound in the cactus *Trichocereus pachanoi* from South America, which has been used since 1300 B.C. as a psychotropic plant in ancient and modern Peru (Schultes and Hofmann, 1987; Mann, 1992).

3.20. *Lupinus albus* (Family Leguminosae)

A small number of lupin species were found around the Mediterranean basin: *L. albus, L. angustifolius, L. luteus, L. hirsutus,* and others. Common to all species are relatively large seeds containing up to 40% protein. As the seeds are also rich in toxic

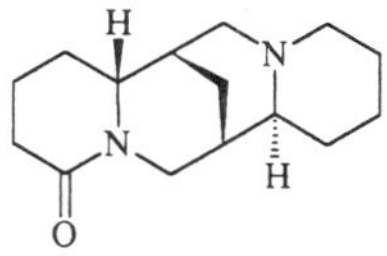

lupanine

40

quinolizidine alkaloids (QA), such as lupanine (**40**), they cannot be eaten directly but only after the alkaloids have been leached out under running water (this debittering technique was invented in both the Old and New World). Lupanine and other QA bind to either the nicotinic or the muscarinic acetylcholine receptor and affect Na^+ and K^+ channels (see Chapter 12). These effects are certainly responsible for toxicity and for the utilization as chemical defense compounds by the lupin plants (Wink, 1988, 1992, 1993a) (see Chapter 11). Intoxications show the following symptoms: After initial nausea and vomiting, convulsions and paralysis occur. Death is caused by respiratory failure.

As lupin seeds have been discovered in Egyptian tombs (in pyramids), *L. albus* (or *L. termis*, a synonym widely used in Egypt) must have been known in antiquity. Dioscorides reports its medicinal utilization: extracts from lupin seeds were used against intestinal worms. Externally they were used against eczema, itches, ulcers, impetigo, and moles. Another application was as an analgesic and for abortion. Visitors to the death oracle at the Acheron had to eat lupin seeds before they were allowed to communicate with the deceased. We can speculate that QA might cause some kind of ecstasy, which made them accessible to visions. Some of these utilizations appear plausible today, because of the activity of QA at AChR and ion channels.

In modern medicine, QA do not play a role, except for sparteine (**41**) which is used as an antiarrhythmic drug. Formerly, it was also used in obstetrics. However, 10–15% of all patients cannot metabolize sparteine, and this application has been abandoned in Europe.

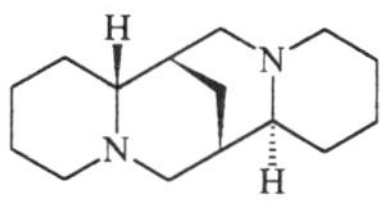

sparteine

41

Lupanine was first isolated from *L. angustifolius* by Hagen in 1885 and from *L. albus* by Soldaini in 1893 and confirmed by Davis in 1897. Its structure was established in 1936/1937 by Winterfeld and colleagues. Sparteine had been isolated much earlier, in 1851, by Stenhouse and its structure was elucidated by Clemo and co-workers in 1928 (citations in Kollmann-Hess, 1988).

3.21. *Lycopodium clavatum, L. annotinum* (Family Lycopodiaceae)

The club moss is a member of the order Lycopodiales (Pteridophyta) and grows on moist grounds in Europe and Asia. The plant is rich in alkaloids such as lycopodine (**42**)

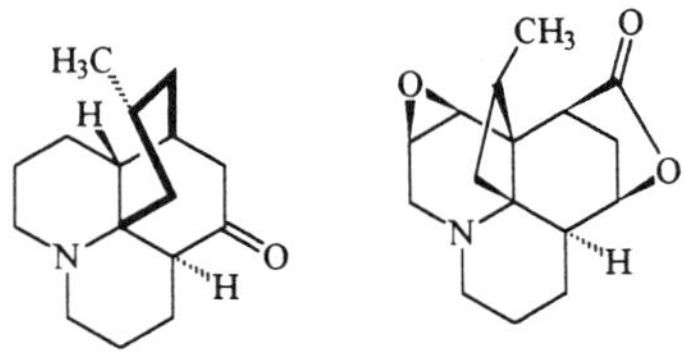

lycopodine annotinine

42 43

and annotinine (**43**) which can be formally considered as QA. These alkaloids are toxic and appear to affect AChR (see Chapter 12). Taken orally, nausea, vomiting, and diarrhea are observed. *Lycopodium* was one of the holy plants (besides mistletoe) of the Druids (the priests of the ancient Celts) and some German ("Druidenkraut") and Welsh ("gras duw" = donation of god) names reflect this old connection. As the wisdom of the Druids was only transferred verbally, we do not know for what purpose extracts of club moss were actually taken.

The club moss produces fine yellow spores in high numbers, which were used as "sulphur vegetabile" as a conspersion in the manufacture of pills. Because the spores are easily inflammable, they were used for fireworks and as photoflash powder. Another application was fingerprint detection.

3.22. *Mandragora officinarum, M. autumnalis* (Family Solanaceae)

The mandrake (Fig. 1) occurs around the Mediterranean and the Near East. Because of massive roots that often have human shapes and the hallucinogenic properties related to tropane alkaloids (mainly L-hyoscyamine but also scopolamine) mandrake has long fueled the fantasies of man and has been mentioned in many old stories.

About 1500 B.C., the Ebers papyrus describes a recipe against intestinal worms, which includes fruits from mandrake. Dioscorides describes its use as a sleep-inducing drug, as a narcotic for surgery, against eye disease, and as an abortifacient (Bellamy and Pfister, 1992). Mandrake was widely known as an aphrodisiac in ancient Greece (the plant was also known as Kirkaia, reflecting its amorous use by Circe?), in pharaonic Egypt (Anchesenamun, the wife of Pharaoh Tutankhamen, is shown to be embedded in mandrake as a symbol of fertility), in the Bible, and even in the *Edda.*

Mandrake became known as a means of warfare: according to the writer Frontinus, the strategist Maharbal (200 B.C.) left his enemy several amphoras of wine that was mixed with extracts from mandrake. When the enemy was drunk and immobilized, Maharbal and his warriors came back and killed their enemy while asleep. During medieval times, extracts from *Mandragora* were used by witches and sorcerers for hallucinogenic purposes (together with *Atropa, Datura* and *Hyoscyamus*).

As hyoscyamine inhibits the muscarinic AChR, most of these activities could be explained rationally (see Chapter 12). More difficult to explain are the various stories concerning the difficulty of harvesting the roots of mandrake. Even Shakespeare refers to this phenomenon in a scene where Juliet takes the sleeping potion. Juliet feared "the shrieks like mandrakes, torn out of the earth, that living mortals, hearing them, run mad."

3.23. *Nicotiana tabacum* (Family Solanaceae)

N. tabacum and related species originate from America and accumulate the pyridine alkaloid nicotine (**44**) and related compounds in all tissues. Nicotine activates the acetylcholine receptor (termed "nicotinic AChR") and is stimulatory for the central and vegetative nervous system at lower doses. Higher doses are rather toxic (50–100 mg of nicotine is lethal for humans), leading to paralysis and death caused by respiratory failure.

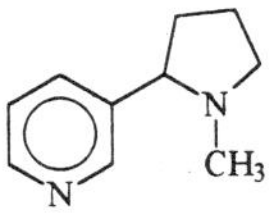

nicotine

44

Leaves of *Nicotiana* were being smoked by Indians for ritual purposes or just for pleasure when Columbus came to South America. Tobacco smoking rapidly became fashionable in western Europe; claims that tobacco had aphrodisiac properties increased its popularity. Despite the obvious dangers of tobacco smoking (heart disease and cancer), smoking is still in vogue in most countries of the world. Although the numbers of smoking men declined in Western countries, increased numbers of smokers who are women can be seen.

Nicotine got its name from Jean Nicot, a scientist of the 16th century; it was first isolated in 1828 and its structure was elucidated in 1893 by Pinner.

Nicotine is an effective insecticide and was widely used for this purpose until the advent of synthetic compounds, such as DDT (Wink, 1993b) (see Chapter 12).

3.24. *Papaver somniferum* (Family Papaveraceae)

Poppy is a Eurasian plant that has played a special role in history. Its latex is rich in benzylisoquinoline and morphinane alkaloids, such as morphine (**45**), codeine (**46**), thebaine, noscapine (**47**), and papaverine (**48**). The dried latex, prepared from unripe fruits, is better known as opium (the alcoholic solution is better known as laudanum). Morphine binds to opioid receptors (targets for endogenous neurotransmitters, such as endorphins), present in the brain and other organs, and has pronounced sleep-inducing, analgesic, and

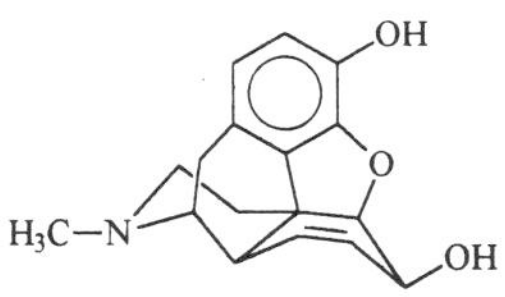

morphine

45

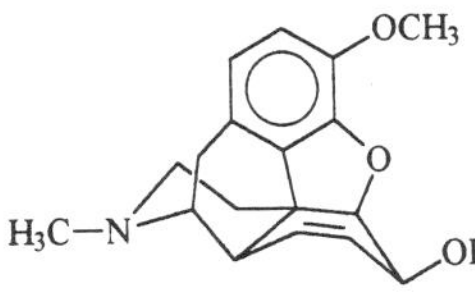

codeine

46

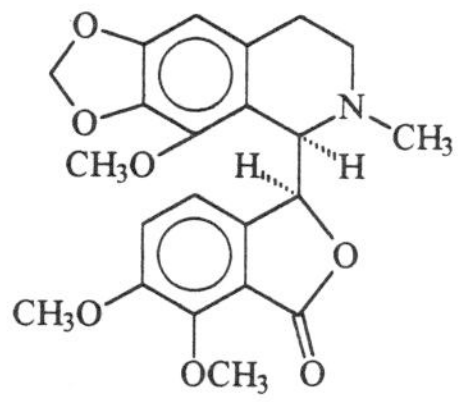

noscapine

47

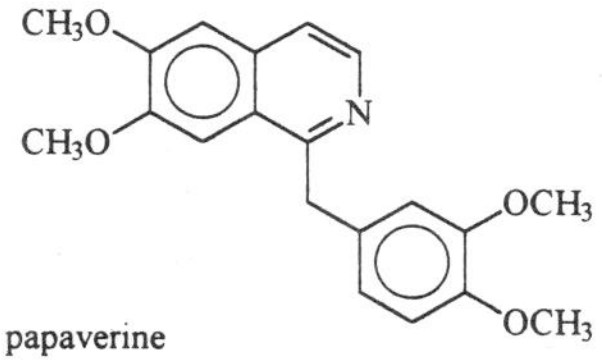

papaverine

48

hallucinogenic properties (see Chapter 12). Morphine is also toxic and 200 mg is lethal for humans. Morphine intoxication shows the following signs: deep sleep and irregular respiration; a pale skin and contracted pupils. Death is observed after 6–8 hours through paralysis, i.e., respiratory failure.

The poppy was known in antiquity and the Sumerians used it as early as 4000 B.C. It was also employed in ancient Egypt, Persia, Arabia, and Greece. An Egyptian recipe of the goddess Isis for the god Ra includes opium ("Xasit") as a remedy against headache. The ancient Greeks and Romans knew the plant well. Poppy was dedicated to the gods of death "Thanatos," of sleep "Hypnos," and of dreams "Morpheus." Diagoras (380 B.C.) and Dioscorides described the harvest of opium by cutting the unripe poppy fruits. The latter gave an excellent description on the medicinal properties of opium (which is more or less even valid today): A pea-sized piece of opium kills pain effectively, induces sleep, activates digestion, silences cough and stomach troubles. Applied on the head (together with rose oil) opium helps against headache, and mixed with almond oil, saffron, and myrrh helps against earache, when brought into the auditory channel. But an overdose of opium will produce extreme lethargy and death. Dioscorides also described the use of poppy seeds (which are rich in oil but almost devoid of alkaloids) for baking bread (as is done today) and the adulteration of opium with latex of *Papaver rhoeas, Glaucium flavum* (this plant was mentioned by Theophrastus as a purgative), or other lactiferous plants. Opium was mixed with hemlock (see *Conium* above) in ancient Greece to execute the condemned (Merk-Schäfer, 1997, Mann, 1992).

More colorful descriptions can be found in the *Odyssey* where opium ("nepenthes") taken with wine made people peaceful and let them forget all sorts of sorrows and pain. When Persephone was kidnaped by Hades to the "underworld," her sorrows were alleviated when Demeter provided her with opium. Virgil reports in the Aeneas that "nepenthes" would even calm the three-headed dog, Cerberus.

The Romans used opium medicinally and for murder in the same way as the Greeks. According to Pliny (50 A.D.) the imperator Nero was an ardent user of toxins and probably murdered his stepbrother Britannicus with a mixture of hemlock and poppy. Although Roman doctors were officially not allowed to dispense opium for people wishing to commit suicide, there are several descriptions from Plautus and other writers suggesting that they did. The Romans were fond of "catapotiae" which had strong effects as a painkiller or sleeping potion because they contained extracts from *Papaver somniferum, Conium maculatum, Mandragora officinarum,* or *Hyoscyamus niger.* According to our present knowledge, the use of these plants was based on pharmacologically active alkaloids, such as morphine, coniine, hyoscyamine, and scopolamine (see Chapters 12 and 18). Hannibal's caring relatives, according to Silius in "de bello punico," regularly helped him to sleep with opium, because he would not sleep otherwise.

Opium in combination with tropane alkaloids (called "theriac") was the main narcotic during the Middle Ages, but was more or less abandoned during the times of the witch-hunt (see *Atropa* and *Datura*, above). It should be mentioned that use of morphine underwent a renaissance in the 20th century. Because of its exceptional pain-killing properties, it is increasingly used to treat cancer patients (see Chapter 18).

Opium smoking was introduced to China in the early 17th century, where it became a major problem. Wars were fought to maintain a free trade in opium. At the beginning of the 20th century, many countries agreed to limit opium production, but only with limited success.

The misuse of opium or morphine is still a problem today, less in the form of "morphinism" (a problem formerly common with doctors) but more with heroinism: As heroin (a diacetyl derivative of morphine, which was first marketed by Bayer Co. in 1898) is more lipophilic, it can pass the blood–brain barrier more easily, and produces stronger hallucinations and addiction (see Chapter 12). Because the market value of opium is still high, poppy is illicitly cultivated in many parts of the world. Opium is processed and distributed through several "dark channels" across the world, from China, Indochina, Afghanistan, and Pakistan to Europe or the United States (Mann, 1992).

In modern pharmacy opium has only restricted application, as a spasmodic and to treat painful diarrhea. The pure compounds morphine (as a painkiller), codeine (to depress cough), and papaverine (to reduce spasm of bile ducts and bronchi) are more important (see Chapter 18.)

Morphine was the first alkaloid that could be isolated as a pure compound by Sertürner in 1806 and obtained its name from the god of dreams, Morpheus. Its structure was elucidated by Gulland and Robinson in 1923 and confirmed by total synthesis in 1952 by Gates and Tschudi (citations in Kollmann-Hess, 1988).

3.25. *Peganum harmala* (Family Zygophyllaceae)

P. harmala is found in southern Europe, the Near East, and the western parts of Asia. *Peganum* seeds and roots are rich in harman alkaloids, such as harmin (**17**), harmaline (**18**), harmalol (**49**), and harman (**50**). As these alkaloids inhibit MAO, an enzyme essential for the metabolism of neurotransmitters, and bind to serotonin receptors, hallucinogenic and psychotropic properties could be expected.

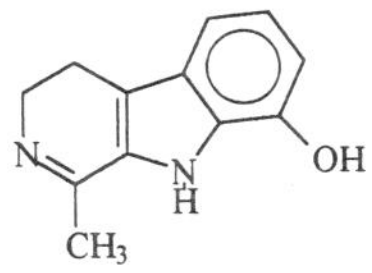

harmalol

49

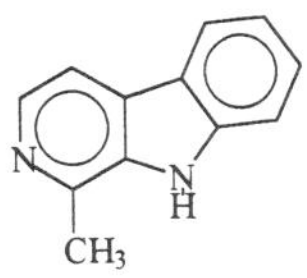

harman

50

Peganum was first used in ancient India and then its utilization as a medicinal plant (e.g., to treat intestinal worms) spread throughout the Mediterranean area. Dioscorides had recognized it and described its different names in Syria, Egypt, and Africa. Interestingly, in Kappadocia this plant was called "moly," the same as the famous antidote mentioned in the *Odyssey*. A clear indication that the hallucinogenic properties of this plant had been explored cannot be found in the old Greek literature. This is surprising because *Banisteriopsis*, a plant from South America that also accumulates β-carboline alkaloids, had been exploited for this purpose by the Indians there (see *Banisteriopsis*).

3.26. *Physostigma venenosum* (Family Leguminosae)

This liana (which can reach up to 15 m in height) occurs in West Africa. Because it is common on the Calabar Coast, it was called the Calabar bean. The seeds contain the alkaloid physostigmine [or eserine (**51**)] which is a potent inhibitor of acetylcholine

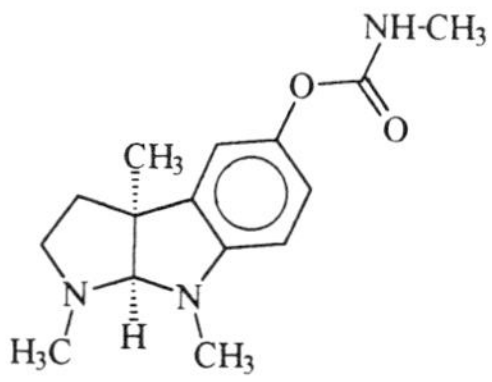

eserine

51

esterase, and thus acts as a parasympathomimetic. It is used medicinally to treat glaucoma, reducing intraocular pressure (see Chapter 18). Another treatment is relevant for patients who produce antibodies against the endogenous AChR, resulting in myasthenia gravis. Physostigmine and its derivative neostigmine extend the biological half-life of ACh (through inhibition of ACh esterase) and the elevated concentrations increase the chance that ACh will bind to the few remaining residual AChR. For treatment of Alzheimer's disease, physostigmine shows positive effects similar to those of galanthamine (another ACh esterase blocker).

In West Africa, the Calabar bean was used in trials by ordeal. For example, the Efik people (according to a report from the missionary W. Daniell in 1840) would feed the condemned person with a certain portion of the Calabar bean ("esere"). Either the person died or was able to vomit and to get rid of the poison. In the latter case, the person was considered innocent and allowed to depart unmolested. The kings of Calabar were usually buried together with hundreds of their people, who were decapitated, poisoned with esere or just buried alive (Mann, 1992). Physostigmine intoxication leads to numbness in the extremities, followed by a prolonged giddiness. Death is caused by paralysis of the heart and by respiratory failure.

Physostigmine was isolated in 1864 by Jobst and Hesse and its structure was elucidated in 1925 by Stedman and Barger (citations in Kollmann-Hess, 1988).

3.27. *Psilocybe mexicana* (Family Agaricaceae)

The mushroom *Psilocybe* occurs in Mexico, although related species have a worldwide distribution, and was known to the Aztecs under the name Teonanacatl. *Psilocybe*

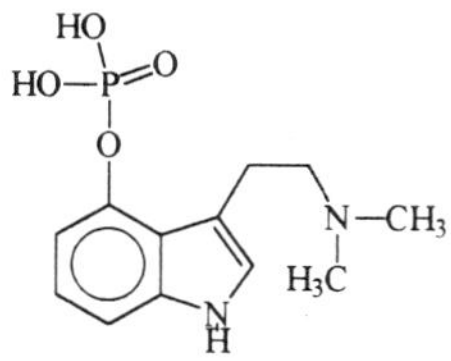

psilocybin

52

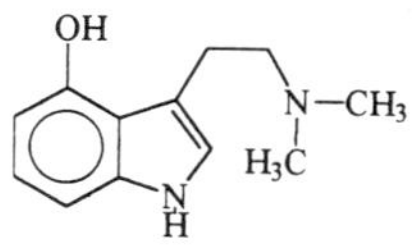

psilocin

53

accumulates the alkaloids psilocybin (**52**) and psilocin (**53**), which can be viewed as analogues to the neurotransmitter serotonin. This small mushroom must have been known to the American Indians for several thousand years, as mushroom-shaped stones (ca. 3000 years old) have been discovered in Guatemala. The alkaloids, which have psychedelic properties similar to those of LSD, have been isolated by A. Hofmann who tested them on himself. Even today, the *Psilocybe* cult is still important in parts of Mexico (Schultes and Hofmann, 1987).

3.28. *Punica granatum* (Family Punicaceae)

Pomegranate, one of the oldest cultivated plants of the Mediterranean basin, originated in south-west Asia. The bark contains simple piperidine alkaloids, such as pseudopelletierine (**54**) and isopelletierine (**55**), which affect AChR (see Chapter 12). Extracts of *Punica* bark can be toxic (< 80 g); symptoms include hypertonia, disturbance of vision, spasms; death is caused by respiratory failure. The garnet-colored fruits are very attractive and obviously stimulated the fantasies of early man.

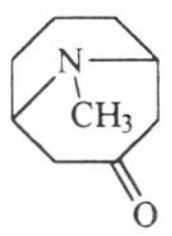

pseudopelletierine

54

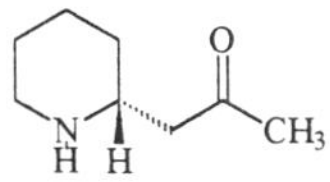

isopelletierine

55

Pomegranate was mentioned in the Ebers papyrus as a drug. Pomegranate was a valuable commodity for Punic traders, and was known to Romans as "Malum punicum." Whereas pomegranate was originally dedicated to the Phoenician mother goddesses Astarte, it was devoted to the goddess Hera and Aphrodite in ancient Greece. The symbols "fertility" and "life" have accompanied the plant to the present. In modern Greece, pomegranate fruits are opened at the entrance of a house on New Year's Day to demonstrate rebirth and fertility (Merk-Schäfer, 1997). It has been speculated that Eve seduced Adam with a pomegranate "apple" and not with *Malus sylvestris*. The kidnaping of Persephone to the underworld became possible after she was seduced by Hades with the aid of a pomegranate.

Dioscorides described several medicinal applications of pomegranate products: Flowers ("Kytinon") had astringent properties whereas the bark "Sidion" was used to kill intestinal worms, such as tapeworm. The latter use was continued until recently when safer remedies became available. The AChR activity of the piperidine alkaloids makes this application plausible.

3.29. *Ruta graveolens* (Family Rutaceae)

R. graveolens occurs throughout the Mediterranean area. It is rich in acridine alkaloids, such as skimmianine (**56**), rutacridine, and graveoline (**57**), and in essential oils,

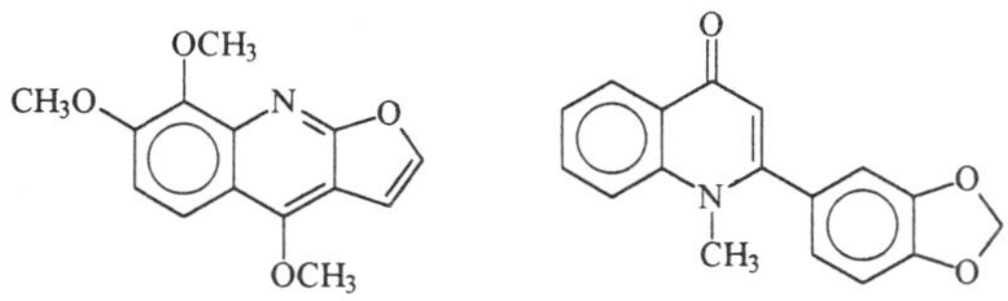

skimmianine graveoline

56 **57**

furanocoumarins, and the flavonoid rutin. Known as "Peganon" in Greece and "Ruta" in Rome, this plant had a special role in ancient mythology. Ruta was dedicated to the goddess Hecate who was recognized as the "daughter of the night" and the "brewer of toxins." In her temples from Colchis to Paphlagonia, she was served by priestesses, called "Pharmakides" who were experienced in toxic and medicinal plants. Dioscorides claims that *Ruta* has the following properties: Seeds and leaves are a general antidote against toxins, including snakebites. Extracts are useful for treating internal infections and in flammations (for example of the bronchi), and external ulcers and eczema. The latter activities can be explained by the cytotoxic and antimicrobial activities of acridone alkaloids (see Chapter 12). In addition, *Ruta* extracts were used for abortion; in this instance both alkaloids and essential oils may be the active ingredients.

The magic touch remained with *Ruta* until medieval times, when the plant was used for exorcism. Together with twigs from the box tree, *Ruta* was employed against witches, the evil eye, and was laid into baby carriages and coffins. *Ruta* was cultivated in every garden, whether those of monasteries or those of simple countrymen. Today *Ruta* has lost its magic touch and is only used for its rutin content to treat edema.

3.30. *Solanum dulcamara* (Family Solanaceae)

Bittersweet woods nightshade is common in Europe and is characterized by its steroidal alkaloids and glycosides such as solanine and soladulcidine (**58**). Whereas they are abundant in green tissue, they are almost absent from red, ripe fruits. Because steroidal alkaloids interact with membrane cholesterol (in the same way as saponins), these alkaloids are cytotoxic and harmful when ingested in higher concentrations. Signs are nausea, vomiting, diarrhea and respiratory problems.

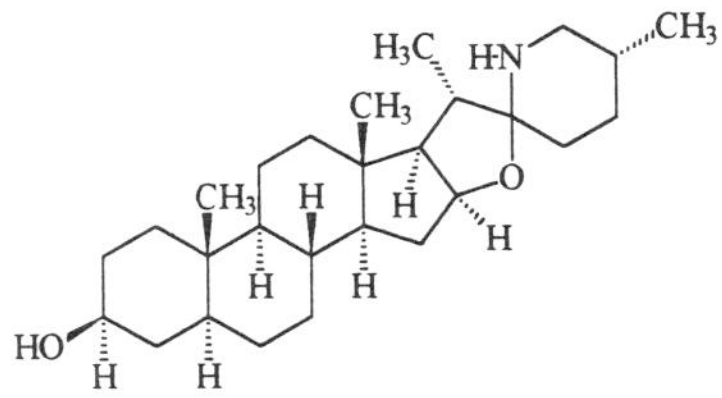

soladulcidine

58

In ancient Greece, a "Strychnos hypnoticon" was recognized, which had red berries; it is suggested that this plant was *S. dulcamara*. Dioscorides considered the fruits as diuretic, the bark as sleep-inducing, and unripe fruits as causing mental disturbance. The diuretic properties are also relevant today: Stipites dulcamarae are used in phytotherapy to treat metabolic disorders accompanied by hindered diuresis.

3.31. *Taxus baccata* (Family Taxaceae)

The yew tree (*T. baccata*) occurs in Eurasia (other *Taxus* species occur in the United States) and produces complex-structured terpene alkaloids, such as taxol (**59**). Taxol and

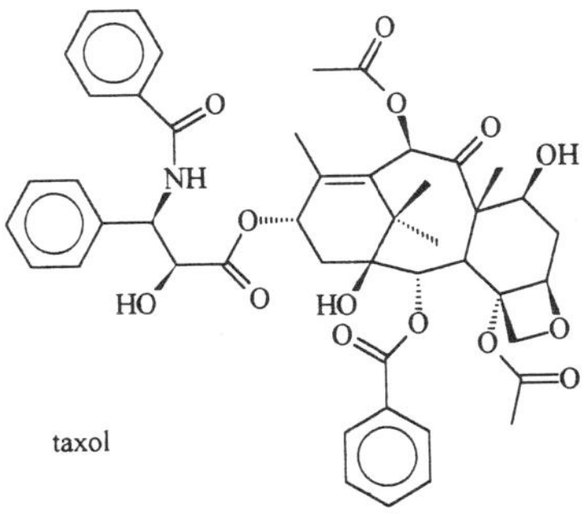

taxol

59

other *Taxus* alkaloids are toxic, e.g., about 500 g of needles (leaves) will kill a fully grown horse. *Taxus* alkaloids bind to microtubules and inhibit their disassembly. Thus, cell division and cell movement are blocked. Because of its antimitotic properties, taxol is presently being introduced in medicine as a potent and promising anticancer drug (see Chapter 12, 18). Intoxication with yew has the following signs: Starting with vomiting, dizziness, and painful diarrhea, respiration is activated at first, but becomes more and more reduced later on. Pulmonary spasms, respiratory and circulatory failure lead to death in coma.

The toxicity of the yew tree was well known: Nikander (200 B.C.) described its use as a poison that renders death very painful because the throat becomes contracted. According to Dioscorides, chicks die by suffocation and humans by diarrhea. Yew toxins were used for suicide; Caesar describes in "de bello gallico" the Celtic chieftain Catuvolcus (53 B.C.)

who preferred to ingest yew rather than become a Roman slave. The ancient Celts employed extracts from bark and needles to prepare potent arrow and spear poisons. Also, the goddess Artemis used arrows poisoned with yew extracts for hunting and murder. According to Pliny, the extracts were first called "taxica" which later developed into "toxica." Dioscorides recognized the plant, which grew in the garden of the goddess Hecate in Colchis, under the name "Smilaxos" and points out that only the Romans call it Taxus. The yew was dedicated to the furies, the Erinyes (the goddesses of vengeance), who pursued evildoers with their anger and punished them with the toxic yew.

Theophrastus stated that animals die when ingesting needles, but that the fruits are not toxic. This is true, the red and fleshy arillus is indeed without toxins, but the seed (which is eliminated intact via the feces) is toxic (see Chapter 12). The wood of the yew tree, which can be up to 2000 years old, has been esteemed for woodworking since ancient times, and is a traditional material for the manufacture of bows.

Another use of extracts made from needles was for abortion, which often had a fatal outcome for both mother and embryo, as 50–100 needles might be lethal.

3.32. *Turbina corymbosa* (Family Convolvulaceae)

Ololiuqui, as *T. corymbosa* (synonym *Rivea corymbosa*) was called by the Aztecs, produces one of the major magical potions and occurs in the American tropics. *Ipomoea violacea* (Family Convolvulaceae) from Central America was used by the Aztecs ("tlitliltzin"), and later by other Indians, in the same way for divinatory activities, in sacrificial ceremonies and even as an aphrodisiac (Schultes and Hofmann, 1987; Mann, 1992; Wasson 1968). Both plants contain ergot alkaloids in their seeds, such as lysergic acid amide (**60**) and *N*-methyl-*N*-hydroxyethyl-lysergic acid amide (**61**). Although these

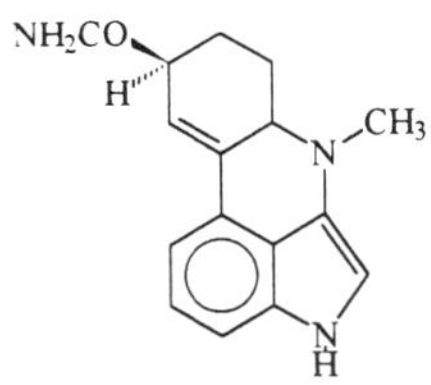

lysergamide

60

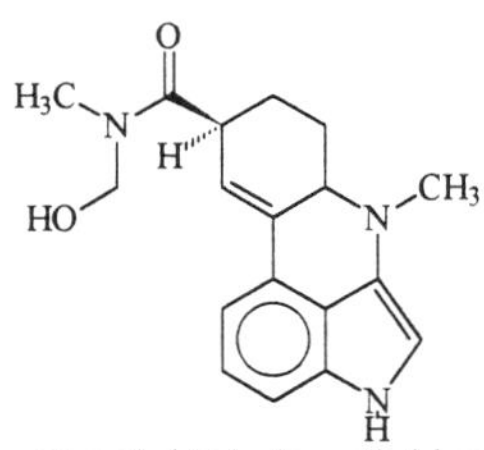

N-methyl-N-hydroxyethyl-lysergamide

61

compounds are less active than LSD, they nevertheless constitute potent hallucinogens, whose activities derive from disturbance of neurotransmission especially at serotonergic, but also adrenergic and dopaminergic receptors (see *Claviceps*, above).

3.33. *Veratrum album* (Family Liliaceae)

V. album occurs in mountainous regions of central and southern Europe and is rich in steroidal alkaloids, such as protoveratrine (**62**), jervine (**63**), or cyclopamine (**64**). These

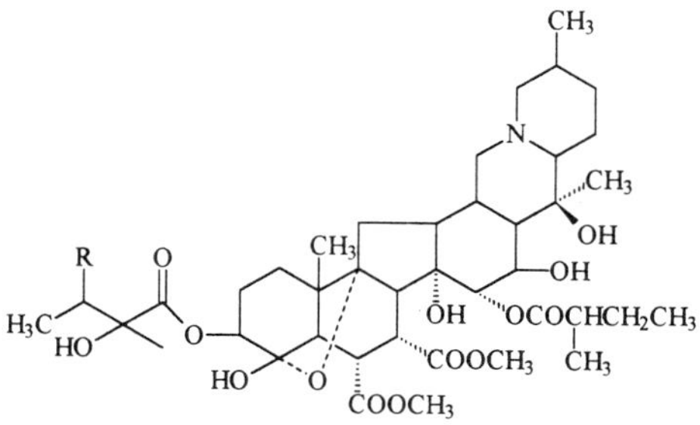

protoveratrine A: R = H
protoveratrine B: R = OH

62

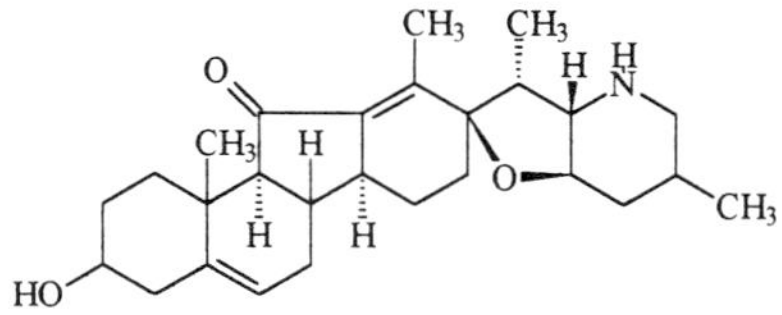

jervine

63

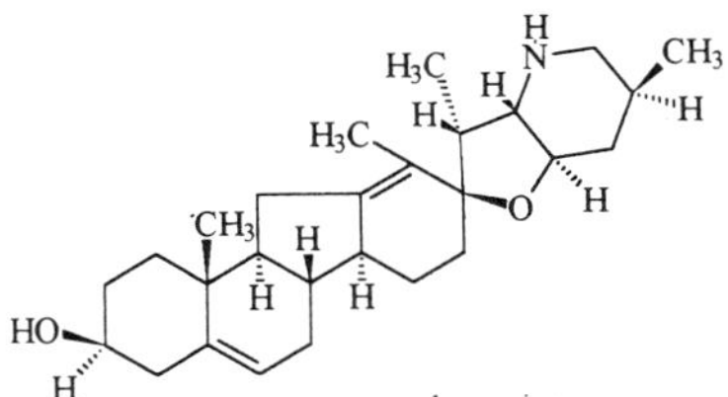

cyclopamine

64

alkaloids activate Na^+ channels and thus cause a reduced signal transduction in neurons. Signs of *Veratrum* intoxication (20 mg of pure alkaloids or 2 g of dried material is lethal) are bradycardia, hypotension, and anesthesia and paralysis. The latter starts with insensitivity of mucosae and of legs and hands.

Known as "Somphia" to the ancient Egyptians, "Veratrum" to the Romans, and "Laginon" to the Celts, veratrum was recognized as a toxic and medicinal plant. In antiquity the plant was used (according to Dioscorides) to initiate emesis, for abortion, and to kill rodents. Powders were known to induce sneezing; and it was still an ingredient of sneezing powder at the beginning of this century (but was later banned because of its toxicity). In folk medicine, extracts from rhizomes were used to treat gout, rheumatism, and trigeminal neuralgia and also against ectoparasites (such as lice). All of these applications are more or less plausible and can be deduced from the biochemical activities of *Veratrum* alkaloids.

The figure of a gigantic Cyclops (having a single central eye) appears in Greek mythology. In the *Odyssey,* for example, Ulysses had to outwit the Cyclops, Polyphemus. Because mythical figures often have a rational base, it is tempting to correlate the ingestion of *Veratrum* alkaloids (for medicinal purposes or via milk from goats that had previously consumed *Veratrum* leaves) with the formation of a cyclopian eye: It can be shown experimentally that the alkaloids jervine and cyclopamine induce this malformation of the embryo if they have been ingested by female animals (e.g., sheep) during the early months of pregnancy (Panter *et al.,* 1990). Of course, another source for this myth could be fossil elephants and mammoths, which were certainly known to the Greeks.

ACKNOWLEDGMENTS

I would like to thank T. Schmeller for drawing the chemical structures and A. Merk-Schäfer for making an unpublished manuscript available to me.

REFERENCES

Major Reviews

Baumann, H., 1986, *Die griechische Pflanzenwelt in Mythos, Kunst und Literatur,* Hirmer-Verlag, Munich.
Bellamy, D., and Pfister, A., 1992, *World Medicine. Plants, Patients and People,* Blackwell, Oxford.
Bisset, N. G., 1989, Arrow and dart poisons, *J. Ethnopharmacol.* **25:**1.
Dioscorides, P., *De Materia Medica libri V. Arzneimittellehre in 5 Bänden,* transl. by J. Berendes, 1970, Wiesbaden.
Gessner, O., 1974, *Gift-und Arzneipflanzen von Mitteleuropa,* Winter Press, Heidelberg.
Hofmann, A., Ruck, C. A. P., Wasson, R. G., 1978, *The Road to Eleusis: Unveiling the Secrets of the Mysteries,* Academic Press, New York.
Joachim, H., (ed.), 1890, *Papyros Ebers,* Berlin.
Kollmann-Hess, M., 1988, *Die "erste Marburger Schule"* (1824–1928), Dtsch. Apoth. Verlag, Stuttgart.
Krafft, F., and Meyer-Abich, A., 1970, *Große Naturwissenschaftler. Biographisches Lexikon,* Fischer, Frankfurt.
Lewin, L., 1984, *Die Gifte in der Weltgeschichte,* 3rd ed., Gerstenberg Verlag, Hildesheim.
Mann, J., 1992, *Murder, Magic and Medicine,* Oxford University Press, London.
Merk-Schäfer, A., 1997, *Heil-, Kult-und Giftpflanzen der Antike: Verwendungen und pharmakologische Wirkungen,* Ph.D. thesis (in preparation).
Pliny, *Naturalis Historia*; *Naturgeschichte,* transl. by G. C. Wittstein, 1987, Nördlingen.
Schmidt, A., 1927, *Drogen und Drogenhandel im Altertum,* Leipzig.
Schultes, R. A., and Hofmann, A., 1980, *The Botany and Chemistry of Hallucinogens,* Thomas, Springfield, IL.
Schultes, R. E., and Hofmann, A., 1987, *Die Pflanzen der Götter,* Hallwag, Bern.
Theophrastus, *Naturgeschichte der Gewächse,* transl. by G. Sprengel, 1871, Darmstadt.
Wasson, R. G., 1968, *Soma: Divine Mushroom of Mortality,* Academic Press, New York.
Wink, M., 1988, Plant breeding: Importance of plant secondary metabolites for protection against pathogens and herbivores, *Theor. Appl. Genet.* **75:**225–233.
Wink, M., 1993a, Allelochemical properties and the raison d'être of alkaloids, in: *The Alkaloids,* Vol. 43 (G. Cordell, ed.), Academic Press,San Diego, pp. 1–118.

Key References

Huxley, A., 1968, *The Doors of Perception,* Chatto & Windus, London.
Panter, K. E., Keeler, R.F., Bunch, T. D., and Callan, R. J., 1990, Congenital skeletal malformations and cleft palate induced in goats by ingestion of Lupinus, Conium and Nicotiana species, *Toxicon* **28:**1377–1385.
Wink, M., 1992, The role of quinolizidine alkaloids in plant–insect interactions, in: *Insect–Plant Interactions,* Vol. IV (E. A. Bernays, ed.), CRC Press, Boca Raton, pp. 133–169
Wink, M., 1993b, Production and application of phytochemicals from an agricultural perspective, in: *Phytochemistry and Agriculture* (T. A. van Beek and H. Breteler, eds.), *Proc. Phytochem. Soc. Eur.* Vol. 34, Oxford University Press, London, pp. 171–213.

Chapter 3

Alkaloids in Arrow Poisons

H. D. Neuwinger

1. INTRODUCTION

A prime example of applied alkaloids—doubtless before medicinal use—are poisons for weapons, especially arrow poisons. The use of poisoned weapons is a fascinating aspect of man's attempts to gain mastery over a hostile environment, to provide food, and to protect himself from his animal and human enemies. Not least important, the biologically active plants yielded drugs and remedies for his diseases. A great deal of anecdotal information has already been lost, but, in spite of their often-prognosticated disappearance, plants containing toxic principles are by no means a thing of the past. Poisons derived from plants continue to be utilized not only in hunting food and against the depredations of wild animals, but also in tribal warfare, especially in Africa. Associated with this, at present, is the extensive use of constituents of the plants providing the hunting poisons as sources of medicines; in Africa these are among the most renowned plants of traditional medicine. The dreaded arrow poisons have provided medicine for effective therapy, or have been applied as tools in research. The best-known examples are ouabain and k-strophanthin for acute cardiac insufficiency, physostigmine for treatment of glaucoma and myasthenia gravis, *d*-tubocurarine as a muscle relaxant in anesthesia, reserpine as an antihypertensive and psychotropic drug, and ajmaline for cases of cardiac rhythm disturbances. Often, however, the therapeutic effects of the plant extract with its complex of compounds may be more beneficial than the effects of an individual compound. Certainly, the highly biologically active arrow poisons are many times more likely to possess therapeutically valuable compounds than are extracts from higher plants selected at random (Spjut and Perdue, 1976).

Sooner or later, arrow poisons will disappear, and therefore there is a real need to research these poisons and to evaluate their potential medicinal use before they disappear completely.

H. D. Neuwinger • Department of Chemistry, University of Heidelberg, D-69120 Heidelberg, Germany (Retired). *Present address:* Hauptstrasse 190, D-68789 St. Leon-Rot, Germany.

Alkaloids: Biochemistry, Ecology, and Medicinal Applications, edited by Roberts and Wink. Plenum Press, New York, 1998.

Two wider reviews exist that deal with arrow poisons, but because of the lack of detailed facts, both are more like general discussions: Perrot and Vogt's book (1913) in France and that of the toxicologist Lewin (1923) in Germany. Both are out-of-date, but they are of historical value in showing the extent to which such poisons have been used. Lewin's work is the more concrete and useful because he investigated the poison activity of many poisoned arrows from Africa, Asia, and South America (Leipzig, Museum of Ethnology). Finally, the literature is vast, very scattered, often incorrect, and of doubtful scientific value. A new, comprehensive review of African arrow poisons is given by Neuwinger (1996).

Peoples in all parts of the world, with the possible exception of Australia and New Zealand, have used poisoned weapons. Early evidence of that use comes from Egypt. In a tomb from the First Intermediate Period (about 2181–2050 B.C.), an arrow was found with largely water-soluble poison, the aqueous extract of which was cardioactive in mice like the cardenolide strophanthin and muscle-relaxant like the alkaloids in curare.

Arrow poisons can be roughly classified as follows:

- *Africa,* with a clear predomination of cardiac poisons (cardenolides)
- *South America,* with almost exclusively muscle-paralyzing (curarizing) poisons (alkaloids)
- *Asia,* mainly with cardiac poisons followed by tetanizing poisons (cardenolides, alkaloids)

With a few exceptions, African and most Asian arrow poisons as cardiac poisons are absolutely deadly; there is no antidote, in contrast to the South American curare poisons, which normally can be survived with the use of antidotes or artificial respiration.

2. AFRICAN ARROW POISONS

The first European killed by an African poisoned arrow was probably Nuno Tristan in 1447, at the mouth of the Gambia river in West Africa. The next centuries showed that poisoned arrows were nowhere else as common or used to such an extent as in Africa. There are poisons containing essentially one plant extract, but more often the poisons consist of a mixture of plant materials with up to a dozen ingredients. Strongholds of such complex poisons are the forest zones. Generally, one or two of the components are the primary source of activity (base poison) and others are added for a variety of reasons: to increase the toxicity and effectiveness of the poison, e.g., by promoting its absorption from the wound; to enable it to adhere better to the arrowhead, but also for magical purposes and perhaps also to mask the real composition of the poison and to divert attention from the real toxic components. Today, most poisons consist of two to four components.

Depending on ethnic origin, there are a great variety of poisons, often with complicated compositions. The unique variety and complexity of African hunting poisons, together with a secrecy that appeared totally insuperable, has made their identification and investigation difficult. Emergence of scientifically useful results started only after the Second World War. In certain areas, even today, the composition and preparation of arrow poison is a closely guarded secret of certain individuals.

Lewin (1923) made a toxicological investigation, using laboratory animals, of many of the poisoned arrows in the Leipzig Museum. He showed for the first time the wide range of cardiac poisons among the African arrow poisons. Little was known, however, about the botanical base and almost nothing about the chemistry of the poisons and their active principles. Since the end of the Second World War, a few detailed and scientifically useful findings on the arrow poisons from certain areas have been published.

In 25 years of research (1968–1993), it has been possible to demonstrate the use of 265 different plants in the preparation of arrow poisons from all over the African continent (Neuwinger, 1996). This comprehensive treatment of the botany, chemistry, pharmacology, and toxicology of the plants also discusses their use in traditional medicine.

2.1. Poisoned Weapons

The bow and arrow is a common weapon. The arrows consist either of hardwood, with the tip hardened in a fire (forest type) (Fig. 1), or they have an iron tip and plant material is wrapped around behind it for better adhesion of the poison (savannah type) (Fig. 2). Experienced hunters never smear the poison on the iron tip but always behind it, so that fresh poison cannot be stripped off when the arrow penetrates the skin. Often the shaft behind the head is notched; when the wounded animal is running through the bush, the shaft easily breaks off and the poisoned tip stays lodged in the wound. The same effect can be achieved by subdividing the arrow; a thin shaft as the holder of the poison is inserted into a hollow in the main shaft (Fig. 3 and 4). When bolting, the animal brushes

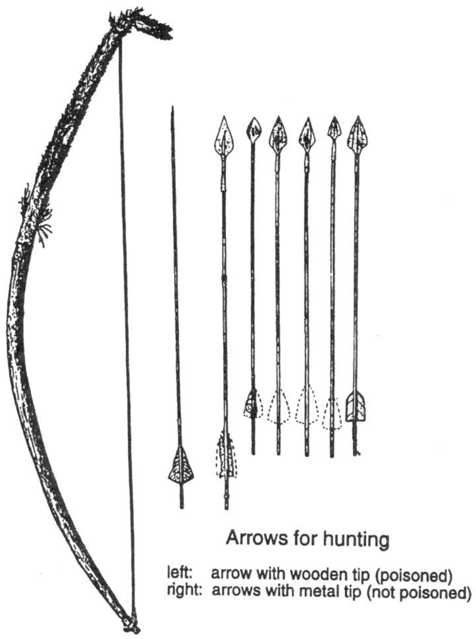

Figure 1. Bow and arrows for hunting of the Mbuti Pygmies in northeast Zaire.

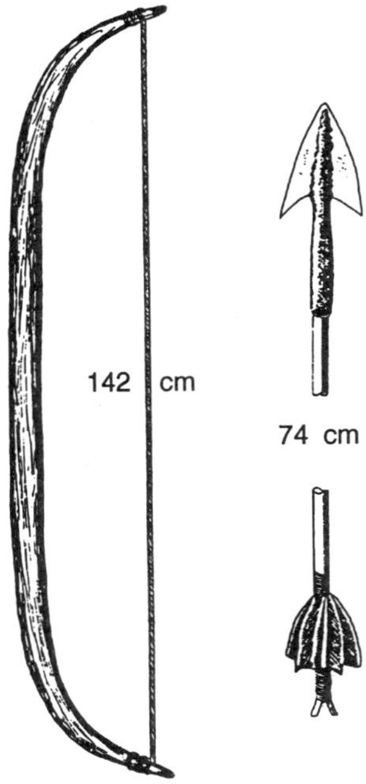

Figure 2. Bow and poisoned arrow of the Tjimba (Namibia/Angola).

off the main shaft which juts out a long way and the thin shaft with the poison stays in the animal and takes effect. In central Africa (southeast Cameroon, southwest Central African Republic, north Congo, north Gabon, and Equatorial Guinea), some tribes, mainly the pygmies, use the crossbow, a very efficient weapon for poisoned arrows (Fig. 5). It was also found among tribes in southeast Asia in former times.

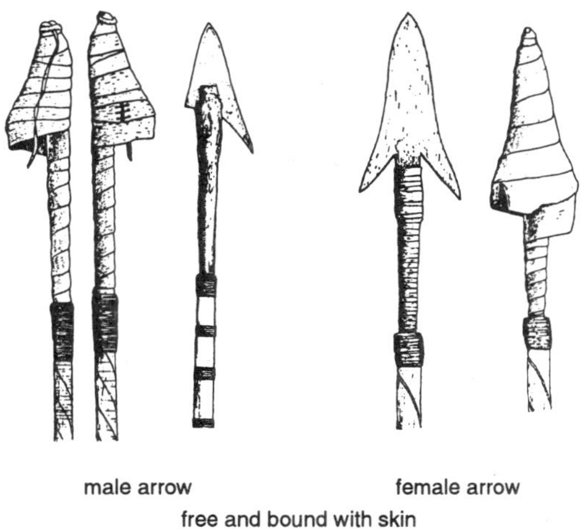

Figure 3. Poisoned arrows of the Hadza (Tanzania, Lake Eyasi).

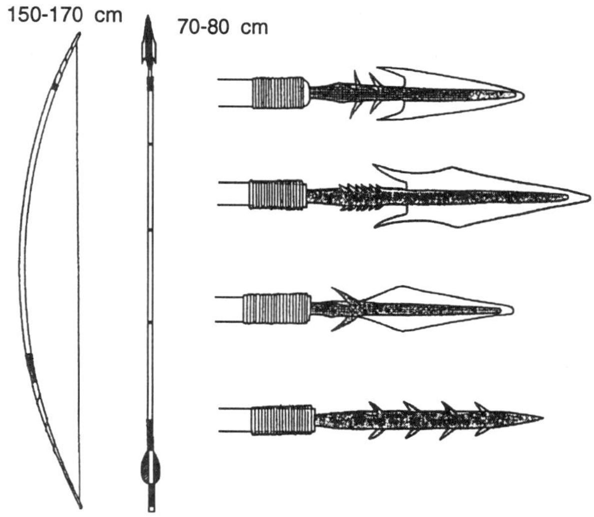

Figure 4. Bow and poisoned arrows of the Bambote (Zaire, Lake Tanganyika).

2.2. Poison Preparation

The secret composition and preparation of poisons is known by the medicine man, magician, herbalist, a certain individual or family and often today by the hunter himself. Poison preparation is considered to be a medicinal-magical art and even today poison-making is sometimes still accompanied by mysterious rituals and taboos. For example, a poison-maker would never prepare poison while ill; in this condition his weakness could be transmitted to the poison and make it weak and ineffective.

Three principal methods of poison preparation are usual:

1. *Boiling in water (aqueous decoction).* The most common method of poison preparation is boiling the ground plant material in water. Often the boiling time is long, from some hours to a few days, with continuous addition of water. After concentration, a tarry mass remains; it is smeared in a thick layer on (wooden tip) or behind (iron tip) the arrowhead. The technique of preparation ranges from primitive boiling in an open pot to a complicated poison extraction using specially built equipment.
2. *Pounding of fresh ingredients and addition of glutinous sap.* This method of poison extraction is mostly restricted to oil-rich plant parts, e.g., seeds and roots. The addition of glutinous sap helps to hold the poisonous mass together and to adhere it to the arrowhead. The adhesive power of the latex of cactiform *Euphorbia,* an almost constant component of arrow poisons, is especially strong. *Euphorbia* species are highly toxic on their own account, and are found almost everywhere.
3. *Squeezing out fresh plant material.* This method for wooden-tipped arrows is typical of the forest tribes, especially the pygmies. There is a large stationary press and a small mobile press for poison-making during the hunt. The plant

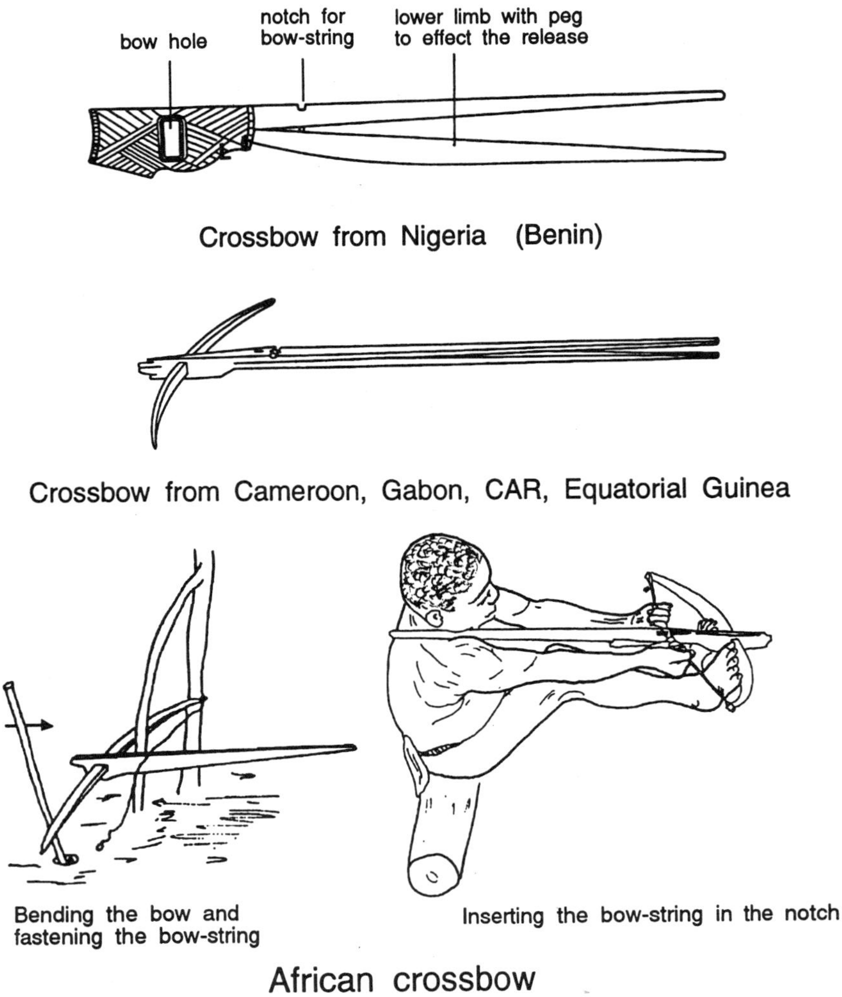

Figure 5. The stock is split laterally throughout the greater part of its length. By bringing the upper and lower limbs together using a squeezing action, the square-sectioned peg fixed to the lower limb forces out the bowstring as it rises in the notch.

material must be fresh and rich in sap: thin twigs, stem bark, and roots of young bushes and trees, lianas, and herbaceous plants.

2.3. Plant Sources and Their Active Principles

Almost all of the primary active ingredients come from plants, only a few from animal sources (toads, snakes). Only the bushmen of the central Kalahari desert use exclusively an animal poison. It is prepared from the highly toxic larvae of several leaf-eating beetles (*Diamphidia, Polyclada*) and the larvae of their flesh-eating parasites

Figure 6. Poison packed in a cigar-shaped parcel (12 × 2 cm) for sale (Kenya).

(*Lebistina*), found deep in the ground beneath certain host bushes and trees (Kao *et al.*, 1989; Neuwinger and Scherer, 1976; Neuwinger, 1996). Bushmen from other areas mix the larvae with plant extracts (Neuwinger, 1996).

In African arrow poisons, cardiac poisons clearly dominate. At least 80% of the poisons are based on cardioactive components, mostly cardiac glycosides. The main plant sources are *Acokanthera, Parquetina, Strophanthus* (APS poisons, Fig. 7), and, from time to time or in special cases, *Adenium, Mansonia, Calotropis, Pergularia, Corchorus,* and *Erythrophleum,* the last an exception in that it contains cardiac alkaloids.

Two other large groups of poisonous principles of African hunting poisons are triterpenoids, especially saponins (many plant genera), and diterpenoids, e.g., the highly irritant esters of diterpenoid alcohols (*Euphorbia, Gnidia, Jatropha*). Most saponins are able to cause hemolysis of erythrocytes using only a few micrograms per milliliter. They also have other toxic properties that do not always run parallel to their hemolytic properties. Saponin-bearing plants have been extensively employed as fishing poisons, e.g., plants with diterpenoid esters (*Euphorbia* species) (Neuwinger, 1994). *Euphorbia* latex is highly esteemed for its glutinous properties. Beyond that, the extremely irritant diterpene esters present may promote the absorption of the toxic principle through stimulation of peripheral blood circulation, in addition to being highly toxic in their own right. *Capsicum frutescens* has similar properties; its fruit or seeds are found in many arrow poisons throughout Africa. Its pungent capsaicinoid complex (related vanillylamides of carboxyl acids with capsaicin as the major compound) not only causes strong irritation externally and inflammation around the arrow wound but is also toxic, parenterally administered (LD_{50} = 1.1 mg kg^{-1} guinea pig i.p.).

Plants containing cyanogens often are used in the preparation of arrow poison; they release volatile hydrogen cyanide when the plant cell is damaged, but it does not survive the poison-making process. The same thing occurs with most animal venom, e.g., snakes, scorpions, or insects; they contribute nothing to the toxic properties of the final product; they may be highly toxic in laboratory tests but they do not survive the usual long-lasting boiling process. Finally, such venoms are few in number. Only toad toxins (bufadienolides, similar to the cardenolides) do survive the boiling process and are active in the finished poison. Many other plant toxins used have a variety of activities, but their toxic effects are long term rather than acute, e.g., sesquiterpene lactones, iridoids, pyrrolizidine alkaloids, tannins.

It is unclear, and difficult to evaluate, to what extent the adjuvant plants contribute to the activity of the poison. It is known that the finished poison is often essentially more active than the concentrated extracts of the corresponding plants; this may be related to possible interactions between the various compounds during the elaboration process of the

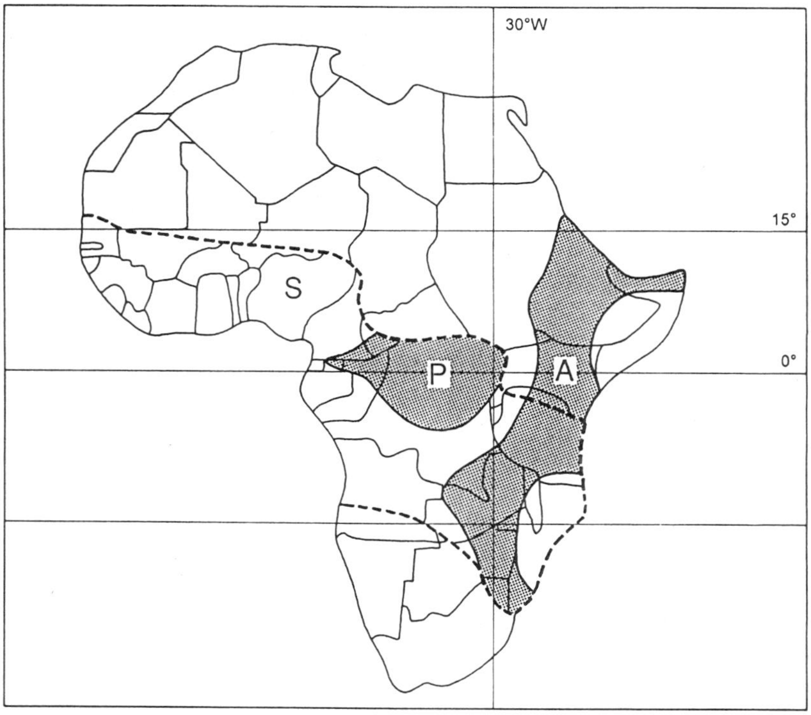

Figure 7. Distribution area of the APS (*Acokanthera–Parquetina–Strophanthus*) hunting poisons in Africa.

poison. The restriction of this review to alkaloid-bearing plant poisons thus gives a fairly one-sided and distorted picture of African arrow poisons. With a few exceptions, alkaloid-bearing plants play a dominant role as adjuvants; several could well be—and some, indeed, are—base plants of a poison.

Strychnos species are a major group of arrow poison adjuvants, their roots being used especially in the equatorial forest regions of central Africa.

The Nyambo, a small tribe of hunters in Rwanda living along the Akagera river which forms the border with Tanzania, use one of the most surprising plants in preparing arrow poison. Their poison is famous because of its powerful action. Base plants are the cardenolide-bearing *Acokanthera schimperi* and the alkaloid-bearing *Strychnos usambarensis* (Angenot, 1971). The root bark of the latter plant, widespread but until that time little noticed, provided a chemical and toxicological surprise: it was found to act like a high-quality curare poison; indeed, alkaloids previously only known in South American calabash curare and *Strychnos* species were obtained from this African *Strychnos* and the poison derived from its roots: C-dihydrotoxiferine, C-curarine, C-calebassine (see structures **50, 53,** and **56,** respectively), and a new one, which was named afrocurarine (**1**) because of its occurrence only in "African curare." The strongly curarizing alkaloids are bis-quaternary dimeric perfectly *symmetrical* indole bases. In contrast, afrocurarine is the first example of a bis-quaternary dimeric *asymmetrical* indole base; it is composed of a strychnine part (type dihydrofluorocurarine) and a corynane part joined via an enamine

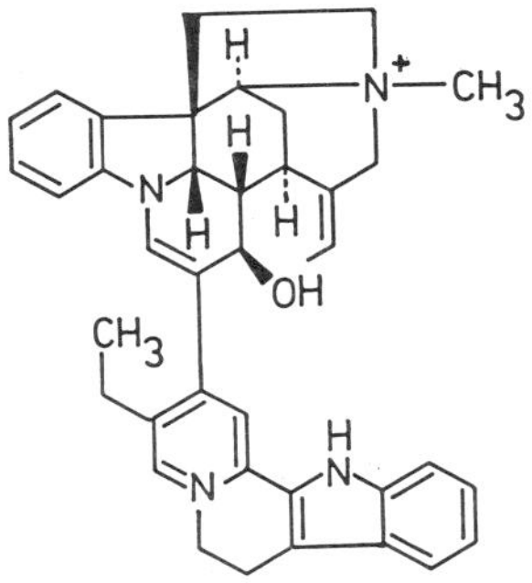

1

bridge. The four bases are potent neuromuscular blocking agents with the final result being complete paralysis of the skeletal or striated muscle apparatus. Afrocurarine was found to be less potent than the other three, probably because of its plane, asymmetrical structure and its stereochemistry. For pharmacology and mechanism of action of these compounds, see Section 3. C-curarine is the most active alkaloid, followed by C-dihydrotoxiferine, C-calebassine, and afrocurarine; the last has only 5% of the activity of (+)-tubocurarine (**2**). In addition to the high toxicity of the four alkaloids as quaternary bases,

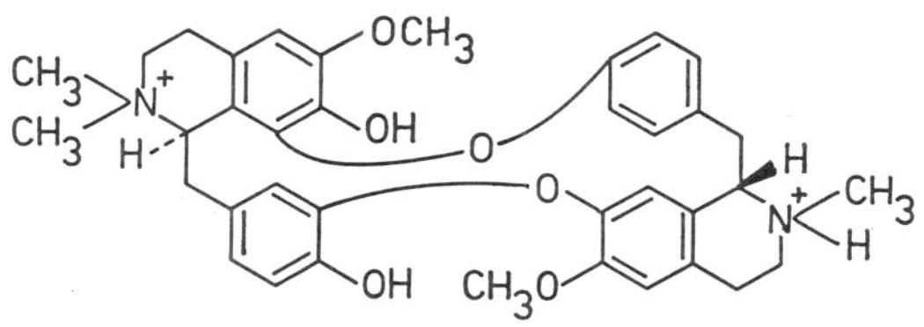

2

they are easily water soluble and therefore the plant is equally suited to the poison-makers as a plant with easily water-extractable cardenolide-glycosides.

Currently, *S. usambarensis* is chemically the best-known *Strychnos* species; more than 60 alkaloids are indicated and 37 characterized. Beyond the 4 curare alkaloids, the roots of the plant contain many other tertiary and quaternary bases with multiple activities. The main representative of the tertiary root alkaloids is usambarensine (**3**) with an atropine-like activity. The other plant parts used in the Nyambo poison, especially the

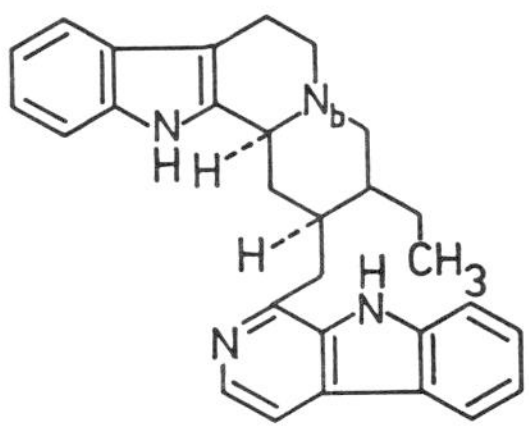

3

leaves, contain other alkaloids which, with a single exception, are tertiary dimeric asymmetrical bisindole alkaloids of the corynanthe type (usambarine and usambarensine derivatives). These alkaloids do not differ markedly in their structures, and have different biological activities than do the root alkaloids. The exception is akagerine (**4**), which

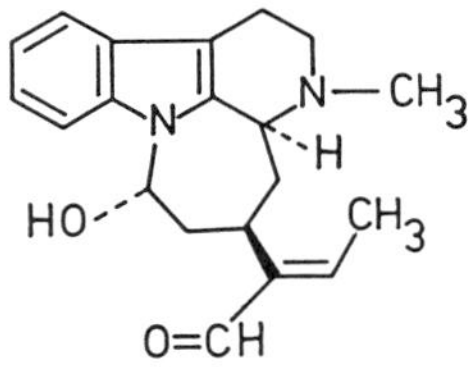

4

belongs to a new type of alkaloid. Only two quaternary bisindoles have been found in the leaves to date; they correspond to the N_b-methyl derivatives of 10- and 11-hydroxyusambarine. In pharmacological studies akagerine was reported to be a convulsant agent, but was 100 times less potent than strychnine (**5**). A number of the leaf alkaloids possess high

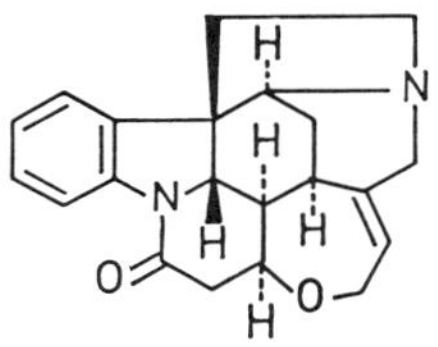

5

in vitro activity against pathogenic protozoa (amoebiasis, malaria) and can exert a strong cytotoxicity on cancer and noncancer cell lines.

Another *Strychnos* species, *S. camptoneura,* whose root bark we have found as an arrow poison in the southwest of the Central African Republic, has toxicological analogy with *S. usambarensis* roots. It is not clear which of the ten compounds isolated to date is responsible for the strong muscle-paralyzing activity, but camptoneurine (**6**) may contribute to the effect.

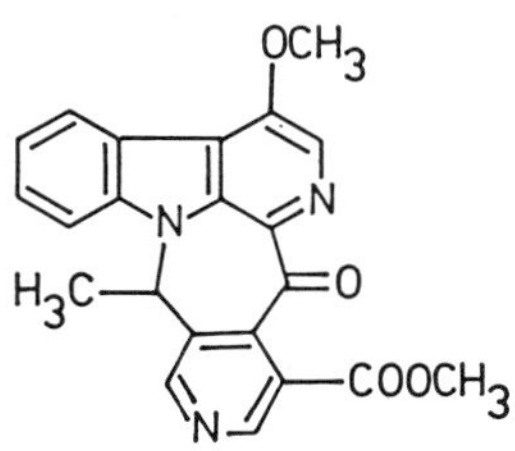

6

Triclisia dictyophylla (Menispermaceae), an arrow poison plant from central Africa, has curarizing activity similar to that of the *Strychnos* species; 25 mg kg^{-1} of raw quaternary alkaloid mixture of the stem bark caused death in rats after 3 min. *Triclisia* species have been found to be excellent sources of bisbenzylisoquinoline alkaloids with the quaternary bisbenzylisoquinoline *N,N'*-dimethylphaeanthine (**7**), the main compound of this type in

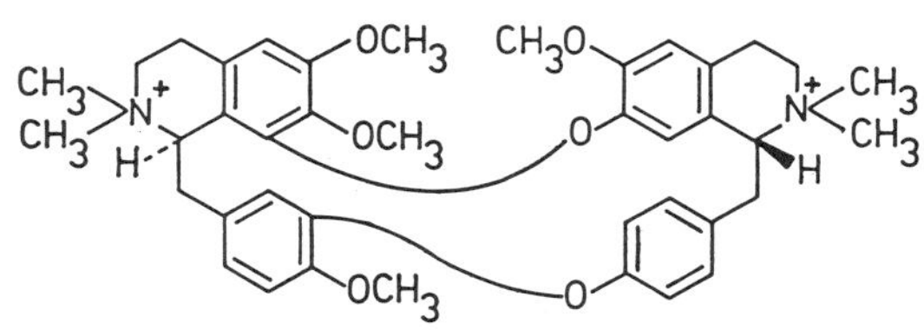

7

T. dictyophylla, responsible for the curarizing effects (LD_{50} = 2.63 mg kg^{-1} mouse i.v.). It has about 9% of the activity of *d*-tubocurarine (**2**) (LD_{50} = 0.14 mg kg^{-1} mouse i.v.); 25 mg kg^{-1} i.p. kills rats in 1 min by asphyxia. Another Menispermaceae, *Cissampelos mucronata,* a thin twiner or trailer, which enters into the preparation of *Strophanthus*-based arrow poison in northern Nigeria, also produces (root extract) curariform paralysis in animal muscle preparations. The root contains a number of tertiary and quaternary bisbenzylisoquinoline and protoberberine alkaloids. *dl*-Hayatine (**8**), *dl*-hayatinine (**9**), *d*-isochondodendrine, and *l*-bebeerine (*l*-curine), the major representatives, have a tubocurarine skeleton. Some quaternary alkaloids are responsible for the curarizing action.

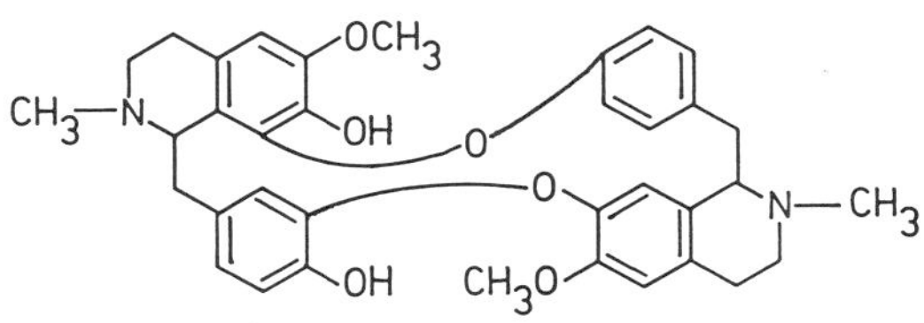

8

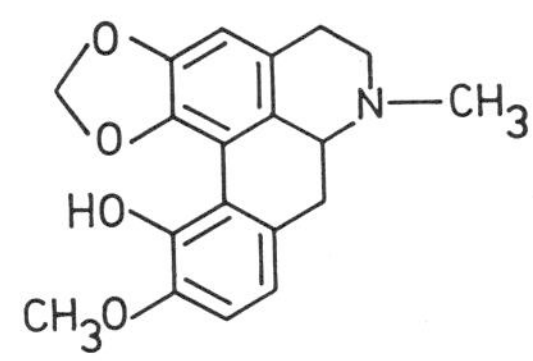

9

In India, the quaternary *N,N'*-dimethylhayatine (hayatine methiodide) was tested clinically in 100 cases as a substitute for *d*-tubocurarine. At the same dosage the duration of effect is about the same while the strength of effect is about half that of *d*-tubocurarine (Pradham *et al.*, 1964).

The aporphine bulbocapnine (**9**), isolated from stem bark and leaves, acts peculiarly on the CNS. It induces catalepsy, the so-called "bulbocapnine rigidity" movement disorder, while maintaining the muscle tonus and the startle reflexes. It also acts hypnotically. Death results from paralysis of the muscles of respiration. It is used therapeutically in the treatment of tremor and of Ménière's disease.

The active principles from *S. icaja,* a *Strychnos* species, are quite different from those of *S. usambarensis.* The red root of this large liana forms an important ingredient of arrow poisons in the Central African Republic and Zaire. *S. icaja* is, however, primarily an ordeal poison and at present is still used in the Central African Republic, Gabon, Congo, and Zaire. Strychnine (**5**) (LD_{50} = 0.47 mg kg^{-1} mouse s.c.) and 12-hydroxystrychnine (LD_{50} = 0.56 mg kg^{-1} mouse s.c.) are responsible for the high toxicity of the roots. Strychnine is always accompanied by minor alkaloids, but with only a weak tetanizing activity, e.g., brucine and α- and β-colubrine.

Strychnine belongs to the analeptics. It acts selectively as a competitive agonist and blocks postsynaptic receptors of the inhibitory transmitter glycine in the spinal cord and motoneurons. It may also have a potent presynaptic action preventing release of the inhibitory transmitter, thus, strychnine (like tetanus toxin) acts by inhibition of inhibitory actions: the excitatory synapses can then exert their unrestrained action, which results in hyperexcitability and tonic convulsions and finally death through respiratory or spinal paralysis. With intravenous injection, the onset of convulsions in the case of brucine is almost instantaneous and death occurs in tonic seizures, but its toxicity is distinctly lower (LD_{50} = 12 mg kg^{-1} mouse i.v.). Recent investigation clearly showed that the aqueous extract of the root bark has, besides the predominantly convulsive effect, a conflicting effect as well, which paralyzes the neuromuscular junction and the cardiac muscle. Indeed, a muscle-relaxing compound has been isolated from the roots: bisnordihydrotoxiferine. This compound is also a sedative, having a marked depressant effect on the CNS. The aqueous quaternary alkaloidal extract of *S. icaja* root bark exhibited a high cardiotoxicity. The leaves of *S. icaja* contain quite different alkaloids with lesser activity (derivatives of icajine and vomitine). Only pseudostrychnine (3-hydroxystrychnine) acts in a similar but weaker way to strychnine.

Another Loganiaceae used as an arrow poison plant in central Africa is *Mostuea* (*brunonis* and *hirsuta*). *M. brunonis* is reported to contain the highly toxic oxindole alkaloid gelsemicine (**10**) (LD_{50} = 0.05 mg kg^{-1} rabbit i.v.) as a major compound of the

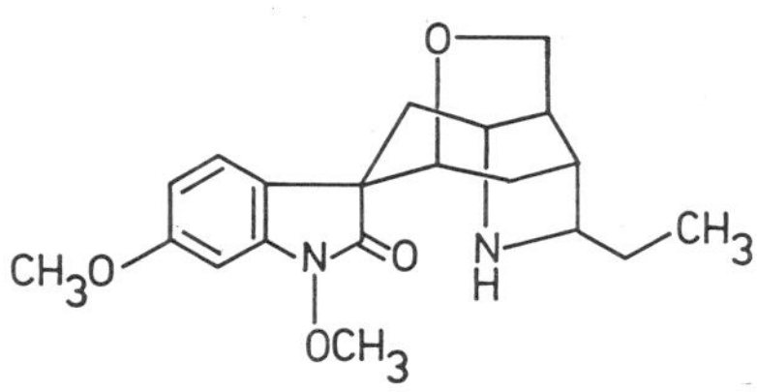

10

leaves and twigs. It affects the respiratory center alone, has no effect on the vagus, but depresses the motoneurons of the brain and spinal cord, resulting in a generalized muscle weakness. Respiratory arrest after fatal doses is not caused by paralysis of the center but is attributable to paralysis of the spinal motoneurons innervating the respiratory muscles. The potent cholinesterase blocker sempervirine, an anhydronium base with the yohimbine nucleus, has been isolated from the roots.

A very interesting ingredient of hunting poisons (Nigeria, Zambia) is *Securidaca longepedunculata.* The root juice, however, is more commonly used as an ordeal poison, especially in the Central African Republic and southeast Zaire. This Polygalaceae is a common suicide poison and probably the best-known African abortifacient. The reputation of its roots as a snakebite antidote is legendary. In 1980, this activity was scientifically confirmed and the active principle was identified as a protein similar to that found in the venom of the spitting cobra (*Naja nigricollis*). The strong-smelling roots contain methylsalicylate and toxic triterpenoid saponins and the stem bark contains alkaloids. The major alkaloid isolated from the bark is the tricyclic securinine (**11r**) with an α,β-unsaturated butenolide ring as in cardenolides. Securinine is highly toxic (LD_{50} = 6.23 mg kg^{-1} mouse i.v.); it has two conspicuous characteristics: an extremely rapid

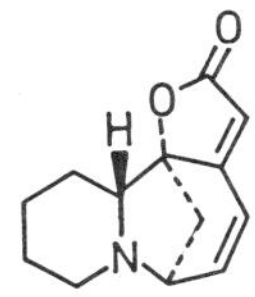

11

initiation of action both orally and by injection and an unusually rapid inactivation in the organism. In mice and rats killed immediately after parenteral administration, 5–30 min after an i.p. injection, no securinine could be detected. It acts primarily on the spinal cord, causing an enhancement of reflex activity and increase in muscular tone in animals; it also influences the functions of the autonomic nervous system. In toxic doses, securinine induces powerful tonic convulsions involving all skeletal muscles, similar to strychnine convulsions. The dosage causing convulsions is very close to that causing death (respiratory arrest). In rats, i.v., it acts 25 times more strongly than strychnine as a CNS stimulant. Surprisingly, ergot alkaloids were found in the roots in 1992, among them elymoclavine (**12**). Such alkaloids are very rare in higher plants. Elymoclavine belongs to the "excitor"

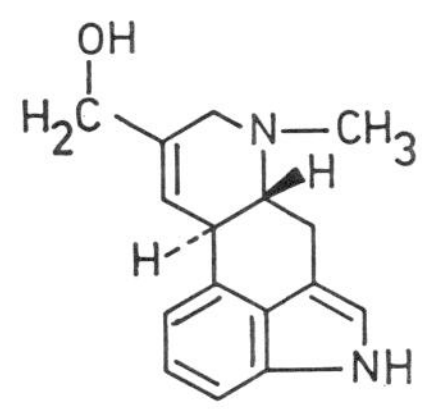

12

group of ergot alkaloids which cause central stimulatory syndromes consisting of sympathetic excitation like those of LSD; it is the most potent stimulant in this group. Dihydrogenation, however, converts excitation to depression: dihydroelymoclavine belongs to the "inhibitor" group and causes mainly central depression.

Derivatives of securinine, e.g., virosecurinine, norsecurinine, allosecurinine, are to be found in a plant of quite another family, the Euphorbiaceae *Flueggea virosa*, an ingredient of a *Strophanthus*-based arrow poison in Nigeria and Benin. The toxicity of allosecurinine, at least, is lower than that of securinine. Some of these derivatives demonstrated significant cytotoxicity in tumor cell lines.

Another commonly used arrow poison plant, which has a good reputation in Nigeria for the treatment of snakebite, is *Schumanniophython magnificum* (Rubiaceae). For this purpose the stem bark juice is applied to a small incision made at the site of the snakebite. Recent studies have demonstrated that polar extracts of the stem bark show activity against cobra cardiotoxin, both *in vivo* and *in vitro*, but not against a curaremimetic neurotoxin component. Crude extracts caused contracture of muscle accompanied by inhibition of the cardiotoxin effect. Ten related chromone alkaloids based on the chromone noreugenin were isolated from root and stem bark. The direct contracture was related to the alkaloid *N*-methyl-schumannificine (**13,** N-Ch_3), while the anticardiotoxin effect was caused by a peptide that is similar in amino acid composition to the cardiotoxins present in snake venom. Another alkaloid, schumannifoside, was also reported to reduce the lethal effect of black cobra venom through a chemical inactivation reaction.

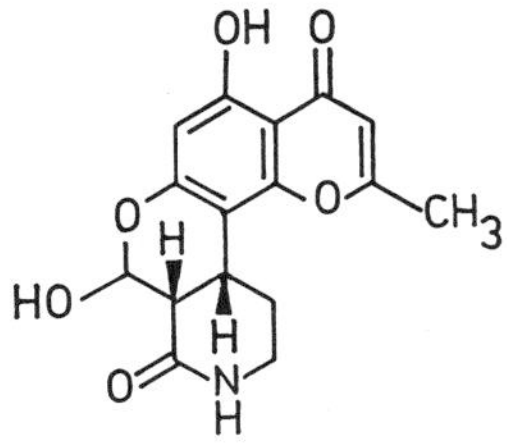

13

In Tanzania, west of Kilimanjaro, Masai-Ndorobo hunters prepare an arrow poison from the herbaceous Compositae *Vernonia hildebrandtii* and the Rutaceae *Zanthoxylum chalybeum*. The former is of special interest because it contains an alkaloid of still-unknown structure which in laboratory animals, especially rabbits s.c., caused a continuous decrease in cardiac contraction, toxic vagal stimulation, profound movement disorders as far as paralysis similar to that of a curarized animal, and death. A narcotic stage lasting several hours was also observed. *Zanthoxylum (Fagara)* species are often found as ingredients of arrow poisons, e.g., *Z. gilletii* in the Ivory Coast and Cameroon, *Z. chalybeum* in Rwanda and Tanzania. *Zanthoxylum* species proved to be rich sources of tertiary and quaternary alkaloids, primarily isoquinolines of various structure types (derivatives of 1-benzyltetrahydroisoquinoline, benzophenanthridines, aporphines, and, to a lesser extent, quinolines). They exhibit a wide range of biological activity and some of them are without doubt highly toxic; the quaternary alkaloids, in particular, are responsible for the toxicity. The simple phenylalkylamine candicine (**14**), perhaps the most toxic *Zanthoxylum* alka-

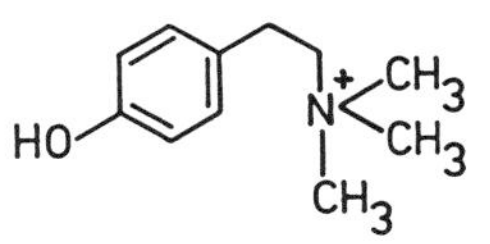

14

loid, causes a curarelike effect in dogs and frogs i.v. at higher doses (6 mg kg^{-1} dog i.v.), with death by asphyxia. The aporphines magnoflorine (**15**), *N*-methylisocorydine (menisperine) (**16**), and *N*-methylcorydine (**17**) also act as neuromuscular blocking agents antagonized by cholinesterase inhibitors such as physostigmine. The stem bark alkaloid

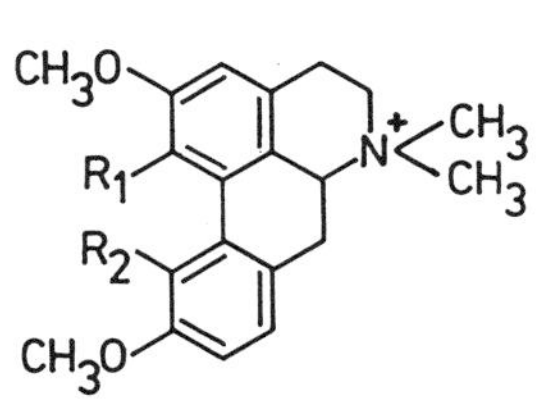

(**15**) $R_1 = R_2 = OH$
(**16**) $R_1 = OCH_3$, $R_2 = OH$
(**17**) $R_1 = OH$, $R_2 = OCH_3$

berberine also antagonizes the curare effect and shows many pharmacological activities. The quinoline alkaloid quinidine exerts exclusively negative actions on the heart.

Many of these tertiary and quaternary isoquinoline and quinoline alkaloids are also found in the Annonaceae *Annickia polycarpa* (*Enantia polycarpa*), an adjuvant to *Mansonia altissima*-based hunting poisons of the Guéré tribes in the southeast Ivory Coast.

Three other Annonaceae also make a significant contribution to African arrow poisons: *Annona senegalensis* in Nigeria (base: *Strophanthus hispidus*), *Pachypodanthium staudtii* in the Ivory Coast (base: *Mansonia altissima*), and *Cleistopholis glauca* in Zaire (base: *Parquetina nigrescens*). Almost all of the alkaloids contained in them possess an isoquinoline-derived structure. In *Cleistopholis* species especially, 1-bisbenzylisoquinolines have been found (Leboeuf *et al.*, 1982). A 95% alcoholic extract of the stem bark of *Annona senegalensis* was fatal to rats with an MLD of 100 mg kg^{-1} i.p., death occurring in 3 hr. The stem bark of *Pachypodanthium staudtii* showed significant toxicity in rats: 500 mg kg^{-1} of crude bark i.p. killed rats within 30 min. Most of its alkaloids are toxicologically unknown.

In the Central African Republic we found a poison hitherto only known as a poison in native judicial procedure: *Physostigma venenosum*. The bean-shaped seeds, known as "Calabar beans" or ésere (both the ordeal poison and the seeds are called ésere in Calabar) is one of the most notorious trial by ordeal poisons in west Africa, although the area of use is relatively small. It is mainly found in southeast Nigeria and neighboring Cameroon. Although strictly forbidden, the trials were still being held occasionally in the 1970s. Presently, the poison is mostly given to an animal as a representative of the accused.

The principle in this type of ordeal was as follows: The poison was given to a suspect to drink or eat. He was assumed to be innocent if he could reject, i.e., vomit, the poison. If the poison was retained, the person was considered to be guilty and was allowed to die from the effects of the poison or was disposed of in some other way. On the other hand, the fetishists who conducted the trials were by no means scrupulous; they had quite a number of "variables" to hand and could control the strength of the poison and therefore could direct the results of "ordeals" to a great extent. (For further information see Neuwinger, 1996.)

In 1952–1953, the anthropologist Simmons (1956) researched in Calabar and produced the following more recent description of the ésere trials:

> The Efik believe that the ésere or Calabar bean possesses the power to uncover witchcraft and to destroy it. A suspect is given 8 ground beans in water to drink. If he is guilty, his mouth trembles and foam flows out of his nose. He is innocent if he raises his right hand and vomits everything up. If the poison still has an effect in spite of his innocence, an antidote is given in the form of a decoction of excrement mixed with water which has been used to wash female genitalia. If someone dies from the ordeal, the eyes are removed, a perforated clay pot is inverted on the head and the corpse is normally thrown into the forest, or it is buried head-down or burned and the ashes buried. By these methods it is hoped to stop the spirit of the witch returning. If the trouble continues in a village, however, suspicion falls on a recently-dead relative as an undiscovered witch. The family exhumes the corpse, in order to determine the extent of the putrefaction. If the corpse appears fresh with very little decay, then the dead person is seen as a witch. The corpse must then be burned.

P. venenosum is a further African poison plant that achieved medicinal importance and is a prime example of the transformation of an extensively used deadly poison to a highly beneficial medicine and an important tool in experimental pharmacology. Up to now, seven indole alkaloids have been reported as occurring in the seeds. They are understood as tricyclic methylcarbaminates. The main alkaloid present in the beans was first isolated in a pure form in 1864 in Germany and called physostigmine and 1 year later in France as eserine.

Physostigmine (**18**), or more correctly (−)-physostigmine, is an indirect-acting para-

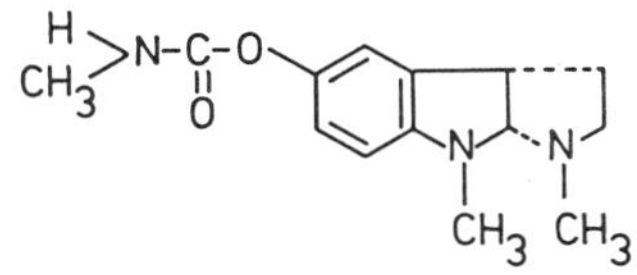

Physostigmine

18

sympathomimetic, a reversible cholinesterase inhibiting agent. It is the strongest cholinesterase blocker known. Alkaloids with such an activity are very rare among plants. As a tertiary amine it can penetrate the blood–brain barrier and thus produce central effects. Death occurs from respiratory arrest, in high doses from heart failure (LD_{50} = 0.45 mg kg^{-1} mouse i.v.).

Physostigmine was the first alkaloid proven to act through inhibition of an enzyme. It has contributed most to our understanding of neurohumoral chemical transmission; it played an important role in understanding the kinetics of the enzyme, cholinesterase, and has contributed to the elucidation of the configuration of the active center of this enzyme. Although replaced by synthetic compounds in many cases, physostigmine has had several uses in clinical medicine: in ophthalmology, as a miotic and in glaucoma; in myasthenia gravis; in testicular paralysis; as an antidote to central poisoning with atropine and other parasympatholytics. Alone or together with atropine, it is protective against organophosphate poisoning. Lastly, it is considered as the prototype of the carbamate insecticides, but is itself a poor insecticide (Holmstedt, 1972). Because of its high toxicity and its very small therapeutic index, LD_{50}/ED_{50}, physostigmine is a dangerous substance. Norphysostigmine and physovenine are similarly toxic.

A plant of still more importance than *Physostigma venenosum* as an ordeal poison for determining innocence or guilt is the Caesalpiniaceae *Erythrophleum*, especially the species *E. suaveolens*, but also *E. ivorense* and *E. africanum*. If macerated or boiled, the red inner stem bark provides a blood-red extract which gives the tree its common name of "red water tree." This legendary extract is the most common and most notorious ordeal poison in Africa. The stem bark or the seeds are used as an arrow poison or poison ingredient mainly by the Mbuti Pygmies in northeast Zaire but also by other tribes in other African countries.

Chemically the *Erythrophleum* alkaloids are tricyclic compounds with a semicyclic, unsaturated side chain as the carrier of the nitrogen, built up from a diterpenic acid which is mostly esterified, more rarely an amide bonded to a secondary or tertiary aminoethanole (**19**). Both series of ester types, *N*-methylaminoethyl esters and *N,N*-dimethylaminoethyl esters, are divided into groups according to different substitutions at C-3, C-4 and C-7 of the tricyclic nucleus.

The *Erythrophleum* alkaloids are quite similar to *Strophanthus* cardenolides in their action as cardiac stimulants (positively inotropic and negatively chronotropic) and in their

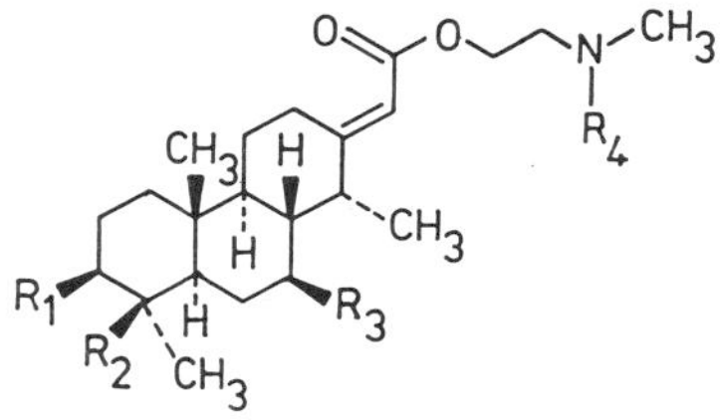

Erythrophleum-Alkaloids

19

high degree of toxicity. Some of these alkaloids, especially coumidine and coumingine (**20**), have almost the same potency as ouabain. The therapeutic index of *Erythrophleum*

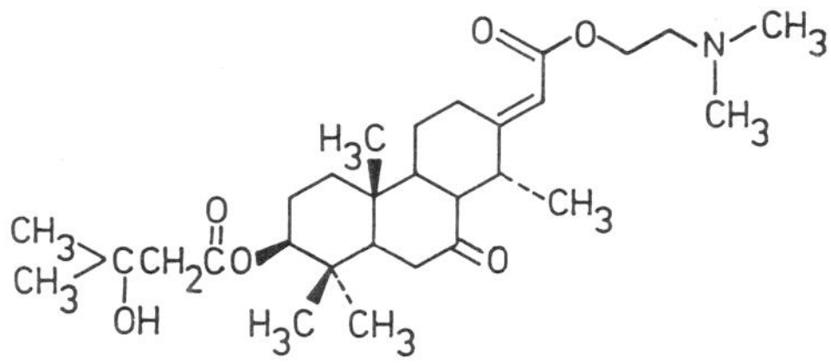

Coumingine

20

alkaloids is higher and the enteral absorption better than that of the cardenolides, but their effect is very brief. Change of the ester to an amide causes an opposite effect: these nor-alkaloids have a negative inotropic action on the heart. Some of the alkaloids are very potent local anesthetics, e.g., cassaine and cassaidine. It is remarkable that the water-soluble extract of the stem bark has been shown to have strong spasmogenic effects on smooth muscle, potentiated by physostigmine and blocked by atropine; in very high doses it causes contracture of skeletal muscle.

One of the most spectacular plants used in central African arrow poison is *Tabernanthe iboga*. The iboga plant is the most important African hallucinogen and up to now the only Apocynaceae to be utilized as such. This psychoactive drug is of great social importance, especially in south Cameroon, Gabon, eastern Equaitorial Guinea, the Congo, and western Zaire. *T. iboga* has the reputation of being a truly divine plant. It is employed by secret societies in initiation rites and messianic-prophetic associations for magical-religious purposes. The most famous of those are the Bwiti cult of the Fang and the Ombudi initiation association of the Mitsogo in Gabon. Sorcerers take the drug before communicating with the spirit world or seeking advice from ancestors. The drug liberates from anxieties and conveys messages from ancestors and spirits to the *iboga*-eaters. Recently, cannibalism has been found in connection with the *iboga* cult in Gabon.

The roots contain indole alkaloids, the so-called iboga alkaloids; they comprise up to 6% of the dried root bark. The major representatives are the alkaloids that mostly occur together—ibogaine, ibogamine, and tabernanthine. The exhaustively studied ibogaine (**21**) mainly exhibits CNS and cardiovascular activities. It acts as a cholinesterase inhibitor, a strong central stimulant, and a hallucinogen; its hallucinogenic effect is similar to

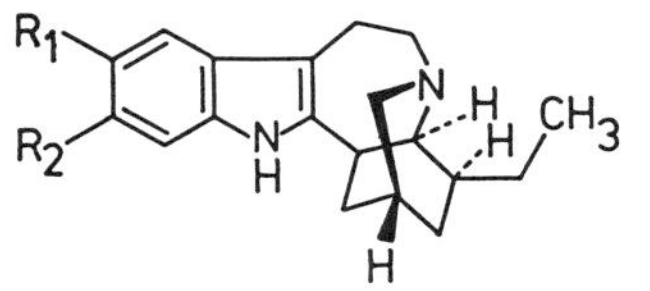

(**21**) $R_1 = H, R_2 = OCH_3$
(**22**) $R_1 = OCH_3, R_2 = H$

that of LSD but weaker. The LD_{50} of tabernanthine (**22**) and ibogaine is 0.38 and 0.42 mg kg^{-1} mouse respectively.

A well-investigated arrow poison adjuvant in Nigeria and Cameroon is *Mucuna pruriens.* The plant is characterized by indole alkaloids and indole-3-alkylamines: *N,N*-dimethyltryptamine and derivatives, bufotenidine (**23**), 6-methoxyharmane, nicotine, and

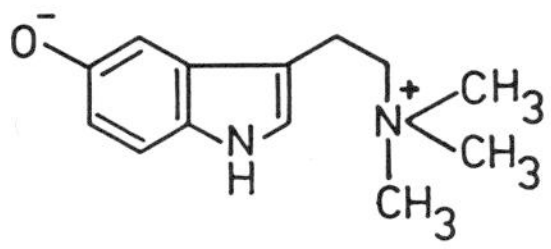

23

choline. *Mucuna* seeds are the richest source of L-dopa known. It is very interesting that Indians on the upper Orinoco river utilize a dart poison with the sole main ingredient of 5-methoxy-*N,N*-dimethyltryptamine (**24**) (up to 8%), as it occurs in *M. pruriens.* The

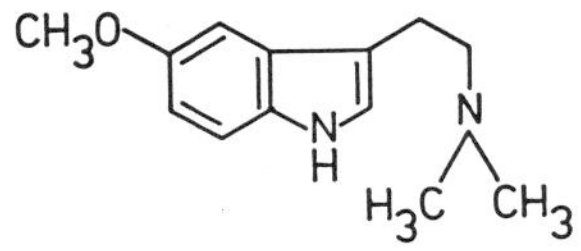

24

toxicity LD_{50} of the most toxic representatives, bufotenidine and 5-hydroxy-*N,N*-dimethyltryptamine, is 1.3 mg kg^{-1} mouse i.v. *N,N*-dimethylated tryptamine derivatives are very potent psychomimetic drugs.

Three types of alkaloidal constituents—β-phenylalkylamines (e.g., ephedrine, β-phenethylamine), quinazolines (e.g., vasicine, vasicinone), and carboxylated tryptamines (e.g., hypaphorine)—in addition to betaine and choline have been isolated from *Sida cordifolia,* a plant that is an ingredient of arrow poisons in northern Nigeria. Their toxicity is moderate. In the last few years, vasicine (**25**) has achieved importance as a

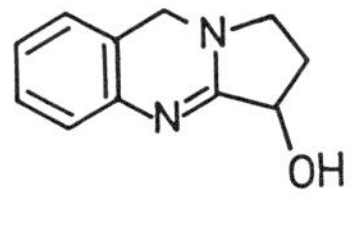

25

powerful uterine stimulant, like oxytocin in activity. It is very selective in its action and accumulation in the uterus. The methylester iodide of hypaphorine (**26**) has curarizing properties.

Phenylalkylamine alkaloids, as well as piperidine alkaloids, contained in the *Aloe* species are often observed as arrow poison adjuvants. In some of them the simple piperidine alkaloids (+)-(*S*)-coniine (**27**) and γ-coniceine (**28**), both highly toxic, were

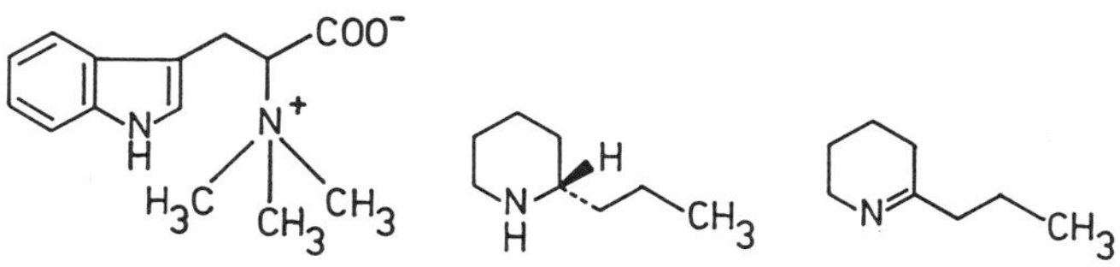

26 **27** **28**

detected. In 399 B.C., Socrates died from poisoning by coniine, which has nicotinic as well as curarising activities leading to a painful death: vomiting followed by paralysis of the striated musculature and nerve centers, beginning at the legs and ascending until finally death occurs from central respiratory arrest while fully conscious. γ-Coniceine is still more powerful than coniine with mainly nicotinic actions and with an oral LD_{50} for mice reported as 12 mg kg^{-1}.

A very toxic plant that enters into the preparation of arrow poisons in the southwest Ivory Coast is the alkaloid-bearing Apocynaceae *Tabernaemontana crassa*. A crude ethanol extract of the stem bark at a dosage of 250 mg kg^{-1} i.p. killed rats in 30 min; similar results were found with a root extract at a dosage of 175 mg kg^{-1}. The stem extract causes reduction of motor activity and muscle paralysis; this is the reason for its use in traditional medicine to calm fits and insanity.

Some of the 34 indole alkaloids from the root bark, stem bark, and seeds have been investigated pharmacologically, but their physiological effects do not match the activity of the general extract. The majority of the bases belong to the iboga type and within that to the coronaridine series. One of the main compounds is ibogaine (see *Tabernanthe iboga*). Most of the alkaloids tested (ibogaine, ibogamine, coronaridine, voacangine, isovoacangine, voacristine, and vobasine) are central stimulants; four of the seven compounds caused catalepsy. All of the bases tested showed a characteristic two-stage hypotensive effect in rats. Perivine, coronaridine, and vobasine caused muscle-relaxant activity in rat muscle preparations. The effect is not nondepolarizing like curare, but like that shown by succinylcholine, caused by a depolarizing neuromuscular blocking mechanism. The three alkaloids have only a moderate grade of toxicity. The leaf decoction is in popular use as a local anesthetic during painful traditional orthopedic procedures, especially bone resetting. Recently the effect was scientifically confirmed. The alkaloids vobasine (**29**) and perivine (**30**) may be responsible; the anesthetic activity of the latter is twice that of cocaine.

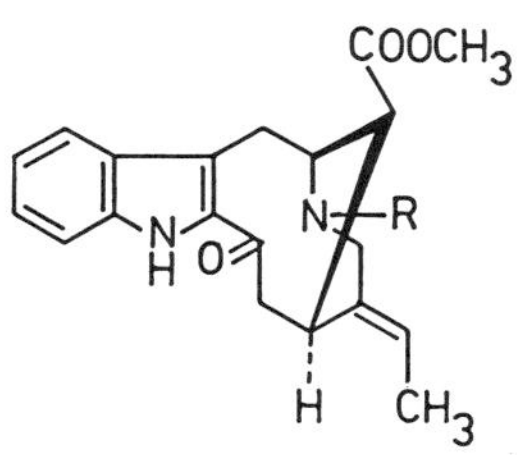

(**29**) R = H
(**30**) R = CH_3

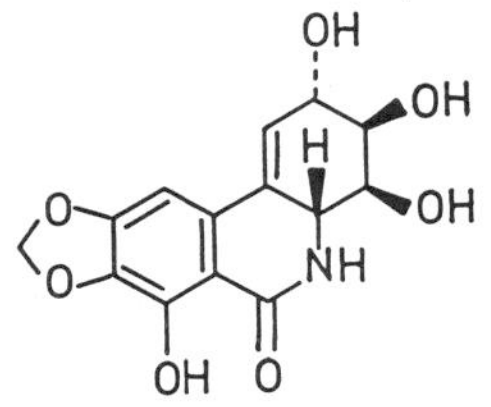

31

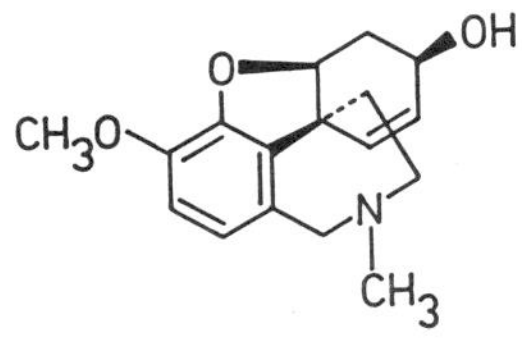

32

Scadoxus, Crinum, and *Boöphone* are arrow poison plants that bear in their tubers the so-called "Amaryllidaceae alkaloids," a unique large group of bases that have so far been found only in that family. A review is given by Martin (1987). The numerous alkaloids with a great variety of structure types are mostly derived from a 5,10β-ethanophenanthridine nucleus. The most common representative is lycorine. They are mostly neurotoxins similar to the tropane alkaloids with high to moderate toxicity. The toxicity of the most toxic alkaloid narciclasine (**31**) (pyrollo[de]phenanthridine type) from *Scadoxus multiflorus* is 5 mg kg^{-1} mouse s.c. Galanthamine (**32**) (dibenzofuran type, LD_{50} = 11.1 mg kg^{-1} mouse s.c.) acts as an antagonist to curare arrow poison and exhibits analgesic activity in mice comparable to that of morphine. *S. multiflorus* and *S. cinnabarinus* as well as *Crinum zeylanicum* are arrow poison ingredients in west and central Africa. *Boöphone disticha,* formerly widely used in southern Africa, is now used only in southern Zaire for this purpose. *Boöphone* poisoning resembles acute drunkenness with impaired vision and hallucinations related to poisoning with tropane alkaloids of *Datura.* The old Boer name "malgif" refers to this mad and confused behavior of animals poisoned with the *Boöphone* bulb.

The tropane alkaloids, another type present in arrow poisons and also in ordeal poisons, come from *Datura* species. Main representatives are (−)-hyoscyamine and (−)-scopolamine (hyoscine) (**33**). They differ in the β-oxido group at position C-6, C-7.

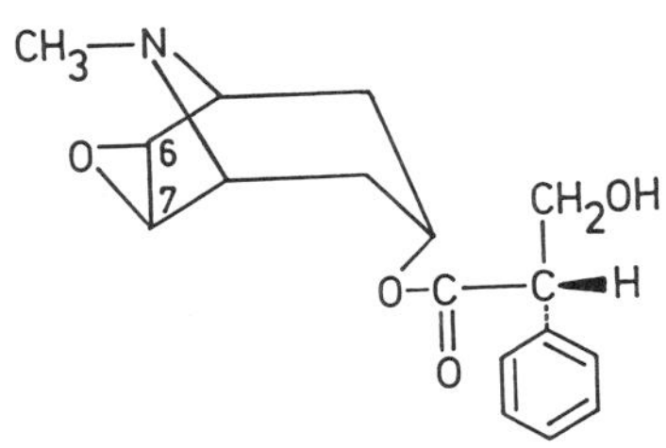

33

Besides these, phenylalkylamines and indole-alkylamines, e.g., noradrenaline, 5-hydroxytryptamine (serotonin), 5-hydroxyindoleacetic acid, were found in different plant parts. Climatic variations have a very considerable influence on the alkaloid content. Arrow poisons based on *Strophanthus hispidus* or *Acokanthera schimperi* with fruit of *D. stramonium* or the very similar *D. metel* have been found in Kenya and Burkina Faso. Tropane alkaloids are widely distributed in the Solanaceae family. Both compounds are isomerized to the racemic forms. Thus, the well-known commercial atropine is racemic *dl*-hyoscyamine. Hyoscyamine and scopolamine are competitive antagonists of the actions of acetylcholine and other muscarine agonists, both have similar peripheral, parasympatholytic properties, but they differ quantitatively in antimuscarinic actions and in action on the CNS. Hyoscyamine and atropine have CNS-stimulating but also -depressing activities, with toxic doses resulting in irritability, disorientation, hallucinations, and delirium; with still larger doses, stimulation is followed by depression, coma, and medullary paralysis. Scopolamine tends to be more CNS-depressant. In both cases, death occurs from respiratory paralysis. Only a moderate grade of toxicity, however, was found for the two alkaloids: $LD_{50} = 95$ mg kg^{-1} for (−)-hyoscyamine and 163 mg kg^{-1} for (−)-scopolamine in mice i.v. On the other hand, the maximum tolerated dose of the alcoholic extract of *D. metel* (whole plant) for mice i.p. is 100 mg kg^{-1}. Its use as an ordeal poison is remarkable in Uganda (Banda tribe) and Tanzania (Pare and Chagga), Senegal, and Botswana. In Uganda, the accused is supposed to divulge his crimes under the influence of the central, psychoactive effects of *Datura* alkaloids (truth or chatter drug). In the various plant parts, both alkaloids are to be found in varying proportions. There is also a direct relationship between plant growth and alkaloid content; in young plant parts, scopolamine predominates, in mature parts hyoscyamine. In flowers and seeds scopolamine is usually the major constituent. When scopolamine predominates, the accused will go into a state of being half-awake and in a trance; he will answer all of the sorcerer's questions about things that he would rather keep secret. This psychic state is especially aspired to by the sorcerers. In Botswana, losing consciousness is a sign of guilt. In Senegal, the leaves are pounded in water and the extract drunk. There have been cases of such poisonings treated in the Dakar neurological clinic in recent times.

Most African wild species of *Dioscorea* (yams) are known to be highly toxic. The tubers must be carefully detoxified by soaking for a couple of days in running or salt water or boiling overnight to degrade the toxic alkaloids, unstable in heat. Five species have been positively proved as ingredients of arrow poisons: *D. dumetorum, D. bulbifera, D. quartiniana, D. sansibarensis,* and *D. smilacifolia*. The *Dioscorea* is best known for the steroid glycosides its species contain. Many of them, however, also contain large amounts of alkaloids based on an isoquinuclidine ring system, such as dioscorine and dihydrodioscorine, the major poisonous principles of such species of *Dioscorea.* Dioscorine (**34**) contains an *N*-methyl group and a lactone ring at C-5. *D. dumetorum* bears dume-

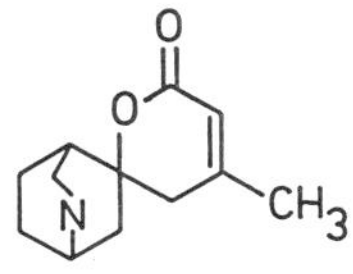

34

torine as a further alkaloid, an α,β-unsaturated γ-lactone joined to a piperidine ring via a CH2 bridge. Dioscorine has a picrotoxin-like convulsant effect and induces a general paralysis of the CNS. The ethanol extract of the tuber exhibited in mice generalized convulsions, followed by tonic–clonic convulsions, and, in lethal doses, death by extensor spasm within 10 min. The alkaloid-containing fraction is hyperglycemic; the crude hydroalcoholic extract, especially the glycosidic fraction, is contrastingly hypoglycemic. In Nigeria, a decoction of the tuber, prepared by steeping the peeled tuber in native gin for 3 days and boiling until the color changes from yellow to brown, is successfully used in the treatment of clinical diabetes mellitus.

Solanum species have a very close relationship with man; many are important food plants. Several wild-growing *Solanum* species are used in preparing arrow poisons, especially the unripe fruit or roots of *S. incanum* (Nigeria, Niger, Rwanda, Kenya), *S. torvum* (Cameroon) and *S. kwebense* (Namibia). Many, if not all, of the *Solanum* species produce mainly poisonous steroidal alkaloids of saponin character ("basic saponins"). Predominant are the glycoalkaloids solasonine (**35**) and solamargine; both are based on the genine

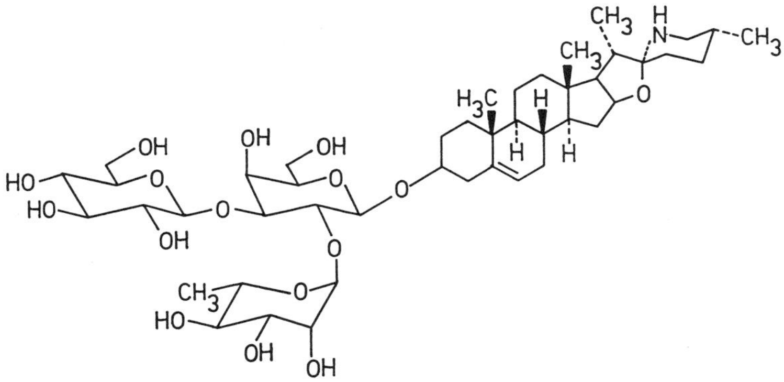

35

solasodine, which is the nitrogen analogue of the sapogenine diosgenine. Both compounds can easily be converted to 16-dehydropregnenolone, a key intermediate in the synthesis of steroid drugs. Besides the steroidal alkaloids, the corresponding steroidal saponins are almost always found in *Solanum*. The fruit contain the highly toxic *N*-dimethylnitrosamine. In basic actions, the *Solanum* steroidal alkaloids are apparently very similar. A number of glycoalkaloids have been shown to produce cardiotonic effects similar to those of cardiac glycosides (cardenolides). Severe hemorrhagic gastroenteritis is generally seen. Death is caused by cardiac or respiratory arrest.

One of the best-known Solanaceae, the tobacco plant *Nicotiana tabacum* and also *N. rustica,* are ingredients of *Strophanthus*-based arrow poisons in Cameroon and Nigeria. The leaves are mostly used. In southeast Cameroon some poison-makers mix pounded *Strophanthus* seeds with juice of the tobacco plant, sometimes hunters smear the thickened juice alone on the arrowhead. *N. tabacum* contains a great number of simple alkaloids, nicotinoids, and others. The most abundant compound by far of *N. tabacum* and *N. rustica* is the pyridine-*N*-methylpyrrollidine nicotine (**36**), one of the few naturally liquid alka-

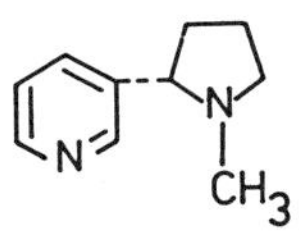

36

loids and an extremely strong poison (LD_{50} = 0.5 mg kg^{-1} mouse i.v.). It is very rapidly absorbed and in fatal doses death ensues within a few minutes. It has great variety and complexity of action. The symptoms of acute nicotine poisoning involve most bodily functions, by actions both at the periphery and on the CNS. It has both stimulant and depressant phases of action. Fatal doses lead to convulsions and respiratory paralysis. Its major mechanism of pharmacological action is an interaction with cholinoceptive receptors at many sites in the body. On the other hand, nicotine is quickly detoxified (by secretion in urine, bile, and feces or by metabolism in the liver). Recently it was recognized that the central actions of nicotine have a considerable therapeutic potential, e.g., in the treatment of Alzheimer's disease and as an anxiolytic. Anabasine (**37**), the major alkaloid in some species (e.g., *N. glauca),* is as toxic as nicotine.

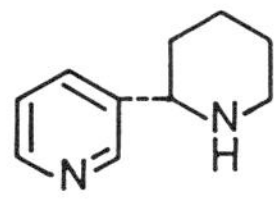

37

One of the plants that is a common ingredient in traditional medicine and also of therapeutic value for modern medicine is *Rauwolfia vomitoria* (Apocynacceae). Its roots often appear in the preparation of arrow poisons in Zaire and the Central African Republic.

R. vomitoria became an important article of commerce as an alternative to the Indian species *R. serpentina* as a source of the indole alkaloid reserpine, therapeutically widely used. It is chemically and pharmacologically one of the most thoroughly investigated African plants. Eighty-six indole alkaloids classified into 20 structure types have been isolated from the roots, stem bark, leaves, and unripe fruit. In 1982, the occurrence of 43 indole alkaloids was indicated in a single batch of stem bark and 39 of them identified. The stem bark and roots mainly yield heteroyohimbine- and yohimbine-type alkaloids (especially reserpiline, reserpine, rescinamine) as well as dihydroindoles (especially ajmaline). The major alkaloid of the leaves is caranaupine, an oxindole. Reserpine (**38**) is

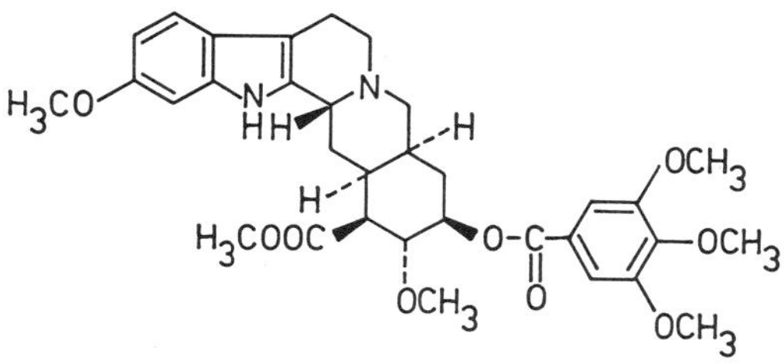

38

a widely used drug for treatment of mild and moderately severe high blood pressure—it has a long-lasting effect—and in psychiatric disorders (neuroleptic). Because of its unpleasant side effects, it is mostly used nowadays in low dosage in combination with other drugs or replaced by drugs with lesser side effects. Decrease of blood pressure occurs as a consequence of decreased sympathetic tone because of a long-lasting depletion of peripheral catecholamine stores. The sedative or tranquillizing activity is in connection with the depletion of serotonin in the stores of brain tissues. Death ultimately results from central respiratory depression (LD_{50} = 0.5 mg kg^{-1} dog i.v., 10 mg kg^{-1} mouse i.v.). Ajmaline (**39**) (LD_{50} = 21 mg kg^{-1} mouse i.v.) exerts no CNS action, it acts like quinidine and is

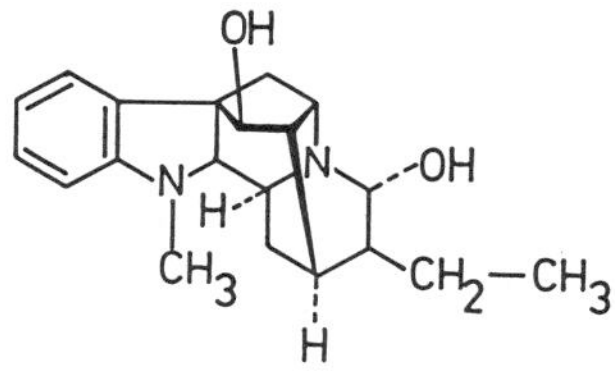

39

widely in use for cardiac disturbances, especially arrhythmia. The major alkaloids are poorly water soluble. The water-soluble root bark extract, as used as an arrow poison, caused severe bradypnea in dogs in high doses, followed by death through respiratory and cardiac arrest. The findings on the interaction between cardiac glycosides and reserpine in the heart and CNS are very remarkable. Ouabain and other cardiac glycosides produce a marked potentiation of the central depressant effects of reserpine: reserpine given simultaneously with ouabain potentiates the toxicity of ouabain and increases mortality from 5 out of 10 dogs (ouabain alone) to 9 out of 10. These synergistic toxic effects and the potentiation of toxicity may be of great interest with regard to arrow poisons: *Rauwolfia* is always used in combination with the cardenolide-bearing *Parquetina nigrescens* or *Strophanthus* species.

Two arrow poison plants that are chemically closely related to *Rauwolfia vomitoria* are *Corynanthe pachyceras* and *Picralima nitida.* The former is an ingredient of *Mansonia altissima*-based arrow poisons in the Ivory Coast; fruit and roots of the latter are often found as adjuvants to arrow poisons in Zaire and the Central African Republic, based on *Parquetina nigrescens* or *Strophanthus. Corynanthe pachyceras* bears yohimbines and oxindoles. The lethal dose of the most poisonous compound, corynantheine (**40**), is 1.4 mg

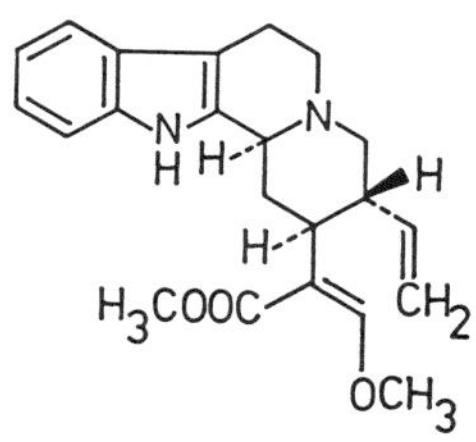

40

kg^{-1} mouse i.v.; it exerts excitatory effects, severe convulsions, dyspnea, and death by respiratory arrest. Most representatives are sympatholytics. The 15 characterized bases of *Picralima nitida,* mostly tertiary, belong to various structure types of the indole alkaloids. In a wide range the principal alkaloid is akuammine (**41**); it is a strong sympathomimetic

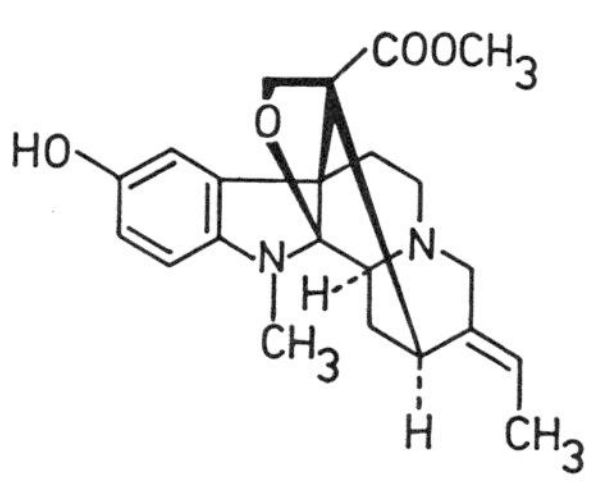

41

with effects similar to those of cocaine. Akuammidine, on the contrary, is a sympatholytic; its local anesthetic action was found to be three times that of cocaine hydrochloride. The only *Picralima* alkaloid of the strychnine type is akuammicine, very closely related to nor-C-fluorocurarine, which exhibits a strychnine-like activity. Pericine and apparicine (**42**)

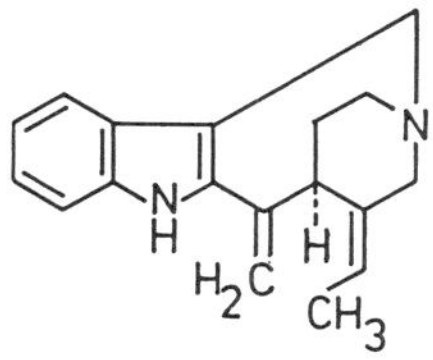

42

(pericalline) are CNS-active and cause tonic–clonic convulsions. The leaves also contain quaternary alkaloids of still unknown structure, melinonine A excepted. *P. nitida* is considered to have potential in drug-resistant malaria chemotherapy; the antimalarial activity at least partly resides in akuammine. The decoction of the stem bark is trypanocidal, similar to diminazine aceturate.

Sarcocephalus esculentus was formerly used as the sole constituent of arrow poisons in the Rio Nunez area of west African Guinea. Nowadays, the very bitter stem bark or the roots are ingredients of *Strophanthus hispidus*-based arrow poisons in northern Nigeria and in the northern part of the Ivory Coast (Senufo tribes).

From the root bark, several indoloquinolizidine (indolopyridine) alkaloids have been isolated: angustine (**43**) and angustoline, known from various *Strychnos* spp., and the six

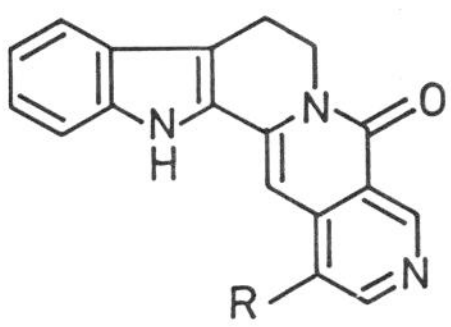

(**43**) R = HC = CH_2
(**44**) R = $COCH_3$

new bases nauclefine, naucletine (**44**), naufoline, descarbomethoxynauclechine, naucleidinale, and epi-19-naucleidinale. Root bark, stem bark, and hardwood contain, moreover, large amounts of the structurally-related alkaloid glycoside strictosamide (**45**) (stric-

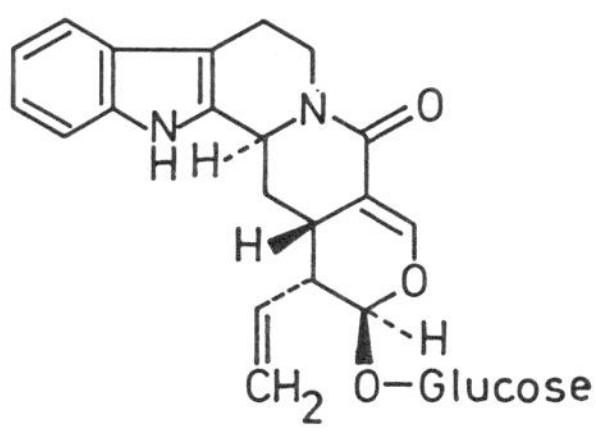

45

tosidine lactam, isovincoside lactam). The aqueous-alcoholic extract of 1 g of stem bark kills a 100-g guinea pig in 15 min. Death occurs from respiratory arrest. In rats, significant central effects, decreased motor activity, and paralysis have been observed.

The tuber of an ornamental known worldwide, the Superb lily, *Gloriosa superba,* is an ingredient of *Strophanthus*-based arrow poisons in northern Nigeria. The individual parts of this plant are highly toxic. From the tuber and seeds, a great number of neutral-phenolic tropolone and some basic nontropolone alkaloids were isolated. The principal alkaloid is mostly colchicine, rarely *N*-formyl-*N*-desacetylcolchicine. Colchicine (**46**), one

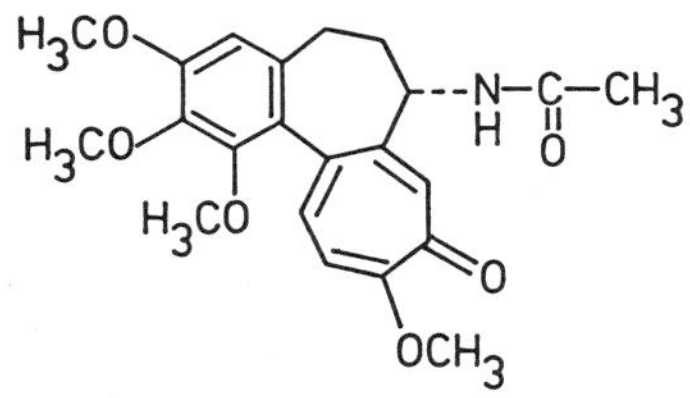

46

of the most fascinating natural compounds, shows a great number of pharmacological activities. It is highly toxic with a small therapeutic index. It is a unique anti-inflammatory agent selective against gouty arthritis, in acute attacks and as a prophylactic. It is also an antimitotic agent and has been widely employed as an experimental tool in the study of cell division and cell function; it can arrest plant and animal cell division. The cytostatic colchicine is probably the oldest cancer chemotherapeutic, but too toxic for the treatment of tumors. Colchicine exhibits a variety of other pharmacological effects: it is one of the strongest poisons (LD_{50} = 0.25 mg kg^{-1} cat i.v., 4.1 mg kg^{-1} mouse i.v.), slow but long-lasting in action. Gastrointestinal disorders, nausea, vomiting, diarrhea, and abdominal pain are the most common effects of colchicine poisoning. In fatal doses, muscular depression is pronounced, an ascending paralysis of the CNS develops, and death results from respiratory arrest.

Another alkaloid that is fundamentally an antimitotic agent is maytansine (**47**),

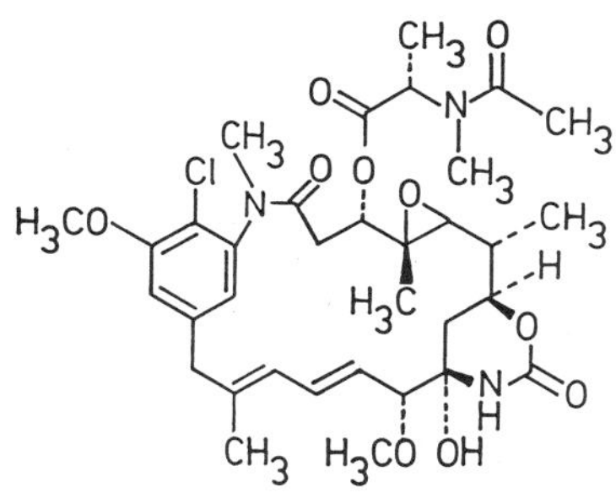

47

isolated from many *Maytenus* species, among them *M. senegalensis.* The stem bark of the twigs of this plant have repeatedly been found as an ingredient of arrow poisons in Burkina Faso, on the base of *Strophanthus hispidus.* It is not clear which compounds are responsible for its toxicity. Wilforine, an alkaloid with a complicated structure, has been isolated, too. The ansa macrolide maytansine (**47**) has been extensively studied; it was found to be an exceptionally interesting antitumor agent, which gave rise to a hopeful anticipation, as it is active at the level of micrograms per kilogram in various tumor test systems. Maytansine binds to tubulin and prevents formation of the microtubules necessary for mitotic spindle formation. This leads to arrest of cell division in metaphase. Phase I and phase II clinical trials, however, finally led to the conclusion that maytansine shows too little activity and is too toxic. The most common toxicities seen were gastrointestinal, nausea, vomiting, and diarrhea; furthermore, neurological toxicities include numbness, jaw and muscle pain, weakness, lethargy, and general malaise (Suffness and Cordell, 1985). For male mice the toxicity with a single dose is $LD_{50} = 1.53$ mg kg^{-1} i.v. Maytansine has, however, a cumulative effect: given for 5 days the LD_{50} is approximately 0.30 mg kg^{-1}.

Alkaloids of unusual structure have been found in the genus *Ziziphus.* The roots or stem bark of *Z. mauritiana* enter into the preparation of *Strophanthus*-based arrow poisons in Niger and Burkina Faso. The fruit are used as a fishing poison in Ethiopia. The stem bark yields a number of cyclopeptide alkaloids. They contain a 14-membered ring system and are based on two structural types: (1) frangulanine type, a ring system formed by the amino acids *p*-hydroxystyrylamine, β-hydroxyleucine and an α-amino acid; the main representative is frangufoline (**48**); (2) Amphibine B system, a ring system formed by

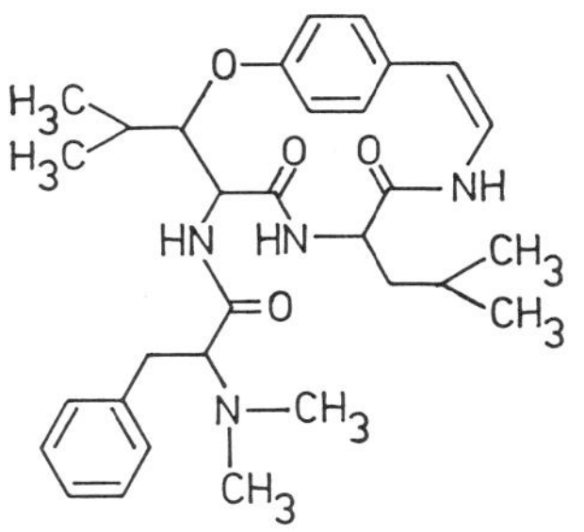

48

p-hydroxystyrylamine, *trans*-3-hydroxypyroline and an α-amino acid; main representatives are amphibine B, D, E, F and mauritine A–F, H. Nothing is known about the pharmacology and toxicology of these compounds (Tschesche and Kaussmann, 1975).

3. SOUTH AMERICAN ARROW POISONS

The northern half of South America has always been a center of many plant derived hunting poisons. By far the most famous and most common of the South American arrow and dart poisons is curare, a name attributed to the large group of poisons with muscle paralysis of a special mechanism. "Curare" and "curarizing" are phenomenological terms describing neuromuscular block of impulse transmission of the motor end plates as a result of inhibition of acetylcholine with the consequence of complete paralysis of skeletal musculature.

Curare is essentially a blowpipe dart poison of the forest tribes living in a roughly semicircular area (about 5000 × 4000 km) stretching from French Guyana to the Mato Grosso and including the northern and western parts of the Amazon basin, the middle and upper parts of the Orinoco basin, the Montana of Ecuador and Peru and northeast Bolivia and adjacent parts of the Brazilian plateau. It has also been prepared for arrows by tribes in the savannah of the Mato Grosso plateau. Curare poisons are still a necessary tool for the survival of a number of Indian tribes in the rain forest of South America.

3.1. Classification and Composition

In 1895, Rudolf Boehm in Germany classified curare poisons, more ethnographically than chemically, into three categories according to the relationship between incineration residues and poison content and the container: tubo-curare (in bamboo tubes), calabash-curare (in gourds), and pot-curare (in earthenware pots) (Fig. 8).

- *Calabash curares* are usually obtained from Loganiaceae (*Strychnos* species) and come principally from the region between the Orinoco in the north and the Negro and lower reaches of the Amazon in the south. A minor type of curare found in use by the Ticuna Indians in Brazil, the so-called bag curare, has similar properties to calabash curare.
- *Tubo curares* are derived from Menispermaceae (chiefly *Chondodendron* and *Curarea* species) and predominate in Peru at the foot of the Andes in the eastern Lowlands ("La Montana"), the region bounded by the rivers Napo, Maranon, and Ucayali.
- *Pot curares* are mixed Loganiaceae/Menispermaceae products and are found mainly in the area covered by the middle reaches of the Amazon and to a lesser extent also in Guyana.

For many decades, all research on curare was based on Boehm's classification. In the course of time, it was found that this classification runs more or less parallel to that based on botanical sources and geographic areas and the term tubo, pot, or calabash curare often does not depend on its content. Many ethnographic observations did not take into consideration the origin of a curare: the poison is merchandise that is transported and marketed

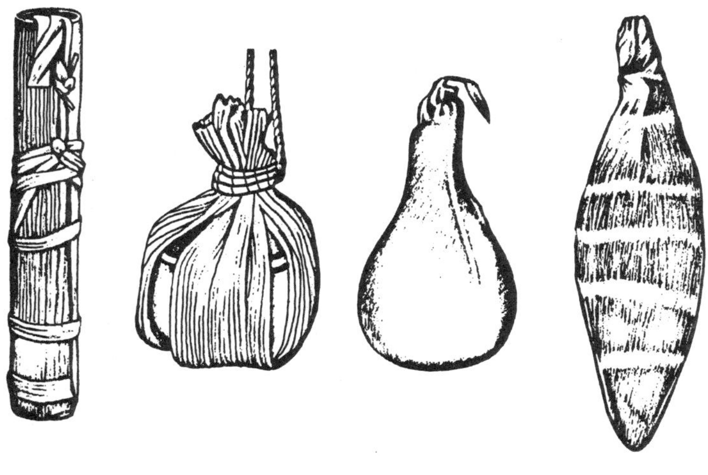

Figure 8. Types of curare poison classified according to their containers (left to right: tubo-curare, pot-curare, calabash-curare, and bag-curare).

Table I
Recipes for Curare

Recipe 1 (Krukoff and Smith, 1937)		Recipe 2 (Schwacke, 1884)	
Strychnos castelnaeana	2 lb outer stem bark	*Strychnos castelnaeana*	stem bark
Curatea tecunarum	5 lb bark	*Anomospermum grandifolium*	leaves
Telitoxicum minutiflorum		*Petiveria alliacea*	leaves
2 *Piper* species	0.25 lb roots	*Dieffenbachia seguine*	stem
Aristolochia species	1 tuber	2 Piperaceae (*Piper*?)	root bark
An undefined Cucurbitaceae	1 tuber	*Markgrafia* species	root bark
Annona ambotay	0.1 lb		
Petiveria species	0.1 lb		

by Indian tribes throughout very large areas and it is often sought by tribes 1000 miles away, who send messengers with gifts to obtain a particularly active curare. The curare markets in Venezuela were well known (Thomas, 1972). Curiously enough, in the 1920s, some of the Ecuadorean Indians smeared their arrow tips with curare imported from the German market, like the curare from the American market. Nowadays, the poison may simply be kept in the nearest handy container (e.g., bottle, tin can) or is smeared directly on the arrow tips and not stored.

Most curares comprise several components, which may be added for various reasons and which differ widely in accordance with local flora and tribal superstitions.

The recipes for curare (Krukoff and Smith, 1937; Schwacke, 1884) used by the Ticuna are given in Table I. The poisons are mostly obtained by squeezing the pounded plant parts, more rarely by long maceration or boiling and concentration. The following procedure is a common one: drip water through a leaf funnel that is filled with the pulverized plant material and concentrate the extract by low heat over the fire.

Among the plants mentioned as adjuvants are *Dioscorea, Nicotiana, Capsicum, Tabernaemontana, Hura, Ficus, Picrasma, Picrolemma, Guadua, Euphorbia, Cocculus, Guatteria, Spigeila,* and *Abuta* (Cheney, 1931; von Reis-Altschul, 1973; von Reis and Lipp, 1982). According to R. E. Schultes (1984, not published, in Bisset, 1989), the Kofan Indians of Colombia/ Ecuador make use of more than 75 different plant species in preparing their various poisons; one-third of them are known to contain highly active compounds, e.g., cardenolides.

3.2. Chemistry

The scientific investigation of the complex curare produced about 70 alkaloids. A great part of them could be separated and characterized, but some of them have only a letter of the alphabet. Curare poisons differ very much in the composition of the alkaloids, but all curares always have in common the presence of dimeric quaternary alkaloids, although of different structures, as the main curarizing principles (Marini-Bettolo, 1975/76; Waser, 1972). A good curare contains up to 12% quaternary alkaloids. The distance between the two protonated N for maximum curare activity is about 0.85–0.9 nm.

In 1970, the structure of (+)-tubocurarine (**2**), the major alkaloid of the tubo or *Chondodendron* curare, was revised to be monoquaternary and not diquaternary as hitherto accepted: it possesses both a tertiary and a quaternary (one ammonium group) basic function, but the second tertiary nitrogen is easily protonated. The main bases of calabash or *Strychnos* curare are easily water-soluble bis-quaternary dimeric indole alkaloids formed by the union of two equal or different monomeric molecules of strychnine type and containing two quaternary ammonium groups.

The bases can be divided into three alkaloid "families" (the prefix C refers to isolation from calabash material):

- The C-toxiferine family (2 × N_b-metho-Wieland-Gumlich aldehyde) (**49,** $N_b - CH_3$); C-toxiferine (**52**), C-alkaloids A (**58**) and E (**55**) as main compounds.
- The C-dihydrotoxiferine family (2 × N_b-metho-18-deoxy-Wieland-Gumlich aldehyde); C-curarine (**53**), C-alkaloid D, C-calebassine (**56**) as main compounds.
- The C-alkaloid-H family (1 × N_b-metho-Wieland-Gumlich aldehyde + 1 × N_b-metho-18-deoxy-Wieland-Gumlich aldehyde); C-alkaloids H (**51**), G (**54**), and F (**57**) as main compounds.

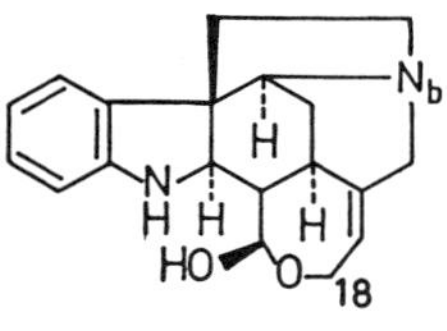

49

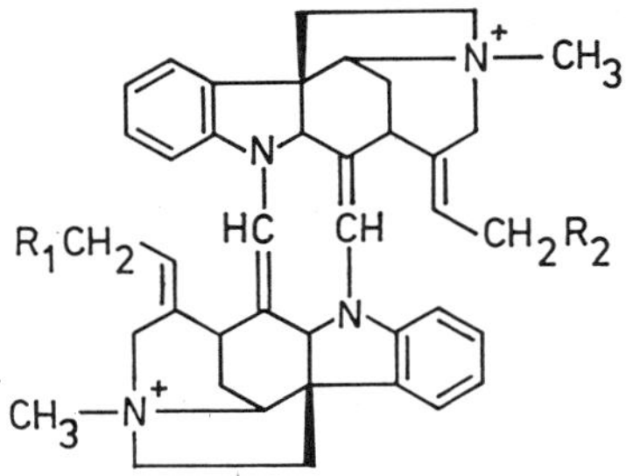

(**50**) C-Dihydrotoxiferine R_1 = H R_2 = H
(**51**) C-Alkaloid H R_1 = OH R_2 = H
(**52**) C-Toxiferine R_1 = OH R_2 = H

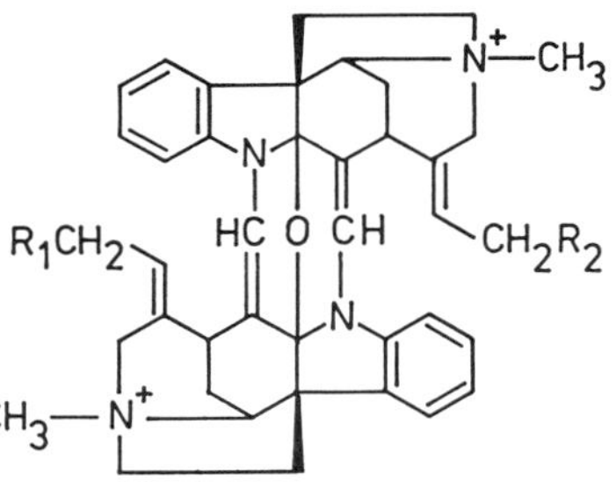

(**53**) C-Curarine R_1 = H R_2 = H
(**54**) C-Alkaloid G R_1 = OH R_2 = H
(**55**) C-Alkaloid E R_1 = OH R_2 = H

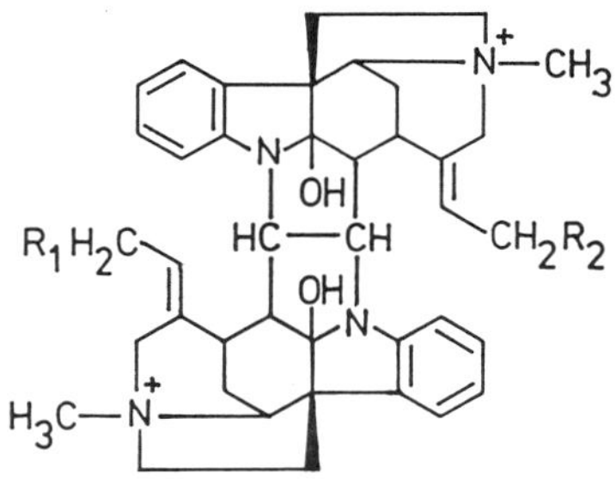

(**56**) C-Calebassine R_1 = H R_2 = H
(**57**) C-Alkaloid F R_1 = OH R_2 = H
(**58**) C-Alkaloid A R_1 = OH R_2 = H

3.3. Pharmacology, Toxicology

When introduced into the bloodstream, all of the dimeric quaternary curare alkaloids are highly active muscle relaxants, acting by a nondepolarizing and competitive mechanism at the neuromuscular junction, competing with ACh for the active surface of the receptor and thus blocking the nerve–muscular transmission. Blockade of the postsynaptic nicotinic acetylcholine receptors causes progressive paralysis of voluntary movement and, as a final result, complete paralysis of the skeletal or striated muscle apparatus. The muscle involvement follows a definite order: eye and ear muscles are the first to be affected, then neck, limb, trunk, intercostal muscles, and diaphragm; death results from anoxia caused by respiratory failure. Artificial respiration prevents death. Physostigmine from the African arrow poison plant *Physostigma venenosum,* immediately given, antagonized all paralytic symptoms.

The toxicities (LD_{50}) of the three curare types for mice i.p. are reported to be 0.8–25 mg kg^{-1} for pot curares, 2–15 mg kg^{-1} for calabash curares, and 5–10 mg kg^{-1} for tubo curares. Calabash or *Strychnos* alkaloids have the greater curarizing potency and moreover, the greater the polarity, the greater the curarizing activity. The most active alkaloids belong to the group with an ether oxygen in the central eight-membered ring: alkaloid E (**55**) (LD_{50} = 0.007–0.012 mg kg^{-1} mice i.v.), alkaloid G (**54**) (LD_{50} = 0.001–0.008 mg kg^{-1} mice i.v.), and C-toxiferine (**52**) (LD_{50} = 0.023 mg kg^{-1} mice i.v.). Some are short-acting, some extremely long-acting. The duration of action of C-toxiferine is outstanding: it takes nearly 2 hr for the muscle to recover after a slow, progressive onset of paralysis; with a 20-fold paralytic dose total paralysis lasts 8 hr. Alkaloid E shows similar properties. The action of C-curarine (**53**) (LD_{50} = 0.050 mg kg^{-1} mouse i.v.) is comparable to that of (+)-tubocurarine (**2**) (LD_{50} = 0.14 mg kg^{-1} mouse i.v.). C-Calebassine (**56**) belongs to the weaker compounds with an LD_{50} = 0.32 mg kg^{-1} mouse i.v.

Curare poisons are very stable, once prepared. Although more than 140 years old, curare samples present in various European museums were shown to have remained very effective. The lethal doses for mice i.p. of such arrow poisons lie between 0.8 and 20 mg kg^{-1}, mostly 2–5 mg kg^{-1} (Bauer and Fondi, 1962; Bauer, 1981; Lewin, 1923). Rather curiously, the oldest arrow poisons (130–140 years old) were the most active. Thus a pot curare of the Mayoruna in Brazil, 130 years old, had a lethal dose of 0.8 mg kg^{-1} mouse i.p. and mainly contained the bases C-alkaloids G (**54**) and A (**58**) and C-toxiferin (**52**). Moisture did not decrease the activity of curare (Bauer and Fondi, 1962). One of the most important unanswered questions concerning all arrow poisons is: especially, what kind of changes or modifications occur in the structure of alkaloids during the elaboration of curare? In the case of Macu curare elaborated with *Chondodendron,* Marini-Bettolo and his "group of Rome" demonstrated that the curarizing activity of the alkaloids obtained from tubo curare are tenfold greater than those extracted from the plant. Experiments indicated that, in the case of the bisbenzylisoquinoline alkaloids, some methylation reactions may occur with the more toxic, curarising quaternary alkaloids during the elaboration of the tubo curare (Marini-Bettolo, 1975/76).

3.4. Medicinal Use

The transformation of curare from deadly poison to a valuable medicinal agent is one of the best examples of the refinement of a crude natural product to an indispensable aid in modern medicine. In 1935, King announced the isolation of a pure crystalline alkaloid, (+)-tubocurarine chloride (**2**), from a museum sample of tubo curare of unknown botanical derivation. Seven years later, the same compound was found in a crude curare known to have been prepared from the single plant *Chondodendron tomentosum.* In 1939, curare was introduced in clinical medicine, first as a preventive of traumatic complications in metrazol convulsive shock therapy and electric shock therapy. Since 1942 the pure alkaloid (+)-tubocurarine has become a useful neuromuscular relaxant in all forms of anesthesia (Bennett, 1968). It continues to be used today. The structure of (+)-tubocurarine was taken as a model for elaborating synthetic curare and semisynthetic curare-like products with fewer sideeffects; it produced the molecular features necessary for muscle-relaxant activity and more exact knowledge of distribution, metabolism, and mechanism of action on the end plates and the transmission of stimulus to the muscles.

Two other curare alkaloids, C-calebassine (**56**) and C-toxiferine (**52**) from calabash curare, are also used in clinical medicine, e.g., in Switzerland, especially for tetanic convulsions; their duration of effect is longer-lasting than that of (+)-tubocurarine with fewer effects on respiration. Most alkaloids of the curare complex lower blood pressure in varying degrees.

It is of interest that the Auca Indians of eastern Ecuador apply their curare (made from *Curarea tecunarum*) directly to bacterial or fungal infections of the skin with good results reported (Davis and Yost, 1983). An extract of the same plant is used by the Deni of the Rio Cunhua in Brazil as a reputedly effective male and female contraceptive (Prance, 1972).

3.4. Other Plant Arrow Poisons

There have also been reports about "curares" based on plant sources other than *Strychnos* and *Chondodendron* (Vellard, 1965), especially *Virola* (Myristicaceae) and *Abuta* (Menispermaceae) species. The use of both species as arrow poisons is ethnologically and pharmacologically remarkable. In 1931 and 1960, the use of *Virola theiodora* and to a lesser extent *V. elongata* (considered by some botanists to be synonymous) as a hallucinogenic snuff by the Yanoama (Waika) of southern Venezuela and northern Brazil was reported, with the remarks that the same plants were applied alternatively as arrow poisons. In 1977, Schultes and Holmstedt established the use of *Virola theiodora* resin as the sole ingredient of the poison made by the Yanoama (Waika) of the Rio Tototobi: the resin of the heated stem bark is applied repeatedly to bamboo arrow tips. An almost identical procedure was observed among the Sanama tribe living more than 300 km north of the Rio Tototobi; both plants were used in the preparation of snuff and arrow poison. The Indians know that these are slow-acting arrow poisons and that the wounded animal must be followed for some time.

Although there was a considerable variation between samples, the indole alkaloids *N*-methyltryptamine, *N,N*-dimethyltryptamine, and 5-methoxy-*N,N*-dimethyltryptamine were the main compounds. It is very interesting that the same components were found as the biologically active constituents of the African arrow poison plant *Mucuna pruriens* (see above). Galeffi *et al.* (1983) confirmed the findings: the poison on Yanoama dart tips in Roraima territory, Brazil, did not contain any curarizing alkaloid, but a single substance in high concentration (8% by weight; 12 mg $dart^{-1}$), which was identified as 5-methoxy-*N,N*-dimethyltryptamine (= *O*-methylbufotenine)(**24**).

These and similar indole alkaloids have a number of peripheral and central activities, foremost their psychomimetic activity, especially their ability to elicit hallucinations. *Virola elongata* bark, at least, does not contain highly toxic constituents which might explain its use as an arrow poison (Macrae and Towers, 1984). Its use for hunting purposes can be explained by the incapacitation of the wounded animal; the predominating 5-methoxy-*N,N*-dimethyltryptamine (**24**), especially, causes much higher behavioral disturbances than other indolalkylamines, in addition to being considerably toxic (LD_{50} = 1.5–2.0 mg kg^{-1} sheep i.v.). It is the most active hallucinogen after LSD. In sheep, guinea pigs, rats, and mice, parenterally injected, it caused death in a short time from cardiac failure.

In 1965, Biocca *et al.* (1965) described a dart-tip curare, which is not stored in particular containers, being used directly for the preparation of dart points. In this curare, two types of active ingredients have been identified: bisindole alkaloids (Yanoama tribes of the Upper Orinoco) and bisbenzylisoquinoline alkaloids (area of Maria de los Guaicas). The same bisbenzylisoquinoline alkaloids, whose sole monoquaternary component is the muscle-relaxant macoline (**59**), have been isolated from *Abuta grisebachii* (Menisper-

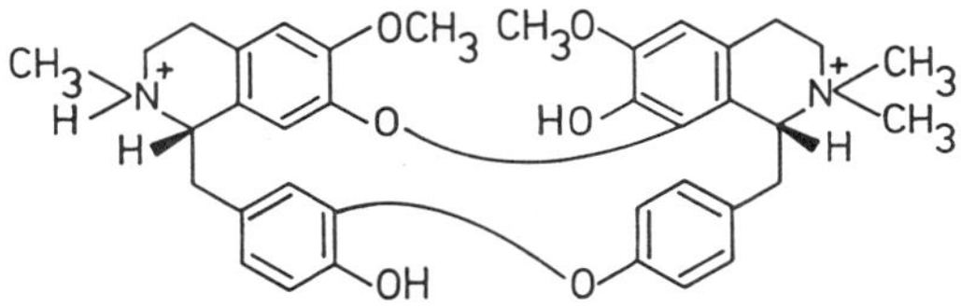

59

maceae), which is the single ingredient of the "macoli" curare of the Sanama Indians.

In Venezuela, arrow poisons derived from *Malouetia* species (*M. nitida, M. schomburghii?*) exert a paralyzing action similar to curare. A related plant from Africa, *M. bequaertiana*, affords the alkaloid malouetine (**60**), an unusual pregnane alkaloid with two

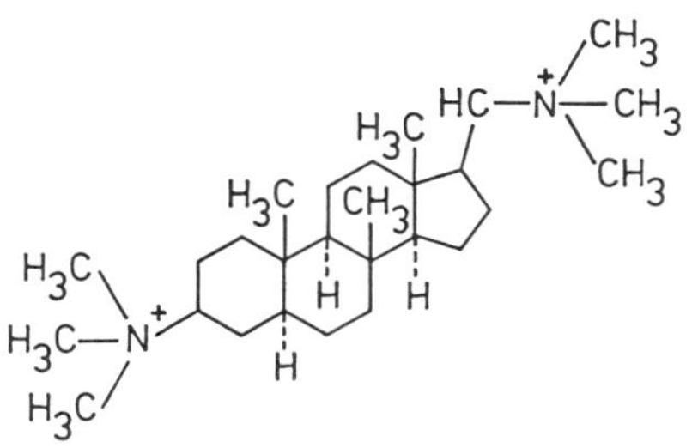

60

quaternary ammonium groups. In frogs, it produces a curarizing effect different from other natural curarizing compounds

4. ASIAN ARROW POISONS

Poisoned weapons have been extensively utilized in southeast Asia. This is the region in which the blowpipe and dart is the principal hunting weapon (Fig. 9). There is a lot of

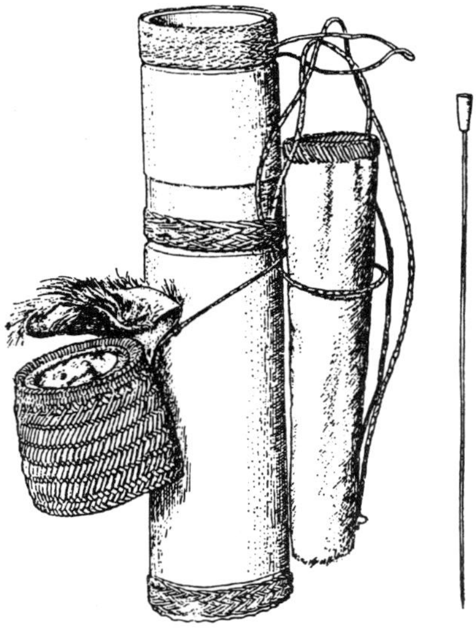

Figure 9. Dart for blowpipe and quiver from Sulawesi (Indonesia).

literature and information on the hunting poisons of this area but, as in Africa, there are two main difficulties regarding information gathered:

1. It is not always clear from the accounts which ethnic group was visited; moreover, changes in ethnological classification and designation make it difficult to relate the findings described in the earlier accounts to those acquired more recently.
2. Almost all of the botanical data and identifications, particularly of *Strychnos*, are incorrect, because in former times the plants collected were generally of poor quality, incomplete, and not well known, but some data have been amended in the course of revisions.

From Burma to China and Indonesia (as far as Timor), the major source is the latex of the highly toxic cardenolide-bearing Moraceae *Antiaris toxicaria*, the "upas" or "ipoh" tree of the Malays and Indonesians. Another important source is the root and stem bark of various species of alkaloid-bearing *Strychnos*, but they have a more restricted use. Their use is known with certainty only in Java, Borneo, Malaya, and Indochina. Arrow and dart poisons based on *Lophopetalum* (Celastraceae), *Strophanthus* (Apocynaceae), *Beaumontia* (Apocynaceae), and *Sapium* (Euphorbiaceae) species are also known in southeast Asia.

4.1. Indonesia, Borneo, Philippines, Hainan, Vietnam, Cambodia

Hunting poisons are still being made, particularly in Borneo and no doubt in other places as well, but generally there is little recent information available. *Strychnos* species (*S. ignatii* and *S. nux-vomica*) have been revealed as important components of the common *Antiaris toxicaria* poisons (Bisset, 1966). Other adjuvants are *Derris, Dioscorea, Sapium,* and *Nicotiana*. In Cambodia, an arrow poison is prepared from the bark of

Sindora cochinchinensis, the sap of a *Xylia* sp., *Capsicum frutescens,* and snake poison; it is still in use (Matras and Martin, 1972).

In the Philippines, a slow-acting arrow poison, formerly also used in India, is made from the seeds of the Papilionaceae *Abrus precatorius*: the seeds are powdered and formed into a paste with which the darts or arrows are smeared. Wounds caused by arrows poisoned with this paste are generally fatal within 24 hr. Boiling the seeds renders the poison ineffective. Local tribes, however, seldom avail themselves of this poison, probably because of its slow effect (Quisumbing, 1947). The seeds contain abrine, a highly toxic glycoprotein, and 0.6–0.9% indole alkaloids, among them *N*-methyltryptamine, *N*-methyltryptophan, and hypaphorine (**26**) (the betaine of tryptophan) as well as saponins. The methylester iodide of hypaphorine has curarizing properties; by itself, it is less toxic for warm-blooded animals. In Zimbabwe (Africa), the plant is one of the most commonly used plants against haematuria caused by *Schistosoma haematobium* infections. A single dose of the powdered seeds is a long-lasting contraceptive, the effect lasting up to 13 menstrual cycles. The seed extracts also have activity against skin cancer. Complexes of human tumor cells with the lectins from *A. precatorius* have been prepared as antitumor vaccines (Tung *et al.,* 1986). Injection of abrine-a into mice bearing Meth-A tumors inhibited the tumor growth by 90%.

Besides *Antiaris toxicaria,* there are very scant data available on arrow poisons or adjuvants in the Philippines. Some other plants that are used in the preparation of hunting poisons are *Diospyros multiflora, Ternstroemia* sp., *Dioscorea hispida, Lophopetalum javanicum, Mallotus floribundus,* and *Strophanthus* sp. (Quisumbing, 1947).

4.2. Burma, Thailand, Malaysia

Malaysia was, and still is, the center of poisoned arrows and darts in southeast Asia. The *Strychnos* species used, solely or as adjuvants, are *S. ignatii, S. nux-vomica, S. vanprukii (S. quadrangularis), S. rufa,* and *S. axillaris*. Less is known about the alkaloids of the latter three species. In *S. axillaris*, the strychnine-type alkaloids spermostrychnine, strychnospermine, and its deacetyl derivative were detected; they are less toxic and cause clonic (but not tonic) convulsions only in high doses. *S. vanprukii*, in contrast to most other Asian *Strychnos* species, has a paralyzing rather than a tetanizing activity. Up to now, only the three common, weakly toxic, pentacyclic indole bases angustine (**43**), angustidine, and angustoline of the vallesiachotaman type have been found.

The alkaloid composition of *S. ignatii* is very variable. The root bark was shown to contain tertiary and also quaternary alkaloids. Both *S. ignatii* and *S. nux-vomica* contain the strong convulsant, strychnine (**5**); *S. nux-vomica* contains the major tertiary base, 12-hydroxystrychnine, as well.

The *seco*-alkaloid diaboline, from *S. ignatii*, is not as "diabolical" as its name suggests; the opening (*seco*) of the amide ring of the basic strychnine skeleton and the introduction of a hydroxyl group at the 17-position produced a marked decrease in toxicity: high doses have a weak curare effect. One of the identified quaternary bases is the monoquaternary C-mavacurine (*S. nux-vomica*), which has a very low curare activity.

Macusine B (**61**) (*S. ignatii*) was found to act on both adrenergic and tryptamine receptors; it also caused clonic convulsions *in vivo*. The occurrence of *N*-cyano-*sec*-

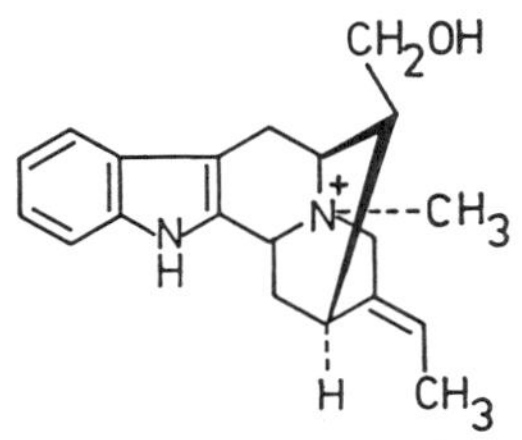

61

pseudostrychnine and -brucine in *S. ignatii* is remarkable. It is noteworthy that one of these mixed poisons from western Malaysia has been shown to act as a weak curare. The poison was most certainly derived from *S. ignatii* (Bisset *et al.,* 1977). Recent examination of nine Malayan dart poisons (1883, 1924) from the Museum of Ethnology in Vienna has confirmed that their main active compounds are cardenolides from *Antiaris toxicaria* and *Strychnos* alkaloids, probably from *S. ignatii*, the most common *Strychnos* in such poisons. Only two poisons contained diaboline as the alkaloidal component (Kopp *et al.,* 1992).

For the other minor alkaloids, see Aimi *et al.* (1989), for the chemistry and pharmacology/toxicology of strychnine and derivatives, see *Strychnos icaja,* above. Alkaloid-bearing plants in Malayan dart poisons are: *Atalantia kwangtungensis, Calamus javensis, Coscinium* sp., *Ervatamia* sp., *Gnetum gnemon*, *Gomphandra* sp., *Micromelum pubescens, Popowia tomentosa, Uvaria* sp., *Glycosmis* sp., *Dioscorea triphylla* (Bisset and Woods, 1966).

The bark and roots of the Apocynaceae *Tabernaemontana corymbosa* have been reportedly used to make arrow poison in Thailand; it also enters into the composition of the dart poison made by the Mantra (Temuan or Belanda) in Malaysia. The plant causes strong alkaloid reactions, but nothing is known about the compounds.

Aconitum is a plant genus that has been widely used for arrow poisons in Asia (Burma, China, Himalaya, Kamchatka, Kuril islands, Sakhalin, Hokkaido, far eastern Siberia) as well as in the United States (Alaska, Aleutian islands) and Europe (Bisset, 1981; Cheney, 1931; Harms, 1927, Lewin, 1923). Among the species so used have been the notorious *A. ferox, A. spicatum*, and *A. laciniatum*. *Aconitum* species have been employed as arrow-poison plants throughout a large part of China, especially in the mountainous southwestern part of the country. The toxicity of these plants has also long been recognized by the Han Chinese (Bisset, 1981). The Ainu of Sakhalin, Hokkaido, and the Kuril islands used predominately aconitum-based poison.

Aconitine (**62**) and mesaconitine (**63**) (both with an LD_{50} of about 0.12 mg kg^{-1} mouse i.v.) are the main toxic constituents of the tubers of the *Aconitum* species used in

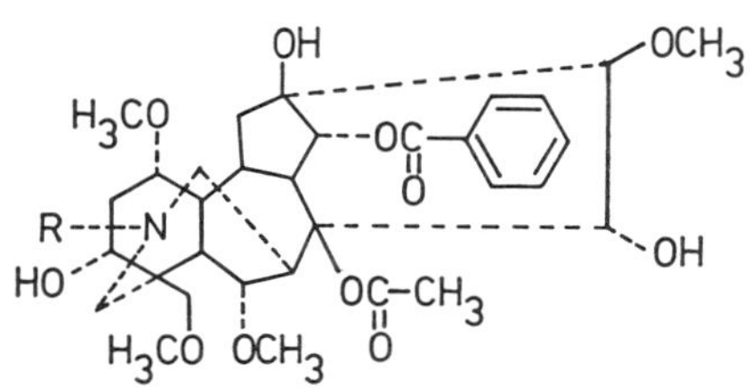

(**62**) R = CH_2–CH_3
(**63**) R = CH_3

arrow poisons. Both are oxygenized C_{19}-diterpenoid ester alkaloids. The ester functions are important for full toxicity. Given orally, 300–400 g of the fresh roots, is lethal to a horse.

Both alkaloids cause prolonged nerve depolarization by binding to the nerve-cell membrane and, through allosteric interaction, preventing normal closure of the Na^+ channels. They have almost the same affinity for the Na^+ channels of brain synaptosomes, heart and skeletal muscle cells. They paralyze the skeletal muscles "curare style," but in contrast to curare and other neuromuscular blocking drugs, they act presynaptically by decreasing the ACh release. The effect is not antagonized by physostigmine.

The principal effect of the two alkaloids is on the cardiovascular and respiratory systems. They act cardiotoxically and neurotoxically, both peripherally and centrally, decreasing blood pressure and body temperature, causing cardiac arrhythmia, giddiness and weakness related to loss of muscle power. Death is from cardiac and respiratory failure.

5. APPENDIX: COMPOSITION OF AFRICAN ARROW POISONS

Nyarwanda (Rwanda)
- *Acokanthera schimperi*
- *Euphorbia grantii*
- *Oreosyce africana*
- *Zehneria scabra*
- *Zanthoxylum chalybeum*
- *Synadenium grantii*
- *Capsicum frutescens*
- *Cucumis aculeata*
- *Rhoicissus tridentata*
- *Solanum incanum*
- *Opuntia ficus-indica*
- *Aloe lateritia*
- *Boöphone species*

Lobi (Burkina Faso)
- *Strophanthus hispidus*
- *Afzelia africana*
- *Capsicum frutescens*
- *Euphorbia kamerunica*
- *Eu. poissonii/unispina*
- *Sapium grahamii*
- *Thevetia peruviana*
- Heads of poisonous snakes

Ewondo (Cameroon)
- *Strophanthus hispidus*
- *Zanthoxylum gilletii*
- *Solanum torvum*
- *Nicotiana tabacum*
- Juice of the bush *etotonga*

Bété (Ivory Coast)
- *Mansonia altissima*
- *Diospyros physocalycina*
- *Funtumia elastica*
- *Harungana madagascariensis*
- *Tabernaemontana crassa*
- *Corynanthe pachyceras*

!Kung bushmen (Namibia)
- *Swartzia madagascariensis*
- *Asparagus exuvialis*
- *Sansevieria aethiopica*
- *Terminalia sericea*
- Larvae of *Diamphidia* beetle

Nyambo (Rwanda)
- *Strychnos usambarensis*
- *Euphorbia tirucalli*
- *Aloe lateritia*
- *Synadenium grantii*
- *Opuntia ficus-indica*
- *Capsicum frutescens*
- *Euclea racemosa subspecies schimperi*
- *Ipomoea wightii*
- *Acokanthera schimperi*

Djerma (Niger)
- *Strophanthus sarmentosus*
- *Pergularia tomentosa*
- *Crinum zeylanicum*
- *Cassia italica*
- *Boscia senegalensis*
- *Ipomoea asarifolia*
- *Ziziphus mauritiana*
- Head of a viper

Guéré of Zanié (Ivory Coast)
- *Mansonia altissima*
- *Erythrophleum suaveolens*
- *Elaeophorbia grandifolia*
- *Zanthoxylum gilletii*
- *Ficus asperifolia*
- *Funtumia elastica*
- *Eriocoelum racemosum*

Dan (Ivory Coast)
- *Mansonia altissima*
- *Tetrapleura tetraptera*
- *Microglossa pyrifolia*
- *Picralima nitida*

Babinga Pygmies (CAR)
- *Parquetina nigrescens*
- *Strychnos camptoneura*
- *Capsicum frutescens*

Fang (Cameroon/Gabon)
Strophanthus gratus
Mucuna pruriens
Dioscorea sansibarensis
Detarium macrocarpum
Toad *Bufo superciliaris*

Kamba (Kenya)
Acokanthera schimperi
Aloe secundiflora
Adenia volkensii
Capparis tomentosa

Socé (Senegal)
Asparagus africanus
Clerodendrum capitatum
Gardenia ternifolia subspecies jovis-tonantis var. goetzei
Acacia macrostachya

Fali (Cameroon)
Strophanthus hispidus
Euphorbia kamerunica
Balanites aegyptiaca
Cochlospermum tinctorium
Elionurus hirtifolius
Aloe species

Guéré of Boo (Ivory Coast)
Mansonia altissima
Diospyros physocalycina
Zanthoxylum gilletii
Annickia polycarpa
Pachypodanthium staudtii
Funtumia africana
Thonningia sanguinea
Dichapetalum toxicarium
Adenia species (lobata?)
Dioscorea species

Mbuti Pygmies (Zaire)
Parquetina nigrescens
Elaeophorbia drupifera
Piptadenistrum africanum
Rauwolfia vomitoria
Anchomanes difformis
Campylospermum elongatum
Alafia grandis
Cleistopholis glauca
Corymborkis corymbis
Capsicum frutescens

REFERENCES

General References

Aimi, N., Sakai, S. I., and Ban, Y., 1989, Alkaloids of Strychnos and Gardneria species, in: *The Alkaloids*, Vol. 36 (A. Brossi, ed.), Academic Press, San Diego, pp. 1–68.

Bauer, W. P., 1981, Curare-Pfeilgiftbereitung—Heute, *Arch. Völkerkunde* **35:**15–30.

Bauer, W. P., and Fondi, M., 1962, Untersuchungen an sehr alten Curare-Pfeilgiften mit besonderer Berücksichtigung ihrer Haltbarkeit, *Sci. Pharm.* **30:**173–182.

Bennett, A. E., 1968, The history of the introduction of curare into medicine, *Anaes. Analg. Curr. Res.* **47:**484–492.

Bisset, N. G., 1966, The arrow and dart poisons of South-East Asia, with particular reference to the Strychnos species used in them, Part 1, Indonesia, Borneo, Philippines and Indo-China, *Lloydia* **29:**1–18.

Bisset, N.G., 1981, Arrow poisons in China. Part II. Aconitum—botany chemistry and pharmacology, *J. Ethnopharmacol.* **4:**247–336.

Bisset, N. G., 1989, Arrow and dart poisons, *J. Ethnopharmacol.* **25:**1–41.

Bisseet, N. G., Woods, M. C., 1966, The arrow and dart poisons of South-East Asia, with particular reference to the Strychnos species used in them, Part II: Burma, Thailand, and Malaya. *Lloydia* **29:**172–195.

Boehm, R., 1895/97, Das südamerikanische Pfeilgift Curare in chemischer und pharmakologischer Beziehung, Abhandl, der mathem-physik, Classe der Königlich-Sächsischen Ges. Wiss., Vol. 22, pp. 200–238 (Part 1); Vol. 24, pp. 1–52 (Part 2).

Cheney, R. H., 1931, Geographic and taxonomic distribution of American plant arrow poisons, *Am. J. Bot.* **2:**136–145.

Davis, E. W., and Yost, J. A., 1983, The ethnomedicine of the Waorani of Amazonian Ecuador, *J. Ethnopharmacol.* **9:**273–297.

Harms, H., 1927, Historisch-kritische Untersuchungen über das Pfeilgift nordostasiatischer Völker, *Arch. Pharm.* 81–90.

Holmstedt, B., 1972, The ordeal bean of Old Calabar. The pageant of Physostigma venenosum in medicine, in: *Plants in the Development of Modern Medicine* (T. Swain, ed.), Harvard University Press, Cambridge, MA, pp. 303–360.

Leboeuf, M., Cavé, A., Bhaumik, P. K., Mukherjee, B., and Mukherjee, R., 1982, The phytochemistry of the Annonaceae, *Phytochemistry* **21**:2783–2813.
Lewin, L., 1923, *Die Pfeilgifte*, Verlag J.A. Barth, Leipzig
Marini-Bettolo, G.B., 1975/76, Advances in the research of curare and Strychnos alkaloids, *Rend. Accad. Naz. 40 quaranta* pp. 61–76.
Martin, S. F., 1987, The Amarylidaceae alkaloids, in: *The Alkaloids*, Vol. 30 (A. Brossi, ed.) Academic Press, San Diego, pp. 251–376.
Matras, J., and Martin, M. A., 1972, Contribution à l'éthnobotanique des Brou (Cambodge-Province de Ratanakiri), *J. Agric. Trop. Bot Appl.* **19**:1–49.
Neuwinger, H.D., 1994, Fish-poisoning plants in Africa, *Bot. Acta* **107**:263–270.
Neuwinger, H. D., 1996, *African Ethnobotany: Poisons and Drugs, Chemistry, Pharmacology, Toxicology*, Chapman & Hall, London.
Perrot, E., and Vogt, E., 1913, *Poisons de flèches et poisons d'épreuves*, Vigot Frères, Paris.
Prance, G. T., 1972, Ethnobotanical from Amazonian Brazil, *Econ. Bot.* **26**:22–37.
Quisumbing, E., 1947, Philippine plants used for arrow and fish poisons, *Philipp. J. Sci.* **77**:127–177.
Schultes, R. E., and Holmstedt, B., 1977, De Plantis toxicariis e mundo novo tropicale commentationes, VIII. Miscellaneous notes on mystericaceous plants of South America, *Lloydia* **34**:61–78.
Simmons, D. C., 1956, Efik divination, ordeals and omens, *S.W.J. Anthropol.* **12**:223–228.
Thomas, D. J., 1972, The indigenous trade system of southeast Estado Bolivar, Venezuela, *Anthropologica* **33**:3–37.
Tschesche, R., and Kaussmann, E. U., 1975, The cyclopeptide alkaloids, in: *The Alkaloids,* Vol. 15 (R. F. Manske, ed.), Academic Press, New York, pp. 165–205.
Tung, T. C., Lin, J. Y., Shionoya, H., Koyanagi, N., and Arai, M., 1986, Lectin tumor cell complexes as antitumor vaccines, Japan Tokyo Koho JP 61 33,008 [86 33,008], 31.7.1986
Vellard, J., 1965, *Histoire du Curare*, Gallimard, Paris
von Reis, S., and Lipp, F. J., 1982, *New Plant Sources for Drugs and Foods from the New York Botanical Garden Herbarium*, Harvard University Press, Cambridge, MA.
von Reis-Altschul, S., 1973, *Drugs and Foods from Little-Known Plants*, Harvard University Press, Cambridge, MA.
Waser, P. G., 1972, Chemistry and pharmacology of natural curare compounds, in: *Neuromuscular Blocking and Stimulating Agents, Intern. Encycl. Pharmacology and Therapeutics* Section 14, Oxford University Press, pp. 205–239.

Specific References

Angenot, L., 1971, De l'éxistence en Afrique Centrale d'un poison de flèche curarisant issu du Strychnos usambarensis Gilg, *Ann. Pharm. F.*, **29**:353–364
Biocca, E., Bovet, C., Galeffi, C., and Marini Bettolol, G.B., 1965, Sui curaro Yanoama. Un nuovo tipo di curaro indigeno "Curaro di torrefazione e percolazione," *Atti Accad. Naz. Lincei Cl. Sci. Fis. Mat. Nat.* **38**:34–38.
Bisset, N. G., Baser, K. H. C., Phillipson, J. D., Bohlin, L., and Sandberg, F., 1977, Muscle-relaxant activity in Asia *Strychnos* species. A re-examination of two western Malaysian dart poisons, *Lloydia* **40**:546–560.
Galeffi, C., Messana, I., and Marini Bettolo, G. B., 1983, *N,N*-dimethyl-5-methoxytryptamine, a component of a dart poison of the Yanoama Indians, *J. Nat Prod.* **46**:586–587.
Kao, C. Y., Salwen, M. J., Hu, S. L., Pitter, H. M., and Woollard, J. M. R., 1989, Diamphidia toxin, the bushmen's arrow poison: Possible mechanism of prey-killing, *Toxicon* **27**:1351–1366.
Kopp, B., Bauer, W. P., and Bernkop-Schnürch, A., 1992, Analysis of some Malaysian dart poisons, *J. Ethnopharmacol.* **36**:57–62.
Krukoff, B. A., and Smith, A. C., 1937, Notes on the botanical components of curare. *Bull. Torrey Club* **64**:401–409.
Macrae, W. D., and Towers, G. H. N., 1984, An ethnopharmacological examination of Virola elongata bark: A South American arrow poison, *J. Ethnopharmacol.* **12**:75–92.
Neuwinger, H. D., and Scherer, G., 1976, Das Larven-Pfeilgift der Buschmänner, *Biol. Unserer Zeit* **6**:75–82.
Pradham, S. N., Pandey, K., and Badola, R. P., 1964, Clinical trial of Hayatin methiodide as relaxant in 100 cases, *Br. J. Anaesth.* **36**:604–611.

Schwacke, W., 1884, Bereitung des Curare-Pfeilgifts bei den Teuna-Indianern, Jahrbuch des Köiglich Bot, Gartens und des Bot. Museuma zu Berlin, Vol. 3, pp. 220–223.
Spjut, R. W., and Perdue, R. E., 1976, Plant folklore: A tool for predicting sources of antitumor activity, *Cancer Treat. Rep.* **60:**979–985.
Suffness, M., and Cordell, G. A., 1985, Antitumor alkaloids. XII. Maytansinoids, in: *The Alkaloids*, Vol. 25 (A. Brossi, ed.), Academic Press, New York, pp. 142–156.

Part II

Biochemistry

Chapter 4

Chemical Taxonomy of Alkaloids

Peter G. Waterman

1. INTRODUCTION

The use of secondary metabolites produced by organisms as systematic markers is as old as the use of morphological characters but has, until this century, been subjective. For example, calling a plant aromatic was invariably an indication of the presence of volatile oils, whereas the sensation of "bitterness," more often than not, denoted the presence of alkaloids.

Relatively recent (post 1950) advances in phytochemical techniques have allowed a proper definition of these chemical characters and with it an assessment of their value as taxonomic markers. Their value was recognized in two seminal publications in chemical taxonomy published in the early 1960s (Alston and Turner, 1963; Swain, 1963). It is probably true to say that the early promise anticipated by the authors in these two books has not always been sustained and that today we recognize quite severe limitations on the use of all types of secondary metabolite in taxonomy.

2. WHAT IS AN ALKALOID?

The definition of an alkaloid has always been rather problematical. In a taxonomic context it is probably best to restrict what is recognized as an alkaloid to the following:

1. A nitrogenous compound in which at least one nitrogen atom is derived directly from an amino acid, and
2. A compound with limited distribution

Thus, nonbasic and noncyclic structures such as simple amines and amides (commonly called *protoalkaloids*) would be considered as alkaloids for taxonomic purposes. A

Peter G. Waterman • Phytochemistry Research Laboratories, University of Strathclyde, Glasgow G1 1XW, Scotland.

Alkaloids: Biochemistry, Ecology, and Medicinal Applications, edited by Roberts and Wink. Plenum Press, New York, 1998.

large group of nitrogenous substances that fail to meet these criteria are nitrogen-containing terpenoids, notably diterpenes and steroids, and polyketides. These are usually called *pseudoalkaloids.*

Such a definition normally works adequately and, as long as the biosynthetic processes involved are understood, allows a compound to be placed in context. Hegnauer (1963a) subclassified alkaloids into biosynthetic families based on recognition of the parent amino acid. To the uninitiated this can be a little perplexing but the value of the approach is seen clearly in the recognition of quinine and related alkaloids as members of the indole–*seco*-loganin family, even though the final structure of quinine has a quinoline skeleton rather than an indole skeleton.

That is not to say that problems do not remain; the biosynthetic origin of nitrogen in alkaloids is not always obvious and there is some dispute over what is meant by an amino acid. For example, Hegnauer (1963a) considered alkaloids with a single nitrogen arising from nicotinic acid to be pseudoalkaloids and yet accepted those derived from anthranilic acid as true alkaloids. Neither nicotinic acid nor anthranilic acid truly qualifies as an essential amino acid, but the nitrogen is part of an established precursor molecule that forms part of the complete alkaloid structure. In that sense they are much closer to true alkaloids than to pseudoalkaloids in which the nitrogen is an adduct into an essentially nonalkaloidal compound.

2.1. A Unifying Theme in the Biosynthesis of True Alkaloids

There is a simple chemical theme running through the mechanism of biosynthesis of the vast majority of alkaloids. This is most easily rationalized chemically (Geissman and Crout, 1969) as (a) Schiff base formation and (b) Mannich condensation (Fig. 1).

Virtually all major groups of true alkaloids can be considered to arise by this route, with the donor of the nitrogen and carbonyl changing from alkaloid group to alkaloid group. The donors for the major alkaloid types are given in Table I.

3. THE IMPACT OF ALKALOIDS ON TAXONOMY

It seems fair to say that the taxonomic insights gained from studying the distribution of alkaloids over the last 40 years have not lived up to the expectations of the 1960s and, with the advent of molecular biology in general and DNA sequencing in particular, their

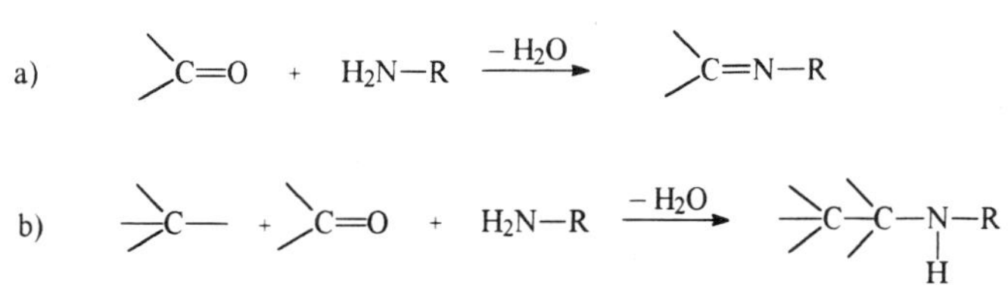

Figure 1. (a) Schiff base formation and (b) Mannich condensation; key reactions in alkaloid formation.

Table I
Precursors of Some Major Alkaloid Classes

Alkaloid type	N Source	C=O source
Amaryllidaceae	Tyrosine/phenylalanine	Phenylbenzaldehyde from tyrosine/phenylalanine
1-Benzylisoquinoline	Tyrosine/phenylalanine	Phenylacetaldehyde from tyrosine/phenylalanine
Betalains	Indole or proline	Tyrosine (betalamic acid)
Emetine type	Tyrosine	*seco*-Loganin
Indole-loganin	Tryptophan	*seco*-Loganin
Tropane	Ornithine	Ornithine(same molecule)
Pyrrolizidine	Ornithine	Ornithine
Quinolizidine	Lysine	Lysine

study for taxonomic purposes has become less of a focus. The relative lack of success can be attributed to a number of factors.

First is the continued lack of an adequate data base from which to work. In the majority of cases, data on alkaloid distribution are gathered from information published for purposes other than systematics; e.g., as an exercise in structure elucidation, or as a result of a search for substances exhibiting a particular type of biological activity. Such data were often far from complete and can lead to a false impression of the biosynthetic capacity of a taxon.

Second, chemical variation is often quite considerable in alkaloid-producing taxa. A typical example of this is seen in the investigation of *Papaver radicatum* in Norway (Folkestad *et al.*, 1988). Of 12 collections studied, all contained two protopine alkaloids and an isopavine alkaloid, but a third isopavine, two further protopines, a promorphinane, and three unidentified alkaloids were found sporadically. While the sporadically distributed alkaloids may have systematic value in identifying subspecies, only the three "common" alkaloids are of value at the specific level. Far too little attention has been paid to this type of problem which can often reveal different taxonomic levels at which alkaloids, appropriately classified (e.g., as individual structures, structural classes), have systematic value.

Third, variation may also be produced by ecological factors. For example, Chandra and Purohit (1980) found that berberine content within and between species growing in the Himalayas diminished with an increase in altitude and concluded that alkaloid content was an adaptive response to altitude as well as a genetic character of the different species studied.

Finally, most of our attempts to use alkaloids (or any other group of metabolites) as systematic markers have suffered from the fact that they have been discussed in terms of either a stand-alone taxonomy or being superimposed on extant classifications. Rarely have they been adequately integrated into a comprehensive synthesis (compare Da Silva *et al.*, 1988; Waterman, 1990). This generally simply reflects the circumstance under which work of this nature is done; that is, by an individual, isolated from those in other disciplines who could contribute to optimizing the value of that individuals findings.

In this chapter discussion has been restricted to typical alkaloids and pseudoalkaloids; several other peripheral groups of compounds have not been included such as nonprotein amino acids, cyanogenic glycosides, and glucosinolates. Taxonomic assessments have been made against just two of the many schemes available, those of Thorne

(1968) and Dahlgren (1977, 1980). Those have been chosen because of their awareness and consideration of secondary metabolites as potential markers.

3.1. Alkaloids Originating from Tyrosine (and Phenylalanine), Usually through Dopa

The "family" of alkaloids formed from the two phenylethylamine amino acids is large (Fig. 2). The most prolific are the 1-benzyltetrahydroisoquinoline group which have the common skeleton C_6-C_2-N-C_2-C_6 (type A in Fig. 2), originating from two amino acid

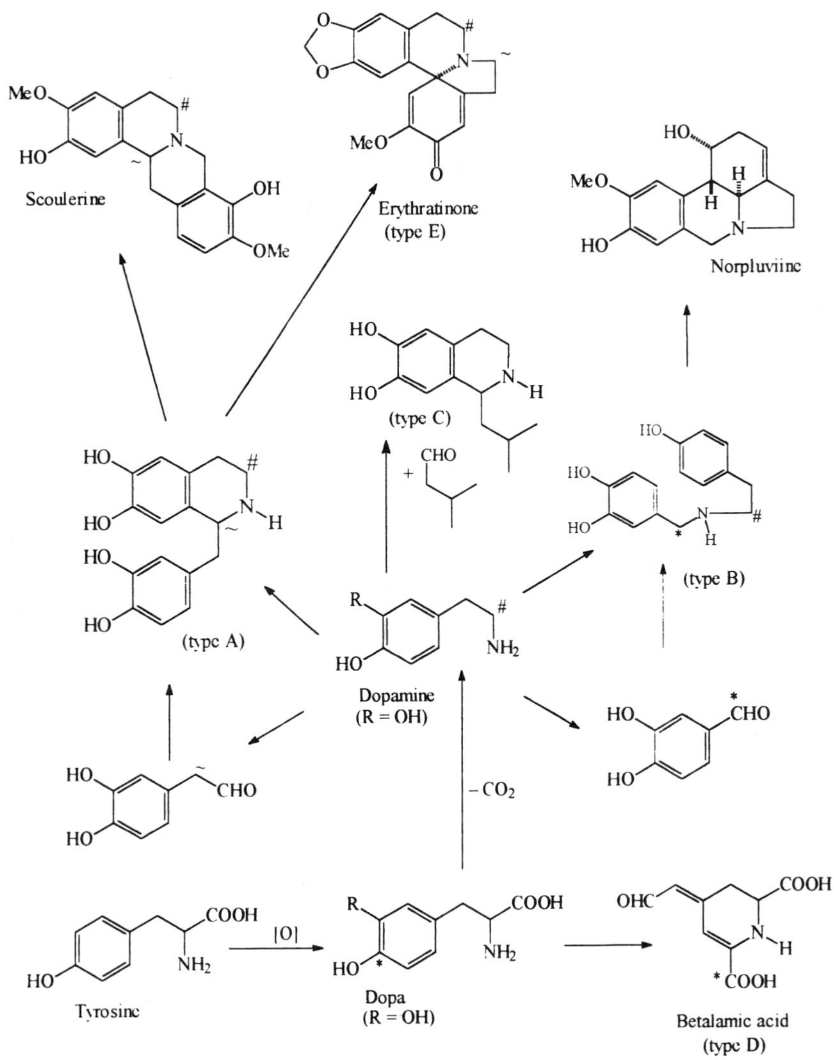

Figure 2. Proliferation of alkaloids from the amino acids tyrosine and phenylalanine (R = H or OH).

units. The second large group, based on C_6-C_2-N-C_1-C_6 (type B in Fig. 2), arise from the same precursors. Simple isoquinoline alkaloids C_6-C_2-N-C_n (type C in Fig. 2) employ a number of different moieties (isoprene or acetate) to donate the carbonyl to the Mannich condensation. The central component of the betalains, betalamic acid (type D in Fig. 2), is a much modified derivative of dopa.

3.1.1. TYPICAL 1-BENZYLISOQUINOLINE ALKALOIDS

The currently known distributions of the major classes of 1-benzyltetrahydroisoquinoline alkaloids are given in Table II, where the families are divided according to whether or not they are considered to be part of the Annoniflorae and Nymphaeiflorae (*sensu* Thorne, 1968). The focus of distribution of these alkaloids within a putatively primitive group of dicotyledonous families was first discussed in detail by Hegnauer (1963a) who identified their occurrence with the major families placed by Wettstein in the Polycarpicae (approximates to the Annoniflorae and Nymphaeiflorae of Thorne). In the same paper Hegnauer drew attention to Wettstein's inclusion of the Papaveraceae in the Rhoedales, and while the Papaveraceae and Fumariaceae were producers of a wide range of benzylisoquinoline alkaloids these were not produced at all by any of the other families of the Rhoedales. His view that the Papaveraceae (including Fumariaceae) was misplaced within the Rhoedales and should be placed in or near the Polycarpicae has been supported in all recent major classifications (see the Papaverineae in Table III).

Yet, despite the obvious value of these alkaloids, their systematic significance must be viewed in the context that their distribution within a family is often sporadic. As pointed out by Hegnauer (1963a) the ability of a taxon to produce alkaloids is not, in an evolutionary sense, a one-way process. In time the number and diversity of alkaloid structures may increase in an evolutionary line; but equally the capacity to form these substances may be gradually or completely lost over evolutionary time. It is to be expected, therefore, that even in alkaloid-rich families like the Annonaceae some species have lost this capacity.

What we see today for the Annoniflorae (Table III) has been interpreted by Kubitzki (1969) as separating this group of families from the mainstream of angiosperm evolution, describing them as "an odd lateral branch of the evolution of the dicotyledons which appears wholly fixed and has remained primitive in different respects, and which has given rise to a limited number of more modern groups like Papaverales (*sensu stricto*) and Rutaceae."

The Papaverales has already been discussed. Hegnauer (1963a) and Price (1963) both drew attention to the capacity of some genera of the Rutaceae to produce benzylisoquinoline alkaloids. While the number of genera with this ability is relatively small (currently 5 out of about 150), the production of these alkaloids is common to almost all species of those genera and they are often produced in large amounts. Fish and Waterman (1973) and Waterman (1975) explored in more detail the similarities between the Berberidales, the Papaverales, and the large rutaceous pantropical genus *Zanthoxylum*, noting in particular the occurrence of unusual benzophenanthridine alkaloids, such as chelerythrine (**1**) and nitidine (**2**).

In his classification of the Rutaceae, Engler placed the benzylisoquinoline producing genera in two subfamilies: *Zanthoxylum* (including *Fagara*) in the Rutoideae, *Toddalia*, *Phellodendron,* and *Fagaropsis* in the Toddalioideae. Hartley (1981) examined the rela-

Table II

Distribution of the Major Classes of 1-Benzylisoquinoline Alkaloids in Angiospermae (Guinaudeau and Bruneton, 1993)

	1	2	3	4	5	6	7	8	9	10	11	12	13	14	15
Annoniflorae and Nymphaeiflorae (*sensu* Thorne, 1968)															
Annonaceae	+	+	++	+	+					+					
Aristolochiaceae	+		+	++	+										
Berberidaceae	++	+	+		++	+		+		+					+
Canellaceae				+											
Eupomatiaceae			+												
Fumariaceae		+	+		++	++	+	++	++		+		+		
Hernandiaceae	+		++												
Lauraceae	+	+	++		+					+					+
Magnoliaceae	+		++												
Menispermaceae	++	+	++	+	+					+	+	+			+
Monimiaceae	+	+	++	+											
Nympheaceae	+	+	++												
Papaveraceae		+	++		++	++		++	++	+			+		+
Piperaceae			+												
Ranunculaceae	++	+	++		+	+		+		+					+
Other taxa															
Alangiaceae					+										
Araceae			+												
Buxaceae	+														
Caprifoliaceae									+						
Euphorbiaceae	+	+	+							+					
Leguminosae			+		+							+			
Liliaceae		+	+												
Rhamnaceae	+		+												
Rutaceae	+		+		+			+	+						
Sapindaceae								+	+						
Umbelliferae	+														

1, benzylisoquinolines and bisbenzylisoquinolines; 2, proaporphines; 3, aporphines; 4, aristolochic acids and aristolactams; 5, protoberberines; 6, phthalideisoquinolines; 7, spiroisoquinolines; 8, protopines; 9, benzophenanthridines; 10, morphinans and morphinanedienones; 11, hasubabans; 12, dibenzazonines; 13, rhoedines; 14, cularines; 15, pavines and isopavines.

Table III
Distribution of Classes of 1-Benzylisoquinoline Alkaloids; Taxa Arranged According to Thorne (1968)

	1	2	3	4	5	6	7	8	9	10	11	12	13	14	15
Annoniflorae															
Annonales															
Winterineae															
Illicinae															
Magnoliineae	+		+												
Annonineae	+	+	+	+	+					+					
Aristolochineae	+		+	+	+										
Laurineae	+	+	+	+	+					+					+
Piperineae			+												
Berberidales															
Berberidineae	+	+	+	+	+	+		+		+	+	+			+
Papaverineae		+	+		+	+	+	+	+	+		+	+	+	+
Nymphaeiflorae															
Nymphaeales	+	+	+												
Rutiflorae															
Rutales	+		+		+			+	+						

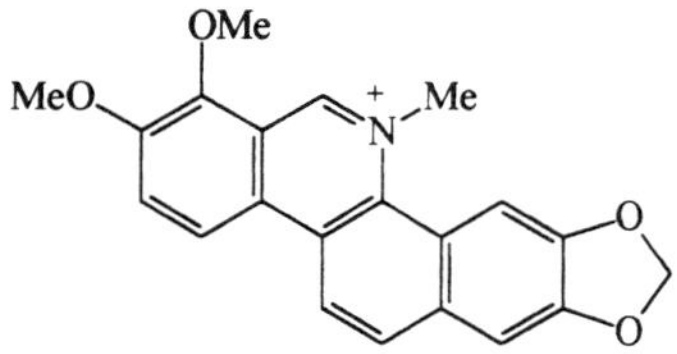

1

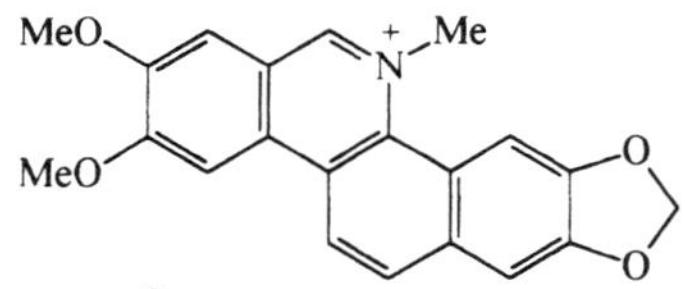

2

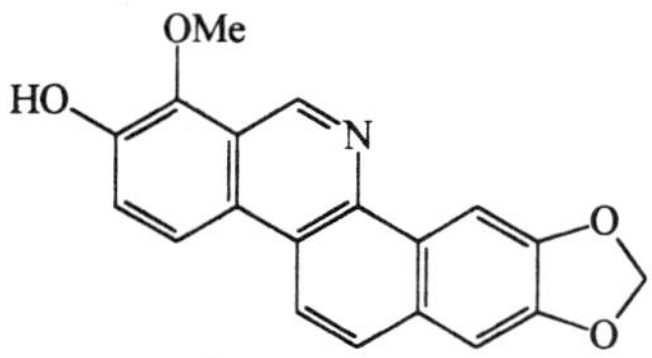

3

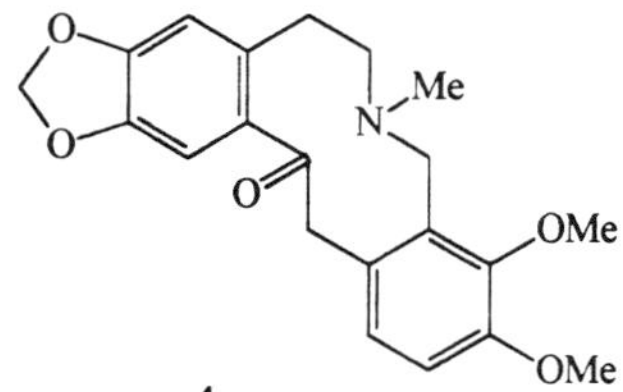

4

tionships between *Zanthoxylum*, *Toddalia,* and *Phellodendron* and concluded that they were close to each other and also linked to a northern subset of species hitherto considered as *Euodia* and for which he raised the generic name *Tetradium*. Hartley's hypothesis raised the possibility that *Tetradium* species could also be benzylisoquinoline alkaloid producers and this has now been confirmed with the isolation of decarine (**3**) from *T. glabrifolium* (Ng *et al.*, 1987) and then α-allocryptopine (**4**) from *T. trichotomum* (Quader *et al.*, 1990). This is a satisfying confirmation of the value of these alkaloids in delineating a group of genera within the Rutaceae. Unfortunately, what the occurrence of these alkaloids means in terms of a relationship between the Rutaceae and the Papaverineae and Berberidineae (Table III) is still far from clear, as is an understanding of what significance can be placed on the occasional occurrence of 1-benzylisoquinoline alkaloids in other apparently unrelated families like the Euphorbiaceae, Buxaceae, and Rhamnaceae (Table II).

3.1.2. ALKALOIDS OF THE AMARYLLIDACEAE

These alkaloids, which are exemplified by norpluviine (Fig. 2), are a feature of the Amaryllidaceae, a monocot taxon (for a detailed account of these alkaloids see Cordell, 1981). Biosynthetically they are distinguished from the 1-benzylisoquinolines only by the contraction of the nonnitrogenous component from C_6C_2 to C_6C_1. Given the origin of these alkaloids from the same precursors by means of the same biosynthetic reactions as for the 1-benzylisoquinoline alkaloids suggests that the two major groups of producers might well share a common origin. This is supported by most taxonomic treatments of the Angiospermae (Hegnauer, 1963b; Dahlgren, 1977) which envisage a monocot line evolving from Polycarpicae-like ancestors. Furthermore, individuals of two monocot taxa, the Liliaceae and Araceae, are reported to contain typical 1-benzylisoquinoline aporphine alkaloids (Table II). It is plausible that these two groups of alkaloids could have evolved separately but, given the weight of other evidence pointing to Polycarpicae-like ancestors for the progenitors of monocots, the distribution of these metabolites does lend strong support to this being the case (Hegnauer, 1963a).

3.1.3. ALKALOIDS OF *ERYTHRINA*

The alkaloids of *Erythrina* (Leguminosae) are characterized by an unusual skeleton that can be regarded as yet a further modification of the Schiff base adduct of dopa and phenylacetaldehyde that probably arises from the same route via a rearrangement of a normal 1-benzylisoquinoline. An example is shown as type E in Fig. 2. A few typical benzylisoquinolines are also known to occur in the Leguminosae (Table II).

3.1.4. THE BETALAINS

The betalains are a group of colored nitrogenous pigments that share a betalamic acid unit (Fig. 2), linked with various other amino acids, notably dopa (betanidin, **5**) and proline (indicaxanthin, **6**). Betalains are often glycosylated. The most interesting feature

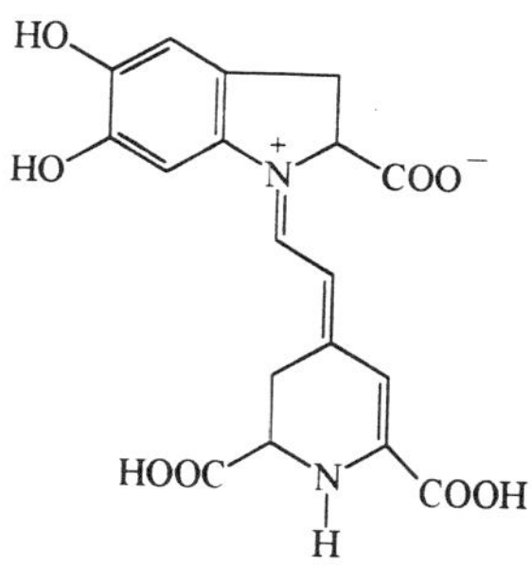

5

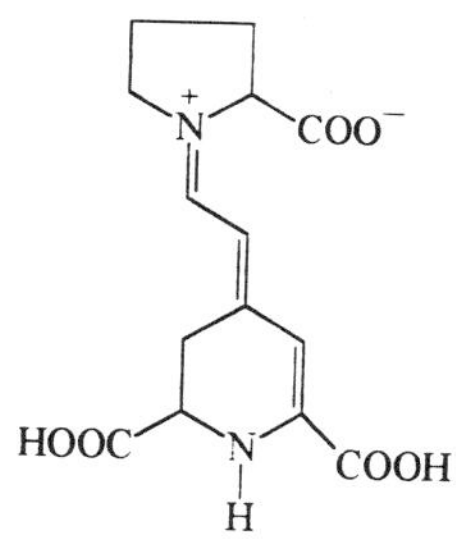

6

of their formation is the production of betalamic acid, which originates through fission of the aromatic ring of dopa, with a subsequent recyclization (Fig. 3).

Within higher plants betalains are confined to families of the Centrospermae (Caryophyllales), where they supplant the flavonoid anthocyanin pigments in flower colors. However, not all taxa of the Caryophyllales produce betalains. They are known from the Aizoaceae, Amaranthaceae, Basellaceae, Cactaceae, Chemopodiaceae, Didiereaceae, Nyctaginaceae, Phytolaccaceae, and Portulacaceae, but not from the Caryophyllaceae or Molluginaceae. The taxonomic significance of these pigments has been reviewed (Harborne and Turner, 1984; Waterman and Gray, 1987; Strack *et al.*, 1993). While their value in attributing taxa of doubtful origin to the Caryophyllales has been very beneficial (Mabry *et al.*, 1978), their phylogenetic importance remains a matter of controversy. Much has been made of their origin from tyrosine and a possible link to the typical benzylisoquinoline-producing taxa.

However, there is considerable and justified resistance to dividing the Caryophyllales into two separate taxa on the basis of presence or absence of betalains as the Caryophyllales is one of the most natural and universally accepted orders on other criteria. Furthermore, betalains have recently been found to occur in Basidomycetes (Strack *et al.*, 1993), an interesting case of chemical convergence.

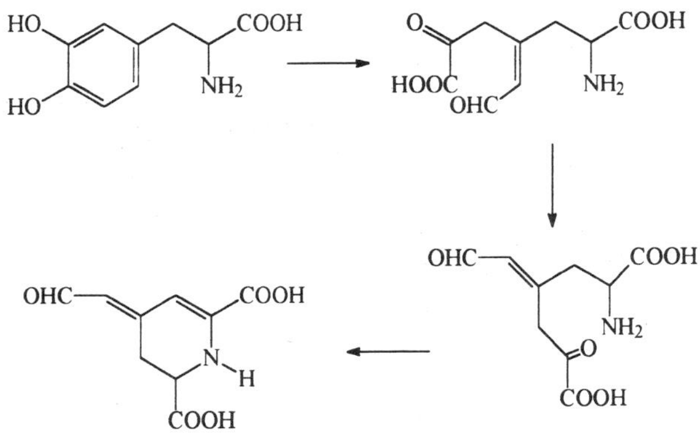

Betalamic acid

Figure 3. Formation of betalamic acid from dopa.

3.1.5. EMETINE-TYPE ALKALOIDS

Emetine (**7**) and related alkaloids occur only in *Urogoga* (*Cephaelis*) *ipecacuanha* and a few allied species in the Rubiaceae and in the Alangiaceae (Cordell, 1981) and are

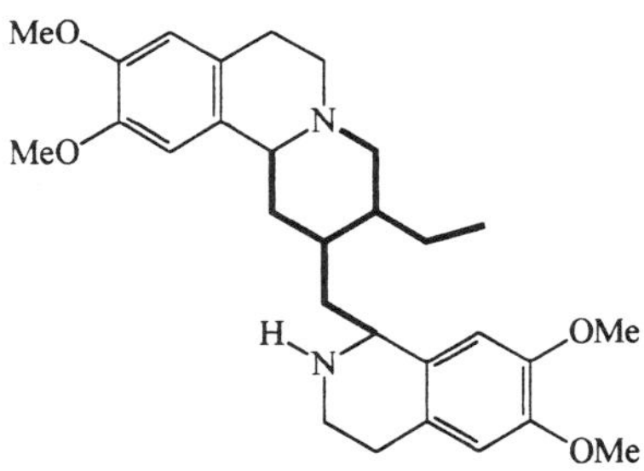

7

of only limited taxonomic value. They are mentioned here because they represent a halfway point between the tyrosine/phenylalanine derived alkaloids discussed so far and the even larger indole–*seco*-loganin group which are dealt with next. The link between the two lies in the presence of a 9- or 10-carbon terpenoid moiety (9 in this case, shown in bold in **7**). This fragment, which arises from *seco*-loganic acid, occurs far more commonly in association with another aromatic amino acid, tryptophan.

3.2. Indole–*seco*-Loganin-Derived Alkaloids

seco-Loganin (**8**) is a monoterpene iridoid that is able to combine with an amine to form alkaloids. The most common amine substrate is tryptamine, thus leading to the numerous indole–loganin alkaloid group. This type of alkaloid occurs most widely in the Gentianiflorae–Gentianales, in the families Loganiaceae, Apocynaceae, and Rubiaceae,

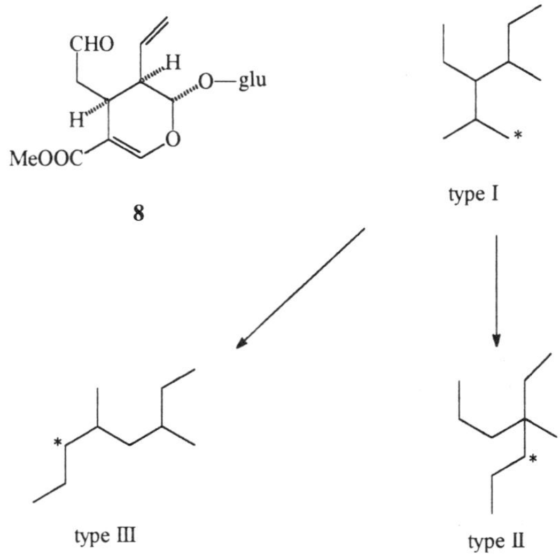

Figure 4. The major structural skeletons derived from *seco*-loganin and found in indole–loganin alkaloids.

and to a lesser extent in the Alangiaceae, Nyssaceae, and Icacinaceae of the Corniflorae–Cornales. These two superorders are adjacent in the systems of both Thorne (1968) and Dahlgren (1980).

The complexity achieved in such indole alkaloids is considerable, particularly in the Gentianales. This is based initially on the various modifications that occur in the *seco*-loganin skeleton (Fig. 4), and then on a vast array of secondary reactions. The *seco*-loganin skeleton has often been considered to take on three major structural types in combination with indole (Tillequin *et al.*, 1993)—type I (e.g., ajmalicine, **9**), type II (e.g., vincadiffor-mine, **10**), and type III (e.g., eglandine, **11**)—while modification of the indole nucleus can lead to quinoline heterocycles in alkaloids such as methyldihydroquinicine (**12**).

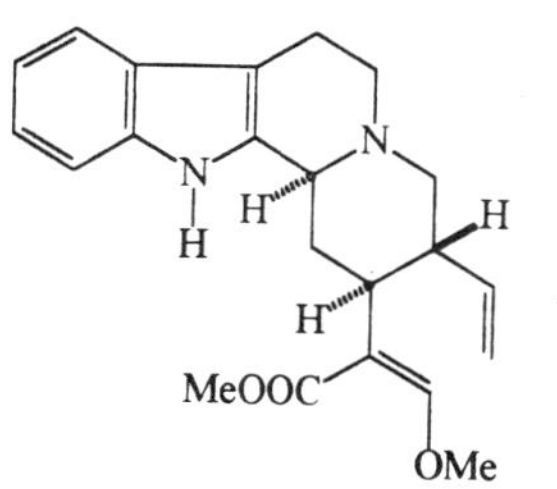

9

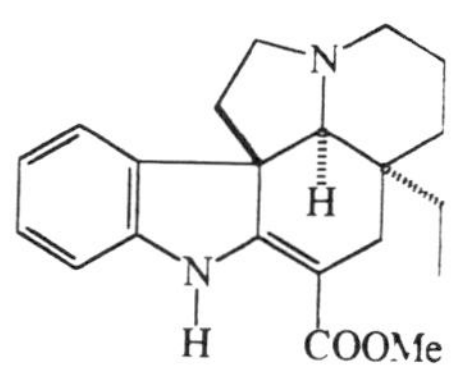

10

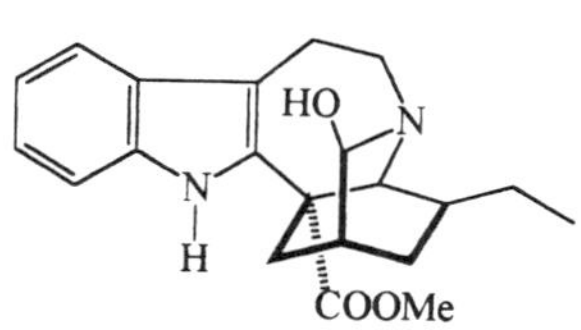

11

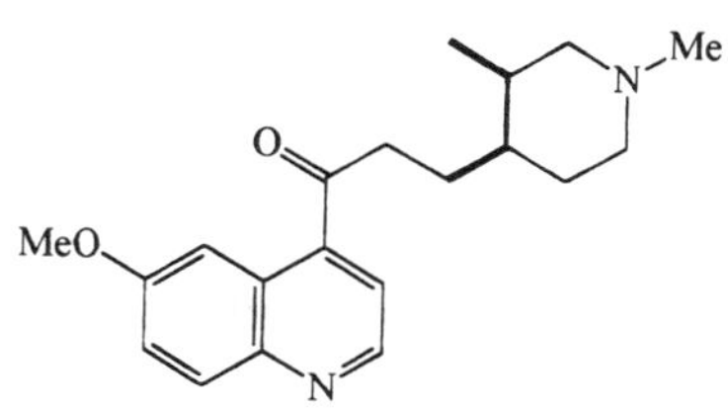

12

Table IV
Distribution of Indole–*seco*-Loganin Alkaloids and *seco*-Iridoids in Families of the Corniflorae–Cornales and Gentianiflorae (*sensu* Dahlgren, 1980)

Taxon	Iridoids	*seco*-Iridoids	Indole–*seco*-loganin alkaloids
Cornales (selected)			
Garryaceae	+		
Alangiaceae	+		+
Nyssaceae	+		+
Cornaceae	+	+	
Davidiaceae	+	+	
Icacinaceae	+	+	+
Hydrangeaceae	+	+	
Gentianiflorae			
Goodeniales			
Goodeniaceae	+	+	
Oleales			
Oleaceae	+	+	
Gentianales			
Loganiaceae	+	+	+
Rubiaceae	+	+	+
Menyanthaceae	+	+	
Gentianaceae	+	+	
Apocynaceae	+	+	+
Asclepiadaceae			

The type III products are, so far, known only from the Apocynaceae and it is possible to associate a number of other particular modifications with singular taxa. But this is the exception rather than the rule and the best taxonomic level for employing these compounds appears to be at the family or higher. But here again, as with the benzylisoquinoline alkaloids in the Polycarpicae, there are striking gaps in the distribution. What does appear to be true is that, not surprisingly, the capacity to produce these alkaloids is tied to the ability to form the iridoid and *seco*-iridoid precursors (Table IV). Both the Corniflorae and the Gentianiflorae commonly produce these monoterpenes (Waterman and Gray, 1987) but the facility to make them form a Schiff base with tryptamine (or dopamine) has either been gained by a few families only, or alternatively more recently lost from many. The Corniflorae and Gentianiflorae clearly share a common metabolic heritage but alkaloid distribution does not add much definition to that central fact.

The one family of the Gentianiflorae that produces neither iridoids nor indole–loganin alkaloids is the Asclepiadaceae. In this case the major secondary metabolites are steroidal cardenolides, a class of compounds also found in some genera of the Apocynaceae.

3.3. Anthranilic Acid as an Alkaloid Substrate

Alkaloids originating from anthranilic acid are largely confined to one family, the Rutaceae (Waterman, 1993). In most, anthranilic acid participates in the formation of a

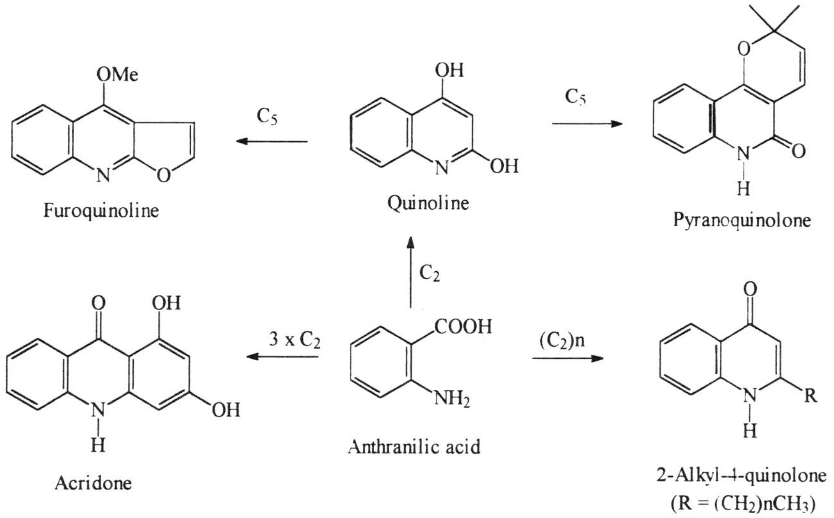

Figure 5. Some typical alkaloids of the Rutaceae formed from anthranilic acid and acetyl coenzyme A.

2,4-dioxygenated quinoline nucleus by combination with a two-carbon unit originating from acetyl coenzyme A. Simple 2-quinolone and 4-quinolone alkaloids are often prenylated leading to furoquinolines and pyranoquinolines. Involvement of three acetyl coenzyme A units leads to acridones whereas a larger number of acetates will give 2-acyl-4-quinolones. Some examples are given in Fig. 5. Anthranilic acid also associates with tryptamine to give the indoloquinazoline nucleus (**13**).

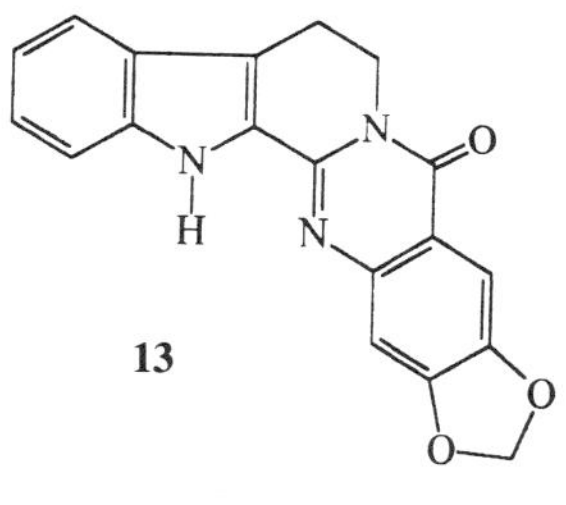

13

The systematic significance of these alkaloids to the Rutaceae has been discussed many times (Price, 1963; Da Silva *et al.,* 1988; Waterman, 1975, 1983, 1990, 1993). The families of the Rutales (Meliaceae, Cneoraceae, Simaroubaceae, Rutaceae) form a homogeneous group with several chemical features in common (Waterman and Grundon, 1983) but these alkaloids appear to be largely, but not quite entirely, confined to the Rutaceae where furoquinolines in particular appear to be very widespread, although often in small amounts. They have proved very useful in establishing the affinities of several small taxa associated with the Rutales (Waterman, 1975, 1983).

Unfortunately, once again there have been occasional reports of furoquinolines being isolated from other taxa, notably the Solanaceae, Asclepiadaceae and Apocynaceae (Gray, 1993). These records all need to be confirmed. To date acridones appear to be unique to the Rutaceae.

3.4. Alkaloids from Ornithine and Lysine

3.4.1. SIMPLE PYRROLIDINE AND TROPANE ALKALOIDS

These alkaloids are formed by a classic Schiff base and Mannich condensation taking place using nitrogen and carbonyl functional groups produced within a single molecule rather than between two molecules. The result is formation of a pyrrolidine ring. The tropane nucleus is derived from the combination of the pyrrolidine with acetate units, the product usually carrying an oxygen at C-3 which is then esterified (Fig. 6).

Tropane alkaloids are common constituents in the Solanaceae, occur quite widely in the Convolvulaceae and Erythroxylaceae, and sporadically in a number of other families including the Cruciferae, Euphorbiaceae, Proteaceae, Rhizophoraceae, and Elaeocarpaceae (Romeike, 1978; Woolley, 1993).

Given their distribution, which places similar alkaloids in a number of superorders that are phylogenetically distant from one another (Table V), the inevitable conclusion is that the tropane nucleus has developed several times during the evolution of the Angiospermae (Hegnauer, 1981). This is perhaps not so surprising given that the events needed to form the pyrrolidine nucleus do not require that two molecules with suitable functional groups are brought together and the steps needed to convert ornithine into the "reactive" intermediate (Fig. 6) are, for the most part, ubiquitous in higher plants.

There is considerable structural diversity in the tropane alkaloids found in different families. This can involve substitution and stereochemistry of those substituents on the tropane nucleus and the form of the esterifying acid on C-3. Some examples are simple oxygenated tropanes like calystegin-B_1 (**14**) from *Calystegia* sp. (Convolvulaceae), cocaine (**15**) from *Erythroxylum coca* (Erythroxylaceae), strobiline (**16**) from *Knightia strobilina* (Proteaceae), brugine (**17**) from *Bruguiera* sp. (Rhizophoraceae), and hyoscine (**18**) from various Solanaceae.

However, while many tropane alkaloids are associated with particular families these

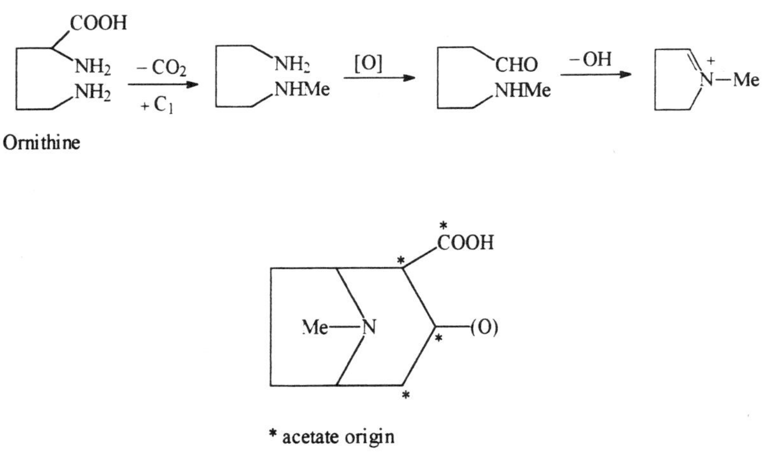

Figure 6. Formation of *N*-methylpyrrolidine from ornithine and addition of acetate to give the tropane nucleus.

Table V
Sources of Tropane Alkaloids and Their Position in the Systems of Thorne (1968) and Dahlgren (1980)

Family	Thorne (1968)	Dahlgren (1980)
Convolvulaceae	Malviflorae–Solanales	Solaniflorae–Solanales
Cruciferae	Cistiflorae–Capparidales	Violiflorae–Capparales
Elaeocarpaceae	Malviflorae–Malvales	Malviflorae–Malvales
Erythroxylaceae	Geraniiflorae–Geraniales	Rutiflorae–Geraniales
Euphorbiaceae	Malviflorae–Euphorbiales	Malviflorae–Euphorbiales
Proteaceae	Proteiflorae–Proteales	Proteiflorae–Proteales
Rhizophoraceae	Corniflorae–Cornales	Myrtiflorae–Rhizophorales
Solanaceae	Malviflorae–Solanales	Solaniflorae–Solanales

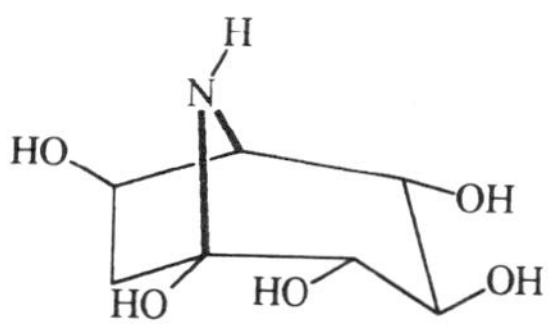

14

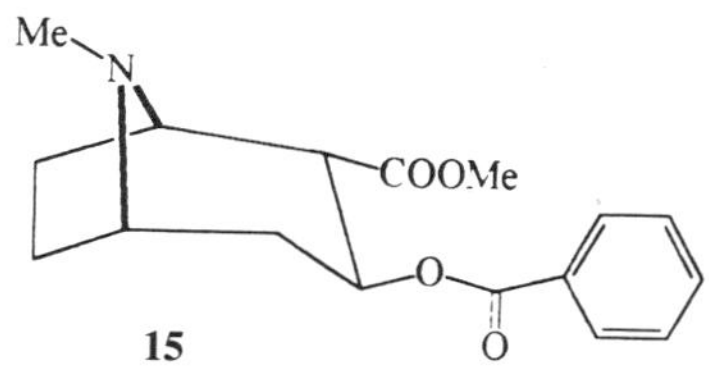

15

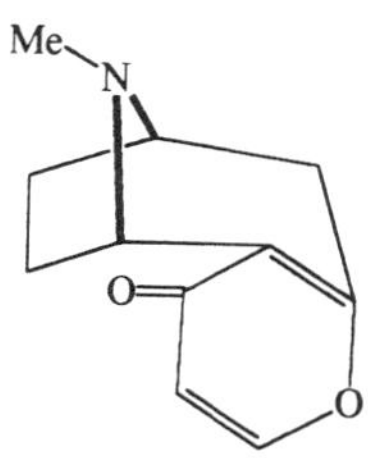

16

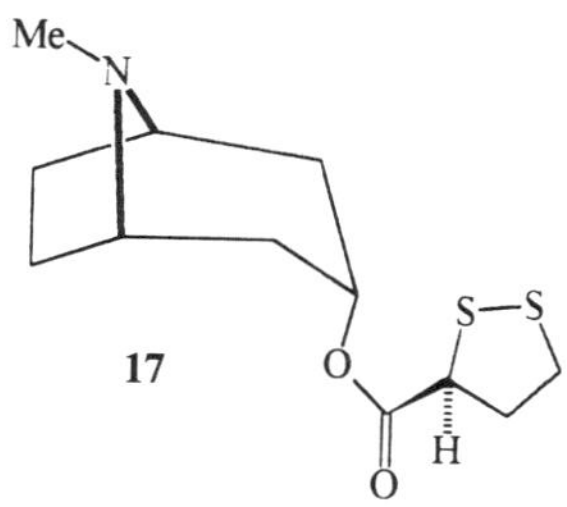

17

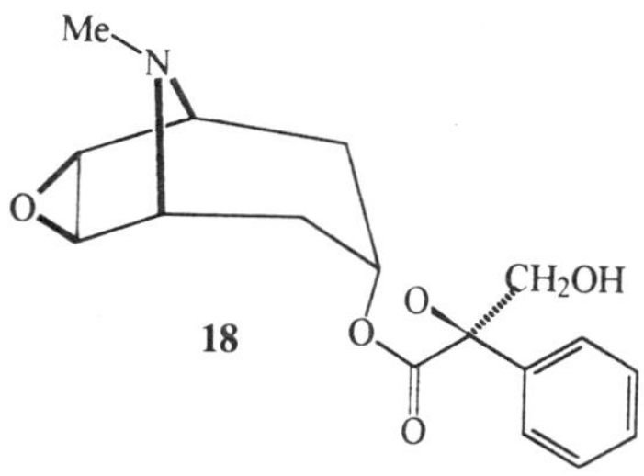

18

generally represent traits rather than valuable taxonomic markers. It seems probable that these alkaloids have little value as phylogenetic markers at higher taxonomic levels and taxonomic implications reside only within the family.

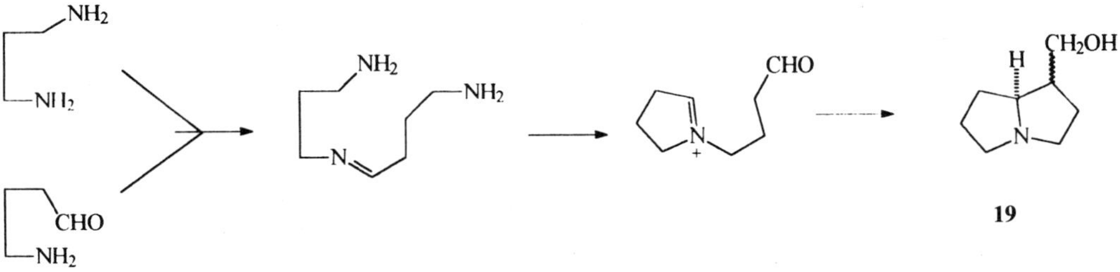

19

Figure 7. Formation of the necine skeleton (**19**) from ornithine.

3.4.2. PYRROLIZIDINE ALKALOIDS

Pyrrolizidines are also derived from ornithine, but in this case the initial cyclization is through a Schiff base involving two molecules originating in ornithine (Fig. 7). The bicyclic necine structure (**19**) occurs quite widely in the Angiospermae; at the last count (Robins, 1993) alkaloids based on this nucleus had been isolated from 15 different plant families (Table VI). As with the tropane alkaloids, the pattern of occurrence is not suggestive of a single biosynthetic event but of several independent occasions on which the ability to produce necines has arisen, leading to the view (Culvenor, 1978) that pyrrolizidines are not taxonomically useful above the family level.

The pyrrolizidine alkaloids may be subdivided according to the type of esterification that occurs in the necine nucleus. Culvenor (1978) recognized four types of esterifying groups (Table VI): (1) monocarboxylic aliphatic acids such as heliotrine (**20**), (2) aryl/aliphatic acids such as malaxine (**21**), (3) macrocyclic diester acids such as monocrotaline (**22**), and (4) 1-amino derivatives like absulin (**23**). The most important source is the Senecioneae tribe in the Asteraceae where macrocyclic pyrrolizidines occur widely.

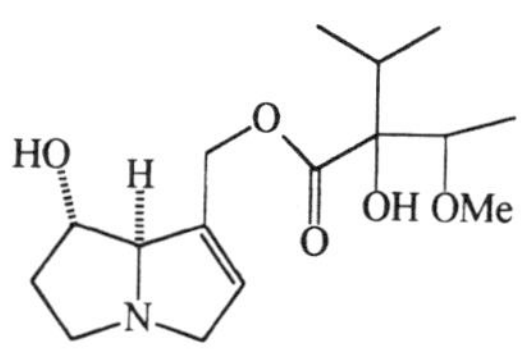

20

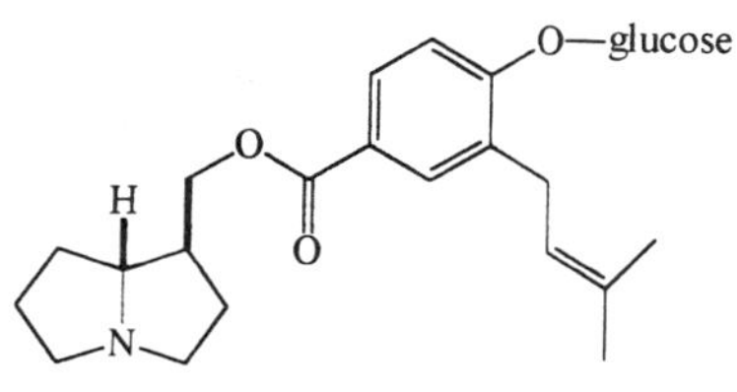

21

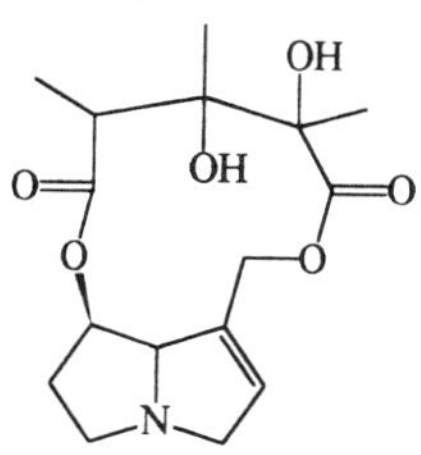

22

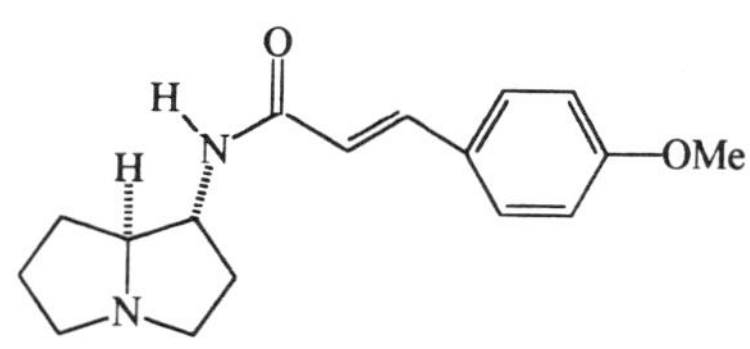

23

Table VI
Sources of Pyrrolizidine Alkaloids (Robins, 1993)—Classification of Dahlgren (1980)

Super order/order/family	Ester type
Monocot	
Liliiflorae/Orchidales/Orchidaceae	Aliphatic and aryl/aralkyl
Commeliniflorae/Poales/Poaceae	1-Amino
Dicot	
Ranunculflorae/Ranunculales/Ranunculaceae	Macrocyclic
Primuliflorae/Ebenales/Sapotaceae	Aliphatic and aryl/aralkyl
Fabiflorae/Fabales/Leguminosae	Macrocyclic and 1-amino
Myrtiflorae/Rhizophorales/Rhizophoraceae	1-Amino
Myrtiflorae/Euphorbiales/Euphorbiaceae	Atypical[a]
Rutiflorae/Geraniales/Linaceae	1-Aminoarylamide
Santaliflorae/Celastrales/Celastraceae	Aliphatic
Santaliflorae/Santalales/Santalaceae[b]	Aryl/aralkyl
Asteriflorae/Asterales/Asteraceae	Aliphatic and macrocyclic
Solaniflorae/Boraginales/Ehretiaceae	Aryl
Solaniflorae/Boraginales/Boraginaceae	Aliphatic macrocyclic
Gentianiflorae/Gentianales/Apocynaceae	Aliphatic and aryl/aralkyl
Lamiiflorae/Scrophulariales/Scrophulariaceae[b]	Macrocyclic

[a]Probably not biosynthetically allied to normal pyrrolizidine alkaloids.
[b]Isolated from parasitic or hemiparasitic species and may be derived from the host.

3.4.3. QUINOLIZIDINE ALKALOIDS

The formation of the quinolizidine nucleus mirrors that of pyrrolizidines but as lysine is the amino acid yields six-membered rings rather than five-membered rings. They may then be extended by the addition of a further lysine-derived piperidine ring to yield alkaloids like lupanine (**24**) and matrine (**25**), or two further piperidines to give piptanthine (**26**).

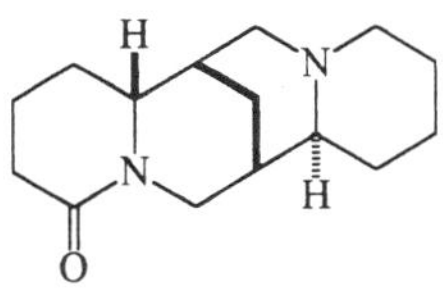

24

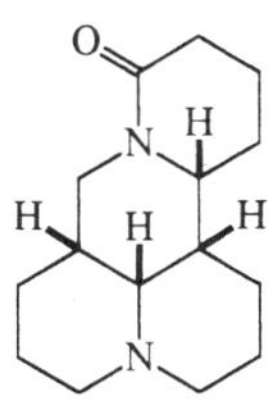

25

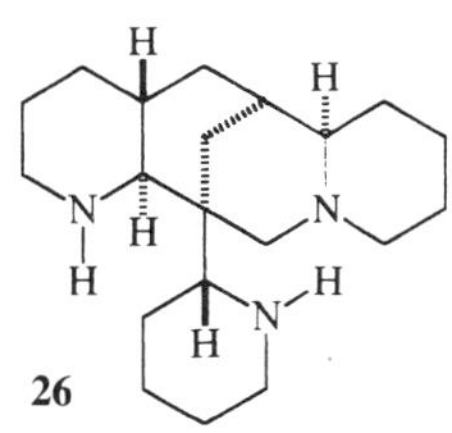

26

The major center of quinolizidine alkaloid production is the Leguminosae–Papilionoideae (Fabaceae) where they are common in a number of tribes (Salatino and Gottlieb, 1980; Wink, 1993). Other families where they are known to be produced are the Berberidaceae, Ranunculaceae, Solanaceae, Chenopodiaceae, and Rubiaceae. Wink (1993) notes that the reported presence of these alkaloids in the Scrophulariaceae, Santalaceae, Loranthaceae, and Cuscutaceae may well reflect their transfer from the sap of the host to parasitic or hemiparasitic species. This suggestion should also be borne in mind with regard to the distribution of pyrrolizidine alkaloids (Table VI).

The distribution of quinolizidine alkaloids in the Papilionoideae has been examined in depth by Salatino and Gottlieb (1980). It is well established that certain tribes contain such alkaloids in most or all species while others are entirely free of them. An attempt to interpret alkaloid structure in phylogenetic terms (Salatino and Gottlieb, 1980) is not convincing, the methodology employed being far too rigid to accommodate the data available. However, there is no doubt that these alkaloids are taxonomically useful and on a number of occasions their presence or absence has been used to assign difficult taxa or to decide between different taxonomic systems. For example, the presence of unusual quinolizidine/indolizidine alkaloids in *Camoensia* strongly supports placing it in the Papilionoideae rather than in the Caesalpinoideae (Waterman and Faulkner, 1978). Attempts to use these alkaloids at lower levels of classification have revealed considerable variation, even within populations (e.g., *Baptisia,* Alston and Turner, 1963).

3.4.4. SIMPLE POLYHYDROXY PYRROLIDINE, PIPERIDINE AND INDOLIZIDINE ALKALOIDS

This group of alkaloids, typified by fagomine (**27**) and castanospermine (**28**), have only recently come to light. As is often the case with a class of natural products that has

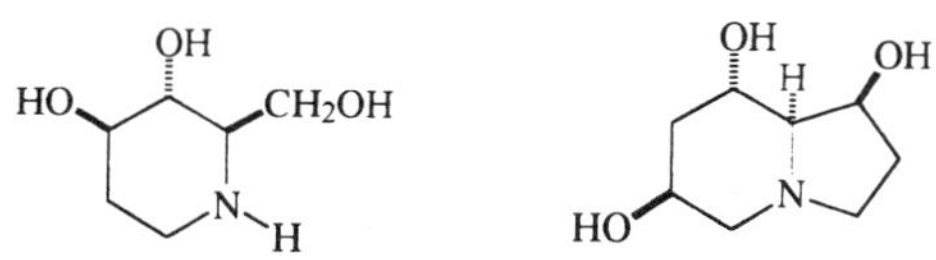

27 **28**

been overlooked, their discovery has led to the rapid identification of a number of related compounds (Molyneux, 1993). The original, and still major, source was the Leguminosae, but such compounds have now been discovered in a number of other families (Aspidiaceae, Euphorbiaceae, Polygonaceae, Moraceae). It is too early to make any claims for the taxonomic value of these alkaloids.

3.5. Pseudoalkaloids

The two major groups of pseudoalkaloids are those based on diterpenes and on steroids. Both parent classes of compound are widespread in the plant kingdom. As their "conversion" into alkaloids requires the insertion of nitrogen into the molecule, often by supplanting an oxygen, there is no general biosynthetic strategy. Thus, there seems little reason to expect their occurrence to reveal any phylogenetic insights at higher taxonomic levels. The value in taxonomy lies only at the level of being an indicator for the placement of taxa at the infrafamily level.

Thus, steroidal alkaloids are found extensively in the Solanaceae, Apocynaceae, Buxaceae, and Liliaceae, while diterpene alkaloids occur within the Ranunculaceae, Garryaceae, Escalloniaceae, and occasionally in the Leguminosae, Rosaceae, and Asteraceae. Within many of these families their presence or absence has obvious taxonomic value.

4. CONCLUDING COMMENTS

Chemical taxonomy is a frustrating business! It is now clear that the ability to produce alkaloids is present in many plants, perhaps many more than was once considered to be the case (Wink and Witte, 1983). At the biosynthetic level the mechanism of alkaloid synthesis has proved to be a very conservative process changing primarily through variation in the substrate employed rather than in the mechanism of formation. Thus, the genetic information for the enzymes governing the mechanism by which alkaloids are formed (Schiff base plus Mannich condensation) are certainly very widespread, and are likely to be present but silent in many non-alkaloid-producing species. These are conditions that lend themselves to the independent development of compounds over evolutionary time, leading to parallelism and convergence.

A further confounding variable is the ecological role that can be attributed to many alkaloids (Harborne, 1988; Hartmann, 1991). The most widely assumed role is in defense of the producer against herbivores, but pollinator attraction, nitrogen reserve, and interaction with soil fungi can also be cited. Alkaloid production can be enhanced by extrinsic factors and may be influenced by altitude (see earlier). Harborne and Turner (1984) draw attention to the finding that the alkaloid content of *Cinchona* bark declines along a high-altitude cline from Colombia, through Ecuador to Peru.

It should be apparent from the earlier sections of this chapter that alkaloids have provided many valuable insights into taxonomic relationships and there is no reason why they should not continue to do so. What is now clear, however, is that they do not provide the unambiguous, easily rationalized, insights that were once anticipated.

REFERENCES

General Reviews

Alston, R. E., and Turner, B. L., 1963, *Biochemical Systematics,* Prentice–Hall, Englewood Cliffs, NJ.

Cordell, G. A., 1981, *Introduction to Alkaloids: A Biogenetic Approach,* Wiley, New York.

Geissman, T. A., and Crout, D. H. G., 1969, *Organic Chemistry of Secondary Plant Metabolism,* Freeman, Cooper, San Francisco

Gray, A. I., 1993, Quinoline alkaloids related to anthranilic acid, in: *Methods in Plant Biochemistry,* Vol. 8 (P. G. Waterman, ed.), Academic Press, San Diego, pp. 271–308.

Guinaudeau, H., and Bruneton, J., 1993, Isoquinoline alkaloids, in: *Methods in Plant Biochemistry,* Vol. 8 (P. G. Waterman, ed.), Academic Press, San Diego, pp. 373–419.

Harborne, J. B., 1988, *Ecological Biochemistry,* 3rd ed., Academic Press, San Diego.

Harborne, J. B., and Turner, B. L., 1984, *Plant Chemosystematics,RR Academic Press, San Diego.*

Hartmann, T., 1991, Alkaloids, in: Herbivores, Their Interactions with Secondary Plant Metabolites, 2nd ed., Vol. 1 (G. A. Rosenthal and M. Berenbaum, eds.), Academic Press, San Diego, pp. 79–121.

Hegnauer, R., 1963a, The taxonomic significance of alkaloids, in: *Chemical Plant Taxonomy* (T. Swain, ed.), Academic Press, San Diego, pp. 389–427.

Hegnauer, R., 1963b, *Chemotaxonomie der Pflanzen,* Vol. 2, Springer-Verlag, Berlin.

Molyneux, R. J., 1993, Polyhydroxy indizolines and related alkaloids, in: *Methods in Plant Biochemistry,* Vol. 8 (P. G. Waterman, ed.), Academic Press, San Diego, pp. 511–530.

Price, J. R., 1963, The distribution of alkaloids in the Rutaceae, in: *Chemical Plant Taxonomy* (T. Swain, ed.), Academic Press, San Diego, pp. 429–452..

Robins, D. J., 1993, Pyrrolizidine alkaloids, in: *Methods in Plant Biochemistry,* Vol. 8, (P. G. Waterman, ed.), Academic Press, San Diego, pp. 175–195.

Strack, D., Steglich, W., and Wray, V., 1993, Betalains, in: *Methods in Plant Biochemistry,* Vol. 8 (P. G. Waterman, ed.), Academic Press, San Diego, pp. 421–450.
Swain, T., (ed.), 1963, *Chemical Plant Taxonomy,* Academic Press, San Diego.
Waterman, P. G., 1983, Phylogenetic implications of the distribution of secondary metabolites within the Rutales, in: *Chemistry and Chemical Taxonomy of the Rutales* (P. G. Waterman, and M. F. Grundon, eds.), Academic Press, San Diego, pp. 377–400.
Wink, M., 1993, Quinolizidine alkaloids, in: *Methods in Plant Biochemistry,* Vol. 8 (P. G. Waterman, ed.), Academic Press, San Diego, pp. 197–239.
Woolley, J. G., 1993, Tropane alkaloids, in: *Methods in Plant Biochemistry,* Vol. 8 (P. G. Waterman, ed.), Academic Press, San Diego, pp. 133–173.

Key References

Chandra, P., and Purohit, A. N., 1980, Berberine contents and alkaloid profile of *Berberis* species from different altitudes, *Biochem. Syst. Ecol.* **8:**379–380.
Culvenor, C. C. J., 1978, Pyrrolizidine alkaloids—occurrence and systematic occurrence in Angiosperms, *Bot. Notiser.* **131:**473–486.
Dahlgren, R. M. T., 1977, A commentary on a diagrammatic presentation of the angiosperms in relation to the distribution of character states, *Plant. Syst. Evol.* Suppl. **1:**253–283.
Dahlgren, R. M. T., 1980, A revised system of classification of the angiosperms, *Bot. J. Linn. Soc.* **90:**91–124.
Da Silva, M.F.G.F., Gottlieb, O. R., and Ehrendorfer, F., 1988, Chemosystematics in the Rutaceae: Suggestions for a more natural taxonomy and evolutionary interpretation of the family, *Plant Syst. Evol.* **161:**97–134.
Fish, F., and Waterman, P. G., 1973, The chemosystematics of the *Zanthoxylum/Fagara* complex, *Taxon* **22:**177–203.
Folkestad, K., Høiland, K., Paulsen, B. S., and Malterud, K. E., 1988, Alkaloid chemotaxonomy of Nordic *Papaver* sect. Scapiflora (Papaveraceae), *Nord. J. Bot.* **8:**139–146.
Hartley, T. G., 1981, A revision of the genus *Tetradium* (Rutaceae), *Gard. Bull. Singapore* **34:**91–131.
Hegnauer, R., 1981, Chemotaxonomy of Erythroxylaceae (including some ethnobotanical notes on Old World species), *J. Ethnopharmacol.* **3:**279–292.
Kubitzki, K., 1969, Chemosystematische betrachtungen zur grossgliederung der Dicotyledon, *Taxon* **18:**360–368.
Mabry, T. G., Neuman, P., and Philipson, W. R., 1978, *Hectorella,* a member of the betalain sub-order Chemopodiineae of the order Centrospermae, *Plant. Syst. Evol.* **130:**163–165.
Ng, K. M., But, P. P-H., Gray, A. I., Hartley, T. G., Kong, Y.-C., and Waterman, P. G., 1987, The biochemical systematics of *Tetradium, Euodia* and *Melicope* and their significance in the Rutaceae, *Biochem. Syst. Ecol.* **15:**587–593.
Quader, A., But, P. P-H., Gray, A. I., Hartley, T. G., Hu, Y.-J., and Waterman, P. G., 1990, Alkaloids and limonoids of *Tetradium trichotomum*: Systematic significance, *Biochem. Syst. Ecol.* **18:**251–252.
Romeike, A., 1978, Tropane alkaloids—Occurrence and systematic importance of angiosperms, *Bot. Notiser.* **131:**85–96.
Salatino, A., and Gottlieb, O. R., 1980, Quinolizidine alkaloids as systematic markers of the Papilionoideae, *Biochem. Syst. Ecol.* **8:**133–147.
Thorne, R. F., 1968, Synopsis of a putatively phylogenetic classification of the flowering plants, *Aliso* **6:**57–66.
Tillequin, F., Michel, S., and Seguin, E., 1993, Tryptamine-derived indole alkaloids, in: *Methods in Plant Biochemistry,* Vol. 8 (P. G. Waterman, ed.), Academic Press, San Diego, pp. 309–371.
Waterman, P. G., 1975, Alkaloids of the Rutaceae: Their distribution and systematic significance, *Biochem. Syst. Ecol.* **3:**149–180.
Waterman, P. G., 1990, Chemosystematics of the Rutaceae: Comments on the interpretation of Da Silva *et al., Plant. Syst. Evol.* **173:**39–48.
Waterman, P. G., 1993, Phytochemical diversity in the order Rutales, *Recent. Adv. Phytochem.* **27:**203–233.
Waterman, P. G., and Faulkner, D. F., 1982, Quinolizidine/indolizidine alkaloids from the seeds of *Camoensia brevicalyx, Phytochemistry* **21:**215–218.
Waterman, P. G., and Gray, A. I., 1987, Chemical systematics. *Nat. Prod. Rep.* **4:**175–203.

Waterman, P. G., and Grundon, M. F., (eds.), 1983, *Chemistry and Chemical Taxonomy of the Rutales,* Academic Press, San Diego.

Wink, M., and Witte, L., 1983, Evidence for a widespread occurrence for the genes of quinolizidine biosynthesis. Induction of alkaloid accumulation in cell suspension cultures of alkaloid 'free' species, *Fed. Eur. Biochem. Soc. Lett.* **159:**196–200.

Chapter 5

Enzymology of Alkaloid Biosynthesis

Margaret F. Roberts

1. INTRODUCTION

The biogenesis of alkaloids has been the cause of much speculation and since the beginning of the century there has been a great effort directed at the elucidation of some of the pathways. The advent of isotopically labeled compounds in the 1950s heralded the beginning of research that has given an insight into many aspects of alkaloid biosynthesis and metabolism.

The majority of alkaloids have been found to be derived from amino acids such as tyrosine, rarely phenylalanine, anthranilic acid, tryptophan/tryptamine, ornithine/arginine, lysine, histidine, and nicotinic acid. However, alkaloids may be derived from purines, i.e., caffeine, "aminated" terpenoids, i.e., aconite, or the steroidal alkaloids such as are found in the Solanaceae and Liliaceae. Alkaloids may also be formed from acetate-derived polyketides where the amino nitrogen is introduced as in the hemlock alkaloid, coniine.

Experiments with labeled precursors enabled the unraveling of many complex biogenetic pathways. Originally alkaloids were thought to be essentially plant products; however, these basic compounds also occur in microorganisms and animals. Although at present the majority of known alkaloids are amino acid derived, an increasing number of alkaloids from insects and marine organisms are being found that are either terpenoid or polyketide in origin.

The interest in growing and manipulating microorganisms and plants in cell culture for commercial purposes gave impetus to the study of alkaloid biosynthesis and in particular to the elucidation of the enzymes involved. It also brought about renewed interest in the regulation of alkaloid synthesis and in the location and means of sequestration of these substances within the plant.

It was not until the early 1970s that the enzymes associated with alkaloid formation were isolated. Now, however, the enzymes of every step of entire pathways, for instance

---

Margaret F. Roberts • The Centre for Pharmacognosy, School of Pharmacy, University of London, London WC1N 1AX, England.

Alkaloids: Biochemistry, Ecology, and Medicinal Applications, edited by Roberts and Wink. Plenum Press, New York, 1998.

from tyrosine to berberine, are known. The enzymes of the relatively few pathways isolated to date clearly indicate that most of the enzymes required are highly specific for a given biosynthetic step. The results of this research have also helped to revise routes to alkaloids hypothesized as a result of feeding radiolabeled precursors to plants. The investigation of enzymes of alkaloid biosynthesis has also helped to answer some of the questions regarding where and at what time during the plant growth cycle the alkaloids are actively made and has provided an insight into the location of enzymes within the plant and the cell.

The present chapter sets out to investigate a limited number of areas where the enzymes of whole pathways have been isolated in an attempt to clarify an area that is important in understanding alkaloid formation, mobilization, and sequestration. As in the enzymes involved in the biosynthesis of the tropane, nicotine, and senecio alkaloids are discussed in the chapter on "hairy" roots (see Chapter 8) they will not be considered here. For a complete overview of alkaloid biosynthesis the reader should consult a number of general texts and review articles on the subject. Some useful texts are:

- The series *The Alkaloids* originally edited by R. H. F. Manske, Academic Press, New York, and currently edited by G. Cordell. This series provides valuable updated information on the chemistry and pharmacology of various classes of alkaloids; 45 volumes were published between 1950 and 1996.
- The series *Alkaloids: Chemical and Biological Perspectives,* edited by S. W. Pelletier, Pergamon Press, Oxford. The first volume appeared in 1983.
- *Secondary Metabolism*, by J. Mann, Clarendon Press, Oxford, 1987.
- *The Biosynthesis of Secondary Metabolites*, by R. B. Herbert, Chapman & Hall, London, 1989.
- *Biochemistry of Alkaloids*, by K. Mothes, H. R. Schutte, and M. Luckner, VEB Deutscher Verlag der Wissenschaften, Berlin, 1985.
- *Indole and Biogenetically Related Alkaloids*, edited by J. D. Phillipson and M. H. Zenk, Academic Press, New York, 1980.
- *The Chemistry and Biology of Isoquinoline Alkaloids*, edited by J. D. Phillipson, M. F. Roberts, and M. H. Zenk, Springer Verlag, Berlin, 1985.
- *Natural Product Reports* published by the Royal Society; this periodical regularly reviews current progress in alkaloid biosynthesis.

2. BIOSYNTHESIS OF ACETATE-DERIVED SIMPLE PIPERIDINE ALKALOIDS

Simple piperidines are found in plants, animals, and microorganisms. In plants, *Conium* species (coniine), *Pinus jeffreyii* (pinidine), *Cassia* sp. (cassine), *Prosopis* sp. (spectaline), and the dimeric alkaloids of *Azima tetracantha* and *Carica papaya* (azimine, azacarpaine, and carpaine) are good examples (Mothes *et al.,* 1985). Coniine has also been found in the pitcher plant, *Sarracenia*, and in a number of *Aloe* species (Dring *et al.,* 1984). In arthropods, simple piperidines may be used as part of chemical defense and communication systems; e.g., in *Aphaenogaster* (ants), anabaseine has been identified as an attractant, and in the venomous fire ant, *Solenopsis saevissma*, simple piperidines appear to be used as part of a defense system. Whereas in plants the piperidine alkaloids

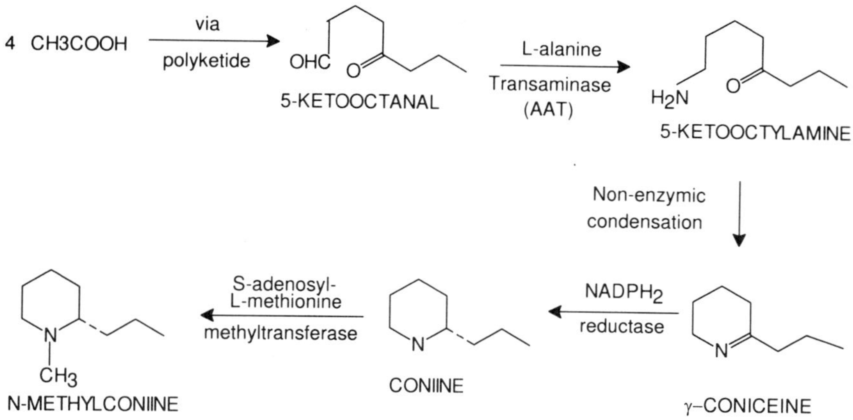

Figure 1. Biosynthesis of the simple piperidines of *Conium maculatum* from an acetate-derived polyketide.

may be formed from acetate or lysine according to the plant species, in microorganisms these alkaloids all appear to be acetate derived. This is also the case for the antibiotics of the cycloheximide type in *Streptomyces noursei* (Mothes *et al.,* 1985).

The piperidine alkaloids of *Conium maculatum*, hemlock, are known traditionally as powerful poisons (Plato, 387 B.C.) with an oral LD_{50} for mice of 12 mg kg^{-1} for γ-coniceine (Bowman and Sanghvi, 1963; see also Chapters 4 and 18).

The biosynthesis of these alkaloids was one of the first to be investigated at the enzyme level (Roberts, 1971). These alkaloids are simple derivatives of piperidine with a three carbon side chain and were originally thought to be derived from lysine. However, it has been conclusively proved by radiolabeled precursor experiments that these alkaloids are derived from acetate (Leete and Olson, 1972). The enzymatic pathway (Fig. 1) which involves L-alanine:5-ketooctanal transaminase (AAT) (Roberts, 1971, 1978), an NADPH-dependant γ-coniceine reductase (CR) (Roberts, 1975), and coniine:*S*-adenosyl-L-methionine methyltransferase (CSAM) (Roberts 1974) is found to occur principally in the aerial parts of the plant. The first committed step in alkaloid formation is the amination of 5-ketooctanal by AAT, of which two isoenzymes with differing substrate requirements occur (Roberts, 1978). It has been suggested that the *Conium* AAT is similar to the amino acid transaminase found in many higher plants, which is involved in amine formation (Hartmann *et al.,* 1972a,b). The operation of this ubiquitous enzyme is dependent on the occurrence of the appropriate aldehydes. The two remaining enzymes in the sequence, the reductase and the methyltransferase, are highly substrate specific.

In *Conium* there is a close link between alkaloid formation and illumination (Fairbairn and Suwal, 1961). The AAT isoenzymes reside in the chloroplast and mitochondria (Roberts, 1981) and it is thought that the chloroplast enzyme, which is not found in spinach (Wink and Hartmann, 1981a), is responsible for alkaloid formation. However, the spinach (mitochondrial) enzyme provided with 5-ketooctanal and L-alanine will produce γ-coniceine. The key to alkaloid production is most likely the availability of 5-ketooctanal. Alkaloids are actively formed in all aerial parts of the plant. In the leaves all three enzymes (AAT, CR, CSAM) are active during leaf expansion but production of these enzymes, and hence alkaloid production, ceases as the leaf reaches maturity. A similar situation is found in the immature fruits, which proved to be the powerhouses of alkaloid production (Roberts, 1985). In the ripened fruits the alkaloids occur in specialist cells (Fairbairn and Suwal,1961) and may well be involved in protection of the seed prior to germination when the alkaloids are totally lost along with the testa. Young seedlings commence generation of γ-coniceine immediately, even at the cotyledon stage in the dark, but activity increases of some 400% with illumination are observed and the plants exhibit a well-expressed diurnal rhythm with respect to alkaloid formation. Plants also appear to lose much of these volatile piperidine alkaloids to the atmosphere and this apparently attracts pollinating insects at flowering (Fairbairn and Suwal, 1961, Roberts, unpublished results).

3. BIOSYNTHESIS OF QUINOLIZIDINE ALKALOIDS

The lysine-derived quinolizidines (QAs) are found in the tribes Genisteae, Podalyrieae, and Sophoreae of the subfamily Papilionoideae of the Fabaceae. Most abundant and best studied are the alkaloids of *Lupinus* species and hence these alkaloids are frequently called the Lupin alkaloids. Approximately 170 QAs are known and these have been divided into ten structural groups (Kinghorn and Balandrin,1984; Aslanov *et al.,* 1987). The most important groups are: the bicyclic lupinine and its esters; the tetracyclic alkaloids of the sparteine/lupanine type and hydroxylupanine esters; tricyclic degradation products of the sparteine/lupanine type (e.g., tetrahydrorhombifoline); tetracyclic α-pyridone alkaloids (e.g., anagyrine); tricyclic α-pyridone alkaloids (e.g., cytisine); and matrine alkaloids (Fig. 2).

Biosynthetic tracer studies established lysine and its decarboxylation product cadaverine as the only precursors of these bi-, tri-, and tetracyclic QAs, (Fraser and Robins, 1987; Golebiewski and Spencer, 1984). The tricyclic QAs are derived from the tetracyclic precursors (Fig. 3) with lupanine the common precursor of this group of alkaloids (Herbert,1989).

Crude enzyme preparations from *Lupinus* cell cultures catalyze the conversion of cadaverine to 17-oxosparteine in the presence of pyruvate as an amino-group acceptor. This specialized transaminase which converts three units of cadaverine into tetracyclic alkaloids deserves further study (Wink and Hartmann, 1981b). This enzyme and lysine decarboxylase which converts lysine to cadaverine were found localized in the chloroplasts of lupin leaves (Wink and Hartmann, 1982). Recent tracer experiments exclude 17-oxosparteine as an intermediate in the pathway to lupanine and sparteine (Fraser and Robins, 1987; Spenser, 1985). It may be assumed that either (1) the production of 17-oxosparteine may be the result of further oxidation of sparteine which is the primary product (Herbert, 1989) or (2) the biosynthetic sequence proceeds via an early tetracyclic intermediate to lupanine/sparteine, which is released and, in cell-free enzymatic preparations, is oxidized to 17-oxosparteine (Hartmann, 1988). The latter suggestion is supported by results with isolated intact chloroplasts which incorporate cadaverine and lysine into lupanine without detectable amounts of 17-oxosparteine (Wink *et al.,* 1980).

Enzymes that methylate cytisine to *N*-methylcytisine in *Laburnum anagyroides, L. alpinum, Cytisus canariensis* (Wink, 1984), and *Thermopsis sp.* (Murakoshi *et al.,* 1977a,b) have been isolated along with enzymes that synthesize QA esters (Murakoshi *et al.,* 1977a,b; Saito *et al.,* 1993).

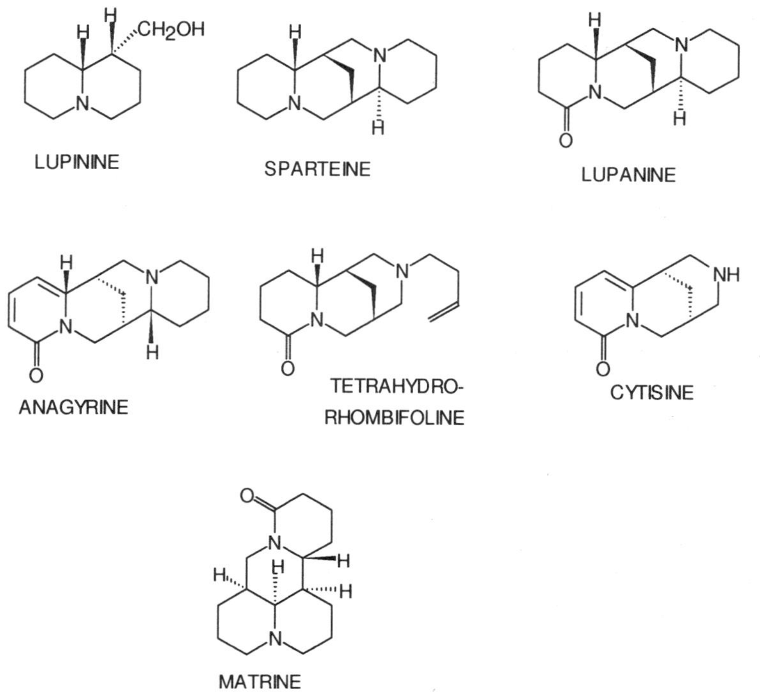

Figure 2. Structural types of quinolizidine alkaloids.

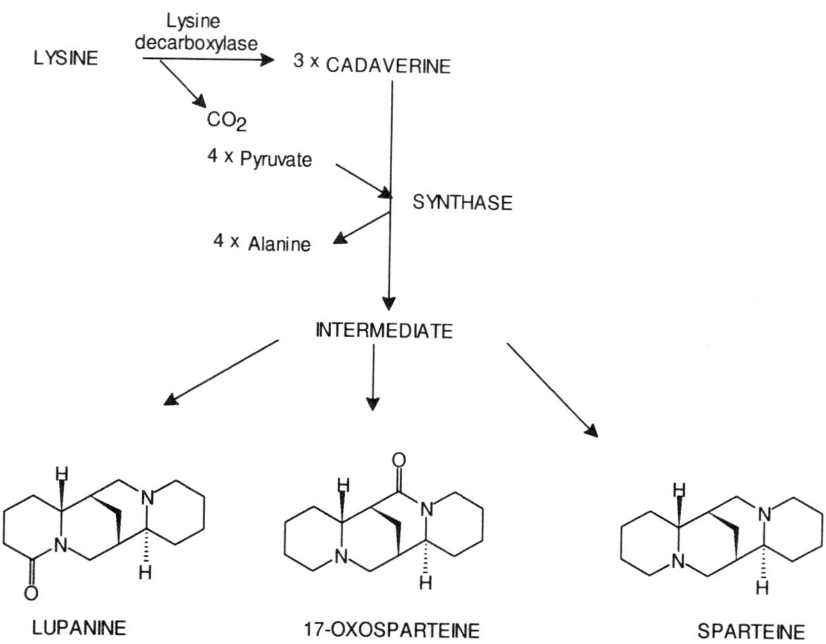

Figure 3. Biosynthesis of *Lupinus* tetracyclic quinolizidines from lysine.

In lupins biosynthesis is restricted to the aerial parts of the plants and this suggests that chloroplasts are the exclusive sites of biosynthesis of QAs. Seeds and fruits are especially rich in QAs and in the stems and leaves the alkaloids are mainly stored in the epidermal cells where concentrations may be as high as 200 mM (Wink, 1986, 1987).

4. BIOSYNTHESIS OF BENZYLISOQUINOLINE ALKALOIDS

Since the discovery and characterization of the medicinal constituents of the opium poppy, *Papaver somniferum* (see Chapters 4 and 18), in the early 1800s, a large number of benzylisoquinoline alkaloids have been isolated. The benzylisoquinolines are found within the families of the superorders Magnoliiflorae (i.e., Annonaceae, Eupomatiaceae, Aristolochiaceae, Magnoliaceae, Lauraceae, Monimiaceae, and Nelumbonaceae) and Ranunculiflorae (i.e., Berberidaceae, Ranunculaceae, Menispermaceae, Fumariaceae, and Papaveraceae). This highly clustered distribution is of interest from a chemotaxonomic point of view as there are few exceptions, the most notable being the erythrans which occur throughout the genus *Erythrina* (Fabaceae).

These alkaloids have common origins in the aromatic amino acid tyrosine. The benzylisoquinolines are formed from two molecules of tyrosine which are elaborated to form (S)-norcoclaurine (Stadler and Zenk, 1990). This alkaloid is an important precursor of a variety of pathways that lead to a series of diverse structures, some of which are shown in Fig. 4.

Plant cell cultures established from various isoquinoline bearing plants have provided most useful systems for the study of biosynthetic pathways at the enzyme level. Excellent progress has therefore been made in unraveling the route to (*S*)-norcoclaurine and the sequences leading to some of the more important groups of isoquinolines. It is only relatively recently, as a result of the investigations of the enzymes of the biosynthetic pathways to morphine and berberine (Zenk *et al.,* 1985), that the early steps of the pathway have been fully elucidated. These studies have also helped to improve our understanding of the localization at the subcellular level of both enzymes and products (Ikuta, 1988; Zenk, 1990).

4.1. The Formation of (*S*)-Norcoclaurine

The first clues on the involvement of (*S*)-norcoclaurine in the synthesis of isoquinoline alkaloids came from experiments with *Annona reticulata* leaves in which ^{14}C label from tyrosine was incorporated into (*S*)-coclaurine and (*S*)-reticuline suggesting that the (*S*)-coclaurine was an intermediate in (*S*)-reticuline formation and inferring that (*S*)-norlaudanosoline (Battersby *et al.,* 1964) was not. Investigation of a number of enzymes involved in tyrosine conversion by Rueffer (Rueffer and Zenk, 1987a) suggested that the first committed step in the biosynthesis of benzylisoquinolines involved the condensation of dopamine with 4-hydroxyphenylacetaldehyde to give (*S*)-norcoclaurine, a compound that proved to be pivotal in the formation of all benzylisoquinoline alkaloids (Fig. 4). (*S*)-Reticuline is readily formed from (*S*)-norcoclaurine as a result of a series of hydroxylations and methylations (Fig. 5). From intermediates observed *in vivo* (Stadler *et al.,* 1989;

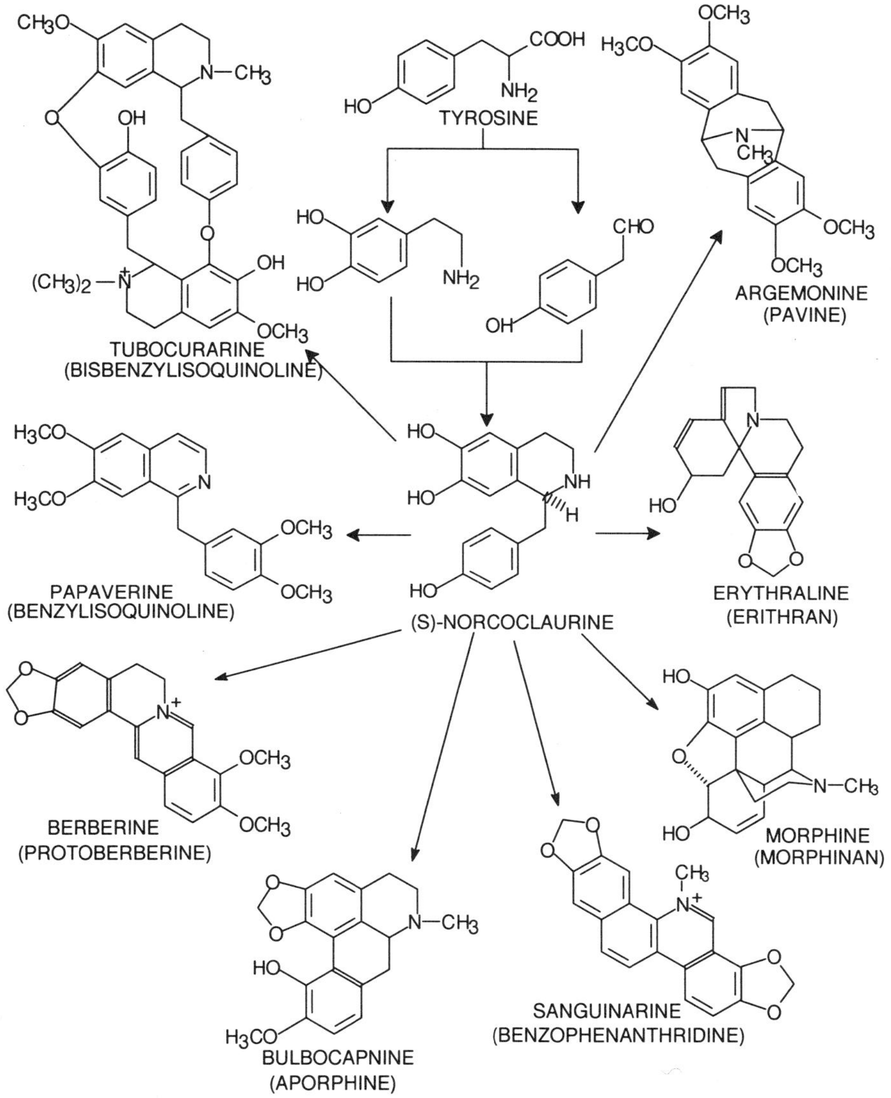

Figure 4. The various groups of benzylisoquinoline alkaloids derived from noncoclaurine.

Stadler and Zenk, 1990; Frenzel and Zenk, 1990; Loeffler and Zenk, 1990) it may be concluded that (*S*)-norcoclaurine is stereospecifically metabolized to (*S*)-reticuline via (*S*)-coclaurine, (*S*)-*N*-methylcoclaurine, and (*S*)-3′-hydroxy-*N*-methylcoclaurine. The order in which the various hydroxylations and methylations occur is, to a degree, substantiated by the distribution of radioactivity in the benzylisoquinoline alkaloids of *Berberis stolonifera* cell cultures after feeding L[U-^{14}C]-tyrosine. The reason that norlaudanosine (previously thought to be the first condensation product from tyrosine) is easily incorporated into reticuline and related alkaloids is the relative nonspecificity of the enzymes of the reticuline pathway (Fig. 5). This contrasts with the high specificity found among the enzymes

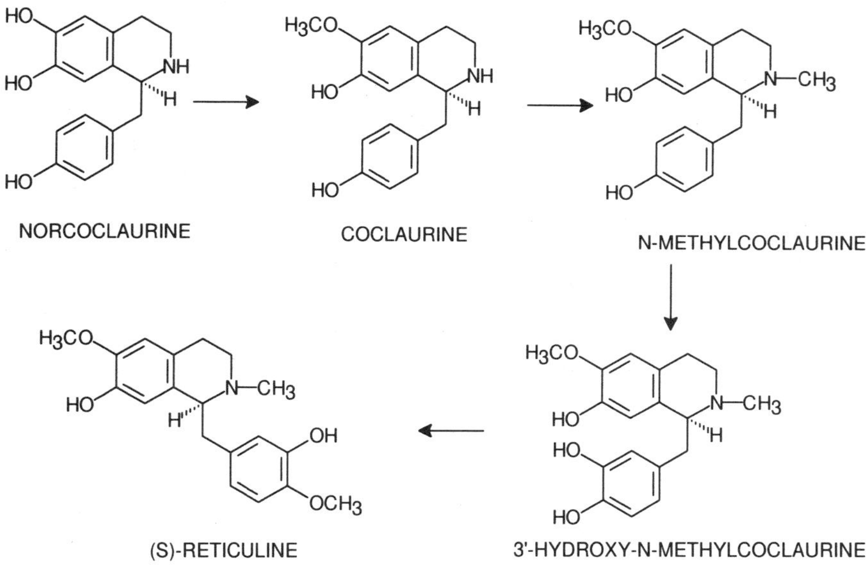

Figure 5. The formulation of (*S*)-reticuline from tyrosine derived tyramine and *p*-hydroxyphenylacetaldehyde.

involved in the synthesis of various more complex structures, e.g., the protoberberines (Fig. 6), the benzophenanthridines (Fig. 7), and the morphinans (Fig. 8).

4.2. Biosynthesis of the Tetrahydroberberine Alkaloids

The enzymatic route to berberine was one of the first to be completely elucidated with all of the participating enzymes isolated and characterized. The conversion of (*S*)-reticuline to (*S*)-scoulerine by the berberine bridge enzyme (Steffens *et al.,* 1984, 1985) may be considered as the first committed step in the production of the tetrahydroprotoberberines and the whole range of alkaloidal types that are derived from this basic skeleton (Fig. 6). It has been shown that the enzyme (*S*)-scoulerine-9-*O*-methyltransferase catalyzes the conversion of (*S*)-scoulerine to (*S*)-tetrahydrocolumbamine (Muemmler *et al.,* 1985; Fujiwara *et al.,* 1993) in which a methylene bridge is formed to yield (*S*)-canadine by (*S*)-canadine synthase. This enzyme is a specific methylenedioxy-bridge forming enzyme (Galaeder *et al.,* 1988; Rueffer and Zenk, 1994). (*S*)-canadine may act as a substrate for the tetrahydroberberine oxidase (STOX) enzyme isolated from *Berberis* and be converted to berberine (Galaeder *et al.,* 1988; Okada *et al.,* 1988). However, the oxidase found in *Coptis japonica* (COX) is specific for (*S*)-canadine (Galaeder *et al.,* 1988; Rueffer and Zenk, 1994). These two oxidases differ in that STOX contains a flavin and produces 1 mole of H_2O_2 and water per mole of substrate consumed, whereas COX has a cofactor requirement for iron and produces 2 moles of H_2O_2 per mole of substrate utilized (Okada *et al.,* 1988) (Fig. 6). Hence, there is a reason not to generalize unless enzymatic steps have been elucidated for each species.

The enzyme activity that was originally thought to represent formation of the methylenedioxy bridge in *Berberis* has been found to be caused by the demethylating activity

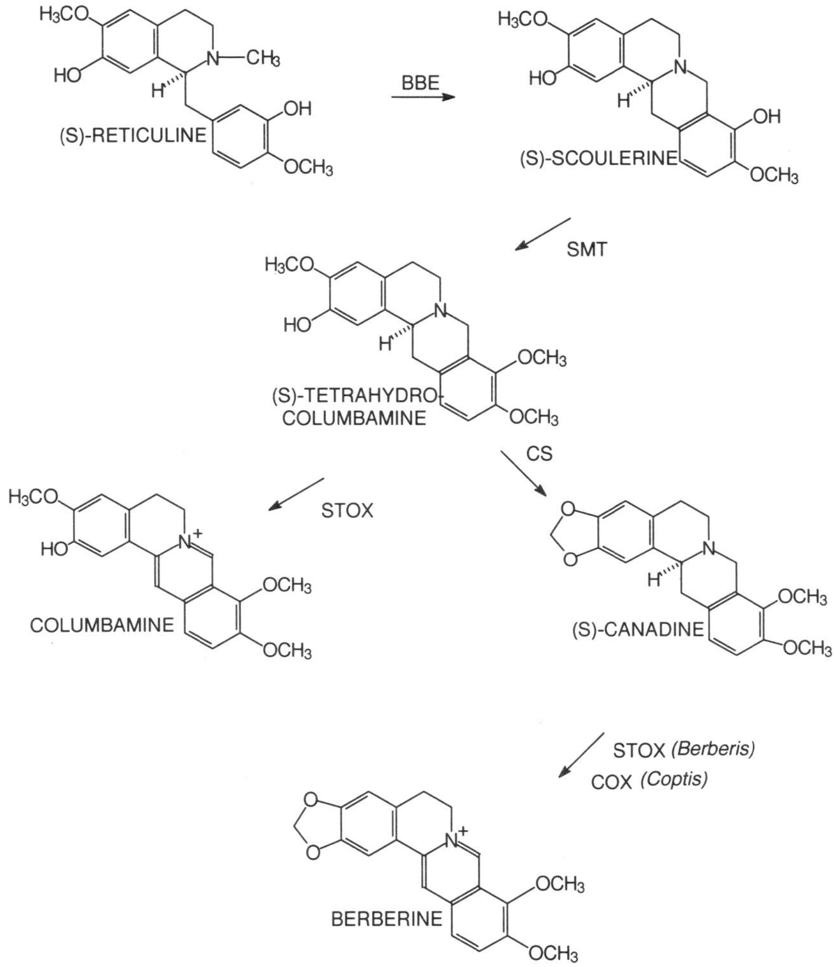

Figure 6. The biosynthesis of berberine from (*S*)-reticuline in *Berberis* species and *Coptis japonica.*

of a peroxidase found within the vesicle. It was also found that the cytochrome P_{450} enzyme (canadine synthase) from microsomes of *Berberis* spp., *Thalictrum* spp., and *Coptis* spp. formed the methylene bridge in (*S*)-tetrahydrocolumbamine but not in the quaternary alkaloid columbamine (Zenk, 1995). Because of this substrate specificity of canadine synthase, the berberine pathway once proposed for *Berberis* spp. (Amann *et al.,* 1984; Rueffer and Zenk, 1985) must be abandoned. Columbamine is, however, converted to palmatine by a specific methyltransferase first isolated from *Berberis wilsoniae* cell cultures (Rueffer and Zenk, 1985).

A unique cytochrome P_{450} enzyme isolated from *B. stolonifera* cell cultures catalyzes the oxidation of three different chiral benzyltetrahydroisoquinolines, namely, (*S*)-coclaurine, (*R*)-*N*-methylcoclaurine, and (*S*)-*N*-methylcoclaurine leading to the formation of three distinct dimeric products, namely, (*R,S*)-berbamunine, (*R,S*)-2′-norberbamunine, and (*R,R*)-guattegaumerine (Stadler and Zenk, 1993).

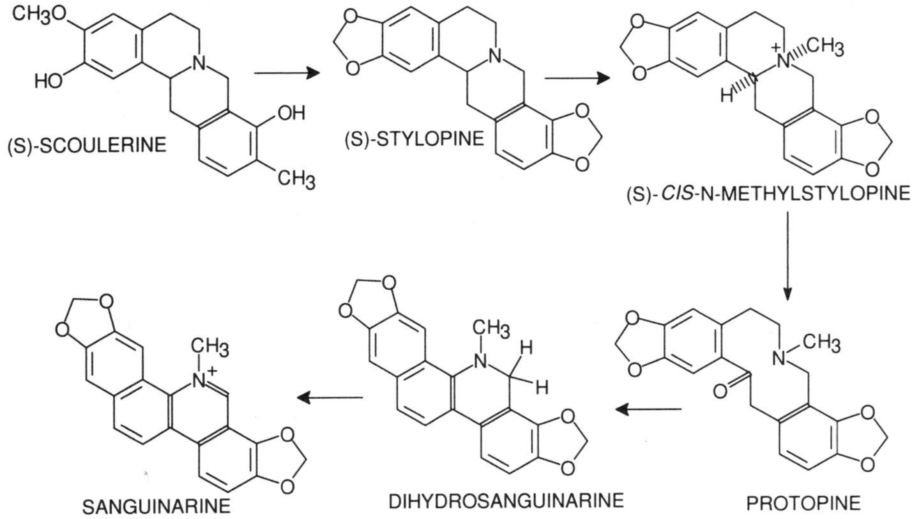

Figure 7. The biosynthesis of protopine and the benzophenanthridine alkaloids from (*S*)-scoulerine.

Important to our understanding of the mechanisms of secondary metabolism was the discovery that all of these enzymes from (*S*)-scoulerine to the production of berberine are firmly associated with vesicles (Amann *et al.,* 1986) which are thought to be derived from the endoplasmic reticulum. These vesicles appear to be specific sites for the formation of quaternary protoberberine alkaloids. Because of their positive charge, the alkaloids are prevented from leaving the vesicles and there is some evidence to suggest that they end up in the vacuole when the vesicle membrane fuses with the tonoplast (Zenk, 1989) (see also Chapter 10). Tertiary tetrahydrobenzylisoquinolines such as (*S*)-scoulerine are able to diffuse freely out of the vesicle to undergo further modifications.

4.3. The Route to the Protopine and Benzophenanthridine Alkaloids

Another important route stems from the formation of the *N*-methylated moieties of the (*S*)-tetrahydroprotoberberines which serve as precursors for the protopine, benzophenanthridine, tetrahydrobenzazepines (rhoadines), and spirobenzylisoquinoline alkaloids (Cordell, 1981) (Fig. 7).

Microsomal, cytochrome P_{450}-dependent enzymes isolated from the cells of *Eschscholtzia californica* convert (*S*)-scoulerine to (*S*)-stylopine by the introduction of methylenedioxy bridges (Bauer and Zenk, 1991). The subsequent *N*-methylation requires *S*-adenosyl-L-methionine:(*S*)-tetrahydro-*cis*-*N*-methyltransferase and this enzyme has been isolated from the cell cultures of a variety of plants found within the Berberidaceae, Fumariaceae, Menispermaceae, Papaveraceae, and Ranunculaceae (Rueffer *et al.,* 1990).

The route to protopine requires oxidation at C-14 of the tetrahydroprotoberberine molecule (Rueffer and Zenk, 1987b). The enzyme responsible for this oxidation is a microsomal cytochrome P_{450}–NADPH-dependent enzyme that hydroxylates, stereo- and regiospecifically, C-14 of (*S*)-*cis*-*N*-methyltetrahydroprotoberberines and has been found

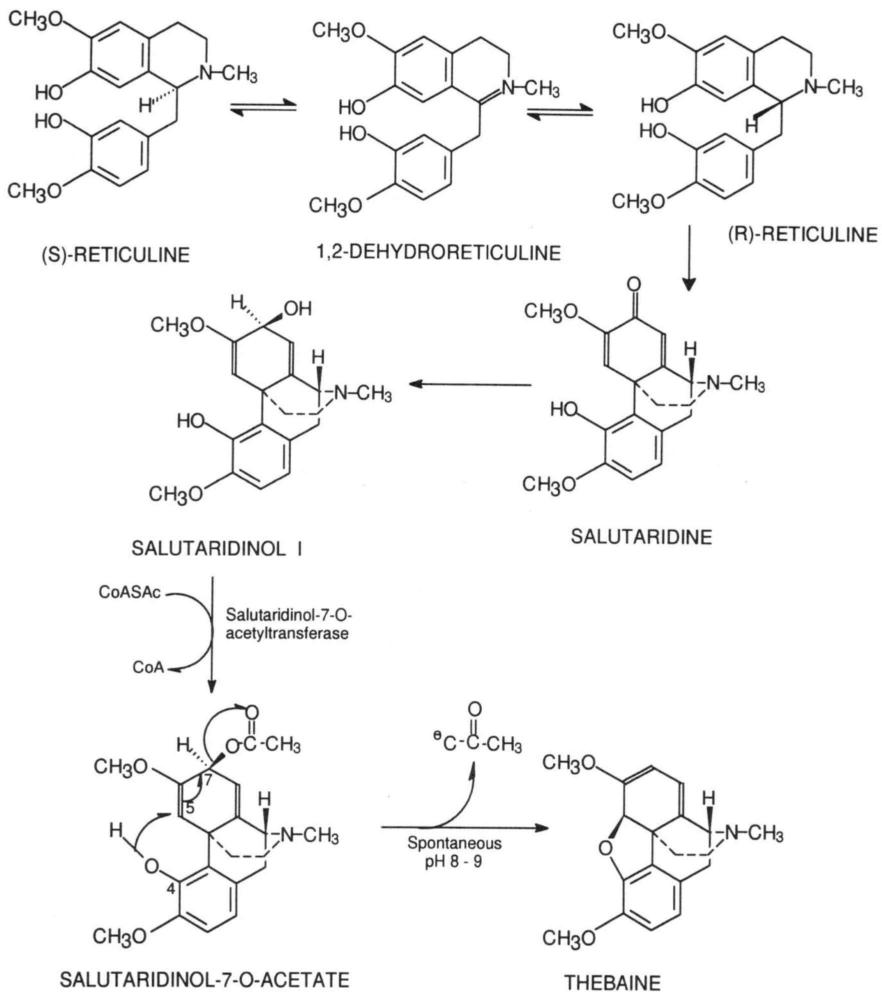

Figure 8. The biosynthesis of morphine via the conversion of (*S*)-reticuline to (*R*)-reticuline in *Papaver somniferum*.

in a number of cell cultures developed from plants of the Fumariaceae and Papaveraceae. Some of the best activity was observed using cell cultures of *Fumaria officinalis* and *F. cordata*. The protopines may be further metabolized to produce benzazepine and benzophenanthridine alkaloids. Protopine has been found to be a central intermediate in the biosynthesis of the benzophenanthridine sanguinarine and also the more highly oxidized alkaloids such as macarpine (Takao *et al.*, 1983; Schumacher and Zenk, 1988). Essential to this conversion is hydroxylation of the tetrahydroprotoberberine skeleton at C-6 and it is this that leads to C-6/N bond fission followed by intramolecular cyclization. Important to these events is the fact that, as acid salts, protopines cannot be represented by a simple *N*-protonated structure. The absence of a carbonyl absorption indicates the closure of the ten-membered ring as shown in Fig. 7. The microsomal enzyme that catalyzes the hydroxylation of protopine with the concomitant formation of dihydrosanguinarine has been isolated from *E. californica* and is strictly dependent on NADPH as a reducing factor and on molecular oxygen. Studies with inhibitors suggested that the enzyme is a cytochrome P_{450}-linked monooxygenase. The enzyme was also found to be specifically present only in plant species that produce benzophenanthridine alkaloids in culture (Tanahashi and Zenk, 1990). The dihydro moieties are readily converted to benzophenanthridine alkaloids by an oxidase (Arakawa *et al.,* 1992). This latter enzyme, together with a 12-*O*-methyltransferase (Kammerer *et al.,* 1994), convert dihydrosanguinarine, dihydrochelirubine, and dihydromaccarpine to sanguinarine, chelirubine, and macarpine, respectively, and hence the route to the benzophenanthridine alkaloids is now clearly defined at the enzyme level. In contrast to the berberine pathway, the enzymes of benzophenanthridine biosynthesis are located in the cytosol.

4.4. Biosynthesis of the Morphinan Alkaloids

4.4.1. (*S*)- AND (*R*)-RETICULINE

The role of reticuline as an intermediate in the biosynthesis of the morphinan alkaloids (Fig. 8) was demonstrated by the isolation of both (*S*)- and (*R*)-reticuline from the opium poppy (Martin *et al.,* 1978). An excess of the (*S*)-reticuline over the (*R*)-isomer was found in opium (poppy latex) obtained from the mature plant, in contrast to the roughly equal amounts of these two isomers to be found in poppy seedlings. Both isomers are found to be incorporated into morphine, the major alkaloid isolated from opium, although incorporation of the (*R*) isomer was slightly more efficient. (*R*)-Reticuline is firmly established in *P. somniferum* as the precursor of the morphinan-type alkaloids (Battersby *et al.,* 1965; Loeffler and Zenk, 1990). (*S*)-Reticuline, however, is the central intermediate in isoquinoline alkaloid biosynthesis (Cordell, 1981). It has been postulated that the (*R*)-reticuline is formed from (*S*)-reticuline by isomerization. This inversion of configuration is most plausibly explained by the intermediate formation of the 1,2-dehydroreticulinium ion originating from (*S*)-reticuline followed by stereospecific reduction to yield the (*R*) counterpart (Battersby *et al.,* 1965). The 1,2-dehydroreticulinium ion is efficiently incorporated into opium alkaloids and its role as a precursor of the morphinan-type alkaloids has been unequivocally established (Borkowski *et al.,* 1978).

The conversion of (*S*)-reticuline to 1,2-dehydroreticuline has been accomplished using a novel oxidase isolated from cell cultures of plants of the Berberidaceae. This enzyme, (*S*)-tetrahydroprotoberberine oxidase, has been previously shown to catalyze, in the presence of oxygen, the dehydrogenation of (*S*)-tetrahydroprotoberberine (Amann *et al.,* 1988). This flavoprotein is compartmentalized in a specific vesicle and can stereospecifically oxidise (*S*)-benzylisoquinolines to their corresponding 1,2-dehydro analogues. Although this enzyme more efficiently oxidizes the tetrahydroprotoberberines, it has been shown to occur in *P. somniferum* roots and leaves (Amann *et al.,* 1986, 1988). The question to be answered is whether, *in vivo,* this is the enzyme primarily responsible for the conversion of (*S*)-reticuline to its iminium ion. The conversion of 1,2-dehydroreticuline to (*R*)-reticuline was brought about by crude cell preparations from young seedlings of *P. somniferum* in the presence of NADPH at pH 8.5. The purified enzyme stereospecifically transfers the pro-*S*-hydride from NADPH to C-1 of the 1,2-dehydroreticuline. The reaction is highly substrate specific with no evidence for the reverse reaction. No activity was found in plants that do not normally synthesize the morphinans and in cell cultures of the genus *Papaver*, i.e., *P. somniferum, P. rhoas, P. bracteatum, P. feddei*, and *P. dubium* (De-Eknamkul and Zenk, 1990, 1992).

The formation of (*R*)-reticuline in this manner enables a narrow range of *Papaver* species to form the morphinandienone alkaloids morphine, codeine, and thebaine which also possess the (*R*) configuration at the chiral center.

The next step in the pathway to morphine, the intramolecular condensation of (*R*)-reticuline in a regio- and stereoselective manner to salutaridine, a morphinandienone, was detected by tracer feeding studies (Barton *et al.*, 1965). The natural occurrence of salutaridine was confirmed by the isolation of the compound from extracts of opium (Hodges and Rapoport, 1982). The enzyme responsible for this reaction, first isolated by Hodges and Rapoport (1982), has recently been found to be a highly selective microsomal bound cytochrome P_{450}-dependent enzyme isolated from young poppy capsules (Zenk *et al.*, 1995; Gerardy and Zenk, 1993). The conversion of salutaridine to salutaridinol with the (7*S*) configuration (Lotter *et al.*, 1992) (Fig. 8) by an NADPH-7-oxidoreductase isolated from *Papaver somniferum* takes the elucidation of the morphinan pathway a step further (Gerardy and Zenk, 1992, 1993). Salutaridinol possesses the correct configuration for an allylic *syn*-displacement of the activated C-7 hydroxyl by the phenolic C-4 hydroxyl to produce thebaine. A highly substrate-specific enzyme that transfers the acetyl moiety from acetyl CoA to the 7-OH group of salutaridinol has been discovered and purified to homogeneity. The salutaridine-7-*O*-acetate that is formed, subsequently spontaneously closes, at a cellular pH of 8–9, to produce the oxide bridge between C-4 and C-5 and thus furnish thebaine (Lenz and Zenk, 1995a).

The sequences from thebaine via various intermediates to morphine, although known from ^{14}C-labeling studies, are as yet poorly understood at the enzyme level (Brochmann-Hanssen, 1984) (Fig. 9). An $NADH/NAD^+$-dependent codeinone reducing enzyme was reported earlier (Furuya *et al.*, 1984; Hodges and Rapoport, 1980) but purification was not accomplished. More recent results suggest that codeinone reductase is not $NADH/NAD^+$ dependent but has a requirement for NADPH. This NADPH-requiring enzyme has now been isolated and characterized from cell cultures of *Papaver somniferum*. Using capsule tissue of differentiated *P. somniferum* plants and applying similar isolation procedures, two isoenzymes were isolated. To determine which isoenzyme is similar to the enzyme isolated from the cell cultures will require the isolation of the appropriate genes. These codeinone reductases ($NADPH/NADP^+$) convert both codeinone to codeine and morphinone to morphine and are thus of prime importance in the biosynthesis of morphine (Lenz and Zenk, 1995b) (Fig. 9). These latest findings take the total elucidation of the biosynthetic route to morphine a step further.

5. INDOLE ALKALOID BIOSYNTHESIS

The monoterpene indole alkaloids are almost as large a group as the benzylisoquinolines and contain a much greater diversity of structures. The majority of these alkaloids have been isolated from three mainly tropical plant families, Loganiaceae, Apocynaceae, and Rubiaceae, all of the Gentianales. The indole alkaloids as a group are rich in biologically active constituents some of which are used as therapeutic agents in medicine (see Chapters 4 and 18). As an example one may cite vinblastine and vincristine,

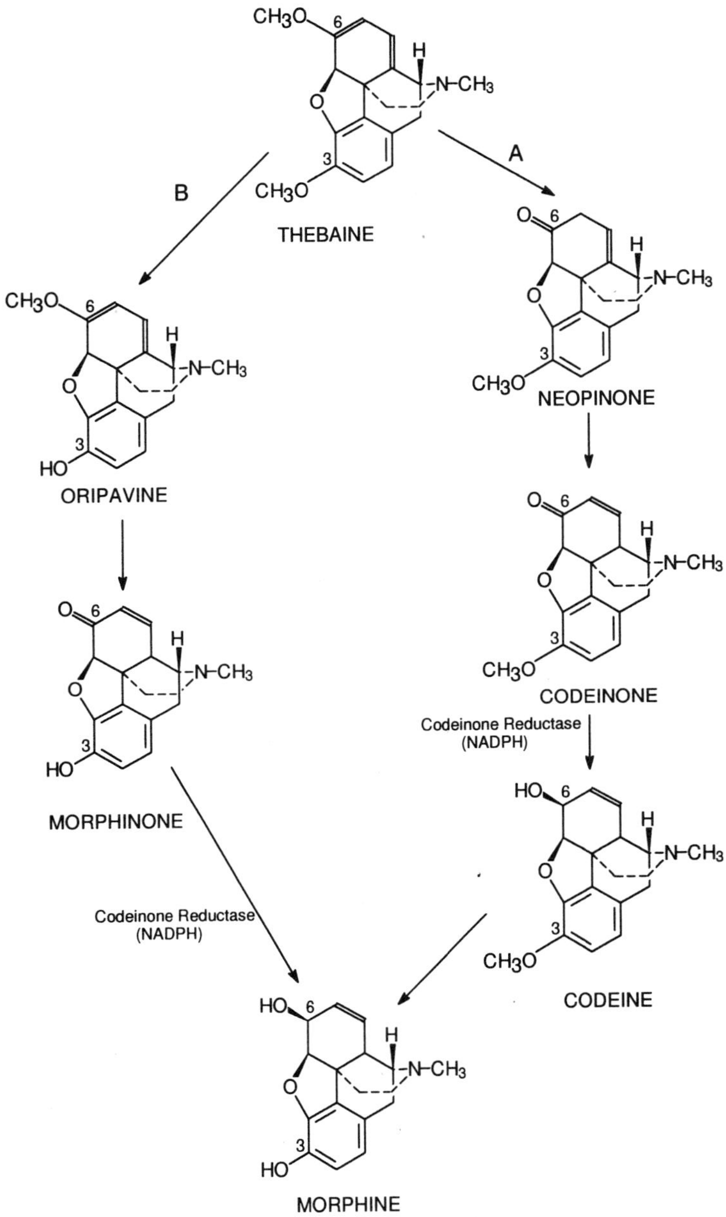

Figure 9. The pathways to morphine from thebaine.

two dimeric alkaloids used in the treatment of leukemia and Hodgkin's disease and present in small amounts in *Catharanthus roseus* (Apocynaceae). These alkaloids have led to extensive investigation of this plant and cell cultures derived from it. However, neither the formation of vincristine or vinblastine nor vindoline, the major alkaloid of *C. roseus,* was unequivocally found in cell cultures (De Luca and Kurz, 1988). Cell cultures of *C. roseus,* however, do produce many other indole alkaloids and have proved to be very useful for biochemical studies. As a result the *Catharanthus* system is one of the best studied *in vitro* with most of the basic studies of indole alkaloid biosynthesis at the enzyme level elucidated using *Catharanthus* cell cultures (Stöckigt, 1984).

Indole alkaloids are derived from tryptophan which, in the case of the terpenoid indoles, is usually first converted to tryptamine by the enzyme tryptophan decarboxylase. This enzyme occurs in the cytosol and has been detected in all parts of the developing seedling and in cell cultures of *C. roseus* (De Luca and Cutler, 1987; De Luca *et al.*, 1988). This enzyme appears to be a pyridoxo-quinoprotein as two molecules of pyridoxal phosphate and two molecules of covalently bound pyrroloquinoline quinone were found per enzyme molecule (Pennings *et al.*, 1989). Tryptamine condenses with the monoterpene, *seco*-loganin, which is derived from geraniol by hydroxylation at C-10. The enzyme responsible for this latter reaction is a cytochrome P_{450}-requiring hydroxylase which was first characterized from *C. roseus* and found in low activity in cell cultures of that plant (Madyastha *et al.*, 1976; Madyastha and Coscia, 1979; Spitsberg *et al.*, 1981). Plant cell cultures have been used to further investigations of this enzyme (Meijer *et al.*, 1993) which appears to have a regulatory effect on alkaloid production; its activity pattern being more closely related to the pattern of indole alkaloid accumulation than that of tryptophan decarboxylase. The intermediate accumulation of tryptamine and its later incorporation into indole alkaloids like ajmalicine indicated that the coordination of the two precursor pathways for monoterpene indole alkaloid formation are not synchronized (Schiel *et al.*, 1987). The most recent studies suggest that loganic acid is synthesized from 10-hydroxynerol via 9,10-deoxygeranial and iridotrial (Balsevich *et al.*, 1982), or via a route involving 10-oxogeranial or 10-oxoneral and iridoidial (Uesato *et al.*, 1986a,b). The latter pathway appears to be the more plausible since a monoterpene cyclase which converts 10-oxongeranial to iridoidial has been isolated. The final step in the formation of *seco*-loganic acid is the methylation of loganic acid. The methyltransferase has been partially purified from young *C. roseus* seedlings (Madyastha *et al.*, 1973).

5.1. Formation of (*S*)-Strictosidine

The stereospecific condensation between tryptamine and *seco*-loganin is carried out by the enzyme (*S*)-strictosidine synthase and results in the formation of the glucoalkaloid (*S*)-strictosidine (Fig. 10) from which most monoterpene indole alkaloids are derived (Fig. 11) to include the quinoline alkaloids of *Cinchona* (Stöckigt and Zenk, 1977; Nagakura *et al.*, 1979).

The isolation of the stereospecific strictosidine synthase and the formation of strictosidine (≡ isovincoside) with the 3α-(*S*) configuration proved conclusively that this was the natural precursor of the terpenoid indole alkaloids. Strictosidine occurs naturally in *Rhazya stricta* and the synthase has been isolated from a number of species: *Amsonia salicifolia, A. tabernaemontana, Catharanthus pusillus, C. roseus, Rauwolfia verticillata, R. vomitoria, R. serpentina, Rhazya orientalis,* and *Voacanga africana*. The enzyme has been purified to homogeneity from *R. serpentina* (Treimer and Zenk, 1979; Hampp and Zenk, 1988). A comparison of the activity of strictosidine synthase from *C. roseus* roots, the only portion of the plant to contain ajmalicine, with that present in plant cell cultures producing the same alkaloid suggested that the plant cell cultures are much more metabolically active.

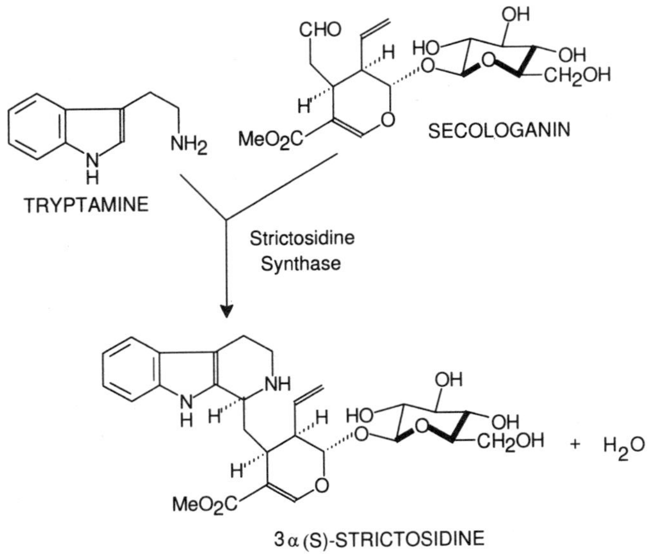

Figure 10. The first committed step in indole alkaloid biosynthesis: the formation of 3α(*S*)-strictosidine from tryptamine and *seco*-loganin by strictosidine synthase.

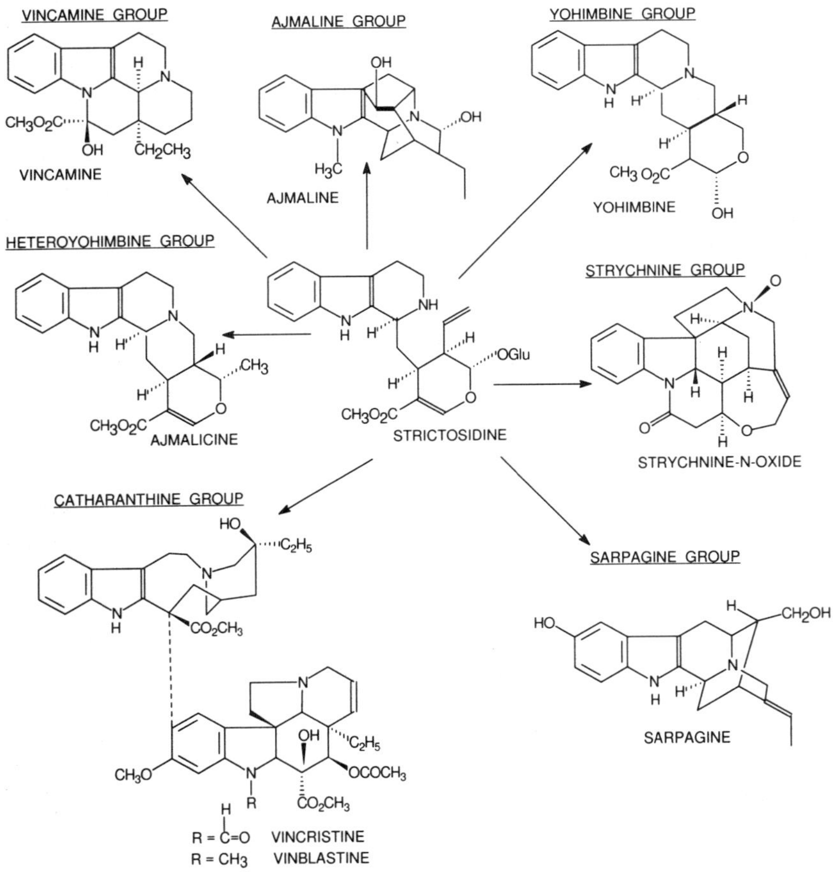

Figure 11. Types of indole alkaloids biosynthesized from strictosidine.

Strictosidine synthase has been immobilized on CNBr-activated Sepharose and this has made possible the production of gram quantities of (*S*)-strictosidine, previously difficult to purify in large quantities from plant cell cultures (Pfitzner and Zenk, 1989).

The cDNA for strictosidine synthase has now been expressed in an enzymatically active form in *Escherichia coli*, *Saccharomyces cereviseae*, and cell cultures of the insect *Spodoptera frugiperda* (Kutchan, 1989; Kutchan *et al.*, 1991). Recently the modified cDNA encoded strictosidine synthase from *Catharanthus roseus* (L.) Don. has been introduced into tobacco plants. Transgenic tobacco plants expressing this construct had from 3 to 22 times greater strictosidine synthase activity than *C. roseus* plants. Ultrastructural immunolocalization demonstrated that strictosidine synthase is a vacuolar protein in *C. roseus* and is correctly targeted to the vacuole in transgenic tobacco (McKnight *et al.*, 1991).

5.2. Deglucosylation of Strictosidine

Deglucosylation of strictosidine, a key reaction in the formation of the many types of indole alkaloids (Fig. 11), is carried out by two highly specific gluco-alkaloid β-glucosidases, strictosidine-β-D glucosidase I and II. They have been isolated from *C. roseus* and a number of other indole alkaloid-containing plants of the Apocynaceae. These specific glucosidases are involved in an essential initial reaction that leads to a complex sequence of events and a series of highly reactive intermediates. Geissoschizine, which is formed from these intermediates (Fig. 12), is converted via geissoschizine dehydrogenase to 4,21-dehydrogeissoschizine. The enzyme that removes the 21α-hydrogen of geissoschizine in an $NADP^+$-dependent reaction has been partially purified from *C. roseus* cell suspension cultures. Compared with other enzymes of the ajmalicine pathway, geissoschizine dehydrogenase shows extremely low specific activity (Pfitzner and Stöckigt, 1982).

5.3. Formation of Corynanthe-Type Alkaloids

Ajmalicine, 19-epi-ajmalicine, and tetrahydroalstonine are formed from 4,21-dehydrogeissoschizine via cathenamine (Fig. 12) (Zenk, 1980). The enzymatic synthesis of these corynanthe-type alkaloids has been investigated using *C. roseus* cell suspension cultures. The soluble enzyme system that catalyzes the formation of these alkaloids has been characterized using a radioimmunoassay with antibodies directed at ajmalicine (Treimer and Zenk, 1978). Using crude enzyme preparations, it has been possible to follow the steric course of hydrogen transfer during the formation of the 3α-heteroyohimbine alkaloids (Fig. 11) (Stöckigt *et al.*, 1983).

5.4. The Formation of Sarpagan-Type Alkaloids

Vinorine and ajmaline and the related alkaloids raucaffrinoline and raucaffricine are also formed via a series of enzymatic steps from 4,21-dehydrogeissoschizine (Fig. 13). The step from 4,21-dehydrogeissoshizine to the sarpagan structure has not been verified at

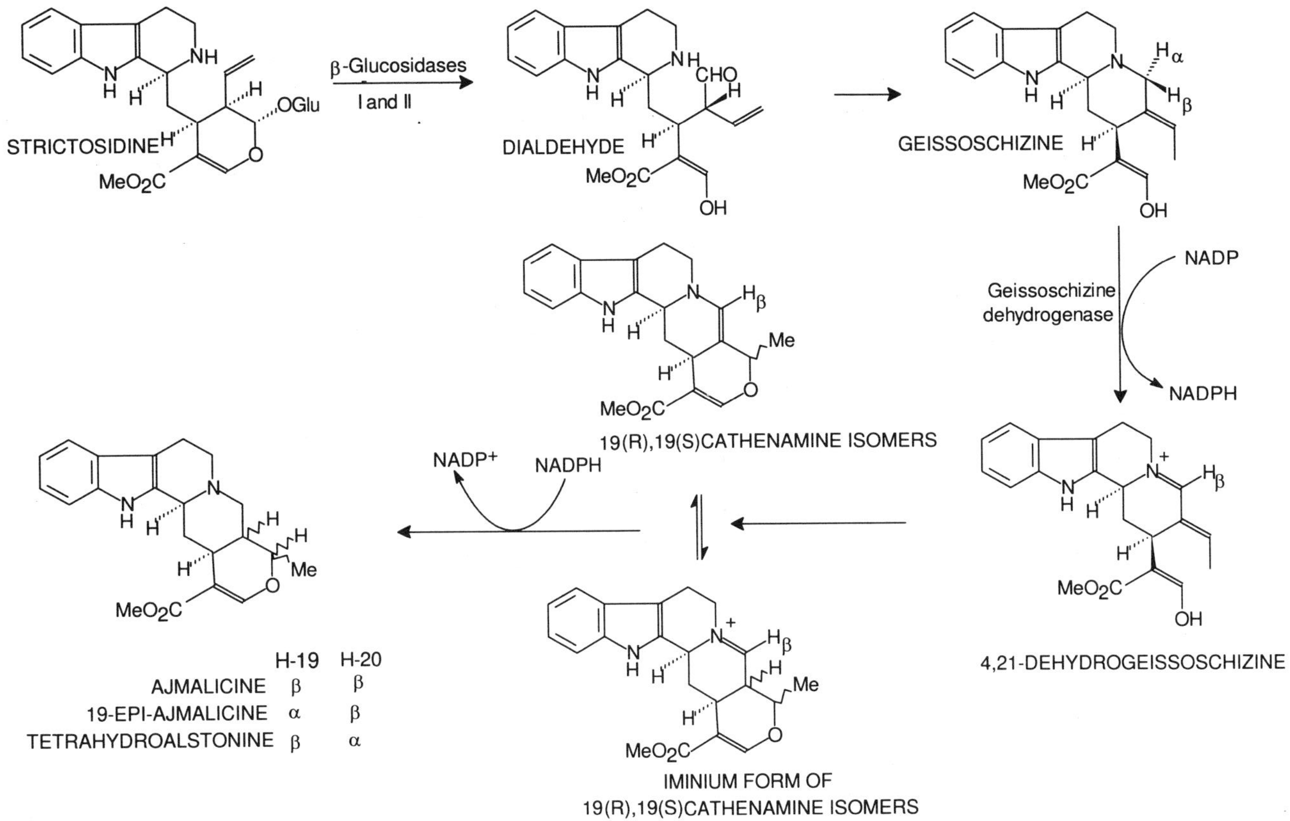

Figure 12. Enzymatic formation of ajmalicine, 19-epi-ajmalicine, and tetrahydroalstonine in *Rauwolfia serpentina*.

the enzyme level. Recent studies have shown that the sarpagine bridge in polyneuridine aldehyde (PNA) (Fig.13) is formed by a microsomal enzyme that requires NADPH and oxygen. Inhibition studies indicate a cytochrome P_{450}-dependent monooxygenase (Schmidt and Stöckigt, 1995). From structural similarities and the next enzyme in the sequence, it has been proved that the alkaloid PNA is one of the stable intermediates at the beginning of this biogenetic route. The enzyme that acts on the aldehyde has been well characterized from *R. serpentina* cells and is the specific methylesterase polyneuridine aldehyde esterase (PNA esterase); the product of the reaction, polyneuridine acid, is very unstable and decarboxylates to give 16-epi-vellosimine (Stöckigt and Schübel, 1988). This is the mechanism by which the monoterpenoid skeleton is transformed into a C-9 unit (Pfitzner and Stöckigt, 1983; Schmidt and Stockigt, 1995). At the same time, 16-epi-vellosimine has the correct stereo-requirement for the formation of the ajmaline-type bond.

The next step in the sequence to ajmaline is catalyzed by the enzyme vinorine synthase (Pfitzner *et al.,* 1986). Vinorine, a constituent of *Rauwolfia* cell cultures, is an acetylated indolenine alkaloid. It transpires that vinorine synthase has a requirement for acetyl CoA as cosubstrate. The acetyl unit has a stabilizing effect (Fig. 13) on the indolenine structure. Vinorine is hydroxylated to vomilenine (21-OH-vinorine) by a cytochrome P_{450}-dependent hydroxylase (Falkenhagen and Stöcktigt, 1995) and this intermediate product is then converted to ajmaline via a series of enzymatic reactions. First is the reduction by an NADPH-requiring reductase to 1,2-dihydrovomilenine followed by further reduction, also NADPH requiring, to acetylnorajmaline. Deacetylation proceeds with the aid of acetylesterase [specific for the 2β (*R*) configuration] to give norajmaline (Polz *et al.,* 1986). This enzyme which catalyzes the hydrolysis of 17-*O*-acetylated alkaloids of the ajmaline group has been purified 27-fold from *R. serpentina.* It has high substrate selectivity and exclusively accepts acetylated ajmaline derivatives with the naturally occurring 2β (*R*)-configuration. The highest enzyme activities were observed in leaves and cell suspension cultures of the tribe Rauvolfieae which are known to synthesize ajmaline and its congeners. Finally, *N*-methylation occurs to complete the biogenetic sequence with the production of ajmaline.

This pathway to ajmaline (Route A, Fig. 13) has other branch points which have an effect on ajmaline production by cell suspension cultures of *R. serpentina.* The first branch point involves 16-epi-vellosimine which, in these cell cultures, may be readily converted to sarpagine. 16-Epi-vellosimine enters this alternative route by spontaneous epimerization to vellosimine which is then reduced by vellosimine reductase to 10-deoxysarpagine followed by hydroxylation to sarpagine.

The second branch route occurs as a result of aldehyde formation (perkakine, Fig. 13) from vomilenine and the conversion to raucaffinoline as a result of an NADPH reductase. However, these two branch routes do not have a significant effect in ajmaline production.

A third branch point occurs (Route B, Fig. 13) with the alternate conversion of vomilenine into its glycoside raucaffricine and has a very significant effect on ajmaline production. In *R. serpentina* cell cultures raucaffricine amounts to 1.2 g liter^{-1} medium whereas ajmaline levels reach only 0.3 g liter^{-1} medium (Schübel *et al.,* 1986). It is interesting to note that raucaffricine is a typical constituent of *R. caffra* but has not been isolated from other *Rauwolfia* species. However, in *Rauwolfia* cell cultures the compound is found in all species tested with a maximum yield in *R. serpentina.* In other words, under this type of growth condition the pathway to ajmaline appears to have become deregulated.

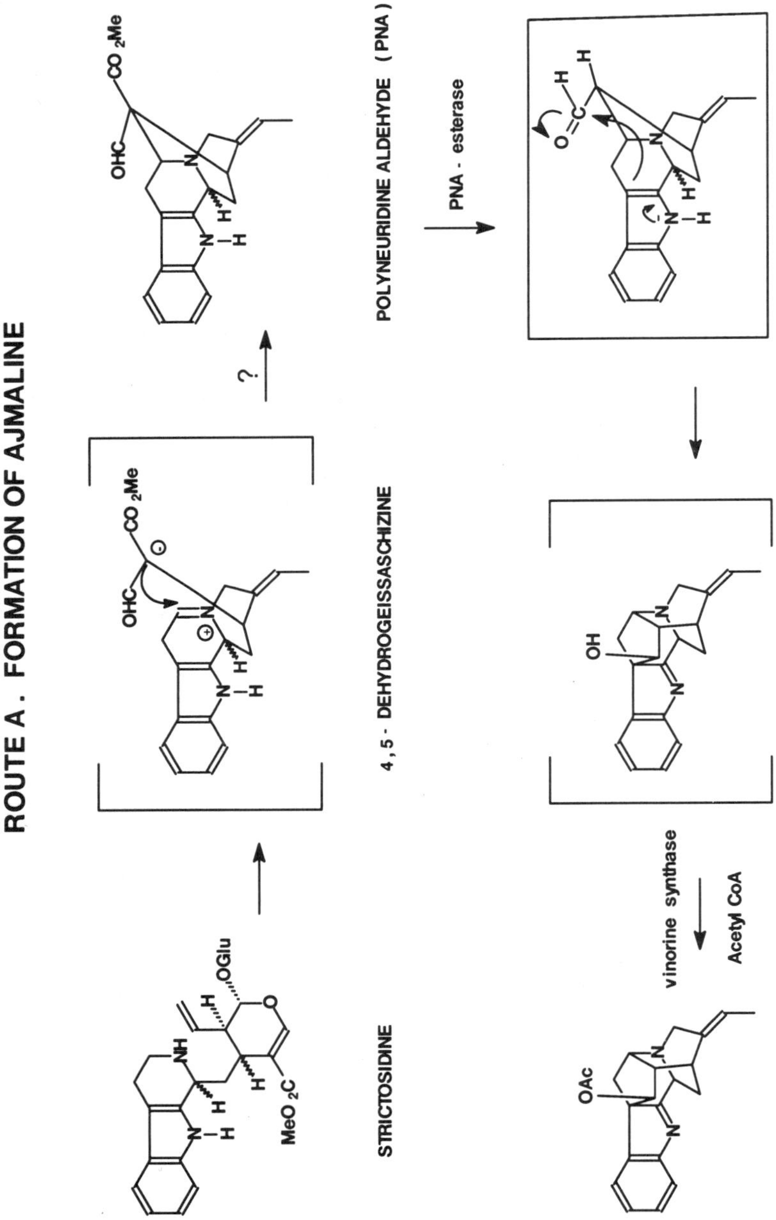
ROUTE A . FORMATION OF AJMALINE
STRICTOSIDINE
4 , 5 - DEHYDROGEISSASCHIZINE
?
POLYNEURIDINE ALDEHYDE (PNA)
PNA - esterase
16 - EPI - VEILOSOMINE
DEACETYLVINORINE
vinorine synthase
Acetyl CoA
VINORINE

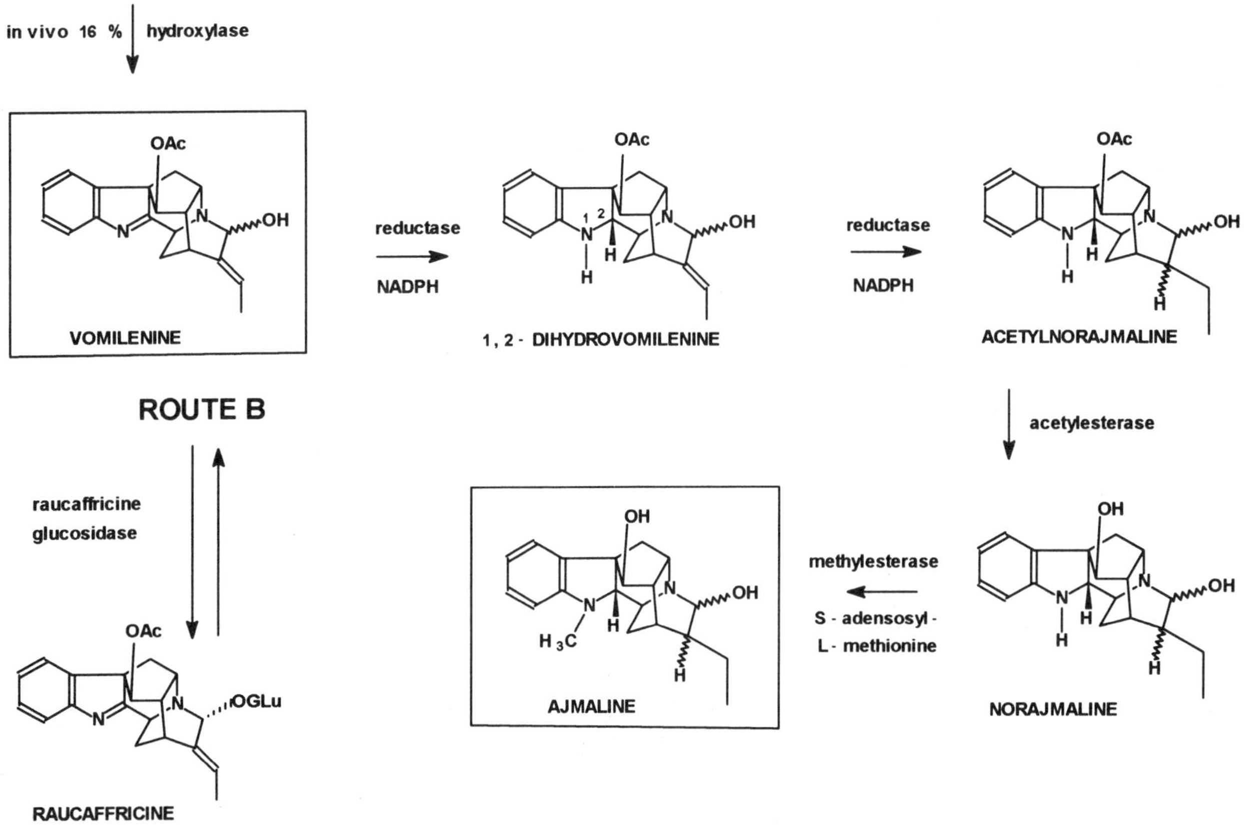

Figure 13. Biosynthesis of ajmaline (route A) and raucaffricine (route B) in *Rauwolfia serpentina* cell cultures.

5.5. Formation of Aspidosperma-Type Alkaloids

It was originally proposed that the biosynthesis of vindoline proceeds from the alkaloid tabersonine by a sequence of three consecutive hydroxylations followed by *N*-methylation, *O*-methylation, and *O*-acetylation. This proposal was based on the results obtained from substrate specificity studies performed with an acetyl CoA-dependent *O*-acetyltransferase (Fahn *et al.*, 1985a) and on the demonstrated presence of an *S*-adenosyl-L-methionine-*O*-methyltransferase activity (Fahn *et al.*, 1985b).

The enzyme-catalyzed acetylation of deacetylvindoline was also thought to be the last step in vindoline synthesis (De Luca *et al.*, 1985). Further studies performed with *C. roseus* seedlings (Balsevich *et al.*, 1982; De Luca *et al.*, 1986) suggested that the route from tabersonine to vindoline was not as proposed by Fahn *et al.* (1985a,b) but rather proceeded by the sequence shown in Fig. 14. The C-4 hydroxylation of 2,3-dihydro-3-hydroxy-*N*(1)-methyltabersonine to the 3,4-dihydroxy derivative utilizes an enzyme that has an absolute requirement for 2-oxyglutarate; enzymatic activity was enhanced by ascorbate, establishing that the enzyme involved is a 2-oxyglutarate-dependent dioxygenase. This enzyme is specific for position 4 of various alkaloid substrates. The appearance of 4-hydroxylase activity was shown to be developmentally regulated and is inducible by light treatment of seedlings. Substrate specificity studies with this enzyme suggested that hydroxylation at C-3 and *N*-methylation are required prior to hydroxylation at position 4. This enzyme, therefore, catalyzes the second to the last step in vindoline synthesis: the conversion of 16-methoxy-3,4-dihydroxy-*N*-methyltabersonine to deacetylvindoline (Fig. 14) (De Carolis *et al.*, 1990). The methyltransferase required for the formation of the substrate for the penultimate step (deacetyloxyvindoline) from 16-methoxy-2,3-dihydro-3-hydroxytabersonine has now been isolated and partially purified (Fahn and Stöckigt, 1990; Dethier and De Luca, 1993).

The final step in the formation of vindoline is the acetylation of 17-*O*-deacetylvindoline by a 17-*O*-acetyltransferase. This enzyme has been purified to homogeneity from *C. roseus* leaves.

Seedlings grown in the dark produced an early accumulation of tabersonine as a major alkaloid. Transfer of 5-day-old seedlings to the light resulted in the rapid loss of vindoline precursors followed by a more gradual disappearance of tabersonine and the subsequent enhancement of vindoline accumulation. Although light enhanced vindoline biosynthesis, it was not essential. The time course of induction indicated that an increase of tryptophan decarboxylase coincided with tabersonine accumulation, whereas increase of acetyl CoA:deacetylvindoline-*O*-acetyltransferase activity coincided with vindoline accumulation. Results with young seedlings suggested that the tabersonine biosynthetic pathway enzymes occur in all plant parts whereas the last five steps in vindoline biosynthesis are restricted to aerial parts of the plant and that the whole pathway to vindoline biosynthesis is developmentally regulated (De Luca *et al.*, 1986). Further investigations of some of the enzymes involved in vindoline production with young seedlings of *C. roseus* showed that while tryptophan decarboxylase (TDC), strictosidine synthase (SS), *N*-methyltransferase (NMT), and *O*-acetyltransferase (OAT) activities appeared early in seedling development, TDC activity was highly regulated and peaked over a 48-hr period achieving a maximum by day 5. Both TDC and SS were present in all tissues of the seedlings. NMT and OAT enzyme activities were induced after TDC and SS had peaked and these

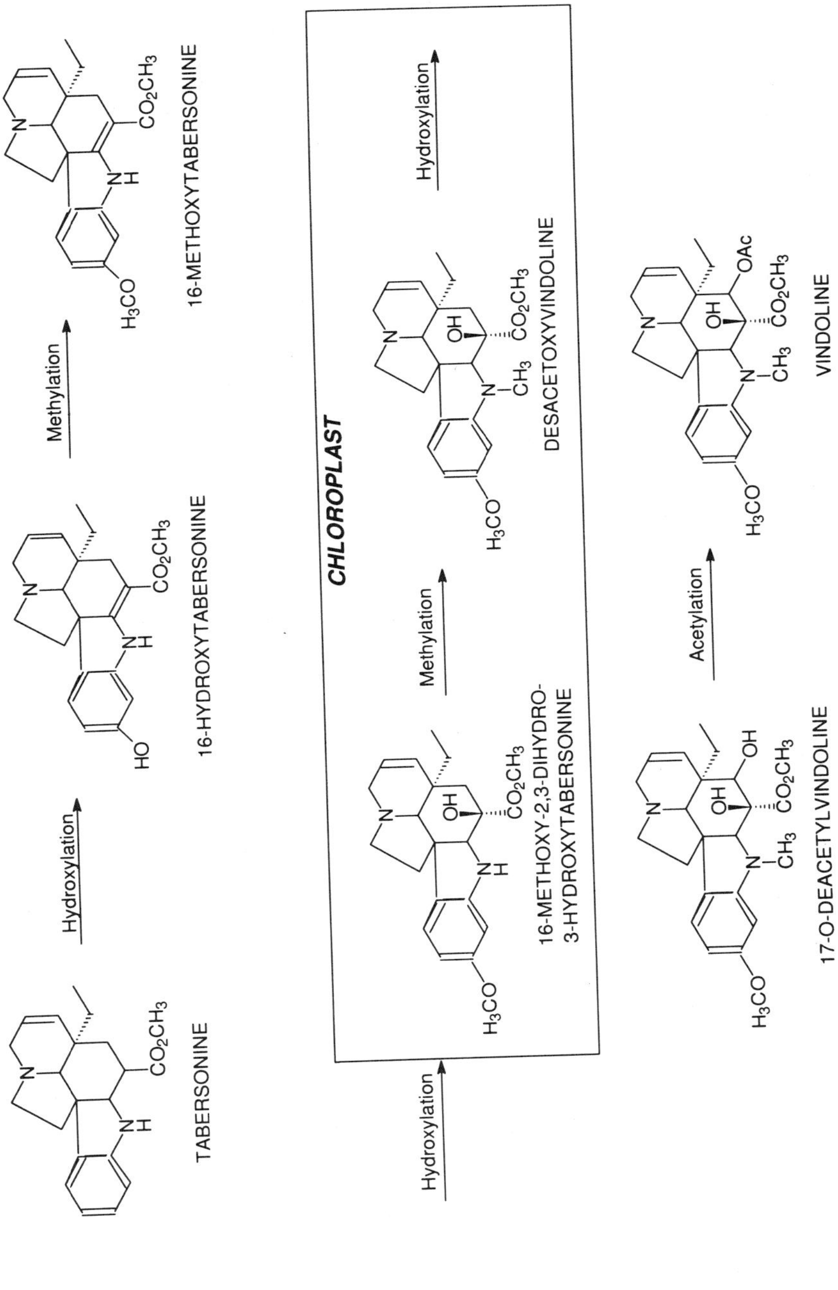

Figure 14. Proposals for the formation of vindoline from tabersonine in *Catharanthus roseus* seedlings.

activities could only be found in hypocotyls and cotyledons. TDC, SS, and NMT did not require light for induction and OAT enzyme activity increased approximately 10-fold after light treatment of dark-grown seedlings (De Luca *et al.,* 1988). TDC, SS and OAT were found to be cytoplasmic enzymes but NMT was found in the chloroplasts associated with the thylakoid. The participation of the chloroplast in this pathway suggests the indole alkaloid intermediates enter and exit the compartment during vindoline synthesis (De Luca and Cutler, 1987).

The methyltransferase that catalyzes the *N*-methylation of 16-methoxy-2,8-dihydro-3-hydroxytabersonine to deacetoxyvindoline has been isolated from young leaves of *C. roseus* plants and has a high degree of specificity for the nitrogen at position 1 of (2*R*,3*R*)-2,3-dihydro-3-hydroxytabersonine. This enzyme has a high degree of specificity for tabersonine derivative substrates containing a reduced double bond at position 2,3 of tabersonine derivatives but the more substituted *N*-desmethyl-deacetylvindoline did not act as a substrate.

Indole alkaloids are generally stored in the vacuole. However, the mechanisms involved are still a matter for discussion (see Chapter 10).

5.6. Biosynthesis of Dimeric Indole Alkaloids

The production in *C. roseus* of the clinically useful alkaloids vincristine and vinblastine has received much attention (Fig. 15). These alkaloids are found, albeit in small amount, in the young leaves but it has not been possible to produce significant amounts of these alkaloids from plant cell cultures. The monomers required for the production of vincristine and vinblastine (vindoline and catharanthine) may be isolated in suitable amounts from the plant (vindoline) or cell cultures (catharanthine). Cell free extracts from *C. roseus* will convert [2-^{14}C]tryptophan and *seco*-loganin into vindoline (Kutney, 1987) and the same cell-free extracts will also couple vindoline and catharanthine to yield the dimeric 3′,4′-anhydrovinblastine, which, in turn, can be converted to the natural dimeric alkaloids leurosine, catharine, and vinblastine (Fig. 15) (Kutney, 1987). More recently the enzyme that couples vindoline and catharanthine to produce 3′,4′-anhydrovinblastine has been partially purified from cell suspension cultures and found to consist of at least five different enzymes which were demonstrably of a peroxidase nature (Endo *et al.,* 1986). Subsequently it was found that this coupling reaction could be achieved with horseradish peroxidase followed by sodium borohydride reduction (Goodbody *et al.,* 1988). It is proposed that the horseradish peroxidase enables condensation of catharanthine and vindoline to an iminium intermediate which is then reduced to 3′,4′-anhydrovinblastine. It has also been possible to convert vinblastine to vincristine with cell suspension cultures of *C. roseus*. An efficient method of obtaining vincristine depends on an efficient method for the conversion of 3′,4′-anhydrovinblastine to vinblastine.

6. ERGOT ALKALOID BIOSYNTHESIS

The fungus *Claviceps purpurea,* normally parasitic on rye and other cereals, has long been known as a mycotoxin producer. Poisoning in man, known as St. Anthony's fire,

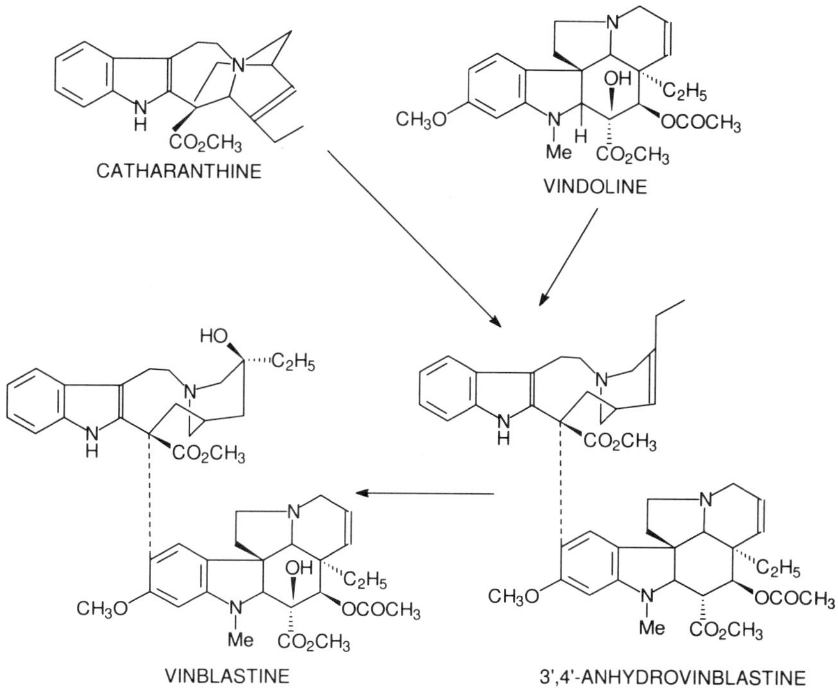

Figure 15. Biosynthesis of the *Catharanthus roseus* dimeric alkaloids, vincristine and vinblastine.

occurred in regular epidemics in Europe in the Middle Ages caused by contamination of the flour used in bread making. This type of mycotoxicosis is now virtually eliminated, unlike others such as those caused by aflatoxins and tricothecenes (Evans, 1996). The toxins responsible are alkaloids of a type that is derived, like other indole alkaloids, from tryptophan and because they are of importance as medicinal agents, they have been extensively studied (see Chapters 1 and 18).

The naturally occurring ergot alkaloids can be divided into two classes on the basis of their chemical structure: the lysergic acid derivatives and the clavine alkaloids. All of them possess the tetracyclic ergoline system. Besides the sclerotia of *Claviceps*, other fungi and several higher plants can contain ergot alkaloids. Some examples of fungi are *Aspergillus fumigatus, Rhizopus arrizus, Penicillium roqueforti, Sclerotium dephinii* and of plants, *Rivea corymbosa* and *Ipomoea tricolor*. The alkaloids from these sources are restricted to low yields and therefore for practical purposes *Claviceps* remains the only commercial source. Until 1960 *Claviceps* growing on rye was the only source of ergot alkaloids; however, it is now possible to grow strains of *C. purpurea* in submerged culture and obtain commercially useful amounts of ergotamine and ergocryptine.

The formation of ergot alkaloids from L-tryptophan is well known (Herbert, 1989; Mothes *et al.*, 1985; Floss and Anderson, 1980).

L-Tryptophan condenses with dimethylallylpyrophosphate (DMAPP) to give γ,γ-dimethylallyltryptophan (DMAT) which is modified via chanoclavine-I to give agroclavine and finally elymoclavine (Fig. 16). From this last alkaloid, lysergic acid may be formed.

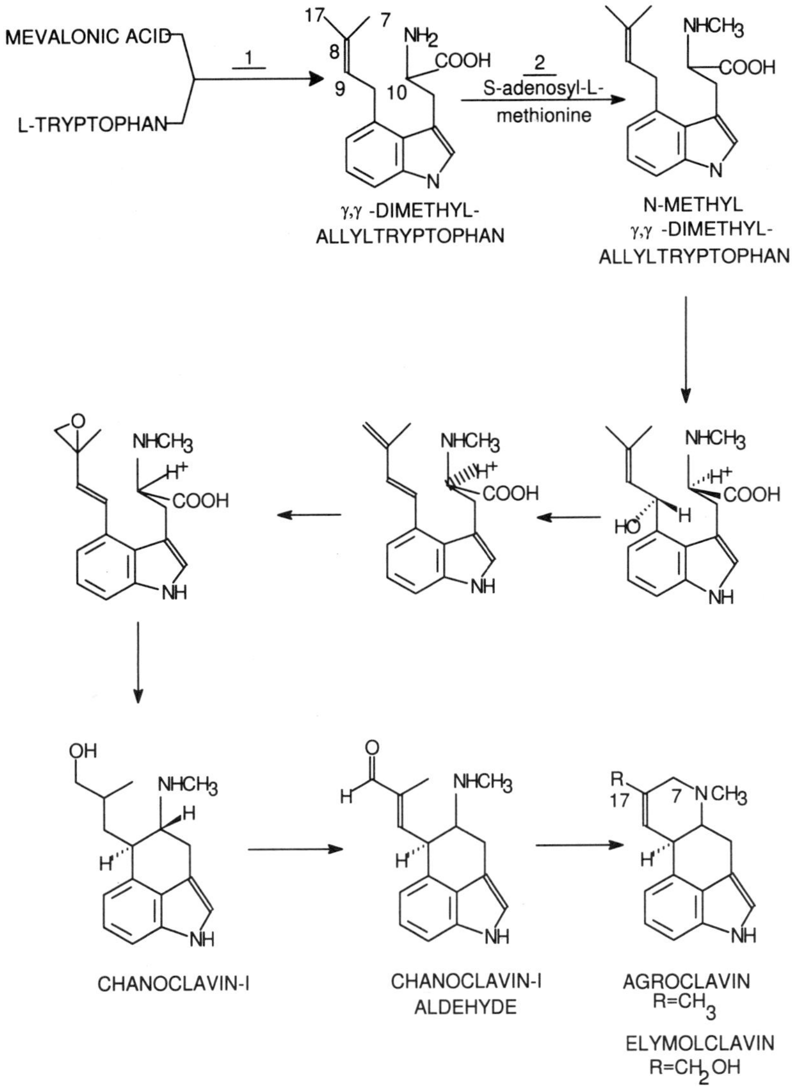

Figure 16. Biosynthesis of agroclavine and elymoclavine by *Claviceps purpurea* cultures.

Pharmaceutically important drugs are derived from lysergic acid which is the precursor of a range of peptidic ergot bases, e.g., ergotamine. The amide portion may be a smaller peptide or simple alkylamide, the basic skeleton being called an ergopeptide. Peptidic ergot bases contain lysergic acid and an amide portion reduced to a tricyclic ring system. In the clavine alkaloids the carboxyl group at C-17 is converted to a group with a lower oxygen state (Fig. 16).

Most of the biosynthetic studies with *C. purpurea* have used submerged or surface cultures. The enzyme responsible for the first step in the biosynthesis of these alkaloids is DMAT synthase. This enzyme which brings about the condensation of L-tryptophan with DMAPP has been isolated from *C. purpurea* cultures, purified, and characterized. The mechanism by which the enzyme works has been probed using a set of analogues of DMAPP and L-tryptophan and it was concluded that the reaction was an electrophilic aromatic substitution similar to that catalyzed by farnesylpyrophosphate synthase. There is a feedback mechanism operative with inhibition of the enzyme by elymoclavine (Shibuya *et al.*, 1990; Gebler and Poulter, 1992; Gebler *et al.*, 1992). The γ,γ-dimethylallyl-L-tryptophan (DMAT) formed is methylated using the methyl group of *S*-adenosylmethionine; the activity of the enzyme in cultures roughly parallels that of other ergoline enzymes and has been isolated, purified, and characterized (Otsuka *et al.*, 1980). This is therefore the second pathway-specific step; however, further reactions are required in the isoprenoid side chain before C-ring formation can take place.

In the conversion of *N*-methyl-DMAT to give chanoclavine-I there is a potential gap in our knowledge. The oxygen atoms of both elymoclavine and chanoclavine-I are derived from molecular oxygen (Kobayashi and Floss, 1987). Chanoclavine aldehyde has been found as a natural constituent of a blocked mutant strain of *C. purpurea* and this strongly suggested that it was an intermediate on the route to the tetracyclic ergolines. Conformation at the enzyme level is required to validate this hypothesis. Elegant experiments with radiolabeled precursors suggest that two *cis–trans* isomerizations occur during the conversion of *N*-methyl-DMAT by ring closure to agroclavine. The most recent experiments on the formation of the C-ring utilizing deuterated intermediates suggest that the incorporation of *N*-methyl-DMAT into chanoclavine-I is by a mechanism that involves C-10 hydroxylation, followed by 1,4-dehydration and epoxidation at C-7, the terminal double bond of the resulting diene (Fig. 16). The epoxide can then cyclize with simultaneous decarboxylation and attack of the resulting C-5 anion on C-10 followed by epoxide ring opening to give chanoclavine-I. Whether the decarboxylation is concerted with ring closure as seems most plausible, or occurs as a separate step remains open (Fig. 17) (Kozikowski *et al.*, 1993).

Chanoclavine-I cyclase catalyzes the conversion of chanoclavine-I and/or chanoclavine-I aldehyde to agroclavine and/or elymoclavine. A requirement for NAD or NADP has been observed. This enzyme's appearance and decline in cultures resembled that of DMAT synthase. The conversion of agroclavine to elymoclavine has been achieved with a cell free preparation. This enzyme, a microsomal hydroxylase, is NADPH requiring and the lack of inhibition by EDTA and cyanide suggests a cytochrome P_{450} monooxygenase. This enzyme had great activity during maximum alkaloid production (Kim *et al.*, 1981).

Elymoclavine is the precursor of lysergic acid though the exact mechanism of formation is still unclear. It is assumed that a double bond shift from $\Delta^{8,9}$ to $\Delta^{9,10}$ occurs at the aldehyde stage and this is confirmed with feeding experiments with the enol acetate of lysergic aldehyde (Fig. 17). To form the amide alkaloids it is suggested that activation of lysergic acid as lysergyl CoA is required but this is still controversial. A particulate fraction isolated from an ergotamine-producing strain of *C. purpurea* converted elymoclavine to paspalic acid. NADPH was required as was cytochrome P_{450}. A particulate system has also been isolated that converts elymoclavine directly to ergotamine (Maier *et al.*, 1981), suggesting that under normal circumstances paspalic acid is not a free intermediate (Fig 17). This is also borne out by the fact that $^{18}O_2$ is incorporated equally into the carbonyl oxygen of lysergic acid and the oxygen attached to the α-carbon of the alanine of ergotamine.

The peptide ergot alkaloids have rather complex structures (Fig. 17). The formation of the modified peptide portion of ergotamine involves the conversion of the α-amino acid

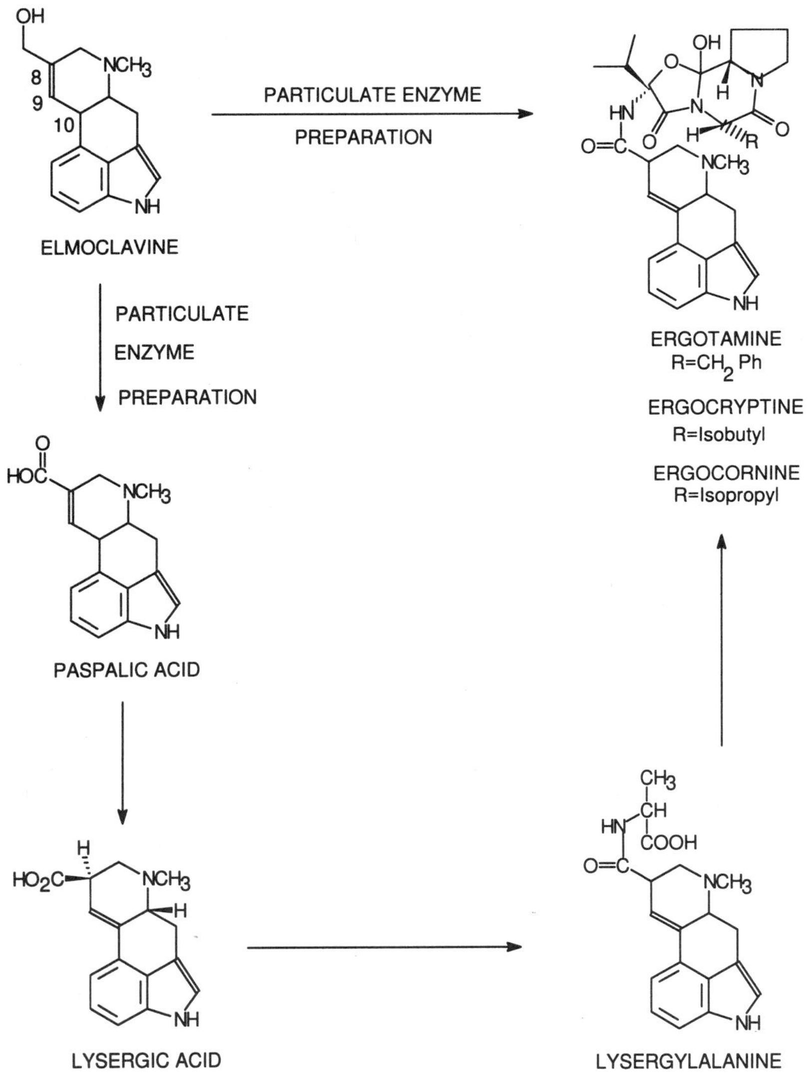

Figure 17. Biosynthesis of ergotamine alkaloids from elymoclavine by *Claviceps purpurea* cultures.

alanine into the corresponding α-hydroxy-α-amino acid moiety which then reacts with the carboxyl group of proline to give the unique cyclol structure (Fig. 17). This transformation is thought to occur after the formation of the entire lysergyl-tripeptide, i.e., ergotamine (lysergyl-alanyl-phenylalanyl-proline), ergocornine (lysergyl-valyl-valyl-proline), and ergocryptine (lysergyl-alanyl-aminobutyryl-proline).

There are many gaps in our knowledge of the biosynthesis of these alkaloids despite much elegant research using radiolabeled precursors and there is now a real need for the isolation of more of the enzymes of these sequences so that improvements to commercial production may be obtained.

7. CYCLOPENIN–VIRIDICATIN ALKALOID BIOSYNTHESIS

The benzodiazepine alkaloids produced by *Penicillium cyclopium* in emerged culture are another important group of alkaloids and along with the quinoline and acridine alkaloids are formed from anthranilic acid. A detailed summary of the biosynthesis of these alkaloids from anthranilic acid may be found in Luckner *et al.* (1977), Mothes *et al.* (1985), and Roos (1990).

P. cyclopium synthesizes cyclopenin alkaloids from anthranilic acid, L-phenylalanine, and L-methionine (Fig. 18). The two oxygen atoms of cyclopenin, cyclopenol, and the oxygen atoms of the hydroxyls of cyclopenol, viridicatin, and viridicatinol are formed from molecular oxygen. Cyclopeptine and dehydrocyclopeptine are intermediates in the biosynthesis of cyclopenin which is converted to cyclopenol. These are converted to viridicatin and viridicatinol, respectively (Fig. 18).

Five enzymes are involved in the synthesis of these alkaloids. The first enzyme in the sequence is a cyclopeptine synthetase system which catalyzes the formation of anthranilic acid AMP, L-phenylalanine AMP, the binding of anthranilic acid and L-phenylalanine as thioesters, formation of the thioester-bound *N*-methyl-L-phenylalanine and *N*-methyl-L-phenylalanineanthranilic acid as well as the cyclization of the latter compound to cyclopeptine. Experiments suggest that cyclopeptine is formed from enzyme bound intermediates by the thiotemplate mechanism of peptide synthesis (Gerlach *et al.,* 1985).

The second enzyme, cyclopeptine dehydrogenase, catalyzes the reversible conversion of cyclopeptine to dehydrocyclopeptine. This enzyme is an NAD(P)-dependent flavoprotein, which is specific for the 3(*S*)-isomer of cyclopeptine. During the reaction the hydrogen atoms are displaced from C-3 and C-10 and the hydride ion is transferred to the *pro*-4(*R*) position of NAD(P).

The third enzyme, dehydrocyclopeptine epoxidase, is an NAD(P)H:O_2 oxidoreductase in which the oxygen atom of the epoxide is derived from molecular oxygen, suggesting that it is introduced by a mixed-function oxidase. Inhibitor experiments indicate that the enzyme is an Fe^{2+}-activated FAD-containing flavoprotein.

Cyclopenin is subsequently hydroxylated at C-3′ by the fourth enzyme in the sequence, cyclopenin-*m*-hydroxylase. This hydroxylation, is carried out by an NADPH:O_2 oxidoreductase. Inhibition of this enzyme by dicoumarol (but not CO) indicates that the enzyme is a flavoprotein.

The last step in the sequence is effected by cyclopenase (cyclopenin methylisocyanate lyase). This enzyme uses either cyclopenin or cyclopenol as substrate to form the quinoline alkaloids viridicatin and viridicatinol, respectively. This occurs as a result of removal of CO_2 which derives from the C-5 carbonyl group and CH_3NH_2 from the *N*-methyl group at position 4 of the benzyldiazepine nucleus (Fig. 18). It is thought that these two groups are extruded as methylisocyanate which is then hydrolyzed. The closure of the heterocyclic ring of the quinoline moiety involves C-5*a* and C-10 of the benzyldiazepene nucleus. The hydroxyl at C-3 of the quinoline moiety is derived from the epoxide oxygen of the benzyldiazepenes.

Cyclopenase, in contrast to the previous enzymes which occur in both hyphae and conidia, occurs only in the conidia. *P. cyclopium* cultures synthesize these alkaloids and conidiospores during a phase of chemical and morphological cell specialization (idiophase) which occurs about 48 hr after culturing and follows the phase of hyphal growth

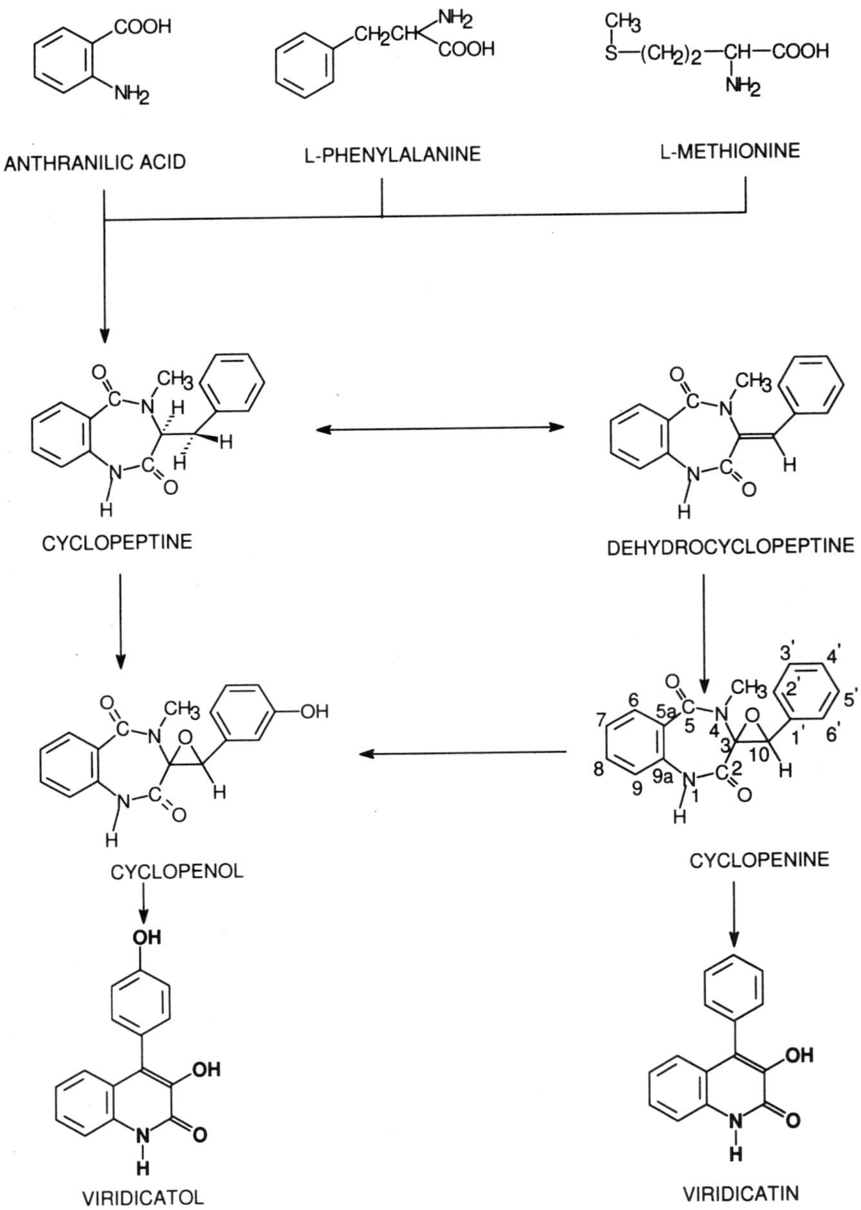

Figure 18. Biosynthesis of cyclopenin/viridicatin alkaloids by *Penicillium cyclopium.*

(trophophase). The quantity of conidia produced equals or surpasses the mycelial dry weight and is followed, after a 12-hr. delay, by the production of benzodiazepine alkaloids. The expression of alkaloid biosynthesis and conidiospore formation is triggered endogenously. A substance (P-factor) which has been isolated but not characterized, is found to accelerate protein synthesis when added to cultures at the stage of early growth and conidiospore germination. The action on alkaloid formation and sporulation is indirect and resembles that of a developmental hormone (Zendem *et al.,* 1982).

8. ACRIDONE ALKALOID BIOSYNTHESIS

Acridone alkaloids comprise a relatively small group of alkaloids that are found solely in some Rutaceae genera. Some 100 examples of this alkaloid group have been isolated and these include the monomeric acridones and the acridone–coumarin dimers (acrimarines) isolated from *Citrus* plants; some binary alkaloids have recently been isolated and described (Takamura *et al.,* 1995).

Robinson (1955) postulated that the monomeric acridone alkaloids were derived from anthranilic acid and acetate via a polyketide. First studies in which [^{13}C]acetate was utilized by cell cultures of *Ruta graveolens* indicated that ring C of the acridone nucleus was acetate derived (Zschunke *et al.,* 1982) Further research revealed that anthranilic acid is specifically incorporated into ring A of rutacridone (Baumert *et al.,* 1982).

Cell-free extracts of *R. graveolens* convert anthranilic acid into *N*-methylanthranilate utilizing *S*-adenosyl-L-methionine and a methyltransferase (Baumert *et al.,* 1983; Maier, *et al.,* 1995). This is the first commited step in the biosynthesis of the rutacridones. The formation of *N*-methylanthraniloyl-CoA from anthranilate utilizing a CoA-ligase (Baumert *et al.,* 1985, 1992) enabled a study of the enzyme that catalyzes the condensation of *N*-methylanthraniloyl-CoA with malonyl-CoA to be made. This enzyme has been purified to homogeneity from *R. graveolens.* cDNA has been isolated from clones harboring acridone synthase and introduced into *E. coli* where high acridone synthase activity was expressed. An insert of roughly 1.4 kb encoded the complete acridone synthase and although this enzyme expressed no chalcone synthase activity, alignments at both DNA and protein levels corroborated a high degree of homology to chalcone synthase (Fig. 19) (Junghanns *et al.,* 1995). The product of this reaction, 1,3-dihydroxy-*N*-methylacridone, leads directly to the more complex acridones such as rutacridone (Baumert *et al.,* 1994). This final step in the biosynthesis of these alkaloids has been hypothesized as requiring mevalonic acid (Baumert *et al.,* 1982). However, the enzyme involved in this step has yet to be elucidated.

9. CONCLUSIONS

These examples of alkaloid biosynthesis serve to indicate how isolation of the enzymes of whole pathways has clarified our hypotheses of alkaloid biosynthesis and enabled investigations to take place at the molecular level (see Chapter 6). There is a need to further investigate systems of vesicular transport and alkaloid sequestration. Little is known, in many instances, about sites of alkaloid synthesis, location of sequestration, and means of translocation (see Chapter 10). Investigations with plant cell cultures have suggested that in many plants, alkaloid production is developmentally regulated and this may account for the lack of production of some alkaloids in cell culture. This should be a major area of study in the future if commercial exploitation is to take place.

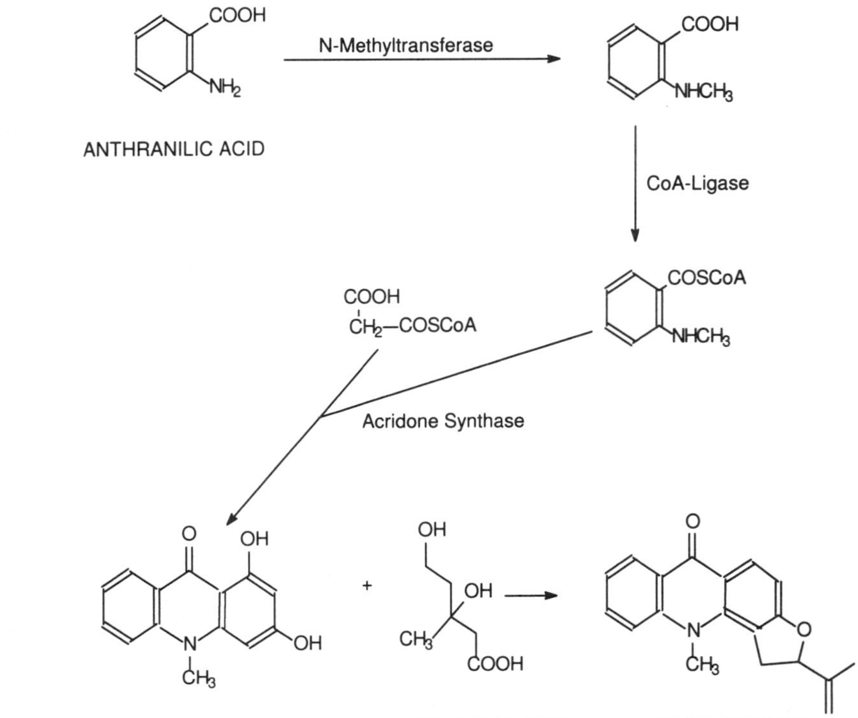

Figure 19. Biosynthesis of acridone alkaloids by *Ruta* cell cultures.

REFERENCES

General Reviews

The Alkaloids, 1950–1995, Vols. 1–47, Academic Press, New York.

Brochmann-Hanssen, E., 1985, Biosynthesis of morphinan alkaloids, in: *The Chemistry and Biology of Isoquinoline Alkaloids* (J. D. Phillipson, M. F. Roberts, and M. H. Zenk, eds.). Springer-Verlag, Berlin, pp. 229–239.

Cordell, G. A., 1981, *Introduction to Alkaloids—A Biogenetic Approach*, Wiley, New York.

Hartmann, T., 1988, Quinolizidines and pyrrolizidines, in: *Cell Culture and Somatic Cell Genetics of Plants,* Vol. 5 (F. Constabel and I. K. Vasil, eds.) Academic Press, San Diego, pp. 277–288.

Herbert, R. B., 1989, *The Biosynthesis of Secondary Metabolites*, 2d ed., Chapman & Hall, London.

Luckner, M., Nover, L., and Bohm, H., 1977, *Molecular Biology, Biochemistry and Biophysics: Secondary Metabolism and Cell Differentiation*, Springer-Verlag, Berlin.

Mann, J., 1987, *Secondary Metabolism*, 2nd ed. Clarendon Press, Oxford.

Mothes, K., Schutte, H. R., and Luckner, M., 1985, *Biochemistry of Alkaloids*, VEB Deutscher Verlag der Wissenschaften, Berlin.

Pelletier, S. W., 1983–1995, *Alkaloids: Chemical and Biological Perspectives*, Vols. 1–8, Pergamon Press, Oxford.

Phillipson, J. D., and Zenk, M. H., (eds.), 1980, *Indole and Biogenetically Related Alkaloids,* Academic Press, New York.

Phillipson, J. D., Roberts, M. F., and Zenk, M. H., (eds.), 1985, *The Chemistry and Biology of Isoquinoline Alkaloids*, Springer-Verlag, Berlin.

Robinson, R., 1955, *The Structural Relationships of Natural Products*, Clarendon Press, Oxford.

Stöckigt, J., 1984, Biosynthesis of tryptophan alkaloids in higher plants: Formation of the sarpagine–ajmaline group, in: *Progress in Tryptophan and Serotonin Research* (H. G. Schlossberger, W. Kochen, B. Linzen, and H. Steinhart, eds.), de Gruyter, Berlin, pp. 777–784.

Zenk, M. H., 1989, Biosynthesis of alkaloids using plant cell culture, *Recent Adv. Phytochem.* **23**:429–457.

Zenk, M. H., 1990, Plant cell cultures: A potential in food and biotechnology, *Food Biotechnol.* **4**:461–470.

Zenk, M. H., 1995, Chasing the enzymes of alkaloid biosynthesis, in: *Organic Reactivity: Physical and Biological Aspects* (B. T. Golding, R. J. Griffin, and H. Maskill, eds.), The Royal Society of Chemistry, Newcastle upon Tyne, U.K., pp. 89–109.

Key References

Amann, M., Nagakura, N., and Zenk, M. H., 1984, (S)-tetrahydroprotoberberine oxidase the final enzyme in protoberberine biosynthesis, *Tetrahedron Lett.* **25**:953–954.

Amann, M., Wanner, G., and Zenk, M. H., 1986, Purification and characterization of (S)-tetrahydroberberine oxidase from cultured *Coptis japonica* cells, *Phytochemistry* **37**:979–982.

Amann, M., Nagakura, N., and Zenk, M. H., 1988, Purification and properties of (S)-tetrahydroprotoberberine oxidase from suspension-cultured cells of *Berberis wilsoniae Eur. J. Biochem.*, **175**:17–25.

Arakawa, H., Clark, W. G., Psenak, M. and Coscia, C. J., 1992, Purification and characterization of dihydrobenzophenanthridine oxidase from elicited *Sanguinaria canadensis* cell cultures, *Arch. Biochem. Biophys.* **299**:1–7.

Aslanov, K. A., Kushmuradov, Y. K., and Sadykov, A. S., 1987, Lupine alkaloids, In: *The Alkaloids*, Vol. 31 (A. Brossi, ed.), Academic Press, New York, pp. 117–192.

Balsevich, J., Constabel, F., and Kurz, W. G. W., 1982, Efficient incorporation of 10-hydroxygeraniol and 10-hydroxynerol into indole alkaloids by a cell suspension culture of *Catharanthus roseus*, *Planta Med.* **44**:231–233.

Barton, D. H. R., Kirby, G. W., Steglich, W., Thomas, G. M., Battersby, A. R., Dobson, T. A., and Ramuz H., 1965, Investigations on the biosynthesis of morphinan alkaloids, *J. Chem. Soc.* 2423–2438.

Battersby, A. R., Binks, R., Francis, R. J., McCaldin, D. J., and Ramuz H., 1964, Alkaloid biosynthesis, Part IV. 1-Benzylisoquinolines as precursors of thebaine, codeine and morphine, *J. Chem. Soc.* 3600–3610.

Battersby, A. R., Foulkes, D. M., and Binks, R., 1965, Alkaloid biosynthesis, Part VIII.Use of optically active precursors for investigation of the biosynthesis of morphine alkaloids, *J. Chem. Soc.* 3323–3332.

Bauer, W., and Zenk, M. H., 1991, Two methylenedioxy bridge forming cytochrome P-450 dependent enzymes are involved in (S)-stylopine biosynthesis, *Phytochemistry* **30**:2953–2961.

Baumert, A., Kuzovkina, N. I., Krauss, G., Hieke, M., and Gröger, D., 1982, Biosynthesis of rutacridone in tissue cultures of *Ruta graveolens* L., *Plant Cell Rep.* **1**:168–171.

Baumert, A., Kuzovkina, N., Hieke, M., and Gröger, D., 1983, N-methylation of anthranilic acid to N-methylanthranilic acid by cell free extracts from *Ruta graveolens* tissue cultures, *Planta Med.* **48**:258–262.

Baumert, A., Kuzovkina, N., and Gröger, D., 1985, Activation of anthranilic acid and N-methylanthranilic acid by cell free extracts from *Ruta graveolens* tissue cultures, *Planta Med.* **50**:125–127.

Baumert, A., Porzel, A., Schmidt, J., and Gröger, D., 1992, Formation of 1,3-dihydroxy-N-methylacridone from N-methylanthranoyl-CoA and malonyl-CoA by cell cultures of *Ruta graveolens*, *Z. Naturforsch.* **47c**:365–368.

Baumert, A., Maier, W., Gröger, D., and Deutzmann, R., 1994, Purification and properties of acridone synthase from cell suspension cultures of *Ruta graveolens* L., *Z. Naturforsch.* **49c**:26–32.

Borkowski, P. R., Horn, J. S., and Rapoport, H., 1978, Role of 1,2-dehydroreticulinium ion in the biosynthetic conversion of reticuline to thebaine, *J. Am. Chem. Soc.* **100**:276–281.

Bowman, W. C., and Sanghvi, I. S., 1963, Pharmacological actions of hemlock (*Conium maculatum*), *J. Pharm. Pharmacol.* **15**:1–15.

De Carolis, E., Chan, F., Balsevich, J., and De Luca, V., 1990, Isolation and characterisation of a 2-oxygluterate dependent dioxygenase involved in the second-to-last step in vindoline biosynthesis, *Plant Physiol.* **94**: 1323–1329.

De-Eknamkul, W., and Zenk, M. H., 1990, Enzymic formation of (R)-reticuline from 1,2-dehydroreticuline in the opium poppy plant, *Tetrahedron Lett.* **34:**4855–4858.

De-Eknamkul, W., and Zenk, M. H., 1992, Purification and properties of 1,2-dehydroreticuline reductase from *Papaver somniferum* seedlings, *Phytochemistry* **31:**813–821.

De Luca, V., and Cutler, A. J., 1987, Subcellular localization of enzymes involved in indole alkaloid biosynthesis in *Catharanthus roseus, Plant Physiol.* **85:**1099–1102.

De Luca, V., and Kurz, W. G. W., 1988, Monoterpene indole alkaloids in: *Cell Culture and Somatic Cell Genetics of Plants*, Vol. 5 (F. Constabel and J. K. Vasil, eds.), Academic Press, San Diego, pp. 385–401.

De Luca, V., Balsevich, J., and Kurz, W. G. W., 1985, Acetyl coenzyme A: deacetylvindoline O-acetyltransferase, a novel enzyme from *Catharanthus roseus, J. Plant Physiol.* **121:**417–428.

De Luca, V., Balsevich, J., Tyler, R. T., Eilert, U., Panchuck, B. D., and Kurz, W. G. W., 1986, Biosynthesis of indole alkaloids: Developmental regulation of the biosynthetic pathway from tabersonine to vindoline in *Catharanthus roseus, J. Plant Physiol.* **125:**147–156.

De Luca, V., Fernandez, J. A., Campbell, D., and Kurz, W. G. W., 1988, Developmental regulation of enzymes of indole alkaloid biosynthesis in *Catharanthus roseus, Plant Physiol.* **86:**447–450.

Dethier, M., and De Luca, V., 1993, Partial purification of a N-methyltransferase involved in vindoline biosynthesis in *Catharanthus roseus*, *Phytochemistry* **31:**663–678.

Dring, J. V., Nash, R. J., Roberts, M. F., and Reynolds, T., 1984, Hemlock alkaloids in aloes. Occurrence and distribution of γ-coniceine, *Planta Med.* **5:**442–443.

Endo, T., Goodbody, A., Vukovic, J., and Misawa, M., 1986, Enzymes from *Catharanthus roseus* cell suspension cultures that couple vindoline and catharanthine to form 3′,4′-anhydrovinblastine, *Phytochemistry* **27:**2147–2149.

Evans, W. C., 1996, *Trease and Evans Pharmacognosy* W. B. Saunders, London, pp. 385–391.

Fahn, W., and Stöckigt, J., 1990, Purification of acetyl-CoA: 17-O-deacetylvindoline 17-O-acetyltransferase from *Catharanthus roseus* leaves, *Plant Cell Rep.* **8:**613–616.

Fahn, W., Gundlach, H., Deus-Neumann, B., and Stöckigt, J., 1985a, Late enzymes of vindoline biosynthesis. Acetyl-CoA:17-O-deacetylvindoline 17-O-acetyl-transferase, *Plant Cell Rep.* **4:**333–336.

Fahn, W., Laussermair, E., Deus-Neumann, B., and Stöckigt, J., 1985b, Late enzymes in vindoline biosynthesis. S-adenosyl-L-methionine:11-O-demethyl-17-O-deacetylvindoline 11-O-methyltransferase and unspecific acetylesterase, *Plant Cell Rep.* **4:**337–340.

Fairbairn, J. W., and Suwal, P. N., 1961, The alkaloids of hemlock (*Conium maculatum* L.) II Evidence for a rapid turnover of the major alkaloids, *Phytochemistry* **1:**38–46.

Falkenhagen, H., and Stöckigt, J., 1995, Enzymic biosynthesis of vomilenine, a key intermediate of the ajmaline pathway, catalysed by a novel cytochrome P_{450}-dependent enzyme from plant cell cultures of *Rauwolfia serpentina*, *Z. Naturforsch.* **50c:**45–53.

Floss, H. G., and Anderson, J. A., 1980, Biosynthesis of ergot toxins, in: The Biosynthesis of Mycotoxins (P. S. Steyn, ed.), Academic Press, New York, pp. 17–67.

Fraser, A. M., and Robins, D. J., 1987, Application of ^{2}H-NMR spectroscopy to the study of the incorporation of enantiomeric ^{2}H-labelled cadaverines into quinolizidine alkaloids, *J. Chem. Soc. Perkin Trans. I* 105–109.

Frenzel, T., and Zenk, M. H., 1990, S-adenosyl-L-methionine:3′-hydroxy-N-methyl-(S)-coclaurine-4′-O-methyltransferase, a regio- and stereoselective enzyme of the (S)-reticuline pathway, *Phytochemistry* **29:**3505–3511.

Fujiwara, H., Takeshita, N., Terano, Y., Fitchen, J. H., Tsujita, T., Katagiri, Y., Sato, F., and Yamada, Y., 1993, Expression of (S)-scoulerine 9-O-methyltransferase in *Coptis japonica* plants, *Phytochemistry* **34:**949–954.

Furuya, T., Yoshikawa, T., and Taira, M., 1984, Biotransformation of codeinone to codeine by immobilised cells of *Papaves somniferum*, Phytochemistry **23:**999–1001.

Galneder, E., Rueffer, M., Wanner, G., Tabata, M., and Zenk, M. H., 1988, Alternative final steps in berberine biosynthesis in *Coptis japonica* cell cultures, *Plant Cell Rep.* **7:**1–4.

Gebler, J. C., and Poulter, C. D., 1992, Purification and characterisation of dimethylallyltryptophan synthase from *Claviceps purpurea*, *Arch. Biochem. Biophys.* **296:**308–313.

Gebler, J. C., Woodside, A. B., and Poulter, C. D., 1992, Dimethylallyltryptophan synthase. An enzyme catalysed electrophilic aromatic substitution, *J. Am. Chem. Soc.* **114:**7354–7360.

Gerardy, R., and Zenk, M.H., 1992, Formation of salutaridine from (R)-reticuline by a membrane-bound cytochrome P-450 enzyme from *Papaver somniferum*, *Phytochemistry* **32:**79–86.

Gerardy, R., and Zenk, M. H., 1993, Purification and characterization of salutaridine : NADPH-7-oxidoreductase, *Phytochemistry* **34:**125–132.

Gerlach, M., Schwelle, N., Lerbs, W., and Luckner, M., 1985, Enzymic synthesis of cyclopentine intermediates in *Penicillium cyclopium*, *Phytochemistry* **24:**1935–1939.

Golebiewski, W. M., and Spenser, I. D., 1984, ^{2}H-NMR spectroscopy as a probe in the stereochemistry of biosynthetic reactions : The biosynthesis of lupanine and sparteine, *J. Am. Chem. Soc.* **106:**7925–7927.

Goodbody, A. E., Endo, T., Vukovic, J., Kutney, J. P., Choi, L. S. L., and Misawa, M., 1988, Enzymic coupling of catharanthine and vindoline to form 3′,4′-anhydrovinblastine by horseradish peroxidase, *Planta Med.* **54:**136–140.

Hampp, N., and Zenk, M. H., 1988, Homogeneous strictosidine synthase from cell suspension cultures of *Rauvolvia serpentina*, *Phytochemistry* **27:**3811–3815.

Hartmann, T., Donges, D., and Steiner, M., 1972a, Biosynthese aliphatischer Monoamine in *Mercuralis perennis* durch Aminosaure-Aldehyd-Transaminierung, *Z. Plant Physiol.* **67:**104–117.

Hartmann, T., Eilert, H. I., and Steiner, M., 1972b, Aldehydaminierung, der biosyntheseweg für primäre aliphatische Monoamine in Blütenpflanzen, *Z. Plant. Physiol.* **68:**11–18.

Hodges, C. C., and Rapoport, H., 1982, Enzymic conversion of reticuline to salutaridine by cell free systems from *Papaver somniferum*, *Biochemistry* **21:**3729–3734.

Ikuta, A., 1988, Isoquinolines, in: *Cell Culture and Somatic Cell Genetics of Plants*, Vol. 5 (F. Constabel and I. K. Vasil, eds.), Academic Press, San Diego, pp. 289–312.

Junghanns, K. T., Kneusel, R. E., Baumert, A., Maier, W., Gröger, D., and Matern, U., 1995, Molecular cloning and heterologous expression of acridone synthase from elicited *Ruta graveolens* L. cell suspension cultures, *Plant Mol. Biol.* **27:**681–692.

Kammerer, L., De-Eknamkul, W., and Zenk, M. H., 1994, Enzymic 12-hydroxylation and 12-O-methylation of dihydrochelirubine in dihydromacarpine formation by *Thalictrum bulgaricum*, *Phytochemistry* **36:**1409–1416.

Kim, I.-S., Kim, S.-U., and Anderson, J. A., 1981, Microsomal agroclavine hydroxylase of *Claviceps* species, *Phytochemistry* **20:**2311–2314.

Kinghorn, A. D., and Balandrin, M. F., 1984, Quinolizidine alkaloids in the Leguminosae: Structural types, analysis, chemotaxonomy and biological activities, *Alkaloids Chem. Biol. Perspect.* **2:**105–148.

Kobayashi, M., and Floss, H. G., 1987, Biosynthesis of ergot alkaloids: Origin of the oxygen atoms in chanoclavine-I and elymoclavine, *J. Org. Chem.* **52:**4350–4352.

Kozikowski, A. P., Chen, C., Wu, J.-P., Shibuya, M., Kim, C.-G., and Floss, H. G., 1993, Probing alkaloid biosynthesis: Intermediates in the formation of ring C, *J. Am. Chem. Soc.* **115:**2482–2488.

Kutchan, T. M., 1989, Expression of enzymatically active cloned strictosidine synthase from the higher plant *Rauvolfia serpentina* in *Escherichia coli*, *FEBS Lett.* **257:**127–130.

Kutchan, T. M., Dittrich, H., Bracher, D., and Zenk, M. H., 1991, Enzymology and molecular biology of alkaloid biosynthesis, *Tetrahedron* **47:**5945–5954.

Kutney, J. P., 1987, Studies in plant tissue culture. The synthesis and biosynthesis of indole alkaloids, *Heterocycles* **25:**617–640.

Leete, E., and Olson, J. O., 1972, Biosynthesis and metabolism of hemlock alkaloids, *J. Am. Chem. Soc.* **94:**5472–5476.

Lenz, R., and Zenk, M. H., 1995a, Acetyl coenzyme A:salutaridinol-7-O-acetyltransferase from *Papaver somniferum* cell cultures, *J. Biol. Chem.* **270:**31091–31096.

Lenz, R., and Zenk, M. H., 1995b, Purification and properties of codeinone reductase (NADPH) from *Papaver somniferum* plant cell cultures and differentiated plants, *Eur. J. Biochem.* **233:**132–139.

Loeffler, S., and Zenk, M. H., 1990, The hydroxylation step in the biosynthetic pathway leading from norcoclaurine to reticuline, *Phytochemistry* **29:**3499–3503.

Lotter, H., Gollsitzer, J., and Zenk, M. H., 1992, Revision of the configuration at C-7 of salutaridinol-I, the natural intermediate in morphine biosynthesis, *Tetrahedron Lett.* **33:**2443–2446.

McKnight, T. D., Bergey, D. R., Burnett, R. J., and Nessler, C. L., 1991, Expression of enzymatically active and correctly targeted strictosidine synthase in transgenic tobacco plants, *Planta* **185:**148–152.

Madyastha, K. M., and Coscia, C. J., 1979, Enzymology of indole alkaloid biosynthesis, *Recent Adv. Phytochem.* **13:**85–129.

Madyastha, K. M., Guarnaccia, R. Baxter, C., and Coscia, C. J., 1973, S-adenosyl-L-methionine:loganic acid methyltransferase: A carboxyl alkylating enzyme from *Vinca rosea*, *J. Biol. Chem.* **248:**2497–2501.

Madyastha, K. M., Meehan, T. D., and Coscia, C. J., 1976, Characterisation of cytochrome P-450 dependent monoterpene hydroxylase from the higher plant *Vinca rosea,Biochemistry* **15:**1097–1102.

Maier, W. D., Baumert, A., and Gröger, D., 1995, Partial purification and characterisation of S-adenosyl-L-methinione:anthranilic acid N-methyltransferase, *J. Plant Physiol.* **145:**1–6.

Maier, W., Schumann, B., and Gröger, D., 1988, Microsomal oxygenases involved in ergoline alkaloid biosynthesis of various Claviceps strains, *J. Basic Microbiol.* **28:**83–93.

Martin, R. O., Warren, M. E., and Rapoport, H., 1978, Reticuline as the tetrahydrobenzylisoquinoline precursor of thebaine in biosynthesis with carbon-14 dioxide, *J. Am. Chem. Soc.* **86:**4726–4727.

Meijer, A. H., De Wall, A., and Verpoorte, R., 1993, Purification of the cytochrome P-450 enzyme geraniol 10-hydroxylase from cell cultures of *Catharanthus roseus,J. Chromatogr.* **635:**237–249.

Muemmler, S., Rueffer, M., Nagakura, N., and Zenk, M. H., 1985, S-adenosyl-L-methionine:(S)-scoulerine 9-O-methyltransferase, a highly stereo- and regio-specific enzyme in tetrahydroprotoberberine biosynthesis, *Plant Cell Rep.* **4:**36–39.

Murakoshi, I., Ogawa, M., Toriisuki, K., Haginiwa, J., Ohmiya, S., and Otomasu, H., 1977a, The enzymic conversion of (−)-lupinine to (−)-(trans-4′-hydroxycinnamoyl)-lupinine by extracts of *Lupinus* seedlings, *Chem. Pharm. Bull.* **25:**527–528.

Murakoshi, I., Sanda, A., Haginiwa, J., Suzuki, N., Ohmiya, S., and Otomasu, H., 1977b, An S-adenosyl-L-methionine:cytisine methyltransferase in *Thermopsis* seedlings, *Chem. Pharm. Bull.* **25:**1970–1973.

Nagakura, N., Rüffer, M., and Zenk, M. H., 1979, The biosynthesis of monoterpenoid indole alkaloids from strictosidine, *J. Chem. Soc. Perkin Trans. I* 2808–2312.

Okada, N., Shinmyo, A., Okada, H., and Yamada, Y., 1988, Purification and characterisation of (S)-tetrahydroberberine oxidase from cultured *Coptis japonica* cells, *Phytochemistry* **27:**979–982.

Otsuka, H., Quigley, F. R., Gröger, D., Anderson, J. A., and Floss, H. G., 1980, *In vivo* and *in vitro* evidence for N-methylation as the second pathway specific step in ergoline biosynthesis, *Planta Med.* **40:**109–119.

Pennings, E. J. M., Groen, B. W., Duine, J. A., and Verpoorte, R., 1989, Tryptophan decarboxylase from *Catharanthus roseus* is a pyridoxo-quinoprotein, *FEBS Lett.* **255:**97–100.

Pfitzner, A., and Stöckigt, J., 1982, Partial purification and characterization of geissoschizine dehydrogenase from suspension cultures of *Catharanthus roseus*, *Phytochemistry* **21:**1585–1588.

Pfitzner, A., and Stöckigt, J., 1983, Characterisation of polyneuridine aldehyde esterase a key enzyme in the biosynthesis of sarpagine, *Planta Med.* **48:**221–227.

Pfitzner, A., Polz, L., and Stöckigt, J., 1986, Properties of vinorine synthase—The *Rauwolfia* enzyme involved in the formation of the ajmaline skeleton, *Z. Naturforsch.* **41c:**103–114.

Pfitzner, U. M., and Zenk, M. H., 1989, Homogeneous strictosidine synthase isoenzymes from cell suspension cultures of *Catharanthus roseus*, *Planta Med.* **55:**525–530.

Plato, 387 B.C., *Phaedo*, translation by R. Hackforth, Cambridge University Press, London, 1955.

Polz, L., Schübel, H., and Stöckigt, J., 1986, Characterisation of 2β(R)-17-O-acetylajmalan:acetylesterase a specific enzyme involved in the biosynthesis of the *Rauwolfia* alkaloid ajmalicine, *Z. Naturforsch.* **42c:**333–342.

Roberts, M. F., 1971, The formation of γ-coniceine from 5-ketooctanal by a transaminase of *Conium maculatum*, *Phytochemistry* **10:**3057–3060.

Roberts, M. F., 1974, An S-adenosyl-L-methionine:coniine methytransferase from *Conium maculatum* L., *Phytochemistry* **13:**1847–1851.

Roberts, M. F., 1975, γ-Coniceine reductase in *Conium maculatum* L., *Phytochemistry* **14:**2393–2397.

Roberts, M. F., 1978, Separation of the formation of γ-coniceine and aliphatic amines from GOT activity in *Conium maculatum*, *Phytochemistry* **17:**107–112.

Roberts, M. F., 1981, Enzymic synthesis of γ-coniceine in *Conium maculatum* chloroplasts and mitochondria, *Plant Cell Rep.* **1:**10–13.

Roberts, M. F., 1985, Alkaloid production in *Conium* fruit, *J. Pharm. Pharmacol.* **37:**141P.

Roos, W., 1990, Benzodiazepine alkaloids, in: *The Alkaloids* (A. Brossi, ed.), Academic Press, San Diego, pp. 63–97.

Rueffer, M., and Zenk, M. H., 1985, Berberine synthase, the methylenedioxy group forming enzyme in berberine synthase, *Tetrahedron Lett.* **26:**201–202.

Rueffer, M., and Zenk, M. H., 1987a, Distant precursors of benzylisoquinoline alkaloids and their enzymatic formation, *Z. Naturforsch.* **42c:**319–332.

Rueffer, M., and Zenk, M. H., 1987b, Enzymatic formation of protopines by a microsomal cytochrome P-450 system of *Cordyalis vaginans*, *Tetrahedron Lett.* **28:**5307–5310.

Rueffer, M., and Zenk, M. H., 1994, Canadine synthase from *Thalictrum tuberosum* cell cultures catalyses the formation of the methylenedioxy bridge in berberine synthesis, *Phytochemistry* **36:**1219–1223.

Rueffer, M., Zumstein, G., and Zenk, M. H., 1990, Partial purification and properties of S-adenosyl-L-methionine:(S)-tetrahydroprotoberberine-cis-N-methyltransferase from suspension-cultured cells of *Eschscholtzia* and *Corydalis*, *Phytochemistry* **29:**3727–3733.

Saito, K., Suzuki, H., Takamatsu, S., and Murakoshi, I., 1993, Acyltransferases for lupin alkaloids in *Lupinus hirsutus*, *Phytochemistry* **32:**87–91.

Schiel, O., Witte, L., and Berlin, J., 1987, Geraniol-10-hydroxylase activity and its relation to monoterpene indole alkaloid accumulation in cell suspension cultures of *Catharanthus roseus*, *Z. Naturforsch.* **42c:**1075–1081.

Schmidt, D., and Stöckigt, J., 1995, Enzymic formation of the sarpagan—bridge: A key step in the biosynthesis of sarpagine-ajmalicine-type alkaloids, *Planta Med.* **61:**254–258.

Schübel, H., Stöckigt, J., Feicht, R., and Simon H., 1986, Partial purification and characterisation of raucaffricine β-D-glucosidase from plant cell suspension cultures of *Rauwolfia serpentina* Benth., *Helv. Chim. Acta* **69:**538–547.

Schumacher, H.-M., and Zenk, M. H., 1988, Partial purification and characterization of dihydrobenzophenanthridine oxidase from *Eschscholtzia californica* cell suspension cultures, *Plant Cell Rep.* **7:**43–46.

Shibuya, M., Chou, H.-M., Fountoulakis, M., Hassam, S., Kim, S.-U., Kobayashi, K., Otsuka, H., Rogalska, E., Cassady, J. M., and Floss, H. G., 1990, Stereochemistry of the isoprenylation of tryptophan catalysed by 4-(γ,γ-dimethylallyl)tryptophan synthase from *Claviceps*, the first pathway specific enzyme in ergot alkaloid biosynthesis, *J. Am. Chem. Soc.* **112:**297–304.

Spenser, I. D., 1985, Stereochemical aspects of the biosynthesis of the pyrrolizidine and the quinolizidine alkaloids, *Pure Appl. Chem.* **57:**453–470.

Stadler, R., and Zenk, M. H., 1990, A revision of the generally accepted pathway for the biosynthesis of the benzyltetrahydroisoquinoline alkaloid reticuline, *Liebigs Ann. Chem.* **6:**555–562.

Spitsberg, V., Coscia, C. J., and Krueger, R. J., 1981, Characterisation of a monoterpene hydroxylase from cell suspension cultures of *Catharanthus roseus* (L.) G. Don, *Plant Cell Reports* **1:**43–47.

Stadler, R., and Zenk, M. H., 1993, The purification and characterization of a unique cytochrome P_{450} enzyme from *Berberis stolonifera* plant cell cultures, *J. Biol. Chem.* **268:**823–831.

Stadler, R., Kutchan, T. M., and Zenk, M. H., 1989, (S)-norcoclaurine is the central intermediate in benzylisoquinoline alkaloid biosynthesis, *Phytochemistry* **28:**1083–1086.

Steffens, P., Nagakura, N., and Zenk, M. H., 1984, The berberine bridge forming enzyme in tetrahydroprotoberberine biosynthesis, *Tetrahedron Lett.* **25:**951–952.

Steffens, P., Nagakura, N., and Zenk, M. H., 1985, Purification and characterization of the berberine bridge enzyme from *Berberis beaniana* cell cultures, *Phytochemistry* **24:**2577–2583.

Stöckigt, J., and Zenk, M. H., 1977, Isovincoside (strictosidine), the key intermediate in the enzymatic formation of indole alkaloids, *FEBS Lett.* **79:**233–237.

Stöckigt, J., Hemschmidt, T., Hofle, G., Heinstein, P., and Formacek, V., 1983, Steric course of hydrogen transfer during enzymic formation of 3-α-heteroyohimbine, *Biochemistry* **22:**3448–3452.

Takamura, Y., Matsushita, Y., Nagareya, N., Abe, M., Takaya, J., Juichi, M., Hashimoto, T., Kan, Y., Takoaka, S., Asakawa, Y., Omura, M., Ito, C., and Furukawa, H., 1995, Citbismine-A, citbismine-B and citbismine-C, new binary acridone alkaloids from Citrus plants, *Chem. Pharm. Bull.* **43:**1340–1345.

Takao, N., Kamigauchi, M., and Okada, M., 1983, Biosynthesis of benzophenanthridine alkaloids, sanguinarine, chelerubine and macarpine, *Helv. Chim. Acta.* **66:**473–484.

Tanahashi, T., and Zenk, M. H., 1990, Elicitor induction and characterization of microsomal protopine-6-hydroxylase, the central enzyme in benzophenanthridine alkaloid biosynthesis, *Phytochemistry* **29:**1113–1122.

Treimer, J. F., and Zenk, M. H., 1978, Enzymic synthesis of corynanthe type alkaloids in cell cultures of *Catharanthus roseus*. Quantitation by radioimmunoassay, *Phytochemistry* **17:**227–231.

Treimer, J. F., and Zenk, M. H., 1979, Purification and properties of strictosidine synthase, the key enzyme in indole alkaloid formation, *Eur. J. Biochem.* **101:**225–233.

Uesato, S., Kanomi, S., Iida, A., Inouye, H., and Zenk, M. H., 1986a, Mechanism for iridane skeleton formation in the biosynthesis of secologanin and indole alkaloids in *Lonicera tartarica, Catharanthus roseus* and suspension cultures of *Rauwolfia serpentina*, *Phytochemistry* **25:**839–842.

Uesato, S., Kanomi, S., Iida, A., Inouye, H., and Zenk, M. H., 1986b, Synthesis of iridoidal by cell free extracts from *Rauwolfia serpentina* cell suspension cultures, *Tetrahedron Lett.* **27:**2893–2896.

Wink, C., and Hartmann, T., 1981a, Properties of subcellular localisation of L-alanine: aldehyde aminotransferase. Concept of an ubiquitous plant enzyme involved in secondary metabolism, *Z. Naturforsch.* **36**:625–632.

Wink, M., 1984, N-methylation of quinolizidine alkaloids: An S-adenosyl-L-methionine: cytisine N-methyltransferase from *Laburnum anagyroides* plant and cell cultures of *L. alpinum* and *Cytisus canariensis*, *Planta* **161**:339–344.

Wink, M., 1986, Cell-free synthesis of the alkaloids ammodendrine and smipine, *Z. Naturforsch.* **41c**:375–380,

Wink, M., 1987, Quinolizidine alkaloids: Biochemistry, metabolism, and function in plants and cell suspension cultures, *Planta Med.* **53**:509–514.

Wink, M., and Hartmann, T., 1981b, Sites of enzymic synthesis of quinolizidine alkaloids and their accumulation in *Lupinus polyphyllus*, *Z. Pflanzenphysiol.* **102**:337–344.

Wink, M., and Hartmann, T., 1982, Localisation of the enzymes of quinolizidine alkaloid biosynthesis in leaf chloroplasts of *Lupinus polyphyllus*, *Plant Physiol.* **70**:74–77.

Wink, M., Hartmann, T., and Witte, L., 1980, Biotransformation of cadaverine and potential biogenetic intermediates of lupanine biosynthesis by plant cell suspension cultures, *Planta Med.* **40**:31–39

Zendem, Z., Khalil, S., and Luckner, M., 1982, P-factor, a developmental hormone of *Penicillium cyclopium*, *Phytochemistry* **21**:839–842.

Zenk, M. H., 1980, Enzymic synthesis of ajmalicine and related indole alkaloids, *J. Nat. Prod.* **43**:438–451.

Zenk, M. H., Rueffer, M., Amann, M., and Deus-Neumann, B., 1985, Benzylisoquinoline biosynthesis by cultivated plant cells and isolated enzymes, *J. Nat. Prod.* **48**:725–735

Zenk, M. H., Gerardy, R., and Stadler, R., 1995, Phenol oxidative coupling of benzylisoquinoline alkaloids is catalyzed by regio- and stereo-selective cytochrome P-450 linked plant enzymes: Salutaridine and berbamunine, *J. Chem. Soc. Chem. Comun.* 1725–1727.

Zschunke, A., Baumert, A., and Gröger, D., 1982, Biosynthesis of rutacridone in *Ruta graveolens*. Incorporation studies with [^{13}C]-acetate, *J. Chem. Soc. Chem. Commun.* 1263–1265.

Chapter 6

Genes in Alkaloid Metabolism

Kazuki Saito and Isamu Murakoshi

1. INTRODUCTION

A number of studies in classical genetics suggest the participation of specific genes in alkaloid metabolism in plants (Böhm, 1985; Waller and Nowacki, 1978). Extensive breeding studies, e.g., screening of mutants and cross-pollination, along with phytochemical investigations on alkaloid-rich ("bitter") and alkaloid-poor ("sweet") forms of *Lupinus* plants revealed several mutant alleles concerned with alkaloid contents in *Lupinus* plants (Saito *et al.,* 1993a; Hondelmann, 1984; Williams and Harrison, 1983; Williams *et al.,* 1984). However, it has been in just the last 5 years that the genes responsible for alkaloid metabolism were isolated as defined DNA fragments of nucleotide sequences. This progress in the molecular biology of alkaloid metabolism has been promoted by recent developments in recombinant DNA technology in higher plants. This technology provides the tools to clarify the regulatory mechanism governing alkaloid metabolism at a molecular level and to express cloned genes in heterologous cells, for mechanistic investigation as well as for biotechnological application.

This chapter focuses on isolation of genes, regulation of gene expression, heterologous expression of cloned genes, and molecular evolution of genes in alkaloid metabolism. Readers should refer to recently published reviews of related topics (Hashimoto and Yamada, 1994; Kutchan, 1993; Saito *et al.,* 1992).

2. ISOLATION OF GENES

Genes involved in the metabolism of several categories of alkaloids or nitrogen-containing compounds have been isolated and characterized (Table I). The first cloned

---

Kazuki Saito and Isamu Murakoshi • Faculty of Pharmaceutical Sciences, Laboratory of Molecular Biology and Biotechnology, Research Center of Medicinal Resources, Chiba University, Inage-ku, Chiba 263, Japan

Alkaloids: Biochemistry, Ecology, and Medicinal Applications, edited by Roberts and Wink. Plenum Press, New York, 1998.

Table I.
Isolated Genes for Biosynthesis of Nitrogen-Containing Secondary Metabolites from Plants

Type of metabolite	Gene/enzyme	Type of DNA	Plant species	Cloning strategy	Reference
Indole alkaloid	Strictosidine synthase	cDNA	*Rauwolfia serpentina*	Oligonucleotide probe	Kutchan *et al.* (1988)
		Genomic DNA (*str1*)	*R. serpentina* *R. mannii*	cDNA probe	Bracher and Kutchan (1992a)
		cDNA	*Catharanthus roseus*	Oligonucleotide probe Antibody screening	McKnight *et al.* (1990) Pasquali *et al.* (1992)
	Tryptophan decarboxylase	cDNA	*C. roseus*	Antibody screening	De Luca *et al.* (1989)
Tropane alkaloid	Hyoscyamine 6β-hydroxylase	cDNA Genomic DNA	*Hyoscyamus niger* *H. niger*	Antibody screening cDNA probe	Matsuda *et al.* (1991) Kanegae *et al.* (1994)
	Tropinone reductase I	cDNA	*Datura stramonium*	Oligonucleotide probe	Nakajima *et al.* (1993a)
	Tropinone reductase II	cDNA	*D. stramonium*	Oligonucleotide probe	Nakajima *et al.* (1993a)
		cDNA	*H. niger*	Oligonucleotide probe	Nakajima *et al.* (1993b)
Isoquinoline alkaloid	Berberine bridge enzyme	cDNA	*Eschscholtzia californica*	Oligonucleotide probe	Dittrich and Kutchan (1991)
	Berbamurine synthase	cDNA	*Berberis stolonifera*	Oligonucleotide probe	Kraus and Kutchan (1995)
	Scoulerine 9-O-methyltransferase	cDNA	*Coptis japonica*	Oligonucleotide probe	Takeshita *et al.* (1995)
Nicotine alkaloid	Putrescine *N*-methyltransferase	cDNA	*Nicotiana tabacum*	Differential screening	Hibi *et al.* (1994)
Heterocyclic amino acid	β-Pyrazolealanine	cDNA	*Citrullus vulgaris*	Genetic complementation	Noji *et al.* (1993, 1994)
General or unknown	Cytochrome P_{450}	cDNA	*C. roseus*	Differential screening Polymerase chain reaction	Vetter *et al.* (1992) Meijer *et al.* (1993b)
	NADPH:cytochrome P_{450} reductase	cDNA Genomic DNA	*C. roseus*	Antibody screening	Meijer *et al.* (1993b)

gene in alkaloid metabolism was the cDNA encoding strictosidine synthase catalyzing the stereospecific condensation of tryptamine and *seco*-loganin to form 3α(*S*)-strictosidine (Kutchan *et al.,* 1988). Since this first molecular cloning, 16 cDNA and genomic DNA encoding 10 different enzymes involved with nitrogen-containing metabolites have been isolated. The genes for cytochrome P_{450} and NADPH:cytochrome P_{450} reductase, both involved in oxygenation of a variety of metabolites, were isolated from periwinkle (*Catharanthus roseus)* (Meijer *et al.,* 1993a,b; Vetter *et al.,* 1992), although the participation of these genes in indole alkaloid metabolism still remains to be confirmed.

Most cDNAs have been isolated by a strategy involving purification of enzyme protein, amino acid sequencing, and screening a library with synthetic oligonucleotide probes or immuno-screening an expression library with an antibody. Although this way is most straightforward and reliable, purification of the enzyme protein normally present in a small amount in plant cells is a time-consuming and tedious task. cDNA clones have also been isolated, without purifying enzymes, by alternative ways, i.e., differential screening of the libraries prepared from mutants (Hibi *et al.,* 1994) or induced cells (Vetter *et al.,* 1992) and genetic complementation in other organisms (Noji *et al.,* 1993, 1994). To verify the identity of isolated clones, it is generally necessary to express the isolated genes as catalytically active proteins in heterologous cells. *Escherichia coli* is often the organism of choice for this purpose.

3. REGULATION

Understanding the regulatory mechanisms operating in alkaloid metabolism is fascinating from the fundamental viewpoint of plant molecular biology and is also a prerequisite for genetic manipulation of alkaloid production in an artificial system, e.g., cell and tissue cultures.

Figure 1 summarizes an accepted general scheme for regulation of alkaloid biosynthesis through specific gene expression.

A variety of internal and external signals, i.e., phytohormones, elicitation, and environmental stress, induce gene expression for alkaloid biosynthesis (Goddijn *et al.,* 1992; Pasquali *et al.*, 1992). These stimuli are transmitted through the cellular signal transduction system to activate the defined protein factor (*trans*-element) present in the nucleus. The activated *trans*-element binds to the particular DNA sequence (*cis*-element), located in the 5′-flanking region of the structural gene encoding a biosynthetic enzyme, and subsequently enhances the transcription of the gene. Another regulatory factor is postulated to explain tissue specific biosynthesis of particular alkaloids, e.g., formation of tropane and nicotine alkaloids in roots or in particular types of root cells: A tissue-specific *trans*-element is present in the cells capable of alkaloid biosynthesis. This tissue-specific *trans*-element binds to a *cis*-element in the 5′-promoter region which is different from that for activation by the cellular signals.

The presence of regulatory regions in the promoter sequence of the *str1* gene for strictosidine synthase and the species-specific binding protein in *Rauwolfia serpentina* are suggested by deletion analyses of the 5′-promoter region and by gel retardation assays (Bracher and Kutchan, 1992a). Species-dependent expression controlled by the promoter of the hyoscyamine 6β-hydroxylase gene was observed in the experiments on transgenic plants using the β-glucuronidase gene as a visible reporter gene (Kanegae *et al.,* 1994).

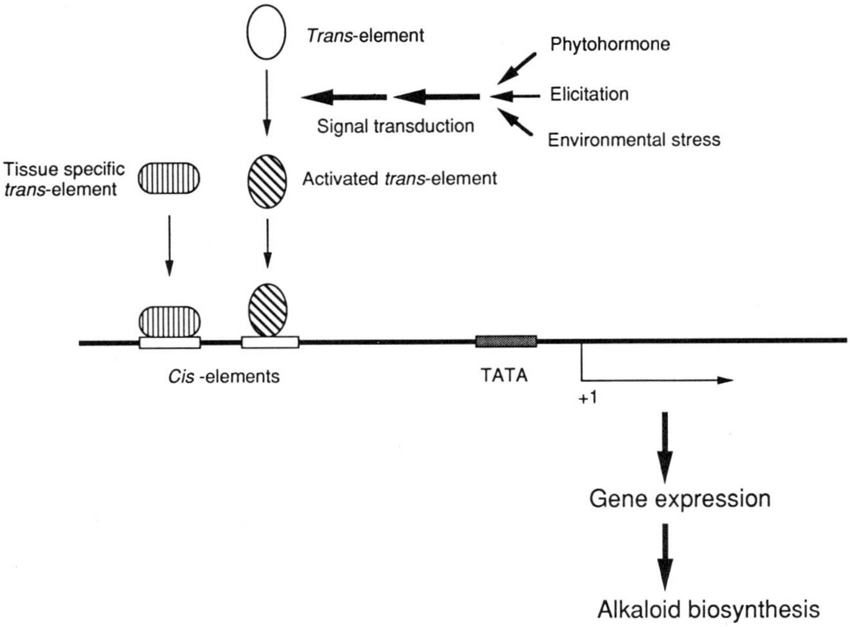

Figure 1. A schematic model for the regulatory mechanism of gene expression in alkaloid biosynthesis.

Pericycle-specific accumulation of hyoscyamine 6β-hydroxylase (Hashimoto *et al.,* 1991) was attributed to the 0.8-kb-length 5′-flanking region of the gene from *Hyoscyamus niger.* The molecular analysis of low-nicotine mutants of *Nicotiana tabacum* suggested the presence of regulatory genes (*Nic1* and *Nic2*) governing the gene expression for nicotine biosynthesis (Hibi *et al.,* 1994). Identification of the "master genes," those controlling the expression of entire biosynthetic genes, as have been isolated and characterized as transcription factors in flavonoid biosynthesis (Dooner and Robbins, 1991), will be a matter of great future interest.

4. EXPRESSION OF RECOMBINANT ENZYMES IN HETEROLOGOUS SYSTEMS

Once a gene is isolated, heterologous expression can be accomplished for several purposes: (1) to identify the biological function of the encoded protein, (2) to overproduce the enzyme supplied for detailed mechanistic study and for biocatalysis of alkaloid production, (3) to change the biosynthetic flow in host plants. Table II lists examples of heterologous expression of the genes involved in alkaloid biosynthesis and of the genes for altering the alkaloid metabolism in host cells.

4.1. Expression in Microorganisms

Prokaryotic expression, usually in *E. coli,* is carried out to obtain catalytically active recombinant proteins. Biotransformation and biosynthesis of particular metabolites were

Table II.
Heterologous Expression of Plant Genes for Nitrogen-Containing Compound Metabolism and of Genes to Alter Alkaloid Metabolism

Host organism	Gene/enzyme (source organism)	Comment	Reference
Microorganism			
Escherichia coli	Strictosidine synthase (*R. serpentina*)	Catalytically active	Kutchan (1989), Kutchan *et al.* (1991)
	Hyoscyamine 6β-hydroxylase (*H. niger*)	Catalytically active, exhibits epoxidase activity	Matsuda *et al.* (1991), Hashimoto *et al.* (1993a)
	Tropinone reductase I (*Datura stramonium*)	Catalytically active	Nakajima *et al.* (1993a)
	Tropinone reductase II (*D. stramonium, H. niger*)	Catalytically active	Nakajima *et al.* (1993a,b)
	Putrescine N-methyltransferase (*Nicotiana tabacum*)	Catalytically active	Hibi *et al.* (1994)
	β-Pyrazolealanine synthase (*Citrullus vulgaris*)	Catalytically active product formation *in vivo*	Noji *et al.* (1993, 1994)
	Scoulerine 9-*O*-methyltransferase (*Coptis japonica*)	Catalytically active	Takeshita *et al.* (1995)
	NADPH:cytochrome P_{450} reductase (*C. roseus*)	Catalytically active	Meijer *et al.* (1993a)
Saccharomyces cereviseae	Strictosidine synthase (*R. serpentina*)	Catalytically active	Kutchan *et al.* (1991)
	Berberine bridge enzyme (*E. californica*)	Catalytically active	Dittrich and Kutchan (1991), Kutchan *et al.* (1991)
	Cytochrome P_{450} (*C. roseus*)		Vetter *et al.* (1992)

(continued)

Table II.
Continued

Host organism	Gene/enzyme (source organism)	Comment	Reference
Plant			
Atropa belladonna	Hyoscyamine 6β-hydroxylase (*H. niger*)	Enhanced scopolamine accumulation	Yun *et al.* (1992), Hashimoto *et al.* (1993b)
Nicotiana tabacum	Tryptophan decarboxylase (*C. roseus*)	Tryptamine accumulation	Songstad *et al.* (1990)
	Cytochrome P_{450} (rabbit)	Altered alkaloid pattern	Saito *et al.* (1991a)
	Lysine decarboxylase (bacterium, *Hafnia alvei*)	Increasing cadaverine accumulation	Herminghaus *et al.* (1991)
	Strictosidine synthase (*C. roseus*)	Targeting to vacuole	McKnight *et al.* (1991)
	Hyoscyamine 6β-hydroxylase (*H. niger*)	Hyoscyamine bioconversion	Yun *et al.* (1993)
	Ornithine decarboxylase (mouse)	Polyamine accumulation	DeScenzo and Minocha (1993)
Nicotiana rustica	Ornithine decarboxylase (yeast)	Enhanced production of nicotine	Hamill *et al.* (1990)
Nicotiana glauca	Lysine decarboxylase (*H. alvei*)	Enhanced production of cadaverine and anabasine	Fecker *et al.* (1992)
Peganum harmala	Tryptophan decarboxylase (*C. roseus*)	Increased production of serotonin	Berlin *et al.* (1993)
Insect cells			
Spodoptera frugiperda	Strictosidine synthase (*R. serpentina*)	Catalytically active Secretion into medium	Kutchan *et al.* (1991, 1994)
	Berberine bridge enzyme (*E. californica*)	Catalytically active Secretion into medium	Kutchan *et al.* (1994)
	Berbamurine synthase (*B. stolonifera*)	Catalytically active Accumulation into microsomes	Kraus and Kutchan (1995)

achieved by feeding biosynthetic precursors to a culture of *E. coli* in which plant genes were overexpressed (Hashimoto *et al.*, 1993a; Noji *et al.*, 1993; Kutchan, 1989). In all of these cases, the reaction products were accumulated in the medium, suggesting free permeability of bacterial cell membrane to the products.

4.2. Expression in Transgenic Plants and Other Higher Organisms

A transgenic plant can be a powerful tool not only for basic research clarifying molecular mechanism but also for genetic improvement of metabolism in crop plants (Saito *et al.*, 1991b, 1992; Weising *et al.*, 1988). Several genes originated from plants, microorganisms and animals have been introduced and expressed in transgenic plants to modify the metabolic pathways. The most successful example is the expression of the hyoscyamine 6β-hydroxylase gene in *Atropa belladonna* (Yun *et al.*, 1992). The only tropane alkaloid accumulated in the transgenic *A. belladonna* expressing the 6β-hydroxylase gene from *H. niger* was scopolamine, while a nontransformed plant contained only a small quantity of scopolamine but a large amount of hyoscyamine. As the 6β-hydroxylase is a bifunctional enzyme catalyzing two consecutive reactions from hyoscyamine to scopolamine, expression of this single gene could change the alkaloid pattern of the host *A. belladonna* plants. Because scopolamine has ten times the commercial value of hyoscyamine, this study was a promising advance in molecular breeding of alkaloid-containing plants.

Cultured insect cells using a vaculovirus-based expression system are also convenient for overproduction of recombinant proteins (Kutchan, 1996). Processing of transit peptide present in premature enzymes took place correctly in the insect cells, although the processed proteins were secreted into the medium, probably because of the apparent lack of alkaloid biosynthetic vesicles and vacuoles (Kutchan *et al.*, 1994).

5. MOLECULAR EVOLUTION

Several recent studies suggest that alkaloids are involved in the defense mechanism of plants that produce them (see Chapter 11). Molecular analysis of the gene encoding strictosidine synthase indicated that the nucleotide sequences of the coding region of this enzyme were highly conserved among *Rauwolfia* plants (Bracher and Kutchan, 1992a,b). This example may imply a strict selective pressure in evolution being caused by the role of alkaloids in the fitness of alkaloid-producing plants.

One can easily postulate that alkaloid metabolism genes have evolved from progenitor genes involved in the basic functions for living cells. The gene encoding hyoscyamine 6β-hydroxylase is assumed to have evolved from an ethylene-forming enzyme, a 2-oxyacid-dependent oxygenase ubiquitously found in plants, based on sequence similarity and strict conservation of exon–intron positions (Hashimoto and Yamada, 1994). The β-pyrazolealanine synthase gene in *Citrullus vulgaris* is identical to the cysteine synthase gene in primary metabolism (Noji *et al.*, 1993, 1994). This may be an example of an ongoing process of evolution of genes from a primary metabolic role to a specified role in secondary metabolism. The traces of gene evolution in alkaloid metabolism have been suggested by the finding that non-alkaloid-producing *Glycine max* may contain a genomic sequence highly homologous to the strictosidine synthase gene (Kutchan, 1993).

Another interesting question is how enzymes have evolved that can catalyze such sophisticated chemical reactions, e.g., regio- and stereospecific reactions. Tropinone reductase I and II, catalyzing stereospecific reduction of the keto group of tropinone to 3α-hydroxy and 3β-hydroxy groups, respectively, were analyzed in detail by dissection of peptides and construction of chimeric enzymes (Nakajima *et al.,* 1993a, 1994). Opposite stereospecificity of the two reductases was ascribed to the carboxy half of the proteins, to which the substrate tropinone is assumed to bind with reverse orientation in the two enzymes. The genes of the two reductases are believed to have evolved from a common ancestor reductase in the Solanaceae. Tigloyltransferases for quinolizidine alkaloids exhibit strict specificity regarding enantiomers of sparteine-type hydroxylated alkaloids but only weak specificity toward the acyl donors (Saito *et al.,* 1993b, Suzuki *et al.,* 1994).

The genes in quinolizidine alkaloid biosynthesis address several other interesting questions concerning evolution. Quinolizidine alkaloids are found mainly in the Leguminosae, but also in unrelated families, e.g., Berberidaceae. To explain the occurrence in different families, there are at least two possibilities: Either the genes for enzymes have evolved independently in both families (convergence) or the genes were selected much earlier but have only been expressed in limited families as a result of selective pressure. The latter possibility was proposed with the experimental evidence for inducible accumulation of quinolizidine alkaloids in several unrelated plant species (Wink and Witte, 1983). The finding of the genomic presence of a homologous sequence to the strictosidine synthase gene in *Glycine max* (Kutchan, 1993) also supports a widespread occurrence of the genes for alkaloid biosynthesis.

Molecular phylogeny of the Papilionoideae was elucidated by the nucleotide sequences of the photosynthetic *rbc*L gene. Only partial correlation was observed between the *rbc*L gene sequence and alkaloid accumulation pattern (Kass and Wink, 1995). However, molecular phylogenetic trees of *Lupinus* plants based on restriction fragment length polymorphism (RFLP) suggested the intra- and interspecies correlations between general RFLP pattern and biosynthetic features of quinolizidine alkaloids including stereospecificity (Yamazaki *et al.,* 1993). It would be premature to conclude the molecular phylogeny of lupins through correlation with alkaloid biosynthesis. We need to accumulate far more extensive data to support molecular genetic evolution of the genes for alkaloid metabolism in plants.

6. PROSPECTS

Several genes involved in alkaloid metabolism have been isolated from higher plants; however, many genes concerned with the metabolism of plant alkaloids still remain to be cloned and characterized. Besides the cloning strategy employing enzyme purification, it is necessary to explore alternative methods, e.g., functional expression in host cells or differential screening with mutant lines, isogenic forms, or cell variants. These cloning strategies, which have been proven to be successful in some cases, may also lead to isolation of the postulated regulatory genes governing the entire biosynthetic pathway of particular alkaloids. The availability of cloned regulatory genes will definitely open a new avenue both for fundamental research and for biotechnological application in the production of pharmaceutically useful alkaloids.

The understanding of molecular evolution of genes in alkaloid metabolism will give us new insights into natural history. Through the study of evolution of genes for secondary metabolism, we will be able to learn how biodiversity in metabolic process has been spread through the plant kingdom as well as the real function of alkaloids in plants.

REFERENCES

General Reviews

Böhm, H., 1985, The biochemical genetics of alkaloids, in: *Biochemistry of Alkaloids* (K. Mothes, H. R. Schütte, and M. Luckner, eds.), VCH, Weinheim, pp. 25–36.

Hashimoto, T., and Yamada, Y., 1994, Alkaloid biogenesis: Molecular aspects, *Annu. Rev. Plant Physiol. Plant Mol. Biol.* **45:**257–285.

Kutchan, T. M., 1993, Strictosidine: From alkaloids to enzyme to gene, *Phytochemistry* **32:**493–506.

Saito, K., Yamazaki, M., and Murakoshi, I., 1992, Transgenic medicinal plants: *Agrobacterium*-mediated foreign gene transfer and production of secondary metabolites, *J. Nat. Prod.* **55:**149–162.

Waller, G. R., and Nowacki, E. K., 1978, *Alkaloid Biology and Metabolism in Plants,* Plenum Press, New York.

Key References

Berlin, J., Rügenhagen, C., Dietze, P., Fecker, L. F., Goddijn, O. J. M., and Hoge, H. C., 1993, Increased production of serotonin by suspension and root cultures of *Peganum harmala* transformed with a tryptophan decarboxylase cDNA clone from *Catharanthus roseus. Transgenic Res.* **2:**336–344.

Bracher, D., and Kutchan, T. M., 1992a, Strictosidine synthase from *Rauvolfia serpentina*: Analysis of a gene involved in indole alkaloid biosynthesis, *Arch. Biochem. Biophys.* **294:**717–723.

Bracher, D., and Kutchan, T. M., 1992b, Polymerase chain reaction comparison of the gene for strictosidine synthase from ten *Rauvolfia* species, *Plant Cell Rep.* **11:**179–182.

De Luca, V., Marineau, C., and Brisson, N., 1989, Molecular cloning and analysis of cDNA encoding a plant tryptophan decarboxylase: Comparison with animal dopa decarboxylases, *Proc. Natl. Acad. Sci. USA* **86:**2582–2586.

DeScenzo, R. A., and Minocha, S., 1993, Modulation of cellular polyamines in tobacco by transfer and expression of mouse ornithine decarboxylase cDNA, *Plant Mol. Biol.* **22:**113–127.

Dittrich, H., and Kutchan, T. M., 1991, Molecular cloning, expression, and induction of berberine bridge enzyme, an enzyme essential to the formation of benzophenanthridine alkaloids in the response of plants to pathogenic attack, *Proc. Natl. Acad. Sci. USA* **88:**9969–9973.

Dooner, H. K., and Robbins, T. P., 1991, Genetic and developmental control of anthocyanin biosynthesis, *Annu. Rev. Genet.* **25:**173–199.

Fecker, L. F., Hillebrandt, S., Rügenhagen, C., Herminghaus, S., Landsmann, J., and Berlin, J., 1992, Metabolic effects of a bacterial lysine decarboxylase gene expressed in a hairy root culture of *Nicotiana glauca, Biotech. Lett.* **14:**1035–1040.

Goddijn, O. J. M., de Kam, R. J., Zanetti, A., Schilperoort, R. A., and Hoge, J. H. C., 1992, Auxin rapidly down-regulates transcription of the tryptophan decarboxylase gene from *Catharanthus roseus, Plant Mol. Biol.* **18:**1113–1120

Hamill, J. D., Robins, R. J., Parr, A. J., Evans, D. M., Furze, J. M., and Rhodes, M. J. C., 1990, Over-expressing a yeast ornithine decarboxylase gene in transgenic roots of *Nicotiana rustica* can lead to enhanced nicotine accumulation, *Plant Mol. Biol.* **15:**27–38.

Hashimoto, T., Hayashi, A., Amano, Y., Kohno, J., Iwanari, H., Usuda, S., and Yamada, Y., 1991, Hyoscyamine 6β-hydroxylase, an enzyme involved in tropane alkaloid biosynthesis, is localized at the pericycle of the root, *J. Biol. Chem.* **266:**4648–4653.

Hashimoto, T., Matsuda, J., and Yamada, Y., 1993a, Two-step epoxidation of hyoscyamine to scopolamine is catalyzed by bifunctional hyoscyamine 6β-hydroxylase, *FEBS Lett.* **329:**35–39.

Hashimoto, T., Yun, D.-J., and Yamada, Y, 1993b, Production of tropane alkaloids in genetically engineered root cultures, *Phytochemistry* **32:**713–718.

Herminghaus, S., Schreier, P. H., McCarthy, J. E. G., Landsmann, J., Botterman, J., and Berlin, J., 1991, Expression of a bacterial lysine decarboxylase gene and transport of the protein into chloroplasts of transgenic tobacco, *Plant Mol. Biol.* **17:**475–486.

Hibi, N., Higashiguchi, S., Hashimoto, T., and Yamada, Y., 1994, Gene expression in tobacco low-nicotine mutants, *Plant Cell* **6:**723–735.

Hondelmann, W., 1984, The lupin–Ancient and modern crop plant, *Theor. Appl. Genet.* **68:**1–9.

Kanegae, T., Kajiya, H., Amano, Y., Hashimoto, T., and Yamada, Y., 1994, Species-dependent expression of the hyoscyamine 6β-hydroxylase gene in the pericycle, *Plant Physiol.* **105:**483–490.

Kass, E., and Wink, M., 1995, Molecular phylogeny of the Papilionoideae (family Leguminosae): rbcL gene sequence versus chemical taxonomy, *Bot. Acta* **108:**149–162.

Kraus, P. F. X., and Kutchan, T. M., 1995, Molecular cloning and heterologous expression of a cDNA encoding berbamunine synthase, a C-O phenol-coupling cytochrome P_{450} from the higher plant *Berberis stolonifera, Proc. Natl. Acad. Sci. USA* **92:**2071–2075.

Kutchan, T. M., 1989, Expression of enzymatically active cloned strictosidine synthase from the higher plant *Rauvolfia serpentina* in *Escherichia coli, FEBS Lett.* **257:**127–130.

Kutchan, T. M., 1996, Heterologous expression of alkaloid biosynthetic genes—a review, *Gene* **179:**73–81.

Kutchan, T. M., Hampp, N., Lottspeich, F., Beyreuther, K., and Zenk, M. H., 1988, The cDNA clone for strictosidine synthase from *Rauvolfia serpentina, FEBS Lett.* **237:**40–44.

Kutchan, T. M., Dittrich, H., Bracher, D., and Zenk, M. H., 1991, Enzymology and molecular biology of alkaloid biosynthesis, *Tetrahedron* **47:**5945–5954.

Kutchan, T. M., Bock, A., and Dittrich, H., 1994, Heterologous expression of the plant proteins strictosidine synthase and berberine bridge enzyme in insect cell culture, *Phytochemistry* **35:**353–360.

McKnight, T. D., Roessner, C. A., Devagupta, R., Scott, A. I., and Nessler, C. L., 1990, Nucleotide sequence of a cDNA encoding the vascular protein strictosidine synthase from *Catharanthus roseus, Nucleic Acids Res.* **18:**4939.

McKnight, T. D., Bergey, D. R., Burnett, R. J., and Nessler, C. L., 1991, Expression of enzymatically active and correctly targeted strictosidine synthase in transgenic tobacco plants, *Planta* **185:**148–152.

Matsuda, J., Okabe, S., Hashimoto, T., and Yamada, Y., 1991, Molecular cloning of hyoscyamine 6β-hydroxylase, a 2-oxoglutarate-dependent dioxygenase, from cultured roots of *Hyoscyamus niger, J. Biol. Chem.* **266:**9460–9464.

Meijer, A. H., Cardoso, M. I. L., Voskuilen, J. T., de Waal, A., Verpoorte, R., and Hoge, J. H. C., 1993a, Isolation and characterization of a cDNA clone from *Catharanthus roseus* encoding NADPH:cytochrome P_{450} reductase, an enzyme essential for reactions catalyzed by cytochrome P_{450} mono-oxygenases in plants, *Plant J.* **4:**47–60.

Meijer, A. H., Souer, E., Verpoorte, R., and Hoge, J. H. C., 1993b, Isolation of cytochrome P_{450} cDNA clones from the higher plant *Catharanthus roseus* by a PCR strategy, *Plant Mol. Biol.* **22:**379–383.

Nakajima, K., Hashimoto, T., and Yamada, Y., 1993a, Two tropinone reductases with different stereospecificities are short-chain dehydrogenase evolved from a common ancestor, *Proc. Natl. Acad. Sci. USA* **90:**9591–9595.

Nakajima, K., Hashimoto, T., and Yamada, Y., 1993b, cDNA encoding tropinone reductase-II from *Hyoscyamus niger, Plant Physiol.* **103:**1465–1466.

Nakajima, K., Hashimoto, T., and Yamada, Y., 1994, Opposite stereospecificity of two tropinone reductases is conferred by the substrate-binding sites, *J. Biol. Chem.* **269:**11695–11698.

Noji, M., Murakoshi, I., and Saito, K., 1993, Evidence for identity of β-pyrazolealanine synthase with cysteine synthase in watermelon: Formation of β-pyrazolealanine by cloned cysteine synthase *in vitro* and *in vivo, Biochem. Biophys. Res. Commun.* **197:**1111–1117.

Noji, M., Murakoshi, I., and Saito, K., 1994, Molecular cloning of a cysteine synthase cDNA from *Citrullus vulgaris* (watermelon) by genetic complementation in an *Escherichia coli* Cys^- auxotroph, *Mol. Gen. Genet.* **244:**57–66.

Pasquali, G., Goddijn, O. J. M., de Waal, A., Verpoorte, R., Schilperoort, R. A., Hoge, J. H. C., and Memelink, J., 1992, Coordinated regulation of two indole alkaloid biosynthetic genes from *Catharanthus roseus* by auxin and elicitors, *Plant Mol. Biol.* **18:**1121–1131.

Saito, K., Noji, M., Ohmori, S., Imai, Y., and Murakoshi, I., 1991a, Integration and expression of a rabbit liver cytochrome P_{450} gene in transgenic *Nicotiana tabacum, Proc. Natl. Acad. Sci. USA* **88:**7041–7045.

Saito, K., Yamazaki, M., Kawaguchi, A., and Murakoshi, I., 1991b, Metabolism of solanaceous alkaloids in transgenic plant teratomas integrated with genetically engineered genes, *Tetrahedron* **47:**5955–5968.

Saito, K., Koike, Y., Suzuki, H., and Murakoshi, I., 1993a, Biogenetic implication of lupin alkaloid biosynthesis in bitter and sweet forms of *Lupinus luteus* and *L. albus, Phytochemistry* **34:**1041–1044.

Saito, K., Suzuki, H., Takamatsu, S., and Murakoshi, I., 1993b, Acyltransferases for lupin alkaloids in *Lupinus hirsutus, Phytochemistry* **32:**87–91.

Songstad, D. D., De Luca, V., Brisson, N., Kurz, W. G. W., and Nessler, C. L., 1990, High levels of tryptamine accumulation in transgenic tobacco expressing tryptophan decarboxylase, *Plant Physiol.* **94:**1410–1413.

Suzuki, H., Murakoshi, I., and Saito, K., 1994, A novel O-tigloyltransferase for alkaloid biosynthesis in plants: Purification, characterization, and distribution in *Lupinus* plants, *J. Biol. Chem.* **269:**15853–15860.

Takeshita, N., Fujisawa, H., Miura, H., Fitchen, J. H., Yamada, Y., and Sato, F., 1995, Molecular cloning and characterization of *S*-adenosyl-L-methionine:scoulerine-9-O-methyltransferase from cultured *Coptis japonica* cells, *Plant Cell Physiol.* **36:**29–36.

Vetter, H.-P., Mangold, U., Schröder, G., Marner, F.-J., Werck-Reichhart, D., and Schröder, J., 1992, Molecular analysis and heterologous expression of an inducible cytochrome P_{450} protein from periwinkle (*Catharanthus roseus* L.), Plant Physiol. **100:**998–1007.

Weising, K., Schell, J., and Kahl, G., 1988, Foreign genes in plants: Transfer, structure, expression, and application, *Annu. Rev. Genet.* **22:**421–477.

Williams, W., and Harrison, J. E. M., 1983, Alkaloid concentration during development in three *Lupinus* species and the expression of genes for alkaloid biosynthesis in seedlings, *Phytochemistry* **22:**85–90.

Williams, W., Harrison, J. E. M., and Jayasekera, S., 1984, Genetical control of alkaloid production in *Lupinus mutabilis* and the effect of a mutant alle *mutal* isolated following chemical mutagenesis, *Euphytica* **33:**811–817.

Wink, M., and Witte, L., 1983, Evidence for a wide-spread occurrence of the genes of quinolizidine alkaloid biosynthesis, *FEBS Lett.* **159:**196–200.

Yamazaki, M., Sato, A., Saito, K., and Murakoshi, I., 1993, Molecular phylogeny based on RFLP and its relation with alkaloid patterns in *Lupinus* plants, *Biol. Pharm. Bull.* **16:**1182–1184.

Yun, D.-J., Hashimoto, T., and Yamada, Y., 1992, Metabolic engineering of medicinal plants: Transgenic *Atropa belladonna* with an improved alkaloid composition, *Proc. Natl. Acad. Sci. USA* **89:**11799–11803.

Yun, D.-J., Hashimoto, T., and Yamada, Y., 1993, Transgenic plants with two consecutive oxidation reactions catalyzed by hyoscyamine 6β-hydroxylase, *Biosci. Biotech. Biochem.* **57:**502–503.

Chapter 7

Production of Alkaloids in Plant Cell Culture

Margaret F. Roberts

1. INTRODUCTION

The destruction of natural habitats and the exploitation of certain plants for drug production has created an interest in alternative sources for plant material. Over the last decade the potential of cultured plant cells as a means of producing important secondary metabolites has been a focus of investigation (Staba, 1985; Rokem and Goldberg, 1985; Fowler, 1986; Yamada and Fujita, 1986; Banthorpe, 1994).

Plant biotechnology may be described (Constabel, 1990) as the enhancement of the formation and accumulation of desirable natural products as well as possible product modification in plants. The methods for tailoring plant resources include plant cell and protoplast culture, manipulation of nuclear and plasmid genes, plant cell and enzyme immobilization, and industrial-scale production or biotransformation. The interest in the production of alkaloids results from their use as pharmaceuticals and from the fact that some products are required in relatively small quantities and are of high value (Deus and Zenk, 1982).

Improvements in plant propagation methods have been developed using plant biotechnology techniques. The regeneration of plants using shoot tip culture and callus culture have proved to be commercially useful techniques. Micropropagation, as this form of culture is usually called today, has a number of advantages over traditional or conventional propagation techniques. It is eminently suited to rapid multiplication of clonally derived (from a single cell) material and can be used to derive and multiply disease-free stock; it may allow for the multiplication of sexually derived nonfertile crosses and, through protoplast fusion and formation of somatic hybrids, may provide an additional

Margaret F. Roberts • The Centre for Pharmacognosy, School of Pharmacy, University of London, London WC1N 1AX, England.

Alkaloids: Biochemistry, Ecology, and Medicinal Applications, edited by Roberts and Wink. Plenum Press, New York, 1998.

means of developing new hybrids with desirable characteristics or traits. These techniques are being used increasingly to investigate a variety of botanical phenomena as well as to improve existing plant stocks and to preserve endangered plant species.

The technique of plant cell culture has proved to be an invaluable tool for the study of some aspects of metabolism. Suspension cultures of cells offer a number of advantages as experimental systems including stability and homogeneity of the culture environment; facile manipulation of the environment, nutrient and hormone conditions; and the rapid regeneration of relatively uniform tissue, on almost any scale, on a nonseasonal basis. The disadvantages that should be borne in mind, which for some studies may cause serious difficulties, stem from the fact that most plant cell cultures lose their phototropic function and cells do not produce morphologically differentiated structures (Ellis, 1984). Compared with plants, they also display an enhanced degree of genome instability (Meins, 1983). As cells in a liquid culture can be considered to be restricted to a very limited range of differentiated states resulting from the level of auxin and cytokinin in the medium, the secondary metabolism phenotype displayed is presumably that which typifies cells of that species at one particular level of differentiation. This leads to the failure of some cultured plant cell systems to accumulate those specific compounds for which the mature plant is noted. Other species, by contrast, synthesize remarkable quantities of taxon-specific metabolites (Table I). These include several alkaloid-containing species such as *Catharanthus roseus* (ajmalicine and serpentine) and *Coptis japonica* (berberine) (Anderson *et al.*, 1987, and references therein).

The fact that cultured plant cells frequently show a marked simplification in the patterns of secondary metabolism has made them ideal candidates for the isolation and study of the enzymes of particular biogenetic sequences. Purified enzymes and active cell-free systems can be prepared easily from cell cultures and this has facilitated the study of individual enzymes. Some caution must be employed, for while many cell cultures do not synthesize many of the alkaloids found in the intact plant, others may form a product via a route somewhat altered from that found in the entire plant and, indeed, differences in this respect have also been observed between cell cultures derived from different plant species

Table I
Examples of Cell Cultures That Produce Alkaloids at Levels Higher Than the Parent Plant

		Yield % dry weight	
Alkaloid	Plant species	Cell culture	Whole plant
Caffeine	*Coffea arabica*	1.60	1.60
Nicotine	*Nicotiana tabacum*	3.40	2.50
Protopine	*Macleaya cordata*	0.40	0.32
Biscoclaurine	*Stephania cepharantha*	2.29	0.92
Ajmalicine	*Catharanthus roseus*	1.30	0.26
Serpentine			
Canthin-6-one	*Ailanthus altissma*	1.27	0.26
	Brucea javanica		
Berberine	*Coptis japonica*	11.40	10.00
	Thalictrum minus		
Jatrorrhizine	*Berberis wilsoniae*	10.00	1.2

that produce the same product (see Chapter 5). The fact that the use of cell cultures has facilitated the isolation of individual enzyme proteins has made possible the isolation of the appropriate genes. For example, isolation of the genes responsible for production of strictosidine synthase, hyoscyamine-6β-hydroxylase, and tetrahydroprotoberberine oxidase, key enzymes of indole tropane and isoquinoline alkaloid biosynthesis, respectively, made possible the further detailed study of these particular enzymes. The genes for these enzymes have been inserted into microorganisms, insect cells in culture, and plants where they have been successfully expressed (Kutchan, 1989; Okada *et al.,* 1989; Kutchan *et al.,* 1991; Matsuda *et al.,* 1991; Yun *et al.,* 1992).

Higher plants are important sources of natural products and are still used commercially to produce a wide range of chemicals used as drugs, flavors, enzymes, perfumes, insecticides, and emulsifying agents. Many of these phytochemicals are of interest as pharmaceuticals, particularly the alkaloids, steroids, hormones, and cardiac glycosides (Anderson *et al.,* 1987; Flores, 1986).

Plants continue to be a major source of medicines (Hay *et al.,* 1988). Among the most important are the physiologically active alkaloids; about 30 of these are used medicinally, including vincristine and vinblastine from *Catharanthus roseus,* codeine and morphine from *Papaver somniferum,* hyoscyamine and scopolamine from *Datura* species, quinine and quinidine from *Cinchona ledgeriana,* and reserpine from *Rauwolfia* species.

These medicinal agents are produced on a commercial scale by large-scale cultivation and solvent extraction of whole plants or isolated plant parts. The production of crude drugs is therefore subject to the vagaries of climate, to crop disease, to varying methods of collection and drying which can influence quality, and to the inherent differences in active constituents arising from genetic variation among plants of the same species. While synthetic routes to many of the alkaloids are well established, economically viable processes exist for only a few of the medicinal alkaloids in common use, e.g., caffeine and ephedrine (Anderson *et al.,* 1987).

The present chapter will discuss the utilization of plant cell cultures for the production of improved plants for cultivation as well as for biologically active compounds for use as drugs *per se* or as templates for the production of potentially new drugs in the future (Heinstein, 1985; Roberts, 1988b; Zenk, 1990). A discussion of the role of plant cell cultures in molecular biology and in the elucidation of alkaloid biosynthetic pathways will be found in Chapters 5, 6, and 8.

2. CELL TECHNOLOGY

2.1. Cell Culture

The feasibility of artificially cultivating plant cells was recognized at the turn of the century; however, it is only some few decades ago that modern development in the cultivation of cells of higher plants as a callus or as a suspension liquid culture really began. In this connection one might mention the following publications: R. J. Gautheret's, *La Culture des Tissues Vegeteaux* (1959) and H. E. Street's *Plant Tissue and Cell Culture* (1977) (Reinert and Yeoman, 1988; Stafford and Warren, 1991).

Cultures of individual cells growing under controlled conditions in a liquid medium

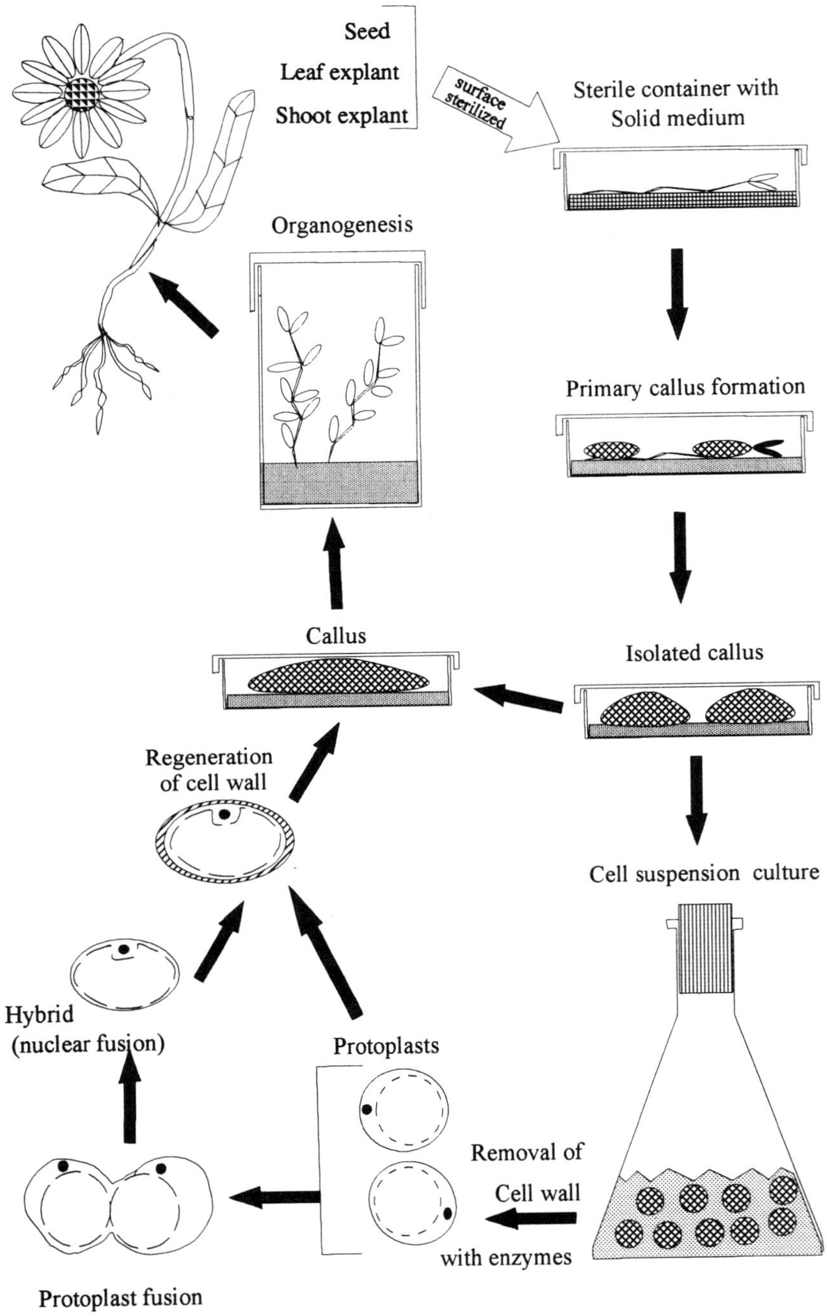

Figure 1. Scheme for the production of cell suspension cultures, the production of protoplasts, and plant regeneration.

or callus cultures consisting of undifferentiated masses of cells developing on a semi-solid medium can be initiated from parenchymatous tissues of shoots, roots and other plant structures (Fig. 1). Maintenance of cultures depends on an adequate supply of nutrients and growth factors and a controlled sterile environment where temperature, pH of the medium, and quality and quantity of light are also important. The cells, although undifferentiated, contain all of the genetic information present in the normal plant, i.e., they are totipotent. Suitable manipulation of the hormone content of the medium makes it possible to initiate the development of roots, shoots, and complete plants from callus cell culture and to encourage the division of cells in suspension culture. Details of these media and hormone requirements can be found in any of a number of texts (e.g., Mantell and Smith, 1983). Environmental factors such as the oxygen content of the medium, light, and temperature may also play a significant role in alkaloid production by plant cell cultures and manipulations of these parameters have improved alkaloid yields.

2.1.1. METHODS OF CULTIVATING CELLS

Batch suspension cultures: In this technique the cells multiply in liquid medium which is being continuously agitated to maintain the cells as small aggregates and to maintain oxygen levels in what is essentially a closed system with no addition or subtraction from the culture, other than gases. Typically the original inoculation of cells into the medium is followed by a lag phase and then, after increasing in mass, the cells undergo a period of rapid division. Finally, a growth phase is reached at which point some component of the medium, essential for growth, has probably been exhausted (Fig. 2). Growth

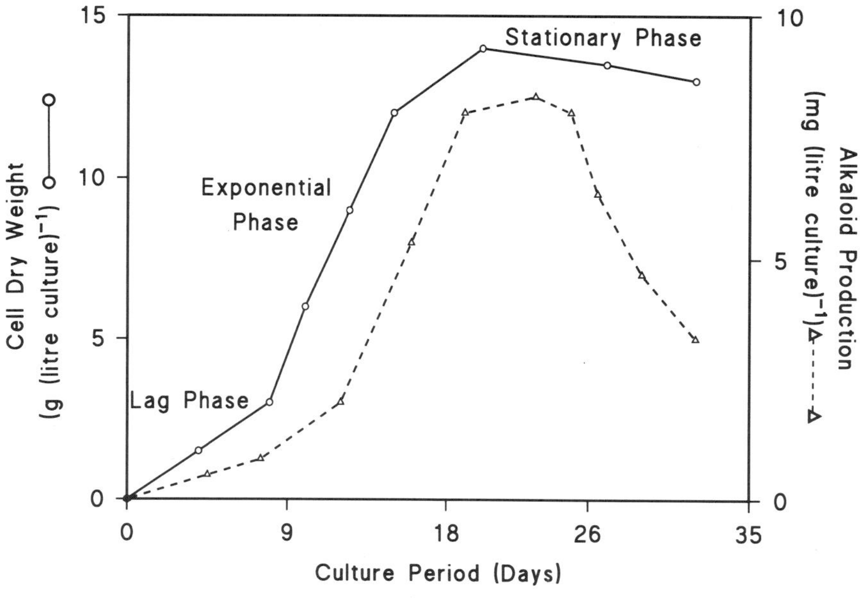

Figure 2. Typical growth curve for plant cells in cell suspension cultures.

will recommence when cells are transferred to a fresh medium or when more medium is added to this original culture.

Semicontinuous cultures: In this instance the system is designed for the periodic removal of culture and the addition of fresh medium by which means the growth of the culture is continuously maintained. This is usually referred to as an open system.

Continuous cultures: Two forms of this open system are the chemostat and turbidostat systems in which the volume of the culture remains constant and fresh medium and culture are, respectively, continuously added and withdrawn. The main feature of these two systems is that cell proliferation takes place under essentially constant conditions. In the chemostat arrangement, a steady state is achieved by adding medium in which a single nutrient is monitored and has been adjusted to be growth-limiting. The turbidostat method depends on maintaining the cultures at a constant cell concentration. This contrasts with the batch culture method, in which the cells find themselves in a continuously transient condition which leads to continuous changes not only in their growth rate but also in their metabolism.

The problems associated with plant cell culture are not identical with those encountered in the fermentation of microorganisms and fungi. The growth rate of plant cells in culture is, for instance, much slower than that of microorganisms and fungi. Plant cells in culture tend to form large aggregates of cells and these aggregates exist in a different environment than do suspended single cells or small groups (three- to four-cell aggregates).

2.2. Establishment of Specific Cell Lines

Critical to the effective production of any secondary metabolite by plants has been the establishment of high-yielding cell lines. This initially requires the selection of appropriate source material and this may mean choosing the appropriate plant species and/or searching for individuals within a particular species that are particularly high producers of a constituent. In some instances, it has been possible to produce stable high-yielding cell lines from selected plants but in other instances results have been disappointing. For the selection of high yielding strains (perhaps a few plants in several thousand), radioimmunoassay (RIA) or enzyme-linked immunosorbent assay (ELISA) has become an important tool as each enables many thousands of plants to be sampled rapidly. Cells from plants selected for the production of cell cultures are usually subject to further techniques of selection.

The variation within a population of cultured cells is related in part to the nature of the explant from which the culture has been initiated but mostly it arises as a result of the act of culture (Yeoman and Forsche, 1980). It is generally accepted that genetic variability arises spontaneously in all types of plant cell cultures, i.e., cell lines derived from a single genotype are phenotypically different (Larkin and Scowcroft, 1981), this phenomenon has been called *somaclonal variation*. An intracultural heterogeneity also occurs (cell-to-cell variation within a single culture), and to suppress or reduce this tendency, cloning with protoplasts, single cells, or small cell aggregates must be used (Fig. 1). Sometimes, however, these procedures can have a destabilizing effect on the genotype of the cell culture, leading to an unstable clone. In some cases, however, the instability can be resolved by repetitive selection until stable production is reached.

Culture systems are best summed up in Fig. 1. Typically, callus cultures may be inspected visually for altered pigmentation or for fluorescence if the compounds of interest are either pigmented or fluoresce; with repeated cloning it may be possible to select stable high yielding cell lines. This procedure has been used with some success in producing stable cell lines for *Thalictrum minus* and *Coptis japonica* yielding berberine but has been less successful in achieving stable cell lines from *Catharanthus roseus* producing ajmalicine, where inherent instability appears to prevail. The plating out of cells from callus or suspension cultures can further assist the selection process. The main problem with this technique is that, unlike microbial cells, plant cells apparently require a minimum inoculum density for growth. Protoplasts may also be prepared from mesophyll cells and plated out for regeneration of callus tissue; however, instances in which protoplasts do not show cell wall regeneration or prolonged cell division and cannot be regenerated into callus or whole plants have been reported. In some works, physical or chemical mutagenic treatments have been used to induce variability prior to selection (Dix, 1990).

From a fundamental and practical point of view the emergence of flow cytometry and cell sorter techniques could improve the exploitation of plant cell cultures (for review see Brown, 1984). On the one hand, it would allow the assessment of a large number of cells according to various parameters (cell size or structure, fluorescent markers), an alternative approach to studying the variability within a cell culture by single cell or protoplast cloning. On the other hand, flow cytometry should permit the sorting of cultures into homogeneous subpopulations of high-producing cells and may therefore avoid the induced variation ascribed to the singleness of isolated cells. Details of the use of cloning techniques and the cell sorter are described for *C. roseus* and *T. minus* by Bariaud-Fontanel *et al.* (1988). It has, on the other hand, been suggested (Hall and Yeoman, 1986) that the number of producing cells within a culture is a function of spatial and temporal heterogeneity, i.e., the proportion of producing cells is determined by cultural conditions. If this is the case, cell sorting would not be particularly useful.

With the cloning techniques described, only about 28% of the *T. minus* clones remained stable in berberine production, while production by the remaining clones either increased or decreased, presumably because the cloning process induced instability. Cloning of *T. minus* produced a strain with a higher production rate than the original culture. Repeated cloning may successfully exploit induced variation to enhance the proportion of producing cells provided these clonings occur before the clones restabilize, i.e., a constant selective pressure must be applied during several cloning cycles and this is where cell sorting and flow cytometry techniques are useful (Brown, 1984; Harkins and Galbraith, 1984). With *C. roseus* (Deus and Zenk, 1982), however, no clone seemed to be more productive than the mother strain of the two cell lines used. In most cases, active selection pressure has to be continuously used to maintain cultures of high-yielding cells. Yields may be further enhanced by variations in media composition or by elicitation and stress; techniques are discussed in Section 2.4.

Cell culture techniques are also used as a route to the production of hybrids. This usually involves the isolation of plant protoplasts from two closely related plants and their fusion using electrofusion techniques followed by individual cloning. The hybrid nature of the cloned cells may be confirmed by cytological investigation and ribosomal-DNA analysis. As an example of this technique, *Duboisia leichhardtii* and *Nicotiana tabacum* have been used to produce a series of hybrids via protoplast fusion techniques. The hybrid plants produced nicotine as the predominant alkaloid, like *N. tabacum,* whereas *D.*

leichhardtii plantlets produced both tropane and nicotine alkaloids. The hybrid plants did, however, have the ability to reduce tropinone and carry out the hydroxylation and epoxidation of hyoscyamine (Endo *et al.,* 1991); see Chapter 8 for tropane alkaloid biosynthesis.

2.3. Growth and Product Accumulation

Secondary metabolites are generally accumulated by cultured cells when entering the stationary phase. This observation indicates an inverse relationship between culture growth (primary metabolism) and accumulation of products (secondary metabolism). Evidence for this inverse relationship can be seen in the finding that actinomycin D and cyclohixamide both inhibit protein synthesis and growth in cultures of *Macleaya cordata*; however, only the former has a stimulatory effect on alkaloid production (von Neumann and Müller, 1974). These results suggest that the metabolic systems responsible for the biosynthesis of specific compounds have to be present within the cell at the time the inhibitors for RNA and protein synthesis are applied and that it is primarily the interrelationship and concentration of nutrients that determine whether growth or biosynthesis of secondary metabolites is promoted. These observations suggest that pathways are present but nonfunctional and that a switching system is responsible for their operation. Two possibilities may be suggested (Yeoman *et al.,* 1980): either (1) the biosynthetic pathways are present but inoperative until switched on by a key enzyme or (2) the biosynthetic pathway for a specific compound is nonexistent and has to be produced when conditions for synthesis of such a compound are favorable. Evidence seems to favor the first hypothesis. Although these schemes are somewhat simplistic, the control of some aspects of differentiation by key enzymes that activate already present pathways would be of selective advantage to stationary organisms like plants, enabling them to respond immediately to environmental changes.

Many plants in cell culture produce their characteristic metabolites but in low yield. Sometimes compounds that have not been found in the original plant occur in cell cultures; examples are new minor alkaloids and anthraquinones from *Cinchona ledgeriana* and *C. pubescens*. Tropane alkaloids not previously obtained from *Atropa belladonna* have been isolated from root suspension cultures. In general, with species of the Solanaceae, root production enhances tropane alkaloid production, and the pattern or profile of alkaloids may undergo a change. For example, canthin-6-one alkaloids in *Ailanthus altissima* occur at much higher levels in cell cultures than they do in the parent plant (Anderson *et al.,* 1985). *Lupinus polyphyllus* cell culture alkaloid levels are only 1/200th those of the source plant leaves and lupanine occurred almost exclusively (95%); in the plant sparteine and other minor alkaloids occurred and lupanine accounted for less than 70% of the total alkaloid (Wink et al., 1983; Constabel, 1990). Although alkaloids from a wide range of plants have been produced satisfactorily in plant cell cultures, in some instances it was apparent that some differentiation is required for production to occur. In *Cinchona ledgeriana* there is a requirement for the production of "rooty" cultures, or transformed roots, in order to obtain quinine and quinidine. The absence or very low production of morphinans in *Papaver somniferum* cell cultures seems related to a lack of developed laticiferous tissue in the unorganized cell culture. *P. bracteatum* cell suspension cultures produced orientalidine and sanguinarine, whereas the root and shoot cultures produced thebaine (Furuya *et al.,* 1972; Zito and Staba, 1982).

Siah and Doran (1991) have shown that *P. somniferum* suspension cultures produce codeine (3.0 mg g^{-1} dw) and morphine (2.5 mg g^{-1} dw) after removal of hormones. These cultures developed without organogenesis or the formation of meristemoids. In this case, improved alkaloid production was correlated with slower overall growth rather than with differentiation. The necessity and effect of morphogenesis (differentiation) in cell cultures for secondary product formation and accumulation remains an open issue. However, the more genes are investigated, the more likely it is that genes of alkaloid synthesis are as tightly regulated as are others. The regulatory proteins that control them are likely to be tissue and developmentally specific and may even be environmentally specific. Suspension cultures of *Catharanthus roseus* do not produce the dimeric alkaloids vincristine and vinblastine, constituents useful in the treatment of leukemia. Leaf organ cultures of *C. roseus* (i.e., cultures developed from leaves, in which differentiation was observed) have been shown to produce these dimeric indole alkaloids. These dimers occur only in cultures that also contained the constituent monomers vindoline and catharanthine (Endo *et al.,* 1987).

2.4. Development of Cultures with Improved Alkaloid Yield

Most secondary metabolites tend to accumulate late in the cell growth cycle, usually at the point where growth has ceased and at the stage where cell differentiation may occur (Lindsey and Yeoman, 1985). It is likely that the switch from primary metabolism to secondary metabolism coincides with medium depletion of one or more nutrients. Competition for common precursors is dependent on the growth rate of the culture (Yeoman, 1987a,b). Nutrient stress has been reported to slow growth and primary metabolism and promote differentiation and secondary metabolism (Collin, 1987). A variety of chemical and physical factors have been investigated in an attempt to optimize the yield of secondary metabolites by plant cell cultures.

Medium composition: Most media developed for plant cell culture generally stimulate cell division and growth but, in most cases, yield only minute amounts of secondary products. It has been found that the amount and form in which key nutrients are supplied to cell cultures may greatly influence the switch mechanisms responsible for activating respective pathways. The effects of medium composition and sucrose concentration on alkaloid production for example have been studied in hairy roots of *Hyoscyamus albus* (Sauerwein and Shimomura, 1991; see also Chapter 8). A large number of carbon sources will support cell growth but the highest growth rates are obtained with sucrose and glucose. Nitrate and ammonia are the main sources of nitrogen used in culture medium and their relative concentration can have a marked effect on both growth and metabolite production. The effect of nitrogen on secondary metabolite production has to be viewed as an interaction with carbon sources available to the cells and even small alterations in the C/N ratio can have a marked influence (Knobloch and Berlin, 1981; Lindsey, 1985; Collin, 1987) .

Phosphorus, while required for growth, has, in excess of 10 mM, an inhibitory effect on secondary metabolism in microorganisms and in indole alkaloid-producing *Catharanthus roseus* cells. It is interesting to note that phosphorus levels in growth medium must be exhausted to below the inhibitory level before induction of tryptophan decarboxylase, the key enzyme of alkaloid biosynthesis, can take place (Kurz and Constabel, 1985).

Growth regulators affect cell growth and differentiation and therefore the quantity and type used are usually critical to secondary metabolite production. Plant growth hormones do not react with the intermediaries of biosynthetic pathways, but appear to shift cytological conditions that favor product formation. Growth regulators employed can be natural, such as indole-3-acetic acid (IAA) and zeatin, or synthetic, such as kinetin, 6-benzyladenine (BA), 2,4-dichlorophenoxyacetic acid (2,4-D), and naphthaleneacetic acid (NAA). In general, IAA seems to favor product synthesis and accumulation more than NAA or 2,4-D. For example, *C. roseus, D. innoxia, H. niger, A. belladonna,* and *P. somniferum* cultures lack alkaloids when exposed to 2,4-D (Kurz and Constabel, 1985; Mantell and Smith, 1983; Payne *et al.,* 1991).

Auxins and cytokinins, while synergists in enhancing growth by cell division, may cause independent responses in secondary metabolism. In cultures of *C. roseus*, BA in the absence of auxins permitted high yields of alkaloids but did not stimulate growth. As both growth and differentiation of cell cultures are necessary for large-scale secondary metabolite production, a two-step procedure where cell mass would be obtained in high level auxin medium, followed by transfer of the cells into medium containing cytokinins for metabolite synthesis, has been advocated (Zenk *et al.,* 1977). Some of these points are best illustrated by the work of Zenk's research group in their investigations of indole alkaloid production in *C. roseus*. Changes in hormones were used to good effect with the development of production media that allowed for the production of serpentine after cell growth had been accomplished. The same technique helped to improve yields of berberine from *Thalictrum minus* (Nakagawa *et al.,* 1986) and serpentine from *Rauwolfia serpentina* (Yamamoto and Yamada, 1987).

Precursors: In some instances alkaloid production may be improved by precursor feeding. For example, tryptophan fed to cultures of *Cinchona ledgeriana* improved production (25%) of quinine and quinidine and similar results were observed with *C. roseus* and the production of ajmalicine and serpentine (Hay *et al.,* 1988). More recently, the various precursors of the tropane alkaloids have been fed to hairy root cultures of a *Datura candida* hybrid where phenylalanine and tropinone stimulated production of hyoscyamine and scopolamine (Robins *et al.,* 1990, 1991a,b). The precursor putrescine and its synthetic analogues considerably improved yields of scopolamine in *Duboisia myoporoides*.

Elicitors: It was found, relatively recently, that fungal infection of some plants elicits the production of secondary metabolites by the plant that are toxic to the fungi. This technique has been applied to plant suspension cultures with some success (Brodelius, 1988). Brodelius showed that *Thalictrum rugosum* (Funk *et al.,* 1987) responded to elicitation by yeast and fungal cell wall glucans to give a 3- to 4-fold increase in berberine yield. More startling has been the more than 100-fold increase in yield of ajmalicine and catharanthine in *C. roseus* cell cultures with *Micromucor isobelina* extracts achieved by DiCosmo's group (DiCosmo *et al.,* 1987). Low concentrations of a *Pythium vexans* extract induced catharanthine synthesis and stimulated ajmalicine production. The elicitor also induces changes in the cell/medium distribution of the alkaloids in that most of the alkaloid is released into the medium. However at high concentrations, extracts of fungal elicitors like *P. vexans* can cause cell death (Nef *et al.,* 1991). It has also been shown that cells may withstand several elicitation treatments with fungal elicitors, e.g., *Botrytis* sp., without the loss of cell viability. In this instance, it was possible to continuously produce sanguinarine and dihydrosanguinarine in *Papaver somniferum* cell suspension cultures. Somewhat advantageously, this technique led to significant release of alkaloid into the medium. Levels of 2 g per 100 ml (73% cell dw) were obtained at each of three successive elicitation treatments (Tyler *et al.*, 1988).This area is discussed in more detail in Chapter 9.

Environmental factors: During scaleup with cell suspension cultures, other environmental factors may also be critical. For example, increasing the dissolved oxygen content of the medium to 50% of air saturation considerably improves the production of berberine and jatrorrhizine by cell cultures of *T. minus* but increasing the oxygen tension by increasing the aeration rates leads to growth reduction, owing presumably to shear stress (Breuling *et al.*, 1985).

Light and temperature may also play a significant role, in that Courtois and Guern (1980) showed increased alkaloid production in *C. roseus* as a result of temperature shifts. Light appears to be a requirement for the production of lupanine in cell cultures of *Lupinus* spp. where alkaloid production is associated with the chloroplast (Wink *et al.*, 1983). However, quinoline alkaloid production, particularly quinine in *Cinchona* cells transformed by *Agrobacterium tumefaciens*, is increased 10- to 12-fold if cells are grown in the dark (Payne *et al.*, 1987b).

2.5. Immobilization of Plant Cells and Enzymes

The development of techniques for plant cell and enzyme immobilization has increased the flexibility of plant cell biotechnology. Immobilization of plant cells or enzymes using such matrices as alginate, agarose, or polyacrylamide has a number of advantages. These include reuse of expensive biocatalysts, stabilization of the biocatalyst, and simplified control with automatic separation of the catalyst from the product. In the case of immobilized whole cells there is a potential for obtaining more stable enzymes in a natural environment with the possibility for using whole biosynthetic pathways or one-step bioconversions. Ideally suited for one-step bioconversions, immobilized cells may be seen as an extension of enzyme immobilization technology that avoids the costs of enzyme purification, and covalent attachment to an enzyme support. Immobilized cells also have a further advantage over immobilized enzyme systems in their high relative activity and extended stability; for example, immobilized cells of *C. roseus* continued to make ajmalicine and serpentine over a 4-month period (Reinhard and Alfermann, 1980; Brodelius and Mosbach, 1982; Rhodes, 1986). Immobilized plant cells are likely to be commercially useful only for reactions not carried out by microorganisms. Examples of reactions efficiently carried out by immobilized cells are: (1) the conversion of (−)codeinone to (−)codeine by immobilized cells of *P. somniferum* with 79% conversion as opposed to 57% for a cell suspension culture (Corchete and Yeoman, 1989; Furnya *et al.*, 1972) and (2) the conversion of theobromine to caffeine by cells of *Coffea arabica* in cell suspension and immobilized cells on polyurethane foam, where the efficiency is usually more than 100% as the best cells for this conversion are also making caffeine *de novo*. In the latter instance, however, both suspension cells and immobilized cells gave similar results (Furuya *et al.*, 1991).

The main area of opportunity with immobilized plant cells, however, undoubtedly rests with the *de novo* synthesis of chemically complex molecules from simple carbon and nitrogen sources, i.e., the production of serpentine and ajmalicine by immobilized cells of *C. roseus* (Lambe *et al.,* 1982).

2.5.1. METHODS FOR THE IMMOBILIZATION OF PLANT CELLS

In general, two types of immobilization methods have been found suitable for plant cells. These are the entrapment of cells within gels and the adsorption of cells on the surface of a polymer. Both methods enable immobilization of cells to be carried out under mild conditions, minimizing damage to cultured plant cells which are extremely sensitive to stress. Two of the materials most widely used are calcium alginate and agarose both of which can be gelled under mild conditions without loss of cell integrity. With any of the gel methods, the biomass concentration in the beads is determined at the time of immobilization as gel beads offer only limited opportunities for cell growth (Kennedy, 1982). If growth occurs the beads rapidly disintegrate and free cells emerge (Morris and Fowler, 1981); thus, gel entrapment methods are generally used only for nongrowing cells in a nutrient- or hormone-limited medium.

The second approach to immobilization of plant cells requires that cells grow in the immobilized state rather than obtaining an immobilized population of cells that are subsequently held in a nongrowing state. This method provides a rigid porous support on which plant cells can grow, enabling the maintenance of cell to-cell contact and gradient formation which is thought to be conducive to the full expression of plant cell potential (Lindsey and Yeoman, 1983). The supports used for this type of work tend to be rigid synthetic polymers formed into a very open pore structure and include such materials as nylons and polyurethanes. For success it is essential that these materials have chemical and microbiological stability, be nonphytotoxic, have enough mechanical stability to withstand agitation in a fermentor, and be autoclavable.

Immobilization of cells separates cells from medium permitting media flow and continuous operation of the cells without loss of biomass; these factors are important with plant cells where slow cell growth and the requirement for complex media make the investment in biomass high. One of the problems with cell immobilization is the intracellular retention of secondary products produced by cells. Some products of interest are, however, released from plant cells, and there is an increasing array of methods for manipulating the permeability barrier therein.

2.5.2. MANIPULATION OF PRODUCT RELEASE

A continuous process is feasible only if the product is released by the cells; where this does not occur naturally, chemical agents have been used to alter membrane structure of the cells. Permeabilization of plant cells has been achieved by the use of chemicals such as DMSO, changing pH and/or increasing the ionic strength of the medium; changing temperature; the use of high and low electric fields and ultrasound (Brodelius *et al.,* 1988; Renaudin *et al.,* 1985; Weathers *et al.,* 1990; Kirby and Hunter, 1990). However, the viability of cells after such treatments varies. Some culture systems, however excrete substances into the medium and this has been observed with a number of root and shoot cultures (Payne *et al.,* 1991; Rhodes *et al.,* 1986; Maranaka *et al.,* 1992); this may also occur with undifferentiated cells in culture, e.g., berberine from *T. minor* (Yamamoto *et al.,* 1987).

2.5.3. BIOREACTOR SYSTEMS

The ultimate goal of the use of immobilized cells is the ability to scale up and utilize cells on a large scale in bioreactors. Two types of system have been tried, namely packed and fluidized beds and hollow fiber systems (Hulst and Tramper, 1989). As an example, the packed bed system has been used for the study of indole alkaloid production in *C. roseus* (Brodelius *et al.,* 1979; Veliky and Jones, 1981) using *C. roseus* cells in alginate packed beads. Growth limiting medium was cycled in a closed loop system and alkaloids were continuously removed by pumping the medium through a chloroform layer. Such a system was capable of maintaining alkaloid production at a steady rate for 60 hr.

Recently a bioreactor has been designed that uses *Thalictrum rugosum* cells immobilized on glass fiber mats (4–5 liters). However, over long periods the system did not produce protoberberine alkaloids as well as shake flask cultures of a similar size (Facchini and DiCosmo, 1991).

2.6. Organ Culture

2.6.1. MULTIPLE SHOOT CULTURES AND ROOT CULTURES

The inherent ability of a large number of plants to form multiple shoots in culture from axillary and shoot tip meristem may be used for the production of secondary metabolites as well as for the preservation and propagation of elite genotypes. Root cultures may also be produced in a similar manner from root tip meristem. Multiple shoots develop when meristem tissue is grown on minimal inorganic basal medium (Murashige and Skoog, 1962) that has been supplemented with benzyladenine, kinetin, and zeatin singly or in combination with low levels of auxin. Cell proliferation development at the basal end of the shoot meristem gives rise to shoot buds (Benjamin *et al.,* 1987). Multiple shoot cultures can be propagated on agar or in liquid medium and have been grown in 2-liter-capacity flasks and multiliter fermentors (Takayama and Misawa, 1981). The type of explant, the hormonal combination, and conditions of growth are all important in the establishment of these cultures.

Shoot and root cultures have been particularly useful for studies of alkaloid biosynthesis where disorganized cultures do not produce the expected product. Tropane alkaloids, scopolamine (hyoscine) and hyoscyamine, were found in only trace amounts in shoot cultures of *A. belladonna* presumably because these alkaloids are normally synthesized in the roots and translocated into the aerial parts. Root cultures are found to produce significant levels of the tropane alkaloids which are very comparable with the levels in the whole plant (Endo and Yamada, 1984; Hashimoto *et al.,* 1986). Shoot and root cultures from high-yielding plants have potential usefulness, particularly for multiplication of elite genotypes, production of alkaloids, and studies of alkaloid formation (Hebel, 1985).

2.6.2. HAIRY ROOTS

One of the major problems that has limited the commercialization of the production of secondary metabolites of unorganized plant cell cultures, root and shoot cultures has been their slow growth and low productivity relative to microorganisms. A recent development has been the genetic transformation of plants with *Agrobacterium rhizogenes* or *A. tumefaciens*. Plants transformed by *A. rhizogenes* produce fast-growing, organized, hairy root cultures which can be both stable and productive (Hamill *et al.,* 1987) (see also Chapter 8).

The introduction into the plant genome of two pieces of transfer DNA (T_1 and T_2) contained on the bacterial plasmid alters the auxin metabolism of the transformed tissue, causing the hairy root phenotype. Production of hairy root cultures is straightforward for some plant genera and there are a number of strains of *A. rhizogenes* that show good virulence characteristics.

Most hairy root cultures demonstrate a high level of productivity of secondary metabolites, similar to or greater than that of the plant from which the hairy roots have been derived. Some of the best examples are found with solanaceous plants of the genera *Nicotiana* and *Datura* (see Chapter 8). Work to date would seem to suggest that production may only be useful where the metabolite is normally synthesized in the root of the plant (Hamill *et al.,* 1986, 1987, 1989; Payne *et al.,* 1987a; Sauerwein *et al.,* 1991).

2.6.3. SHOOTY AND ROOTY TERATOMAS

The production of transgenic plants has been accomplished using a number of different methods to transfer genetic material. In the production of shooty and rooty teratomas, *A. tumefaciens* has been used as a method for the insertion of the Ti plasmid mutants into gene 1 and/or 2 (Shi, shoot inhibition) or gene 4 (Roi, root inhibition). *Nicotiana* hairy roots and rooty teratomas were able to produce nicotine, nornicotine, and anatabine; these alkaloids were not produced by the shooty teratomas, although the latter have the ability to accumulate the alkaloids and to demethylate nicotine to nornicotine (Saito *et al.,* 1989).

Whether genetic engineering of plants is a success depends on the efficiency with which a particular species or cultivar can be engineered and on the identification and isolation of genes conferring traits of economic interest.

2.7. Plant Propagation by Tissue Culture Techniques

Vegetative propagation of plants has been practised and the methods employed improved for centuries; tissue culture techniques have considerably expanded both its scope and potential. The tissue culture technique used is called *micropropagation* and it has proved to be a powerful tool for extending regenerative processes to a wider range of cells and tissues in diverse species. Micropropagation involves the *in vitro*, sterile multiplication of plants on a precisely defined growth medium incorporating specific growth regulators known to elicit specific growth responses in plant tissues. The growth regulators can be used to stimulate either normal shoot growth or the development of precocious axillary branching or the rooting of the small shoots developed. Then by moving plant tissue from one medium to another, it can be grown, multiplied, and rooted prior to its transfer to the greenhouse or field. A particular example is the rapid shoot proliferation that has been established by adventitious shoot formation on internodal segments of *Cephaelis ipecacuanha* using media containing benzyladenine and kinetin and a 16-hr

Table II
Examples of Pharmaceutically Important Alkaloids Extracted from Plants

Natural product	Pharmacological activity
Reserpine	Central nervous system
Caffeine	Autonomic nervous system
Physostigmine, pilocarpine	Cholinergic
Atropine, hyoscyamine, scopolamine	Cholinergic blocking
Ephedrine	Adrenergic
Nicotine	Ganglion blocking
Vincristine, vinblastine, harringtonine, camptothecin	Chemotherapeutics, anticancer
Emetine	Antiamoebic
Quinine	Antimalarial
Berberine	Antibacterial
Papaverine, theophylline, vincamine	Vasodilator
Reserpine, rescinamine	Hypotensive
Quincine, ajmalicine	Antiarrhythmic
Morphine, codeine	Analgesics
Glaucine, noscapine	Antitussive
Colchicine	Anti-inflammatory
Tubocurarine	Muscle relaxant
Cocaine	Local anesthetic

light regime. These shoots were relatively easily rooted on hormone-free Woody Plant or Gamborg B5 solid media. One-year-old regenerated plants cultivated in the greenhouse demonstrated the same alkaloid content (roots 0.82% dw emetine, and 2.16% dw cephaeline) as the parental plant (Yoshimatsu and Shimomura, 1991).

Micropropagation can lead to very rapid multiplication of elite individuals with rates of from 2 to 12 times over a 3- to 4-week period. This allows the introduction of new varieties much more rapidly than is otherwise possible as well as the production of disease free material.

The production of stable regenerating callus (2*n* meristematic cells) and, with many species, the development of mass somatic embryogenesis within the callus is now possible. These embryos have no food reserves as do those in seeds, however, and therefore this rather special problem has to be overcome before successful plant regeneration can be accomplished.

Micropropagation is particularly useful for production of high-yielding strains of a plant on a commercial scale in a short time. Plants with improved characteristics can be identified and cloned and plantation planning made easier. This technique has been put to particularly good use for the production of many young plants of *Datura sanguinea*, a commercial source of scopolamine, in the Ecuadorean Andes where the growing season is restricted to 90 days (October to December) (Evans, 1989).

2.8. Cell Cultures for the Production of Alkaloids

Plant cell cultures have been studied for their potential as an alternative to the whole plant for producing commercially important secondary metabolites. Investigations have centered on plants that produce the more pharmacologically important alkaloids currently still in use as drugs today (Table II). Despite initial optimism, cell cultures were found to produce inferior yields compared with the whole plant. In a few instances, however, the reverse situation occurs; for example, ajmalicine and serpentine from *C. roseus* and berberine from *Coptis japonica*.

Although few high-yielding cultures have been developed, more than 200 alkaloids, including several novel compounds, have been identified in plant cell cultures. These are tabulated and documented in a comprehensive treatise by Hay *et al.* (1988). The major interest has focused on those alkaloids that contain tropine, indole and isoquinoline moieties as well as medicinally topical alkaloids like camptothecin and taxol, both of which are proving useful in cancer therapy. In the following sections, alkaloids will be classified biogenetically according to their origins.

2.8.1. TRYPTOPHAN-DERIVED ALKALOIDS

This group of alkaloids includes the antimalarial quinine from the root cultures of *Cinchona ledgeriana,* and the anticancer compound camptothecin, from *Camptotheca accuminata* (see also Chapter 18). More than 100 tryptophan-derived alkaloids, commonly known as indole alkaloids, have been identified from plant cell cultures. More than 90% of them are derived biosynthetically from both tryptophan and *seco*-loganin. Others, the non-iridoid types, e.g., canthin-6-ones and β-carbolines, lack the *seco*-loganin component. Representatives of most of the main structural types have been isolated and examples are given in Table III. This large number of alkaloids has resulted from studies of relatively few species. Some of the most investigated genera are *Alstonia, Camptotheca, Catharanthus, Cinchona, Ochrosia, Picralima, Rauwolfia, Rhazya, Stemmadenia, Tabernaemontana,* and *Voacanga* which, apart from *Cinchona* (Rubiaceae) and *Camptotheca* (Nyssaceae) belong to the Apocynaceae. A more detailed account is found in Hay *et al.* (1988). The Apocynaceae contain many of the indole alkaloids with useful pharmacological activities; for example, the antileukemic indole alkaloids vincristine and vinblastine occur in small amounts in *Catharanthus roseus* (see Chapter 18). Despite considerable scientific effort, these alkaloids have been found, in *C. roseus* cell cultures, in only small amounts (Miura *et al.,* 1987, 1988).

Cell-free extracts from *C. roseus* suspension cultures have been found to convert anhydrovinblastine to vinblastine; however, the conversion was only about 12% (Endo *et al.,* 1987). Similarly, cell free extracts will convert vindoline and catharanthine to 3′,4′-anhydrovinblastine (Endo *et al.,* 1987); this latter reaction may also be accomplished using a horseradish peroxidase preparation (Goodbody *et al.,* 1988).

Cell cultures of *C. roseus* have proved to be an excellent source of monomeric indole alkaloids, although ajmalicine and serpentine are usually the major products.

Shoot cultures of *C. roseus* produce vindoline and catharanthine, the monomers of the anticancer agent vinblastine, as major constituents as well as strictosidine, ajmalicine, hoeshammericine, and vindolinine. The levels of alkaloid produced suggested that a shoot structure favored vindoline formation. It has also been possible to isolate low levels of dimeric alkaloids in shoot cultures and from callus cultures with differentiated roots (Miura *et al.,* 1987).

Despite the lack of success in producing significant amounts of vincristine and

Table III
Some Representative Examples of Tryptophan-Derived Alkaloids from Plant Cell Culture

Indoles		
Noniridoid		
1. β-Carbolines		
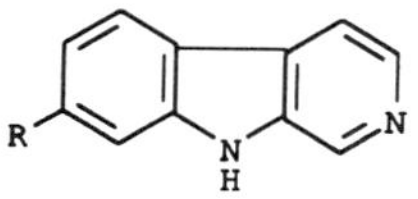	Harman R=H Harmaline R=OCH_3	*Peganum harmala*
2. Canthin-6-ones		
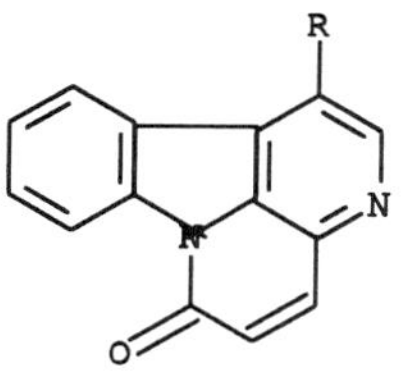	Canthin-6-one, R=H 1-MeO-canthin-6-one R=OCH_3 1-OH-canthin-6-one R=OH	*Brucea javanica* *Ailanthus altissima*
Iridoids		
1. Coryntheans		
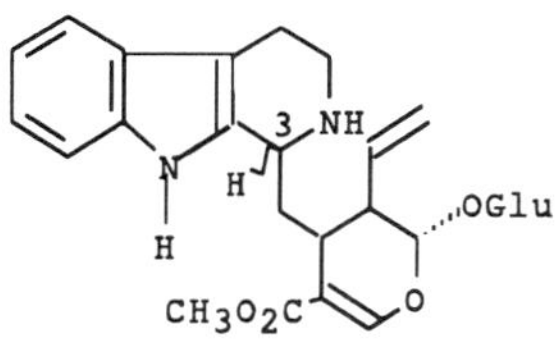	Strictosidine	*Rauwolfia serpentina* *Catharanthus roseus*
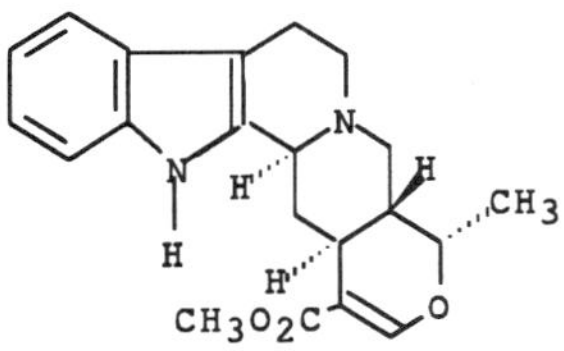	Ajmalicine	*R. serpentina* *C. roseus*
2. Aspidospermatans		
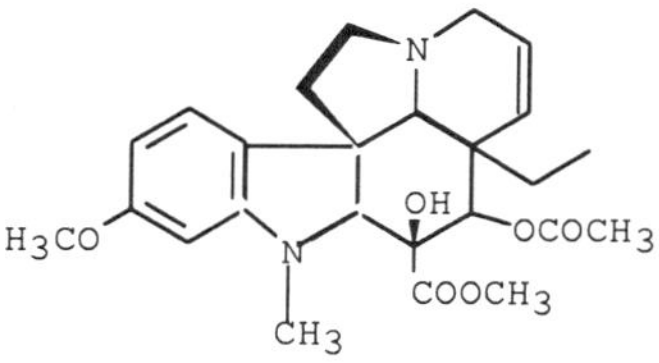	Vindoline	*C. roseus*
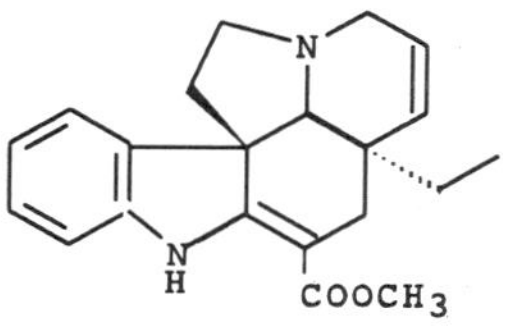	(−)-Tabersonine	*C. ovalis* *Stemmademia tormentosum* *Voacanga africana*
3. Ibogans		
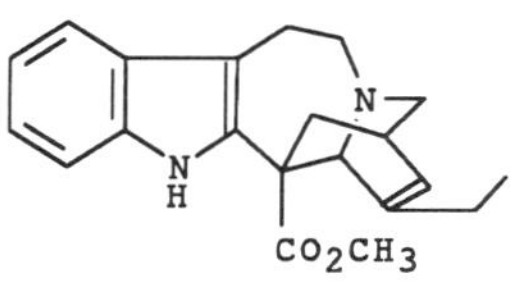	Catharanthine	*C. roseus* *Tabernaemontana divaricata*

(*continued*)

Table III
(Continued)

4. Strychnans 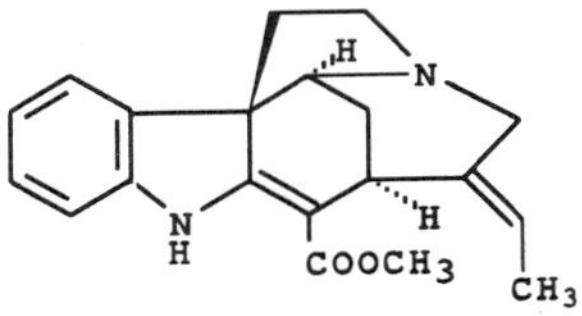	Akuammicine	*C. roseus* *C. ovalis* *Rhazya stricta*
5. Plumerans 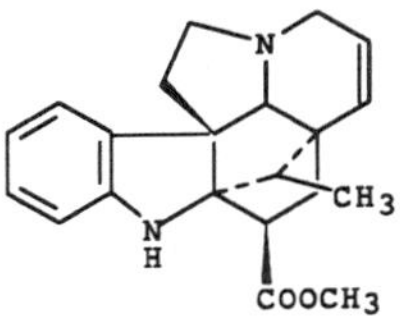	Vindoline	*C. roseus* *C. ovalis*
6. Eburnans 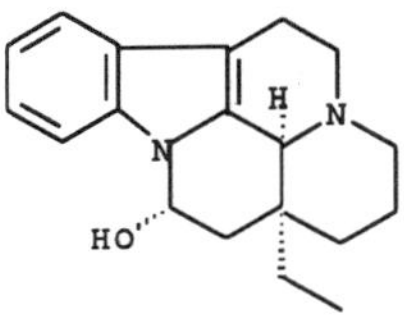	(+)-Eburnamine	*R. stricta*
7. Vallesiachotamans 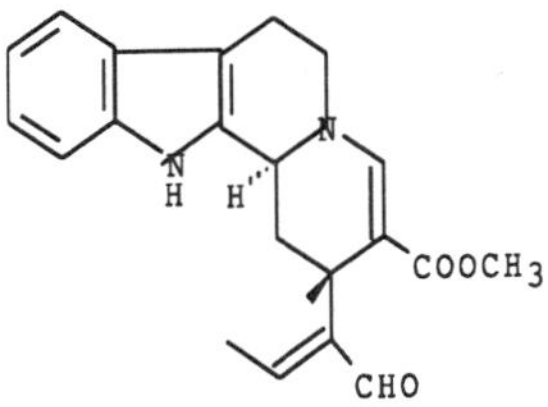	Vallesiachotamin	*C. roseus* *R. stricta*
8. Apparicines 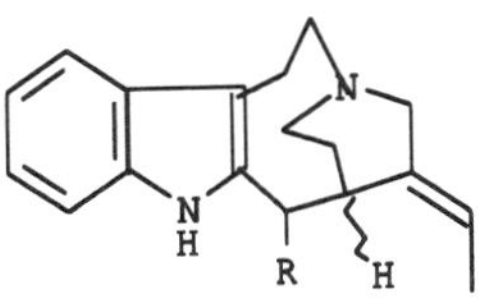	Apparicine R=CH_2	*C. roseus* *Picralima nitida*
Indoleisoquinolines 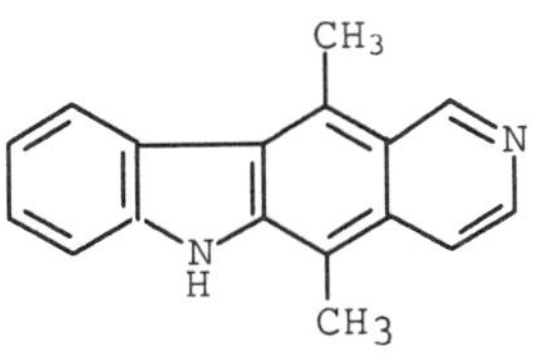	Ellipticine	*Ochrosia elliptica*

Table III
(Continued)

Dimeric indoles		
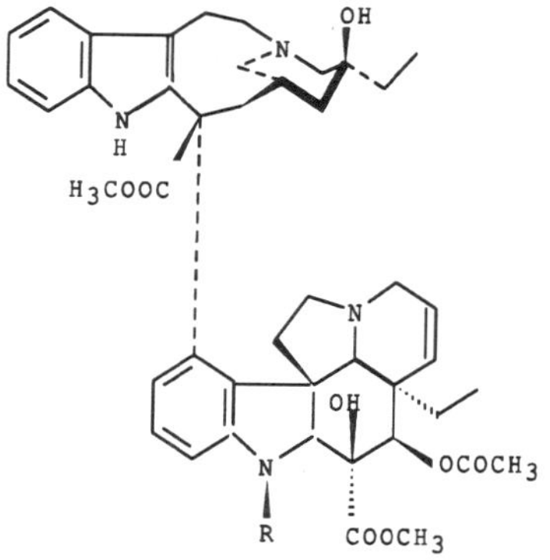	Vinscristine R=CHO Vinblastine $R=CH_3$	*C. roseus*
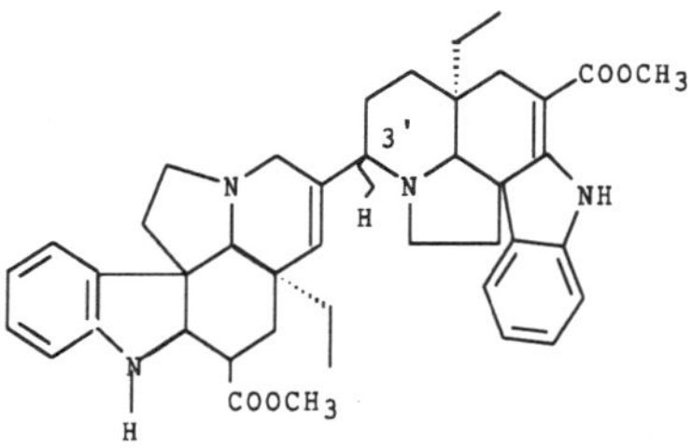	Voafrine A C-3′H α-configuration Voafrine B C-3′H β-configuration	*Voacanga africana*
Quinolines		
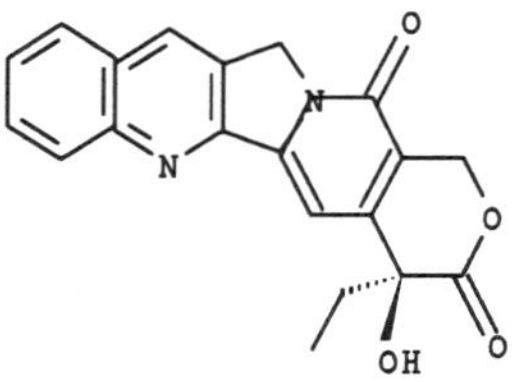	Camptothecin	*Camptotheca accuminata*
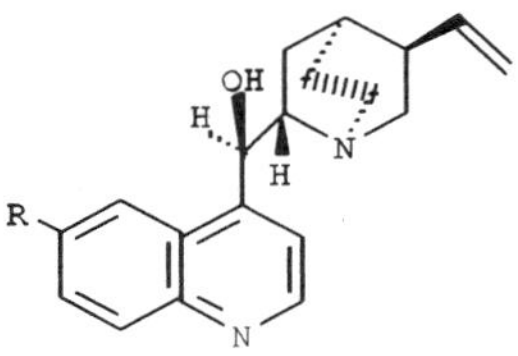	Cinchonidine R=H Quinine $R=OCH_3$	*Cinchona ledgeriana* *C. pubescens*

vinblastine from cell cultures, other dimeric indoles have been found: isoleurosine from *C. roseus* and voafrin A and B from the related *Voacanga africana*.

Studies with indole alkaloids demonstrate the potential of plant cell cultures for secondary metabolite production. Compounds not yet detected in the whole plant and, in some instances, compounds not previously described as natural products have been isolated from cell cultures. For example, novel monomeric indole alkaloids have been reported from a number of species, including 21-hydroxycyclolochnerin from *C. roseus* and epchro-

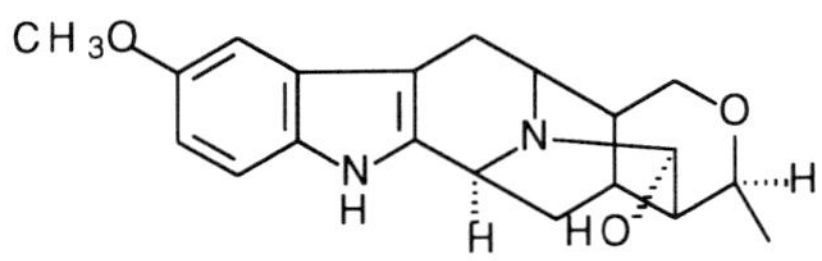

21-hydroxycyclolochnerin

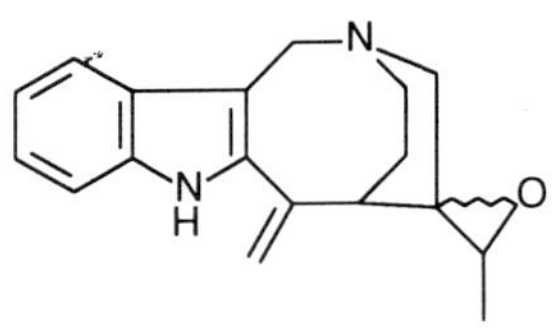

Epchrosine

Figure 3. Novel indole alkaloids from cell suspension cultures.

sine from *Ochrosia elliptica* (Fig. 3). The use of opiate binding studies in conjunction with the isolation of active compounds has led to the identification of two alkaloids with opioid activity from *Picralima nitida* cell cultures. One, pericine (Han *et al.*, 1988), proved to be a novel alkaloid . This type of approach has use for the discovery of potentially useful new drugs or prospective templates for new drugs (see also Chapter 12).

Rauwolfia serpentina cell cultures have been a rich source of alkaloids, including ajmalicine, reserpine, serpentine, vomilenine, and yohimbine and the glycoalkaloid raucaffricine, which proved to be the major alkaloid in *R. serpentina* cell suspension cultures. This glycoalkaloid was previously reported from plants of *R. caffra* but not from *R. serpentina*. The major interest lies in the fact that in these cultures all of the strictosidine is channeled into the production of the glycoalkaloid rather than into reserpine. Shoot cultures of *R. serpentina* contained yohimbine, ajmalicine, ajmalidine, and other dihydroindoles as major components . The alkaloid content of these cultures was greater than in the parent shoots; however, in the plant, most of the alkaloids are found in the roots and the aerial parts have a simplified alkaloid profile.

Cell cultures of *Rhazya stricta* are a rich source of indole alkaloids with five of the major structural types of indoles, namely, corynanthean, aspidospermatan, strychnan, eburnan, and vallesiachotaman represented by the alkaloids isolated (Table III). With the exception of akuammicine, all are typical of the genus *Rhazya*. Examples of cultures producing alkaloids that are not typical of the parent plant include 3-oxoisovoacangine from *Tabernaemontana elegans* cultures and reserpiline from *Ochrosia elliptica* cultures. Although reserpiline has not been found in differentiated plants of *O. elliptica*, it has been detected as a minor alkaloid of *O. balansae*.

Transformed root cultures of *Amsonia elliptica* have also been found to produce the same indole alkaloids, 17α-*O*-methylyohimbine, pleiocarpine, and vallesiachotamine as the normal roots in culture. Although yields were considerably enhanced by the inclusion of naphthyleneacetic acid (NAA) in the medium for both normal and hairy root cultures, levels of alkaloid in the normal root culture were always in excess of the levels found in the hairy roots (Sauerwein *et al.,* 1991).

Indole alkaloids obtained from cell cultures of *Cinchona* species (Rubiaceae) and *Camptotheca* (Nyssaceae) may also be considered as quinolines (Table III). Alkaloids from *Cinchona* species include cinchonamine and 10-methoxycinchonamine from *C. pubescens* and quinine, quinamine, and quinidine from *C. pubescens* and *C. ledgeriana* (Hay *et al.,* 1988) but the levels were low in comparison to the bark of the whole plant and in some instances the nonmethoxylated analogues, cinchonidine and cinchonine, have proved to be the major alkaloids in the cultures.

In some cultures there is degradation of the alkaloids, for example, in some clones of *Catharanthus roseus,* ajmalicine is rapidly produced and degraded at the beginning of the growth phase while the degradation rate was also high in the stationary phase (Dos Santos *et al.,* 1994). In *Tabernaemontana divaricata* coronaridine was converted to heyneanine, conopharyngine to 19-hydroxyconopharyngine, vobasine to dregamine and tabernaemontanine, and tabersonine to voaphylline and an unknown chromophore of the tabersonine type (Dagnino *et al.,* 1994).

Multiple shoot cultures established from seedlings of *Cinchona ledgeriana* or *C. succirubra* had growth characteristics that were significantly better than the root cultures (Hay *et al.,* 1988). Shoot cultures produced comparatively high yields of alkaloids, e.g., 10% of the bark alkaloids of an 8- to 20-year-old tree and 50% of the bark alkaloid content of a 1-year-old tree.

Rauwolfia plants regenerated from root and stem callus when compared with the parental stock had a similar alkaloid content, although it appeared to be slightly higher in the roots of regenerated plants. Ajmalicine, serpentine, and reserpine are lower than in the roots of the parental stock. The glucoalkaloid raucaffricine was identified as a constituent of all samples, thus providing the first evidence for its occurrence in the roots of *R. serpentina* (Ruyter *et al.,* 1991).

C. ledgeriana cultures transformed with *Agrobacterium tumefaciens* (Payne *et al.,* 1987b) were also capable of producing these quinoline alkaloids without the addition of phytohormones; however, the type and amount of alkaloid did not seem to differ from those reported for the untransformed lines of *C. ledgeriana.* Quinine and quinidine are of commercial importance: quinine is used as an antimalarial drug and as a bitter flavoring in the soft drink industry, while quinidine is used to treat cardiac arrhythmias. *R. serpentina* and *R. vomitoria* do not produce the alkaloids typical of the plant bark but synthesize the alkaloids vinorine and perakine as well as two alkaloids found in cell suspension cultures, vomilenine and ajmaline (Falkenhagen *et al.,* 1993).

Camptothecin (Table III) produced by cell cultures of *Camptotheca accuminata* is another example of a tryptophan derived quinoline alkaloid. This alkaloid has marked antitumor activity, and is currently being reevaluated as an anticancer drug (see also Chapter 18).

2.8.2. TYROSINE-DERIVED ALKALOIDS

The production of isoquinoline alkaloids by plant cell cultures has been the subject of a number of comprehensive reviews (Rueffer, 1985; Roberts, 1988a; Hay *et al.,* 1988;

Zenk, 1990). Isoquinoline alkaloids of the following structural types have been identified from plant cell cultures: benzylisoquinolines, bisbenzylisoquinolines, proaporphines, aporphines, protoberberines, protopines, benzophenanthridines, phthalidoisoquinolines, morphinans, aristolactams, and *Cephalotaxus* alkaloids. To date a relatively small number of genera and species within the Berberidaceae, Cephalotaxaceae, Fumariaceae, Menispermaceae, Papaveraceae, and Ranunculaceae have been studied. Some typical examples are given in Table IV.

Some of the major yields in the production of secondary metabolites from cell cultures have been isoquinoline alkaloids. Yields of 2.7 g $liter^{-1}$ of jatrorrhizine from *Berberis stolonifera* cultures and more than 3 g $liter^{-1}$ from *B. wilsoniae* grown in 20-liter airlift bioreactors have been obtained. High-berberine-producing cultures have been derived from *Coptis japonica* with selected lines reaching 13.2% berberine on a dry weight basis (1.39 g $liter^{-1}$).

The protoberberine group of alkaloids shows the greatest variety in its aromatic substitution pattern. Three species that produce an extremely wide range of protoberberine alkaloids are *Coptis japonica*, *Nandina domestica*, and *Thalictrum minus*. In addition, cultures of these three species also produce the aporphine alkaloid magnoflorine. A number of species from the genera *Chasmanthera, Chelidonium, Corydalis, Dicentra, Eschscholtzia* and *Tinospora* have been shown to produce up to three different structural types of isoquinolines in cell cultures while *Stephania, Papaver*, and *Fumaria* species produce an even wider range. *Stephania* cell cultures, for example, produce aporphine, bisbenzylisoquinoline, and protoberberine alkaloids and are the only cultures reported to produce aristolactam-type alkaloids (cepharanones I and II) (Table IV).

Papaver species in cell culture also produce an extensive range of isoquinoline alkaloids including aporphine, protoberberine, protopine, benzophenanthridine, phthalide-isoquinoline, and morphinan types. Many of these alkaloids are not typical of the parent plants. Despite the demonstrated ability of *Papaver* cultures to produce significant amounts of the isoquinoline alkaloids, codeine and morphine—the most important from a pharmaceutical viewpoint—have proved extremely difficult to produce at a level approaching that found in the plant (Siah and Doran, 1991).

Although significant metabolic blocks in the pathway to morphine from (*S*)-reticuline must occur, cells of *P. somniferum* immobilized using calcium alginate or reticulate polyurethane are capable of converting codeinone to codeine with rates of 70.4 and 86% quoted. Most of the codeine (87.5%) was excreted from the cells. These bioconversions were significantly better than conversions with cell suspensions which had a bioconversion ratio of 57–60% (Corchete and Yeoman, 1989). Other groups have established cell cultures that will convert thebaine to neopine but these cultures would not metabolize neopine, codeine, papaverine, or DL-norlaudanosoline and no trace of morphine was found. Bioconversion of *Papaver* alkaloids has not been confined to *P. somniferum* as papaverine (a benzylisoquinoline) may be converted to a series of demethylated products, i.e., 6-monodemethylpapaverine and 4′-monodemethylpapaverine, by the non-alkaloid producing plant *Silene alba*. Yields, however, were of the order of 11%. Interestingly, these two demethyl products are the two major compounds formed by demethylation in man and similar reactions occur in guinea pig, rabbit, and microorganisms.

The greatest range of isoquinoline-type alkaloids reported from a cell suspension culture has been obtained from *Fumaria capreolata*. Ten isoquinoline alkaloids representing six structural classes have been found in this species, i.e., benzylisoquinoline (methyl-

Table IV
Representative Tyrosine-Derived Alkaloids from Plant Cell Cultures

Isoquinolines		
1. Benzylisoquinolines		
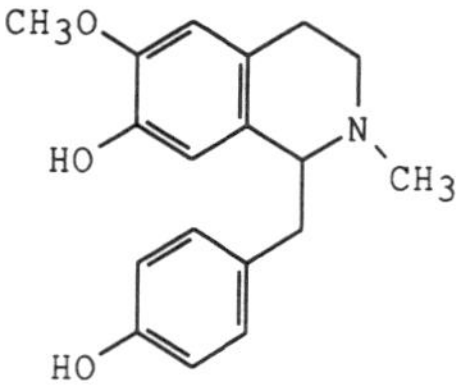	*N*-Methylcoclaurine	*Fumaria capreolata*
2. Bisbenzylisoquinolines		
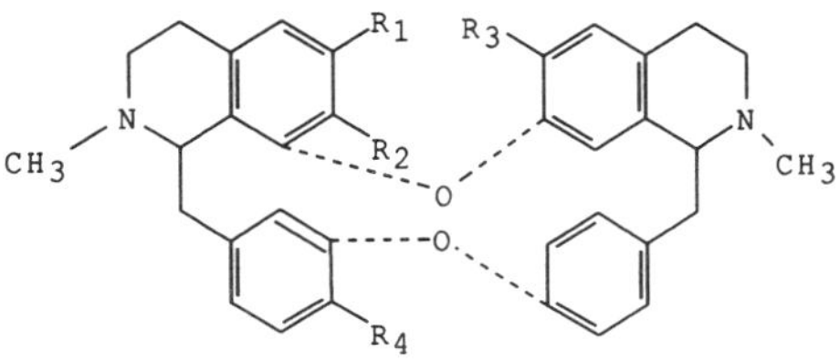	Berbamine $R_1=R_2=R_3=R_4=H$ Biscoclaurine $R_1=R_3=OCH_3$ $R_2=R_4=OH$	*Stephania cepharantha*
3. Proaporphines		
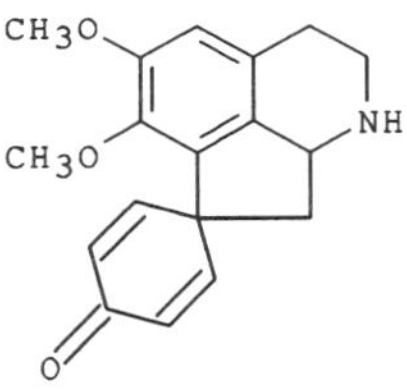	Stephanine	*Tinospora caffea* *T. cordifolia*
4. Aporphines		
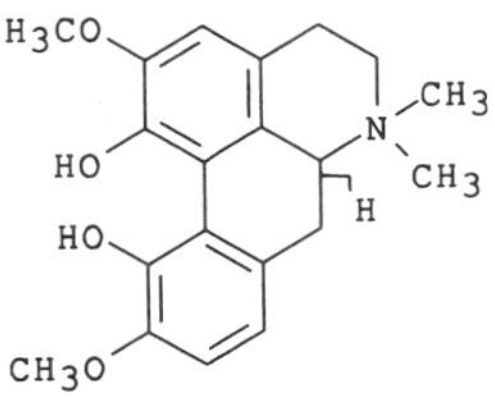	Magnoflorine	*Coptis japonica* *Corydalis incisa* *Eschscholtzia californica* *Fumaria capreolata* *Papaver bracteatum* *P. somniferum* *Thalictrum minus*
5. Berberines		
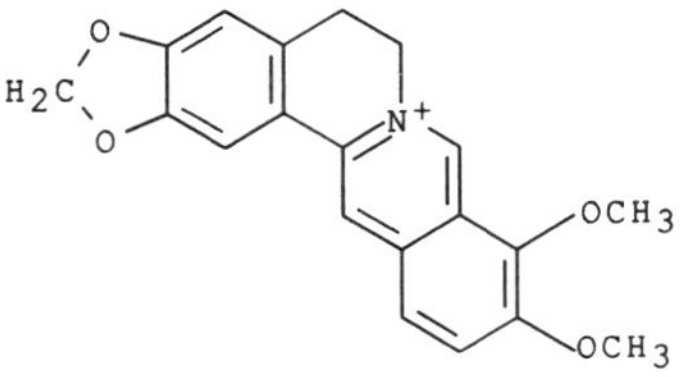	Berberine	*Berberis wilsoniae* *Coptis japonica* *Thalictrum minus*

(*continued*)

Table IV
(Continued)

6. Protopines		
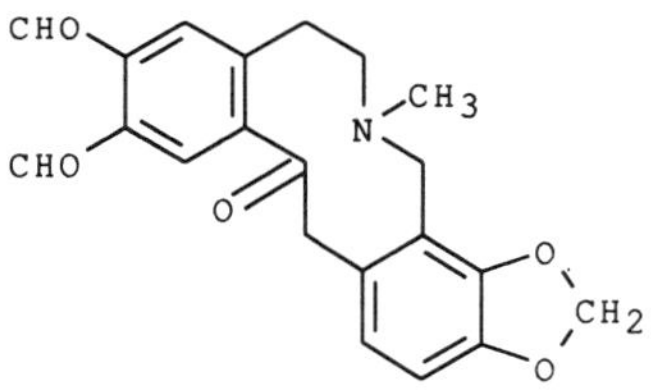	Cryptopine	*Papaver somniferum*
7. Benzophenanthridines		
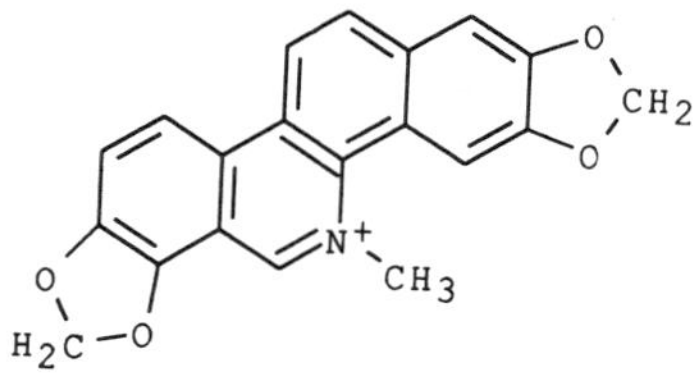	Sanguinarine	*Chelidonium majus* *Corydalis ophiocarpa* *Eschscholtzia californica* *Fumaria cordata* *Macleaya cordatga*
8. Phthalideisoquinolines		
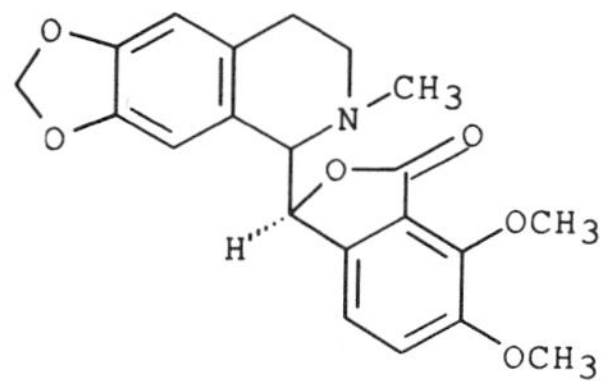	Noscapine	*Papaver rhoas* *P. somniferum*
9. Morphinans		
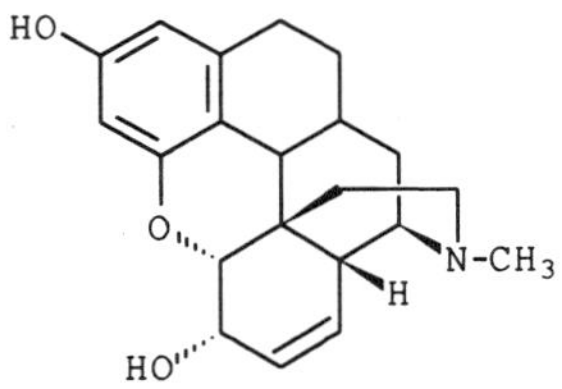	Morphine	*Papaver somniferum* *P. setigerum*
10. Aristolactams		
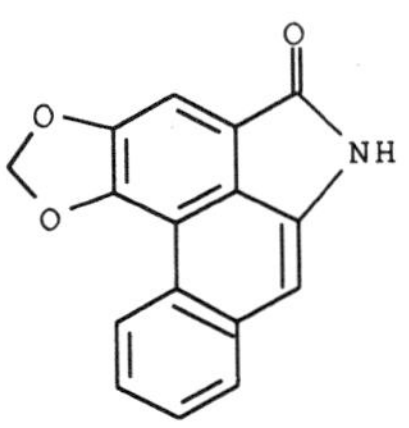	Cepharanone-A	*Stephania cepharantha*

Table IV
(Continued)

11. Ipecac alkaloids

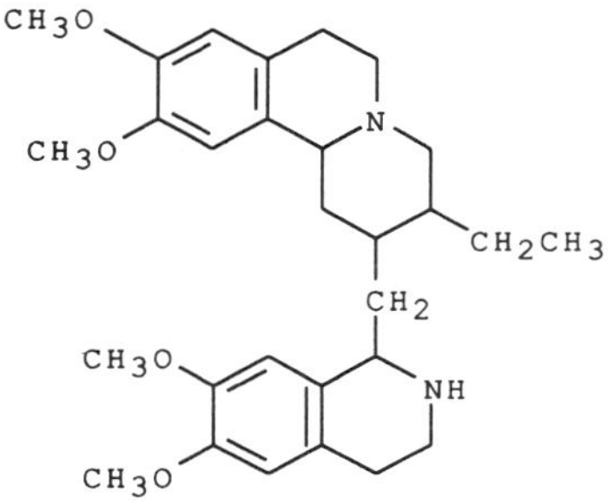

Emetine — *Cephaelis ipecacuanha*

12. Cephalataxus alkaloids

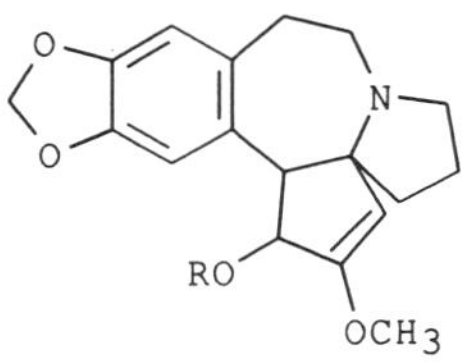

Cephalotaxine R=H — *Cephalotaxus harringtonia*
Harringtonine R=

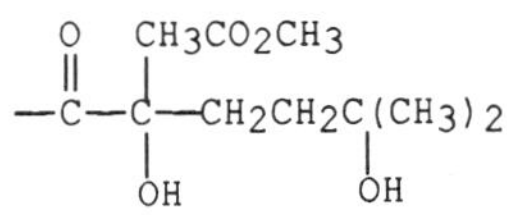

coclaurine, reticuline), aporphine (isoboldine, magnoflorine), protoberberine (coptisine, dehydrocheilanthifoline, scoulerine), protopine (protopine), benzophenanthridine (sanguinarine), and morphinan (pallidine) (Table IV).

In addition to the extensive range of known isoquinoline alkaloids that have been identified from plant cell cultures, several novel compounds have also been found. The novel oxoaporphine, norcepharadione, has been isolated from callus tissues of *Stephania cepharantha* and norsanguinarine was located as a natural product for the first time from callus cultures of *Papaver somniferum*. Fungal elicitors have been found to stimulate sanguinarine production in *P. somniferum* where cell selection has produced cell lines high yielding in this alkaloid (currently being used for removal of dental plaque) (Eilert and Constabel, 1985; Eilert *et al.*, 1985). Plant cell cultures have also been shown to produce isoquinoline alkaloids that are not typical of the parent plant and in some instances new to a particular genus. Many of the alkaloids from *P. somniferum* cell cultures, e.g., stylopine and protopine, have not been identified from the parent plant. Aromoline, a bisbenzylisoquinoline found in *Stephania* cultures, had not been previously isolated from the genus.

Adventitious root cultures of *Cephaelis ipecacuanha* established from leaf segments of shoot cultures *in vitro* have produced emetine and cephaeline at levels that were equivalent to those found in the roots of 1-year-old greenhouse grown plants and thus may be considered to have potential commercial usefulness (Yoshimatsu and Shimomura, 1991).

Table V
Examples of Lysine-Derived Quinolizidine Alkaloids from Plant Cell Cultures

Tricyclics		
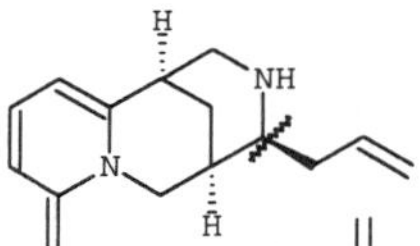	11-Allylcytisine	*Cytisus canariensis*
	(−)-Tinctorine	
Tetracyclics		
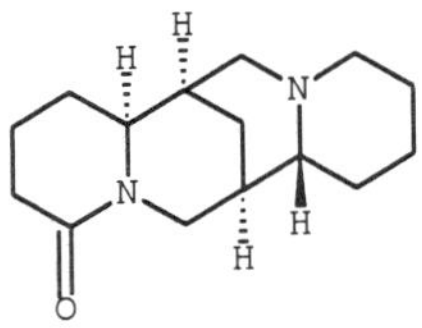	(−)-Lupanine	*Lupinus luteus* *L. hartwegii* *Cytisus scoparius* *Genista pilosa* *Baptisia australis* *Laburnum alpinum* *Sophora japonica*

2.9. Lysine-Derived Quinolizidine Alkaloids

Plants producing quinolizidine alkaloids occur in a number of families but they are particularly abundant in the Leguminosae (Fabaceae). Ten species of the Leguminosae representing six genera, i.e., *Baptisia, Cytisus, Genista, Laburnum, Lupinus*, and *Sophora*, have been intensively investigated in cell culture. In general, the alkaloids produced by cell cultures differ both quantitatively and qualitatively from the parent plant. Overall the level of alkaloid produced by the cultures was several orders of magnitude lower than that of the parent. The alkaloid pattern of the cell suspension cultures was much simpler than that of the whole plant; a marked difference was absence from the cell culture of the α-pyridone alkaloids such as cytisine and *N*-methylcytisine, which predominate in the plant. Twelve quinolizidine alkaloids were, however, identified in cell cultures of the species investigated. Lupanine proved to be the major alkaloid in cultures of all ten species and was accompanied by sparteine, tetrahydrorhombifoline, 17-oxosparteine, 13-hydroxylupanine, 4-hydroxylupanine, 17-oxolupanine, and 13-hydroxylupanine esters as minor alkaloids in some species (Table V) (Wink *et al.*, 1983).

2.10. Ornithine-Derived Alkaloids

Tropane alkaloids derived from ornithine, such as atropine (± hyoscyamine), (−)-hyoscyamine, and scopolamine (hyoscine) (Table VI), are medicinally important secondary plant products because of their action on the autonomic nervous system. These alkaloids are synthesized mainly in the roots of some Solanaceous plants, and studies with *in vitro* cultures indicate that the production of significant levels of tropane alkaloids depends on the degree of organogenesis of the culture. Tropane alkaloids have been

Table VI
Examples of Ornithine-Derived Tropane Alkaloids from Plant Cell Cultures

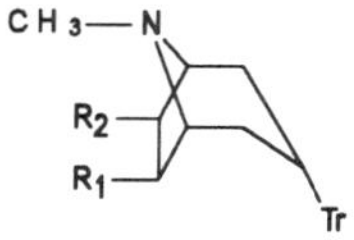	Hyoscyamine $R_1=R_2=H$	*Datura innoxia*
		Duboisia hopwoodii
		D. leichhardtii
	6β-Hyoscyamine $R_1=OH$, $R_2=H$	*D. myoporoides*
		Hyoscyamus niger
		H. muticus
Tr=tropic acid		*H. albus*
	Scopolamine	*Scopolia paviflora*
	R_1, R_2= 	*S. acutangula*

reported from undifferentiated callus cultures of various genera of the Solanaceae, although the majority of reports involve highly differentiated cultures, in particular root cultures (Hay *et al.,* 1988).

Recent studies with *Duboisia* have shown that, in general, callus cultures produce only low levels of the desired tropanes, in contrast to differentiated roots obtained from callus which produce a range of alkaloids typical of the parent plant.

Yamada and co-workers have initiated callus cultures of *Atropa belladona, Datura stramonium*, and *Hyoscyamus niger*, but only *H. niger* was found to contain significant levels of alkaloids. The main alkaloid, hyoscyamine, could be increased to a level comparable with the whole plant by means of cell selection and optimization of cultural conditions. The scopolamine content of the callus, however, remained low, being about one-tenth that of the parent plant. This content was increased significantly by inducing root formation in the cultures, indicating that differentiation is required for scopolamine biosynthesis in *H. niger.*

Where the levels of alkaloid production are increased relative to the parent plant, this may reflect the different state of development of the transformed root systems. In normal root cultures, tropane alkaloid content is known to vary markedly with maturity. Normally, both root cultures and hairy root cultures do not develop secondary xylem, as the transfer of growing tips to fresh medium occurs before that stage of growth.

This has been substantiated by studies using immunohistochemical techniques with monoclonal antibodies and immunogold/silver enhancement to locate the enzyme hyoscyamine-6β-hydroxylase, responsible for scopolamine production in the pericycle of young roots. During secondary thickening and production of new root primordia the pericycle cells are lost or disrupted and enzyme activity is lost (Hashimoto *et al.,* 1991). Shoot cultures established from *Duboisia* species failed to produce tropane alkaloids, and these results support the association of tropane alkaloid production with root organogenesis. Studies with *Scopolia paviflora* cultures have also shown a close relationship between organization of the tissue and alkaloid formation. On the whole, undifferentiated cell cultures have low alkaloid content while root cultures contain the same alkaloids as the parent plant with some yields in excess of 1%. Hairy root cultures may produce hyoscyamine and scopolamine at levels in excess of that of the parent plant in *Hy-*

oscyamus albus and *Datura candida* hybrid, respectively (Sauerwein and Shimomura, 1991; Christen *et al.,* 1990).

Although cell cultures derived from plants containing tropane alkaloids frequently do not produce these alkaloids in large amounts, some parts of the pathway to the tropanes are operable and allow the efficient conversion of some intermediates. These studies have helped to elucidate factors that regulate production of tropanes in plants and cell cultures. Various species of *Datura,* i.e., *D. metel, D. ferox,* and *D. stramonium,* esterify exogenously supplied tropine with acetic acid to produce acetyltropine rather than carry out esterification with tropic acid (Alfermann and Reinhard, 1988). Indeed, hairy root cultures developed from hybrids of *Datura candida* with *D. aureus* and *D. candida* with a *D. candida* cultivar produce a multiplicity of esters alternative to those of tropic acid (Christen *et al.,* 1990; Robins *et al.,* 1990). It has also been found with some suspension cultures, notably those of *D. innoxia,* that (*RS*)-tropic acid was converted into its glucose esters (*RS*)-3-hydroxy-2-phenylpropionyl-β-D-glucopyranoside and (*RS*)-2-*O*-(3-hydroxy-2-phenylpropionyl)-D-glucose. These bioconversions of tropic acid are observed with other species; for example, a culture of *Eucalyptus persianiana* converts (*RS*)-tropic acid into its glucoside (*RS*)-3-*O*-β-D-glucopyranosyl-2-phenylpropionic acid in addition to its glucose esters. Suspension cultures of *Coffea arabica* convert (*RS*)-tropic acid into its glucose, sucrose, and isotrehalose esters and a small amount of its glucoside; (*RS*)-2-(4-hydroxyphenyl)propionic acid into its glucose and sucrose esters and a small amount of its glucoside and (*RS*)-ethyl-2-(4-hydroxyphenyl)propionate into its gentiobioside. The formation of sucrose esters and the linkage of the aglycone to the C-6 position of glucose are characteristic of the biotransformation of carboxylic acids by suspension cultures of *C. arabica.* The suspension cultures of *C. arabica* selectively converted (*R*)-tropic acid into its isotrehalose ester on administration of (*RS*)-tropic acid (Ushiyama and Furuya, 1989).

Tropic acid may also be observed to form hyoscyamine but not scopolamine when NAA rather than 2,4-D is used for cell cultures of *D. innoxia.* The formation of both acetyltropine and hyoscyamine have been reported for *D. stramonium* (Hay *et al.,* 1988). At a feeding concentration of 100 mg liter^{-1}, ± hyoscyamine was converted to 6β-hydroxyhyoscyamine (10 mg liter^{-1}) in 4 weeks and 4 mg liter^{-1} of scopolamine was formed by week 5 in cell cultures of *Anisodus tanguticus.* Immobilized cells were found to behave in a similar manner (Cheng *et al.,* 1987, 1989).

In organ culture, using roots of *Hyoscyamus niger* and *Duboisia* hybrid, it has been found that 10 to 20% of hyoscyamine added to the medium was converted to scopolamine. A similar conversion of hyoscyamine to scopolamine may be obtained with transformed roots of a *D. candida* × *D. aurea* hybrid (Robins *et al.,* 1990).

2.11. Nicotinic Acid-Derived and Purine Alkaloids

Nicotinic acid derived alkaloids are produced by cell cultures of *Duboisia* and *Nicotiana* species (Table VII). Cultured roots of *Duboisia* species yield nicotine and nornicotine and traces of anabasine occur in *D. myoporoides* callus. *Nicotiana tabacum* callus yields 3.75% total alkaloid calculated on a cell dry weight basis with nornicotine as the major alkaloid. Hairy root cultures of *N. rustica* developed by transformation with *Agrobacterium rhizogenes* release nicotine into the medium to a maximum concentration

Table VII
Some Examples of (a) Nicotine-Derived Pyridines and (b) Purine Alkaloids Found in Plant Cell Cultures

a. Pyridines			
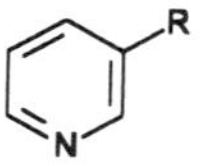	Nicotine: R=	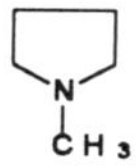	*Duboisia hopwoodii* *D. myoporoides* *Nicotiana tabacum* *N. rustica*
	Anabasine: R=	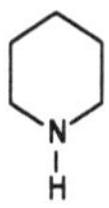	
b. Purines			
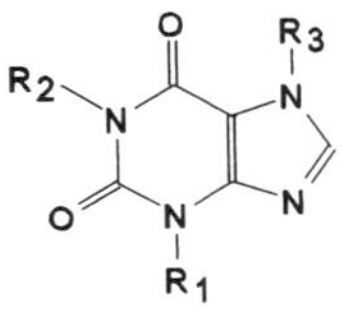	Caffeine $R_1=R_2=R_3=CH_3$ Theobromine $R_1=R_2=CH_3$; $R_3=H$		*Coffea arabica*

of 10 mg liter^{-1}. Suspension cultures of *N. tabacum* produce other tobacco alkaloids, including anatabine, anabasine, myosmine, anatelline, and nicotelline.

Cell suspension cultures of *Nicotiana plumbaginifolia* which do not accumulate tobacco alkaloids were found to maintain their ability to demethylate nicotine to nornicotime. The highest bioconversion yield was of the order of 53.2%. Light appeared to enhance this catalytic activity of the cells suggesting that this particular metabolic step may be bound to photodependent systems (Manceau *et al.,* 1989). Hairy roots, rooty teratomas, and shooty teratomas of *N. tabacum* also have the ability to convert nicotine to nornicotine but only the hairy roots and rooty teratomas produce nicotine and anabasine (Saito *et al.,* 1989).

The purine alkaloids caffeine and theobromine (Table VII) are produced by callus and suspension cultures of *Coffea arabica* and *Camellia sinensis*. Caffeine production of up to 128 mg liter^{-1} has been observed with some suspension cultures of *Coffea arabica*; however, it should be noted that these cultures produced large amounts of chlorogenic acid (480 mg liter^{-1}) and some correlation between acid and alkaloid in the cell vacuole was observed (Baumann and Rohrig, 1989). Immobilized cells also produced caffeine (80 mg liter^{-1}) with 50% in the medium (Furuya *et al.,* 1990).

2.12. Anthranilic Acid-Derived Alkaloids

Quinolone, and furoquinoline alkaloids (Table VIII) are typical secondary metabolites of the family Rutaceae. Two genera of the family have been studied in cell culture, viz., *Ruta* and *Choisya*. Callus and cell suspension cultures of *Ruta graveolens* have been shown to produce a diverse range of alkaloids, including furoquinolines, skimmianine,

Table VIII
Representative Examples of Anthranilic Acid-Derived Alkaloids from Plant Cell Cultures

Furoquinolines		
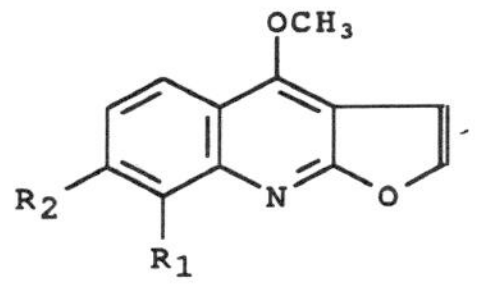	Skimmianine $R_1=R_2=OCH_3$	*Choisya ternata*
	6-Methoxydictamine $R_1=H$; $R_2=OCH_3$	*Ruta graveolens*
Acridones		
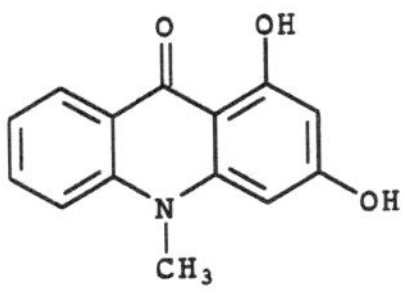	1,3-Dihydroxy-*N*-methylacridone	*R. graveolens*
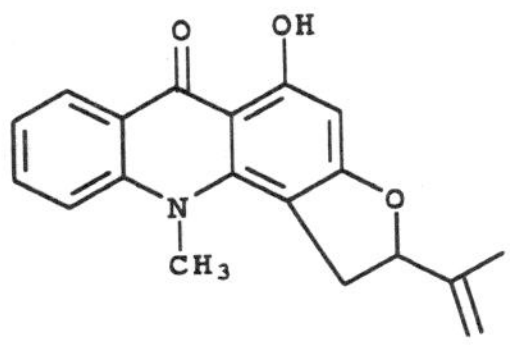	Rutacridone	*R. graveolens*

6-hydroxydictamine, acridones, 1,3-dihydroxy-*N*-methylacridine, rutacridone, and one quinolone, edulinine (Table VIII). Two new alkaloids termed furacridone, have been isolated for the first time from callus cultures of *R. graveolens*; they are 1-hydroxyrutacridone epoxide and rutagravin. These represent a new pentacyclic system of acridone alkaloids with a difuranoid moiety (Gröger and Baumert, 1994). Callus and suspension cultures of other *Ruta* species—*R. macrophylla, R. corsica, R. chalepensis*, and *R. bracteosa*—also produce rutacridone epoxide and hydroxyrutacridone epoxide. Two dihydrofuroquinoline alkaloids, platydesmine and balfourodine, have been identified from cultures of *Choisya ternata.*

2.13. Acetate-Derived Alkaloids

Eight steroidal alkaloids have been reported from cultures of ten *Solanum* species (Table IX) (Hay *et al.*, 1988). Total alkaloid concentrations of 0.1 mg g^{-1} dry weight of green suspension cultures of *S. dulcamara* have been obtained in which soladulcidine and solasodine are the major alkaloids. Factors that affect the production of solasodine in cultures of *S. laciniatum* and *S. nigrum* have been investigated. Tomatidine has been obtained from cultures of *Lycopersicon esculentum.*

Taxol (structure 59, page 41) is a diterpene pseudoalkaloid isolated from the Pacific yew (*Taxus brevifolia* Nutt.). In 1964 it was found that the bark of *T. brevifolia* is toxic to

Table IX
Acetate-Derived Alkaloids from Plant Cell Cultures

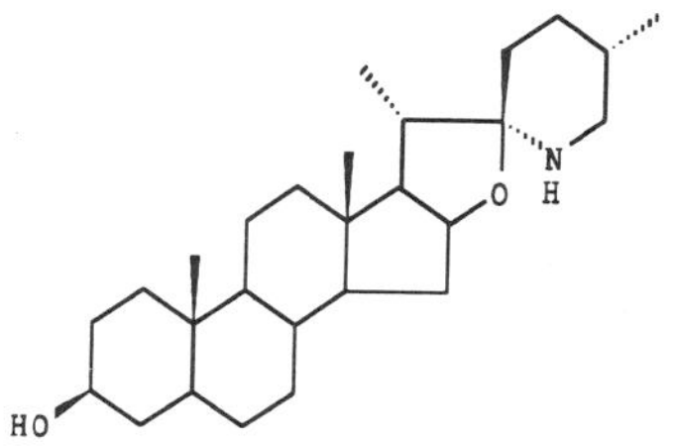	Solasodine	*Solanum dulcamara* *S. nigrum* *S. laciniatum*
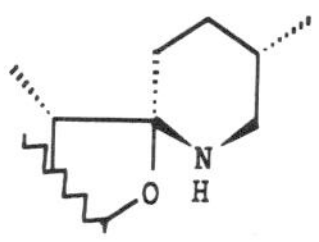	Tomatidine	*Lycopersicon esculentum*

murine leukemia cells. Taxol was obtained in the laboratory of Munro Wall in 1969 and the structure was fully elucidated by 1970. The concentration in the bark is very low (100 mg kg^{-1}). Yew bark is thin and yields an average 3 kg from a century old tree. The trees are slow growing and are found mainly in the forests of northwestern Canada (British Columbia), and the United States (southern Alaska, northern California, Washington, Oregon, Montana, and Idaho). Hence, alternative sources have been sought. Synthesis, partial synthesis, extraction from yew needles, and plant cell culture have all been investigated.

Plant cell cultures have been generated from needles, stems, and fruits. The most successful cultures have been produced from stem tissue but require the constant addition of substances to complex with the phenolics produced by these cultures and thus prevent browning. Taxol yields of 0.0014% were obtained from *T. cuspidata* cell cultures and this has to be compared with 0.035% extracted from the needles and 0.022% from the stems of the same plant (Fett-Netto *et al.*, 1992). Other related constituents have been isolated from *T. baccata* cell cultures which also promote tubulin assembly (Ma *et al.*, 1994).

Recently, genetic transformation of mature *Taxus* species has proved successful and this may ultimately lead to control of taxol production in cell cultures. However, to date partial synthesis using 10-deacetylbaccatin III from *Taxus* needles seems to represent the best way forward (Han *et al.*, 1994).

The foregoing sections have presented a brief account of the diversity of alkaloids that have been found in plant cell and organ cultures. Most of the original papers are referenced in Hay *et al.* (1988). Those subsequent to 1986 are given in the text.

3. INDUSTRIAL INTEREST

3.1. The Problem of Cost-Effectiveness

Many classes of compounds have been produced via plant cell cultures, including steroids, alkaloids, terpenoids, quinones, and enzymes, which has created interest in this technology as a possible alternative for the production of commercially useful constituents of plant origin. Unfortunately, factors that limit the success of plant cell culture techniques include slow growth of the plant cells and lack of efficient productivity. When chemicals of interest are produced, accumulation is usually low and the desired secondary metabolites are often retained within the cells. Commercial production is currently economically feasible only when the compounds produced are unique and not readily synthesized through modern chemical techniques. Because the development of this new technology is of necessity expensive, interest has been limited to a small number of candidates: those of high value that are required in relatively small amounts. This interest has centered around food colors, flavors and fragrances, pesticides, and pharmaceuticals. Of the pharmaceuticals, berberine, codeine, morphine, scopolamine, atropine, vinblastine, L-dopa, hyoscyamine, diosgenin, digoxin, quinine, vincristine, vinblastine, shikonin, ajmalicine, and serpentine have been the most investigated. Research on the production of alkaloids by plant cell cultures accounts for some 67% of all research on pharmaceuticals from cell cultures while pharmaceuticals in general account for 72.5% of the effort in plant cell culture phytochemical research.

The most valuable of the alkaloids used are the antileukemics vinblastine and vincristine at \$U.S. 5,000,000 kg^{-1} with a worldwide requirement on the order of 5 kg $year^{-1}$. These alkaloids only occur in small quantities in the plant *Catharanthus roseus*. The cheapest is probably quinine at \$100 kg^{-1} and an annual requirement of about 500,000 kg. If one assumes that 10% of the market share would be required at first, a typical volume would be 300 kg $year^{-1}$. Current costs of a product produced by plant tissue culture range from \$500 kg^{-1} to \$1500 kg^{-1} (Fowler and Scragg, 1988).

Successful biotechnological development of plant cells, leading to metabolite levels that compare favorably with microbial fermentations (antibiotics, vitamins), requires the following strategy:

1. High-producing plants for the establishment of callus
2. The selection of fast growing, high yielding stable cell lines
3. The establishment of specific production medium different in composition from growth medium
4. Yields of product in excess of that produced by the plant and at levels in cell culture that are in excess of 1 g $liter^{-1}$ of culture medium (Zenk, 1990)

One of the major problems is that, at present, many of the plants with biotechnologically interesting compounds cannot be induced to produce these compounds in cell cultures in economically justifiable amounts (Fowler *et al.*, 1990). Particular examples in the alkaloid area are *Papaver* (morphinan alkaloids) and *Datura* (tropane alkaloids). Industrial application is still very restricted.

3.2. The Problems Associated with Scaleup

Like conventional microbial processes, industrial bioprocesses based on plant cells would consist of two major sets of processing operations, a fermentation section and a downstream processing section.

In the fermentation section, plant cells are grown in appropriate bioreactors to the desired densities and the product is accumulated either intracellularly or excreted into the medium.

Plant cells are now cultivated in volumes up to 75×10^3 liters in fermentor systems specifically developed for plant cell cultures which are much more susceptible to shear damage than are cultured microbial cells. Plant cells are larger than microbial cells, tend to form aggregates of up to 200 cells, and grow slowly compared with microbes. Consequently, there is a preproduction phase before the cells start producing. It takes 20–40 days to establish a large-scale culture and this time-consuming start adds significantly to production costs. Airlift fermentors have been developed that are characterized by low shear levels and while good growth and productivity appear achievable up to biomass levels of about 30 g dw $liter^{-1}$, with biomass levels above this mixing and aeration are incomplete and culture performance tends to fall off. For commercial viability, biomass levels of 100 g $liter^{-1}$ may be required and this level of cell productivity is not achievable with present airlift fermentors (Fowler *et al.*, 1990). In some instances it is possible to use the conventional stir tank as is used by Mitsui Petrochemicals for shikonin production (Curtin, 1983). The problems associated with high cell density with shikonin production by *Lithospermum erythrorhizon* and berberine from *Coptis japonica* have to some extent been overcome by Mitsui Petrochemicals by the use of growing and production media. Another critical element is the supply of nutrients. With current production methods it is possible to increase nutrients as the number of cells increases but if cell density gets too high, cell growth and berberine production in *C. japonica* cease (Fujita, 1988). At the same time an adequate supply of oxygen has to be provided to maintain production in high density cultures. At high densities, cell secretions (apparently superannuated cell substances) often inhibit both cell growth and berberine production. To overcome this handicap a perfusion method is used in which part of the medium is renewed continuously. This technique, with *C. japonica*, permitted adequate production of cells, 70 g dw $liter^{-1}$, with no decrease in berberine content.

Downstream processing operations for the harvesting of compounds may typically involve cell disruption by physical or chemical means, solid–liquid separation by centrifugation or filtration, and product recovery via adsorption, liquid–liquid extraction, or distillation. Thus, unit operations currently used for recovery of microbial products would be directly applicable to production of secondary metabolites from plant cell culture with only minor modifications. At present, the fermentation process for plant cells is the major contributor to capital and production costs (Kreis and Reinhard, 1989).

3.3. Areas of Potential Industrial Interest

Although there has been potential industrial interest in a number of alkaloids of pharmaceutical interest, a large part of this effort has centered on the production of indole alkaloids in *C. roseus* and *Rauwolfia serpentina* and the production of berberine from *C. japonica* and *T. minor*. Mitsui Petrochemicals has well documented processes (Fujita, 1988) for the large-scale production of berberine from *C. japonica* and production of ajmalicine from callus cultures of *R. serpentina* on an industrial basis has been reported (Vukovic *et al.*, 1988). Vincristine and vinblastine preparation has involved the utilization of catharanthine from cell cultures and vindoline from the *C. roseus* plant in the formation of the dimer 3′,4′-anhydrovinblastine, either by chemical means, utilizing ferric ions in acid medium (Vukovic *et al.*, 1988), or by the use of a peroxidase isolated from *C. roseus* cell suspension cultures (Smith *et al.*, 1988) and subsequently immobilized. It is likely in the future that these useful antileukemic compounds will be synthesized in part or in whole using cell cultures and/or isolated enzymes. Although these dimeric alkaloids have now been found in very low levels in cell suspension cultures of *C. roseus*, it may take further exploration of this condensing enzyme to determine the feasibility of producing genetically engineered cultures that are able to form the dimers vincristine and vinblastine.

4. SUMMARY

Plant cell culture is proving to be a useful tool in the study of alkaloid biology at the cellular level. Use of cell cultures has facilitated studies on the biosynthesis of alkaloids at the enzyme level and on their metabolism. Studies on the regulation and location of alkaloids within the cell have been helped by the availability of cells in culture. These matters are be discussed in Chapters 5, 8, and 10.

The culturing of plant cells has yielded improved methods for the propagation of elite, essentially genetically identical, plants for agricultural and horticultural purposes as well as for the conservation of endangered species. New methods for the production of hybrids, for example by protoplast fusion, have proved useful for the production of genetically altered plants and this along with micropropagation techniques may revolutionize agricultural, horticultural, and industrial processes. From a commercial point of view, cell cultures have been of interest as an alternative method for the production of commercially useful metabolites, in particular medicinal compounds and flavors. Cell cultures are slow growing compared with microorganisms but techniques are already available and proven for the transfer of genes for particular parts of biosynthetic pathways from plants to microorganisms where they have been efficiently expressed. Although currently few industrial processes involving plant cells are in operation, the techniques and industrial experience on problems of scaleup are at a stage where commercial exploitation is a reality. These techniques may serve best for the production of new drugs or templates for new drugs.

Plant cell culture on a commercial scale is most likely to be economically feasible in instances where the molecules required are not readily synthesized abiotically and where the source plant is slow growing and likely to become an endangered species in the event of commercialization. This is particularly true of the current interest in the anticancer substance taxol, where alternatives to the trees *Taxus brevifolia* or *T. chinensis* will have to be found if commercial exploitation is to proceed.

REFERENCES

General Reviews

Alfermann, A. W., and Reinhard, E., 1988, Biotransformation of synthetic and natural compounds by plant cell cultures, in: *Plant Cell Biotechnology*, NATO ASI Series, Vol H18 (M. S. S. Pais, F. Mavituna, and J. M. Novais, eds.), Springer-Verlag, Berlin, pp. 275–283.

Anderson, L. A., Phillipson, J. D., and Roberts, M. F., 1985, Biosynthesis of secondary products by cell cultures of higher plants, in: *Advances in Biochemical Engineering/Biotechnology, Plant Cell Culture*, Vol. 31 (A. Fiechter, ed.), Springer-Verlag, Berlin, pp. 1–36.

Anderson, L. A., Phillipson, J. D., and Roberts, M. F., 1987, Alkaloid production by plant cells, in: *Plant and Animal Cell Cultures, Process Possibilities* (C. Webb and F. Mavituna, eds.), Ellis Horwood, Chichester, pp. 172–192.

Banthorpe, D. V., 1994, Secondary metabolism in plant tissue culture: Scope and limitations, *J. Chem. Soc. Nat. Prod. Rep.* 303–328.

Brodelius, P., 1988, Stress induced secondary metabolism in plant cell cultures, in: *Plant Cell Biotechnology,* NATO ASI Series, H18 (M. S. S. Pais, F. Mavituna, and J. M. Novais, eds.), Springer-Verlag, Berlin, pp. 195–209.

Brodelius, P., and Mosbach, K., 1982, Immobilised plant cells in: *Advances in Applied Microbiology,* Vol. 28 (A. I. Laskin, ed.), Academic Press, New York, pp. 1–26.

Curtin, M. E., 1983, Harvesting profitable products from plant cell culture, *Biotechnology*, **1:**649–657.

Dix, P. J., 1990, *Plant Cell Line Selection: Procedures and Applications*, VCH, Weinheim.

Ellis, B. E., 1984, Probing secondary metabolism in plant cell tissue cultures, *Can. J. Bot.* **62**:2912–2917.

Evans, W. C., 1989, Cell and organ culture and clonal propagation as sources of drugs, in: *Trease & Evans' Pharmacognosy*, 13th ed., Baillière–Tindell, London, pp. 68–78.

Flores, H. E., 1986, Use of plant cells and organ culture in the production of biological chemicals, in: *Applications of Biotechnology to Agricultural Chemistry* (H. Lebaron, R. O. Mumma, R. C. Honeycutt, and J. H. Dresing, eds.), ACS Symp. Ser. 190, pp. 66–86.

Fowler, M. W., 1986, Industrial applications of plant cell culture, in: *Plant Cell Culture Technology Botanical Monograph*, Vol. 23 (M. M. Yeoman, ed.), Blackwell, Oxford, pp. 202–227.

Fowler, M. W., and Scragg, A. H., 1988, Natural products from higher plants and plant cell cultures, in: *Plant Cell Biotechnology, NATO ASI Series*, Vol. H18 (M. S. S. Pais, F. Mavituna, and J. M. Novais, eds.), Springer-Verlag, Berlin, pp. 166–177.

Fowler, M., Cresswell, R. C., and Stafford, A. M., 1990, An economic and technical assessment of the use of plant cell cultures for natural product synthesis on an industrial scale, *Ciba Found. Symp.* **154:**157–174.

Gautheret, R. J., 1959, *La Culture des Tissues Vegeteaux*, Masson, Paris.

Hamill, J. D., Parr, A. J., Rhodes, M. J. C., Robins, R. J., and Walton, N. J., 1987, New routes to plant secondary products, *Biotechnology* **5:**800–804.

Hay, C. A., Anderson, L. A., Roberts, M. F., and Phillipson, J. D. 1988, Alkaloid production by plant cell cultures, in: *Biotechnology in Agriculture* (A. Misrahi and A. L. van Wezel, eds.), Liss, New York, pp. 97–140.

Heble, M. R., 1985, Multiple shoot cultures: A viable alternative *in vitro* system for the production of known and new biologically active plant constituents, in: *Primary and Secondary Metabolism of Plant Cell Cultures* (K. H. Newmann, W. Barz, and E. Reinhard, eds.), Springer-Verlag, Berlin, pp. 281–289.

Heinstein, P. F., 1985, Future approaches to the formation of secondary natural products in plant cell suspension cultures, *J. Nat. Prod.* **48:**1–9.

Hulst, A. C., and Tramper, J., 1989, Immobilised plant cells: A literature survey, *Enzyme Microb. Technol.* **11:**546–558.

Kennedy, J. F., 1982, The future of immobilised cell technology, *Nature* **299:**777–778.

Kurz, W. G. W., and Constabel, F., 1985, Aspects affecting biosynthesis and biotransformation of secondary metabolites in plant cell cultures, in: *CRC Critical Reviews in Biotechnology*, Vol. 2, Issue 2 (G. Stewart and I. Russell, eds.), CRC, Boca Raton, pp. 105–118.

Lindsey, K., and Yeoman, M. M., 1983, Novel experimental systems for studying the production of secondary metabolites by plant tissue cultures, in: *Plant Biotechnology* (S. H. Mantell and H. Smith, eds.), Cambridge University Press, London, pp. 39–66.

Lindsey, K., and Yeoman, M. M., 1985, Dynamic aspects of plant cell culture, in: *Cell Culture and Somatic Cell Genetics of Plants: A Comprehensive Treatise* (I. Vasil, ed.), Academic Press, New York, pp. 61–101.

Mantell, S. H., and Smith, H., (eds.), 1983, *Plant Biotechnology,* Cambridge University Press, London.

Meins, F., 1983, Heritable variation in plant cell culture, *Annu. Rev. Plant Physiol.* **34:**327–346.

Payne, G., Bringi, V., Prince, C., and Shuler, M. L., (eds.), 1991, *Plant Cell and Tissue Culture in Liquid Systems*, Hanser, Munich.

Reinert, J., and Yeoman, M., (eds.), 1982, *Plant Cell and Tissue Culture, a Laboratory Manual*, Springer-Verlag, Berlin.

Reinhard, E., and Alfermann, A. W., 1980, Biotransformation by plant cells, in: *Advances in Biochemical Engineering 16, Plant Cell Cultures I* (A. Fiechter, ed.), Springer-Verlag, Berlin, pp. 49–83.

Rhodes, M. J. C., 1986, Immobilised plant cells, in: *Topics in Enzyme and Fermentation Biotechnology*, Vol. 10 (A. Weisman, ed.), Ellis Horwood, Chichester, pp. 51–87.

Roberts, M. F., 1988b, Medicinal products through plant biotechnology, in: *Manipulating Secondary Metabolism in Culture* (R. J. Robins and M. J. C. Rhodes, eds.), Cambridge University Press, London, pp. 201–216.

Stafford, A., and Warren, G., 1991, *Plant Cell and Tissue Culture,* Open University Press, Biotechnology Series.

Street, H., (ed.), 1977, *Plant Tissue and Cell Culture*, Blackwell, Oxford.

Yamada, Y., and Fujita, Y., 1986, Production of useful compounds from culture, in: *Handbook of Plant Cell Culture* (D. A. Evans, W. R. Sharp, P. V. Ammirator, and Y. Yamada, eds.), Macmillan Co., New York, pp. 717–728.

Yeoman, M. M., 1987a, Techniques, characteristics, properties and commercial potential of immobilised cell, in: *Cell Culture and Somatic Genetics,* Vol. 4: (F. Constabel and I. K. Vasil, eds.), Academic Press, San Diego, pp. 197–215.

Yeoman, M. M., 1987b, Bypassing the plant, *Ann. Bot.* **60**(Suppl.):157–174.

Yeoman, M. M., Lindsey, K., Miedzybrodzka, H. B., and McLauchlan, W. R., 1980, Accumulation of secondary products as a facet of differentiation in plant cell cultures, in: *Differentiation in Vitro* (M. M. Yeoman and D. E. S. Truman, eds.), Cambridge University Press, London, pp. 65–82.

Zenk, M. H., 1990, Plant cell cultures: A potential in food and biotechnology, *Food Biotechnol.* **4**:461–470.

Key References

Bariaud-Fontanel, A., Julien, M., Coutos-Thevenot, P., Brown, S., Courtois, D., and Petiard, V., 1988, Cloning and cell sorter, in: *Plant Cell Biotechnology*, NATO ASI Series, Vol. H18 (M. S. M. Pais, F. Mavituna, and J. M. Novais, eds.), Springer-Verlag, Berlin, pp. 403–420.

Baumann, T. W., and Röhrig, L., 1989, Formation and intracellular accumulation of caffeine and chlorogenic acid in suspension cultures of *Coffea arabica*, *Phytochemistry* **28:**2667–2669.

Benjamin, B. D., Roja, P. C., Heble, M. R., and Chadha, M. S., 1987, Multiple shoot cultures of *Atropa belladonna*: Effect of physicochemical factors on growth and alkaloid formation, *J. Plant Physiol.* **129:**129–135.

Breuling, M., Alfermann, A. W., and Reinhard, E., 1985, Cultivation of cell cultures of *Berberis wilsoniae* in 20-l airlift bioreactors, *Plant Cell Rep.* **4:**220–223.

Brodelius, P., Deus, B., Mosbach, K., and Zenk, M. H., 1979, Immobilised plant cells in the production and transformation of natural products, *FEBS Lett.* **103:**93–97.

Brodelius, P., Funk, C., and Shillito, R. D., 1988, Permeabilisation of cultivated plant cells by electroporation for the release of intracellularly stored secondary products, *Plant Cell Rep.* **7:**186–188.

Brown, S., 1984, Analysis and sorting of plant material by flow-cytometry, *Physiol. Veg.* **22:**341–345.

Cheng, K.-D., Zhu, W.-H., Li, X.-L, Mang, C., Sun, Z.-M., and Yang, D.-H., 1987, Biotransformation of hyoscyamine by suspension cultures of *Anisodus tanguticus*, *Planta Med.* **53:**212–213.

Cheng, K.-D., Fang, H.-J., Zhu, W.-H., Mang, C., Yang, D.-H., and Li, L., 1989, The biotransformation of 6 hydroxyhyoscyamine to scopolamine by suspension and immobilised cells of *Anisodus tanguticus*, *Planta Med.* **55:**391–392.

Christen, P., Roberts, M. F., Phillipson, J. D., and Evans, W. C., 1990, Alkaloids of hairy root cultures of a *Datura candida* hybrid, *Plant Cell Rep.* **9:**101–104.

Collin, H. A., 1987, Determinants of yield of secondary products in plant tissue cultures, *Adv. Bot. Res.* **113:**146–187.

Constabel, F., 1990, Medicinal plant biotechnology, *Planta Med.* **56:**421–425.

Corchete, P., and Yeoman, M. M., 1989, Biotransformation of (−)-codeinone to (−)-codeine by *Papaver somniferum* cells immobilized in reticulate polyurethane foam, *Plant Cell Rep.* **8:**128–131.

Courtois, D., and Guern, J., 1980, Temperature response of *Catharanthus roseus* cells cultivated in liquid medium, *Plant Sci. Lett.* **17:**473–482.

Dagnino, D., Schripsema, J., and Verpoorte, R., 1994, Terpenoid indole alkaloid biotransformation capacity of *Tabernaemontana divaricata*, *Phytochemistry* **35:**671–676.

Deus, B., and Zenk, M. H., 1982, Exploitation of plant cells for the production of natural compounds, *Biotechol. Bioeng.* **24:**1965–1974.

DiCosmo, F., Quesnel, A., Misawa, M., and Tallevi, S. G., 1987, Increased synthesis of ajmalicine and catharanthine by cell suspension cultures of *Catharanthus roseus* in response to fungal culture-filtrates, *Appl. Biochem. Biotechnol.* **14:**101–106.

Dos Santos, R. I., Schripsema, J., and Verpoorte, R., 1994, Ajmalicine metabolism in *Catharanthus roseus* cell cultures, *Phytochemistry* **35:**677–681.

Eilert, U., and Constabel, F., 1985, Ultrastructure of *Papaver somniferum* cells cultured *in vitro* and treated with fungal homogenate eliciting alkaloid production, *Protoplasma* **128:**38–42.

Eilert, U., Kurz, W. G. W., and Constabel, F., 1985, Stimulation of sanguinarine accumulation in *Papaver somniferum* cell cultures by fungal elicitors, *J. Plant Physiol.* **119:**65–76.

Endo, T., and Yamada, Y., 1985, Alkaloid production in cultured roots of three species of *Duboisia*, *Phytochemistry* **24:**1233–1236.

Endo, T., Goodbody, A., Vukovic, J., and Misawa, M., 1987, Biotransformation of anhydrovinblastine to vinblastine by a cell free extract of *Catharanthus roseus* cell suspension cultures, *Phytochemistry* **26:**32–34.

Endo, T., Hamaguchi, H., Eriksson, T., and Yamada, Y., 1991, Alkaloid biosynthesis in somatic hybrids of *Duboisia leichhardtii* F. Muell., and *Nicotiana tabacum* L., *Planta* **183:**505–510.

Facchini, P. J., and DiCosmo, F., 1991, Plant cell bioreactor for the production of protoberberine alkaloids from immobilised *Thalictrum rugosum* cultures, *Biotechnol. Bioeng.* **37:**397–403.

Falkenhagen, H., Kuzovkina, I. N., Alterman, I. E., Nikolaeva, L. A., and Stöckigt, J., 1993, Alkaloid formation hairy roots and cell suspension cultures of *Rauvolvia serpentina* Benth., *Nat. Prod. Lett.* **3:**107–112.

FettNeto, A. G., DiCosmo, F., Reynolds, W. E., and Sakata, K., 1992, Cell cultures of *Taxus* as a source of the antineoplastic drug taxol and related taxanes, *Biotechnology* **10:**1572–1575.

Fujita, Y., 1988, Industrial production of shikonin and berberine, *Ciba Found. Symp.* **137:**228–238.

Funk, C., Gügler, K., and Brodelius, P. E., 1987, Increased secondary product formation in plant cell suspension cultures after treatment with a yeast carbohydrate preparation (elicitor), *Phytochemistry* **26:**401–405.

Furuya, T., Ikuta, A., and Syono, K., 1972, Alkaloids from callus tissue of *Papaver somniferum*, *Phytochemistry* **11:**3041–3044.

Furuya, T., Yoshikawa, T., and Taira, M., 1984, Biotransformation of codeinone to codeine by immobilised cells of *Papaver somniferum*, *Phytochemistry* **23:**999–1001.

Furuya, T., Koge, K., and Orihara, Y., 1990, Long term culture and caffeine production of immobilised coffee (*Coffea arabica* L.) cells in polyurethane foam, *Plant Cell Rep.* **9:**125–128.

Furuya, T., Orihara, Y., and Koge, K., 1991, Biotransformation of theobromine and caffeine in suspension and polyurethane foam immobilised coffee (*Coffea arabica* L.) cells, *Plant Cell Rep.* **9:**659–662.

Goodbody, A. E., Endo, T., Vukovic, J., Kutney, J. P., Choi, L., and Misawa, M., 1988, Enzymic coupling of catharanthine and vindoline to form 3′-4′-anhydrovinblastine by horseradish peroxidase, *Planta Med* **52:**136–140.

Gröger, D., and Baumert, A., 1984, Acridone alkaloid biosynthesis in tissue cultures of *Ruta graveolens*, in: *Progress in Tryptophan and Serotonin Research*, de Gruyter, Berlin, pp. 785–788.

Hall, R. D., and Yeoman, M. M., 1986, Temporal and spatial heterogeneity in the accumulation of anthocyane in cell cultures of *Catharanthus roseus* (L.) G. Don., *J. Exp. Bot.* **37:**48–60.

Hamill, J. D., Parr, A. J., Robins, R. J., and Rhodes, M. J. C., 1986, Secondary product formation by cultures of *Beta vulgaris* and *Nicotiana rustica* transformed with *Agrobacterium rhizogenes*, *Plant Cell Rep.* **5:**111–114.

Hamill, J. D., Robins, R. J., and Rhodes, M. C. J., 1989, Alkaloid production by transformed root cultures of *Cinchona ledgeriana*, *Planta Med.* **55:**354–357.

Han, K.-H., Fleming, P., Walker, K., Loper, M., Chilton, W. S., Mocek, U., Gordon, M. P., and Floss, H., 1994, Genetic transformation of mature *Taxus*: An approach to genetically control the in vitro production of the anticancer drug taxol, *Plant Sci.* **95:**187–196.

Harkins, K. R., and Galbraith, D. W., 1984, Flow sorting and the culture of plant protoplasts, *Physiol. Plant.* **60:**43–52.

Hashimoto, T., Yukimune, Y., Yamada, Y., 1986, Tropane alkaloid production in *Hyoscyamus* root cultures, *J. Plant Physiol.* **124:**61–75.

Hashimoto, T., Hayashi, A., Amano, Y., Kohno, J., Iwanari, H., Usuda, S, and Yamada, Y., 1991, Hyoscyamine 6β-hydroxylase, an enzyme involved in tropane alkaloid biosynthesis, is localised at the pericycle of the root, *J. Biol. Chem.* **266:**4648–4653.

Kirby, N. J., and Hunter, C. S., 1990, Repeated harvest of vacuole-located secondary product from *in vitro* grown plant cells using 1. 02 MHz ultrasound, *Appl. Microbiol. Biotechnol.* **33:**448–451.

Knobloch, K.-H., and Berlin, J., 1981, Phosphate mediated regulation of putrescine biosynthesis in cell suspension cultures of *Nicotiana tabacum*, *Planta Med.* **42:**167–142.

Kreis, W., and Reinhard, E., 1989, The production of secondary metabolites by plant cells cultivated in bioreactors, *Planta Med.* **55:**409–416.

Kutchan, T. M., 1989, Expression of enzymatically active cloned strictosidine synthase from the higher plant *Rauvolfia serpentina* in *Escherichia coli*, *FEBS Lett.* **257:**127–130.

Kutchan, T. M., Dittrich, H., Bracher, D., and Zenk, M. H., 1991, Enzymology and molecular biology of alkaloid biosynthesis, *Tetrahedron* **47:**5945–5954.

Lambe, C. A., Reading, A., Roe, S., and Rosvear, A., 1982, Production of alkaloids with immobilised cells of *Catharanthus roseus*, *Enzyme Eng.* **6:**137–138.

Larkin, P. J., and Scowcroft, W. R. 1981, Somaclonal variation—A novel source of variability from plant cell culture for plant improvement, *Theor., Appl. Genet.* **60:**197–214.

Lindsey, K., 1985, Manipulation by nutrient limitation, of the biosynthetic activity of immobilized cells of *Capsicum frutescens*, Mill. cv. annuum, *Planta* **165:**126–133.

Ma, W. W., Park, G. L., Gomez, G. A., Nieder, M. H., Adams, T. L., Ansley, G. S., Sahai, O. P., Smith, R. J., Stalhut, R. W., Hylands, P. J., Bitsch, F., and Shackelton, C., 1994, New bioactive taxoids from cell cultures of *Taxus baccata*, *J. Nat. Prod.* **57:**116–122.

Manceau, F., Fliniaux, M. A., and Dubreuil, A. J., 1989, Ability of *Nicotiana plumbaginifolia* cell suspension to demethylate nicotine to nornicotine, *Phytochemistry* **28:**2671–2674.

Matsuda J., Okabe, S., Hashimoto, T., and Yamada, Y., 1991, Molecular cloning of hyoscyamine 6-hydroxylase a 2-oxoglutarate-dependent dioxygenase from cultured roots of *Hyoscyamus niger*, *J. Biol. Chem.* **266:**9460–9464.

Miura, Y., Hirata, K., and Kurano, N., 1987, Isolation of vinblastine in callus culture with differentiated roots of *Catharanthus roseus* (L.) G. Don., *Agric. Biol. Chem.* **51:**611–614.

Miura, Y., Hirata, K., Kurano, N., Miyamoto, K., and Uchida, K., 1988, Formation of vinblastine in multiple shoot culture of *Catharanthus roseus*, *Plant Med.* **54:**18–20.

Morris, P., and Fowler, M. W., 1981, A new method for the production of fine plant cell suspension cultures, *Plant Cell Tissie Org. Cult.* **1:**15–24.

Muranaka, T., Ohkawa, H., and Yamada, Y., 1992, Scopolamine release into the media by *Dubiosia leichhardtii* hairy roots, *Appl. Microbiol. Biotechnol.* **37:**554–559.

Murashige, T., and Skoog, F., 1962, A revised medium for rapid growth and bioassays with tobacco tissue cultures, *Physiol. Plant.* **15:**473–497.

Nakagawa, K., Fukui, H., and Tabata, M., 1986, Hormonal regulation of berberine production in cell suspension cultures of *Thalictrum minus*, *Plant Cell Rep.* **5:**69–71.

Nef, C., Rio, B., and Chrestin, H., 1991, Induction of catharanthine synthesis and stimulation of major indole alkaloid production by *Catharanthus roseus* cells under non-growth-altering treatment with *Pythium vexans* extracts, *Plant Cell Rep.* **10:**26–29.

Okada, N., Koizuma, N., Tanaka, T., Ohkubo, H., Nakanishi, S., and Yamada, Y., 1989, Isolation, sequence, and bacterial expression of a cDNA for (S)-tetrahydroberberine oxidase from cultured berberine-producing *Coptis japonica* cells, *Proc. Natl. Acad. Sci. USA* **86:**534–538.

Payne, J., Hamill, J. D., Robins, R. J., and Rhodes, M. J. C., 1987a, Production of hyoscyamine by "hairy root" cultures of *Datura stramonium*. Planta Med. **53:**474–478.

Payne, J., Rhodes, M. J. C., and Robins, R. J., 1987b, Quinoline alkaloid production by transformed cultures of *Cinchona ledgeriana*. Planta Med. **53:**368–372.

Rhodes, M. J. C., Hilton, M., Parr, A. J., Hamill, J. D., and Robins, R. J., 1986, Nicotine production by hairy root cultures of *Nicotiana rustica*. Fermentation and product recovery, *Biotechnol. Lett.* **8:**415–420.

Roberts, M. F., 1988a, Isoquinolines (*Papaver* alkaloids), in: *Cell Culture and Somatic Cell Genetics of Plants*, Vol. 5, (F. Constabel and I. K. Vasil, eds.), Academic Press, San Diego, pp. 315–330.

Robins, R. J., Parr, A. J., Payne, J., Walton, N. J., and Rhodes, M. J. C., 1990. Factors regulating tropane alkaloid production in a transformed root culture of a *Datura candida* × *D. aurea* hybrid, *Planta* **181:**414–422.

Robins, R. J., Parr, A. J., Bent, E. G., and Rhodes, M. J. C., 1991a, Studies on the biosynthesis of tropane alkaloids in *Datura stramonium* L. transformed root cultures. 1. The kinetics of alkaloid production and the influence of feeding intermediate metabolites, *Planta* **183:**185–195.

Robins, R. J., Parr, A. J., and Walton, N. J., 1991b, Studies on the biosynthesis of tropane alkaloids in *Datura stramonium* L. transformed root cultures. 2. On the relative contributions of L-arginine and L-ornithine to the formation of the tropane ring, *Planta* **183:**196–201.

Rokem, J. S., and Goldberg, I., 1985, Secondary metabolites from plant cell suspension cultures. Methods for yield improvement, in: *Advances in Biotechnological Processes*, Vol. 4 (A. Mizrahi and A. L. van Wezel, eds.), Liss, New York, pp. 241–274.

Rueffer, M., 1985, The production of isoquinoline alkaloids by plant cell cultures, in: *The Chemistry and Biology of Isoquinoline Alkaloids* (J. D. Phillipson, M. F. Roberts, and M. H. Zenk, eds.), Springer-Verlag, Berlin, pp. 265–280.

Ruyter, C. M., Akram, M., Illahi, I., and Stöckigt, J., 1991, Investigation of the alkaloid content of *Rauvolfia serpentina* roots from regenerated plants, *Planta Med.* **57**:328–330.

Saito, K., Murakoshi, I., Inze, D., and Van Montagu, M., 1989, Biotransformation of nicotine alkaloids by tobacco shooty teratomas induced by a Ti plasmid mutant, *Plant Cell Rep.* **7**:607–610.

Sauerwein, M., and Shimomura, K., 1991, Alkaloid production in hairy roots of *Hyoscyamus albus* transformed with *Agrobacterium rhizogenes*, *Phytochemistry* **30**:3277–3280.

Sauerwein, M., Ishimaru, K., and Shimomura, K., 1991, Indole alkaloids in hairy roots of *Amsonia elliptica*, *Phytochemistry* **30**:1153–1155.

Siah, C. L., and Doran, P. M., 1991, Enhanced codeine and morphine production in suspended *Papaver somniferum* cultures after removal of exogenous hormones, *Plant Cell Rep.* **10**:349–353.

Smith, J. I., Amouzou, E., Yamaguchi, A., McLean, S., and DiCosmo, F., 1988, Peroxidase from bioreactor cultivated *Catharanthus roseus* cell cultures mediates biosynthesis of α-3′4′-anhydrovinblastine, *Biotechnol. Appl. Biochem.* **10**:568–574.

Staba, E. J., 1985, Milestone in plant tissue culture systems for production of secondary products, *J. Nat. Prod.* **48**:203–209.

Takayama, S., and Misawa, M., 1981, Mass propagation of *Begonia* and *Hiemalis* plantlets by shake cultures, *Plant Cell Physiol.* **22**:461–467.

Tyler, R. T., Eilert, U., Rijnders, C. O. M., Roewer, I. A., and Kurz, W. G. W., 1988, Semicontinuous production of sanguinarine and dihydrosanguinarine by *Papaver somniferum* L. cell suspension cultures treated with fungal homogenate, *Plant Cell Rep.* **7**:410–413.

Ushiyama, M., and Furuya, T., 1989, Biotransformation of (RS)-tropic acid in suspension cultures of *Coffea arabica, Datura innoxia, Eucalyptus perriniana*, and *Nicotiana tabacum*, *Phytochemistry* **28**:2333–2339.

Veliky, I. A., and Jones, A., 1981, Bioconversion of gitoxigenin by immobilised plant cells in a column bioreactor, *Biotechnol. Lett.* **3**:551–554.

von Neumann, D., and Müller, E., 1974, Beiträge zur Physiologie der Alkaloide IV. Alkaloidbildung in Kallus Kulturen von *Macleaya*, *Biochem. Physiol. Pflanz.* **165**:271–282.

Vukovic, J., Goodbody, A. E., Kutney, J. P., and Misawa, M., 1988, Production of 3′,4′-anhydrovinblastine: A unique chemical synthesis, *Tetrahedron* **44**:325–331.

Weathers, P. J., Diiorio, A., Cheetham, R., and O'Leary, M., 1990, Recovery of secondary metabolites with minimal loss of cell viability, in: *Current Plant Science and Biotechnology in Agriculture, Progress in Plant Cellular and Molecular Biology* (H. J. J. Nijkamp, L. H. W. Van der Plas, and J. Van Aatrijk, eds.), Kluwer, Dordrecht, pp. 582–586.

Wink, M., Witte, L., Hartmann, T., Theuring, C., and Volz, V., 1983, Accumulation of quinolizidine alkaloids in plants and cell suspension cultures: Genera *Lupinus, Cytisus, Baptisia, Genista, Laburnum*, and *Sophora*, *Planta Med.* **48**:253–257.

Yamamoto, O., and Yamada, Y., 1987, Selection of reserpine producing cell strain using UV light and optimisation of reserpine production in the selected cell strain, *Plant Cell Tissue Org. Cult.* **8**:125–133.

Yamamoto, O., Suzuki, M., Suga, Y., Fukui, H., and Tabata, M., 1987, Participation of an active transport system in berberine secreting cells of *Thalictrum minus*, *Plant Cell Rep.* **6**:356–359.

Yeoman, M. M., and Forche, E., 1980, Cell proliferation and growth in callus cultures, in: *International Review of Cytology* Suppl. **11A**:1–24.

Yoshimatsu, K., and Shimomura, K., 1991, Efficient shoot formation on internodal segments and alkaloid formation in the regenerates of *Cephaelis ipecacuanha* A. Richard, *Plant Cell Rep.* **9**:567–570.

Yun, D.-J., Hashimoto, T., and Yamada, Y., 1992, Metabolic engineering of medicinal plants: Transgenic *Atropa belladonna* with improved alkaloid composition, *Proc. Natl. Acad. Sci. USA* **89**:11790–11803.

Zenk, M. H., El-Shagi, H., Arens, H., Stöckigt, J., Weiler, E. W., and Deus, B., 1977, Formation of the indole alkaloids serpentine and ajmalicine in cell suspension cultures of *Catharanthus*, in: *Plant Tissue Culture and Its Biotechnological Application* (W. Barz, E. Reinhard, and M. H. Zenk, eds.), Springer-Verlag, Berlin, pp. 27–43.

Zito, S. W., and Staba, E. J., 1982, Thebaine from root cultures of *Papaver bracteatum*, *Planta Med.* **45**:53–54.

Chapter 8

The Biosynthesis of Alkaloids in Root Cultures

Richard J. Robins

1. INTRODUCTION

Alkaloid production by *in vitro* cultures has a long history. Root cultures of alkaloid-producing species were established more than 50 years ago (Butcher and Street, 1964) but it was not until the 1950s that alkaloid production was demonstrated in such systems (Mitra, 1972; Solt, 1957). Research was limited, however, because of the difficulty of establishing and maintaining root cultures. The advent of dispersed suspension cultures stimulated a period of intense activity and major advances were made in the biochemistry of some alkaloid classes, notably the indole and bisisoquinoline types (Zenk, 1989). These examples proved, unfortunately, to be exceptions rather than the rule and several other important classes, such as the tropanes, quinolines, and pyrrolizidines, were accumulated at only very low levels or not at all. Several groups in the early 1980s reexamined the possibility of using root cultures for alkaloid production and the establishment of cultures of *Hyoscyamus niger* (Hashimoto and Yamada, 1983) and *Senecio vulgaris* (Hartmann and Toppel, 1987; Toppel *et al.,* 1987) were reported. In 1986, a further step forward was made when Hamill *et al.* (1986) reported the use of transformed root cultures for nicotine formation. Four years later, the same authors demonstrated that nicotine accumulation in *Nicotiana rustica* transformed roots could be increased by genetic engineering (Hamill *et al.,* 1990). The advantage of transformation is that, by engineering the delivery system, genes can be inserted into the plant (Jones *et al.,* 1992). Furthermore, transformed roots have no phytohormone requirement and show good stability in culture.

Since these early reports, a wide range of root cultures, both transformed and untransformed, have been generated that can accumulate alkaloids (Rhodes *et al.,* 1990). This chapter is not intended to summarize all of these reports (see also Chapter 5). Rather,

Richard J. Robins • Laboratoire d' Analyse Isotopique et Électrochimique de Métabolismes, CNRS UPRES-A 6006, University of Nantes, BP 92208, F-44322 Nantes Cedex 03, France.

Alkaloids: Biochemistry, Ecology, and Medicinal Applications, edited by Roberts and Wink. Plenum Press, New York, 1998.

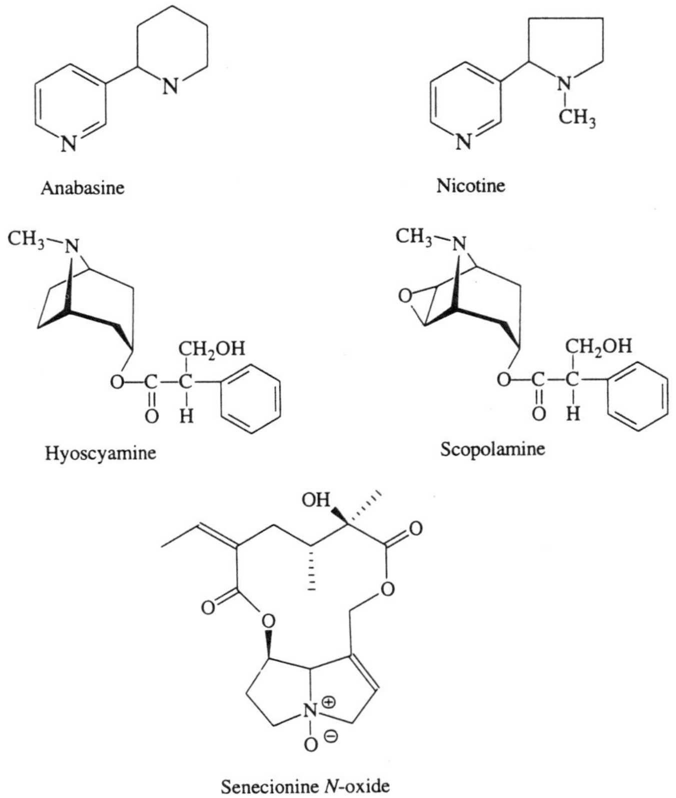

Figure 1. Structures of the principal alkaloids referred to in the text.

it will concentrate on three classes of alkaloids for which substantial progress has been made in our understanding of the biochemistry and regulation of their formation primarily because of work in root cultures. These are the pyrrolidine alkaloids of *Nicotiana*, the tropane alkaloids of *Datura* and *Hyoscyamus*, and the pyrrolizidine alkaloids of *Senecio* and *Eupatorium* (Fig. 1). It will further consider the contribution made by genetic engineering in helping unravel the complexities of these pathways.

2. THE PYRROLIDINE ALKALOIDS

Nicotine is made in suspension cultures of, for example, *N. tabacum* grown in suitable hormonal conditions (Röper *et al.*, 1985). Such cultures, however, lack stability and often require different conditions for growth and nicotine production. Stable transformed root cultures of *Nicotiana* making and accumulating nicotine can, in contrast, easily be generated, cultured, and used to study alkaloid formation. *N. rustica* and *N. tabacum* root cultures principally contain nicotine, which is made from putrescine and nicotinic acid (Fig. 2). Putrescine is a primary metabolite, produced by the decarboxylation of either arginine or ornithine, and is used for the biosynthesis of the polyamines spermine and spermidine, which play a central role in the control of cell function. It is also

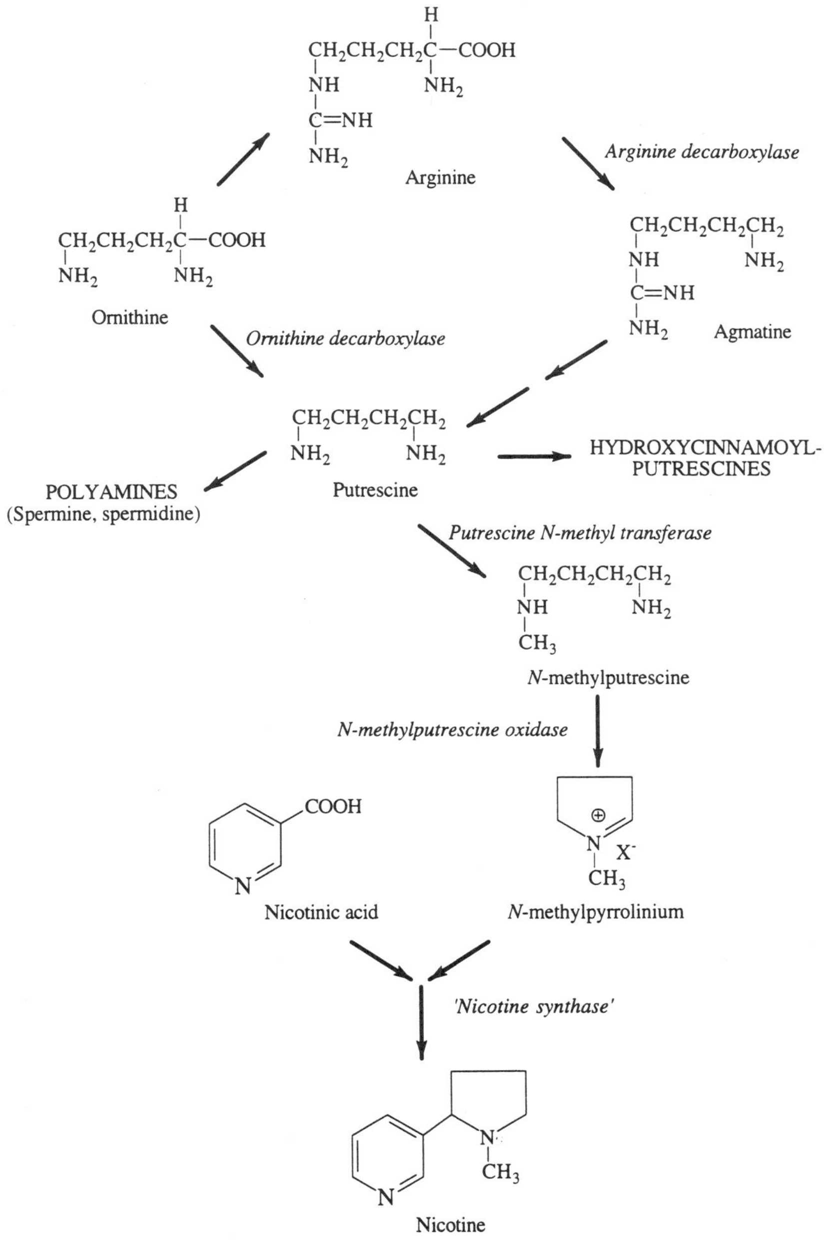

Figure 2. Scheme showing the pathway of biosynthesis of nicotine from ornithine/arginine and nicotinic acid.

involved in the formation of another group of secondary compounds, the hydroxycinnamoylputrescines. The conversion of putrescine to *N*-methylputrescine is, therefore, the first committed step of the alkaloidal pathway. *N*-Methylpyrrolinium, formed by the oxidative deamination of *N*-methylputrescine, is condensed with a putative intermediate derived by the decarboxylation of nicotinic acid.

By most standards, the pathway from putrescine is short, involving possibly only three specific enzymes, namely, putrescine *N*-methyltransferase (PMT), *N*-methylputrescine oxidase (MPO) (Mizusaki *et al.,* 1973), and "nicotine synthase" (Friesen and Leete, 1990). The regulation of these enzymes, however, and the control of flux into the pathway are still not well understood. Furthermore, nicotine formation also requires nicotinic acid, itself the product of a multistep pathway from aspartate.

2.1. Regulation of the Pyrrolidine Alkaloid Pathway in *Nicotiana*

Feeding experiments in root cultures of *N. rustica* indicate that a major limitation in accumulation occurs subsequent to *N*-methylpyrrolinium formation. However, small enhancements are seen with putrescine or agmatine but not with ornithine or arginine, indicating a possible limitation in the supply of putrescine (Walton and McLauchlan, 1990; Walton *et al.,* 1988). If this is so, then regulation may occur at the level of the availability of putrescine. The difficulty is in deciding the extent to which this step influences the final accumulation of product. Which decarboxylase enzyme is involved is indicated by evidence obtained from callus cultures of *Nicotiana.* From feeding and inhibition experiments with substrate-activated "suicide" inhibitors of ODC and ADC, namely, α-difluoromethylornithine (DFMO) and α-difluoromethylarginine (DFMA), Tiburcio and Galston (1986) showed that arginine is probably the preferred origin of the putrescine incorporated into nicotine. In addition, PMT is closely regulated and is important in the overall regulation of the pathway. In suspension cultures of *N. tabacum,* productive lines show high PMT activity while nonproductive lines have only trace levels of this enzyme (Ohta and Yatazawa, 1980). In root cultures, nicotine production and PMT activity are lost if roots are subcultured in media containing phytohormones (Rhodes *et al.,* 1989). This effect is reversible: integral roots competent in nicotine production are obtained when cells are passaged into phytohormone-free medium. Therefore, PMT, rather than ADC or ODC, is identified as a good target for genetic engineering.

An added complication is that nicotine biosynthesis involves the incorporation of nicotinic acid (Fig. 2). It was found that when nicotinic acid or nicotinamide was fed to root cultures of *N. rustica,* nicotine accumulation was enhanced (Robins *et al.,* 1987). Therefore, it might well be that the availability of this moiety can be as important in nicotine accumulation as is the availability of the putrescine-derived portion.

2.1.1. THE ENZYMES OF PYRROLIDINE ALKALOID FORMATION

2.1.1a. Putrescine *N*-Methyltransferase

This enzyme is extremely unstable but has been partially purified from cultures of *Hyoscyamus niger* and fully from *D. stramonium* (see below). A clone has been isolated by differential screening of cDNA libraries from high and low nicotine-yielding *N. tabacum* plants (Hibi *et al.,* 1994). The enzyme shows considerable sequence homology to spermidine synthase but is distinct as it only shows PMT activity when expressed in *E. coli.*

2.1.1b. *N*-Methylputrescine Oxidase

This activity appears to be more stable than PMT and has been isolated in pure form from *N. tabacum* transformed root cultures (McLauchlan *et al.,* 1993). It is quite widely distributed in the Solanaceae, as shown by Western blotting, and is apparently both immunologically (McLauchlan *et al.,* 1993) and kinetically (Hashimoto *et al.,* 1990; Walton and McLauchlan, 1990) related to a wide range of diamine oxidases in plants.

2.2. Studying the Pyrrolidine Alkaloid Pathway by Genetic Engineering

2.2.1. INCREASING THE AMOUNT OF ALKALOID ACCUMULATED

While PMT is clearly important in determining the overall extent to which cultures can make pyrrolidine alkaloids, the level of activity normally found in transformed root cultures of *N. rustica* does not limit the ability of the cultures to accumulate nicotine. As feeding putrescine, however, did have some effect on nicotine levels, experiments were conducted to try to enhance nicotine formation by engineering the supply of this metabolite (Hamill *et al.,* 1990).

When *odc* from *Saccharomyces cereviseae* was expressed with the enhanced cauliflower mosaic virus 35 S protein promoter in transgenic roots of *N. rustica*, the level of ODC was enhanced in several root clones (Fig. 3A). Furthermore, the level of ODC in *odc* lines remained elevated even in the late stationary phase of the culture, in contrast to control lines. Other enzymes (ADC, PMT, and MPO) were not enhanced. This indicated that the introduced gene was being expressed in a deregulated manner and this was confirmed by showing that ODC mRNA was also present at a high level throughout the growth cycle. Some of the *odc*-expressing clones had increased levels of putrescine and, in particular, *N*-methylputrescine. In addition, the mean nicotine content of the 14-day-old cultures was increased from 2.28 ± 0.22 to 4.04 ± 0.48 μmol g^{-1} fresh mass (Fig. 3B).

The increments in both ODC activity and nicotine were small but significant. Why, then, was a larger increase in nicotine not found? Presumably, once the supply of putrescine was enhanced, other enzymes contributed more than before to limiting nicotine accumulation. As MPO is present at, typically, two- to fivefold higher levels than PMT, it could be that PMT becomes limiting. Now that *pmt* has been cloned (Hibi *et al.,* 1994), this possibility can be tested directly. However, the enzyme responsible for the condensation of *N*-methylpyrrolinium with decarboxylated nicotinic acid, nicotine synthase (Friesen and Leete, 1990), was only measured at a very low level of activity, quite inadequate to account for the rates of nicotine accumulation observed in cultures. Thus, the mechanism and regulation of this important step in nicotine formation requires further investigation in order to estimate the extent to which nicotine accumulation is controlled by the condensation step. Feeding nicotinic acid to the ODC-overproducing cultures did not increase nicotine accumulation, suggesting that the supply of this moiety was not limiting.

2.2.2. ALTERING THE MIXTURE OF ALKALOIDS ACCUMULATED

Genetic engineering has also been used to show that it is possible *in vivo* to manipulate the mixture of alkaloids accumulated. Anabasine, a minor alkaloid in some *Nicotiana* species, is derived from lysine, via cadaverine, in a pathway parallel to that for the biosynthesis of nicotine (Fig. 4). Root cultures of *N. rustica* (Walton *et al.,* 1988) and *N. hesperis* (Walton and Belshaw, 1988) accumulate anabasine when fed cadaverine, the

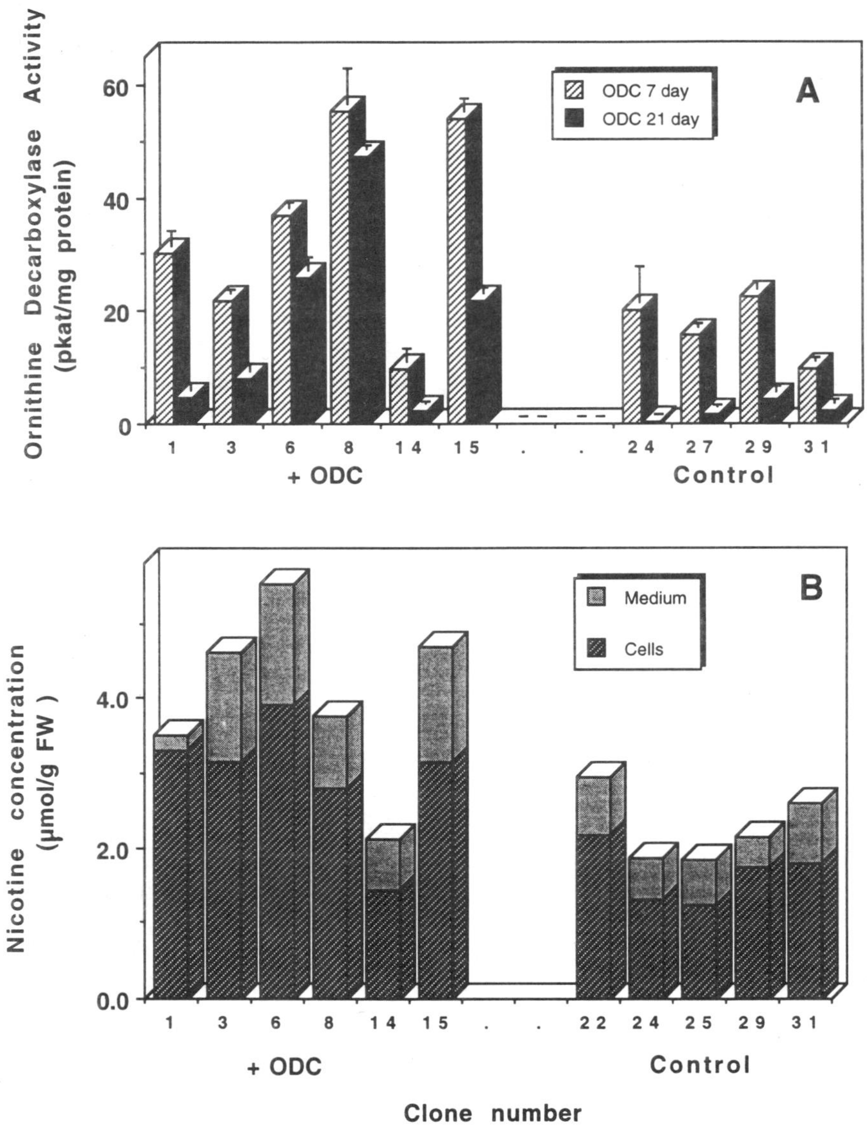

Figure 3. The effect of expressing the *S. cereviseae odc* gene in transgenic *N. rustica* root cultures. (A) Levels of ornithine decarboxylase activity in roots expressing the *odc* gene (+ODC) or the *cat* gene (Control). ODC activity ± SE is shown in clones at 7 and 21 days following subculture. (B) Nicotine accumulation in 14-day-old roots.

nicotine/anabasine ratio in the former changing from 10 : 1 to 1 : 5. Thus, the enhanced anabasine formation is at the expense of nicotine, indicating that the two pathways may be competing for nicotinic acid. Alternatively, and even more interestingly, it may be that some steps in each pathway are catalyzed by the same enzyme. It has been shown that MPO from *N. tabacum* catalyzes the oxidation of both *N*-methylputrescine and cadaverine (Walton and McLauchlan, 1990). Although the enzyme has a 34-fold higher affinity for

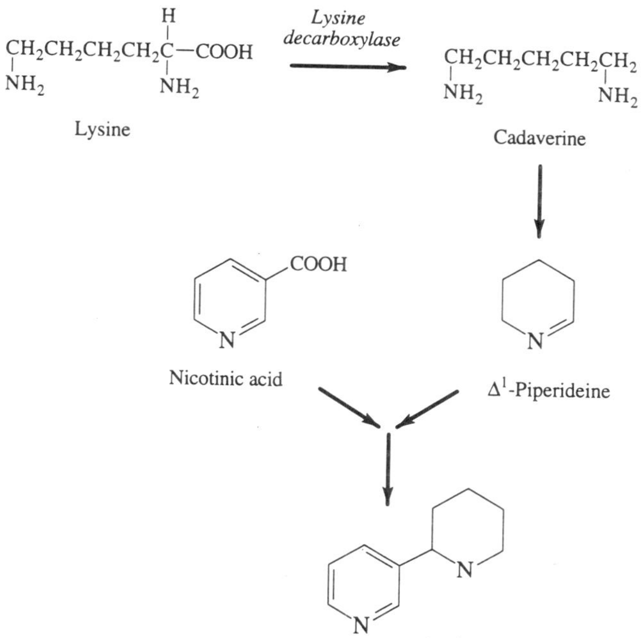

Figure 4. Scheme showing the pathway of biosynthesis of anabasine from lysine and nicotinic acid.

N-methylputrescine, the capacity of the enzyme to oxidize cadaverine is 3-fold greater. Hence, when supplied in excess, cadaverine might be expected to dominate the reaction, leading to more anabasine and less nicotine, as observed.

Walton *et al.* (1988) found that feeding lysine hardly affected the alkaloid ratio, indicating that a deficiency in lysine decarboxylase (LDC) might be responsible for the low natural levels of anabasine. In order to test this, Berlin and co-workers inserted the *ldc* from *Hafnia alvei* into transgenic root cultures of *N. glauca* under the control of the cauliflower mosaic virus 35 S promoter (Fecker *et al.*, 1992). *N. glauca* root cultures only contain low LDC activity, even though anabasine is accumulated as a major product. Two clones were isolated showing about a sixfold increase of LDC activity. This was accompanied by a tenfold rise in cadaverine, a twofold rise in anabasine, and a change in the nicotine/anabasine ratio from 75 : 25 in controls to 60 : 40 in *ldc* transgenic roots. In terms of the effect on anabasine production, this experiment was similar in outcome to that of Hamill *et al.* (1990) in *N. rustica.* Much greater enhancement of both the gene product and the diamine were found, however. The experiment clearly showed that anabasine production is partially limited by cadaverine supply and provided further evidence for at least one common step in the pathways of nicotine and anabasine production.

3. THE TROPANE ALKALOIDS

In contrast to *Nicotiana*, tropane-alkaloid-producing genera were found intransigent to forming alkaloids in dispersed cultures (Robins and Walton, 1993). Thus, only a very limited understanding of the biochemistry and regulation of the tropane alkaloid pathway was available prior to studies made in root cultures. Both untransformed (Hashimoto *et al.,* 1986) and transformed root cultures (Payne *et al.,* 1987; Robins *et al.,* 1990, 1991a) accumulate high levels of tropane alkaloids.

The alkaloids hyoscyamine and scopolamine (Fig. 1) have been known for many years to have sympathomimetic activity and currently are widely used as muscle relaxants. Thus, the pathway of tropane alkaloid production presents not only an interesting biochemical problem but also a realistic target for the genetic manipulation of a useful plant-derived product.

3.1. The Biosynthetic Route to the Tropane Alkaloids

Root cultures of *Datura, Atropa, Hyoscyamus,* and related genera contain predominantly hyoscyamine and scopolamine (Fig. 1). Tropane and pyrrolidine alkaloids have a common biosynthetic pathway to *N*-methylpyrrolinium (Fig. 2), the first unique step toward the tropanes being the condensation of *N*-methylpyrrolinium with a C_3 unit to form the bicyclo[3.2.1] of tropinone (Fig. 5). This is stereospecifically reduced to form tropine (tropan-3α-ol), which is esterified with a moiety of phenyllactic acid to form littorine. Phenyllactic acid is derived from phenylalanine via phenylpyruvic acid. The rearrangement of the phenyllactoyl group in littorine to form a tropoyl residue completes hyoscyamine biosynthesis. Further metabolism of hyoscyamine, involving the introduction of a 7β-hydroxyl group followed by oxidation to the 6β,7β-epoxide, results in scopolamine formation.

As can be seen, the pathway by which scopolamine is made is more complex than that for nicotine. Not only is a C_3 unit, derived from acetate, required, but a further pathway from phenylalanine that provides the tropic acid moiety is involved. Root cultures have been used extensively to identify the pathway of synthesis by the feeding of labeled intermediates, to study the regulation of the flux into the alkaloids and to provide a source from which to isolate enzymes and genes.

3.2. Regulation of the Tropane Alkaloid Pathway in *Datura, Hyoscyamus,* and *Atropa*

This pathway presents a much greater challenge to the biochemist than that for the biosynthesis of nicotine. As well as having a greater number of steps, it is branched and root cultures can contain a range of products derived from tropinone (Fig. 6). Investigations have tended to focus on understanding the enzymes involved at the branch points and in investigating the role these play in regulating the flux into the different groups of products.

3.2.1. STUDIES OF THE FLUX OF METABOLITES THROUGH THE PATHWAY

The enzymes of hyoscyamine and scopolamine biosynthesis are present throughout much of the growth cycle of both *Datura* (Robins *et al.,* 1991a) and *Hyoscyamus* (Hashi-

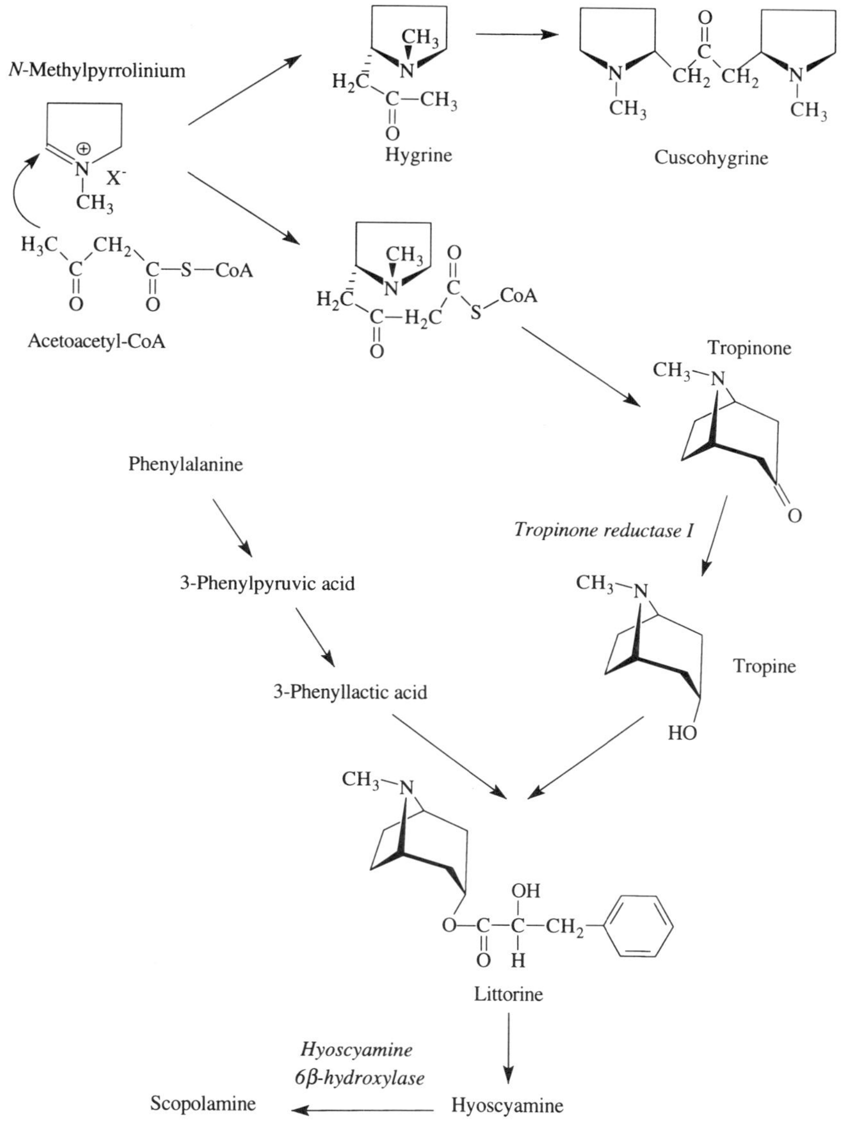

Figure 5. Scheme showing the pathway of biosynthesis of hyoscyamine and scopolamine from *N*-methylpyrolinium, acetoacetyl-CoA, and phenylalanine. The pathway to *N*-methylpyrrolinium is illustrated in Fig. 2.

moto *et al.,* 1989) root cultures. The level of activity present is maximal in rapidly growing tissue, indicating that the pathway is regulated to some degree. Furthermore, the amounts of ODC, ADC, and PMT determined *in vitro* in extracts from *D. stramonium* roots, do not greatly exceed the minimum required to synthesize the amounts of alkaloid accumulated *in vivo* (Rhodes *et al.,* 1989). In contrast, the amounts of tropinone reductases are much higher (Portsteffen *et al.,* 1992, 1994). Whether any of these activities actually limit flux was investigated by feeding ornithine, arginine, putrescine, agmatine, *N*-methylputrescine, tropinone, and tropine to root cultures of *D. stramonium* (Robins *et al.,* 1991a). None of

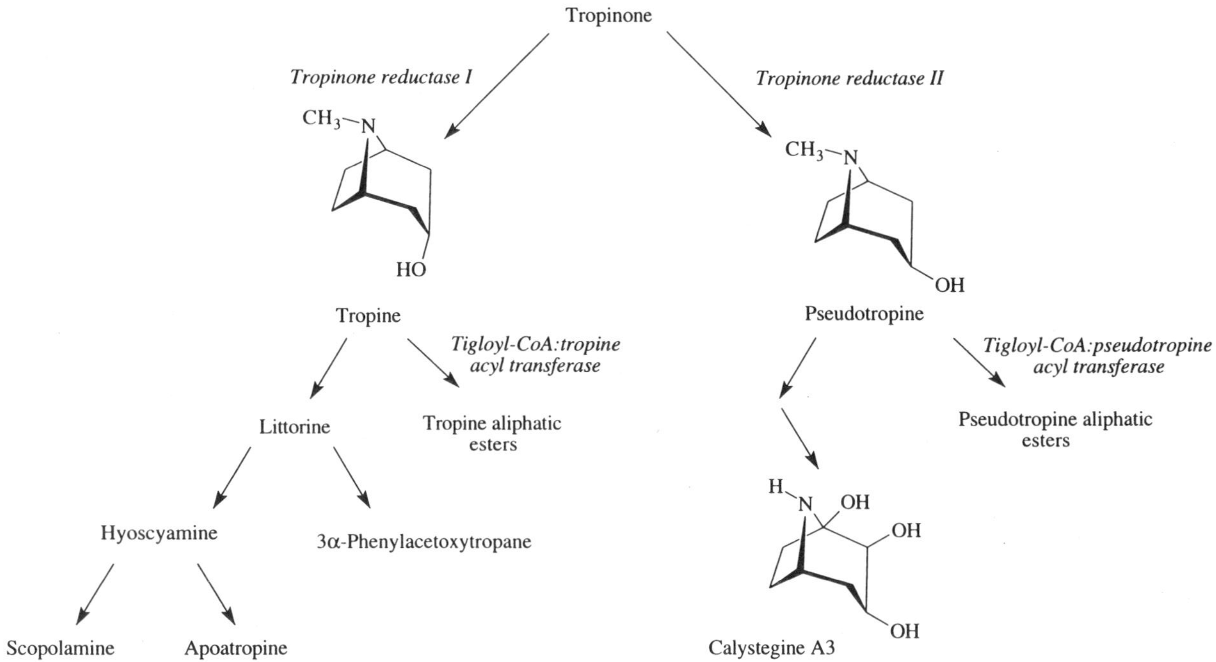

Figure 6. The diverse groups of products that originate from tropinone in species expressing the tropane alkaloid pathway.

these precursors caused more hyoscyamine to accumulate. Rather, tropine and acetyltropine levels rose. From these findings, it was deduced that, in these root cultures, the esterification of tropine may be crucial in limiting hyoscyamine accumulation. In order to ensure that this was not just related to a lack of phenyllactic acid, both tropic and phenyllactic acids were fed. Neither enhanced hyoscyamine accumulation (Robins *et al.,* 1991a).

Some root cultures, notably of *Hyoscyamus* and *Duboisia*, accumulate scopolamine as a major product (Robins and Walton, 1993). In contrast, only traces of scopolamine were found in *D. stramonium* roots. This implies that the expression of hyoscyamine 6β-hydroxylase (H6H) that forms the 6,7-epoxide is variable and a good target for genetic engineering (see below and Chapter 6).

3.2.2. STUDIES THAT DEFINE THE METABOLIC PATHWAY

Numerous labeling experiments have been performed in intact plants aimed at delineating the tropane pathway (Leete, 1990). Nevertheless, with the advent of studies in root cultures a number of steps remained to be unraveled. In addition, some inconsistencies had been reported which were resolved by studying the biochemistry of the pathway.

The first of these concerned the formation of *N*-methylputrescine. On evidence obtained from feeding experiments with specifically labeled putrescines, Leete and McDonell (1981) suggested that hyoscyamine formation in *Datura* might proceed via, 4-*N*-methylornithine, not *N*-methylputrescine. Walton *et al.* (1990) were able to show, however, that this alternative pathway was not active as neither of the enzymes required could be detected, whereas ODC and PMT were present in excess. Thus, it was confirmed that PMT, as in *Nicotiana*, is the first enzyme of the pathway. Further support for this conclusion is given in Section 3.2.3.

Again as in *Nicotiana*, it was debatable whether ADC or ODC might provide the putrescine incorporated. This possibility was tested by growing roots in the presence of DFMO and DFMA (Robins *et al.,* 1991b) in experiments analogous to those done by Tiburcio and Galston (1986) in *Nicotiana*. Inhibiting specifically ADC, depressed hyoscyamine accumulation and the pools of intermediates, indicating that ADC might be more important for the tropane alkaloid pathway also.

Another area of uncertainty concerned the route by which the tropic acid moiety is incorporated. Although early reports claimed to synthesize hyoscyamine from tropine and tropic acid or tropoyl-CoA, these findings were not readily substantiated (Robins and Walton, 1993). Furthermore, while good incorporation of label from phenylalanine was possible, tropic acid was hardly incorporated at all. A series of experiments in which labeled phenyllactic acids were fed to plants of *D. innoxia* or root cultures of *D. stramonium* confirmed unequivocally that this compound was an intermediate of the pathway (Chesters *et al.,* 1994, 1995; Robins *et al.,* 1994a; Ansarin and Woolley, 1993, 1994) and that it is the (*R*)-isomer that is incorporated (Chesters *et al.,* 1995). Tropic acid was found unable to dilute out the incorporation of ^{13}C from phenyllactic acid, indicating that *free* tropic acid was probably not an intermediate in the pathway (Robins *et al.,* 1994a). This deduction was strengthened by a subsequent experiment in which it was demonstrated that littorine (the phenyllactoyl ester of tropine) rearranges *in vivo* to hyoscyamine (Robins *et al.,* 1994b). Direct rearrangement was demonstrated unequivocally by incorporating three ^{2}H nuclei in the *N*-methyl of the tropinyl portion and two ^{13}C nuclei in the phenyllactoyl moiety.

During the incorporation experiments, it was noticed that ^{13}C from phenyl-[1,3-^{13}C]-

lactate was incorporated into two other alkaloids, 3α-phenylacetoxytropane and 3α-(2′-hydroxyacetoxy)tropane with retention of the nuclei. Furthermore, both nuclei of (*R*)-phenyl-[2-^{2}H, 2-^{13}C]lactate were retained in 3α-(2′-hydroxyacetoxy)tropane but neither was incorporated into 3α-phenylacetoxytropane. These unexpected results led to the proposal that the rearrangement of littorine to hyoscyamine might occur by a P_{450}-catalyzed reaction and that these minor alkaloids are adventitious side-products of the mechanism (Robins *et al.,* 1995).

The calystegines were first shown to belong biosynthetically to the tropane alkaloids by Goldmann *et al.* (1992), who showed that [^{14}C]ornithine was incorporated. This finding has been confirmed by Dräger *et al.* (1994), who got good incorporation of ^{15}N-labeled tropinone by root cultures of *Atropa belladonna.* By contrast, feeding experiments with plants of *D. innoxia* (Abraham *et al.,* 1995) and root cultures of *D. stramonium* (Robins *et al.,* 1997) have shown that hygrine is not on the direct pathway to hyoscyamine. When ^{13}C-labeled hygrine was fed, no incorporation into hyoscyamine or scopolamine could be detected, although it was efficiently incorporated into cuscohygrine. Instead, these authors have demonstrated, by feeding ^{13}C-labeled substrate, that 4-(1-methyl-2-pyrrolidinyl)-3-oxobutanoate is probably an intermediate.

3.2.3. STUDIES OF THE REGULATION OF THE PATHWAY

In attempting to understand such a pathway at the biochemical level, it is important to identify those enzymes that can be expected to make a strong contribution to overall flux control and/or that are sensitive to endogenous or exogenous effectors. As in *Nicotiana,* PMT appears to play a key role in affecting the overall flux into the pathway. Treating *D. stramonium* root cultures with phytohormones causes dispersion of the cultures, an active degradation of tropine and hyoscyamine, and a loss of PMT activity (Robins *et al.,* 1991c). PMT is completely absent in dispersed cultures. Root integrity, alkaloid production, and PMT activity are fully restored following the removal of phytohormones. Now that a clone for PMT is available, it will be most interesting to examine this phenomenon at the molecular level. By examining levels of transcript and comparing these with PMT activity and protein (determined with antibodies), it should be possible to determine the turnover of PMT and relate its rapid disappearance to changes in synthesis or degradation. Hibi *et al.* (1994) clearly demonstrate that *pmt* expression in *N. tabacum* plants is downregulated by auxin, in agreement with the observed effect of auxins on PMT activity in root cultures (Robins *et al.,* 1991c; Rhodes *et al.,* 1989).

The important regulatory role of PMT is also indicated by its apparent sensitivity to feed-forward activation by agmatine. In roots grown under normal conditions, the level of PMT is not influenced by adding exogenous agmatine. Roots treated with DFMA, however, contained decreased PMT activity (Robins *et al.,* 1991b) but normal levels were restored by adding agmatine to the cultures. As DFMA treatment results in a loss of agmatine from the system, this effect was interpreted as a possible stimulation of PMT expression by agmatine. Again, this can now be tested at the molecular level.

3.2.4. STUDIES OF THE ENZYMES OF TROPANE ALKALOID BIOSYNTHESIS

Several enzymes of tropane alkaloid biosynthesis have been purified and characterized from root cultures. This work will only be briefly summarized here.

3.2.4a. Putrescine *N*-Methyltransferase

This enzyme proved to be very difficult to purify but, after 3 years' work, Walton *et al.* (1994) were able to report its purification from *D. stramonium*. The key to this success was the use of affinity chromatography on an L-ornithine–Sepharose matrix. The properties of a partially purified *H. niger* protein (Hibi *et al.*, 1992) are very similar.

3.2.4b. Tropinone Reductases I and II

The tropine-forming (TR I) and pseudotropine-forming (TR II) tropinone reductases, respectively, have been purified from *A. belladonna* (Dräger and Schaal, 1994), *D. stramonium* (Portsteffen *et al.*, 1992, 1994), and *H. niger* (Hashimoto *et al.*, 1992). The crucial separation of the two activities was achieved either by the sequential use of Butyl-Fractogel and TMAE-Fractogel or by chromatography on DEAE-Sepharose.

TR I and TR II have quite distinct properties. The recent cloning of the genes for these enzymes (Nakajima *et al.*, 1993), however, indicates that they are structurally very similar. The major difference, it is suggested, probably lies in the orientation in which the substrate enters the active site.

3.2.4c. Tigloyl-CoA: Pseudotropine Acyltransferase

This enzyme has been purified from roots of *D. stramonium* (Rabot *et al.*, 1995). It catalyzes the transfer to pseudotropine of an acyl group from a range of acyl-CoA thioesters. A similar enzyme, to date only partially purified, catalyzes the equivalent reaction with tropine (Robins *et al.*, 1991d). The purpose of this pseudotropine-utilizing enzyme is not apparent. Esters of pseudotropine do not accumulate significantly in *D. stramonium* roots, although they do appear under abnormal metabolic conditions (Dräger *et al.*, 1992).

3.2.4d. Hyoscyamine 6β-Hydroxylase

This was the first enzyme of tropane alkaloid metabolism to be purified and remains the most rigorously studied. It was obtained in pure form from *H. niger* root cultures and the preparation showed activity both as the 7β-hydroxylase and as the 6,7β-epoxidase (Hashimoto and Yamada, 1987). A clone for H6H was obtained following the purification of enzyme activity. The gene shows some similarity to other hydroxylases, including those involved in oxidative reactions resulting in the formation of ethylene and anthocyanins (Matsuda *et al.*, 1991).

3.3. Genetic Engineering of Tropane Alkaloid Biosynthesis

3.3.1. GENETIC MANIPULATION OF SCOPOLAMINE FORMATION

Typically, the alkaloid spectrum of transformed root cultures of *A. belladonna* contains hyoscyamine and scopolamine in a ratio between 10: 1 and 5: 1 (Robins and Walton,

1993). Following the introduction of the *h6h* gene into cultures of transformed roots of *A. belladonna*, an engineered root line was isolated (Hashimoto *et al.,* 1993) that showed increased H6H activity and about a 2-fold higher accumulation of 7β-hydroxyhyoscyamine and scopolamine. This experiment effectively demonstrated that the ability of these cultures to accumulate hyoscyamine was limited by H6H activity and that, by engineering the appropriate gene, this deficiency could be rectified. *A. belladonna* plants, expressing the *h6h* gene, almost exclusively contained scopolamine, in contrast to controls (Yun *et al.,* 1992).

3.3.2. USE OF GENETIC ENGINEERING TO DEFINE THE PATHWAY

As indicated above, purified H6H showed both hyoscyamine 6β-hydroxylase and hyoscyamine 6,7β-epoxidase activities. Whether or not these two reactions were catalyzed by the same polypeptide had been debated for some time until the experiments of Yun *et al.* (1993) showed that only the single polypeptide is required to carry out both reactions. The *h6h* was inserted into transgenic *N. tabacum* plants. These plants, as a result of this single insertion, acquired the ability to biotransform hyoscyamine into scopolamine, showing unequivocally that a single gene product was responsible for both the hydroxylation and epoxidation steps.

4. THE PYRROLIZIDINE ALKALOIDS

This group of alkaloids is found in a wide range of families, centered around the Asteraceae and Boraginaceae (Hartmann and Witte, 1995). Their occurrence in many *Senecio* species accounts for the high toxicity of these plants, which are commonly responsible for the loss of livestock and, occasionally, for the poisoning of humans. Curiously, the bases are not themselves toxic but are activated by the intestinal flora following ingestion.

4.1. The Biosynthetic Route to the Pyrrolizidine Alkaloids

The biosynthesis of pyrrolizidine alkaloids has been studied mostly in *Senecio* species (Robins, 1989, 1991). These alkaloids are esters between a necine base, derived from arginine or ornithine via homospermidine (Fig. 7), and a necic acid moiety, frequently derived from isoleucine. The formation of homospermidine from two molecules of putrescine is the first committed step in the pathway and pyrroline is not an intermediate (Böttcher *et al.,* 1993). Thus, there is a close parallel between this pathway and that described for the tropane alkaloids with two pathways starting from amino acids that provide acidic and alkamine moieties which are condensed by esterification late in the pathway.

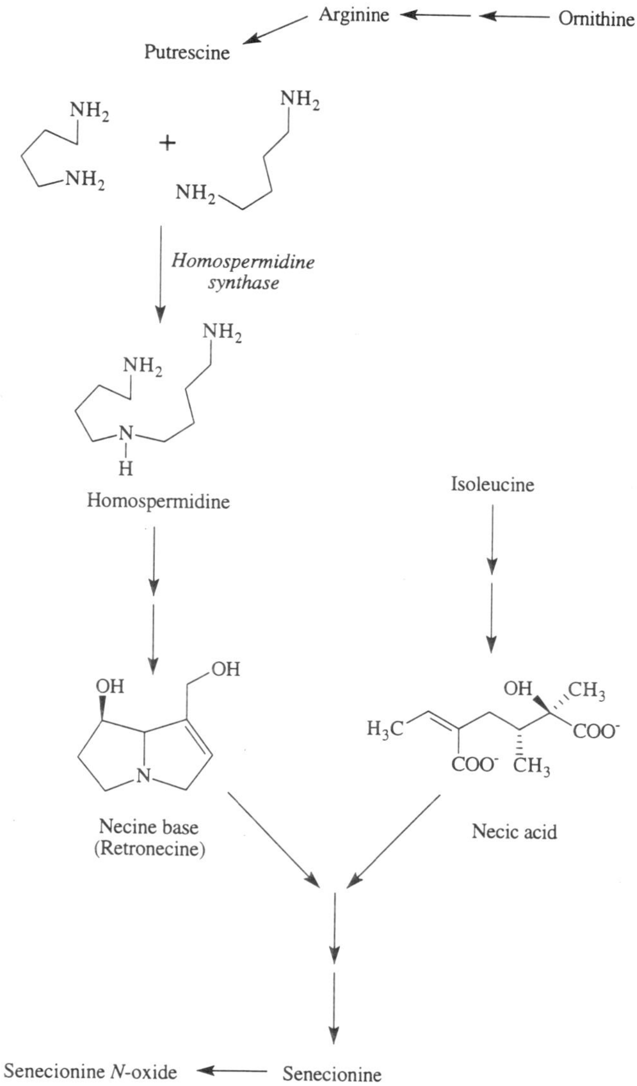

Figure 7. Scheme showing the pathway of biosynthesis of senecionine-*N*-oxide from ornithine/arginine and isoleucine.

4.2. Regulation of the Biosynthetic Pathway

4.2.1. STUDIES OF THE FLUX THROUGH THE PATHWAY

The advent of root cultures of *Senecio* competent in pyrrolizidine alkaloid biosynthesis (Toppel *et al.,* 1987) was an important step in studying their biosynthesis. The major product accumulated is senecionine *N*-oxide (Fig. 1). Neither suspension cultures nor shoot cultures of *Senecio* will make pyrrolizidine alkaloids, indicating that the root is the sole site of biosynthesis (Hartmann, 1994). Other genera, however, have proved less capable of making pyrrolizidine alkaloids in culture, even as roots.

Feeding experiments in root cultures of *S. vulgaris* with a range of ^{14}C-labeled precursors and inhibitors of metabolism showed both ornithine and arginine to be incorporated into senecionine *N*-oxide (Hartmann *et al.,* 1988). Unexpectedly, DFMA prevented labeling from both precursors, whereas DFMO had no effect. Thus it would appear that, in contrast to *Nicotiana* and *Datura*, label from ornithine is incorporated via arginine. The mechanism for this is not clear. Interestingly, spermidine was as well incorporated as putrescine, indicating a rapid interconversion of these two metabolites. Both spermine and spermidine reduced the incorporation of label from arginine, suggesting that there is feedback control of agmatine biosynthesis which, adventitiously, leads to a depression of alkaloid formation. This interaction between alkaloid and polyamine formation was not apparent in *D. stramonium* roots (Robins *et al.,* 1991b). Possibly a higher degree of regulation is required in this pathway because of the greater demand for putrescine—homospermidine requiring 2 moles per mole of alkaloid made.

The senecionine *N*-oxide that accumulates is apparently made only in the actively growing parts of the root cultures. It does not significantly turn over but is slowly transported throughout the root mass, as shown by pulse-feeding experiments (Sander and Hartmann, 1989). Some limited metabolism occurs during this process, primarily oxidation and acetlylation.

The biosynthesis of the necic acid moiety has, in contrast, received relatively little attention. Label from [^{14}C]isoleucine is effectively incorporated into senecionine *N*-oxide by root cultures of *S. vulgaris* (Hartmann and Toppel, 1987).

4.2.2. STUDIES OF THE ENZYMES OF PYRROLIZIDINE ALKALOID FORMATION

To date, only one enzyme has been partially purified and characterized for this pathway.

4.2.2a. Homospermidine Synthase

This labile activity was successfully purified about 60-fold from root cultures of *Eupatorium cannabinum* after it was found that spermidine and NAD^+ would stabilize the protein (Böttcher *et al.,* 1993). The key to purification was a step involving affinity chromatography on a spermidine–Sepharose matrix, a solution similar to that used by Walton *et al.* (1994) for another labile putrescine-metabolizing enzyme, PMT. The activity is present in all pyrrolizidine-alkaloid-forming species so far examined and its presence confirms that the biosynthetic route is not via free pyrroline.

The enzyme is intriguing in that it carries out two sequential steps the first of which is a deaminative oxidation that generates NADH and the second of which is a reduction, utilizing NADH. This reaction sequence was clearly demonstrated by using chirally C1-^2H labeled putrescines (Böttcher *et al.,* 1994). These authors also suggest that spermidine may act, at least in part, as a cosubstrate with putrescine in homospermidine formation. The apparently rapid interconversion of putrescine and spermidine in these cultures, however, makes this a difficult problem to unravel. Unlike many enzymes of secondary metabolism, homospermidine synthase is inactive with $NADP^+$, perhaps indicating an origin in primary metabolism.

5. CONCLUSIONS

Our understanding of alkaloid biosynthesis has been greatly improved by the use of culture systems. Root cultures have played, and will continue to play, an important role in this field of investigation. They are not, however, a solution to all of the problems of alkaloid production in culture. First, as indicate above, disorganized suspension cultures can be highly productive for some classes of alkaloid. Secondly, some important genera have proved intransigent to culture as roots. In some cases, even when cultures have been generated, the desired products are absent. For example, Williams and Ellis (1993) could not detect morphinan alkaloids in transformed root cultures of *Papaver somniferum.* Nevertheless, as in the examples given above, the presence in root culture of organized tissue appears beneficial to the formation of natural products. Thirdly, some products are not biosynthesized in roots but require shooty cultures. The essential oils of *Mentha* species have only been made *in vitro* in shooty cultures (Spencer *et al.,* 1993). Root cultures do have the advantage that they can be grown reproducibly on a large scale to provide kilogram amounts of homogeneous material without losing their biosynthetic capacity (Hilton and Rhodes, 1990). Thus, they offer excellent material for the isolation of both enzymes and mRNA. Whether the early predictions of their value for the biotechnological production of natural products (Hamill *et al.,* 1987) will be substantiated remains to be seen but they will continue to provide the research scientist with a useful experimental system for many years to come.

REFERENCES

General Reviews

Butcher, D. N., and Street, H. E., 1964, Excised root culture, *Bot. Rev.* **30:**513–586.

Hamill, J. D., Parr, A. J., Rhodes, M. J. C., Robins, R. J., and Walton, N. J., 1987, New routes to secondary products, *Bio/Technology* **5:**800–804.

Hartmann, T., 1994, *Senecio* species: Biochemistry of the formation of pyrrolizidine alkaloids in root cultures, in: *Biotechnology in Agriculture and Forestry,* Vol. 26 (Y. P. S. Bajaj, ed.), Springer-Verlag, Berlin, pp. 339–355.

Hartmann, T., and Witte, L., 1995, Chemistry, biology and chemoecology of the pyrrolizidine alkaloids, in: *Alkaloids: Chemical and Biological Perspectives,* Vol. 9 (S. W. Pelletier, ed.), Pergamon Press, Oxford, pp. 155–233.

Leete E., 1990, Recent developments in the biosynthesis of tropane alkaloids, *Planta Med.* **56:**339–352.

Rhodes, M. J. C., Robins, R. J., Aird, E. L. H., Payne, J., Parr, A. J., and Walton, N. J., 1989, Regulation of secondary metabolism in transformed root cultures, in: *Primary and Secondary Metabolism of Plant Cell Cultures II* (W. G. W. Kurz, ed.), Springer-Verlag, Berlin, pp. 58–72.

Rhodes, M. J. C., Robins, R. J., Parr, A. J., and Walton, N. J., 1990, Secondary metabolism in transformed root cultures, in: *Secondary Products from Plant Tissue Culture* (B. V. Charlwood and M. J. C. Rhodes, eds.), Oxford University Press, Oxford, pp. 201–225.

Robins, D. J., 1989, Biosynthesis of pyrrolizidine alkaloids, *Chem. Soc. Rev.* **18:**375–408.

Robins, D. J., 1991, Stereochemistry of enzymic processes in the biosynthesis of pyrrolizidine alkaloids, *Experientia* **47:**1118–1122.

Robins, R. J., and Walton, N. J., 1993, The biosynthesis of tropane alkaloids, in: *The Alkaloids*, Vol. 44 (G. A. Cordell, ed.), Academic Press, San Diego, pp. 115–187.

Zenk, M. H., 1989, Biosynthesis of alkaloids using plant cell cultures, *Recent Adv. Phytochem.* **23:**429–457.

Key References

Abraham, T. W., and Leete, E., 1995, New intermediate in the biosynthesis of tropane alkaloids in *Datura innoxia. J. Amer. Chem. Soc.* **117:**8100–8105.

Ansarin, M., and Woolley, J. G., 1993, The obligatory role of phenyllactate in the biosynthesis of tropic acid, *Phytochemistry* **32:**1183–1187.

Ansarin, M., and Woolley, J. G., 1994, The rearrangement of phenyllactic acid in the biosynthesis of tropic acid, *Phytochemistry* **35:**935–939.

Böttcher, F., Adolph, R.-D., and Hartmann, T., 1993, Homospermidine synthase, the first pathway-specific enzyme in pyrrolizidine alkaloid biosynthesis, *Phytochemistry* **32:**679–689.

Böttcher, F., Ober, D., and Hartmann, T., 1994, Biosynthesis of pyrrolizidine alkaloids: Putrescine and spermidine are essential substrates of enzymatic homospermidine formation, *Can. J. Chem.* **72:**80–85.

Chesters, N. C. J. E., O'Hagan, D., and Robins, R. J., 1994, The biosynthesis of tropic acid in plants. Evidence for the direct rearrangement of phenyllactate to tropate, *J. Chem. Soc. Perkin Trans. I* 1159–1162.

Chesters, N. C. J. E., O'Hagan, D., and Robins, R. J., 1995, The biosynthesis of tropic acid: The (*R*)-D-phenyllactyl moiety is processed by the mutase involved in hyoscyamine biosynthesis in *Datura stramonium, J. Chem. Soc. Chem. Commun.* 127–128.

Dräger, B., and Schaal, A., 1994, Tropinone reduction in *Atropa belladonna* root cultures, *Phytochemistry* **35:**1441–1447.

Dräger, B., Portsteffen, A., Schaal, A, McCabe, P. H., Peerless, A. C. J., and Robins, R. J., 1992, Levels of tropinone-reductase activities influence the spectrum of tropane esters found in transformed root cultures of *Datura stramonium, Planta* **188:**581–586.

Dräger, B., Funck, C., Höhler, A., Mrachatz, G., Portsteffen, A., Schaal, A., and Schmidt, R., 1994, Calystegines as a new group of tropane alkaloids in Solanaceae, *Plant Cell Tissue Org. Cult.* **38:**235–240.

Fecker, L. F., Hillebrandt, S., Rügenhagen, C., Herminghaus, S., Landsmann, J., and Berlin, J., 1992, Metabolic effects of a bacterial lysine decarboxylase gene expressed in hairy root culture of *Nicotiana glauca, Biotechnol Lett.* **14:**1035–1040.

Friesen, J. B., and Leete, E., 1990, Nicotine synthase—An enzyme from *Nicotiana* species which catalyses the formation of *(S)*-nicotine from nicotinic acid and 1-methyl-Δ'-pyrrolinium chloride, *Tetrahedron Lett.* 6295–6298.

Goldmann, A., Milat, A.-L., Ducrot, P.-H., Lallemand, J.-Y., Maille, M., Lepingle, A., Charpin, I., and Tepfer, D., 1992, Tropane derivatives from *Calystegia sepium. Phytochemistry* **29:**2125–2127.

Hamill, J. D., Parr, A. J., Robins, R. J., and Rhodes, M. J. C., 1986, Secondary product formation by cultures of *Beta vulgaris* and *Nicotiana rustica* transformed with *Agrobacterium rhizogenes, Plant Cell Rep.* **5:**111–114.

Hamill, J. D., Robins, R. J., Parr, A. J., Evans, D. M., Furze, J. M., and Rhodes, M. J. C., 1990, Over-expressing a yeast ornithine decarboxylase gene in transgenic roots of *Nicotiana rustica* can lead to enhanced nicotine formation, *Plant Mol. Biol.* **15:**27–38.

Hartmann, T., and Toppel, G., 1987, Senecionine *N*-oxide, the primary product of pyrrolizidine alkaloid biosynthesis in root cultures of *Senecio vulgaris, Phytochemistry* **26:**1639–1643.

Hartmann, T., Sander, H., Adolph, R., and Toppel, G., 1988, Metabolic links between the biosynthesis of pyrrolizidine alkaloids and polyamines in root cultures of *Senecio* species, *Planta* **175:**82–90.

Hashimoto, T., and Yamada, Y., 1983, Scopolamine production in suspension cultures and redifferentiated roots of *Hyoscyamus niger, Planta Med.* **47:**195–199.

Hashimoto, T., and Yamada, Y., 1987, Purification and characterization of hyoscyamine 6β-hydroxylase from root cultures of *Hyoscyamus niger* L.: Hydroxylase and epoxidase activities in the preparation, *Eur. J. Biochem.* **164:**277–285.

Hashimoto, T., Yukimune, Y., and Yamada, Y., 1986, Tropane alkaloid production in *Hyoscyamus* root cultures, *J. Plant Physiol.* **124:**61–75.

Hashimoto, T., Yukimune, Y., and Yamada, Y., 1989, Putrescine and putrescine *N*-methyltransferase in the biosynthesis of tropane alkaloids in cultured roots of *Hyoscyamus albus,* I. Biochemical studies, *Planta* **178:**123–130.

Hashimoto, T., Mitani, A., and Yamada, Y., 1990, Diamine oxidase from cultured roots of *Hyoscyamus niger, Plant Physiol.* **93:**216–221.

Hashimoto, T., Nakajema, K., Ongena, G., and Yamada, Y., 1992, Two tropinone reductases with distinct stereospecificities from cultured roots of *Hyoscyamus niger, Plant Physiol.* **100:**836–845.

Hashimoto, T., Yun, D.-J., and Yamada, Y., 1993, Production of tropane alkaloids in genetically engineered root cultures, *Phytochemistry* **32:**713–718.

Hibi, N., Fujita, T., Hatano, M., Hashimoto, T., and Yamada. Y., 1992, Putrescine *N*-methyltransferase in cultured roots of *Hyoscyamus albus, Plant Physiol.* **100:**826–835.

Hibi, N., Higashiguchi, S., Hashimoto, T., and Yamada. Y., 1994, Gene expression in tobacco low-nicotine mutants, *Plant Cell* **6:**723–735.

Hilton M., and Rhodes, M. J. C., 1990, Growth and hyoscyamine production of 'hairy root' cultures of *Datura stramonium* in a modified stirred tank reactor, *Appl. Microbiol. Biotechnol.* **33:**132–138.

Jones, J. D. G., Shlumukov, L., Carland, F., English, J., Scofield, S. R., Bishop, G. J., and Harrison, K., 1992, Effective vectors for transformation, expression of heterologous genes, and assaying transposon excision in plants, *Transgenic Res.* **1:**285–297.

Leete, E., and McDonell, J. A., 1981, The incorporation of [1-^{13}C,^{14}C, *methylamino*-^{15}N]-*N*-methylputrescine into nicotine and scopolamine established by means of carbon-13 nuclear magnetic resonance, *J. Am. Chem Soc.* **103:**658–662.

Matsuda, J., Okabe, S., Hashimoto, T., and Yamada, Y., 1991, Molecular cloning of hyoscyamine 6β-hydroxylase, a 2-oxoglutarate-dependent dioxygenase, from cultured roots of *Hyoscyamus niger, J. Biol. Chem.* **266:**9460–9464.

McLauchlan, W. R., McKee, R. A., and Evans, D. M., 1993, The purification and immunocharacterisation of *N*-methylputrescine oxidase from transformed root cultures of *Nicotiana tabacum* L. cv SC58, *Planta* **191:**440–445.

Mitra, G. C., 1972, Atropine production and growth of excised Belladonna root culture, *Indian J. Exp. Biol.* **10:**217–218.

Mizusaki, S., Tanabe, Y., Noguchi, M., and Tamaki, E., 1973, Changes in the activities of ornithine decarboxylase, putrescine *N*-methyltransferase and *N*-methylputrescine oxidase in tobacco roots in relation to nicotine biosynthesis, *Plant Cell Physiol.* **14:**103–110.

Nakajima, K., Hashimoto, T., and Yamada, Y., 1993, Two tropinone reductases with different stereospecificities are short-chain dehydrogenases evolved from a common ancestor, *Proc. Natl. Acad. Sci. USA* **90:**9591–9595.

Ohta, S., and Yatazawa, M., 1980, Metabolic key step discriminating nicotine producing tobacco callus strain from ineffective one, *Biochem. Physiol. Pflanz.* **175:**382–385.

Payne, J., Hamill, J. D., Robins, R. J., and Rhodes, M. J. C., 1987, Production of hyoscyamine by 'hairy root' cultures of *Datura stramonium, Planta Med.* **53:**474–478.

Portsteffen, A., Dräger, B., and Nahrstedt, A., 1992, Two tropinone reducing enzymes from *Datura stramonium* transformed root cultures, *Phytochemistry* **31:**1135–1138.

Portsteffen, A., Dräger, B., and Nahrstedt, A., 1994, The reduction of tropinone in *Datura stramonium* root cultures by two specific reductases, *Phytochemistry* **37:**391–400.

Rabot, S., Peerless, A. C. J., and Robins, R. J., 1995, Tigloyl-CoA:pseudotropine acyl transferase—a novel enzyme of tropane alkaloid biosynthesis, *Phytochemistry* **39:**315–322.

Robins, R. J., Hamill, J. D., Parr, A. J., Smith, K., Walton, N. J., and Rhodes, M. J. C., 1987, Potential for use of nicotinic acid as a selective agent for isolation of high nicotine-producing lines of *Nicotiana rustica* hairy root cultures, *Plant Cell Rep.* **6:**122–126.

Robins, R. J., Parr, A. J., Payne, J., Walton, N. J., and Rhodes, M. J. C., 1990, Factors regulating tropane-alkaloid production in a transformed root culture of a *Datura candida* × *D. aurea* hybrid, *Planta* **181:**414–422.

Robins, R. J., Parr, A. J., Bent, E. G., and Rhodes, M. J. C., 1991a, Studies on the formation of tropane alkaloids in *Datura stramonium* L. transformed root cultures, 1. The kinetics of alkaloid production and the influence of feeding intermediate metabolites, *Planta* **183:**185–195.

Robins R. J., Parr, A. J., and Walton, N. J., 1991b, Studies on the formation of tropane alkaloids in *Datura stramonium* L. transformed root cultures, 2. On the relative contributions of L-arginine and L-ornithine to the formation of the tropane ring, *Planta* **183:**196–201.

Robins R. J., Bent, E. G., and Rhodes, M. J. C., 1991c, Studies on the formation of tropane alkaloids in *Datura stramonium* L. transformed root cultures, 3. The relationship between morphological integrity and alkaloid biosynthesis, *Planta* **185:**385–390.

Robins, R. J., Bachmann, P., Robinson, T., Yamada, Y., and Rhodes, M. J. C., 1991d, The formation of 3α and 3β-acetoxytropanes by *Datura stramonium* transformed root cultures involves two acetyl-CoA-dependent acyltransferases, *FEBS Lett.* **292:**293—297.

Robins, R. J., Woolley, J. G., Ansarin, M., Eagles, J., and Goodfellow, B. J., 1994a, Phenyllactic acid but not tropic acid is an intermediate in the biosynthesis of tropane alkaloids in *Datura* and *Brugmansia* transformed root cultures, *Planta* **194:**86–94

Robins, R. J., Bachmann, P., and Woolley, J. G., 1994b, Biosynthesis of hyoscyamine involves an intramolecular rearrangement of littorine, *J. Chem. Soc. Perkin Trans. I* 615–619.

Robins, R. J., Chesters, N. C. J. E., O'Hagan, D., Parr, A. J., Walton, N. J., and Woolley, J. G., 1995, The biosynthesis of hyoscyamine: The process by which littorine rearranges to hyoscyamine, *J. Chem. Soc. Perkin Trans. I* 481–485.

Robins, R. J., Abraham, T. W., Parr, A. J., Eagles, J., and Walton, N. J., 1997, The biosynthesis of tropane alkaloids in *Datura stramonium:* The identity of the intermediates between *N*-methylpyrrolinium salt and tropinone. *J. Amer. Chem. Soc.* (in press).

Röper, W., Schulz, M., Chaouiche, E., and Meloh, K. A., 1985, Nicotine production by tissue cultures of tobacco as influenced by various culture parameters, *J. Plant Physiol.* **118:**463–470.

Sander, H., and Hartmann, T., 1989, Site of synthesis, metabolism and translocation of senecionine *N*-oxide in cultured roots of *Senecio erucifolius, Plant Cell Tissue Org. Cult.* **18:**19–31.

Solt, M. L., 1957, Nicotine production and growth of excised tobacco root culture, *Plant Physiol.* **32:**480–484.

Spencer, A., Hamill, J. D., and Rhodes, M. J. C., 1993, *In vitro* biosynthesis of monoterpenes by *Agrobacterium* transformed shoot cultures of two *Mentha* species, *Phytochemistry* **32:**911–919.

Tiburcio, A. F., and Galston, A. W., 1986, Arginine decarboxylase as the source of putrescine for tobacco alkaloids, *Phytochemistry* **25:**107–110.

Toppel, G., Witte, L., Riebesehl, B., van Borstel, K., and Hartmann, T., 1987, Alkaloid patterns and biosynthetic capacity of root cultures of some pyrrolizidine alkaloid producing *Senecio* species, *Plant Cell Rep.* **6:**466–469.

Walton, N. J., and Belshaw, N. J., 1988. The effect of cadaverine on the formation of anabasine from lysine in hairy root cultures of *Nicotiana hesperis, Plant Cell Rep.* **7:**115–118.

Walton, N. J., and McLauchlan, W. R., 1990, Diamine oxidation and alkaloid production in transformed root cultures of *Nicotiana tabacum, Phytochemistry* **29:**1455–1457.

Walton, N. J., Robins, R. J., and Rhodes, M. J. C., 1988, Perturbation of alkaloid production by cadaverine in hairy root cultures of *Nicotiana rustica, Plant Sci.* **54:**125–131.

Walton, N. J., Robins, R. J., and Peerless, A. C. J., 1990, Enzymes of *N*-methylputrescine biosynthesis in relation to hyoscyamine formation in transformed root cultures of *Datura stramonium* and *Atropa belladonna, Planta* **182:**16–141.

Walton, N. J., Peerless, A. C. J., Robins, R. J., Rhodes, M. J. C., Boswell, H. D., and Robins, D. J., 1994, Purification and properties of putrescine *N*-methyltransferase from transformed roots of *Datura stramonium* L, *Planta* **193:**9–15.

Williams, R. D., and Ellis, B. E., 1993, Alkaloids from *Agrobacterium rhizogenes*-transformed *Papaver somniferum* cultures, *Phytochemistry* **32:**719–723.

Yun, D.-J., Hashimoto, T., and Yamada, Y., 1992, Metabolic engineering of medicinal plants: Transgenic *Atropa belladonna* with improved alkaloid composition, *Proc. Natl. Acad. Sci. USA* **89:**11799–11803.

Yun, D.-J., Hashimoto, T., and Yamada, Y., 1993, Expression of hyoscyamine 6β-hydroxylase gene in transgenic tobacco, *Biosci. Biotechnol. Biochem.* **57:**502–503.

Chapter 9

Induction of Alkaloid Biosynthesis and Accumulation in Plants and *in Vitro* Cultures in Response to Elicitation

U. Eilert

1. INTRODUCTION

This chapter reviews the area of induced rapid alkaloid accumulation in plants and focuses on the regulatory mechanisms involved. The majority of known alkaloids are constitutively expressed secondary metabolites and are considered to be part of a constitutive defense system. However, alkaloid biosynthesis is affected by environmental factors such as light/irradiation, temperature, and water supply. Changes in one or more of these factors exert direct effects on primary metabolism and thus influence the direction of secondary metabolism as well. The resulting alterations in alkaloid content are generally not expected to occur rapidly, i.e., within a few hours or 1–2 days. Alternatively, alkaloid biosynthesis may be directly regulated by environmental factors, if, for example, light receptors are involved and the pathway is localized partially or fully in plastids/chloroplasts. Thus, both direct and indirect regulatory effects resulting from photosynthesis would be responsible for changes in alkaloid accumulation. Overall alkaloid induction by environmental factors is difficult to interpret with respect to the physiological role of the products. As for the numerous data on the effect of culture conditions on alkaloid accumulation *in vitro*, only a description of the phenomena observed can presently be given. This topic is reviewed in related chapters of this book (Chapters 5, 7, and 8) and interesting comparisons of the effects of culture conditions, as well as the inducing factors, on the accumulation of secondary metabolites are made.

Typical events causing rapid induction of secondary metabolism are physical wounding or microbial infection. Although most alkaloid pathways are constitutively expressed,

U. Eilert • Institute for Pharmaceutical Biology, Technical University of Braunschweig, D38106 Braunschweig, Germany.

Alkaloids: Biochemistry, Ecology, and Medicinal Applications, edited by Roberts and Wink. Plenum Press, New York, 1998.

some have been shown to respond to such stimuli. The main emphasis in this review is on the analysis of the effect of microbial induction on alkaloid biosynthesis and accumulation, where significant progress has been made in the past decade. Rather than provide a detailed literature review, more emphasis will be placed on the presentation and discussion of well-studied systems. The best-characterized systems are compared in order to relate the underlying processes and regulatory mechanisms that control the induction and accumulation of specific secondary metabolites as far as they are presently understood. A clear separation has to be made between induction of alkaloid accumulation in soil grown plants and the responses of aseptically grown *in vitro* systems. Induced accumulation in the plant is of particular phytopathological interest as it certainly forms part of the plant's defense machinery. However, reports on induced alkaloid accumulation in plants are quite rare. *In vitro* induction, in contrast, has been reported for a number of different classes of alkaloids, which are expressed constitutively in the plant. Thus, the physiological and phytopathological significance of *in vitro* induction is not clear and subject to discussion. Cell culture systems are, however, ideal tools to study the biosynthesis of alkaloids at the transcriptional and translational level under controlled and reproducible conditions. The progress that has been made in these areas was largely possible because of such inducible *in vitro* systems.

2. INDUCTION OF ALKALOID PRODUCTION BY CLIMATIC AND SOIL CONDITIONS IN FIELD-GROWN PLANTS

Plants producing several pharmaceutically important alkaloids such as tropanes, poppy alkaloids, or various monoterpene indole alkaloids are cultivated for commercial production. Variation of alkaloid yield in these plants is of economic importance, thus the effects of climatic conditions and soil quality have been studied. The literature on this topic has been thoroughly reviewed by Waller and Nowacki (1978) and consists mostly of case reports describing specific phenomena without elucidation of the relevant biochemical background. In addition, in field experiments multifactorial changes occur rendering the results even more difficult to interpret with respect to the regulating factors. Relatively little new data have been reported concerning the effect of environmental factors on the alkaloid content of field-grown plants.

2.1. Induction by Heat Stress

An interesting observation was made by Hanson *et al.* (1983) in barley plants exposed to heat stress. Normally, seedlings accumulate *N,N*-dimethyltryptamine, known as gramine, only during the first days of development. When grown at high temperatures, 30°C by day and 25°C at night, plants maintained a high level of gramine, and maintained the initial rate of gramine synthesis over 14 days. Gramine is considered a defense compound with many different deterrent activities (Sepulveda and Corcuera, 1990; see Chapter 11).

2.2. Induction by Light

Light, as the source of energy in photosynthesis, affects primary and secondary metabolism of plants in a very complex way. Plants possess several types of photoreceptors for perception of light. These include the red/far-red phytochrome photoreceptors, the blue/UV-A light-absorbing cryptochrome, and one or more UV-B light-absorbing photoreceptors (Thompson, 1991) which may control expression of genes. Although light-induced alkaloid accumulation is difficult to interpret, it may be a consequence of a shift in primary metabolism but also steps in the alkaloid biosynthetic pathways may be directly light regulated.

The best known example for light inducible alkaloids is the accumulation of steroidal alkaloids in potato tubers. Conner (1937) found blue light and in particular UV light to possess the strongest inducing activity. A clear correlation between light intensity and alkaloid content was not established (Gull and Isenberg, 1960). Similar studies by Dale *et al.* (1993) compared light-induced alkaloid accumulation in relation to greening. Measurements were performed over 7 days. For each of six potato varieties tested, a good correlation between the relative increase in chlorophyll and glycoalkaloid content was established. However, a quantitative comparison of the induced levels of total chlorophyll and alkaloids of different varieties did not show this correlation. These results indicated that other factors, in addition to light, are contributing to the regulation of glycoalkaloid accumulation. Because of our lack of knowledge about the enzymes of steroidal alkaloid biosynthesis, it is not clear whether transcriptional, translational, and/or posttranslational modifications are involved in this light regulation.

Direct light regulation was demonstrated for monoterpene indole alkaloid accumulation in periwinkle (*Catharanthus roseus*) (DeLuca *et al.,* 1986; Balsevich *et al.,* 1986). When seedlings were germinated in the dark, they accumulated high levels of the indole alkaloid tabersonine and tabersonine derived intermediates of the pathway to vindoline. In contrast to light-grown controls, vindoline itself accumulated only in trace amounts. Aerts and DeLuca (1992) studied the physiological and biochemical background in detail. Exposure to white light induced a rapid increase in 2-oxoglutarate-dependent *N*-(1)-methyl-16-methoxy-2,3-dihydro-3-hydroxytabersonine-4-hydroxylase and acetyl CoA:deacetylvindoline 4-*O*-acetyltransferase, the two enzymes that catalyze the last two steps in vindoline biosynthesis (DeLuca *et al.,* 1988; DeCarolis *et al.,* 1990). The light-induced increase of the *O*-acetyltransferase persisted for 2–3 days followed by a decline. A 30-min exposure of dark-grown seedlings to red light was sufficient to trigger the induction process (Aerts and DeLuca, 1992). Posttreatment with far red light partially or completely prevented the induction of the two enzymes as well as vindoline accumulation. This indicates that phytochrome is at least partially involved in the light induced vindoline accumulation in *C. roseus* seedlings. A close correlation to photosynthesis may exist as a thylakoid membrane-associated *N*-methyltransferase is involved in the late stages of vindoline biosynthesis. In *C. roseus* regulatory effects are not limited to red light. In cell suspension cultures blue light irradiation caused an induction of serpentine accumulation and a concomitant decline of the level of its precursor alkaloid ajmalicine (Knobloch *et al.,* 1982). However, monoterpene indole alkaloid accumulation is not exclusively controlled by light as it is also developmentally regulated and tissue specific.

3. INDUCTION OF ALKALOID PRODUCTION *IN VITRO*

The relatively few publications on induction of alkaloid accumulation in field-grown plants do not mean a diminishing interest in the topic as a multitude of papers have been published on the effect of culturing conditions on secondary metabolism in plant cell cultures. In the 1970s plant cell cultures from almost any plant became widely available. They were recognized as ideal systems for the study of secondary metabolism under conditions more controlled than possible with potted or field-grown plants. Furthermore they offered a novel potential for biotechnological production of secondary metabolites. This potential remains to be realized despite many attempts to increase production. The two most popular approaches were to establish a series of cultures in hopes of finding high-producing cell lines followed by an optimization of culture conditions (see Chapter 7). The investigations of several cell lines derived from one plant (Constabel *et al.,* 1981) revealed that results obtained with one cell line alone were not always representative as other lines may behave differently. The importance of this fact is frequently ignored. Numerous studies reported the use of only one particular line to perform extensive and time-consuming physiological, biochemical, and/or molecular investigations. Contradictory results may simply be explained by culture-specific traits.

The extremely low success rate in finding highly productive cultures led to the realization that expression of many pathways is negatively correlated with growth and positively correlated with differentiation. These findings promoted experimentation with varied hormonal regimes as well as growth-limiting nutrients. Studies were also expanded to include monitoring of enzymes of respective pathways. In the course of such experiments with suspension cultures of *C. roseus* a production medium was developed that induced the accumulation of monoterpene indole alkaloids (Zenk *et al.,* 1977). The medium contained an elevated sucrose and a reduced nitrogen concentration, an altered hormone regime, and lacked phosphate. It would not sustain growth when used for continued subculturing. Alkaloid accumulation would begin about 5 days after transfer to this medium and reach maximum accumulation by 10–15 days after transfer. At the biochemical level a transient induction of tryptophan decarboxylase, the enzyme channeling tryptophan into the pathway, was measured (Noé and Berlin, 1985) and the highest specific activity was reached around 24 hours after transfer to production medium. In contrast, strictosidine synthase, which catalyzes the formation of the first monoterpene indole alkaloid, was not induced. The pattern and amount of accumulated alkaloids varied widely among strains. Several different medium components were identified to affect alkaloid accumulation and a "super production medium" was expected. However, optimum concentrations of each component did not result in an additive or synergistic effect. Because one induction factor could even counteract another, multiple modes of action appear to be involved. For example, the induction of alkaloid biosynthesis by elevated sucrose concentration may not only be the consequence of the change in carbon nutrition but may also occur as a consequence of osmotic stress. Rapid alkaloid induction by osmotic stress is well documented for different classes of alkaloids. In *C. roseus* cell cultures alkaloid accumulation was induced by adding 8% (w/v) sucrose to the medium (Knobloch and Berlin, 1980). Sugar alcohols (e.g., sorbitol, 100–200 mM) or high salt concentrations (e.g., 25–50 mM NaCl or 20–40 mM KCl) also induced a 1.5- to 2-fold increase of alkaloid level within 3 days after application (Smith *et al.*, 1987a). Effects of other nutrients and hormones will not be discussed here (see Chapter 7).

The development of production media for other cell cultures and for the production of other secondary metabolites has also been reported, e.g., benzophenanthridine accumulation in *Eschscholtzia californica* cultures (Berlin *et al.*, 1983). These data are in contrast to results by Hösel *et al.* (1985); their suspension cultures of *E. californica* formed 1-(4′-hydroxyphenyl)-2-nitroethane, a HCN-releasing compound, when cultured in normal growth medium supplemented with 6% sorbitol. Maximum content was reached within 3 days. Neither use of a complex production medium nor addition of fungal elicitors (see below) induced accumulation of benzophenanthridine alkaloids.

Purine alkaloid accumulation in suspension cultures of *Coffea arabica* grown under salt stress (125 mM NaCl) was dependent on aggregate size. Salt inhibited alkaloid accumulation in small cell aggregates but enhanced it in large ones. Growth rate remained largely unaffected (Frischknecht and Baumann, 1985). Purine alkaloid content was increased when low producing cultures were exposed to high light intensity. Additive effects were observed when cultures were exposed to NaCl stress and high light intensity simultaneously. Under these conditions a shift in alkaloid profile from theobromine to caffeine also occurred.

These examples for induction of alkaloid accumulation by osmotically stressed cultures raise the question of the physiological function. Generally, secondary metabolite levels are higher in plants growing under stress conditions. Increased secondary metabolite level contributes to the protection of precious tissue. In the natural environment saline soils present a permanent osmotic stress to which plants must adapt from germination to maturity. Thus, a special, rapid inducibility of secondary metabolism in response to osmotic stress is not likely to exist in the plant except in young seedlings. Cell cultures, however, are artificial systems and are very sensitive to their environment as their cells, particularly in suspension cultures, are much more exposed to their environment than are cells of a differentiated, soil-grown plant. Responses to osmotic stress in cell cultures may be markedly different from those in the plant and difficult to interpret with respect to their physiological significance. However, the activated mechanisms are possibly related to wound reactions.

4. INDUCTION BY WOUNDING

Wounding may be caused by several different types of injuries: e.g., mechanical wounding via physical forces such as wind, rain, hail, or frost; wounding by feeding organisms like mammals, mollusks, insects, or nematodes and wounding by cell-attacking microorganisms such as fungi, bacteria, or viruses. Wounds, however, may also occur during the normal course of development of a plant, e.g., growth cracks. In each case a set of reactions is triggered to ensure proper closure of the wound (Bostock and Stermer, 1989). Some reactions will be common to all types of wounds, whereas others will be induced as a specific response to a certain type of wounding. In both cases secondary metabolites may be involved and induction of secondary metabolite accumulation is a particularly common response to microbial infection. The remainder of this chapter is divided into two parts: a short presentation of the effects of physical wounding on alkaloid content of plants and a detailed review of systems accumulating alkaloids in response to microbial stress.

4.1. Effects of Mechanical Wounding

Alkaloid levels are affected by wounding in a number of plants. Solanaceous plants, in particular, show marked effects. Potato tubers have long been known for rapid accumulation of steroidal alkaloid glycosides when sliced (McKee, 1955). Temperature has a modulating effect on this response (Currier and Kuc, 1975). On the molecular and biochemical level a rapid increase of mRNA for HMG-CoA-reductase with a maximum as early as 30 min after wounding followed by an increase of the respective enzyme activity was found (Yang *et al.,* 1991). This enzyme is considered to be the rate limiting step in the biosynthetic pathway. Detailed analysis revealed the presence of a differentially expressed HMG-CoA-reductase gene family. Expression of one gene was induced by arachidonic acid and its induction was correlated to sesquiterpenoid phytoalexin accumulation (see below). Two other genes were wound and methyljasmonate inducible but were inhibited by arachidonic acid (Choi and Bostock, 1994). These results deviate in detail from observations made in a previous study by Choi *et al.* (1992) and have been explained by the authors as effects resulting from a strain-specific variation. The complex regulation shows that precise mechanisms regulate the flow of substrates into sesquiterpenoid phytoalexin or steroidal alkaloid biosynthesis, depending on the triggering event.

Wounding also affects alkaloid accumulation in Solanaceae, producing other types of alkaloids. *Atropa acuminata* responds with a 1.5-fold increase of tropane alkaloid content to wounding of the leaves. Repeated mechanical wounding even caused a doubling of the alkaloid level after 8 days. As feeding mollusks also caused a similar increase, the response was interpreted to be a defense against herbivores (Khan and Harborne, 1990). Wound inducible alkaloid accumulation seems to be a trait of the whole genus *Nicotiana.* In species predominantly accumulating anabasine or nornicotine, the level of their main alkaloids increased (Baldwin and Ohnmeiss, 1993). Detailed studies of the effect of mechanical wounding on nicotine alkaloid content were performed with plants of *Nicotiana sylvestris.* Baldwin (1988a) observed a 5.5-fold increase in nicotine content when greenhouse grown plants were wounded. However, alkaloid accumulation was not induced when potted plants kept outdoors under the same conditions as the controls were wounded (Baldwin, 1988b). A convincing explanation is lacking.

Because both tropane alkaloids and nicotine are produced in roots, a signal must be transmitted from the wound site to the site of alkaloid biosynthesis. In the case of nicotine, Baldwin (1989) and Baldwin *et al.* (1994a) presented evidence that such alkaloid biosynthesis activating signals are sent to the roots by phloem transport, which in turn triggers transport of nicotine to the leaves via the xylem. The signal may be jasmonic acid (Baldwin *et al.,* 1994b). Its potential role as a messenger in induction processes will be discussed later.

Wound induced alkaloid accumulation has also been reported in plants from families other than Solanaceae. When leaves of *Catharanthus roseus* plants were cut, alkaloid levels increased by up to 100%, but only in actively growing regions (Frischknecht *et al.,* 1987). Enzyme activities were not measured and it cannot be completely ruled out that the observed effect was caused by reduced sink space, as roots are known to synthesize alkaloids quite actively. The 1.5- to 4-fold increases of quinolizidine alkaloid content occurring in lupines in response to wounding (Wink, 1983) were believed to result from an induction of alkaloid biosynthesis. Lupin leaves are known to be the site of alkaloid synthesis. Alkaloid accumulation is likely to be inducible in many more plants by mechanical injuries. The growing interest in the role of alkaloids in defense will stimulate research in this relatively neglected field.

4.2. Elicitor Induction

Numerous plants respond to attack by microorganisms with a rapid and frequently transient accumulation of secondary metabolites. Only if they are synthesized *de novo* and if they were found to confer resistance is the term *phytoalexin* applicable to these compounds. Phytoalexins belong to various structural classes depending on biosynthetic origin, e.g., isoflavonoids, furanocoumarins or sesquiterpenoids, and are often specific to one species or are a trait of a whole plant family. The intensity of the response usually varies among cultivars of a plant species and is strongly affected by the microorganism involved. In contrast, the pattern of induced compounds is little affected (Bailey, 1987).

The process of phytoalexin induction requires recognition of the microbial attack. Detailed information on the presently understood sequence of events can be found in reviews by Dixon and Lamb (1990), Scheel and Parker (1990), Ryan and Farmer (1991), Yoshikawa *et al.* (1993), Dixon *et al.* (1994), and Choi and Bostock (1994). Briefly, cell wall fragments of the attacking microorganism and sometimes of the attacked plant itself have been shown to fulfill a signaling function. They induce the same response as the microorganism itself. Typically they are carbohydrates or glycopeptides, which are either released directly or via the activity of microbial or plant enzymes. These cell wall fragments and liberating enzymes are termed *elicitors* or more precisely *biotic elicitors*. Their mode of action remains an active area of investigation. Binding sites located in the cell membrane are postulated and were demonstrated in at least one case for oligosaccharide elicitors. The succeeding steps in the signaling chain are not yet fully understood and are quite controversial, as a number of different molecules have been shown to be closely linked to the process. The order of events might be as follows: (1) rapid changes of cytoplasmic Ca^{2+} and of K^+ ion concentrations take place, combined with liberation of H_2O_2, referred to as an *oxidative burst*; (2) membrane lipids become oxidized to yield jasmonic acid and/or methyljasmonate, which may induce changes in the intracellular cAMP level and/or inositolphosphate level as well as in the concentrations of Ca^{2+} and/or K^+ ions. Nitric oxide, salicylate, ethylene, and other hormones are also involved in modifying the response. The multitude of factors and processes involved suggest that possibly no universal signaling chain exists in all plants but rather that modifications have evolved allowing for a differential and species-specific response.

Elicitation of plant secondary metabolism has received increased attention during the past 15 years not only from a phytopathological point of view but also as an ideal tool to study enzymology and regulation of secondary metabolism. Concomitantly, plant cell cultures became popular systems for such studies, because they are easy to manipulate and frequently offer a much better starting material than the plant itself for the isolation of enzymes (Zenk, 1991). Thus, elicitor induction has been attempted with a large number of cell culture systems and success was reported for many classes of secondary metabolites

(Eilert, 1987). The response to elicitation not only is determined by the type and concentration of elicitor but also is exceedingly dependent on the cell culture line, the medium conditions, and the physiological state of the cells when elicited.

4.2.1. INDUCTION OF BENZYLISOQUINOLINE ALKALOID ACCUMULATION

Eilert *et al.* (1985) reported elicitor induced accumulation of the benzophenanthridine alkaloid sanguinarine in suspension cultures of *Papaver somniferum*. Autoclaved culture homogenates of different fungi served as elicitors. Within 72 hr after treatment the sanguinarine level reached 2.9% of the dry weight of the cells and as much as 40% of the total sanguinarine formed was present in the medium. Feeding experiments with ^{14}C-labeled tyrosine and phenylalanine demonstrated the *de novo* synthesis of sanguinarine (Eilert and Constabel, 1986). Cline *et al.* (1990) successfully reproduced the elicitation of *P. somniferum* cell cultures by treatment with a culture homogenate of *Verticillium dahliae* and obtained the same response in treated *P. bracteatum* suspension cultures. The same elicitor also induced benzophenanthridine alkaloid accumulation in cell cultures of another member of the Papaveraceae, *Sanguinaria canadensis*. The predominant alkaloid that accumulated was chelirubine, which is only a minor alkaloid in the plant. The level of sanguinarine, the main alkaloid (1–3% dw) of the plant, was little affected (Cline *et al.*, 1990). Following the concept of alkaloid production media, Cline *et al.* (1993) subjected *S. canadensis* cells to hormonal deprivation for various periods and observed accumulation of benzophenanthridines. When cells were simultaneously subjected to elicitation, only cells in the growth phase responded with additional benzophenanthridine accumulation.

The response of suspension cultures of California poppy, *Eschscholtzia californica*, was studied by several groups. The plant contains dihydro derivatives of sanguinarine: chelirubine, macarpine, and chelerythrine (Berlin *et al.*, 1983). Addition of cell wall components of *Penicillium* sp. or *Saccharomyces* sp. induced accumulation of chelirubine, chelerythrine, sanguinarine, chelilutine, and macarpine. Antibiotics such as colistin, bacitracin, or polymyxin B also induced alkaloid accumulation while an elicitor compound from *Phytophthora* sp. was inactive (Schumacher *et al.*, 1987). Notable accumulation began about 2 days postelicitation and reached its maximum about 10 days later. During this period total benzophenanthridine content increased from 20 to 160 mg liter^{-1} culture. A different time course was reported by Collinge and Brodelius (1989) using a carbohydrate fraction of yeast as elicitor. Only 6–8 hr after elicitation, maximum levels of sanguinarine were reached. Macarpine continued to accumulate while the level of its precursor sanguinarine declined, indicating turnover. Similar to elicited *P. somniferum* cultures, a substantial release of alkaloids into the medium was observed.

Elicitor-inducible benzophenanthridine accumulation was shown to occur as a common trait in cell cultures of Papaveraceae. Not only *Eschscholtzia* sp., *Papaver* sp., and *Sanguinaria canadensis* but also cultured cells of *Chelidonium majus*, *Glaucium* sp., *Corydalis* sp. (Schumacher *et al.*, 1987), and *Thalictrum rugosum* (Funk *et al.*, 1987), a member of the Ranunculaceae, responded with accumulation of quaternary benzophenanthridines. However, the main alkaloid varies. Berberine, for instance, accumulated in elicitor-treated cell cultures of *T. rugosum* and in particular in those lines that showed a low constitutive accumulation. The response was often dependent on the growth stage. In the differentiated plants where high constitutive levels are present, no induced accumulation in response to elicitor treatment was shown. Thus, although the strong antimicrobial activity of berberine and sanguinarine points to a function as defense compounds, their role as phytoalexins remains to be proven.

Elicitor-induced benzophenanthridine accumulation is based on induction of enzymes involved in the specific biosynthetic pathways. Several enzymes are affected and results obtained in different systems are somewhat controversial with respect to which are the key regulatory steps.

In *T. rugosum* cultures Collinge and Brodelius (1989) observed a rapid induction of tyrosine decarboxylase (maximum activity 3 hr after elicitation), which catalyzes the first step in biosynthesis of benzophenanthridines. The authors assigned a key regulatory role to this enzyme, as a good correlation existed between its induction and induced alkaloid accumulation.

In *E. californica* cultures Tanahashi and Zenk (1990) demonstrated the induction of a cytochrome P_{450}-dependent monooxygenase, protopine-6-hydroxylase, and of the berberine bridge-forming enzyme (Dittrich and Kutchan, 1991). The latter enzyme is encoded by a single gene whose mRNA level increased transiently to a peak 6 hr after elicitor treatment. Its maximum enzyme activity was reached after 20 hr and preceded maximum alkaloid accumulation.

Induction of dihydrosanguinarine oxidase (benzo[*c*]phenanthridine oxidase) was observed in *P. bracteatum* and *S. canadensis* (Cline *et al.*, 1990) as well as in *E. californica*, *Dicentra* sp., *Fumaria* sp., and *Chelidonium* sp. with a maximum of 24 hours after elicitation (Schumacher and Zenk, 1988). Induction of this oxidative enzyme reflects the trend of cell cultures to accumulate highly oxidized compounds after elicitation (Schumacher and Zenk, 1988; Cline *et al.*, 1990). Key regulatory properties were proposed for this enzyme, as its induction was much better correlated with the time course of alkaloid accumulation than was the transient induction of tyrosine decarboxylase.

In *P. somniferum* the possible regulatory role of tyrosine decarboxylase has been demonstrated (Facchini and DeLuca, 1994). In the same cell line, previously studied by Eilert *et al.* (1985) and Eilert and Constabel (1986), a correlation was found between tyrosine decarboxylase induction and alkaloid accumulation (Facchini and DeLuca, personal communication). Molecular analysis of tyrosine/dopa decarboxylase revealed a gene family of 10–14 members, which was divided into two subclasses based on amino acid sequence similarity. One of the subclasses showed rapid and very transient induction by elicitors. Maximum mRNA level is reached 1 hr after induction. The second class shows a much slower but more lasting increase. Analysis in the plant showed a differential and tissue-specific expression of this gene family.

In addition to elicitors, compounds thought to be involved in the cell signaling chain were also tested. Ethylene may be released in elicitor-treated *P. somniferum* (Songstad *et al.*, 1989) and *E. californica* (Piatti *et al.*, 1991) cultures but is apparently not required for benzophenanthridine accumulation. Jasmonic acid was shown to induce benzophenanthridine biosynthesis in *E. californica* cultures (Gundlach *et al.*, 1992) and its involvement in signal transduction was suggested based on its endogenous formation from linolenic acid and its strict correlation to elicitor-induced product accumulation (Müller *et al.*, 1993).

4.2.2. ANTHRANILIC ACID-DERIVED ALKALOIDS

Acridone and furoquinoline alkaloids are both derived from anthranilic acid and occur only in members of the Rutaceae. Callus cultures of *Ruta graveolens* respond to elicitation by homogenates of different fungi with accumulation of acridone epoxides (Wolters and Eilert, 1982; Eilert *et al.,* 1984). The response of *R. graveolens* to a culture homogenate of the yeast *Rhodotorula rubra* was studied in detail and found to be inversely proportional to the degree of differentiation of the *in vitro* system. Dedifferentiated suspension cultures showed an increased accumulation not only of acridone epoxides but also of furoquinoline alkaloids and of furanocoumarins. In shoot cultures and sterile hydroponic cultures the constitutive levels of these metabolites were up to tenfold greater than in dedifferentiated cultures, thus exceeding their maximum inducible levels. Acridone epoxide accumulation, in contrast, was inducible (Eilert, 1989).

In their studies on the role of jasmonic acid in signal transduction, Gundlach *et al.* (1992) claimed an induction of rutacridone in cell cultures of *Ruta chalepensis.* Because *Rhodotorula* elicitor did not affect the accumulation of this compound in *R. graveolens,* these results suggest a species-specific modification of the response, or additional factors that are involved in signal transduction.

Studies of pathway-specific enzymes involved in the biosynthesis of elicitor-inducible alkaloids were also performed. 4-Hydroxy-3-(3-methyl-2-butenyl)-2-quinolinone *O*-methyltransferase is the only furoquinoline pathway-specific enzyme identified and it proved to be inducible in suspension cultures of *R. graveolens* (Chang and Eilert, 1993). This enzyme demonstrates the usefulness of elicitation in cell culture, as this technique contributed significantly to the discovery of the enzyme. The existence of this OMT provides biochemical evidence that the biosynthetic pathway proposed by Boulanger *et al.* (1973) from incorporation studies is correct.

In the case of acridone biosynthesis several enzymes were studied. Elicitor induction was demonstrated for anthranilic acid *N*-methyltransferase (Eilert and Wolters, 1989) and acridone synthase (Baumert *et al.,* 1990), whereas 1,3-dihydroxy-*N*-methylacridone 3-*O*-methyltransferase, a branch point enzyme leading to simple acridones and thus not involved in the biosynthesis of the inducible acridone epoxides, was not affected (Eilert *et al.,* 1984; Guibraltarskkaya and Eilert, 1993). Anthranilate synthase could also play a key regulatory role in the elicitor-induced accumulation of both the furoquinolines and the acridone epoxides. Elicitor treatment induced up to a 3-fold increase in specific anthranilate synthase activity and the induction rate of anthranilate synthase was correlated with induced acridone epoxide accumulation (Bohlmann and Eilert, 1994). Further evidence for the crucial role played by anthranilate synthase in the flow of substrate to alkaloids was gained from molecular analysis. *R. graveolens* contains two genes that encode the alpha subunit of anthranilate synthase, both being constitutively expressed in seedlings and cell cultures. However, elicitation of cell suspension cultures resulted in a 100-fold increase in the transcription of only one of the two genes (Bohlmann *et al.,* 1995). Both alpha subunits are synthesized as cytosolic precursor proteins, which are transported into plastids. When studying the effect of tryptophan, a potent allosteric feedback inhibitor of anthranilate synthase, the gene for the inducible alpha subunit was found to be much less sensitive than the noninducible isogene (Bohlmann *et al.,* 1995). These results, showing the potential role of each alpha subunit gene, helped to identify the regulation of the flow of chorismate into the different pathways; one feeding anthranilate into tryptophan biosynthesis and the other supplying substrate for alkaloid formation.

Benzoxazinones and their precursor amides are a third class of inducible, anthranilic acid derived alkaloids found in members of the monocotyledonous family Poaceae and of the dicotyledonous family Caryophyllaceae. In both cases the alkaloids are involved in plant defense and are true phytoalexins. In Poaceae they may also appear as preformed phytochemicals. The avenalumins I–III, the first N-containing phytoalexins discovered, accumulate in oats (*Avena sativa*) in response to infection with the rust fungus *Puccinia coronata* f. *avenae* or with rice blast fungus (Mayama *et al.,* 1986). In some oat lines avenalumin accumulation was also induced by treatment with victorin, a toxin isolated from *Helminthosporium victoriae*. Inhibitors of transcription were able to repress the accumulation of avenalumin suggesting the involvement of pathway specific enzyme induction (Mayama *et al.,* 1986). In wheat the benzoxazinones are constitutively present as glucosides. On infection with rust fungus the content of two benzoxazinones decreases while the level of another highly methylated derivative increases (Bücker and Grambow, 1990). In rye Barnes and Putnam (1987) discuss a function as allelochemicals. For a more detailed review of the effects see Chapter 14.

Dianthramides, amides of anthranilic acid with benzoic acid varying in substitution pattern, and the derived benzoxazinone compounds called dianthalexins as well as dianthanilides, decarboxylated amides, are found in *Dianthus caryophyllus*. Increase of benzoxazinone levels was found in elicitor-treated plants (Niemann and Baayen, 1988) as well as in treated cuttings (Ponchet *et al.,* 1982) and in elicitor-treated cell suspensions (Gay, 1985). In cell cultures the response proved to be dependent on the time point of treatment during the culturing period (Gay 1985). Differences in response to *Fusarium* homogenate by *D. caryophyllus* cultivars were correlated to their resistance to *Fusarium oxysporum* f. sp. *dianthi* (Baayen and Niemann, 1989). Such modified responses were not noted when other elicitors such as *Phytophthora* elicitor or even Hg^{2+} ions were used (Buiatti *et al.,* 1985). The amount and profile of products, however, were found to vary depending on the type of elicitor (Niemann, 1993). The modified response was reflected in the enzyme level. A glycoprotein elicitor from *Phytophthora* as well as yeast cell wall glucans induced similar increases in benzoyl-CoA:anthranilate *N*-benzoyltransferase, the first enzyme of the pathway (Reinhard and Matern, 1989). The induction of the hydroxylases, which introduce the 4- or 5-OH group into the molecule, exhibited marked differences (Reinhard and Matern, 1991). Niemann *et al.* (1990) observed by pyrolysis-MS analysis that accumulation occurred in the immediate vicinity of the elicitor applied. In the affected part of the xylem dianthalexins and dianthramides made up 1% of the fresh weight.

4.2.3. MONOTERPENE INDOLE ALKALOIDS

The first report on elicitor induction of monoterpene indole alkaloid accumulation was published by Eilert *et al.* (1986). Different well characterized cell lines of periwinkle (*Catharanthus roseus*) were treated with culture homogenates of various fungi. Autoclaved homogenates of *Rhodotorula rubra* and of *Pythium aphanidermatum* were most effective and distinct responses were noted. One line showed no effect, whereas several lines exhibited a rapid accumulation of tryptamine and strictosidine lactam and one line responded with a rapid accumulation of strictosidine lactam, ajmalicine, tabersonine, and catharanthine, the same monoterpene indole alkaloid pattern produced by this line when grown on production medium. The response to elicitation was also strongly affected by the time point of treatment. Alkaloid accumulation occurred in particular when cells were treated with elicitor during the late growth stages. In contrast to cells cultured in alkaloid production medium, where alkaloids accumulated within the cells, elicitor treatment resulted in considerable amounts of alkaloids accumulating in the medium. These results have been observed with most elicitor-inducible compounds, but the physiological basis is poorly understood. Cell leakage, cell death, and active excretion are considered to contribute to this effect to various degrees but detailed investigations of the biochemistry of the induced excretion process are still missing. The effects of elicitation on the product level were confirmed by other groups. DiCosmo *et al.* (1987) also found a cell line responding with ajmalicine and catharanthine accumulation. Berlin *et al.* (1987) compared different induction strategies and found a fungal elicitor to cause only tryptamine accumulation while hormone deprivation resulted in alkaloid accumulation. Both fungal homogenates and abiotic elicitors such as vanadate induced indole alkaloid accumulation in some of the cell lines tested (Smith *et al.,* 1987b). Cline *et al.* (1990) found an increased accumulation of serpentine on elicitation and the maximum level was reached 5 days after treatment. Maximum levels of other alkaloids such as ajmalicine were reached between 16 and 72 hr after treatment. These results suggested that serpentine accumulated probably as a result of the turnover of ajmalicine by oxidation as was previously observed with benzophenanthridines.

Browning of cells is a response that generally accompanies elicitor-induced alkaloid formation (Eilert, 1987). However, Nef *et al.* (1991) found a *C. roseus* cell line that showed no phenol production after elicitation with *Pythium vexans* homogenate but that accumulates indole alkaloids. Although both responses are generally tightly linked, modified control appears to be possible.

Moreno *et al.* (1993) studied the effect of feeding of terpenoid precursors in a cell line showing only an increased tryptamine accumulation when elicited and found that alkaloid accumulation was not triggered. Thus, the failure of this cell line to produce alkaloids is not likely to be caused by a block of the monoterpenoid pathway as was assumed.

The effect of elicitation on the activities of monoterpene indole alkaloid biosynthesis enzymes has been extensively studied. A rapid transient induction of tryptophan decarboxylase and strictosidine synthase (Eilert *et al.,* 1987) was preceded by mRNA induction for both enzymes (Roewer *et al.,* 1992) in an alkaloid-accumulating cell line suggesting control primarily at the level of transcription. Maxima for both mRNA levels were observed 6–10 hr after elicitation and preceded maximum enzyme activities. Both alkaloid accumulation and the induction kinetics of both enzymes were dependent on the physiological stage of the elicited cells. Pasquali *et al.* (1992) confirmed the coordinated increase of mRNA levels of these two enzymes after treatment with elicitor from *Pythium aphanidermatum* using a tryptamine-accumulating cell line. A direct downregulation by auxin on the mRNA levels of tryptophan decarboxylase and strictosidine synthase was found. Elicitor was shown to override the inhibition caused by auxins, which matches very well with the effects observed on the product level.

Moreno (1994) studied the effect of elicitation on different metabolic pathways in the same tryptamine-accumulating *C. roseus* cell suspension culture. An induction of tryptophan decarboxylase, but not strictosidine synthase or strictosidine glucosidase activities was observed. In addition, geraniol-10-hydroxylase and isopentenyl diphosphate isomerase were inhibited. Anthranilate synthase activity was induced, but chorismate mutase, an enzyme competing for the same substrate, was not induced. Isochorismate synthase, a third chorismate-utilizing enzyme, was induced in suspension cultures and in shoot cultures. It is thought to be involved in the biosynthesis of 2,3-dihydroxy benzoic acid, a major compound of the inducible phenolics (Moreno *et al.,* 1994).

Aerts *et al.* (1994) studied the effect of methyljasmonate on monoterpene indole alkaloid metabolism in *C. roseus* and *Cinchona* sp. The young seedlings used represent a system with more physiological relevance than dedifferentiated suspension cultures. Exposure to 1 ppm methyljasmonate vapor induced a doubling of alkaloid content. However, the response of *C. roseus* seedlings was confined to the narrow time interval where the normal, developmentally regulated, onset of alkaloid synthesis occurred. Interestingly, the tabersonine–vindoline pathway was stimulated slightly more (2- to 3-fold) than the catharanthine branch (1.7-fold). An approximately 10 fold increase in specific activity of tryptophan decarboxylase (TDC), strictosidine synthase, deacetoxyvindoline hydroxylase, and deacetoxyvindoline acetyltransferase, all pathway-specific enzymes, was shown. The induction kinetics were not tightly coordinated, indicating additional regulatory mechanisms to be present. In *Cinchona* seedlings the level of cinchonine and dihydrocinchonine doubled under the treatment with methyljasmonate and the induced alkaloid accumulation was preceded by a transient increase in TDC activity. A similar induction of TDC by methyljasmonate was reported using suspension cultures of *Camptotheca acuminata*, an indole alkaloid producing member of the Nyssaceae (Burnett *et al.*, 1993a). The same group also studied the expression of 3-hydroxy-3-methylglutaryl CoA reductase using promoter fusions with the GUS reporter gene in transgenic tobacco. Wounding caused a strong induction of this enzyme while methyljasmonate treatment resulted in a complete suppression of its expression (Burnett *et al.*, 1993b). The effect of fungal elicitors on the secondary metabolism of suspension cultures of *Cinchona* is quite different. Alkaloid formation was not induced but an accumulation of anthraquinones was noted with unusually slow accumulation kinetics (Wijnsma *et al.*, 1985) and maximum levels were reached 10 days after elicitation. The different responses observed illustrate the problems that occur when comparing the response of different systems. These findings also suggest that methyljasmonate does not always induce the same response as elicitors. Other or additional signaling factors are likely to be involved at least in some systems.

Canthin-6-one alkaloids are also derived from tryptophan and produced in *Ailanthus altissima*, a member of the Simaroubaceae. In suspension cultures of this plant canthin-6-one alkaloid accumulation was inducible with cell wall preparations of *Phytophthora* species or yeast (Krauss *et al.*, 1989).

4.2.4. QUINOLIZIDINE ALKALOIDS

Alkaloids, polyamines, cAMP, and HCl can induce quinolizidine alkaloid accumulation whereas chitosan was only a poor inducer (Wink, 1985). The increase is transient, rapid, and takes place in low-producing cultures only.

4.2.5. STEROIDAL ALKALOIDS, TROPANES, NICOTINE

In members of the Solanaceae the typical phytoalexins are sesquiterpenoids. While wounding induces accumulation of steroidal alkaloids in potato, as discussed above, fungal infections redirect the metabolism from triterpenoid synthesis to sesquiterpenoid phytoalexin synthesis. In experiments where slices of potato tubers were infected with *Phytophthora infestans* or *Helminthosporium carboneum*, wound-induced glycoalkaloid accumulation was inhibited (Shih *et al.*, 1973). Arachidonic acid and eicosapentaenoic acid, both considered to be key elicitor active compounds in *Phytophthora*, also inhibited glycoalkaloid accumulation (Tjamos and Kuc, 1982). A key regulatory role in the flow of substrate seems to be exercised by the gene family of HMG-CoA-reductase (see Section 4.1).

Sesquiterpenoid accumulation was also induced when cell cultures of nicotine- or tropane alkaloid-producing plants were elicited. Generally, alkaloid accumulation remained unaffected. Chappell *et al.* (1987) as well as Threlfall and Whitehead (1988) reported this for *Nicotiana* cell cultures, Miguel and Barroso (1994) for *Hyoscyamus albus* cultures, Brachet and Cosson (1986) for salt stressed *Datura stramonium* suspensions, and Furze *et al.* (1991) for root cultures of *D. stramonium* after treatment with 1 mM Cu^{2+} or Cd^{2+} ion solutions. There are, however, reports on successful induction of tropane alkaloid accumulation. Sevón *et al.* (1992) measured a doubling of the hyoscyamine content in a high-producing and a quadrupling in a low-producing hairy root strain of *Hyoscyamus muticus* after treatment with 10–100 μM Cu^{2+} ions. In a screening for elicitor activity, most fungal homogenates did not affect the hyoscyamine level, some inhibited the accumulation, and only in a few cases was a weak increase, accompanied by reduction of growth, observed (Sevón *et al.,* 1993). Chitosan showed a poor induction activity (Sevon *et al.,* 1992). Ballica *et al.* (1994) successfully used *Phytophthora megasperma* cell wall preparations to induce tropane alkaloid accumulation in suspension cultures of *D. stramonium.* When applied at the beginning of the stationary phase, alkaloid levels increased as much as fivefold after an additional culturing period of 15 days and reached levels of 4 mg g^{-1} dw. This is an unusually long period for elicitation effects and no data are given to describe the accumulation kinetics. In neither system with inducible tropane alkaloid accumulation was the occurrence of additional sesquiterpenoid accumulation investigated. Considering the high constitutive levels and the low antimicrobial activity of tropane alkaloids, they do not fulfill the criteria of a true phytoalexin. Thus, their inducibility by elicitors is not to be expected. Yet experiments with cell cultures, physiologically deviant in many aspects from the differentiated plant, show that some potential for inducibility may exist. For better classification of the phenomena, knowledge about the biochemical background of the induced accumulation is required.

5. CONCLUDING REMARKS

Various factors stimulate rapid induction of alkaloid accumulation but our knowledge of the underlying regulatory mechanisms is as yet rudimentary. Little is known of the effect of salt stress. Progress has been made in understanding light regulation of primary metabolism and its effect on secondary metabolism. Investigations of the processes involved in wound and elicitor induction of secondary metabolism are most advanced as a result of the attractive commercial applications with respect to disease resistance.

The number of systems discovered that possess elicitor- or wound-inducible alkaloid metabolism is quite low. Most induction effects are limited to cell cultures with very low constitutive alkaloid levels. Thus, the response may not be representative for the differentiated, intact plant (Eilert, 1987), which usually accumulates high amounts of alkaloids constitutively. The physiological significance of such induction phenomena must be assessed critically. The existence of inducible pathways and their complex regulation on different levels indicate their probable physiological role. Alkaloid levels are normally determined by extraction of complete organs. If only some cells respond, as was shown for *Dianthus* benzoxazinones (Niemann *et al.,* 1990), this will confer resistance but the induction may not be seen as a result of high background levels of alkaloids in other cells. Tissue specific differences in constitutive or inducible expression of the same compound are known from other systems (Eilert, 1987). In combination with molecular tools the potential exists to address these questions.

A largely neglected field in alkaloid biochemistry is turnover, which is quite controversially discussed. As the induced alkaloid accumulation is frequently transient, such systems offer the interesting opportunity to study turnover. Results might foster studies on this important but largely neglected field.

Elicitor induction has been suggested as a possible way to boost production of alkaloids in cell cultures to commercially applicable levels (Eilert and Wolters, 1989). Development of a semicontinuous process with repeated elicitations to produce sanguinarine (Tyler *et al.,* 1988) demonstrated this potential. Recently, Byun and Pedersen (1991) combined elicitation and two phase extraction to further optimize sanguinarine yields in *E. californica* cultures. Nevertheless, costs and risks involved with plant cell culture still appear to deter industry from commercial application. This situation could change if production of extremely valuable compounds is achieved. Considering the market prices of secondary metabolites and the production levels achieved, these hopes, however, are not likely to be fulfilled in the near future.

REFERENCES

General Reviews

Bailey, J. A., 1987, Phytoalexins: a genetic view of their significance in: *Genetics and Plant Pathogenesis* (P. R. Day and G. J. Jellis, eds.), Blackwell, Oxford, pp. 233–234.

Bostock, R. M., and Stermer, B. A., 1989, Perspectives on wound healing in resistance to pathogens, *Annu. Rev. Phytopathol.* **27:**343–371.

Cline, S. D., Psenak, M., McHale, R. J., Krueger, R. J., and Coscia, C. J., 1990, Preferential induction of biological oxidation systems in alkaloid metabolism of plant tissue cultures subjected to fungal elicitation, in: *Biological Oxidation Systems*, Vol. I (C. C. Reddy, G. A. Hamilton, and K. M. Madyyastha, eds.), Academic Press, San Diego, pp. 99–113.

Dixon, R. A., and Lamb, C. J., 1990, Molecular communication in interactions between plants and microbial pathogens, *Annu. Rev. Plant Physiol. Plant Mol. Biol.* **41:**339–367.

Dixon, R. A., Harrison, M. J., and Lamb, C. J. 1994, Early events in the activation of plant defense responses, *Annu. Rev. Phytopathol.* **32:**479–501.

Eilert, U., 1987, Elicitation: Methodology and aspects of application, in: *Cell Culture and Somatic Cell Genetics of Plants*, Vol. 4 (I. Vasil and F. Constabel, eds.), Academic Press, San Diego, pp. 153–196.

Gundlach, H., Müller, M. J., Kutchan, T. M., and Zenk, M. H., 1992, Jasmonic acid is a signal transducer in elicitor-induced plant cell cultures, *Proc. Natl. Acad. Sci. USA* **89:**2389–2393.

Müller, M. J., Brodschelm, W., Spannagl, E., and Zenk, M. H., 1993, Signalling in the elicitation process is mediated through the octadecanoid pathway leading into jasmonic acid, *Proc. Natl. Acad. Sci. USA* **90:**7490–7494.

Ryan, C. A., and Farmer, E. E., 1991, Oligosaccharide signals in plants: A current assessment, *Annu. Rev. Plant Physiol. Plant Mol. Biol.* **42:**651–674.

Scheel, D., and Parker, J. E., 1990, Elicitor recognition and signal transduction in plant defense gene activation, *Z. Naturforsch.* **45c:**569–575.

Thompson, W. F., 1991, Physiological and molecular studies of light-regulated nuclear genes in higher plants, *Annu. Rev. Plant Physiol. Plant Mol. Biol.* **42:**423–466.

Waller, G. R., and Nowacki, E. K., 1978, *Alkaloid Biology and Metabolism in Plants*, Plenum Press, New York, pp. 85–119.

Wink, M., 1985, Metabolism of quinolizidine alkaloids in plants and cell suspension cultures: Induction and degradation, in: *Primary and Secondary Metabolism of Plant Cell Cultures* (K. H. Neumann, W. Barz, and E. Reinhard, eds.), Springer-Verlag, Berlin, pp. 107–116.

Yoshikawa, M., Yamaoka, N., and Takeuchi, Y., 1993, Elicitors: Their significance and primary modes of action in the induction of plant defense reactions, *Plant Cell Physiol.* **34:**1163–1173.

Zenk, M. H., 1991, Chasing the enzymes of secondary metabolism: Plant cell cultures as a pot of gold, *Phytochemistry* **30:**3861–3863.

Zenk, M. H., El-Shagi, H., Arens, H., Stöckigt, J., Weiler, E. W., and Deus, B., 1977, Formation of the indole alkaloids serpentine and ajmalicine in cell suspension cultures of *Catharanthus roseus*, in: *Plant Tissue Culture and Its Biotechnological Application* (W. Barz, E. Reinhard, and M. H. Zenk, eds.), Springer-Verlag, Berlin, pp. 27–42.

Key References

Aerts, R. J., and DeLuca, V. 1992, Phytochrome is involved in the light-regulation of vindoline biosynthesis in *Catharanthus*, *Plant Physiol.* **100:**1029–1032.

Aerts, R. J., Gisi, D., DeCarolis, E., DeLuca, V., and Baumann, T. W., 1994, Methyl jasmonate vapor increases the developmentally controlled synthesis of alkaloids in *Catharanthus* and *Cinchona* seedlings, *Plant J.* **5:**635–643.

Baayen, R. P., and Niemann, G. J., 1989, Correlations between accumulation of dianthramides, dianthalexin and unknown compounds, and partial resistance to *Fusarium oxysporum* f. sp. *dianthi* in eleven carnation cultivars, *J. Phytopathol.* **126:**281–292.

Baldwin, I. T., 1988a, The alkaloidal responses of wild tobacco to real and simulated herbivory, *Oecologia* **77:**378–381.

Baldwin, I. T., 1988b, Damage-induced alkaloids in tobacco: Pot-bound plants are not inducible, *J. Chem. Ecol.* **4:**1113–1120.

Baldwin, I. T., 1989, The mechanism of damage-induced alkaloids in wild tobacco, *J. Chem. Ecol.* **15:**1661–1680.

Baldwin, I. T., and Ohnmeiss, T. E., 1993, Alkaloidal responses to damage in *Nicotiana* native to North America, *J. Chem. Ecol.* **19:**1143–1153.

Baldwin, I. T., Karb, M. J., and Ohnmeiss, T. E., 1994a, Allocation of ^{15}N from nitrate to nicotine: Production and turnover of an induced defense, *Ecology* **75:**995–1002.

Baldwin, I. T., Schmelz, C. A., and Ohnmeiss, T. E., 1994b, Wound induced changes in root and shoot jasmonic acid positively correlate with induced nicotine synthesis in *Nicotiana sylvestris* Spegazzini and Comes, *J. Chem. Ecol.* **20:**2139–2157.

Ballica, R., Ryu, D. D. Y., and Kado, C. I., 1994, Tropane alkaloid production in *Datura stramonium* suspension cultures: Elicitor and precursor effects, *Biotechnol. Bioeng.* **41:**1075–1081.

Balsevich, J., DeLuca, V., and Kurz, W. G. W., 1986, Altered alkaloid pattern in dark grown seedlings of *Catharanthus roseus*. The isolation and characterisation of 4-desacetoxyvindoline, a novel indole alkaloid and proposed precursor of vindoline, *Heterocycles* **24:**2415–2421.

Barnes, J. P., and Putnam, A. R., 1987, Role of benzoxazinones in allelopathy by rye, *J. Chem. Ecol.* **13:**889–906.

Baumert, A., Maier, W., and Gröger, D., 1990, Characterization of acridone-specific enzymes and elicitor-stimulation of acridone formation in cell cultures of *Ruta graveolens*, *Planta Med.* **56:**494.

Berlin, J., Forche, E., Wray, V., Hammer, J., and Hösel, W., 1983, Formation of benzophenanthridine alkaloids by suspension cultures of *Eschscholtzia californica*, *Z. Naturforsch.* **38c:**346–352.

Berlin, J., Mollenschott, C., and DiCosmo, F., 1987, Comparison of various strategies designed to optimize indole alkaloid accumulation of a cell suspension culture of *Catharanthus roseus*, *Z. Naturforsch.* **42c:**1101–1108.

Bohlmann, J., and Eilert, U., 1994, Elicitor induced secondary metabolism in *Ruta graveolens* L.: Role of chorismate utilizing enzymes, *Plant Cell Tissue Org. Cult.* **38:**191–200.

Bohlmann, J., DeLuca, V., Eilert, U., and Martin, W., 1995, Purification and cDNA cloning of anthranilate synthase from *Ruta graveolens*: Expression and properties of *in vivo* and recombinant enzymes, *Plant J.* **7:**491–501.

Boulanger, D., Bailey, K. B., and Steck, W., 1973, Formation of edulinine and furoquinoline alkaloids from quinoline derivates by cell suspension cultures of *Ruta graveolens*, *Phytochemistry* **12:**2399–2405.

Brachet, J., and Cosson, L., 1986, Changes in the total alkaloid content of *Datura innoxia* Mill subjected to salt stress, *J. Exp. Bot.* **37:**650–656.

Bücker, C., and Grambow, H. J., 1990, Alterations in 1,4-benzoxazinone levels following inoculation with stem rust in wheat leaves carrying various alleles for resistance and their possible role as phytoalexins in moderately resistant leaves, *Z. Naturforsch.* **45c:**1151–1155.

Buiatti, M., Scala, A., Bettini, P., Nascari, G., Morpurgo, R., Bogani, P., Pellegrini, G., Gimelli, F., and Venturo, R., 1985, Correlations between *in vivo* resistance to *Fusarium* and *in vitro* response to fungal elicitors and toxic substance in carnation, *Theor. Appl. Genet.* **70:**42–47.

Burnett, R. J., McKnight, T. D., and Nessler, C. L., 1993a, Isolation and expression of a tryptophan decarboxylase gene from *Camptotheca acuminata*, *Plant Physiol.* **102**(Suppl.), Poster 87.

Burnett, R. J., Maldonado-Mendoza, I. E., McKnight, T. D., and Nessler, C. L., 1993b, Expression of a 3-hydroxy-3-methylglutaryl coenzyme A reductase gene from *Camptotheca acuminata* is differentially regulated by wounding and methyljasmonate, *Plant Physiol.* **103:**41–48.

Byun, S. Y., and Pedersen, H., 1991, Elicitation and *in situ* recovery of alkaloids in suspension cultures of California poppy, *J. Microbiol. Biotechnol.* **1:**220–226.

Chang, A., and Eilert, U., 1993, A first specific enzyme of furoquinoline alkaloid biosynthesis: S-adenosyl-L-methionine: 4-hydroxy-3-(3-methyl-2-butenyl)-2-quinolinone O-methyltransferase, *Planta Med.* **59:**585.

Chappell, J., Nable, R., Fleming, P., Andersen, R. A., and Burton, H. R., 1987, Accumulation of capsidiol in tobacco cell cultures treated with fungal elicitor, *Phytochemistry* **26:**2259–2260.

Choi, D., and Bostock, R. M., 1994, Involvement of *de novo* protein synthesis, protein kinase, extracellular Ca^{2+}, and lipoxygenase in arachidonic acid induction of 3-hydroxy-3-methylglutaryl coenzyme A reductase genes and isoprenoid accumulation in potato (*Solanum tuberosum* L.), *Plant Physiol.* **104:**1237–1244.

Choi, D., Ward, B. L., and Bostock, R. M., 1992, Differential induction and suppression of potato 3-hydroxy-3-methylglutaryl coenzyme A reductase genes in response to *Phytophthora infestans* and to its elicitor arachidonic acid, *Plant Cell* **4:**1333–1344.

Cline, S. D., McHale, R. J., and Coscia, C. J., 1993, Differential enhancement of benzophenanthridine alkaloid content in cell suspension cultures of *Sanguinaria canadensis* under conditions of combined hormonal deprivation and fungal elicitation, *J. Nat. Prod.* **56:**1219–1228.

Collinge, M. A., and Brodelius, P. E., 1989, Dynamics of benzophenanthridine alkaloid production in suspension cultures of *Eschscholtzia californica* after treatment with a yeast elicitor, *Phytochemistry* **28:**1101–1104.

Conner, H. W.,1937, Effect of light on solanine synthesis in the potato tuber, *Plant Physiol.* **12:**79–98.

Constabel, F., Rambold, S., Chatson, K. B., Kurz, W. G. W., and Kutney, J. P., 1981, Alkaloid production in *Catharanthus roseus* (L.) G. Don. VI. Variation in alkaloid spectra of cell lines derived from one single leaf, *Plant Cell Rep.* **1:**3–5.

Currier, W. W., and Kuc, J., 1975, Effect of temperature on rishitin and steroid glycoalkaloid accumulation in potato tuber, *Phytopathology* **6:**1194–1197.

Dale, M. F. B., Griffiths, D. W., Bain, H., and Todd, D., 1993, Glycoalkaloid increase in *Solanum tuberosum* on exposure to light, *Ann. Appl. Biol.* **123:**411–418.

DeCarolis, E., Chan, F., Balsevich, J., and DeLuca, V., 1990, Isolation and characterization of a 2-oxoglutarate-dependent dioxygenase involved in the second-to-last step in vindoline biosynthesis, *Plant Physiol.* **94:**1323–1329.

DeLuca, V., Balsevich, J., Tyler, R. T., Eilert, U., Panchuk, B. D., and Kurz, W. G. W., 1986, Biosynthesis of indole alkaloids: Developmental regulation of the biosynthetic pathway from tabersonine to vindoline in *Catharanthus roseus*, *J. Plant Physiol.* **125:**147–156.

DeLuca, V., Fernandez, J. A., Campbell, D., and Kurz, W. G. W., 1988, Developmental regulation of enzymes of indole alkaloid biosynthesis in *Catharanthus roseus*, *Plant Physiol.* **86:**447–450.

DiCosmo, F., Quesnel, A., Misawa, M., and Tallevi, S. G., 1987, Increased synthesis of ajmalicine and catharanthine by cell suspension cultures of *Catharanthus roseus* in response to fungal culture-filtrates, *Appl. Biochem. Biotechnol.* **14:**101–106.

Dittrich, H., and Kutchan, T. M., 1991, Molecular cloning, expression, and induction of berberine bridge enzyme, an enzyme essential to the formation of benzophenanthridine alkaloids in the response of plants to pathogenic attack, *Proc. Natl. Acad. Sci USA* **88:**9969–9973.

Eilert, U., 1989, Elicitor induction of secondary metabolism in dedifferentiated and differentiated *in vitro* systems of *Ruta graveolens* L. in: *Primary and Secondary Metabolism of Plant Cell Cultures* (W. G. W. Kurz, ed.). Springer-Verlag, Berlin, pp. 219–228.

Eilert U., and Constabel, F., 1986, Elicitation of sanguinarine accumulation in *Papaver somniferum* cells by fungal homogenates—an induction process, *J. Plant Physiol.* **125:**167–172.

Eilert U., and Wolters, B., 1989, Elicitor induction of S-adenosyl-L-methionine:anthranilic acid N-methyltransferase in cell suspension and organ cultures of *Ruta graveolens* L., *Plant Cell Tissue Org. Cult.* **18:**1–18.

Eilert, U., Ehmke A., and Wolters, B., 1984, Elicitor-induced accumulation of acridone alkaloid epoxides in *Ruta graveolens* suspension cultures, *Planta Med.*, 508–512.

Eilert, U., Kurz, W. G. W., and Constabel, F., 1985, Stimulation of sanguinarine accumulation in *Papaver somniferum* cell cultures by fungal elicitors, *J. Plant Physiol.* **119:**65–76.

Eilert, U., Constabel, F., and Kurz, W. G. W., 1986, Elicitor stimulation of monoterpene indole alkaloid formation in suspension cultures of *Catharanthus roseus*, *J. Plant Physiol.* **126:**11–22.

Eilert, U., DeLuca, V., Constabel, F., and Kurz, W. G. W., 1987, Elicitor-mediated induction of tryptophan decarboxylase and strictosidine synthase activities in cell suspension cultures of *Catharanthus roseus*, *Arch. Biochem. Biophys.* **254:**491–497.

Facchini, F., and DeLuca, V., 1994, Differential and tissue specific expression of a gene family for tyrosine, dopa decarboxylase in opium poppy, *J. Biol. Chem.* **269:**26684–26690.

Frischknecht, P. M., and Baumann, T. W., 1985, Stress-induced formation of purine alkaloids in plant tissue culture of *Coffea arabica*, *Phytochemistry* **24:**2255–2257.

Frischknecht, P. M., Bättig, M., and Baumann, T. W., 1987, Effect of drought and wounding stress on indole alkaloid formation in *Catharanthus roseus*, *Phytochemistry* **26:**707–710.

Funk, C., Gügler, K., and Brodelius, P. 1987, Increased secondary product formation in plant cell suspension cultures after treatment with a yeast carbohydrate preparation (elicitor), *Phytochemistry* **26:**401–405.

Furze, J. M., Rhode, M. J. C., Parr, A. J., Robins, J. R., Withehead, I. M., and Threlfall, D. R., 1991, Abiotic factors elicit sesquiterpenoid phytoalexin production but not alkaloid production in transformed root cultures of *Datura stramonium*, *Plant Cell Rep.* **10:**111–114.

Gay, L., 1985, Phytoalexin formation in cell cultures of *Dianthus caryophyllus* treated by an extract from the culture medium of *Phytophthora parasitica*, *Physiol. Plant Pathol.* **26:**143–150.

Guibraltarskaya, E., and Eilert, U., 1993, S-adenosyl-L-methionine: 1,3-dihydroxy-N-methylacridone 3-O-methyltransferase a new enzyme from *Ruta graveolens*, *Planta Med.* A647.

Gull, D. D., and Isenberg, F. M., 1960, Chlorophyll and solanine content and distribution in four varieties of potato tubers, *Amer. Soc. Hortic. Sci.* **75:**545–556.

Hanson, A. D., Ditz, K. M., Singletary, G. W., and Leland, T. J., 1983, Gramine accumulation in leaves of barley grown under high temperature stress, *Plant Physiol.* **71:**896–904.

Hösel, W., Berlin, J., Hanzlik, T. N., and Conn, E. E., 1985, *In-vitro* biosynthesis of 1-(4′-hydroxyphenyl)-2-nitroethane and production of cyanogenic compounds in osmotically stressed cell suspension cultures of *Eschscholtzia californica* Cham., *Planta* **166:**176–181.

Khan, M. B., and Harborne, J. B., 1990, Induced alkaloid defence in *Atropa acuminata* in response to mechanical and herbivore leaf damage, *Chemoecology* **1:**77–80.

Knobloch, K. H., and Berlin, J., 1980, Influence of medium composition on the formation of secondary compounds in cell suspension cultures of *Catharanthus roseus* (L.) Don., *Z. Naturforsch.* **35c:**551–556.

Knobloch, K. H., Bast, G., and Berlin, J. 1982, Medium- and light-induced formation of serpentine and anthocyanins in cell suspension cultures of *Catharanthus roseus*, *Phytochemistry* **21:**591–594.

Krauss, G., Kraus, G.-J., Baumbach, R., and Gröger, D., 1989, Elicitation of canthin-6-one alkaloid accumulation in cell suspension cultures of *Ailanthus altissima*, *Z. Naturforsch.* **44c:**712–714.

McKee, R. K., 1955, Host–parasite relationships in the dry rot disease of potatoes, *Ann. Appl. Biol.* **43:**147–148.

Mayama, S., Tani, T., Ueno, T., Midland, S. L., Sims, J. J., and Keen, N. T., 1986, The purification of victorin and its phytoalexin elicitor activity in oat leaves, *Physiol. Mol. Plant Pathol.* **29:**1–18.

Miguel, M., and Barroso, J. G., 1994, Accumulation of stress metabolites in cell suspension cultures of *Hyoscyamus albus*, *Phytochemistry* **35:**371–375.

Moreno, P. R. H., 1994, Influence of stress factors on the secondary metabolism of cell cultured *Catharanthus roseus* (L.) G. Don, Ph.D. thesis, University of Leiden.

Moreno, P. R. H., van der Heijden, R., and Verpoorte, R., 1993, Effect of terpenoid precursor feeding and elicitation on formation of indole alkaloids in cell suspension cultures of *Catharanthus roseus*, *Plant Cell Rep.* **12:**702–705.

Moreno, P. R. H., van der Heijden, R., and Verpoorte, R., 1994, Elicitor-mediated induction of isochorismate synthase and accumulation of 2,3-dihydroxy benzoic acid in *Catharanthus roseus* cell suspension and shoot cultures, *Plant Cell Rep.* **14:**188–191.

Nef, C., Rio, B., and Chrestin, H., 1991, Induction of catharanthine synthesis and stimulation of major indole alkaloids production by *Catharanthus roseus* cells under non-growth-altering treatment with *Pythium vexans* extracts, *Plant Cell Rep.* **10:**26–29.

Niemann, G. J., 1993, The anthranilamide phytoalexins of the Caryophyllaceae and related compounds, *Phytochemistry* **34:**319–328.

Niemann, G. J., and Baayen, R. P., 1988, Involvement of phenol metabolism in resistance of *Dianthus caryophyllus* to *Fusarium oxysporum* f. sp. *dianthi*, *Neth. J. Plant Pathol.* **94:**289–301.

Niemann, G. J., Baayen, R. P., and Boon, J. J., 1990, Localization of phytoalexin accumulation and determination of changes in lignin and carbohydrate composition in carnation (*Dianthus caryophyllus* L.) xylem as a consequence of infection with *Fusarium oxysporum* f. so. *dianthi*, by pyrolysis-mass spectrometry, *Neth. J. Plant Pathol.* **96:**133–153.

Noé, W., and Berlin, J., 1985, Induction of *de-novo* synthesis of tryptophan decarboxylase in cell suspensions of *Catharanthus roseus, Planta* **166:**500–504.

Pasquali, G., Goddijn, O. J. M., de Waal, A., Verporte, R., Schilperoort, R. A., Hoge, J. H. C., and Memelink, J., 1992, Coordinated regulation of two indole alkaloid biosynthetic genes from *Catharanthus roseus* by auxin and elicitors, *Plant Mol. Biol.* **18:**1121–1131.

Piatti, T., Boller, T., and Brodelius, P. E., 1991, Induction of ethylene biosynthesis is correlated with but not required for induction of alkaloid accumulation in elicitor-treated *Eschscholtzia* cells, *Phytochemistry* **30:**2151–2154.

Ponchet, M., Martin-Tanguy, J., Andréoli, C., and Martin, C., 1982, Apparition de substances de type phénolamide lors de l'interaction *Dianthus caryophyllus* L. var. Scania–*Phytophthora parasitica* Dastur., *Agronomie* **2:**37–44.

Reinhard, K., and Matern, U., 1989, The biosynthesis of phytoalexins in *Dianthus caryophyllus* L. cell cultures: Induction of benzoyl-CoA:anthranilate N-benzoyltransferase activity, *Arch. Biochem. Biophys.* **275:**295–301.

Reinhard, K., and Matern U., 1991, Different types of microsomal enzymes catalyze ortho- or para-hydroxylation in the biosynthesis of carnation phytoalexins, *FEBS Lett.* **294:**67–72.

Roewer, I., Cloutier, N., Nessler, C. L., and DeLuca, V., 1992, Transient induction of tryptophan decarboxylase and strictosidine synthase genes in cell suspension cultures of *Catharanthus roseus*, *Plant Cell Rep.* **11:**86–89.

Schumacher, H.-M., and Zenk, M. H., 1988, Partial purification and characterization of dihydrobenzophenanthridine oxidase from *Eschscholtzia californica* cell suspension cultures, *Plant Cell Rep.* **7:**43–46.

Schumacher, H.-M., Gundlach, H., Fiedler, F., and Zenk, M. H., 1987, Elicitation of benzophenanthridine alkaloid synthesis in *Eschscholtzia* cell cultures, *Plant Cell Rep.* **6:**410–413.

Sepulveda, B. A., and Corcuera, L. J., 1990, Effect of gramine on the susceptibility of barley leaves to *Pseudomonas syringae*, *Phytochemistry* **29:**465–467.

Sevón, N., Varjonen, T., Hiltunen, R., and Oksman-Caldentey, K.-M., 1992, Effect of sucrose, nitrogen, and copper on the growth and alkaloid production of transformed root cultures of *Hyoscyamus muticus*, *Planta Med.* **58:**A 609.

Sevón, N., Soumalainen, M., Hiltunen, R., and Oksman-Caldentey, K.-M., 1993, Effect of fungal elicitors on hyoscyamine content in hairy root cultures of *Hyoscyamus muticus*, *Planta Med.* **59:**A 661

Shih, M., Kuc, J., and Williams, E. B., 1973, Suppression of steroid glycoalkaloid accumulation as related to rishitin accumulation in potato tubers, *Phytopathology* **63:**821–826.

Smith, J. I., Smart, N. G., Kurz, W. G. W., and Misawa, M., 1987a, The use of organic and inorganic compounds to increase the accumulation of indole alkaloids in *Catharanthus roseus* (L.) G. Don cell suspension cultures, *J. Exp. Bot.* **38:**1501–1506.

Smith, J. I., Smart, N. G., Misawa, M., Kurz, W. G. W., Tallevi, S. G., and DiCosmo, F., 1987b, Increased accumulation of indole alkaloids by some cell lines of *Catharanthus roseus* in response to addition of vanadyl sulphate, *Plant Cell Rep.* **6:**142–145.

Songstad, D. D., Giles, K. L., Park, J., Novakovski, D., Epp, D., Friesen, L., and Roewer, I., 1989, Effect of ethylene on sanguinarine production from *Papaver somniferum* cell cultures, *Plant Cell Rep.* **8:**463–466.

Tanahashi, T., and Zenk,M. H., 1990, Elicitor induction and characterization of microsomal protopine-6-hydroxylase, the central enzyme in benzophenanthridine alkaloid biosynthesis, *Phytochemistry* **29:**1113–1122.

Threlfall, D. R., and Whitehead, I. M., 1988, Coordinated inhibition of squalene synthase and induction of enzymes of sesquiterpenoid phytoalexin biosynthesis in cultures of *Nicotiana tabacum*, *Phytochemistry* **27**:2567–2580.

Tjamos, E. C., and Kuc, J. A., 1982, Inhibition of steroid glycoalkaloid accumulation by arachidonic and eicosapentaenoic acids in potato, *Science* **217**:542–543.

Tyler, R. T., Eilert, U., Rijnders, C. O. M., Roewer, I. A., and Kurz, W. G. W., 1988, Semi-continuous production of sanguinarine and dihydrosanguinarine by *Papaver somniferum* L. cell suspension cultures treated with fungal homogenate, *Plant Cell Rep.* **7**:410–413.

Wijnsma, R., Go, J. T. K. A., van Weerden, I. N., Harkes, P. A. A., Verpoorte, R., and Baerheim Svendsen, A., 1985, Anthraquinones as phytoalexins in cell and tissue cultures of *Cinchona* species, *Plant Cell Rep.* **4**:241–244.

Wink, M., 1983, Wounding-induced increase of quinolizidine alkaloid accumulation in lupin leaves, *Z. Naturforsch.* **38c**:905–909.

Wolters, B., and Eilert, U., 1982, Acridonepoxidgehalte in Kallus kulturen von *Ruta graveolens* und ihre Steigerung durch Misch kultur mit Pilzen, *Z. Naturforsch.* **37c**:575–583.

Yang, Z., Park, H., Lacy, G. H., and Cramer, C. L., 1991, Differential activation of potato 3-hydroxy-3-methylglutaryl coenzyme A reductase genes by wounding and pathogen challenge, *Plant Cell* **3**:397–405.

Chapter 10

Compartmentation of Alkaloid Synthesis, Transport, and Storage

Michael Wink and Margaret F. Roberts

1. INTRODUCTION

The formation and sequestration of secondary metabolites is often regulated in space and time and correlated with differentiation (Wiermann, 1981). The importance of this correlation has been encountered in undifferentiated plant cell cultures, which either do not produce an alkaloid that is typically present in the intact plant or they accumulate the compounds in very low yields. The comparably few systems that produce high amounts of alkaloids are unfortunately rather the exception (Chapter 7). On the other hand, differentiated organ cultures, such as root or shoot cultures, are very reliable production systems, indicating that it is not the *in vivo* condition that is responsible for low yields in cell suspension cultures, but rather the degree of differentiation, i.e., differential gene expression of the enzymes involved in alkaloid biosynthesis and alkaloid storage.

This chapter focuses on the compartmentation of alkaloid formation, transport and sequestration in intact plants, as several aspects of the biosynthesis of alkaloids in cell suspension and organ cultures have already been dealt with in Chapters 7–9.

2. ALKALOID FORMATION

2.1. Sites and Compartmentation of Alkaloid Formation

Theoretically, we can distinguish between the following situations for alkaloid production in plants:

Michael Wink • Institute for Pharmaceutical Biology, University of Heidelberg, D-69120 Heidelberg, Germany. *Margaret F. Roberts* • The Centre for Pharmacognosy, School of Pharmacy, University of London, London WC1N 1AX, England.

Alkaloids: Biochemistry, Ecology, and Medicinal Applications, edited by Roberts and Wink. Plenum Press, New York, 1998.

1. Biosynthesis and accumulation by all cells of a plant
2. Production and storage in a particular organ or tissue
3. Biosynthesis in a specific organ or tissue, followed by translocation to other plant parts where storage occurs

Although these questions have been analyzed for only a limited number of plants, it can be generalized that the second and third possibilities are the ones most often encountered in nature (Table I). Additional information can be drawn from experiments with root and shoot cultures (see Chapters 7 and 8): If root cultures produce a certain type of alkaloid, it suggests that the roots are a site of alkaloid formation in this species; it does not, however, rule out the possibility that other organs, such as shoots or leaves, have a similar capacity. If roots are cultured in the light, they sometimes differentiate into green, photosynthetically active roots, which can express enzyme activities normally absent from roots but present in shoots. For example, quinolizidine alkaloids (QA) are synthesized in mesophyll chloroplasts and in consequence cannot be detected in root cultures grown in the dark. If roots were cultured in the light and became green, some of them were able to produce alkaloids, albeit at a low level (Wink, 1987b). Therefore, growth conditions can influence results substantially and may even lead to wrong conclusions.

In *Coffea arabica* purine alkaloids, particularly caffeine, are synthesized and stored in large quantities in the seed. Directly after germination caffeine remains in the cotyledons surrounding the endosperm and does not migrate to the hypocotyl and root. In older seedlings caffeine accumulation continues during leaf expansion and in the mature plant the fruits actively synthesize purine alkaloids as they mature (Aerts and Baumann, 1994;

Table I
Examples for Site-Specific Alkaloid Formation and Storage in Plants[a]

Genera	Alkaloid	Roots	Shoots/leaves	Fruits	Latex
Lupinus, Cytisus, Laburnum	Lupanine, cytisine	S	B, S	S	
Conium	Coniine	?	B, S	B, S	
Papaver	Morphine	?	B, S	S	S
Chelidonium	Chelidonine, sanguinarine	B, S	B, S	S, B?	S
Berberis, Mahonia	Berberine	B, S	B, S	B?, S	
Atropa, Datura, Hyoscyamus	Scopolamine				
Mandragora	Hyoscyamine	B, S	S	S	
Erythroxylum	Cocaine	S	B, S	S	
Nicotiana	Nicotine	B, S	S	S	
Senecio	Senecionine	B, S	S	S	
Heliotropium	Heliotrine	S	B, S	S	
Peganum	β-Carbolines	B, S	S	S	
Sanguinaria	Sanguinarine	B, S	S	?	S
Cephaelis	Emetine, cephaeline	B, S	B, S		
Coffea	Caffeine	S?	B, S	B, S	
Solanum	Solanine	B, S	B, S	B?, S	
Cinchona	Quinine	B, S	B, S		
Catharanthus	Vindoline	P	B, S	?	S
Ailanthus	Canthin-6-one	B, S	B, S	?	
Ruta	Acridones	B, S	S, B	P	

[a]Abbreviations used: B, biosynthesis; S, storage; P, present (but B and S not known).

Roberts and Waller, 1978). In *Cinchona* species the synthesis of alkaloids, e.g., quinine, rapidly increases on germination but stops after about 4 days; it is therefore assumed that the alkaloids are autotoxic. At this stage they occur principally in the root of the seedlings (Aerts *et al.,* 1991). In the mature tree large quantities are found in the bark, synthesis presumably having been initiated in the periderm region of the bark. Stem and branch barks contain a higher alkaloid content than the root bark (McHale, 1986). However, between the seedling and the mature tree there is a huge gap in our knowledge of the biosynthesis and transportation of these alkaloids. *Cephaelis ipecacuanha,* 2 years old, contains alkaloids (emetine, cephaeline) throughout the plant with the highest levels found in the roots and young leaves as they expand. The experimental results suggest that biosynthesis takes place in all parts of the plant (Ideda *et al.,* 1988). In *Catharanthus roseus*, vindoline, catharanthine and ajmalicine are found in the leaves of young plants and there is evidence that synthesis occurs mainly in this part of the plant and is light regulated (Aerts and DeLuca, 1992; Endo *et al.,* 1987). However, in mature plants, mRNA levels for tryptophan decarboxylase and strictosidine synthase were much higher in the roots than in the leaves and stems (Pasquali *et al.,* 1992). As geraniol-10-hydroxylase activity was also higher in the roots than in the aerial parts of the plant, it is tempting to suggest that the roots may be the main site for the biosynthesis of alkaloid precursors (Meijer *et al.,* 1993) . Accumulation of alkaloids in the leaves occurs in the laticifers and specialized cells of the cortex (Eilert *et al.,* 1985; Yoder and Marlberg, 1976).

Much less is known about the intracellular compartmentation of alkaloid biosynthesis. It is likely that their usual site of synthesis is the cytoplasm, but a few compounds are known to be synthesized in organelles. Enzymes of QA biosynthesis and of the precursor lysine are localized in the chloroplast (Wink and Hartmann, 1982a) and in consequence enzyme activities and alkaloid formation are governed by light (Fig. 9 in Chapter 11). Thus, QA accumulation in lupin leaves and the subsequent translocation in the phloem follow a diurnal rhythm (Wink and Witte, 1984). The transaminase responsible for the biosynthesis of the piperidine alkaloid γ-coniceine has been detected in both chloroplasts and mitochondria (Roberts, 1981). The fact that the enzyme activity is maximal at pH 8.0 and is stimulated in the light and by greening of etiolated seedlings suggests that the chloroplast enzyme is the most important in alkaloid synthesis (see Chapter 5). Isoguinoline alkaloids such as berberine are synthesized in small vesicles, which later fuse with the tonoplast to release their alkaloidal content into the vacuole (Sato *et al.,* 1990, 1993, 1994; Zenk, 1989; Amann *et al.,* 1986).

2.2. Long-Distance Transport

In case the site of biosynthesis differs from the site of alkaloid storage, we must assume a long distance transport, which can be via the phloem, xylem, or apoplastic space. Early experiments with the fluorescence microscope and *Ruta graveolens* showed heavy deposits of acridone alkaloids in the xylem (Verzar-petri *et al.,* 1976) In a few instances it has been possible to tap xylem and phloem fluid directly (e.g., for quinolizidines, pyrrolizidines, tropanes, and nicotine) whereas for most other plant examples the evidence is indirect. If aphids, which feed entirely on the phloem sap, sequester dietary alkaloids or discard them in the honeydew, it can be assumed that the alkaloids are present in the phloem. Data from this type of experiment are shown in Table II

Table II
Examples for Long-Distance Transport of Alkaloids in Plants[a]

Alkaloid	Genera	Phloem	Xylem
Lupanine and other QA	*Lupinus, Genista, Cytisus, Sophora, Laburnum, Spartium*	Yes	No
Senecionine and other PAs	*Senecio*	Yes	No
Aconitine	*Aconitum*	Yes	?
Swainsonine and other polyhydroxy alkaloids	*Astragalus*	Yes	No
Nicotine	*Nicotiana*	No	Yes
Hyoscyamine and other tropanes	*Atropa, Datura, Hyoscyamus*	No	Yes
Rutacridones	*Ruta*	No	Yes

[a]Data from Wink (1987b).

3. ALKALOID STORAGE

3.1. Vacuolar Sequestration of Alkaloids

A prerequisite for the defense function is the sequestration of critical amounts of the respective active alkaloid, otherwise the necessary effect (which is usually dose dependent) cannot be achieved. Whereas lipophilic material (e.g., many terpenoids) is stored in resin ducts, oil containers, dead cells, or trichomes (Wiermann, 1981), hydrophilic compounds (e.g., alkaloids, nonprotein amino acids, organic acids, and glycosides) are usually confined to idioblasts as in *Ruta graveolens* (Eilert *et al.,* 1986) or the vacuole which seems to be especially adapted to the bulk storage of allelochemicals [the term *defense and signal compartment* was coined to emphasize this vacuolar function (Wink, 1993a)]. The vacuoles employed for defense or signaling can be expected in cells that are positioned in strategically favorable positions, such as epidermal tissues, or in flowers or fruits, i.e., vacuolar sequestration of secondary metabolites shows a high degree of cell and tissue specificity.

It had been observed by light microscopy in the last century that many of the colored flower pigments (e.g., anthocyanins, flavonol glycosides) or the red betalaines (which could be classified as alkaloids) of *Beta* roots are exclusively sequestered in the vacuole. Modern microspectrometric investigations (Gonnet and Hieu, 1992; Schnabl *et al.,* 1986) and direct determinations in isolated vacuoles (Table III) have confirmed these early observations. The fate of the uncolored metabolites remained almost unknown for many decades. Progress came when it was possible to prepare protoplasts from a wide number of plants in the 1970s (Ruesink, 1980) and to produce intact vacuoles without a significant efflux of vacuolar contents. Several protocols, which employ the lysis of protoplasts in slightly hypotonic media and gradient centrifugations (i.e., sucrose or Ficoll gradients), often result in pure and intact vacuoles (for reviews see Willenbrink, 1987; Mäder, 1984; Leigh, 1983; Ryan and Walker-Simmons, 1983; Wagner, 1982; Marty *et al.,* 1980). Centrifugation of the vacuoles through a silicone layer, in order to separate the vacuoles from the surrounding medium, was another important step in the production of intact vacuoles (see Mende and Wink, 1987; Wagner, 1982).

Table III
Vacuolar Sequestration of Defense and Signal Compounds[a]

Compounds	References
Phenolics	
Anthocyanins	Hrazdina *et al.* (1982), Ishikura (1981), Wiermann (1981)
Bergenin	Taneyama (1992)
Coumaroyl glycosides (esculin)	Werner and Matile (1985), Oba *et al.* (1981)
Flavonol glycosides	Hopp and Seitz (1987), Matern (1987), Van Genderen and Van Hemert (1986), Hrazdina *et al.* (1982), Wiermann (1981)
Gallic acid	Taneyama (1992)
7-Glucosyl-pleurostimin	Harborne *et al.* (1993)
Isoflavanone malonyl glycosides	Mackenbrock *et al.* (1992)
Isoflavone malonyl glycosides	Mackenbrock *et al.* (1992)
Kaempherol 3,7-*O*-glycoside	Schnabl *et al.* (1986)
Orientin-*C*-glycosides	Harborne *et al.* (1993)
Pterocarpan malonyl glycosides	Mackenbrock *et al.* (1992)
Quercetin 3-triglucoside	Weißenböck *et al.* (1986)
7-Rhamnosyl-6-hydroxyluteolin	Harborne *et al.* (1993)
Shikimic acid	Holländer-Czytko and Amrhein (1983)
Tricin 5-glucoside	Harborne *et al.* (1993)
Terpenoids	
Convallatoxin and other cardenolides	Löffelhardt *et al.* (1979)
Gentiopicroside	Keller (1986)
Oleanolic acid (3-*O*-glucoside)	Szakiel and Janiszowska (1993)
Oleanolic acid (3-*O*-glucuronide)	Szakiel and Janiszowska (1993)
Primary cardiac glycosides (lanatoside A, C; purpureaglycoside A)	Christmann *et al.* (1993), Kreis and Reinhard (1987)
Saponins (avenacosides)	Urban *et al.* (1983)
Oligosaccharides	
Gentianose	Keller and Wiemken (1982)
Gentiobiose	Keller and Wiemken (1982)
Stachyose	Keller and Matile (1985)
Nitrogen-containing compounds (excluding alkaloids)	
Cyanogenic glycosides (linamarin)	Saunders and Conn (1978)
Glucosinolates	Lüthy and Matile (1984), Wei *et al.* (1981)
Sinapoylglucosides	Sharma and Strack (1985)
Alkaloids	
Ajmalicine	Renaudin (1989), Deus-Neumann and Zenk (1984)
Atropine	Mende and Wink (1987)
Berberine	Sato *et al.* (1993, 1994)
Betaine	Matoh *et al.* (1987)
Betalaines	Leigh (1983)
Catharanthine	Deus-Neumann and Zenk (1984)
Codeine	Pham and Roberts (1991), Roberts (1987)
Dopamine	Roberts (1987), Homeyer and Roberts (1984)
Lupanine	Mende and Wink (1987)
Morphine	Pham and Roberts (1991), Roberts (1987)
Nicotine	Renaudin and Guern (1987), Saunders (1979)
Noscapine	Pham and Roberts (1991)
Papaverine	Pham and Roberts (1991), Roberts (1987)
Polyamines	Pistocchi *et al.* (1988)
S-Reticuline	Deus-Neumann and Zenk (1986)

(*continued*)

Table III (*Continued*)

Compounds	References
Sanguinarine	Matile *et al.* (1970)
Scopolamine	Deus-Neumann and Zenk (1984)
S-Scoulerine	Deus-Neumann and Zenk (1986)
Senecionine *N*-oxide	Ehmke *et al.* (1987, 1988)
Serpentine	Blom *et al.* (1991b), Deus-Neumann and Zenk (1984)
Solanidine	Han *et al.* (1989)
Thebaine	Pham and Roberts (1991), Roberts (1987)
Vindoline	Brisson *et al.* (1992), Deus-Neumann and Zenk (1984)
Defense proteins	
Phytohemagglutinin	Chrispeels and Raikhel (1992), Chrispeels (1991)
Protease inhibitors	Ryan and Walker-Simmons (1983)
Chitinase	Mackenbrock *et al.* (1992)
β-1,3-Glucanase	Mackenbrock *et al.* (1992)

[a]Data from Wink (1997a).

High concentrations of secondary metabolites are sometimes sequestered in vacuoles. For example, in latex vacuoles (vesicles) of *Chelidonium majus*, the concentrations of sanguinarine, chelidonine, and berberine were 500–1000 mM (Hauser and Wink, 1990), and latex vacuoles of *Papaver somniferum* sequester up to 500 mM morphine (Pham and Roberts, 1991). As many of these compounds are toxic and could be dangerous to the plant producing them, Matile (1984) considered storage in the vacuole as a means of detoxification and coined the term *toxic compartment*. This is a plausible description from the point of view of plant biochemistry, but as explained above, *defense and signal compartment* (Wink, 1993a) better describes the functional role and corresponding ecological importance.

3.2. Mechanisms Underlying Vacuolar Sequestration

3.2.1. UPTAKE ACROSS THE TONOPLAST

Biomembranes are semipermeable. Small and lipophilic molecules, such as O_2, CO_2, N_2, or benzene, can pass through membranes rapidly by diffusion. Polar metabolites, such as sugars, ions and charged molecules, such as acids, and amino acids will diffuse freely at a very low rate. The tonoplast does not appear to differ in its semipermeability from other membranes, such as the plasmalemma. Lipophilic synthetic dye probes, such as methyl red, methylene blue, auramine O, and acridine orange, pass through both membranes readily by diffusion (Wink, 1990). Dyes that carry charged sulfate groups or are glycosides, such as procion blue, carminic acid, alizarine red, and indigo carmine, cannot pass through both membranes by diffusion (Wink, 1990). For the charged or polar molecules, the tonoplast and the plasmalemma thus form penetration barriers. Ions, sugars, and amino acids cannot pass the tonoplast by simple diffusion. Instead, it is more or less certain that the uptake is achieved with the aid of special protein-based channels, pores, or transporters. Channels are characterized by turnover numbers in the range of 10^7–10^8 sec^{-1}, whereas carriers are rather slow with turnover rates of 10^2–10^4 sec^{-1} (Sanders,

1990). Using isolated vacuoles, uptake kinetics and mechanisms have been studied for a number of molecules, including Na^+, K^+, Ca^{++}, NO_3^-, Cl^-, SO_4^{2-}, malate, citrate, glucose, sucrose, arginine, lysine, phenylalanine, other amino acids and dopamine (Hediger, 1994; Wink, 1993a; Taiz, 1992; Hedrich and Schröder, 1989; Blumwald, 1987; Martinoia *et al.,* 1985; Boller *et al.,* 1975). Uptake of these solutes usually depends on time, hydrogen ion concentrations, and temperature. It shows saturation kinetics that are often multiphasic (Nissen 1991) and is regulated by SH reagents, metabolic and competitive inhibitors. All of these data imply the presence of specific transport proteins (Hediger, 1994) some of which have been purified and reconstituted in artificial membranes, such as the carriers for glutamine, arginine (Thume and Dietz, 1991), and malate (Martinoia *et al.,* 1991). In a few cases the corresponding genes have been isolated and expressed (Hsu *et al.,* 1993; Tsay *et al.,* 1993; Riesmeier *et al.,* 1992; Sauer and Tanner, 1989).

Most of the defense and signal compounds found in the vacuole (Table III) are hydrophilic, polar (such as sugars and glycosides), or even charged molecules, such as nonprotein amino acids, alkaloids, some glycosides or proteins. Because these compounds have not been synthesized in the vacuole, they have to pass the tonoplast first in order to be accumulated in the vacuole.

Considering the structural diversity of secondary metabolites, we cannot assume that the biochemical mechanisms that lead to their sequestration in vacuoles are identical. Most lipophilic metabolites will pass the tonoplast by *simple diffusion*, as was shown for several weakly basic alkaloids, such as nicotine, ajmalicine, colchicine, vinblastine, ergotamine, sanguinarine, vindoline, quinine, and cinchonamine (Blom *et al.,* 1991a; Hauser and Wink, 1990; Renaudin, 1989; McCaskill *et al.,* 1988; Renaudin and Guern, 1987; Kurkdjian, 1982).

Other alkaloids, which are charged species under cytosolic pH conditions, or polar glycosides appear to pass the tonoplast via a *carrier mechanism.* A few examples are given in Table IV. It is remarkable that for *S*- and *R*-reticuline or *S*- and R-scoulerine, these carriers discriminate for the naturally occurring *S*-configured compounds and were thus stereoselective (Deus-Neumann and Zenk, 1986).

In the case of latex vacuoles of *Papaver somniferum*, morphine uptake was stimulated by ATP, but the authors suggest the presence of an alkaloid channel instead of a specific morphine transporter (Roberts *et al.,* 1991).

Table IV
Examples of Carrier-Mediated Uptake Mechanisms of Alkaloids

Alkaloid	Plant	Mechanism[a]	Reference
S-Scoulerine	*Fumaria*	H^+ AP	Deus-Neumann and Zenk (1986)
S-Reticuline	*Fumaria*	H^+ AP	Deus-Neumann and Zenk (1986)
Catharanthine	*Catharanthus*		Deus-Neumann and Zenk (1984)
Atropine	*Atropa*	H^+ AP	Mende and Wink (1987)
Lupanine	*Lupinus*	H^+ AP	Mende and Wink (1987)
Sparteine	*Lupinus*	H^+ AP	Mende and Wink (1987)
13-Hydroxylupanine	*Lupinus*	?	Hauser and Wink (1990)
Senecionine *N*-oxide	*Senecio*	?	Ehmke *et al.* (1987, 1988)
Dopamine	*Papaver*	?	Homeyer and Roberts (1984)

[a]H^+ AP, H^+–alkaloid antiport mechanism.

In addition to the above-mentioned alkaloid carriers, similar mechanisms have been proposed for glycosides and other polar metabolites: primary cardiac glycosides (Kreis and Hölz, 1991; Kreis and Reinhard, 1987), coumaroyl glycosides (Alibert *et al.*, 1985; Rataboul *et al.*, 1985; Werner and Matile, 1985), acylated anthocyanins (Hopp and Seitz, 1987), flavonoids such as apigenin-7-*O*-(β-D-malonylglucoside) (Matern, 1987; Matern *et al.*, 1986), 1-(malonylamino)-cyclopropane-1-carboxylic acid (Bouzayen *et al.*, 1989) and glucosides and glucuronides of the triterpene oleanolic acid (Szakiel and Janiszowska, 1991, 1992, 1993; Janiszowska and Szakiel, 1992; Szakiel and Kasprzyk, 1989).

Whether particular carriers exist for all of these allelochemicals or whether they can hijack existing carriers for primary metabolites is not clear at present. The transport system for 1-aminocyclopropane-1-carboxylic acid (ACC) can be classified as a neutral L-amino acid carrier with a high affinity for ACC and other nonpolar aminoacids (Saftner, 1994), and this would support the possibility of the use of existing carriers.

3.2.2. VACUOLAR SEQUESTRATION AGAINST A CONCENTRATION GRADIENT

The concentrations of ions, sugars, acids, signal and defense chemicals are often remarkably high in vacuoles and orders of magnitude lower in the cytoplasm (Boller and Wiemken, 1986; Leigh *et al.*, 1981). Thus, all of these molecules have to be sequestered in the vacuole against a concentration gradient. The question to be considered therefore concerns the driving force for the uphill accumulation of substances in the vacuoles.

A major tonoplast protein is H^+-ATPase (Bremberger *et al.*, 1988; Sze, 1985; Leigh and Walker, 1980) which along with pyrophosphatase (Bremberger *et al.*, 1988; Rea and Sanders, 1987; Leigh and Walker, 1980) is responsible for the transport of protons into the vacuole (Taiz, 1992; Hedrich *et al.*, 1989; Rea and Sanders, 1987; Sze, 1985; Thom and Komor, 1984). As a consequence, the vacuolar hydrogen ion concentration is orders of magnitude higher than that of the cytoplasm. This proton gradient (proton motive force) can be utilized for secondary active transport systems (Hedrich and Schröder, 1989; Kurkdjian and Guern, 1989; Blumwald, 1987; Reinhold and Kaplan, 1984), such as substrate–proton antiport mechanisms (Getz, 1991; Blackford *et al.*, 1990; Blumwald, 1987; Blumwald and Poole, 1985 a,b; Briskin *et al.*, 1985; Thom and Komor, 1984, 1985; Hager and Hermsdorf, 1981; Lüttge *et al.*, 1981).

There is experimental evidence that some secondary metabolites are transported into the vacuole by a H^+ antiport mechanism (Table IV), such as lupanine (Fig. 1; Mende and Wink, 1987), *S*-reticuline, *S*-scoulerine (Deus-Neumann and Zenk, 1986), and 1-(malonylamino)-cyclopropane-1-carboxylic acid (Bouzayen *et al.*, 1989). In these cases, vacuolar uptake was dependent on Mg/ATP and could be inhibited by reagents that break down proton gradients. In addition, lupanine transport was enhanced by K^+, indicating either that the membrane potential could be additionally involved (Mende and Wink, 1987) or that K^+ activates the H^+-translocating pyrophosphatase as was shown for beet vacuoles (Leigh, 1983). The transport of glutathione S-conjugates of xenobiotics into the vacuole was also Mg/ATP dependent, but was not inhibited by compounds that disrupt secondary activated uptake processes; therefore, proton antiport mechanisms cannot be the driving force in this case. Instead, the carrier appears to be a specific ATPase, similar to the export pumps in mammalian liver. A connection to (multidrug resistance proteins) MDR-like ATPases appears possible (Li *et al.*, 1995; Martinoia *et al.*, 1993; Tommasini *et al.*, 1993). Recently, it has been shown that the protein encoded by the Bronze-2 gene in maize, which is responsible for the deposition of anthocyanins in the vacuole, is a glutathione S-transferase (Marrs *et al.*, 1995). Thus, anthocyanins, e.g., cyanidin-3-glucoside, like xenobiotics may be conjugated with glutathione. Many herbicide conjugates are metabolized to cysteine conjugates and then acylated with malonic acid. Because malonylated cyanidin-3-glucoside is the major maize anthocyanin, Marrs *et al.* (1995) postulate in analogy that anthocyanin is transported into the vacuole as the glutathione S-conjugate with the aid of the ATP-dependent glutathione S-conjugate "export" pump. Once in the vacuole anthocyanins are further converted to the malonyl derivative. However, Hopp and Seitz (1987) have shown that a carrier system exists for acylated anthocyanins at the tonoplast, which would contradict the assumption of Marrs *et al.* (1995). Whether the glutathione conjugate pump also transports other secondary metabolites, such as alkaloids, remains to be shown (Wink 1997).

It has been emphasized that an apparent uphill transport can also be achieved by certain "trapping" reactions (Boller and Wiemken, 1986; Matile 1978, 1984). Normally, passive transport or diffusion will come to a standstill when equal concentrations are reached on both sides of the tonoplast. If we assume that the molecule that enters the vacuole is trapped or otherwise changed, then it is removed from the equilibrium and the uphill process can go on until all trapping molecules are exhausted and equal concentrations of "free" molecules are reached on both sides.

The trapping reactions discussed (Table V) include "ion traps," which could be relevant for nitrogenous compounds, e.g., alkaloids, which become protonated in the acidic vacuole (Renaudin, 1989; Guern *et al.*, 1987; Boller and Wiemken, 1986; Nishimura, 1982), and which do not pass membranes as protonated molecules, except when highly lipophilic (Hauser and Wink, 1990; Renaudin, 1989; Kurkdjian, 1982). The monoterpene indole alkaloids ajmalicine and vindoline appear to cross the tonoplast by simple diffusion (Blom *et al.*, 1991b; Renaudin, 1989; McCaskill *et al.*, 1988; Renaudin and Guern, 1987) and not by carrier-mediated transport, as reported earlier (Deus-Neumann and Zenk, 1984, 1986). In the vacuole ajmalicine is effectively converted into the more polar serpentine by nonspecific peroxidases, so serpentine cannot leak out of the vacuole (Blom *et al.*, 1991a). It has been suggested that intravacuolar serpentine binds to the tonoplast (Pradier *et al.*, 1988). In other instances conformational changes with solvent

Table V
Mechanisms for Uphill Accumulation of Defense/Signal Compounds in Vacuoles

Active transport
ATPase coupled "pumps"
Secondary active transport
Proton gradients generated by H^+-ATPase, PP_iase (proton motive force) used by proton antiport carriers
Trapping mechanisms
Protonation of alkaloids in acidic vacuole
Binding of secondary compounds to tannin, other phenolics or polyphosphates
Complexation of alkaloids with chelidonic acid, meconic acid
Binding of secondary metabolites to proteins ?
Crystallization (Ca-oxalate)
Conjugation
Change of conformation or configuration

composition and proton concentration have been cited, e.g., apigenin-7-*O*-(6-*O*-malonyl)glucoside in the vacuoles of *Petroselinum hortense* (Matern *et al.,* 1983)

Another trapping procedure that has been suggested is the binding of molecules to complexing compounds such as tannins or polyphosphates. This type of complexing could occur with alkaloids but there are as yet no examples that have been investigated in detail (Matile, 1978). Better studied are the complexing reagents chelidonic and meconic acid. Isolated latex vacuoles of *Chelidonium majus* take up sanguinarine and various other lipophilic alkaloids (Hauser and Wink, 1990; Matile, 1978). It could be shown that these vesicles contain about 660 mM (range 200–1300 mM) chelidonic acid, which readily complexes alkaloids (Hauser and Wink, 1990). The reaction of these alkaloids with chelidonic acid obviously provides the apparent "energy" of accumulation against a concentration gradient. We do not know, however, the mechanism for concentrating chelidonic acid in vesicles. In the case of latex vacuoles of *Papaver somniferum*, the concentration of meconic acid can be as high as 250 mM and was considered as relevant for trapping morphine, codeine, papaverine, and other poppy alkaloids. Also in these vesicles high levels of sulfate and chloride were recorded and salt formation with these ions must occur as well (Pham and Roberts, 1991). In cells of *Coptis japonica*, berberine content of the vacuoles was correlated with their malic acid accumulation; the resulting malate–berberine complex might trap berberine in the vacuole (Sato *et al.,* 1992). The accumulation of lupanine in epidermal vacuoles was also favored by malate, suggesting a similar mechanism (Wink and Mende, 1987). In *Coffea arabica* compartmentation of purine alkaloids, e.g., caffeine, depends exclusively on the physical chemistry of their vacuolar complexation with chlorogenic acid (Waldhauser and Baumann, 1996).

Crystallization, which has been observed for the alkaloids berberine and sanguinarine and for calcium oxalate (Renaudin and Guern, 1990; Lange and Kohlenbach, 1982; Franceschi and Horner, 1980), is another, albeit rare, trapping mechanism.

Summarizing, it is apparent that the uptake of metabolites into the vacuole can be directly or secondarily energy-dependent (ATP-driven proton antiports or direct export "pumps") or can be accomplished by various trapping or binding reactions. The trapping mechanisms are compatible with both transport and diffusion processes.

4. TURNOVER OF ALKALOIDS

For a long time, alkaloids and other secondary metabolites had been considered as stable end products or waste products. This concept changed over the last decades and as outlined in Chapter 11, alkaloids and other secondary metabolites are today regarded as signal and/or defense compounds, i.e., molecules that are important for the fitness of the plants producing them, but certainly no longer as waste products. Following the fate of alkaloids in legumes, it became evident that also the "end product" idea needs to be corrected: Alkaloids are usually found in considerable amounts in photosynthetically active leaves of legumes, whereas senescing leaves (especially of perennial species) are devoid of them, indicating that alkaloids are mobilized (as are other N-containing metabolites) and translocated to the stems. If alkaloids were waste products, we should have seen the opposite trend. Members of the Papilionoideae (family Leguminosae) mainly store quinolizidine and dipiperidine alkaloids and occasionally pyrrolizidine alkaloids (PAs) (in the genus *Crotalaria*) in their seeds. As alkaloid contents are usually high (variation between 1 and

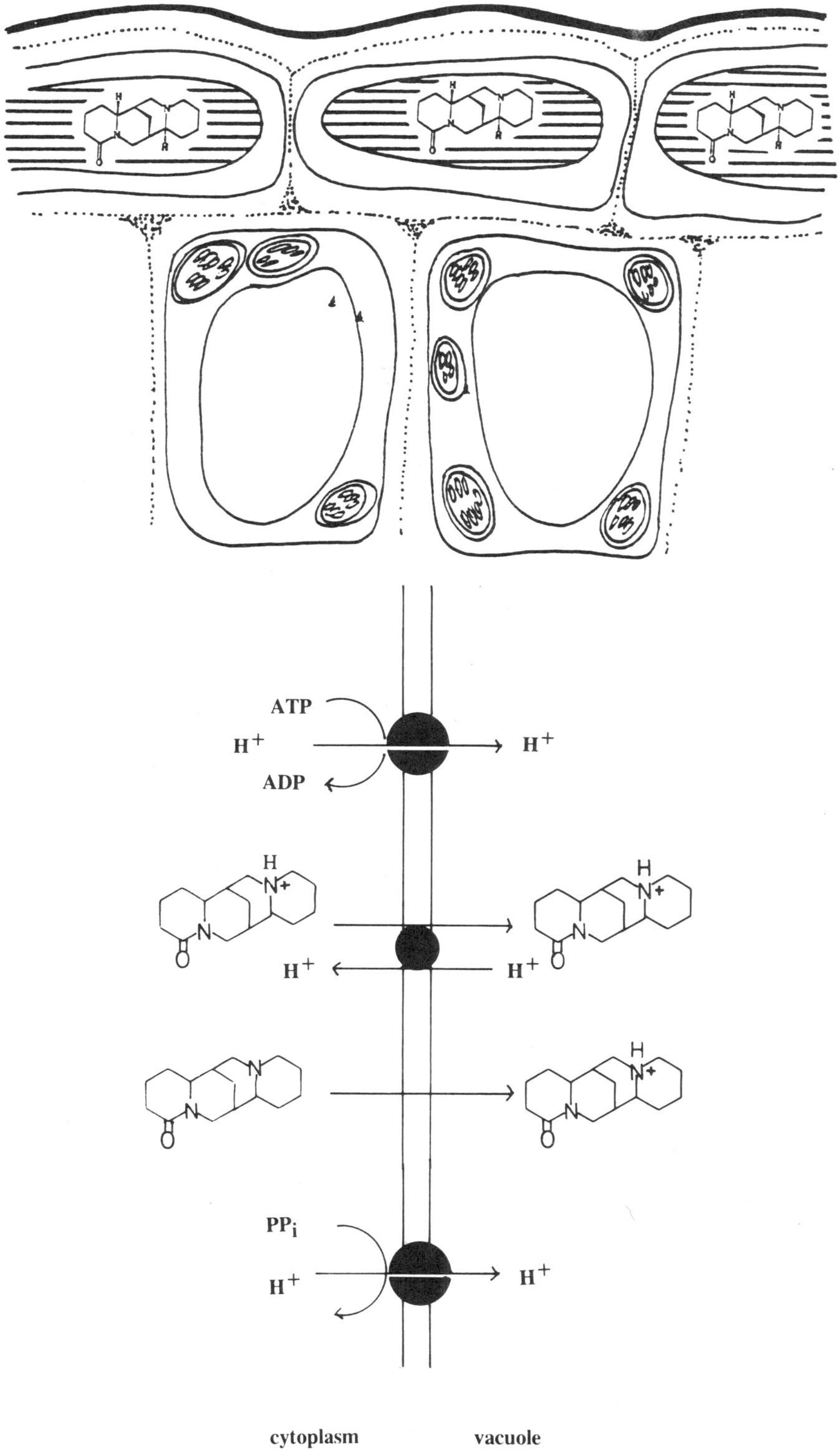

Figure. 1. Compartmentation of quinolizidine alkaloids (QAs) in lupins. Above: Synthesis of QAs in mesophyll chloroplast, but storage predominantly in the epidermal vacuoles. Below: Free QA base can diffuse across the tonoplast, whereas the protonated alkaloid, which is present under physiological conditions, enters the vacuole by a proton-antiport fueled carrier. From Wink, M., 1993a, The plant vacuole: A multifunctional compartment, *J. Exp. Bot.* **44**(Suppl.):231–246; by permission of Oxford University Press.

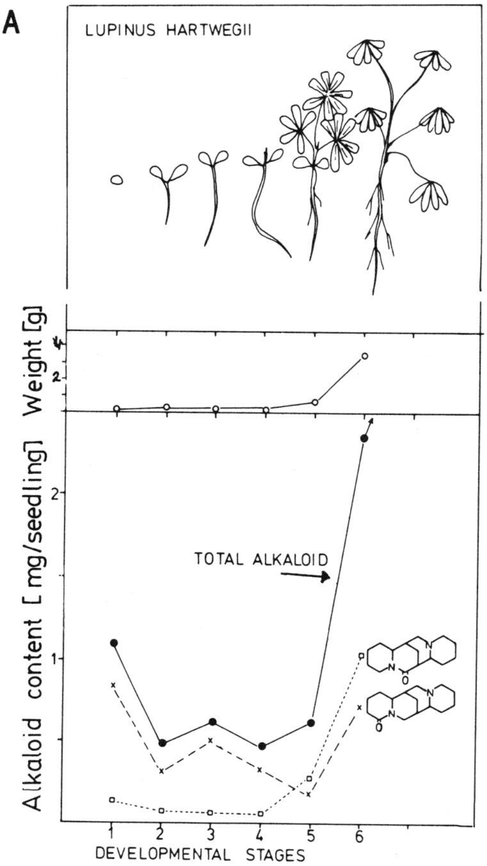

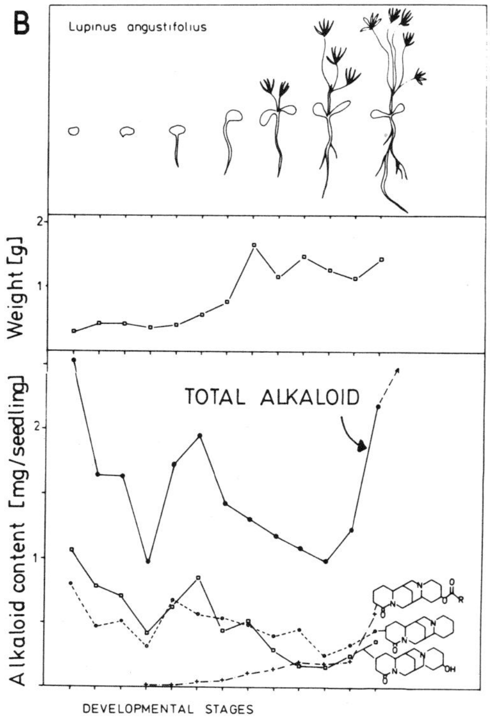

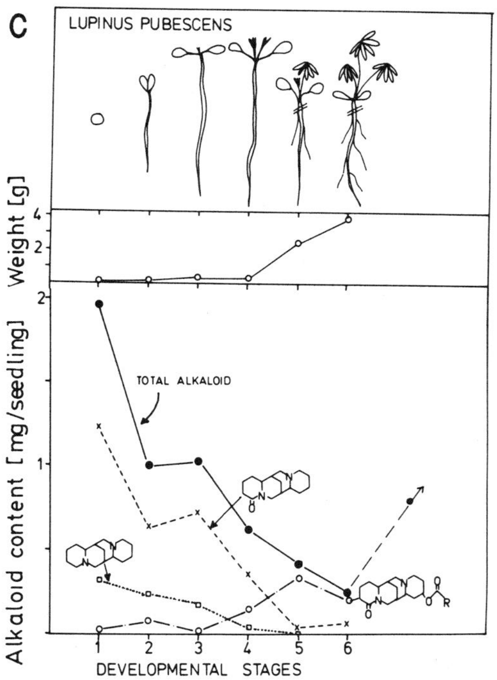

Figure. 2. Kinetics of alkaloid contents and alkaloid patterns during the germination of lupins (A–C) *Lupinus hartwegii, L. angustifolius, L. pubescens*, (D) *Cytisus scoparius*, (E) *Spartium junceum*, (F) *Baptisia australis*, and (G) *Laburnum anagyroides*. From Wink and Witte (1985); by permission of *Zeitschrift der Naturforschung*.

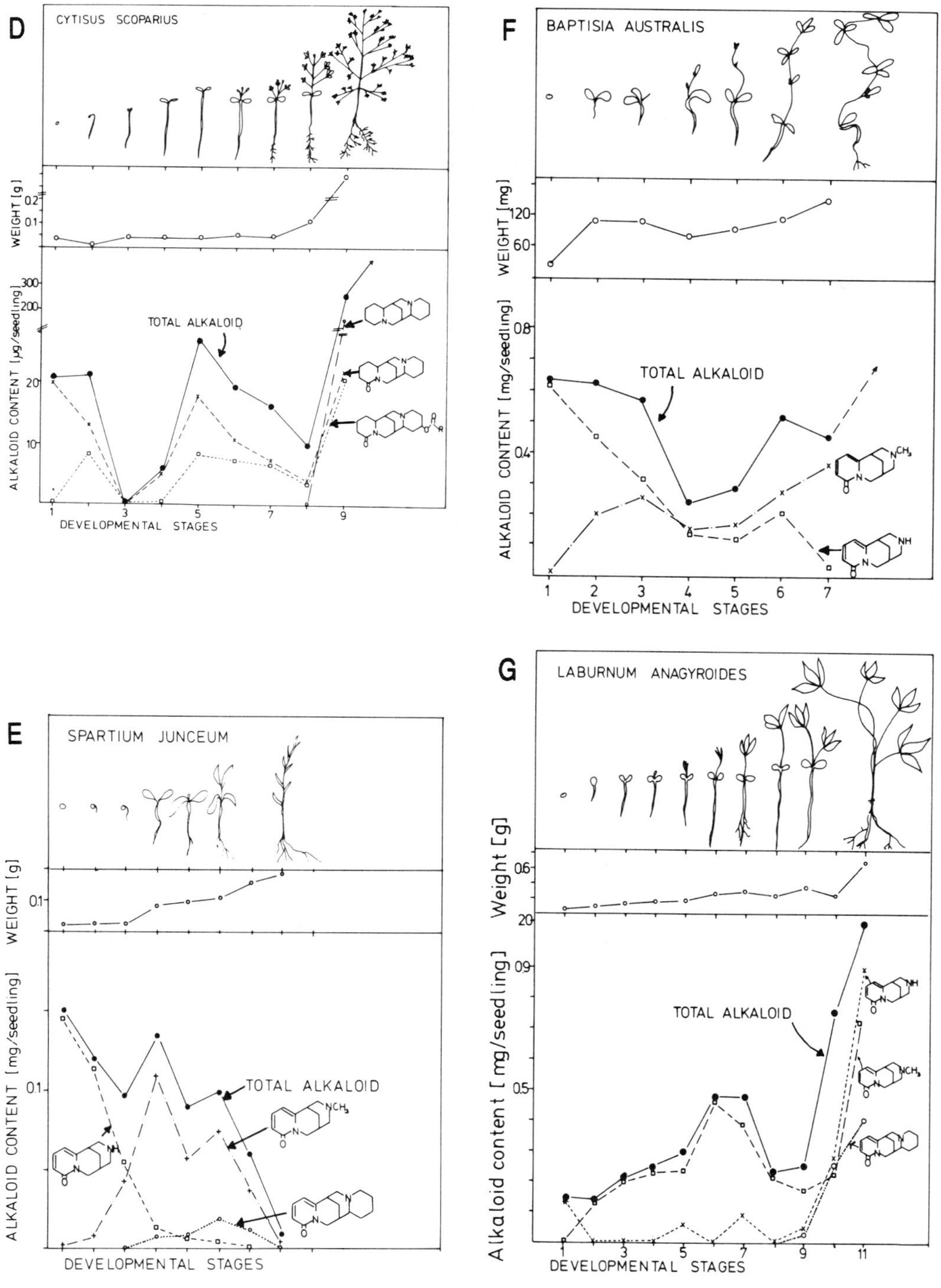

Figure. 2. *(Continued)*

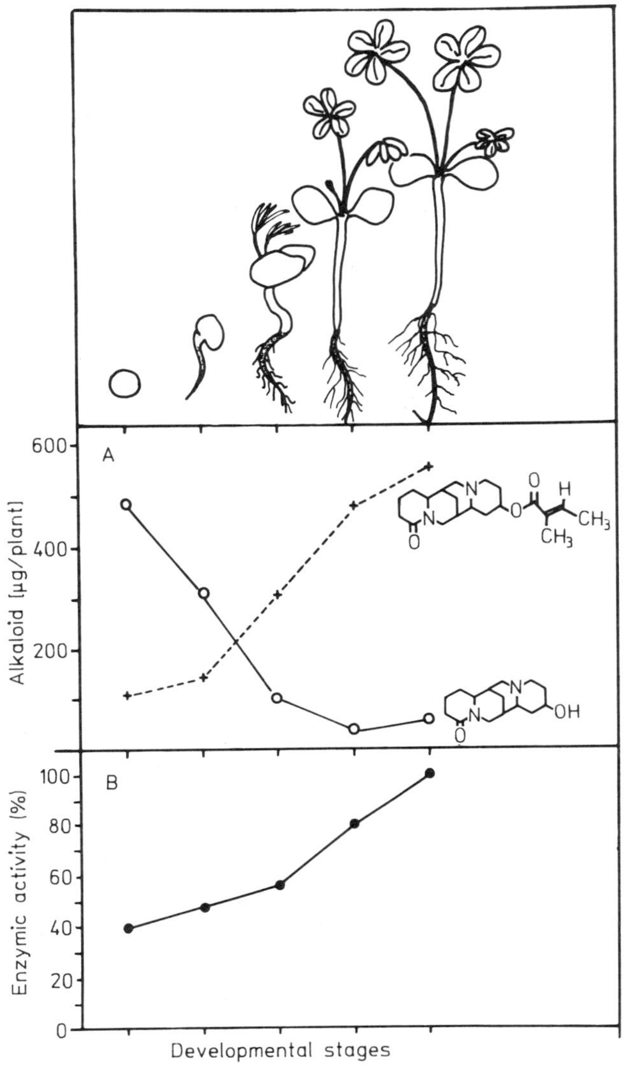

Figure 3. Formation of ester alkaloids during the growth of *Lupinus albus* seedlings and relation to the activity of tigloyl-CoA:13-hydroxylupanine tigloyltransferase. From Wink and Hartmann, (1982b).

8% (dry weight)], and as QAs and PAs are toxic to herbivores and feeding deterrents, it is very likely that these alkaloids serve as chemical defense compounds in the seeds (see Chapter 11). On a closer look, a dual function became obvious. During germination and seedling growth a marked decrease of alkaloids takes place (Fig. 2), suggesting that the alkaloids are degraded and their nitrogen incorporated into the amino acids/proteins of the developing seedling (similar to the situation of storage proteins or nonprotein amino acids, which show similar kinetics). When lupin seedlings have formed their first leaves, then a *de novo* synthesis of alkaloids sets in (Wink and Witte, 1985). Experimental evidence corroborates that lupins are able to degrade and utilize the alkaloidal N as a nitrogen source (Wink and Witte, 1985): Callus cultures of *Lupinus polyphyllus, L. hartwegii,*

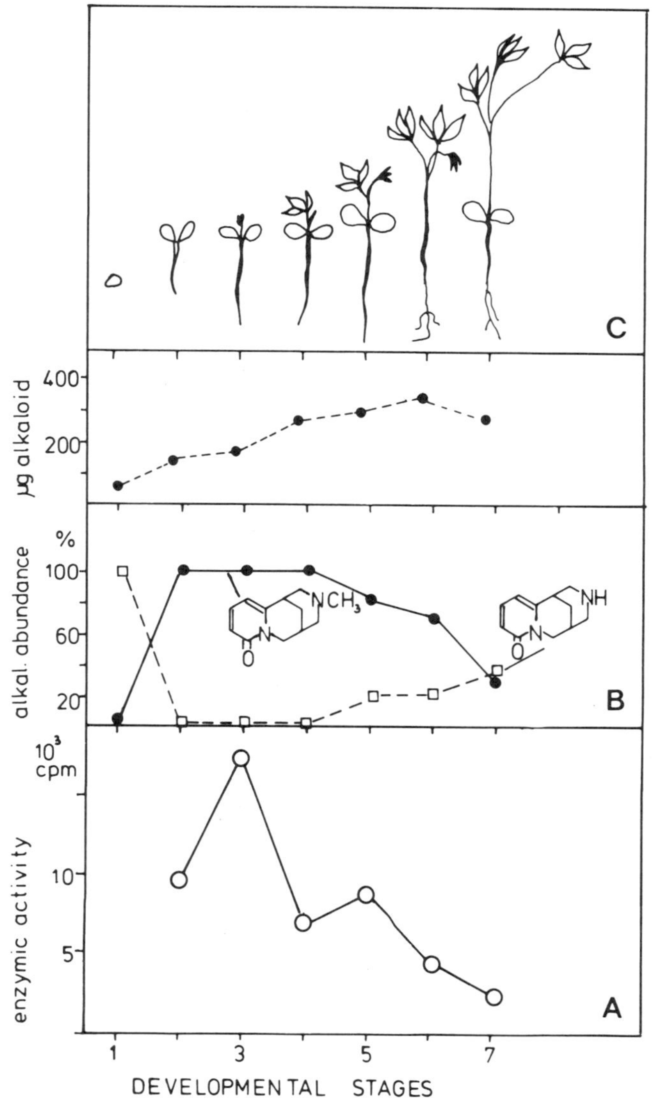

Figure 4. Formation of *N*-methylcytisine in seedlings of *Laburnum anagyroides* in relation to the apparent activity of SAM:cytisine *N*-methyltransferase. Stage 1 represents mature seeds, the following numbers the subsequent development stages. About 20 seedlings per stage were analyzed for methyltransferase activity (A), and quinolizidine alkaloids (C). In B the ratio of the main alkaloids cytisine and *N*-methylcytisine is recorded. Alkaloids were determined by capillary GLC, the enzyme activity under standard assay conditions.

Cytisus scoparius, and *Chamaecytisus purpureus* survived on media that contained sparteine as a sole nitrogen source for over 6 weeks, whereas controls on N-free media had died.

During the early stages of germination two types of enzymatic reactions are typical (Figs. 3 and 4): In seeds that are rich in tetracyclic QA of the lupanine type, hydroxylated alkaloids, such as 13-hydroxylupanine, are relatively abundant. A conversion of these hydroxylated alkaloids into ester alkaloids, such as 13-tigloyloxylupanine or 13-cinnamoyloxylupanine, takes place (Fig. 3); also, the activity of the corresponding enzyme, a tigloyl-CoA:13-hydroxylupanine *O*-tigloyltransferase, is increased (Fig. 4) (Wink and Hartmann, 1982b). As the ester alkaloids have insecticidal and allelopathic properties, this conversion would be plausible in ecological terms. Plants of the genera *Laburnum, Genista, Baptisia, Thermopsis, Spartium,* which produce cytisine-type QAs, store cytisine as the main component in seeds. During the early stages of germination, cytisine is converted into *N*-methylcytisine (Fig. 4), a reaction catalyzed by *S*-adenosyl-L-methionine: cytisine *N*-methyltransferase. Apparent enzyme activities are highest during these early developmental stages (Wink, 1984).

These observations suggest that alkaloids are not inert end products of metabolism. Instead turnover appears to be a regular feature, which could be explained from ecological and physiological requirements.

A similar situation does not occur in the fruits of *Conium maculatum* where the alkaloids present in the fruit are lost at germination with the testa and young seedlings have to synthesize alkaloids *de novo*. Here alkaloids must only have a protective role (see Chapter 5).

5. SEQUESTRATION AS AN INTEGRAL PART OF THE DEFENSE CONCEPT OF PLANTS

The close integration of vacuolar sequestration into the defense strategy of a plant is exemplified in the following (Fig. 7):

QAs, which figure as the characteristic secondary metabolites of many legumes, especially within the Papilionoideae, are produced by the chloroplast of leaf mesophyll cells (Wink and Hartmann, 1982a). After synthesis they are exported to the vacuole, obviously by means of a tonoplast proton antiport carrier (Mende and Wink, 1987). Part of the alkaloids are transported via the phloem (Wink and Witte, 1984) to all other parts of a lupin plant, especially the stems and fruits. In stems and petioles, the alkaloids enter the epidermal cells and are taken up into the vacuole against a concentration gradient by a transport system (Wink and Mende, 1987). Concentrations of QAs in epidermal vacuoles can be as high as 200 mM. As these alkaloids serve as defense compounds against herbivores (e.g., insects) the epidermal localization can be interpreted as a means of placing them at a strategically important site where they can ward off intruders as they start nibbling at a plant (Wink, 1985, 1987, 1988, 1992).

For comparison, the compartmentation of two other nitrogenous compounds, e.g., cyanogenic glycosides and glucosinolates, are shown in Figs. 5 and 6, as the integration appears to be even more intricate:

In *Sorghum*, the cyanogenic glycosides are stored in epidermal vacuoles, which lack, however, β-glucosidase which is localized in the chloroplast of adjacent cells (Wajant *et al.,* 1995; Kojima *et al.,* 1979; Saunders and Conn, 1978). On wounding, for example by a herbivore, the cellular integrity breaks down and both vacuolar contents and β-glucosidase come into contact (Fig. 5). The cyanogenic glycosides are hydrolyzed. Hydroxynitrile lyase, which releases HCN, is found in the cytoplasm of both epidermal and mesophyll cells. The HCN generated is a strong inhibitor of mitochondrial respiration and thus a strong toxin. In *Hevea brasiliensis* leaves, the cyanogenic glycoside linamarin is also exclusively stored in the vacuole and the linamarase in the apoplastic space (Gruhnert

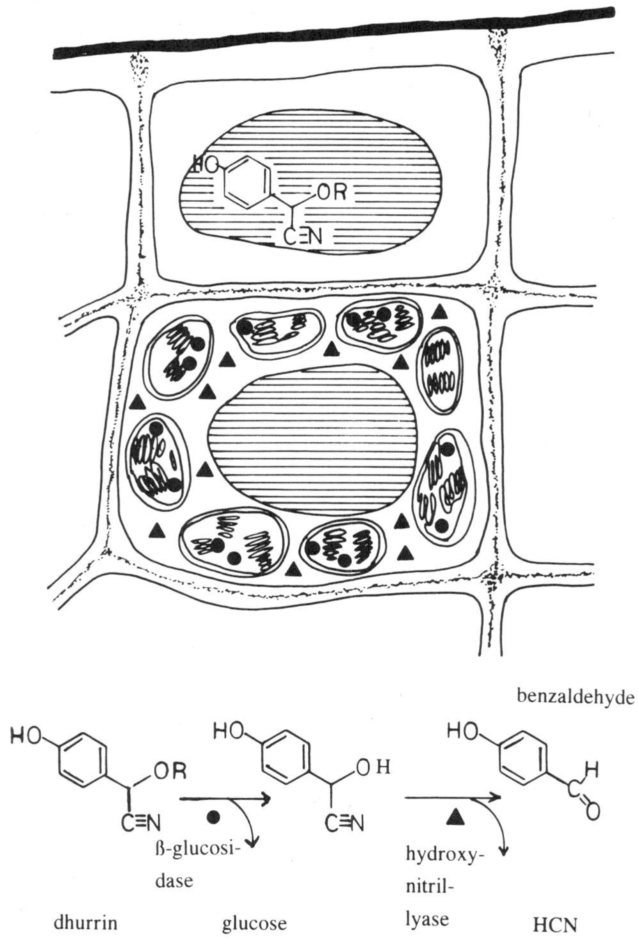

Figure 5. Schematic illustration of the compartmentation of cyanogenic glycosides. Shown below are the enzymatic breakdown of dhurrin and HCN production. From Wink, M., 1993a, The plant vacuole: A multifunctional compartment, *J. Exp. Bot.* **44**(Suppl):231–246; by permission of Oxford University Press.

et al., 1994), indicating that the spatial compartmentation of cyanogenic glycosides and their metabolizing enzymes have evolved several times during evolution.

In root tissues of horseradish (*Armoracia rusticana*), the glucosinolates, in addition to ascorbate (an activator of myrosinase), are also stored in vacuoles (Fig. 6), but the hydrolyzing enzyme, the myrosinase, is localized in the cytoplasmic membranes and in cell walls (Lüthy and Matile, 1984; Matile, 1980, 1984). On wounding, all components come into contact and mustard oil, a strong animal deterrent, membrane destabilizing agent, and antibiotic, is released.

How can we explain the cell- and tissue-specific vacuolar sequestration of alkaloids and other secondary metabolites? If we accept the existence of specific channels, pores, and carriers in the tonoplast, we must assume that corresponding genes exist. Thus, by cell- and tissue-specific expression of these transporter genes and the respective biosynthetic enzymes, the different vacuolar types can be generated; whether a vacuole

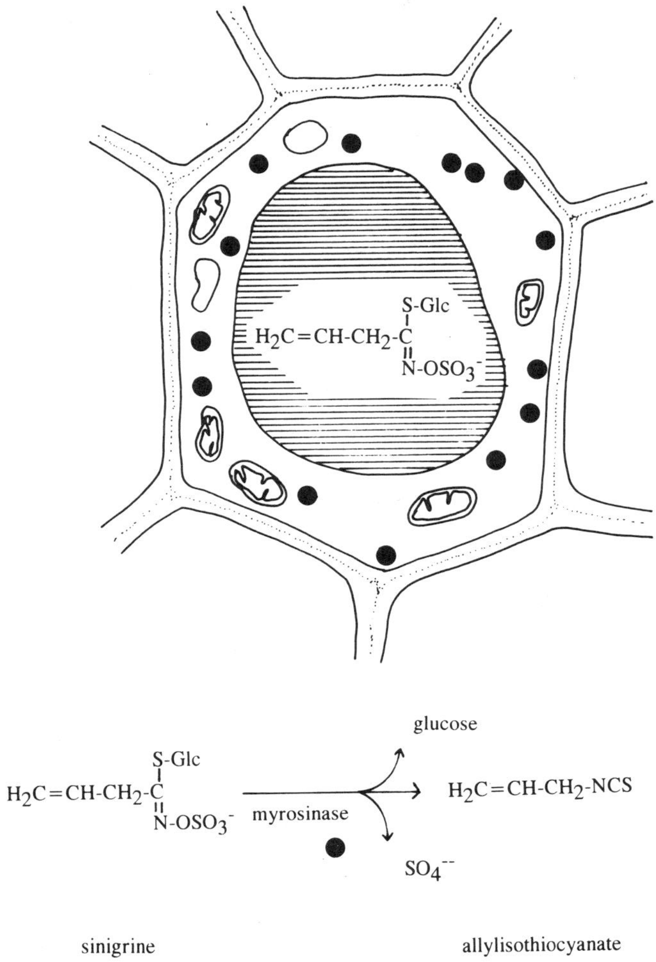

Figure 6. Schematic illustration of the compartmentation of glucosinolates in horse raddish. From Wink, M., 1993a, The plant vacuole: A multifunctional compartment, *J. Exp. Bot.* **44**(Suppl.):231–246; by permission of Oxford University Press.

functions in malate or in alkaloid storage is indirectly gene dependent. These matters will certainly be resolved when genes of carriers and channels become available, hopefully in the near future.

REFERENCES

General Reviews

Alibert, G., Boudet, A. M., Canut, H., and Rataboul, P., 1985, Protoplasts in studies of vacuolar storage compounds, in: *Physiological Properties of Plant Protoplasts* (P. E. Pilet, ed.), Springer-Verlag, Berlin, pp. 105–115.

Blumwald, E., 1987, Tonoplast vesicles as a tool in the study of ion transport at the plant vacuole, *Physiol. Plant.* **69:**731–734.

Boller, T., and Wiemken, A., 1986, Dynamics of vacuolar compartmentation, *Ann. Rev. Plant Physiol.* **37:**137–164.

Boller, T., and Wiemken, A., 1987, Dynamics of lysosomal functions in plant vacuoles, in: *Plant Vacuoles* (B. Marin, ed.), Plenum Press, New York, pp. 361–368.

Chrispeels, M. J., 1991, Sorting of proteins in the secretory system, *Annu. Rev. Plant Physiol. Plant Mol. Biol.* **42:**21–53.

Guern, J., Renaudin, J. P., and Brown, S. C., 1987, The compartmentation of secondary metabolites in plant cell cultures, in: *Cell Culture and Somatic Cell Genetics of Plants*, Vol. 4 (F. Constabel and I. Vasil, eds.), Academic Press, San Diego, pp. 43–76.

Hager, A., and Hermsdorf, P., 1981, A H/Ca-antiporter in membranes of microsomal vesicles from maize coleoptiles, a secondary energized Ca pump, *Z. Naturforsch.* **36c:**1009–1012.

Hediger, M. A., 1994, Structure, function and evolution of solute transporters in procaryotes and eucaryotes, *J. Exp. Biol.* **196:**15–49.

Hedrich, R., and Schröder, J., 1989, Physiology of ion channels and ion pumps of higher plant cells, *Annu. Rev. Plant Physiol. Plant Mol. Biol.* **40:**539–569.

Kreis, W., and Hölz, H., 1991, Zellulärer Transport und Speicherung von Naturstoffen, *Naturwiss. Rundsch.* **44:**463–470.

Kurkdjian, A., and Guern, J., 1989, Intracellular pH—Measurement and importance in cell activity, *Annu. Rev. Plant Physiol. Plant Mol. Biol.* **40:**271–303.

Leigh, R. A., 1983, Methods, progress and potential for use of isolated vacuoles of solute transport in higher plants, *Physiol. Plant.* **57:**390–396.

Mäder, M., 1984, Die Pflanzliche Vakuole aus biochemischer Sicht, *Biol. Unserer Zeit* **14:**171–176.

Marty, F., Branton, D., and Leigh, R. A., 1980, Plant vacuoles, in: *The Biochemistry of Plants: A Comprehensive Treatise,* Vol. 1 (N. E. Tolbert, ed.), Macmillan Co., New York, pp. 625–658.

Matern, U., 1987, Die Isomerenfalle für Sekundärmetabolite, eine Alternative zum Ionenfallen-Modell, *Biol. Unserer Zeit* **17:**148–152.

Matile, P., 1978, Biochemistry and function of vacuoles, *Annu. Rev. Plant Physiol.* **29:**193–213.

Matile, P., 1984, Das Toxische Kompartiment der Pflanzenzelle, *Naturwissenschaften* **71:**18–24.

Meijer, A. H., Verpoorte, R., and Hoge, J. H. C., 1993, Regulation of enzymes and genes involved in terpenoid indole alkaloid biosynthesis in *Catharanthus roseus*, *J. Plant Res. Spec. Iss.* **3:**145–164.

Reinhold, L., and Kaplan, A., 1984, Membrane transport of sugars and amino acids, *Annu. Rev. Plant Physiol.* **35:**45–83.

Renaudin, J. P., and Guern, J., 1990, Transport and vacuolar storage of secondary metabolites in plant cell cultures, in: *Secondary Products from Plant Tissue Culture* (B. C. Charwood and M. J. C. Rhodes, eds.), *Proc. Phytochem. Soc. Eur.* **30:**59–78.

Ruesink, A., 1980, Protoplasts of plant cells, *Methods Enzymol.* **69:**69–84.

Ryan, C. A., and Walker-Simmons, M., 1983, Plant vacuoles, *Methods Enzymol.* **96:**580–589.

Sanders, D., 1990, Kinetic modeling of plant and fungal membrane transport systems, *Annu. Rev. Plant Physiol. Plant Mol. Biol.* **41:**77–107.

Sze, H., 1985, H-translocating ATPase: Advances using membrane vesicles, *Annu. Rev. Plant Physiol.* **36:**175–208.

Taiz, L., 1992, The plant vacuole, *J. Exp. Biol.* **172:**113–122.

Wagner, G. J., 1982, Compartmentation in plant cells: The role of the vacuole, in: *Recent Advances in Phytochemistry*, Vol. 16 (L. Creasy and G. Hrazdina, eds.), Plenum Press, New York, pp. 1–45.

Wiermann, R., 1981, Secondary plant products and cell and tissue differentiation, in: *The Biochemistry of Plants* Vol. 7, Academic Press, San Diego, pp. 85–116.

Willenbrink, J., 1987, Die pflanzliche Vacuole als Speicher, *Naturwissenschaften* **74:**22–29.

Wink, M., 1987a, Quinolizidine alkaloids: Biochemistry, metabolism, and function in plants and cell suspension cultures, *Planta Med.* **53:**509–514.

Wink, M., 1987b, Physiology of the accumulation of secondary metabolites with special reference to alkaloids, in: *Cell Culture and Somatic Cell Genetics of Plants*, Vol. 4, (F. Constabel and I. Vasil, eds.), Academic Press, San Diego, pp. 17–41.

Wink, M., 1988, Plant breeding: Importance of plant secondary metabolites for protection against pathogens and herbivores, *Theor. Appl. Genet.* **75:**225–233.

Wink, M., 1990, Physiology of secondary product formation in plants, in: *Secondary Products from Plant Tissue Culture* (B. C. Charwood and M. J. C. Rhodes, eds.), *Proc. Phytochem. Soc. Eur.* **30:**23–41.

Wink, M., 1993a, The plant vacuole: A multifunctional compartment, *J. Exp. Bot.* **44**(Suppl.):231–246.

Wink, M., 1993b, Allelochemical properties or the raison d'être of alkaloids, in: *The Alkaloids*, Vol. 43, (G. A. Cordell, ed.), Academic Press, San Diego, pp. 1–118.

Wink, M., 1997, Compartmentation of secondary metabolites and xenobiotics in plant vacuoles, *Adv. Bot. Res.* **25:**141–169.

Zenk, M. H., 1989, Biosynthesis of alkaloids using tissue culture, *Recent Adv. Phytochem.* **23:**429–457.

Special References

Aerts, R. J., and Baumann, T. W., 1994, Distribution and utilization of chlorogenic acid in *Coffea* seedlings, *J. Exp. Bot.* **45:**457–503.

Aerts, R. J., and DeLuca, V., 1992, Phytochrome is involved in the light-regulation of vindoline biosynthesis in *Catharanthus*, *Plant Physiol.* **100:**1029–1032.

Aerts, R. J., Snoeijer, W., Aerts-Teerlink, O., van der Meijden, E., and Verpoorte, R., 1991, Control and biological implications of alkaloid synthesis in *Cinchona* seedlings, *Phytochemistry* **30:**3571–3577.

Alibert, G., Carrasco, A., and Boudet, A. M., 1982, Changes in biochemical composition of vacuoles isolated from *Acer pseudoplatanus* L during cell culture, *Biochim. Biophys. Acta* **721:**22–29.

Amann, M., Wanner, G., and Zenk, M. H., 1986, Purification and characterisation of (S)-tetrahydroberberine oxidase from cultured *Coptis japonica* cells, *Phytochemistry* **37:**979–982.

Blackford, S., Rea, P. A., and Sanders, D., 1990, Voltage sensitivity of H/Ca antiport in higher plant tonoplast suggests a role in vacuolar calcium accumulation, *J. Biol. Chem.* **265:**9617–9620.

Blom, T. J. M., Van Vliet, T. B, Schripsema, J., Val, J., Van Iren, F., Verpoorte, R., and Libbenga, K. R., 1991a, Uptake and accumulation of the alkaloids quinine and cinchonamine in cultured cells of *Cinchona robusta* and *Catharanthus roseus*, *J. Plant Physiol.* **138:**436–442.

Blom, T. J. M., Sierra, M., Van Vliet, T. B., Franke-van Dijk, M. E., De Koning, P., Van Iren, F., Verpoorte, R., and Libbenga, K. R., 1991b, Uptake and accumulation of ajmalicine into isolated vacuoles of cultured cells of *Catharanthus roseus* (L.) G. Don., and its conversion into serpentine, *Planta* **183:**170–177.

Blumwald, E., and Poole, R. J., 1985a, Na/H-antiport in isolated tonoplast vesicles from storage tissue of *Beta vulgaris*, *Plant Physiol.* **78:**163–167.

Blumwald, E., and Poole, J., 1985b, Nitrate storage and retrieval in *Beta vulgaris*—Effects of nitrate and chloride on proton gradients in tonoplast vesicles, *Proc. Natl. Acad. Sci. USA* **82:**3683–3687.

Boller, T., Dürr, M., and Wiemken, A., 1975, Characterization of a specific transport system for arginine in isolated yeast vacuoles, *Eur. J. Biochem.* **54:**81–91.

Bouzayen, M., Ltche, A., Pech, J.-C., and Marigo, G., 1989). Carrier-mediated uptake of 1-(malonylamino)-cyclopropane-1-carboxylic acid in vacuoles isolated from *Catharanthus roseus* cells, *Plant Physiol.* **91:**1317–1322.

Bremberger, C., Haschke, H.-P., and Lüttge, U., 1988, Separation and purification of the tonoplast ATPase and pyrophosphatase from plants with constitutive and inducible crassulacean acid metabolism, *Planta* **175:** 465–470.

Briskin, D. P., Thornley, W. R., and Wyse, R. E., 1985, Membrane transport in isolated vesicles from sugarbeet taproot. II. Evidence for a sucrose/K-antiport, *Plant Physiol.* **78:**871–875.

Brisson, L., Charest, P. M., De Luca, V., and Ibrahim, R. K., 1992, Immunocytochemical localization of vindoline in mesophyll protoplasts of *Catharanthus roseus*, *Phytochemistry* **31:**465–470.

Chrispeels, M. J., and Raikhel, N. V., 1992, Short peptide domains target proteins to vacuoles, *Cell* **68:**613–616.

Christmann, J., Kreis, W., and Reinhard, E., 1993, Cardenolide sinks and occurrence of cardenolides in the sieve tubes of *Digitalis lanata*, *Bot. Acta* **106:**419–427.

Deus-Neumann, B., and Zenk, M. H., 1984, A highly selective alkaloid uptake system in vacuoles of higher plants, *Planta* **162:**250–260.

Deus-Neumann, B., and Zenk, M. H., 1986, Accumulation of alkaloids in plant vacuoles does not involve an ion-trap mechanism, *Planta* **166:**44–53.

Ehmke, A., von Borstel, K., and Hartmann, T., 1987, Specific uptake of the N-oxides of pyrrolizidine alkaloids by cells, protoplasts and vacuoles from *Senecio* cell cultures, in: *Plant Vacuoles*, Vol. 134, NATO ASI Series (B. Marin, ed.), Plenum Press, New York, pp. 301–304.

Ehmke, A., von Borstel, K., and Hartmann, T., 1988, Alkaloid-N-oxides as transport and vacuolar storage compounds of pyrrolizidine alkaloids in *Senecio vulgaris*, *Planta* **176:**83–90.

Eilert, U., Wolters, B., and Constabel, F., 1985, Ultrastructure of acridone alkaloid idioblasts in roots and cell cultures of *Ruta graveolens*, *J. Can. Bot.* 64:1089–1096.

Endo, T., Goodbody, A., and Misawa, M., 1987, Alkaloid production in root and shoot cultures of *Catharanthus roseus*, *Planta Med.* 479–482.

Franceschi, V. R., and Horner, H. T., 1980, Calcium oxalate cristals in plants, *Bot. Rev.* **46:**361–427.

Getz, H. P., 1991, Sucrose transport in tonoplast vesicles of red beet roots is linked to ATP hydrolysis, *Planta* **185:**261–268.

Gonnet, J. F., and Hieu, H., 1992, *In situ* micro-spectrophotometric and micro-spectrocolorimetric investigation of vacuolar pigments in flowers of cultivars of carnation (*Dianthus caryophyllus*), *J. Hortic. Sci.* **67:**663–676.

Gruhnert, C., Biehl, B., and Selmar, D., 1994, Compartmentation of cyanogenic glucoside and their degrading enzymes, *Planta* **195:**36–42.

Han, S. R., Campbell, W. F., and Salunkhe, D. K., 1989, Ultrastructural localization of solanidine in potato tubers, *J. Food Biochem.* **13:**377–389.

Harborne, J. B., Greenham, J., Williams, C. A., Eagles, J., and Markham, K. R., 1993, Ten isoprenylated and C-methylated flavonoids from the leaves of three *Vellozia* species, *Phytochemistry* **34:**219–226.

Hauser, M.-T., and and Wink, M., 1990, Uptake of alkaloids by latex vesicles and isolated mesophyll vacuoles of *Chelidonium majus* (Papaveraceae), *Z. Naturforsch.* **45c:**949–957.

Hedrich, R., Kurkdjian, A., Guern, J., and Flügge, U. I., 1989, Comparative studies on the electrical properties of the H translocating ATPase and pyrophosphatase of the vacuolar/lysosomal compartment, *EMBO J.* **8:**2835–2841.

Holländer-Czytko, H., and Amrhein, N., 1983, Subcellular compartmentation of shikimic acid and phenylalanine in buckwheat cell suspension cultures grown in the presence of shikimate pathway inhibitors, *Plant Sci. Lett.* **29:**89–96.

Homeyer, B. C., and Roberts, M. F., 1984, Dopamine accumulation in *Papaver somniferum* latex, *Z. Naturforsch.* **39c:**1034–1037.

Hopp, W., and Seitz, H. U., 1987, The uptake of acylated anthocyanin into isolated vacuoles from a cell suspension culture of *Daucus carota*, *Planta* **170:**74–85.

Hrazdina, G., Marx, G. A., and Hoch, H. C., 1982, Distribution of secondary plant metabolites and their biosynthetic enzymes in pea (*Pisum sativum*) leaves. Anthocyanins and flavonol glycosides, *Plant Physiol.* **70:**745–748.

Hsu, L.-C., Chiou, T.-J., Chen, L., and Bush, D. R., 1993, Cloning a plant amino acid transporter by functional complementation of a yeast amino acid transporter mutant, *Proc. Natl. Acad. Sci. USA* **90:**7441–7445.

Ideda, K., Teshima, D., Aoyama, T., Satake, M., and Shimomura, K., 1988, Clonal propagation of *Cephaelis ipecacuanha*, *Plant Cell Rep.* **7:**288–291.

Ishikura, N., 1981, Flavonoids in the petal cells of *Anagallis arvensis* f. *coerulea* containing a blue crystalline anthocyanin, *Z. Pflanzenphysiol.* **103:**469–473.

Janiszowska, W., and Szakiel, A., 1992, The transport of 3H-oleanolic acid and its monoglycosides to isolated vacuoles of protoplasts from *Calendula officinalis* leaves, *Phytochemistry* **31:**2993–2997.

Keller, F., 1986, Gentiopicroside is located in the vacuoles of root protoplasts of *Gentiana lutea*, *J. Plant Physiol.* **122:**473–476.

Keller, F., and Matile, P., 1985, The role of the vacuole in storage and mobilization of stachyose in tubers of *Stachys sieboldii*, *J. Plant Physiol.* **119:**369–380.

Keller, F., and Wiemken, A., 1982, Differential compartmentation of sucrose and gentianose in the cytosol and vacuoles of storage root protoplasts from *Gentiana lutea*, *Plant Cell Rep.* **1:**274–277.

Kojima, M., Poulton, J. E., Thayer, S., and Conn, E. E., 1979, Tissue distribution of dhurrin and of enzymes involved in its metabolism in leaves of *Sorghum bicolor*, *Plant Physiol.* **63:**1022–1028.

Kreis, W., and Reinhard, E., 1987, Selective uptake and vacuolar storage of primary cardiac glycosides by suspension-cultured *Digitalis lanata* cells, *J. Plant Physiol.* **128:**311–326.

Kurkdjian, A., 1982, Absorption and accumulation of nicotine by *Acer pseudoplatanus and Nicotiana tabacum* cells, *Physiol. Veg.* **20:**73–83.

Lange, H., and Kohlenbach, H. W., 1982, Differentiation of alkaloid cells in cultures of *Macleaya* mesophyll protoplasts, *Planta Med.* **46:**78–81.

Leigh, R. A., and Walker, R. R., 1980, ATPase and acid phosphatase activities associated with vacuoles isolated from storage roots of red beet, *Planta* **150:**222–229.

Leigh, R. A., Ahmad, N., and Wyn Jones, R. G., 1981, Assessment of glycine, betaine and proline compartmentation by analysis of isolated beet vacuoles, *Planta* **153:**34–41.

Li, Z.-S., Zhao, Y., and Rea, P.A., 1995, Magnesium adenosine 5′-triphosphate-energized transport of glutathione-S-conjugates by plant vacuolar membrane vesicles, *Plant Physiol.* **107:**1257–1268.

Löffelhardt, W., Kopp, B., and Kubleka, W., 1979, Intracellular distribution of cardiac glycosides in leaves of *Convallaria majalis*, *Phytochemistry* **18:**1289–1291.

Lüthy, B., and Matile, P., 1984, The mustard oil bomb: Rectified analysis of the subcellular organization of the myrosinase system, *Biochem. Physiol. Pflanz.* **179:**5–12.

Lüttge, U., Smith, J. A. C., Marigo, G., and Osmond, C. B., 1981, Energetics of malate accumulation in the vacuoles of *Kalanchoe tubiflora* cells, *FEBS Lett.* **126:**81–84.

McCaskill, D. G., Martin, D. L., and Scott, A. I., 1988, Characterization of alkaloid uptake by *Catharanthus roseus* (L.) G. Don protoplasts, *Plant Physiol.* **87:**402–408.

McHale, D., 1986, The cinchona tree, *Biologist* **33:**45–53.

Mackenbrock, U., Vogelsang, R., and Barz, W., 1992, Isoflavone and pterocarpan malonylglucosides and β-1,3-glucan- and chitin-hydrolases are vacuolar constituents in chickpea (*Cicer arietinum* L.), *Z. Naturforsch.* **47c:**815–822.

Marrs, K. A., Alfenito, M. R., Lloyd, A. M., and Walbot, V., 1995, A glutathione S-transferase involved in vacuolar transfer encoded by the maize gene Bronze-2, *Nature* **375:**397–400.

Martinoia, E., Flügge, U. I., Kaiser, G., Heber, U., and Heldt, H. W., 1985, Energy-dependent uptake of malate into vacuoles isolated from barley mesophyll protoplasts, *Biochim. Biophys. Acta* **806:**311–319.

Martinoia, E., Vogt, E., Rentsch, D., and Amrhein, N., 1991, Functional reconstitution of the malate carrier of barley mesophyll vacuoles in liposomes, *Biochim. Biophys. Acta* **1062:**271–278.

Martinoia, E., Grill, E., Tommasini, R., Kreuz, K., and Amrhein, N., 1993, ATP-dependent glutathione-S conjugate export pump in the vacuolar membrane of plants, *Nature* **364:**247–249.

Matern, U., Heller, W., and Himmelspach, K., 1983, Conformational changes of apigenin-7-O-(6-O-malonylglycoside), a vacuolar pigment from parsley, with solvent composition and proton concentration, *Eur. J. Biochem.* **133:**439–448.

Matern, U., Reichenbach, C., and Heller, W., 1986, Efficient uptake of flavonoids into parsley (*Petroselinum hortense*) vacuoles requires acylated glycosides, *Planta* **167:**183–189.

Matile, P., 1980, The mustard oil bomb: Compartmentation of myrosinase systems, *Biochem. Physiol. Pflanz.* **175:**722–731.

Matile, P., Jans, B., and Rickenbacher, R., 1970, Vacuoles of *Chelidonium* latex: lysosomal property and accumulation of alkaloids, *Biochem. Physiol. Pflanz.* **161:**447–458.

Matoh, T., Watanabe, J., and Takahashi, E., 1987, Sodium, potassium, chloride, and betaine concentrations in isolated vacuoles from salt-grown *Atriplex gmelini* leaves, *Plant Physiol.* **84:**173–177.

Mende, P., and Wink, M., 1987, Uptake of the quinolizidine alkaloid lupanine by protoplasts and vacuoles of *Lupinus polyphyllus* cell suspension cultures, *J. Plant Physiol.* **129:**229–242.

Nishimura, M., 1982, pH in vacuoles isolated from castor bean endosperm, *Plant Physiol.* **70:**742–744.

Nissen, P., 1991, Multiphasic uptake mechanisms in plants, *Int. Rev. Cytol.* **126:**89–134.

Oba, K., Conn, E., Canut, H., and Boudet, A. M., 1981, Subcellular localization of 2-(βD-glucosyloxy-)cinnamic acids and the related β-glucosidase in leaves of *Melilotus alba*, *Plant Physiol.* **68:**1359–1363.

Pasquali, G., Goddijn, O. J. M., DeWaal, A., Verpoorte, R., Schilperoort, R. A., Hoge, J. H. C., and Memlelink, J., 1992, Coordinated regulation of two indole alkaloid biosynthetic genes from *Catharanthus roseus* by auxin and elicitors, *Plant Mol. Biol.* 18:1121–1131.

Pham, T. D. T., and Roberts, M. F., 1991, Quantitative characterization of the contents of *Papaver somniferum* latex vacuoles, *Phytochem. Anal.* **2:**68–73.

Pistocchi, R., Keller, F., Bagni, N., and Matile, P., 1988, Transport and subcellular localisation of polyamines in carrot protoplasts and vacuoles, *Plant Physiol.* **87:**514–518.

Pradier, J. M., Barbier-Brygoo, H., Ephitikhine, G., and Guern, J., 1988, Interaction of an alkaloid, serpentine, with tonoplast vacuoles from *Catharanthus roseus* G. Don., *C. R. Acad. Sci. Ser. 3*, **306:**283–289.

Rataboul, P., Alibert, G., Boller, T., and Boudet, A.M., 1985, Intracellular transport and vacuolar accumulation of o-coumaric acid glucoside in *Melilotus alba* mesophyll cell protoplasts, *Biochim. Biophys. Acta* **816:**25–36.

Rea, P., and Sanders, D., 1987, Tonoplast energizations: Two H-pumps, one membrane, *Physiol. Plant.* **71:**131–141.

Renaudin, J.P., 1989, Different mechanisms control the vacuolar compartmentation of ajmalicine in *Catharanthus roseus* cell cultures, *Plant Physiol. Biochem. (Paris)* **27:**613–621.

Renaudin, J. P., and Guern, J., 1987, Ajmalicine transport into vacuoles isolated from *Catharanthus roseus*, in: *Plant Vacuoles*, Vol. 134, NATO ASI Series (B. Marin, ed.), Plenum Press, New York, pp. 339–347.

Riesmeier, J. W., Willmitzer, L., and Frommer, W. B., 1992, Isolation and characterisation of a sucrose carrier cDNA from spinach by functional expression in yeast, *EMBO J.* **11:**4705–4713.

Roberts, M. F., 1981, Enzymic synthesis of coniceine in *Conium maculatum* chloroplasts and mitochondria, *Plant Cell Rep.* **1:**10–13.

Roberts, M. F., 1987, *Papaver* latex and alkaloid storage vacuoles, in: *Plant Vacuoles*, Vol. 134, NATO ASI Series (B. Marin, ed.), Plenum Press, New York, pp. 513–528.

Roberts, M. F., and Waller, G. R., 1978, N-methyl transferases and 7-methyl-N-nucleoside hydrolase activity in *Coffea arabica* and the synthesis of caffeine, *Phytochemistry* **18:**451–455.

Roberts, M. F., Homeyer, B. C., and Pham, T. D. T., 1991, Further studies of sequestration of alkaloids in *Papaver somniferum* latex vacuoles, *Z. Naturforsch.* **46c:**377–388.

Saftner, R. A., 1994, Stereoselectivity and structural determinants in molecular recognition by the ACC transport system in isolated maize mesophyll vacuoles, *Physiol. Plant.* **92:**543–554.

Sato, H., Kobayashi, Y., Fukui, H., and Tabata, M., 1990, Specific differences in tolerance to exogenous berberine among plant cell cultures, *Plant Cell Rep.* **9:**133–136.

Sato, H., Taguchi, G., Fukui, H., and Tabata, M., 1992, Role of malic acid in solubilizing excess berberine accumulating in vacuoles of *Coptis japonica*, *Phytochemistry* **31:**3451–3454.

Sato, H., Tanaka, S., and Tabata, M., 1993, Kinetics of alkaloid uptake by cultured cells of *Coptis japonica*, *Phytochemistry* **34:**697–701.

Sato, H., Tanaka, T., Tanaka, S., and Tabata, M., 1994, Binding of berberine to a membrane fraction from *Coptis* cells, *Phytochemistry* **36:**1363–1367.

Sauer, N., and Tanner, W., 1989, The hexose carrier from *Chlorella*: cDNA cloning of a eukaryotic H-cotransporter, *FEBS Lett.* **259:**43–46.

Saunders, G. A., 1979, Investigations of vacuoles isolated from tobacco. I. Quantification of nicotine, *Plant Physiol.* **64:**74–78.

Saunders, G. A., and Conn, E. E., 1978, Presence of the cyanogenic glycoside dhurrin in isolated vacuoles from *Sorghum*, *Plant Physiol.* **61:**154–157.

Schnabl, H., Weissenböck, G., and Scharf, H., 1986, *In vivo* microspectrophotometric characterization of flavonol glycosides in *Vicia faba* guard and epidermal cells, *J. Exp. Bot.* **37:**61–72.

Sharma, V., and Strack, D., 1985, Vacuolar localisation of 1-sinapoyl glucose:L-malate sinapoyltransferase in protoplasts from cotyledons of *Raphanus sativus*, *Planta* **163:**563–568.

Szakiel, A., and Janiszowska, W., 1991, The effect of pH and ATP on the transport of oleanolic acid monoglycosides into isolated vacuoles of *Calendula officinalis* leaves, *Acta Biochim. Pol.* **38:**47–51.

Szakiel, A., and Janiszowska, W., 1992, Competition between oleanolic acid glycosides in their transport to isolated vacuoles from *Calendula officinalis* leaf protoplasts, *Acta Biochim. Pol.* **39:**107–112.

Szakiel, A., and Janiszowska, W., 1993, The kinetics of transport of oleanolic acid monoglycosides into vacuoles isolated from *Calendula officinalis* leaf protoplasts, *Acta Biochim. Pol.* **40:**136–138.

Szakiel, A., and Kasprzyk, Z., 1989, Distribution of oleanolic acid glycosides in vacuoles and cell walls isolated from protoplasts and cells of *Calendula officinalis* leaves, *Steroids* **53:**501–511.

Taneyama, M., 1992, Studies on C-glycosides in higher plants. IV. Intracellular distribution of bergenin and related compounds in *Saxifraga stolonifera* leaves, *Bot. Mag.RR, Tokyo* **105:**565–571.

Thom, M., and Komor, E., 1984, H-sugar antiport as the mechanism of sugar uptake by sugar cane vacuoles, *FEBS Lett.* **173:**1–4.

Thom, M., and Komor, E., 1985, Electrogenic proton translocation by the ATPase of sugarcane vacuoles, *Plant Physiol.* **77:**329–334.

Thume, M., and Dietz, K.-H., 1991, Reconstitution of the tonoplast amino-acid carrier into liposomes, *Planta* **185:**569–575.

Tommasini, R., Martinoia, E., Grill, E., Dietz, K., J., and Amrhein, N., 1993, Transport of oxidised glutathione into barley vacuoles: Evidence for the involvement of the glutathione-S-conjugate ATPase, *Z. Naturforsch.* **48c:**868–871.

Tsay, Y.-F., Schroeder, J., Feldmann, K. A., and Crawford, N. M., 1993, The herbicide sensitivity gene CHL1 of Aribidopsis encodes a nitrate—inducible nitrate transporter, *Cell* **72:**705–713.

Urban, B., Laudenbach, U., and Kesselmeier, J., 1983, Saponin distribution in the etiolated leaf tissue and subcellular localization of steroidal saponins in etiolated protoplasts of oat, *Avena sativa*, *Protoplasma* **118:**121–123.

Van Genderen, H. H., and Van Hemert, J., 1986, Sites of flavone accumulation in some genotypes of *Silene pratensis* (Rafn.) Godron et Gren. (Caryophyllaceae), *New Phytol.* **104:**33–39.

Verzar-Petri, G. K., Czedo, H., Mollmann, K., Szendrei, K., and Reisch, J., 1976, Fluoreszenzmikroskopische Untersuchungen ueber die Lokalisierung von Acridon-Alkaloiden in Geweben von *Ruta graveolens*, *Planta Med.* **29:**370–375.

Wajant, H., Riedel, D., Benz, S., and Mundry, K.-W., 1995, Immunocytological localization of hydroxynitrile lyases from *Sorghum bicolor* L., and *Linum usitatissimum* L. *Plant Sci.* **103:**145–154.

Waldhauser, S. S. M., and Baumann, T. W., 1996, Compartmentation of caffeine and related purine alkaloids depends exclusively on the physical chemistry of their vacuoles complex formation with chlorogenic acid, *Phytochemistry* **42:**985–996.

Wei, X., Roomans, G. M., Seveus, L., and Pihakaski, K., 1981, Localization of glucosinolates in roots of *Sinapis alba* using x-ray microanalysis, *Scanning Electron Microsc.* **2:**481–488.

Weißenböeck, G., Hedrich, R., and Sachs, G., 1986, Secondary phenolic products in isolated guard cell, epidermal cell and mesophyll cell protoplasts from pea (*Pisum sativum* L.) leaves: Distribution and determination, *Protoplasma* **134:**141–148.

Werner, C., and Matile, P., 1985, Accumulation of coumaroylglucosides in vacuoles of barley mesophyll protoplasts, *J. Plant Physiol.* **118:**237–249.

Wink, M., 1984, N-Methylation of quinolizidine alkaloids: An S-adenosyl-L-methionine: cytisine N-methyltransferase from *Laburnum anagyroides* plants and cell cultures of *L. alpinum* and *Cytisus canariensis*, *Planta* **161:**339–344.

Wink, M., 1985, Chemische Verteidigung der Lupinen: Zur biologischen Bedeutung der Chinolizidinalkaloide, *Plant Sys. Evol.* **150:**65–81.

Wink, M., 1987b, Site of lupanine and sparteine biosynthesis in intact plants and *in vitro* organ cultures, *Z. Naturforsch.* **42c:**868–872.

Wink, M., 1992, The role of quinolizidine alkaloids in plant–insect interactions, in: *Insect–Plant Interactions*, Vol. 4 (E. A. Bernays, ed.), CRC Press, Boca Raton, pp. 131–166.

Wink, M., and Hartmann, T., 1982a, Localisation of the enzymes of quinolizidine alkaloid biosynthesis in leaf chloroplasts of *Lupinus polyphyllus*, *Plant Physiol.* **70:**74–77.

Wink, M., and Hartmann, T., 1982b, Enzymatic synthesis of quinolizidine alkaloid esters: A tigloyl-CoA: 13-hydroxylupanine O-tigloyltransferase from *Lupinus albus* L., *Planta* **156:**560–565.

Wink, M., and Mende, P., 1987). Uptake of lupanine by alkaloid-storing epidermal cells of *Lupinus polyphyllus*, *Planta Med.* **53:**465–469.

Wink, M., and Witte, L., 1984, Turnover and transport of quinolizidine alkaloids: Diurnal variation of lupanine in the phloem sap, leaves and fruits of *Lupinus albus* L., *Planta* **161:**519–524.

Wink, M., and Witte, L., 1985, Quinolizidine alkaloids as nitrogen source for lupin seedlings and cell cultures, *Z. Naturforsch.* **40c:**767–775,

Yoder, L. R., and Marlberg, P. G., 1976, Reactions of alkaloid and histochemical indicators in laticifers and specialised parenchyma cells of *Catharanthus roseus* (Apocynaceae), *Am. J. Bot.* **63:**1167–1173.

Part III

Ecology

Chapter 11

Chemical Ecology of Alkaloids

Michael Wink

1. INTRODUCTION

In previous chapters the authors considered the distribution of alkaloids within the plant kingdom and how the plants synthesize, transport, and store them. In this chapter I shall concentrate on the question of the role and function of alkaloids, to what purpose they are produced.

Let us consider the food chain in our ecosystem first (Fig. 1). Plants constitute the major group of photoautotrophic organisms which utilize solar energy to fix CO_2 into sugar and produce ATP as fuel and $NADPH_2$ as reduction equivalents, which serve to build up all of the other essential components of a cell. Animals and most microorganisms (except the chemo- or photoautotrophic bacteria) are heterotrophic organisms, which rely on complex, plant-made organic molecules for their energy requirement or other metabolic functions. Thus, plants serve as a major and the ultimate source of food for animals and microorganisms (Fig. 1). Plants struggle for life as do other organisms and have evolved strategies against herbivorous animals, microorganisms, and viruses. Plants also compete with other plants (of the same or of different species) for space, light, water, and nutrients.

How plants defend themselves against microorganisms (bacteria and fungi), viruses, herbivores and other plants is an old but still controversial topic in botany, ecology, and evolutionary studies. As plants do rather well in nature, the need for defense is not obvious. We are well aware of the defense strategies of higher animals (including man) against microbes and predators (Table I). The complex immune system with its cellular and humoral components is a well-studied and well-documented area in the context of animal–microbe interactions. Against predators animals evolved weapons, armor, crypsis, thanatosis, deimatic behavior, flight, or defense chemicals (usually called "poisons") (see Edmunds, 1974, for an overview).

Michael Wink • Institute for Pharmaceutical Biology, University of Heidelberg, D-69120 Heidelberg, Germany.

Alkaloids: Biochemistry, Ecology, and Medicinal Applications, edited by Roberts and Wink. Plenum Press, New York, 1998.

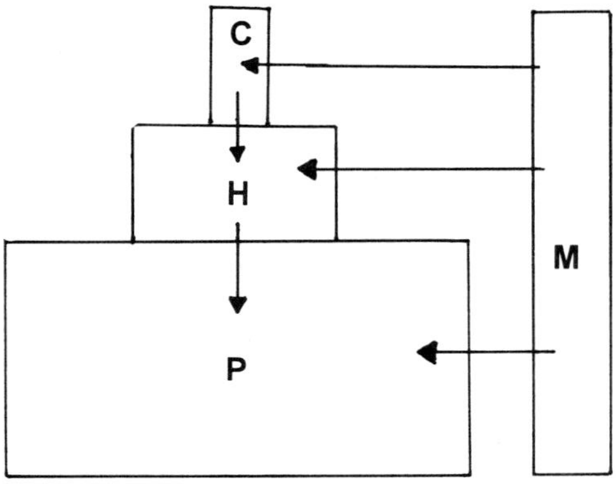

Figure 1. Schematic illustration of the terrestrial food pyramid (biomass). Main trophic levels include plants (P), herbivores (H), carnivores (C), and microorganisms (M). Arrows indicate the direction of attack or predation.

Table I
Defense Strategies of Animals versus Those of Plants and Fungi

	Defense strategies					
Measure	Mammals	Birds	Amphibia	Insects	Plants	Fungi
Carnivores/herbivores						
Flight	++	++	++	++	–	–
Weapons	++	+	*	+	*	–
Armor	*	–	–	*	++	–
Anachoresis	*	*	*	*	–	–
Crypsis	*	*	*	*	–	–
Aposematism	*	*	*	*	?	?
Thanatosis	*	*	*	*	–	–
Deimatic behavior	*	*	*	*	–	–
Group defense	*	*	–	*	–	–
Open growth	–	–	(*)	–	++	++
Toxins	(*)	(*)	+	++	++	++
Microorganisms						
Antibodies	++	++	++	–	–	–
Macrophages	++	++	++	++	–	–
Lytic enzymes	++	++	++	++	++	++
Toxins	–	–	*	?	++	++

++ = major strategy
+ = used by many but not all organisms
– = not used
* = restricted for few groups of organisms
(*) = very restricted to special cases

Table II
Selected Defense Strategies of Plants

Open growth	Replacement of leaves or branches on wounding
Physical protection	
"Armour"	Indigestible cell walls containing cellulose, lignin, suberin, callose
	Presence of hydrophobic cuticular layers (as a penetration barrier directed against microbes)
	Formation of a thick bark of roots and stems
"Weapons"	Mechanical weapons
	Spines, thorns, hooks, trichomes
	Glandular and stinging hairs (sometimes filled with defense chemicals)
	Laticifers and resin ducts (often filled with defense chemicals)
Chemical protection	Inhibitory or toxic proteins
	Lectins
	Protease inhibitors
	Toxalbumins
	Chitinase (against fungal cell walls)
	β-1,3-Glucanase (against bacteria)
	Peroxidase and phenolase (to degrade microbial toxins)
	PR proteins
	Secondary metabolites
	Constitutive defense chemicals
	Preformed chemicals activated by attack or wounding (cyanogenic glycosides, glucosinolates, coumaroylglycosides, alliin, ranunculin)
	"Phytoalexins" induced by elicitors or wounding

It is obvious that sessile plants cannot rely on many strategies that animals apply for defense. Table II shows some of the defense strategies of plants (Harborne, 1993; Wink, 1988, 1992, 1993a–d).

Natural products (or allelochemicals) are the major means involved in the overall defense strategy of plants. For many years secondary metabolites, of which more than 50,000 (including 12,000 alkaloids) have been described, were considered to be waste products or otherwise functionless molecules, illustrating the biochemical virtuosity of nature (Mothes *et al.*, 1985).

Toward the end of the 19th century, Stahl (1888) had advocated that natural products are used by plants for chemical defense against herbivores. The plant physiologists of that period had not accepted the Darwinian view of evolution and therefore were not inclined to accept the defense concept. As a consequence, the chemical defense hypothesis was ignored and remained forgotten for 60–70 years until the debate was reopened by Fraenkel in 1959. Although the defense hypothesis has presently become more and more accepted, critics from chemistry and botany can be heard regularly. While we don't know the function of each secondary metabolite, we can summarize the present argument as follows.

Most secondary metabolites are important for the fitness of the plant producing them, in that they serve as defense compounds against microbes and viruses, competing plants, and/or herbivores. In some instances, other functions, e.g., as signal compounds, include the attraction of pollinating or seed-dispersing animals, e.g., by colored compounds [betalains (within the Centrospermae), anthocyanins, carotenoids, flavonoids], fragrances (terpenes, amines, and aldehydes), or sweet substances (sugars). Physiological roles have often been discussed, such as UV protection, N transport or N storage (Wink and Witte, 1984, 1985), or pigments (carotenoids). As secondary metabolites have rather a group-specific distribution, these physiological functions cannot be universal but special and additional. Furthermore, many natural products have multiple functions, a fact that is easily overlooked as most scientists usually specialize within a very narrow range, i.e., a microbiologist is not usually curious as to whether an antibiotic alkaloid also deters the feeding of caterpillars. To understand all of the interactions we need to adopt a holistic and interdisciplinary approach. Although the defense and signal hypotheses are probably valid for most compounds, it is likely that other compounds exist for which we have no identified function, especially if they occur in minor quantities.

Overviews on the field of chemical ecology are found in Swain (1977), Levin (1976), Harborne (1993), Schlee (1992), Rosenthal and Janzen (1979), Rosenthal and Berenbaum (1991), and Fritz and Simms (1992).

2. FUNCTION OF ALKALOIDS

About 20–30% of higher plants accumulate alkaloids (see Chapter 4). Within systematic alkaloid-accumulating groups, most members produce alkaloids, e.g., 60–70% of the Solanaceae and Apocynaceae species produce alkaloids. Some alkaloids have a wide distribution in nature: caffeine occurs in the largest number of families, lycorine in the largest number of genera, and berberine in the largest number of species. Alkaloids are not restricted to higher plants (although they are here most numerous), but are also present in club mosses (*Lycopodium*), horsetail (*Equisetum*), fungi, and animals, such as marine sponges, worms (e.g., Nereidae), bryozoa, snails, insects (e.g., Coccinellidae, Solenopsidae), amphibia (toads, frogs, salamanders), fishes, and even birds and mammals (see Chapters 15 and 16).

Alkaloids represent one of the largest group of natural products and were considered to be waste products for a long time (even by eminent alkaloid researchers such as W. O. James and Kurt Mothes). The waste product argument probably derived from animal physiology: Animals take up relatively large amounts of proteins and nucleic acids, containing more nitrogen than is needed for metabolism which is eliminated via uric acid or urea (Urich, 1990). As nitrogen is a limiting nutrient for most plants, a nitrogenous waste product would be *a priori* unlikely. If alkaloids were waste products, we would expect an accumulation in old organs being shed; in contrast, many plants (especially perennial species) remobilize their nitrogenous natural products (including alkaloids) from senescing organs such as old leaves or aerial parts. It is noteworthy that the alkaloids stored by animals were never considered to be waste products by zoologists but were always regarded as toxins, i.e., defense chemicals (see Chapter 15).

Although more than 12,000 alkaloids have been described so far, only about 600 have been partly analyzed for their biochemical properties and even fewer for their ecophysiological roles (Wink, 1993a; Brown and Trigo, 1995). This does not imply that the other alkaloids are inactive or without a function; they just have not been studied in detail. Research projects have addressed the question of whether a given alkaloid was a potential candidate for treating bacterial, fungal, or viral diseases, or killing parasites and cancer cells. Although not intended as ecological studies, the results can be interpreted and extrapolated in an ecological sense. We can safely assume that many more functions and activities will become evident if alkaloids are analyzed with a more ecological or biological perspective, i.e., using ecologically relevant test organisms and employing adequate alkaloid concentrations. The last point needs to be stressed: because of the toxic properties of most alkaloids, many pharmacological experiments apply much lower doses than are found in the plant. As a consequence, ecologically relevant properties may be easily overlooked.

In conclusion, we can definitely accept the entirely plausible hypothesis that most alkaloids of plants, microorganisms, and animals serve, like other allelochemicals, as defense and signal compounds. This idea is intuitively straightforward, because many alkaloids are known to be human and animal poisons.

In this chapter, the functions of alkaloids in plant–herbivore interactions will be discussed in more detail. Plant–plant and plant–microbe interactions and the distribution and function of alkaloids in animals will only be treated briefly, as these aspects are extended in Chapters 13–17.

3. PLANT–HERBIVORE INTERACTIONS

3.1. Invertebrates

Insects are extremely successful from the evolutionary point of view and represent the largest class of organisms; they comprise at least 1 million and perhaps as many as 20–30 million species.

Most insects are herbivores and their adaptation to host plants and their chemistry is often very close and complex (Bernays, 1982; Bernays and Chapman, 1994). As many plants need insects for pollination and seed dispersal, but try to avoid insect herbivory, the interplay between attraction and deterrence can be very complicated. In the latter context it can be observed that plants attract insects by chemical means (colors, fragrances, sugar, amino acids) and at the same time other secondary metabolites are employed to prevent herbivory on flowers and seeds. The close association between plants, especially the Angiospermae, and insects evolved during the last 200 million years (Swain, 1977). Some scientists have called this phenomenon a "coevolutionary" process (Ehrlich and Raven, 1964), but it has to be recalled that the associations seen today are not necessarily those in which the chemical interactions originally evolved, i.e., the current associations may be quite recent.

Insect herbivores can be divided in two large groups, whose strategies with respect to the plant's defense chemistry differ substantially (Bernays and Chapman, 1994): The polyphagous species exploit a wide range of host plants, whereas the mono-/oligophagous insects often specialize on one or a small number of host plants which are often systematically related and accumulate the same class of secondary compounds. For these "specialists" the originally noxious defense compounds are no longer toxic but often attractive feeding and oviposition stimulants (Duffey, 1980; Schoonhoven, 1972; Bernays and Chapman, 1994).

Insects cope with dietary allelochemicals using one of several strategies:

1. Insects are commonly endowed with fantastic and powerful olfactory receptors and can select between plants with high or low amounts of "toxins" and also can ascertain

the food quality present, such as lipid, protein, or carbohydrate contents (Bernays and Chapman, 1994). The polyphagous "generalists" are usually deterred from feeding on plants sequestering high amounts of toxic allelochemicals and either select those with less active ones (Schoonhoven, 1972) or change host plants rapidly, consuming small portions of a particular poison at a given time and thus avoid severe intoxication.

2. A species "learns"; or more accurately, during evolution variants have been favored by natural selection which can tolerate a noxious defense compound: (a) by developing a mechanism to avoid toxin resorption in the gut. (b) If resorption cannot be prevented, to eliminate the toxin quickly via the Malpighian tubules or degrade it by detoxifying microsomal and other enzymes. Most polyphagous species have evolved active detoxification mechanisms, such as microsomal oxidases, glutathione transferase, and peroxidase, which promote rapid detoxification and elimination of dietary secondary products (Ahmad, 1983; Brattsten, 1988; Brattsten and Ahmad, 1986). (c) By developing a target site that is resistant to the toxin, i.e., a receptor that does not bind the exogenous ligand any longer. For example, in the monarch butterfly (*Danaus plexippus*) which stores dietary cardiac glycosides, the ouabain binding site of Na^+/K^+-ATPase has been made ouabain insensitive by a single point mutation (Holzinger *et al.*, 1992; Holzinger and Wink, 1996).

3. Alternatively, a species not only tolerates a plant's defense compound, but it also exploits it for its own chemical protection or for other purposes, such as pheromones (see Section 3.2.)

3.1.1. INSECT FEEDING DETERRENCE AND ALKALOID TOXICITY

In general, we would expect that alkaloids are active feeding deterrents against most insects. The examples compiled in Table III indicate that this assumption is valid, i.e., many alkaloids can act as feeding deterrents at higher concentrations (0.5%), although only a limited number of alkaloids have been assessed in this context. Given the choice, polyphagous insects tend to select a diet with no or only a small dose of alkaloids. Also the specialists avoid most "toxins" except those of their host plants. These findings indicate that under natural conditions, plants with a high load of alkaloids should be safe from most herbivorous insects (which is indeed the case) with the exception of particular monophagous species or a few very resistant polyphagous ones.

If animals have no choice or if they are very hungry, the deterrency threshold value is much reduced and they often feed on a diet containing alkaloids that they would normally avoid. In this case the toxicity of an ingested alkaloid can be assessed. Alternatively, alkaloid toxicity can be determined to some degree by topical application, although such data are of limited ecological relevance.

A substantial number of alkaloids display significant insect toxicity: examples include nicotine, piperine, lupin alkaloids, caffeine, and rayanodine. The toxic effects of alkaloids on insects (Table III) can be caused by their interference with a diversity of cellular and intracellular targets. As most mechanisms have not yet been elucidated for insects, we can only extrapolate from the mechanisms outlined in Chapter 12.

3.1.2. SEQUESTRATION OF ALKALOIDS BY INSECTS

Plants that defend themselves effectively constitute an ecological niche, almost devoid of herbivores and pathogens. It is not surprising that during evolution a number of

Table III
Activity of Alkaloids Toward Insects

Alkaloid	Effect[a]	ED_{50}
Ajmalicine	FD for polyphagous *Syntomis* (Lep.)	1%
Ajmaline	FD for *Syntomis*	0.1%
Anabasine	FD for *Syntomis*	0.1%
Arecoline	FD for *Syntomis*	<0.1%
	FD for *Phormia*	10 mM
Aristolochic acid	FD for *Spodoptera, Lymantria*	0.25–0.5%
	I for *Papilio* (Lep.)	0.5%
Atropine	FD for *Phormia*	0.6 mM
	I for *Bruchidius*	0.1%
Berberine	I for *Euxoa* (Lep.)	0.3%
	FD for *Phormia*	0.6 mM
	FD for bees	0.01%
	I for bees	0.003%
Boldine	FD for *Syntomis*	0.01%
Brucine	FD for *Syntomis*	1%
	FD for bees	0.05%
	I for bees	0.02%
Caffeine	FD for *Phormia*	2.5 mM
	FD for *Syntomis*	<0.1%
	FD for bees	0.03%
	I for *Bruchidius*	1%
	I for bees	0.2%
Castanospermine	FD for aphids	0.1 mM
Chaconine	FD for *Choristoneura*	0.1 mM
Chelidonine	FD for *Syntomis*	0.01%
Cinchonidine	FD for *Syntomis*	0.1%
	FD for bees	0.04%
Cinchonine	FD for bees	0.007%
Codeine	FD for *Phormia*	10 mM
Colchicine	FD for *Locusta* (Orth.)	0.001%
	I for *Bruchidius*	0.1%
	FD for *Syntomis*	<0.01%
	FD for bees	0.2%
	I for bees	0.03%
Coniine	FD for *Phormia*	5 mM
Cytisine	FD for *Syntomis*	0.01%
	FD for *Acyrthosiphon* (Hom.)	0.02%
	FD for *Formica* (Hym.)	0.01%
Deoxynojirimycine	FD for aphids	2.5 mM
Emetine	FD for *Syntomis*	<0.1%
L-Ephedrine	I for *Bruchidius*	0.1%
	FD for *Syntomis*	0.01%
	FD for bees	0.09%
Ergometrine	FD for *Syntomis*	1%
Ergotamine	FD for *Syntomis*	<0.1%
Glaucine	FD for *Spodoptera, Lymantria*	0.25-0.5%
Gramine	FD for aphids	<1 mM
	I for *Schizaphis* (Hom.)	0.01%
Harmaline	FD for *Syntomis*	<1%

(continued)

Table III
(Continued)

Alkaloid	Effect[a]	ED_{50}
Harmine	FD for *Syntomis*	<1%
	FD for bees	0.08%
	phototoxic for *Trichoplusia*	
Heliotrine	FD for *Choristoneura*	1.6 mM
	FD for bees	0.09%
	I for bees	0.1%
Hyoscyamine	FD for *Syntomis*	0.01%
	FD for bees	0.005%
	I for bees	0.1%
Jacobine	FD for *Locusta*	0.001%
Lasiocarpine	FD for *Choristoneura*	1.2 mM
Lobeline	FD for *Syntomis*	<1%
	FD for bees	0.008%
Lupanine	FD for *Syntomis*	0.1%
	I for *Plutella* (Lep.)	6 mM
	I for *Dysdercus* (Hom.)	12 mM
	I for *Ceratitis* (Dipt.)	3 mM
	I for *Phaedon* (Col.)	12 mM
Lupinine	FD for *Acyrthosiphon*	0.08%
Nicotine	FD for *Syntomis*	<0.1%
	FD for bees	0.03%
	I for *Bruchidius*	0.1%
	I for bees	0.2%
Papaverine	FD for *Spodoptera, Lymantria*	0.25–0.5%
	FD for *Phormia*	10 mM
	FD for *Syntomis*	<0.1%
Perloline	FD for *Locusta*	0.1%
Physostigmine	FD for *Syntomis*	0.01%
Pilocarpine	FD for *Phormia*	2.5 mM
	FD for *Syntomis*	0.1%
Protoveratrine B	FD for *Syntomis*	0.01%
Quinine	FD for bees	0.01%
	I for bees	0.02%
	FD for *Phormia* (Dipt.)	0.6 mM
	FD for *Locusta* (Orth.)	0.01%
Reserpine	I for *Bruchidius* (Col.)	0.1%
	FD for *Syntomis*	1%
Sanguinarine	FD for *Spodoptera, Lymantria*	0.25–0.5%
	FD for *Syntomis*	<1%
Scopolamine	FD for *Syntomis*	0.01%
	FD for bees	0.03%
Senecionine	FD for *Choristoneura*	1.6 mM
Solanidine	FD for *Choristoneura*	0.1 mM
Solanine	FD for *Choristoneura*	1 mM
Sparteine	FD for *Syntomis*	0.1%
	FD for bees	0.03%
	FD for *Acyrthosiphon*	0.01%
	I for bees	0.05%
	I for *Plutella*	50 mM
	I for *Dysdercus*	50 mM
	I for *Ceratitis*	9 mM

Table III
(Continued)

Alkaloid	Effect[a]	ED_{50}
Strychnine	I for Bruchidius	0.1%
	FD for *Phormia*	10 mM
	FD for *Syntomis*	<1%
	FD for bees	0.02%
	I for bees	0.2%
13-Tigloyloxylupanine	FD for *Choristoneura* (Lep.)	1.4 mM
	I for *Plutella*	12 mM
	I for *Dysdercus*	6 mM
	I for *Phaedon*	6 mM
	I for *Ceratitis*	6 mM
Tomatidine	FD for *Choristoneura*	1 mM
	FD for *Syntomis*	1%
Tomatine	FD for *Locusta*	0.1%
	FD for *Choristoneura*	0.1 mM
	FD for *Phormia*	10 mM
Vincamine	FD for *Syntomis*	0.01%
	FD for bees	0.08%
Yohimbine	FD for *Phormia*	2.5 mM
	FD for bees	0.008%

[a]FD, feeding deterrent; I, insecticidal (after Detzel and Wink, 1993; Wink and Schneider, 1990; Wink, 1993b).

organisms were selected that have specialized on a particular host plant species and found ways to tolerate or even to exploit the defense chemistry of their hosts. As compared to the huge number of potential enemies, the number of adapted specialists is usually small and in general a "status quo" or equilibrium can be observed between specialists (or parasites) and their hosts. A specialist is well advised not to kill its host, for to do so would destroy its own resources; a mutualism is more productive.

Superficially, these observations seem to contradict the working hypothesis, that secondary metabolites are primarily defense compounds. But these specialists are only the exceptions to the general rule. In this context we should recall that our immune system is fantastic in warding off bacteria, fungi, viruses, and parasites. We usually take notice of its existence only when it fails, i.e., when a specialized pathogen has found a way to undergo the immune response. Nobody would call the immune system ineffective because of this! Considering the specialized herbivores that have overcome the chemical defense barrier of plants a similar logic applies.

On a basic evolutionary level we find insects that can tolerate the defense chemistry of their host plants. One such example is *Manduca sexta,* whose larvae live on *Nicotiana* and other solanaceous plants. The tobacco hornworm can even grow on a diet with more than 1% nicotine without any adverse effects. The alkaloids present, such as nicotine or hyoscyamine, are not stored by the insects but degraded or directly eliminated with the feces. In order to avoid toxicity it has been postulated either that nicotine may not diffuse into nerve cells or that the ACH receptor no longer binds nicotine, as in "normal" animals. Recent experiments from my laboratory have shown that *Manduca* has ACh receptors that can bind nicotine. Furthermore, we have sequenced the alpha subunit of the receptor which does not show a substantial target site modification as compared to other moths

Table IV
Examples of the Storage of Dietary Quinolizidine and Pyrrolizidine Alkaloids by Insects[a,b]

Insects	Host plant	Acquisition: By larvae	Acquisition: By adults	Utilized as pheromone
PA (natural sources)				
Lepidoptera				
Euploea treitschkei	*Parsonia*	+		+
Gnophaela latipennis	*Hackelia*	+		
Hyalurga syma	*Heliotropium*	+		
Idea leuconoe	*Parsonia*	+		+
Mechanitis polymnia	Eupatorieae, Boraginaceae	+	+	?
Thitorea spp.	*Prestonia*	+		+
Tyria jacobaeae	*Senecio*	+		
	Adenostyles	+		
	Petasites	−		
Utetheisa ornatrix	*Crotalaria*	+		+
U. pulchelloides	*Heliotropium*	+		+
Coleoptera				
Oreina cacaliae	*Adenostyles*	+	+	
O. speciosissima	*Adenostyles*	+	+	
Coccinella spp.	*Senecio/Aphis*	+	+	
Homoptera				
Aphis jacobaeae	*Senecio*	+	+	
Aphis cacaliaster	*Senecio*	+	+	
PA (experimental feeding)				
Lepidoptera				
Amauris sp.	P		+	+
Arctia caja	P, S	+		
Callimorpha dominula	S	+		
Creatonotos transiens	P, S	+		+
Danaus spp.	P		+	+
Diacrisia sannio	S	+		
Euploea spp.	P		+	+
Nyctemera coleta	P, S	+	+	
Phragmatobia fuliginosa	S	+		+
Spilosoma lubricipeda	S	+		
Tyria jacobaeae	S	+		
Orthoptera				
Zonocercus variegatus	P, S	+	+	
QA (natural sources)				
Lepidoptera				
Uresiphita reversalis	*Teline*	+		
Homoptera				
Macrosiphumn albifrons	*Lupinus*	+	+	
Aphis cytisorum	*Laburnum*	+	+	
	Petteria	+	+	
	Cytisus	+	+	
Aphis genistae	*Spartium*	+	+	
	Sophora	+	+	
	Genista	+	+	

[a]After Hartmann and Witte (1995), Nickisch-Rosenegk and Wink (1993a), Wink (1992, 1993b), Brown and Trigo (1995).
[b]P, PA-rich plant material; S, isolated PA (mostly heliotrine, senecionine, or monocrotaline).

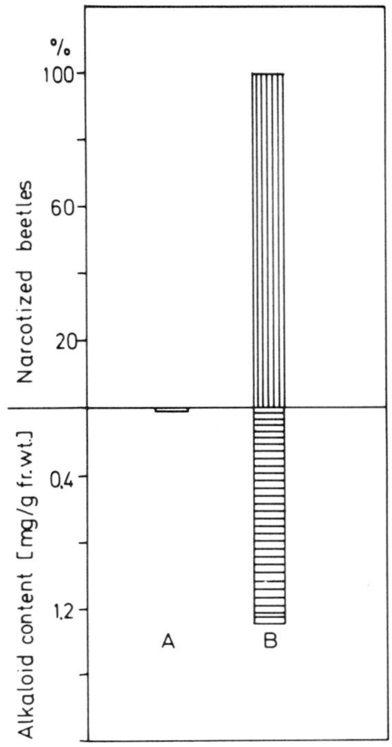

Figure 2. Effects of quinolizidine alkaloids stored by the lupin aphid, *Macrosiphum albifrons*, on a carnivorous beetle (*Carabus problemataicus*) (after Wink and Römer, 1986). About 12 individually kept beetles were given aphids without QA (control: A) or aphids with a QA content of 1.3 mg g^{-1} fresh weight (B). Experiments were evaluated after 16 hr; in B all beetles lay on their backs and remained narcotized for more than 48 hr, whereas the control beetles showed no symptoms.

(V. Theile, T. Schmeller, and M. Wink, unpublished). The potato beetle (*Leptinotarsa decemlineata*) lives on *Solanum* species containing steroidal alkaloids, which are tolerated but not stored by this species. The bruchid beetle, *Bruchidius villosus,* predates seeds of quinolizidine alkaloid (QA)-rich plants, such as *Laburnum anagyroides.* This beetle eliminates most of the dietary cytisine with the feces (Szentesi and Wink, 1991).

Insect herbivores, which not only feed on alkaloidal plants but also sequester the dietary alkaloids and exploit them for their own defense (Blum, 1981; Duffey, 1980; Rosenthal and Janzen, 1979; Rosenthal and Berenbaum, 1991; Brown and Trigo, 1995), have further evolved.

In a number of plants alkaloids are translocated via the phloem (e.g., quinolizidine, pyrrolizidine, and polyhydroxy alkaloids; Wink, 1987a, 1990; Dreyer *et al.,* 1985; Hartmann and Witte, 1995; Vrieling, 1991). If aphids live on these plants, they come in direct contact with the alkaloids present. A few adapted aphids can store the dietary alkaloids (Table IV). Examples are QA in *Aphis cytisorum, A. genistae,* and *Macrosiphum albifrons* (Wink and Witte, 1991) and pyrrolizidine alkaloids (PA) in *Aphis jacobaeae* and *A. cacaliaster* (Hartmann and Witte, 1995; Wink, unpublished). For alkaloid-storing *M. albifrons* it was shown experimentally that the QA stored provide protection against carnivorous beetles, such as *Carabus problematicus* (Fig. 2), or *Coccinella septempunctata* or syrphids (*Episyrphus balteatus*). *Acyrthosiphon spartii* prefers sparteine-rich *Cytisus scoparius* plants; it is likely that this species also stores QA.

A further (tritrophic) interaction has also become evident: *Aphis cytisorum* and *A. genistae* colonies are regularly visited by ants, which collect honey dew. Regarding *Lasius niger* collected from an *A. cytisorum* colony, the ants contained about 45 μg cytisine g^{-1}

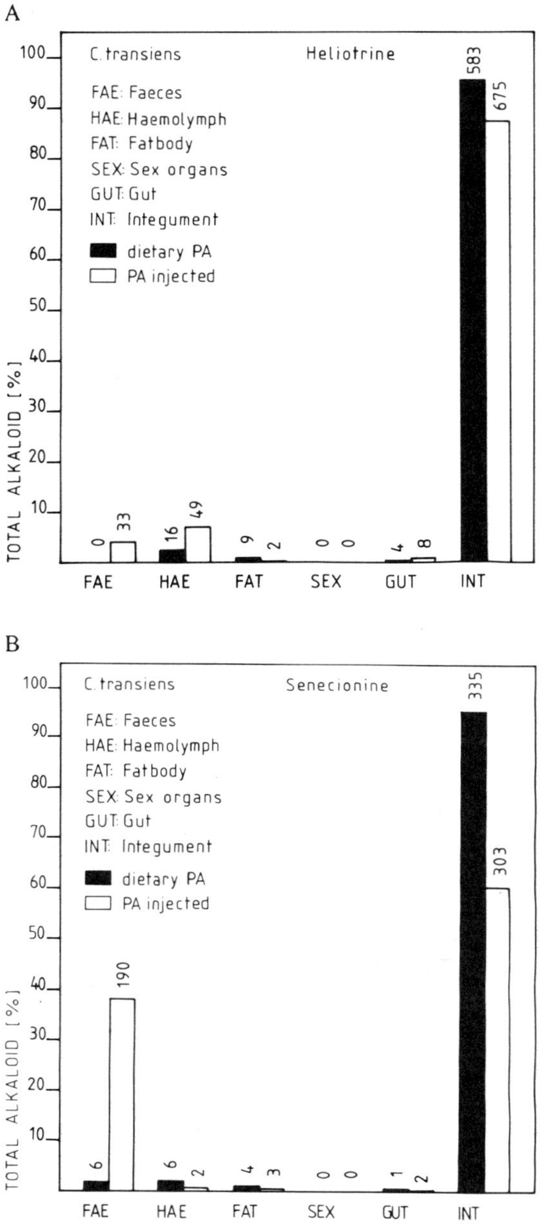

Figure 3. Distribution of pyrrolizidine alkaloid in larvae of the arctiid moth, *Creatonotos transiens* (after Nickisch-Rosenegk *et al.*, 1990). (A, B) PAs [heliotrine (A) or senecionine (B)] were administered orally or via injection into the hemolymph. After 48 hr PA distributions in the different parts of the insects were determined. (C, D) Kinetics of heliotrine within the first 48 hr after feeding, differentiated for males (C) and females (D). (By permission of *Zeitschrift der Naturforschung*.)

fresh weight (Szentesi and Wink, 1991). As cytisine is a very toxic alkaloid, it would be interesting to find out whether the ants gain protection from the alkaloids obtained from aphids. Regarding PA a tritrophic interaction has also been reported (Hartmann and Witte, 1995): Ladybirds (*Coccinella*) sequestered PA from *Aphis jacobaeae* feeding on PA-rich *Senecio jacobaea.* It is likely that, in ladybirds, besides the endogenously produced coccinellines the PA also serve as chemical protectants

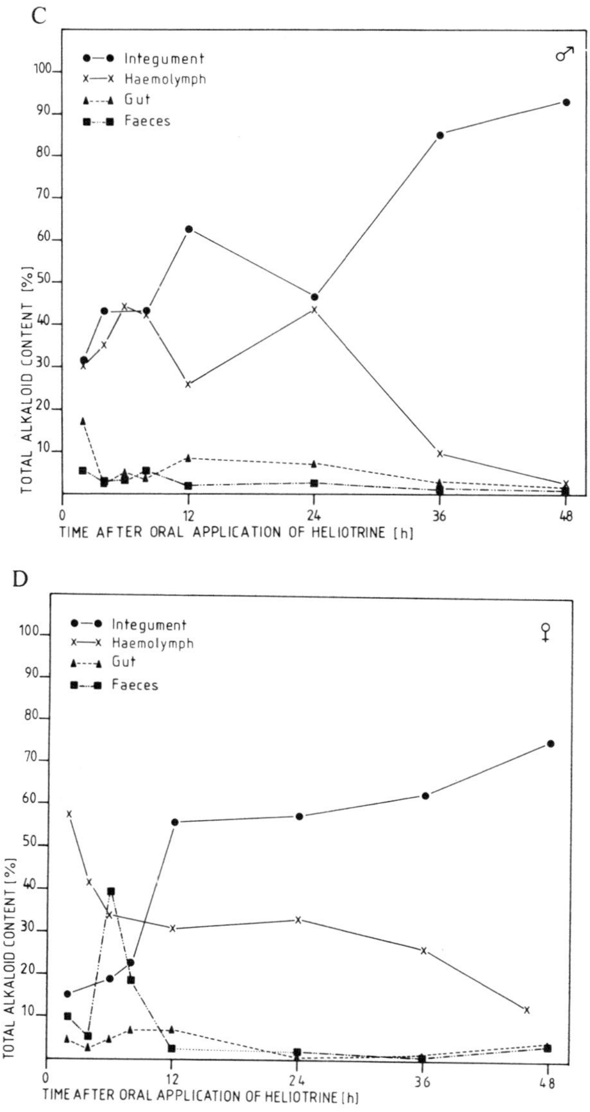

Figure 3 (*Continued*)

Larvae of the pyralid moth, *Uresiphita reversalis,* live on QA-producing plants, such as *Teline monspessulana* (Table IV). The larvae store some of the dietary alkaloids, especially in the integument and silk glands. The uptake is both specific and selective and achieved by a carrier mechanism: Whereas alkaloids of the 10-oxosparteine type dominate in the plant, it is the more toxic cytisine that is accumulated by the larvae and the 10-oxosparteines are eliminated with the feces (Wink *et al.,* 1991). These larvae gain some protection from storing QA as was shown in experiments with predatory ants (*Iridomyrmex humilis*) and the paper wasp (*Mischocyttarus flavitarsus*). When the larvae pupate, most of their stored alkaloids are used to impregnate the silk of the cocoon, thus providing defense for this critical developmental stage. The emerging moth lives cryptically and has no aposematic coloring and does not contain alkaloids. In contrast, the alkaloid-rich larvae were aposematically colored and live openly on the plants (Bernays and Montllor, 1989; Montllor *et al.,* 1990, 1991).

Especially among lepidopteran larvae, many examples have been reported of insects sequestering PA of their host plants (Table IV) and probably making use of them for their own defense (Schneider, 1987; Hartmann and Witte, 1995; Brown and Trigo, 1995). PA are almost always stored as their *N*-oxides and not as a free base. Because PA *N*-oxides cannot diffuse freely across biomembranes, they can be easily stored and retained in specific organs or tissues, in general, the integuments (Nickisch-Rosenegk *et al.,* 1990; Nickisch-Rosenegk and Wink, 1993). In addition, because they also affect muscarinic ACh and serotonergic receptors, the *N*-oxides probably contribute to the deterrence and toxicity of PA (Table III) (Schmeller *et al.,* 1997). Following resorption from the midgut (probably with the aid of a carrier mechanism; Wink and Schneider, 1988), PA remain transiently in the hemolymph before they are transferred to the integuments (Fig. 3). In larvae of the arctiid moth, *Creatonotos transiens,* we could show that during metamorphosis the PA are partly transferred to the ovary in females and to the spermatophore in males. During copulation, females obtain the PA-rich spermatophore (as a "nuptial gift") and transfer the alkaloids to the eggs. Thus, both sexes contribute their PA to the clutch which may benefit from the PA as a chemical protectant (Nickisch-Rosenegk *et al.,* 1990). A similar phenomenon has been reported for *Utetheisa ornatrix* (Dussourd *et al.,* 1988; Conner *et al.,* 1981).

Besides their use as chemical defense compounds, some insects exploit PA as precursors for pheromones (Fig. 5). In *C. transiens* PA are converted into hydroxydanaidal (the hydroxyl group at C-7 is *R*-configured), which is dissipated via the inflatable scent organs of the male (Fig. 4) (Schneider *et al.,* 1982). Because the pheromone content depends on the storage of dietary PA during the larval stages, PA-rich males should be especially attractive to females. If males with a high content of hydroxydanaidal are selected, this behavior would ensure that females can obtain a PA-rich nuptial gift during copulation. Normally, PA are 7*R*-configured in nature. In the event that PA are present in the 7*S* configuration, larvae of *C. transiens* and a few other species can invert the configuration to the correct 7*R* form (Wink *et al.,* 1988, 1990; Schulz *et al.,* 1993; Nickisch-Rosenegk *et al.,* 1993). In androconial organs of Danainae and Ithomiinae, PA-derived male courtship pheromones have been found, such as hydroxydanaidal, danaidal, danaidone, and ithominae lactone. In these butterflies PA are often acquired as adults while feeding on nectar or wilting PA-containing plants (Boppré, 1990; Hartmann and Witte, 1995).

In *C. transiens* the PA adaptation even went one step further: Only if the larvae feed on a PA-rich diet are the males able to develop their big inflatable scent organs (corema) (Fig. 4). In this case, PA serve as a morphogen which triggers the morphological development of the coremata (Schneider *et al.,* 1982; Boppré, 1990; Schneider, 1993).

A number of other alkaloids (e.g., aconitine, cinchonine, aristolochic acid, cocaine, polyhydroxy alkaloids, β-carbolines, cycasin) have also been reported to be sequestered by insects (extensive review in Brown and Trago, 1995).

It is worth recalling that a number of animals (sponges, nudibranchs, worms, insects, toads, frogs, and salamanders) are able to synthesize their own defense compounds, among which are several alkaloids (see Chapters 15 and 16). The endogenously produced and the acquired alkaloids appear to serve as chemical defense compounds, in analogy to the situation found in plants. In animals we observe the trend that sessile species, such as

Figure 4. Corema development in *Creatonotos transiens* and alkaloid feeding. (A) Adult moth; (B) larva; (C) male corema (larva had PA as a diet); (D) corema of an insect raised without PA.

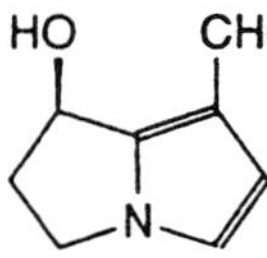

1

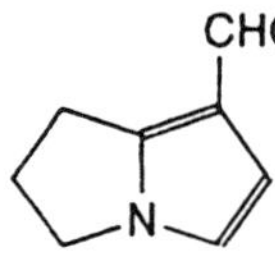

2

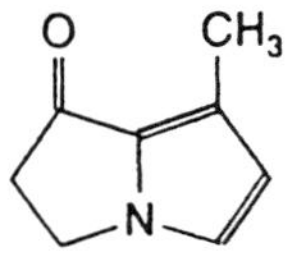

3

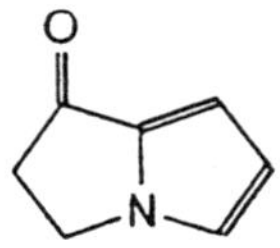

4

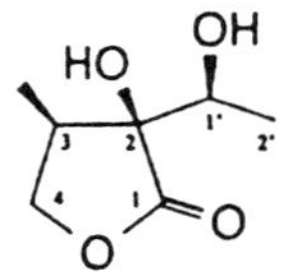

5

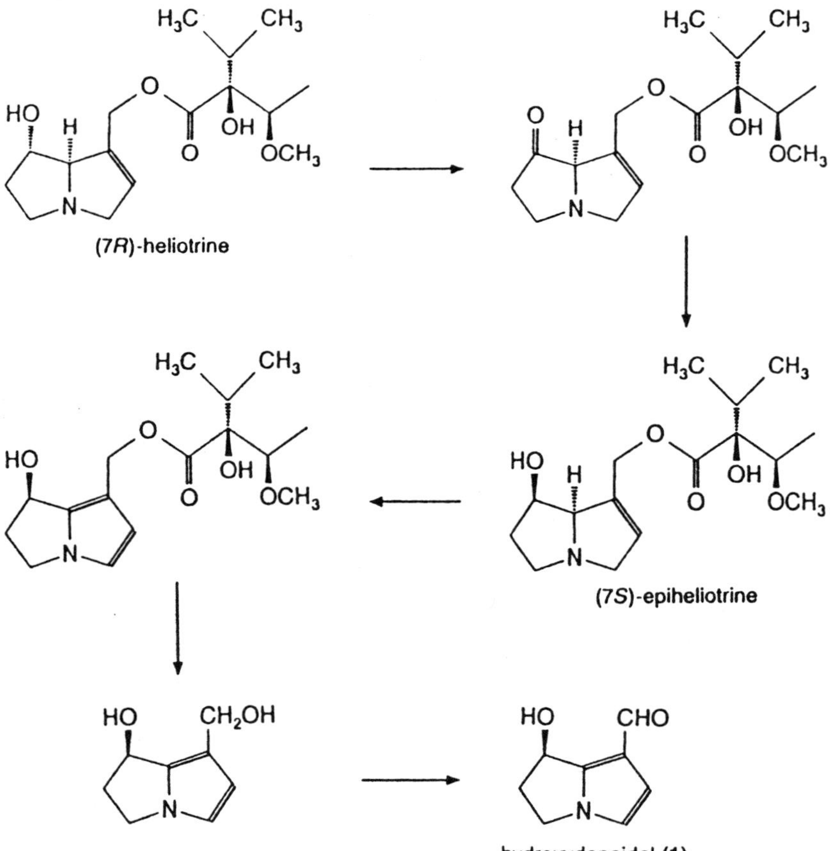

Figure 5. Structures of PA-derived pheromones [**1** = (7*R*)-hydroxydanaidal, **2** = danaidal, **3** = danaidone, **4** = nordanaidone, **5** = ithominae lactone] and pathway of heliotrine transformation in *Creatonotos transiens* and other arctiid moths (after Wink *et al.*, 1988; Schulz *et al.*, 1993; Hartmann and Witte, 1995).

bryozoa, or slow moving ones without armor, such as worms, nudibranchs, frogs, toads, and salamanders, produce active allelochemicals but not so animals with weapons, armor, or the possibility of escape by flight. Plants just share their strategy with sessile animal species (cf. Table I). In this context it seems amazing that hardly anyone has doubted the defense role of animal alkaloids, whereas people did and still do doubt the defense role regarding alkaloids in plants.

3.2. Vertebrate Herbivores

As man and his livestock are herbivores rather than carnivores, a large body of information on adverse effects of alkaloids and other dietary secondary metabolites has been accumulated over the centuries (see also Chapters 1, 12, and 17).

Many alkaloids exhibit a bitter or pungent taste for vertebrates and a bitter or pungent diet is normally instinctively avoided. Examples of bitter-tasting alkaloids (at least in man) are quinine, strychnine, brucine, emetine, and sparteine and for pungent alkaloids, capsaicine and piperine. It should be recalled that these taste properties are not identical for all animals. For example, geese, which are obligate herbivores, hardly avoid food with alkaloids or smelly compounds (amines, mercaptoethanol) which are strong repellents for humans. On the other hand, fragrances that are attractive to us, are highly repellent to geese (Wink *et al.,* 1993). Even within a population taste can differ significantly: a substantial proportion of humans cannot detect the smell of HCN, whereas others are highly sensitive. Furthermore, olfactory sensitivity can differ with age, sex, and hormonal cycles.

The bitterness varies with the chemical structure: in the case of QA the following scale was determined for man: Mean detection levels are 0.00085% for sparteine, 0.0021% for lupanine, and 0.017% for hydroxylupanine (Wink, 1992). Whereas we know most parameters of olfactory qualities in *Homo sapiens,* much less or hardly anything is known for most other vertebrates.

Alkaloids are infamous for their toxic properties in vertebrates (see Chapters 2 and 3) and plants that produce alkaloids are often classified as poisonous or toxic. For a number of alkaloids the respective LD_{50} values have been determined with laboratory animals, especially mice, but also with rats, guinea pigs, cats, rabbits, dogs, or pigeons (Table X in Chapter 12). As rodents are herbivores (and thus adapted to allelochemicals), they are not especially sensitive to alkaloids as toxins and some of the data may be misleading. The toxic effects observed with complete animals have their counterpart in the cytotoxic effect determined for some alkaloids (Table V). Most of these data have been obtained by screening many natural products for anticancer activity. But an alkaloid that can kill a cancer cell is usually also toxic for "normal" cells and the complete animal. Therefore, the data shown in Table V are another indication of the general toxicity of alkaloids toward animals. The mechanisms underlying the toxic effects have been elucidated in some detail. Often molecular targets and processes are involved that are important for all cells, such as DNA, RNA, proteins, replication, transcription, protein biosynthesis, membrane assembly and stability, electron chains, or metabolically important enzymes or proteins, such as receptors, hormones, signal compounds (see Chapter 12 for a more detailed discussion).

Whereas many insect herbivores are "specialists," vertebrate herbivores are rather polyphagous, although some specialization may occur. For example, grouse (*Lagopus*

Table V
Cytotoxic Properties of Selected Alkaloids[a]

Alkaloid	Target cells
Acronycine	L1210 cells, *Plasmodium*
Arecoline	*Trypanosoma*
Berberine	*Trypanosoma, Plasmodium*
Boldine	Human epidermoid carcinoma cells
Camptothecine	L1210 Walker sarcoma cells, KB and P388 leukemia cells
Chelerythrine	Tumor cells
Cinchonine	*Plasmodium*
Colchicine	General cytotoxicity
Coptisine	Cytotoxic
Echinatine *N*-oxide	P388 leukemia cells
Ellipticine	L1210 sarcoma cells
Emetine	Trypanosoma
Fagaronine	KB, L1210, P388 leukemia cells
Harmaline	*Trypanosoma*
Harmine	*Trypanosoma*
Harringtonine	Tumor cells
Heliotrine	Tumor cells
Indicine *N*-oxide	P388 cells
Jatrorrhizine	*Plasmodium*
Liriodenine	A-549, HCT-8, KB, P388 leukemia cells
Lycorine	NIH/3T3 cells
Matrine	Ascites tumor, mouse sarcoma 180
Olivacine	*Trypanosoma,* L1210, KB cells
Palmatine	*Plasmodium*
Quinine	*Plasmodium, Trypanosoma*
Sanguinarine	Tumor cells
Senecionine	Tumor cells
Solamargine	PLC, PRF cells
Tabernamine	P388 leukemia cells
Vinblastine	*Trypanosoma,* Wilms' tumor, lymphoma cells

[a]More data are found in Wink (1993a).

lagopus) or capercaillies (*Tetrao urogallus*) prefer plants of the Ericaceae or Coniferae; crossbills, the seeds of *Picea* and *Abies,* which are rich in terpenes. The Australian koala is oligophagous and consumes certain terpene-rich species of the genus *Eucalyptus.* While a single plant can be a host for hundreds of insect larvae, hundreds of plants comprise a daily menu for a larger grazing mammal.

Vertebrates share a few strategies with insects in coping with allelochemicals. But a sequestration and storage of dietary alkaloids has hardly been reported (as opposed to its regular occurrence in insects; see above); the storage of quinolizidine-type alkaloids in castoreum (derived from food plants) is an exception rather than a rule. Strategies of vertebrates include:

- Avoidance of alkaloid-rich plants (usually labeled toxic or poisonous by man) which is facilitated by the bitter or pungent taste of most alkaloids.
- Sampling of food from a wide variety of sources and thus minimizing the ingestion of high amounts of a single toxin.

- Detoxification of dietary alkaloids, which can be achieved by symbiotic bacteria or protozoa (Dowd, 1992), living in the rumen or intestines, or by liver enzymes, which are specialized for the chemical modification of xenobiotics. Carnivorous animals, such as cats, are known to be much more sensitive than herbivores toward plant poisons. It has been suggested that animals that do not face the problem of toxic food are not adapted to the handling of allelochemicals.
- Some animals such as ungulates, monkeys, parrots, or geese ingest soil (so-called "geophagy"). For geese and chimpanzees it was shown that the ingested soil binds dietary allelochemicals, especially alkaloids (Table VI) (Wink *et al.,* 1993; Zippin *et al.,* 1998).
- Animals are intelligent organisms, able to learn. The role of learning in food and toxin avoidance is rather important but is not understood in most species.

How most vertebrate herbivores manage to avoid, tolerate, or detoxify their dietary allelochemicals has not been explored. Sometimes, only domesticated animals were used in experiments, but they tend to make more mistakes in food choice than the wild ones. More evidence is available for humans; as with most other herbivores, alkaloid-rich diets are avoided, but if toxins have been ingested, they are normally detoxified in the liver. This evolutionary trait is very helpful for *Homo sapiens,* as it endowed us with a way of coping with man-made chemicals that pollute the environment. Besides these biological adaptations, man has evolved a number of "tricks," some of them obviously not anticipated by evolution:

Many fruits or vegetables are peeled before consumption. Because many alkaloids and other natural products are stored in the epidermis, such as steroidal alkaloids in potato tubers, or cucurbitacins in cucumbers, peeling eliminates some of the compounds.

Most food is boiled in water. This leads to the thermal destruction of a number of toxic allelochemicals, such as lectins, protease inhibitors, and some esters and glycosides.

Table VI
Absorption of Alkaloids to Soil and Charcoal[a]

	Adsorption (%)				
	Soil			Charcoal	
Alkaloid	1	2	3	1	2
Cytisine	75	74	n.d.[b]	n.d.	100
Harmine	99	99	n.d.	n.d.	100
Heliotrine	92	87	n.d.	n.d.	100
Nicotine	93	92	n.d.	n.d.	100
Quinine	99	99	n.d.	n.d.	100
Scopolamine	73	71	n.d.	n.d.	100
Sparteine	93	92	74	85	100
Thebaine	94	93	n.d.	n.d.	100

[a]Soils 1–3 and charcoal 1 were collected from places where the geese had ingested similar material. Charcoal 2 was activated charcoal, known for optimal binding capacities. Alkaloids were dissolved in aqueous solutions. After adding 100 mg of soil or charcoal, solutions were centrifuged and the alkaloid content of the supernatant was determined (after Wink *et al.,* 1993).
[b]n.d. = not determined.

Furthermore, many water-soluble compounds are leached out into the cooking water and are discarded after cooking (e.g., steroidal and polyhydroxy alkaloids in potatoes, quinolizidines in lupins, or cyanogenic glycosides in yams).

South American native Indians ingest clay when alkaloid-rich potato tubers are on the menu. As clay binds steroidal alkaloids, geophagy is an ingenious way to detoxify toxins (Johns, 1990).

Man has modified the allelochemical composition of his crop plants, in that he has reduced unpleasant taste components by plant breeding. From the point of view of avoidance this strategy is plausible, but from the point of view of chemical ecology, as will be discussed later (Section 6), not so. These plants often lose their resistance vis-à-vis herbivores and pathogens in this way, resistance that must be replaced by man-made pesticides.

In general, only a few plants are exploited by man as food, out of the more than 300,000 existing species. Thus, despite all of our intelligence, we have only achieved a rather moderate success in the utilization of plants as a staple diet, indicating the importance and power of chemical plant defenses.

4. PLANT–MICROBE INTERACTIONS

It is a common observation that dead plants rot easily under the action of bacteria and fungi, whereas metabolically active, intact plants usually resist a microbial attack. Thus, plants (despite lacking an immune system) must have a means of defending themselves against microorganisms. Plants have an epidermis that is covered by a more or less thick cuticle consisting of waxes, alkanes, and other lipophilic natural products. This cuticle layer is water repellent and chemically rather inert and thus constitutes an important penetration barrier against most bacteria and fungi. In perennial plants and in roots we find another variation of this principle in that they often form resistant bark tissues (Table II).

The only way for microbes to enter a healthy plant is via the stomata or at sites of injury inflicted by herbivory, wind, or other accidents. Immediately after wounding, most plants start to accumulate suberin, lignin, callose, gums, or other resinous substances which seal the injured areas. In addition, antimicrobial agents are produced, such as 1,3-glucanase and chitinase, i.e., lytic enzymes stored in the vacuole which can degrade bacterial and fungal cell walls, protease inhibitors which can inhibit microbial proteases or just secondary metabolites with antimicrobial activity (Table II) (compare Chapter 9 and 17). These antimicrobial natural products, which can be either constitutive or inducible, often also interfere with herbivores; i.e., they often exhibit multiple functions (an observation often overlooked in the phytopathology literature).

Secondary compounds with antimicrobial activity include many phenolics (such as flavonoids, isoflavones, tannins, and simple phenolics), glucosinolates, nonproteinogenic amino acids, cyanogenic glycosides, acids, aldehydes, saponins, triterpenes, sesquiterpenes and last but not least alkaloids (see Chapter 17). It is likely (although not determined in many instances) that a substantial number of the 12,000 alkaloids have antimicrobial properties, directed against ubiquitous and generalist microbes, which have not specialized on a particular host plant.

Most plants are known to be parasitized or infected by at least a few specialized bacteria or fungi which often form close associations. In these circumstances the antimicrobial effect expected from the secondary metabolites present in the plant is rarely observed. We suggest that these specialists have adapted to the chemistry of their host plants and found a way to handle it. Mechanisms (which are yet to be determined in most instances) can be the inhibition of biosynthesis or the degradation of the respective secondary compounds. Many phytopathogenic bacteria and fungi produce their own secondary metabolites, which are often toxic to plants. It is assumed that these phytotoxins serve to weaken the host plants' defense, but this may not be the whole story.

Many grasses are infected with fungi that produce ergot or pyrrolizidine (loline-type) alkaloids. It has been assumed that these fungi are proper parasites. But recent experimental evidence suggests that the interaction between grasses and fungi may be of a symbiotic nature (Clay, 1990): For example, ergot alkaloids are strong vertebrate toxins, mimicking the activity of neurotransmitters such as dopamine, serotonin, and noradrenaline (Table I of Chapter 12). In fact, herbivorous impact on populations that were highly infected by fungi was much less than those without. These fungi exploit the nutrients of their host plants but supply them with strong poisons, which are not produced by the plants themselves. As these fungi do not kill their hosts, this commensalism seems to be of mutual benefit.

Some other associations between plants and fungi are symbiotic in nature, such as *rhizobia* in root nodules of legumes or microrhizal fungi in many species. In lupins, N-fixing *rhizobia* are present in both alkaloid-rich and alkaloid-free plants; they must thus be able to tolerate the alkaloids that are also present in the root.

Like animals, plants are hosts for a substantial number of viruses which are often transmitted by sucking insects, such as aphids and bugs (Heteroptera). Resistance to viral infection can be achieved either by biochemical mechanisms that inhibit viral development and multiplication or by warding off vectors, such as aphids, in the first place.

In two instances it was directly shown that alkaloids, such as quinine and sparteine, can inhibit the multiplication of a plant virus, here the potato X virus (Wink, 1987b, 1993a–d). All other evidence for antiviral activities (Table VII) of alkaloids comes from

Table VII
Examples of Antiviral Activities[a]

Alkaloid	Virus
Acronycine	Herpes simplex virus
Camptothecine	Herpes and others
Castanospermine	Cytomegalovirus and retroviruses
Cinchonidine	Potato X virus
Citracridone I	Herpes simplex virus
Crytopleurine	Herpes simplex virus
Didemnin	Herpes simplex virus
Grandisine	Herpes simplex virus
Harmine	Sindbis and murine cytomegalovirus
Hippeastrine	Herpes simplex virus
Lycorine	Herpes simplex and Rauscher virus
Maytansine	Murine sarcoma virus
Narciclasine	Rauscher virus
Norharman	Herpes simplex virus
Pretazettine	Herpes simplex and Rauscher virus
Sparteine	Potato X virus

[a]After Wink (1993a).

experiments with animal viruses. As viral life strategies are related in plants and animals, we suggest that a wider number of plant viruses may be controlled by alkaloids in nature than our limited data imply.

Fungal and bacterial multiplication can be controlled at the level of replication, transcription, protein biosynthesis, posttranslational protein modification, and membrane/ cell wall integrity, whereas viruses are more difficult to inhibit (see Chapters 12 and 18).

5. PLANT–PLANT INTERACTIONS

Plants often compete with other plants, either of the same or of a different species, for space, light, water, and nutrients. This phenomenon becomes especially evident in deserts or semideserts where resources are limited and thus competition intensive. A number of biological mechanisms to avoid competition have been described, such as temporal spacing of the vegetation period, i.e., some species flower when others are still dormant or ungerminated.

Plants can inhibit each other also by their secondary metabolites (so-called "allelopathy") (Rice, 1984; Waller, 1987; Inderjit *et al.,* 1995). Secondary products are often excreted by the root or rhizophore or they are leached from the surface of intact leaves or from decaying dead leaves by rain. Both processes will increase the concentration of allelochemicals in the soil surrounding a plant, where the germination of a potential competitor may occur. The area of allelopathy is well documented at the level of controlled *in vitro* experiments but how it works in the ecosystem is still often a matter of controversy (Waller, 1987; Inderjit *et al.,* 1995). Allelopathic natural products have been recorded in all classes of secondary metabolites; however, the alkaloid group was somewhat neglected probably because their remarkable animal toxicities were more obvious.

As discussed in more detail in Chapter 14, nearly all structural types of alkaloids can exhibit allelopathic activities. At higher concentrations, a marked reduction in the germination rate of seeds can be recorded. More sensitive, however, are radicles and hypocotyls. They respond to alkaloids at a much lower level, usually showing growth inhibition or overstimulation. Both effects reduce the fitness of a seedling. The inhibitory effect can be absent for the endogenously produced alkaloid, as was reported for quinolizidine alkaloids in lupins (Wink, 1983a) and colchicine in *Colchicum autumnale.* Some of the cellular targets, discussed under plant–animal interactions, may also be affected by alkaloids in allelopathy.

A special case of plant–plant interactions can be seen in parasitic or hemiparasitic plants. The role of alkaloids in these interactions is discussed in Chapter 13.

6. ECOLOGICAL RELEVANCE OF ALKALOIDS

Because many of the allelochemical properties discussed above were determined in *in vitro* systems, it might be argued that they are not relevant under field conditions. In the following, evidence that supports the ecological relevance of alkaloids is discussed in more detail. For an alkaloid to serve as a chemical defense compound, the following criteria should be met:

1. The alkaloid or alkaloid mixture should have significant effects against microbes and/or animals
2. The compounds should be present in the plant in concentrations that are of the same order or even higher than those determined in the bioassays or animal experiments.
3. The compound should be present in the plant at the right time and right place.
4. Experimental evidence should show that an alkaloid or alkaloidal mixture promotes the fitness of a plant.

Many examples have been given concerning the first criterion; the second, third, and fourth will be explained below.

6.1. Are Alkaloid Concentrations in Plants Sufficiently High?

The accumulation of secondary metabolites only makes sense in view of their ecological function as defense compounds; however, they can fulfill these functions only if the amount stored is appropriate. The synthesis and maintenance of high levels of a defense compound is very demanding from the point of view of physiology and biochemistry. In view of the activities of these compounds, it can be assumed that many natural products would probably interfere with the metabolism of the producing plant if these metabolites were accumulated in the same compartments where they were made. Whereas biosynthesis takes place in the cytoplasm, in vesicles (berberine) or organelles such as chloroplasts [QA (Wink and Hartmann, 1982), coniceine (Roberts, 1981)], the site of accumulation of water-soluble alkaloids is the central vacuole; that of lipophilic compounds, the latex, resin ducts, or glandular hairs (see Chapter 10).

In general, all parts of an alkaloidal plant accumulate alkaloids (Table VIII). Although the site of synthesis is often restricted to a particular organ, such as roots or leaves, a translocation via the phloem, xylem, or apoplastically must occur (Wink, 1987a, 1993c) (see Chapter 10). Alkaloid levels can vary with respect to organ and development and even diurnal fluctuations have been observed for quinolizidine (see Fig. 9) and tropane alkaloids (Waller and Nowacki, 1978; Wink and Witte, 1984; Sporer *et al.,* 1993). Alkaloid levels are usually highest during the time of flowering and fruit/seed formation. In annual species the leaves, flowers, and seeds are often alkaloid-rich, whereas in perennial ones, like shrubs and trees, we find alkaloid-rich stem and root barks as well. All of these plant parts and organs have in common that they are important for fitness or reproduction and thus for long-term survival of the species. Spiny species, which invest in mechanical defense as well, accumulate less alkaloid than soft-bodied ones; examples are isoquinoline alkaloids in cacti or QA in legumes (Fig. 6). If a plant produces few and large seeds, their alkaloid levels tends to be higher than in species with many and small seeds (Fig. 7).

Summarizing the relevant phytochemical literature we find that alkaloid levels are between 0.1 and 15% (dw), which is equivalent to 0.01–1.5% fresh weight or 0.1–15 mg g^{-1} fw. As a representative example, Table VIII lists QA concentrations in different legume species and in some of their tissues. As compared with the inhibitory concentrations determined in plant–plant, plant–microbe, and plant–herbivore interactions, plant alkaloid levels are of the same order or one order of magnitude greater (Wink, 1992, 1993a–d).

Table VIII
Organ-Specific QA Concentrations in Selected Legumes

Species	Organ/Tissue	Total alkaloid (per g fresh wt)
Cytisus scoparius	Stem epidermis	46 mg/g; 200 mM
	Shoots	2 mg
	Leaves	0.2–1 mg
	Seeds	2 mg[a]
	Roots	0.03 mg
Lupinus polyphyllus	Petiole epidermis	1.7-10 mg
	Stem epidermis	6.3 mg
	Leaves	1–4 mg
	Stems	1–2 mg
	Flower	
	Pollen	1.8 mg
	Carpels	1.3 mg
	Petals	0.4 mg
	Fruits	1.6 mg
	Seeds	30–40 mg[a]
	Roots	0.2 mg
L. mutabilis	Stem epidermis	5.3 mg
L. albus	Stem epidermis	6.3 mg
	Phloem sap	0.5–1.2 mg/ml
	Leaves	2.8 mg
	Stem	0.7 mg
	Flower	4.1 mg
	Fruit	3.1 mg
	Seed	43.0 mg[a]
	Roots	0.5 mg
L. angustifolius	Phloem sap	0.8 mg/ml
	Xylem sap	0.05 mg/ml
Laburnum anagyroides	Leaves	0.3 mg
	Twigs	
	Bark	11.1 mg
	Wood	0.5 mg
	Flower	0.4 mg
	Fruit	0.5 mg
	Seed	10–30 mg[a]
	Endosperm	21 mg
	Testa	2 mg

[a]Dry weight.

The specific characteristics of individual molecular species, as well as the absolute amounts accumulated are important, and need to be addressed. Usually one to five main alkaloids dominate in a plant, but are accompanied by several (up to 80) minor alkaloids. As the alkaloids present in a particular plant usually share a biosynthetic pathway, their basic skeletons are often similar. To facilitate phytochemical investigations these compounds are usually classified as belonging to a particular alkaloid group, such as pyrrolizidine or quinolizidine alkaloids. This does not mean, however, that the biological activities of individual alkaloids are identical, i.e., that they are addressed to the same

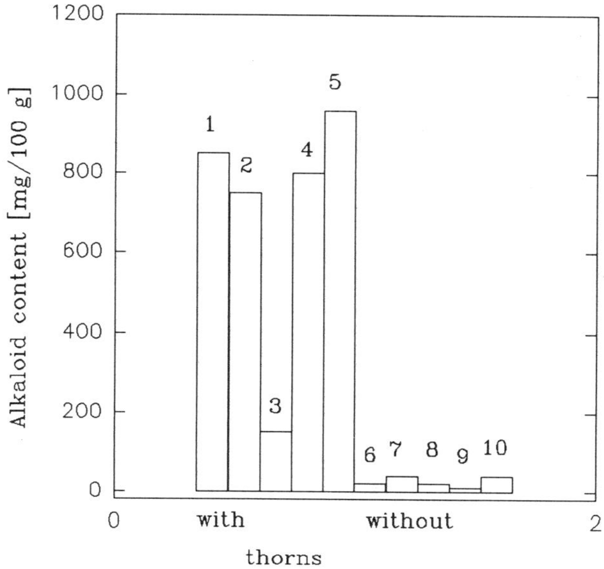

Figure 6. Relationship between mechanical and chemical defense: alkaloid content versus thorns in brooms. 1, *Genista radiata;* 2, *G. lydia;* 3, *G. sagittalis;* 4, *Teline monspessulana;* 5, *T. linifolia;* 6, *G. silvestris;* 7, *G. hispida;* 8, *Callicotome spinosa;* 9, *Ulex europaeus;* 10, *Echinospartum horridum.* After R. Greinwald, 1988, Untersuchungen zur chemotaxonomischen Bedeutung von Leguminosenalkaloiden und zum Alkaloidstoffwechsel in transformierten Geweben und Zellkulturen, Ph.D. dissertation, Universität Würzburg.

molecular target (Chapter 12, Fig. 9). On the contrary, the addition of small substituents to a molecule, such as a lipophilic side chain, while it seems to be a small and insignificant variation from a phytochemical point of view, may render the compound more lipophilic and thus more resorbable. In consequence, its toxicity may be higher.

The qualitative patterns are not constant but usually differ between organs, developmental stages, individuals, populations and, species. For a herbivore or pathogen the variable alkaloid profiles are very demanding, as these organisms not only have to adapt to one group of chemicals but also to most of the individual compounds (at least the major alkaloids) present. Because the composition of these chemicals also change quantitatively, it is even more difficult for them to cope. Thus, we suggest that structural diversity and its continuous variation is a means of counteracting the selection of adapted specialists.

In medicine we apply a related strategy to control microbial infections: In order to prevent bacteria developing resistance toward a particular antibiotic, mixtures of structurally different antibiotics are often applied which are oriented to various molecular targets. If only one antibiotic were given to all patients, the development of resistance would be much favored.

6.2. Occurrence of Alkaloids at the Right Site and Right Time

Intuitively, a valuable plant organ must be more protected than others. The preferential storage of alkaloids in very actively growing young tissue, seeds, petals, carpels, or

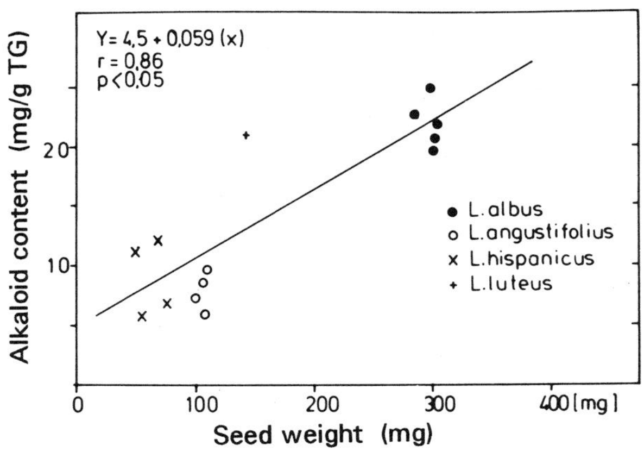

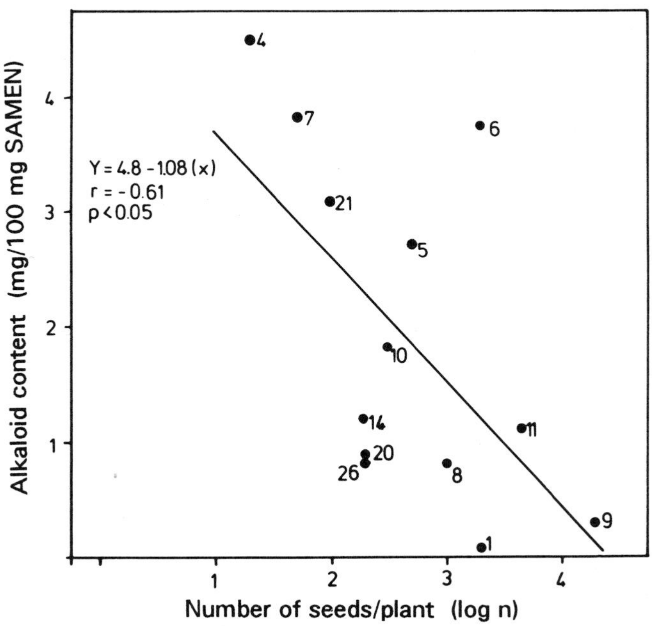

Figure 7. Correlation between alkaloid content and seed size and seed numbers in legumes (after Wink, 1985). (Top) Relation between seed weight and alkaloid content: bigger seeds contain proportionally more QA than smaller seeds. (Bottom) Relation between alkaloid content and seed numbers per plant: seeds from plants with few seeds are alkaloid-rich, whereas plants having many small seeds contain rather low levels of QA. Numbers refer to different legume species (Wink, 1985). By permission of Springer-Verlag.

pollen (Table VIII) provides evidence that alkaloids are stored at the right site, as these organs are important for reproduction and thus species survival.

At the tissue level we can also see these constraints: As discussed in Chapter 10, alkaloids are often accumulated in a cell- and tissue-specific fashion. For example, isoquinoline alkaloids in the Papaveraceae are abundant in the latex, where they are sequestered in many small latex vesicles. In latex vesicles of *Chelidonium majus* the concentration of protoberberine and benzophenanthridine alkaloids can be in the range of 0.6–1.2 M,

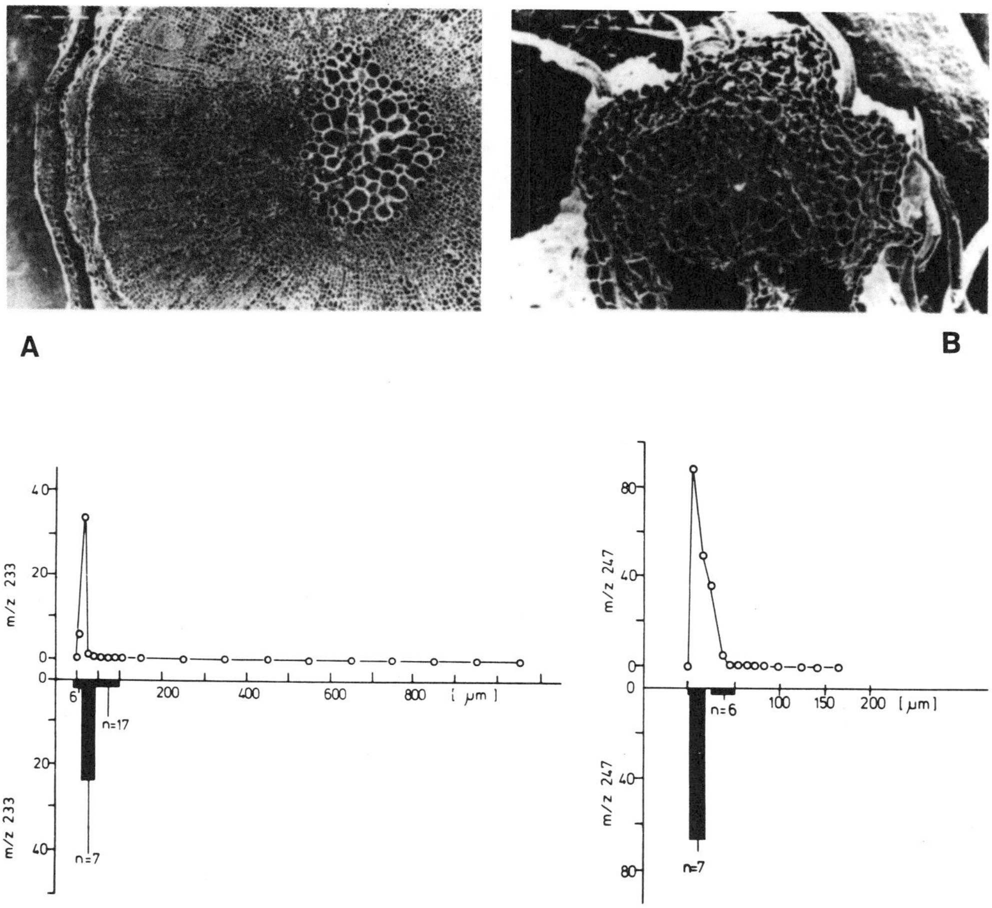

Figure 8. Epidermal storage of quinolizidine alkaloids in *Lupinus polyphyllus* and *Cytisus scoparius* (after Wink *et al.*, 1984). Distribution of sparteine (A) and lupanine (B) in cross sections of *C. scoparius* stems and *L. polyphyllus* petioles analyzed by laser desorption mass spectrometry. SEM photos show the cross sections analyzed by LAMMA 1000. The graphs demonstrate the occurrence of the molecular ions of sparteine (*m/z* 233) and lupanine (*m/z* 247) in scans carried out from epidermal to central cell layers. The lower panel presents a statistical analysis of several LAMMA measurements (means ± SD).

M, which is achieved by their complexation with equal amounts of chelidonic acid. If a herbivore wounds such a plant, the latex spills out immediately. Besides gluing the mandibles of an insect, the high concentration of deterrent and toxic alkaloids will usually do the rest. Indeed, *Chelidonium* plants are hardly attacked by herbivores. In addition, as these alkaloids are also highly antimicrobial, the site of wounding is quickly sealed and impregnated with natural antibiotics.

Lupin alkaloids are preferentially stored in the epidermal, subepidermal, and hypodermal tissues of stems and leaves, reaching local concentrations between 20 and 200 mM (Fig. 8) (Wink *et al.*, 1984; Wink, 1983a,b). A small herbivore or pathogen will enter the plant by injuring the epidermis first, where it is stopped by a chemical barrier. Epidermal storage of alkaloids does not seem to be restricted to legume species, but is probably a common phenomenon, although hardly studied with modern techniques. The accumulation of many alkaloids in root or stem bark, such as berberine, cinchonine, and quinine, can be interpreted in a similar way, i.e., that alkaloids are deposited at strategically important sites, where they can ward off an intruder at the first opportunity.

The biosynthesis and accumulation of alkaloids can be regarded as constitutive processes, ensuring that alkaloids are present when needed for defense. The diurnal cycles (Fig. 9) and alkaloid turnover rates (see Chapter 10), however, seem to contradict this observation. To explain this phenomenon, in the case of alkaloids that have been selected as analogues of neurotransmitters (see examples in Chapter 12), it is essential that the structure and configuration of the molecule does not change. For L-hyoscyamine, which binds to mAChR, a racemization to atropine can occur under physiological conditions, which would halve its activity. Other alkaloids, which are stored in vacuoles, could be oxidized by peroxidase and thus would lose their intrinsic activity. A steady turnover and/or a diurnal fluctuation would ensure that the level of the correctly configured alkaloid is always present. It should be recalled that other molecules that are important for the correct function of a cell (e.g., enzymes and receptors) exhibit a pronounced turnover.

Plants appear to be reactive with respect to challenges by pathogens and herbivores. When infected by microorganisms, many plants begin to synthesize a series of antimicrobial compounds, ranging from proteins to secondary products including alkaloids (Table II). Chapter 9 considered the role of alkaloids in this context.

But what is the situation after wounding by a herbivore? When eaten by large herbivores, a *de novo* synthesis of defense compounds would be almost useless for a plant (except perhaps for trees), for the reaction would not be quick enough. But the situation is different for small herbivores such as insects or worms, which may feed on a particular plant for days or weeks. Here the *de novo* production of an allelochemical would be worthwhile. There are, indeed, experimental data supporting this view: In *Liriodendron tulipa* several aporphine alkaloids accumulate after wounding which are otherwise not present. In tobacco the production of nicotine, in lupins that of QA and in *Atropa belladonna* that of hyoscyamine is induced by wounding, thus increasing the already high levels of alkaloids by factors between 1.2 and 5. While the response was seen after 2–4 hr in lupins (Wink, 1983b), it took days in *Nicotiana* (Baldwin, 1989) and in *Atropa* (Harborne, 1993). We suggest that the wound-induced stimulation of alkaloid formation is not an isolated phenomenon, but part of the chemical defense system (see examples in Tallamy and Raupp, 1991).

Summarizing, it can be assumed that alkaloids are present at the right place, at the right time, and in the right concentrations to fulfill their ecological defense functions.

6.3. Evidence for Alkaloid-Mediated Fitness

The arguments and circumstantial evidence discussed in the last paragraphs appear convincing and support the hypothesis that alkaloids serve as defense compounds. It is, however, difficult to prove that alkaloids are important for the fitness and survival of the plants producing them.

For one group of alkaloids, though, critical data are available. As mentioned before, QA constitute the main secondary products of many Leguminosae, especially in the genera *Lupinus, Genista, Cytisus, Baptisia, Thermopsis, Sophora, Ormosia,* and others

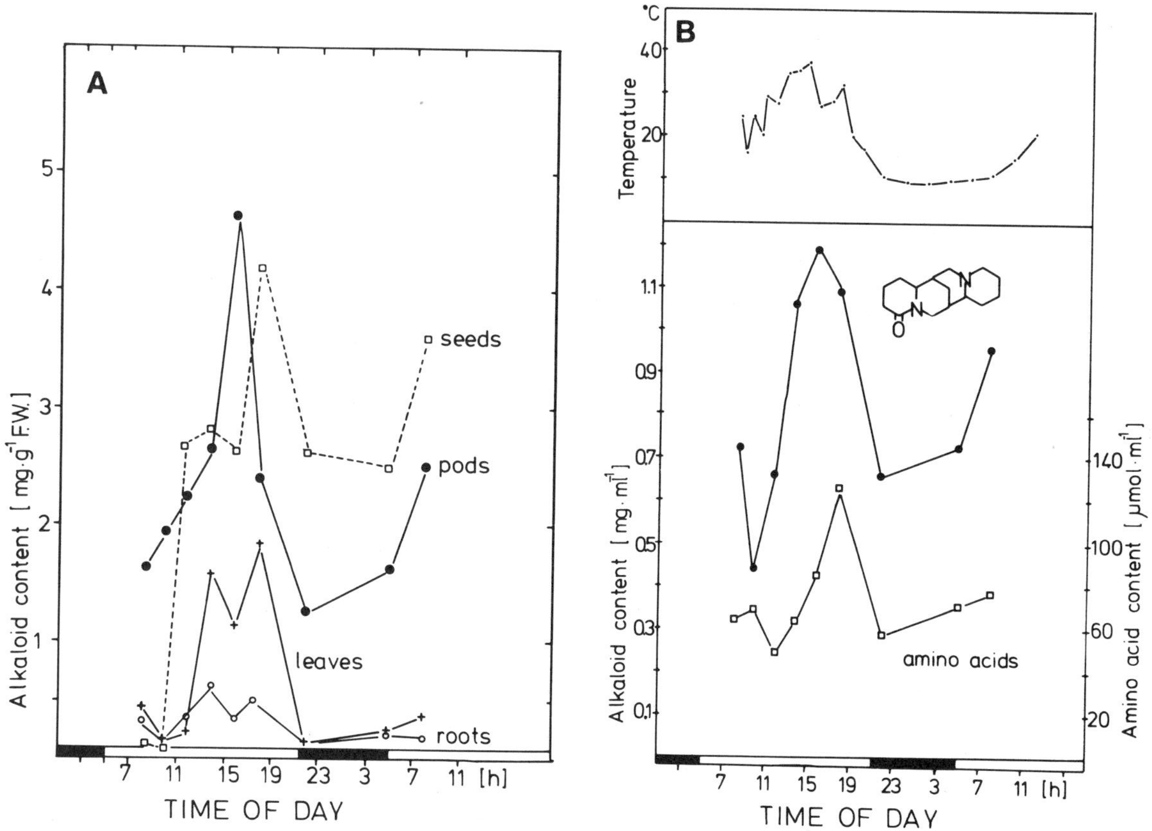

Figure 9. Diurnal fluctuation of lupanine and other QA in the leaves, phloem sap, and maturing fruits of *Lupinus albus* plants kept under natural day–night conditions (after Wink and Witte, 1984). (A) QA fluctuations in leaves, pods, seeds, and roots. (B) Lupanine and amino acids in the phloem sap.

(Kinghorn and Balandrin, 1984; Wink, 1993a). Lupins have relatively large seeds that contain up to 40–50% protein, up to 20% lipids, and 2–8% alkaloids. In order to use lupin seed for animal or human nutrition, our ancestors cooked the seeds and leached out the alkaloids in running water. This practice has been reported from Egypt and Greece in the Old World and for the native Indians and Incas of the New World and is still in use today. The treated seeds taste sweet, in contrast to the alkaloid-rich ones, which are very bitter.

At the turn of this century, German plant breeders set out to grow alkaloid-free lupins, the so-called "sweet lupins." Although extremely rare in nature (frequency $<10^{-4}$), the efforts were successful and at present, "sweet" varieties with an alkaloid content lower than 0.01% exist for *Lupinus albus, L. mutabilis, L. luteus, L. angustifolius,* and *L. polyphyllus.* As far as it has been determined, the sweet varieties differ chemically from their original bitter wild forms, only in their degree of alkaloid accumulation. This offers the chance to test whether bitter lupins have a higher ecological fitness than sweet ones.

The results of these experiments were clear cut: When lupins were grown without being fenced in and without protection from man-made chemicals, a dramatic effect was regularly observed, especially with regard to herbivores: Rabbits (*Cuniculus europaeus*) (Fig. 10A) and hares (*Lepus europaeus*) clearly prefer the sweet plants and leave the bitter plants almost untouched, at least as long as there was an alternative food source. A similar result was seen for a number of insect species, such as aphids, beetles, thrips, and leaf-mining flies (Fig. 10B, Table IX), i.e., the sweet forms are attacked, whereas the alkaloid-rich ones were largely protected. For the specialized and adapted aphid (*Macrosiphum albifrous*) the opposite behavior was observed as expected.

A sweet variety of *L. luteus* was infested by *Acyrthosiphon pisii* in Poland. The invasion of the aphids became a serious problem not only because the aphid enfeebled the plant by sucking its phloem sap, but also because it transferred a viral disease (lupin narrow leafness). In a mixed population of sweet and bitter lupins, sweet plants are at a disadvantage and after a few generations will disappear. The infestation by the aphid and the following viral infection accelerated the elimination of alkaloid-poor plants, which, even without infection, are already inferior in seed production. This observation again stresses the importance of alkaloid for the fitness of lupins.

Plant breeders have observed that bacterial, fungal, and viral diseases are more abundant in the sweet forms, but this effect has not been documented in necessary detail. These experiments and observations clearly prove the importance of alkaloids for lupins, but it should be recalled that other secondary metabolites, such as phenolics, isoflavones, terpenes, saponins, stachyose, erucic acid, and phytic acid, are present in lupins which may have additional or even synergistic effects.

The lupin example also illustrates an intrinsic problem of traditional plant breeding. Knowing the ecological importance of QA for the fitness of lupins, it seems doubtful whether the selection of sweet lupins was a wise strategy. In order to grow sweet lupins, fences are needed and worse, man-made chemical pesticides (which have a number of well-documented disadvantages) are needed to substitute for the alkaloids no longer present. It can be assumed that similar strategies, i.e., breeding away unwanted chemical traits, have been chosen with our other agricultural crops, with the consequent reduction in overall fitness. We can easily observe their reduced fitness by trying to leave crop species to themselves in the wild: they will quickly disappear and not colonize new habitats. There are, however, alternatives: Taking lupins as an example, we could devise large-scale

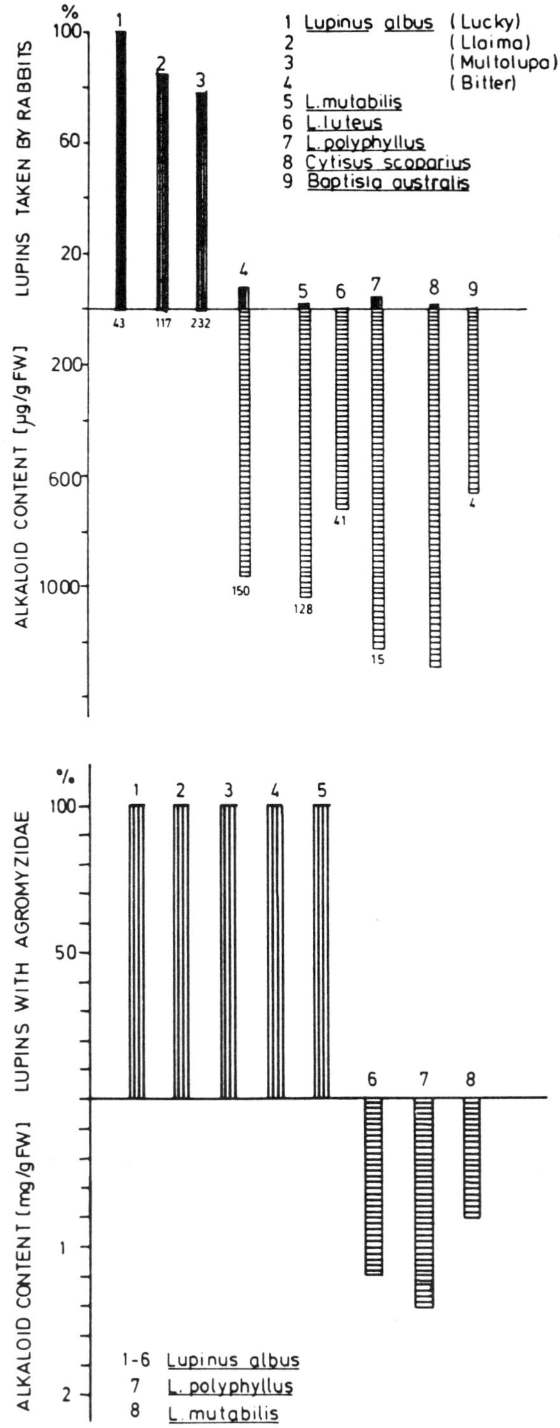

Figure 10. Examples of alkaloid-mediated fitness in alkaloid-rich and alkaloid-poor (bitter versus sweet) lupins (after Wink, 1985, 1987b, 1988). (A) Differential grazing by rabbits; (B) differential infestation by leaf-mining flies (Agromyzidae). Upper panels indicate the degree of herbivory, lower panels the alkaloid content of the respective plants.

Table IX
Correlation between Alkaloid Content and Herbivory in Polyphagous and Specialized Herbivores[a]

Species	Lupin	Alkaloid content	Effect
Nonadapted herbivores			
Vertebrates			
Sheep		Sweet	Sweet lupins are preferred, bitter
		Bitter	discriminated
Lepus europaeus		Sweet	Sweet lupins are preferred, bitter
		Bitter	discriminated
Oryctolagus europaeus	*L. albus*	0.01 mg/g	Herbivory almost 100%
		2.0 mg/g	Herbivory <10%
Insects			
Agromyzidae	*L. albus*	0.01 mg/g	Heavy infestation, 100% incidence
		2.0 mg/g	Infestation < 1%
Sitona lineatus	*L. albus*	<0.02 mg/g	100% herbivory
		1500 mg/g	Low or no herbivory
	L. mutabilis	2500 mg/g	Low or no herbivory
Myzus sp.	*L. luteus*	0.01 mg/g	Infestation 100%
		>0.7 mg/g	Infestation < 1%
Aphis fabae	*L. polyphyllus*	Sweet	Infestation
		Bitter	No infestation
Frankliniella tritici	*Lupinus*	Sweet	Heavy infestation
		Bitter	No infestation
F. bispinosa	*Lupinus*	Sweet	Heavy infestation
		Bitter	No infestation
Adapted herbivores			
Macrosiphum albifrons	*L. albus*	0.01 mg/g	Infestation < 10%
		2.0 mg/g	Infestation 100%
	L. polyphyllus	>1 mg/g	Infestation 80%
	L. angustifolius	1.5 mg/g	Infestation 100%
	L. mutabilis	2.5 mg/g	Infestation 30%

[a]Note that generalists are deterred by alkaloids, whereas specialists are attracted.
Sweet = low-alkaloid variety; bitter = high-alkaloid variety (>0.5%) (after Wink and Römer, 1986; Wink, 1988, 1992).

procedures to remove alkaloids from the seeds after harvest (similar to refining sugar from sugar beets) and to economically produce more valuable products, such as pure protein, lipid, and dietary fibers from bitter seeds. A spin-off product would be alkaloids, which could be used either in medicine (sparteine is exploited as a drug to treat heart antiarrhythmias) or in agriculture as a natural biorational plant protective (Wink, 1993d).

7. CONCLUSION

Although the biological activities of many alkaloids have not yet been studied and their ecological functions remain to be elucidated or proven in most instances, we can, nevertheless, safely conclude that alkaloids are neither waste nor functionless molecules, but that they are important fitness components, probably primarily antiherbivore compounds. As nature obviously favors multitasking, additional activities such as allelopathic or antimicrobial activities are plausible.

As pointed out above, many features of alkaloid physiology and biochemistry only become plausible if this concept is applied. The role of alkaloids in human history (Chapters 2, 3) and their present application in agriculture and medicine (Chapters 17, 18) can be regarded as a benefit that evolved in an ecological context.

REFERENCES

Major Reviews

Bernays, E., 1982, The insect on a plant—A closer look, in: *Insect–Plant Relationships* (J. H. Visser and A. K. Minks, eds.), Wageningen, pp. 3–17.

Bernays, E. A., and Chapman, R. F. 1994, *Host-Plant Selection by Phytophagous Insects,* Chapman & Hall, London.

Blum, M. S., 1981, *Chemical Defenses of Arthropods,* Academic Press, New York.

Boppré, M., 1990, Lepidoptera and pyrrolizidine alkaloids, *J. Chem. Ecol.* **16:**165–186.

Brattsten, L. B., and Ahmad, S., 1986, *Molecular Aspects of Insect–Plant Associations,* Plenum Press, New York.

Brown, K. S., and Trigo, J. L., 1995, The ecological activity of alkaloids, in: *The Alkaloids,* Vol. 47 (G. Cordell, ed.), Academic Press, San Diego, pp. 227–354.

Clay, K., 1990, Fungal endophytes of grasses, *Annu. Rev. Ecol. Syst.* **21:**275–297.

Duffey, J., 1980, Sequestration of plant natural products by insects, *Annu. Rev. Entomol.* **25:**447–477.

Edmunds, M., 1974, *Defense in Animals,* Longman, Harlow.

Fritz, R. S., and Simms, E. L., (eds.), 1992, *Plant Resistance to Herbivores and Pathogens: Ecology, Evolution and Genetics,* University of Chicago Press, Chicago.

Harborne, J. B., 1993, *Introduction to Ecological Biochemistry,* 4th ed., Academic Press, San Diego.

Hartmann, T., and Witte, L., 1995, Chemistry, biology and chemoecology of the pyrrolizidine alkaloids, in: *Alkaloids: Chemical and Biological Perspectives* (S. W. Pelletier, ed.), Vol. 9, Pergamon Press, Oxford, 155–233.

Inderjit, Dakshini, K. M. M., and Einhellig, F. A., 1995, Allelopathy: Organisms, processes and applications, *ACS Symp. Ser.* 582, pp. 117–126.

Johns, T., 1990, *With Bitter Herbs They Shall Eat It,* University of Arizona Press, Tucson.

Kinghorn, A. D., and Balandrin, M. F., 1984, Quinolizidine alkaloids of the Leguminosae: Structural types, analysis, chemotaxonomy, and biological activities, in: *Alkaloids: Chemical and Biological Perspectives* (S. W. Pelletier, ed.), Vol. 2, Wiley, New York, pp. 105–148.

Levin, D. A., 1976, The chemical defenses of plants to pathogens and herbivores, *Annu. Rev. Ecol. Syst.* **7:**121–159.

Mothes, K., Schütte, H. R., and Luckner, M., 1985, *Biochemistry of Alkaloids,* Verlag Chemie, Weinheim.

Rice, E. L., 1984, *Allelopathy,* Academic Press, San Diego.

Robinson, T., 1974, Metabolism and function of alkaloids in plants, *Science* **184:**430–435.

Rosenthal, J., and Berenbaum, M. R., 1991, *Herbivores: Their Interaction with Secondary Plant Metabolites,* 2nd ed., Academic Press, San Diego.

Rosenthal, J., and Janzen, D., 1979, *Herbivores: Their Interactions with Plant Secondary Metabolites,* Academic Press, San Diego.

Schlee, D., 1992, *Ökologische Biochemie,* 2nd ed., Fischer Verlag, Stuttgart.

Schneider, D., 1987, The strange fate of pyrrolizidine alkaloids, in: *Perspectives in Chemoreception Behaviour* (R. F. Chapman, E. A. Bernays, and J. G. Stoffolano, eds.), Springer, Berlin, pp. 123–142.

Schoonhoven, L. M., 1972, Secondary plant substances and insects, in: *Structural and Functional Aspects of Phytochemistry* (C. V. Runekless and T. C. Tso, eds.), *Recent Adv. Phytochem.* **5:**197–224.

Swain, T., 1977, Secondary compounds as protective agents, *Annu. Rev. Plant Physiol.* **28:**479–501.

Tallamy, D. W., and Raupp, M. J., (eds.), 1991, *Phytochemical Induction by herbivores,* Wiley, New York.

Urich, K., 1990, *Vergleichende Biochemie der Tiere,* Fischer Verlag, Stuttgart.
Waller, G. R., 1987, *Allelochemicals. Role in Agriculture and Forestry,* ACS Symp. Ser., 330, American Chemical Society, Washington, DC.
Waller, G. R., and Nowacki, E., 1978, *Alkaloid Biology and Metabolism in Plants,* Plenum Press, New York.
Wink, M., 1987a, Physiology of the accumulation of secondary metabolites with special reference to alkaloids, in: *Cell Culture and Somatic Cell Genetics of Plants,* Vol. 4 (F. Constabel and I. Vasil, eds.), Academic Press, San Diego, pp. 17–41.
Wink, M., 1988, Plant breeding: Importance of plant secondary metabolites for protection against pathogens and herbivores, *Theor. Appl. Genet.* **75:**225–233.
Wink, M., 1990, Physiology of secondary product formation in plants, in: *Secondary Products from Plant Tissue Culture* (B. V. Charlwood and M. J. C. Rhodes, eds.), Clarendon Press, Oxford, pp. 23–41.
Wink, M., 1992, The role of quinolizidine alkaloids in plant–insect interactions, in: *Insect–Plant Interactions* (E. A. Bernays, ed.), Vol. IV, CRC Press, Boca Raton, pp. 133–169.
Wink, M., 1993a, Quinolizidine alkaloids, in: *Methods in Plant Biochemistry,* Vol. 8 (P. Waterman, ed.), Academic Press, San Diego, pp. 197–239.
Wink, M., 1993b, Allelochemical properties and the raison d'être of alkaloids, in: *The Alkaloids,* Vol. 43 (G. Cordell, ed.), Academic Press, San Diego, pp. 1–118.
Wink, M., 1993c, The plant vacuole: A multifunctional compartment, *J. Exp. Bot.* **44**(Suppl.):231–246.
Wink, M., 1993d, Production and application of phytochemicals from an agricultural perspective, in: *Phytochemistry and Agriculture,* Proc. Phytochem. Soc. Eur. Vol. 34 (T. A. van Beek and H. Breteler, eds.), Oxford University Press, London, pp. 171–213.

Key References

Ahmad, S., 1983, Mixed-function oxidase activity in a generalist herbivore in relation to its biology, food plants and feeding history, *Ecology* **64:**235–243.
Baldwin, I. T., 1989, Mechanism of damage-induced alkaloid production in wild tobacco, *J. Chem. Ecol.* **15:**1661–1680.
Bernays, C. B., and Montllor, J., 1989, Aposematism of *Uresiphita reversalis* larvae (Pyralidae), *J. Lepid. Soc.* **43:**261–273.
Brattsten, L. B., 1988, Enzymatic adaptations in leaf-feeding insects to host-plant allelochemicals, *J. Chem. Ecol.* **14:**1919–1939.
Conner, W. E., Eisner, T., van der Meer, R. K., Guerrero, A., and Meinwald, J., 1981, Precopulatory sexual interactions in an arctiid moth (*Utetheisa ornatrix*): Role of a pheromone derived from dietary alkaloids, *Behav. Ecol. Sociobiol.* **9:**227–235.
Detzel, A., and Wink, M., 1993, Attraction, deterrence or intoxication of bees (*Apis mellifera*) by plant allelochemicals, *Chemoecology* **4:**8–18.
Dowd, P. F., 1992, Insect fungal symbionts: A promising source of detoxifying enzymes, *J Ind. Microbiol.* **9:**149–161.
Dreyer, D., Jones, K. C., and Molyneux, R. J., 1985, Feeding deterrency of some pyrrolizidine, indolizidine, and quinolizidine alkaloids towards pea aphid *(Acyrthosiphon pisum)* and evidence for phloem transport of the indolizidine alkaloid swainsonine, *J. Chem. Ecol.* **11:**1045–1051.
Dussourd, D. E., Ubik, K., Harvis, C., Resch, J., Meinwald, J., and Eisner, T., 1988, Biparental endowment of eggs with acquired plant alkaloid in the moth *Utetheisa ornatrix, Proc. Natl. Acad. Sci. USA* **85:**5992–5996.
Ehrlich, P. R., and Raven, P. H., 1964, Butterflies and plants: A study of coevolution, *Evolution* **18:**586–608.
Fraenkel, G., 1959, The raison d'être of secondary substances, *Science* **129:**1466–1470.
Hauser, M.-T., and Wink, M., 1990, Uptake of alkaloids by latex vesicles and isolated mesophyll vacuoles of *Chelidonium majus* (Papaveraceae)., *Z. Naturforsch.* **45c:**949–957.
Holzinger, F., and Wink, M., 1996, Mediation of cardiac glycoside insensitivity in the monarch (*Danaus plexippus*): Role of an amino acid substitution in the ouabain binding site of Na^+, K^+-ATPase. *J. Chem. Ecol.* **22,** 1921–1937.
Holzinger, F., Frick, C., and Wink, M., 1992, Molecular base for the insensitivity of the monarch (*Danaus plexippus*) to cardiac glycosides, *FEBS Lett.* **314:**477–480.
Montllor, C. B., Bernays, E. A., and Barbehenn, R. V., 1990, Importance of quinolizidine alkaloids in the relationship between larvae of *Uresiphita reversalis* (Lepidoptera: Pyralidae) and a host plant, *Genista monspessulana, J. Chem. Ecol.* **16:**1853–1865.

Montllor, C. B., Bernays, E. A., and Cornelius, M. L., 1991, Responses of two hymenopteran predators to surface chemistry of their prey: Significance for an alkaloid-sequestering caterpillar, *J. Chem. Ecol.* **17**:391.

Nickisch-Rosenegk, E. von, and Wink, M., 1993, Sequestration of pyrrolizidine alkaloids in several arctiid moths (Lepidoptera: Arctiidae), *J. Chem. Ecol.* **19**:1889–1903.

Nickisch-Rosenegk, E. von, Schneider, D., and Wink, M., 1990, Time-course of pyrrolizidine alkaloid processing in the alkaloid exploiting arctiid moth, *Creatonotos transiens, Z. Naturforsch.* **45c**:881–894.

Roberts, M. F., 1981, Enzymic synthesis of γ-coniceine in *Conium maculatum* chloroplasts and mitochondria, *Plant Cell Rep.* **1**:10–13.

Schmeller, T., El-Shazly, A. and Wink, M., 1997, Allelochemical activities of pyrrolizidine alkaloids: Interactions with neuroreceptors and acetylcholine related enzymes. *J. Chem. Ecol.* **23**:399–416.

Schneider, D., 1993, Danaine butterflies a didactic story about chemical ecology. *Nat. Hist. Mus. Los Angeles Cty. Contrib. Sci. Ser.,* 19–28.

Schneider, D., Boppre, M., Zweig, I., Horsley, S. B., Bell, T. W., Meinwald, J., Hansen, K., and Diehl, E. W., 1982, Scent organ development in *Creatonotos* moths: Regulation by pyrrolizidine alkaloids, *Science* **215**:1264–1265.

Schulz, S., Francke, W., Boppre, M., Eisner, T., and Meinwald, J., 1993, Insect pheromone biosynthesis: Sterochemical pathways of hydroxydanaidal production from alkaloidal precursors in Creatonotos transiens (Lepidoptera, Arctii.), *Proc. Natl. Acad. Sci. USA,* **90**:6834–6838.

Sporer, F., Sauerwein, M., and Wink, M., 1993, Diurnal and developmental variation of alkaloid accumulation in *Atropa belladonna, Acta Hortic.* **331**:381–386.

Stahl, E., 1888, Pflanzen und Schnecken. Jenaer, *Z. Naturweis.* **22**:557.

Szentesi, A., and Wink, M., 1991, Fate of quinolizdine alkaloids through three trophic levels: *Laburnum anagyroides* (Leguminosae) and associated organisms, *J. Chem. Ecol.* **17**:1557–1573.

Vrieling, K., Smit, W., and Meijden, E. van der, 1991, Tritrophic interactions between aphids (*Aphis jacobaeae* Schrank), ant species, *Tyria jacobaeae* L. and *Senecio jacobaea* L. lead to maintenance of genetic variation in pyrrolizidine alkaloid concentration, *Oecologia* **86**:177–182.

Wink, M., 1983a, Inhibition of seed germination by quinolizidine alkaloids. Aspects of allelopathy in *Lupinus albus* L, *Planta* **158**:365–368.

Wink, M., 1983b, Wounding-induced increase of quinolizidine alkaloid accumulation in lupin leaves, *Z. Naturforsch.* **38c**:905–909.

Wink, M., 1985, Chemische Verteidigung der Lupinen: Zur biologischen Bedeutung der Chinolizidinalkaloide, *Plant Syst. Evol.* **150**:65–81.

Wink, M., 1987b, Chemical ecology of quinolizidine alkaloids in: *Allelochemicals. Role in Agriculture and Forestry,* ACS Symp. Ser. 330 (G. R. Waller, ed.), American Chemical Society, Washington, DC, pp. 524–533.

Wink, M., and Hartmann, T., 1982, Localization of the enzymes of quinolizidine alkaloid biosynthesis in leaf chloroplasts of *Lupinus polyphyllus, Plant Physiol.* **70**:74–77.

Wink, M., and Römer, P., 1986, Acquired toxicity—The advantages of specializing on alkaloid-rich lupins to *Macrosiphum albifrons* (Aphidae), *Naturwissenschaften* **73**:210–212.

Wink, M., and Schneider, D., 1988, Carrier-mediated uptake of pyrrolizidine alkaloids in larvae of the aposematic and alkaloid-exploiting moth, *Creatonotos, Naturwissenschaften* **75**:524–525.

Wink, M., and Witte, L., 1984, Turnover and transport of quinolizidine alkaloids: Diurnal variation of lupanine in the phloem sap, leaves and fruits of *Lupinus albus* L., *Planta* **161**:519–524.

Wink, M., and Witte, L., 1985, Quinolizidine alkaloids as nitrogen source for lupin seedlings and cell suspension cultures, *Z. Naturforsch.* **40c**:767–775.

Wink, M., and Witte, L., 1991, Storage of quinolizidine alkaloids in *Macrosiphum albifrons* and *Aphis genistae* (Homoptera: Aphididae), *Entomol. Gener.* **15**:237–254.

Wink, M., Heinen, H. J., Vogt, H., and Schiebel, H. M., 1984, Cellular localization of quinolizidine alkaloids by laser desorption mass spectrometry (LAMMA 1000), *Plant Cell Rep.* **3**:230–233.

Wink, M., Schneider, D., and Witte, L., 1988, Biosynthesis of pyrrolizidine alkaloid-derived pheromones in the arctiid moth, *Creatonotos transiens:* Stereochemical conversion of heliotrine, *Z. Naturforsch.* **43c**:737–741.

Wink, M., Nickisch-Rosenegk, E. von, and Schneider, D., 1990, Processing of pyrrolizidine alkaloids and cardenolides in three moths, *Syntomis mogadorensis, Syntomeida epilais* and *Creatonotos transiens, Symp. Biol. Hung.* **39**:53–61.

Wink, M., Montplor, C., Bernays, E. A., and Witte, L., 1991, *Uresiphita reversalis* (Lepidoptera: Pyralidae): Carrier-mediated uptake and sequestration of quinolizidine alkaloids obtained from the host plant *Teline monspessulana, Z. Naturforsch.* **46c**:1080–1088.

Wink, M., Hofer, A., Bilfinger, M., Englert, E., Martin, M., and Schneider, D., 1993, Geese and plant dietary allelochemicals—Food palatability and geophagy, *Chemoecology* **4**:93–107.

Zippin, J., Mahaney, W. C., Milner, M. W., Sanmugadas, K., Hancock, R. G. V., Aufreiter, S., Campbell, S., Huffman, M. A., and Wink, M., Geochemistry and mineralogy of termite mound soil eaten by the chimpanzees of the Mahale mountains, Tanzania (to be published).

Chapter 12

Modes of Action of Alkaloids

Michael Wink

1. INTRODUCTION

We can safely assume that most alkaloids play an important role in the ecology of plants. In general, alkaloids serve as defense chemicals against herbivores and to a lesser degree against bacteria, fungi, and viruses or provide a means of interaction with other plants (see Chapters 13 and 14). A protective function has also been attributed to those alkaloids that are produced or sequestered by animals (see Chapters 15 and 16). In order to fulfill this function, alkaloids must closely interact with specific targets in herbivores, predators, microorganisms, or competing plants, i.e., they must either inhibit or otherwise deregulate important processes that are vital for these organisms. A thorough understanding of how these capabilities are effected is important for a comprehension of the evolutionary and ecological implications of alkaloids and their rational use in medicine or as natural pesticides in agriculture.

Whereas we know the structures of more than 12,000 individual alkaloids, our knowledge of their biological activities and functions is still rather limited. In this chapter I have tried to summarize and discuss the modes of action of the better known alkaloids, considering molecular targets first and then more complex interactions with organs or complete organisms. For overviews see Habermehl (1983), Harborne (1993), Luckner (1990), Mann (1992), Mothes *et al.* (1985), Mutschler (1981), Rimpler (1990), Robinson (1981), Rosenthal and Berenbaum (1991, 1992), Roth *et al.* (1994), Teuscher and Lindequist (1994), Wagner (1993), Waller (1987), and Wink (1992a,b, 1993a).

2. MOLECULAR TARGETS OF ALKALOIDS

In the following I have identified a number of important cellular molecular targets that are often affected by allelochemicals.

Michael Wink • Institute for Pharmaceutical Biology, University of Heidelberg, D-69120, Heidelberg, Germany.

Alkaloids: Biochemistry, Ecology, and Medicinal Applications, edited by Roberts and Wink. Plenum Press, New York, 1998.

2.1. Biomembranes

Cells can only operate if they are enclosed by an intact biomembrane and eukaryotic cells must be substructured by an internal membrane system providing a complex compartmentation, i.e., separated reaction chambers. As biomembranes are almost impermeable for ions and polar molecules, cells can prevent the uncontrolled flux or migration of essential metabolites. The controlled flux of these compounds across biomembranes is achieved by specific transport proteins, which can be ion channels, pores, or carrier proteins. If the transport of metabolites has to proceed against a concentration gradient, transport processes must be either directly or indirectly energized (Alberts *et al.,* 1994; Lodish *et al.,* 1995). These complex transport systems are targets of many natural products.

Steroidal alkaloids, such as solanine and tomatine which are present in many members of the Solanaceae, can form complexes with the cholesterol present in biomembranes. Important for this interaction is the presence of a lipophilic portion of the molecule (given by the steroidal moiety) and a hydrophilic portion (provided by the sugar side chain). Whereas the lipophilic moiety "dives" into the lipophilic interior of the membrane and interacts with the structurally similar cholesterol, the hydrophilic side chain remains outside and binds to external sugar receptors. As phospholipids are in continuous motion, a tension easily builds up which leads to membrane disruption, i.e., transient "holes" occur in the biomembrane rendering the cell leaky. A similar mechanism has been postulated for saponins, a widely distributed group of natural products, to which the steroidal alkaloids may be assigned (Fig. 1). Steroidal alkaloids can also interact with other targets, such as receptors.

The alkaloids tetrandrine, berbamine, and cepharanthine have also been reported to interfere with membrane integrity (Wink, 1993a).

2.2. Signal Transduction at Biomembranes

Cells carefully control the homeostasis of their ion concentrations through the action of specific ion channels (e.g., Na^+, K^+, Ca^{2+}, and Cl^- channels) and of active Na^+, K^+, or Ca^{2+} pumps, such as Na^+/K^+-ATPase and Ca^{2+}-ATPase (for an overview see Alberts *et al.,* 1994). Ion gradients and ion fluxes mediated by these channels and pumps are the main elements in active transport processes, and in neuronal and neuromuscular signaling.

2.2.1. Na^+/K^+-ATPase

Cardiac glycosides, found in plants, some insects, and in the skin of toads (Bufonidae), are potent and well-known inhibitors of Na^+/K^+-ATPase. A few alkaloids—harmaline, nitidine, sanguinarine, capsaicin, cassaine, and solenopsine (from ants)—exhibit Na^+/K^+-ATPase inhibition.

2.2.2. NEUROTRANSMITTER RECEPTORS

Signal transduction in the central nervous system and in neuromuscular junctions is mediated by receptor proteins residing in the membrane which are directly or indirectly

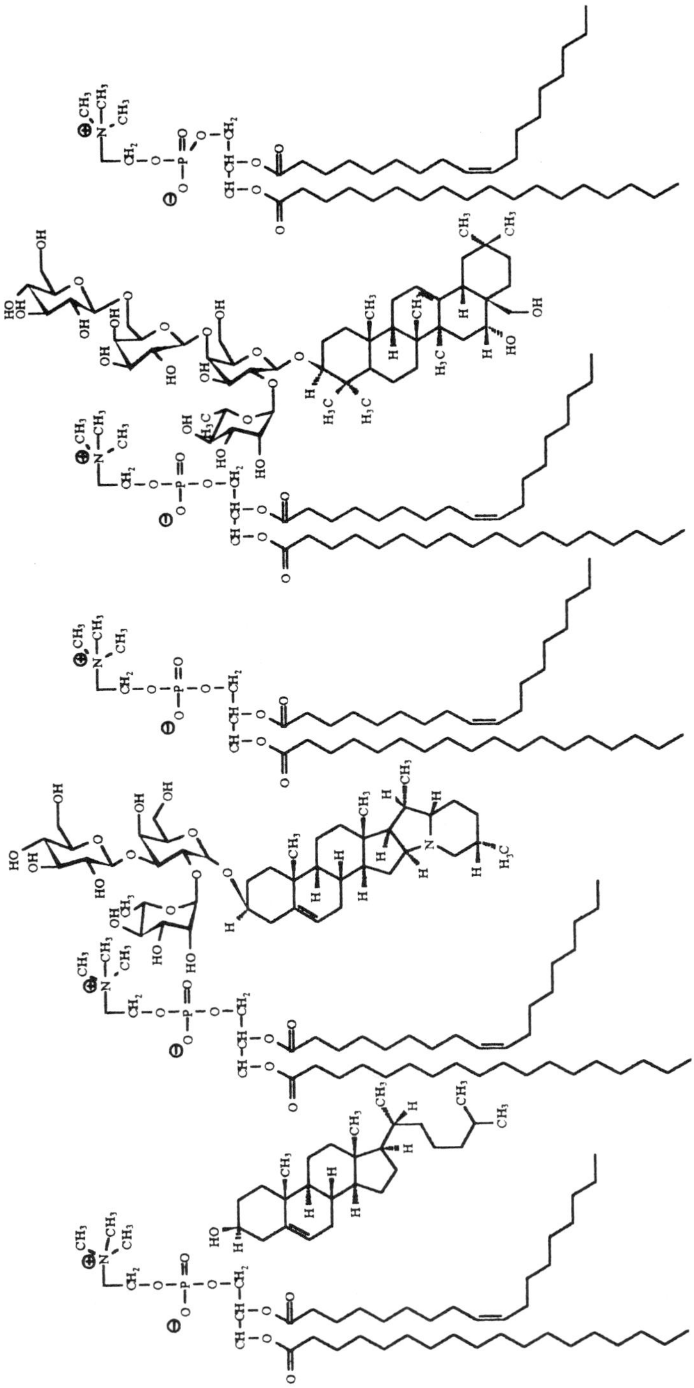

Figure 1. Possible interaction of saponins and steroidal alkaloids with membrane cholesterol.

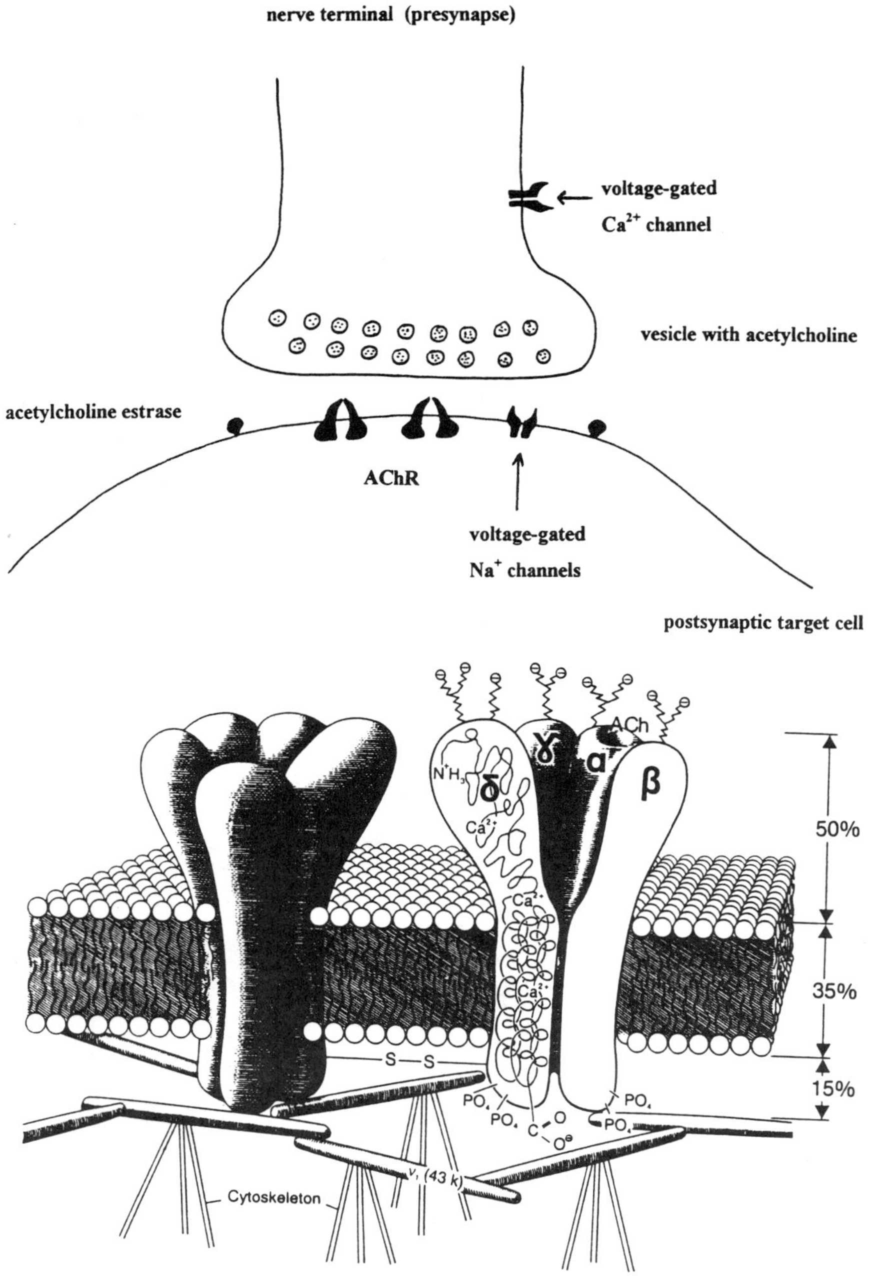

Figure 2. Signal transduction in excitable synapses. (Bottom) Nicotinic ACh receptor; (top) schematic view of a chemical synapse. Events during neurotransmission: When a nerve impulse arrives and depolarizes the plasma membrane, voltage-gated Ca^{2+} channels open and Ca^{2+} enters the presynapse. Increased Ca^{2+} concentrations activate the synaptic vesicles (by a complicated interaction of several proteins) so that they can fuse with the presynaptic membrane by exocytosis. Thus, acetylcholine is released into the synaptic cleft and binds to AChR on the postsynaptic membrane. The ligand-activated Na^{+} channel transiently opens. Inflow of Na^{+} depolarizes locally the membrane and opens voltage-gated Na^{+} channels. This in turn opens more Na^{+} channels, allowing more Na^{+} to enter, which further polarizes the membrane. In nerve cells this mechanism produces a self-propagating depolarization (action potential), whereas at the neuromuscular junction, the depolarization induces Ca^{2+} channels of the sarcoplasmatic reticulum to open transiently and to release Ca^{2+} into the cytoplasm. The increase of Ca^{2+} causes the myofibrils in the muscle cell to contract.

coupled with ion channels. The neurotransmitters involved include, among others, noradrenaline (NA), serotonin, dopamine, histamine, glycine, GABA, and acetylcholine (ACh). The following mechanisms have been deduced for the direct and indirect channel systems, which are very similar across a wide range of animals (Fig. 2):

Type I is a ligand-gated channel, i.e., a receptor that is part of an ion-channel complex. The well-studied nicotinic ACh receptor consists of five subunits, two of them binding ACh. When the neurotransmitter binds, a conformational change induces the opening of a Na^+/K^+ channel for microseconds, allowing Na^+ ions (the external concentration is about 145 mM) to enter the cell following a concentration gradient (the internal Na^+ concentration is between 5 and 15 mM). The ligand quickly dissociates from the receptor and in the case of ACh is hydrolyzed by ACh esterase.

In Type II the muscarinic ACh receptor is an integral protein. When ACh binds, the receptor changes its conformation, inducing a conformational change in an adjacent G-protein molecule, consisting of three subunits. The α-subunit dissociates and then activates the enzyme adenylyl cyclase, which in turn produces cAMP from ATP. The cAMP molecule, a second messenger, activates protein kinases or Ca^{2+} channels directly. Whereas ACh is degraded in the synaptic gap, the biogenic amines are taken up by the presynaptic membrane and consequently by the synaptic vesicles. These uptake processes involve transport proteins.

Nicotinic ACh receptors belong to Type I (Fig. 2), while the muscarinic ACh receptor, noradrenaline, serotonin, and dopamine receptors belong to Type II. The family of GABA receptors, some of which are important for memory storage in the brain, resemble Type I systems.

Quite a number of alkaloids are known whose structures are more or less similar to those of endogenous neurotransmitters and can function as structural analogues (review in Wink, 1993a). In addition, several plants produce compounds that are identical to animal neurotransmitters, such as acetylcholine and histamine in stinging hairs of *Urtica,* or serotonin and dopamine in several species. Targets can be:

1. The receptor itself (Table I) through inhibition or overstimulation
2. The enzymes that deactivate neurotransmitters after they have bound to a receptor (Table II)
3. Transport processes (Table III), which are important for the storage of the neurotransmitters in synaptic vesicles
4. Enzymes involved in the biosynthesis of a neurotransmitter

2.2.3. VOLTAGE-GATED Na^+/K^+ CHANNELS

The stimulation of neurotransmitter-activated ion channels leads to a rapid influx of Na^+ ions which in turn activates voltage-gated Na^+ and K^+ channels essential for further signal transduction. These Na^+ and K^+ channels constitute another important target for alkaloids (Table IV).

2.2.4. KEY ENZYMES OF SIGNAL PATHWAYS

Whereas receptor/ion channel interactions represent the initial part of many signal pathways, key enzymes that produce or inactivate second messengers or amplify the

Table I
Examples of Alkaloids That Bind to Neurotransmitter Receptors, Either Activating or Inhibiting Them[a]

Receptor	Ligand	Alkaloid	Occurrence
1. Acetylcholine receptor			
Nicotinic receptor	Acetylcholine	Nicotine	*Nicotiana*, many other plants
		C-toxiferine	*Strychnos*
		Tubocurarine	*Chondodendron*
		Coniine	*Conium*
		Cytisine and other QA	Several legumes
		Lobeline	*Lobelia*
		Anabasine	*Anabasis, Nicotiana*
		Histrionicotoxin	Frogs
Muscarinic receptor	Acetylcholine	Hyoscyamine (atropine)	*Atropa, Hyoscyamus, Datura, Mandragora,* etc.
		Scopolamine	Several Solanaceae
		Arecoline	*Areca*
		Pilocarpine	*Pilocarpus*
		Muscarine	*Amanita, Inocybe, Clitocybe,* and other fungi
		Sparteine and other QA	Several legumes
2. Adrenergic receptors	Noradrenaline/adrenaline	Ergot alkaloids	*Claviceps*
		Yohimbine	*Pausinystalia*
		Yohimba	*Aspidosperma*
		Rauwolscine	*Rauwolfia*
		Corynanthine	*Rauwolfia*
		Norlaudanosoline	Papaveraceae
		Ephedrine, norephedrine	*Ephedra*

3. Serotonin receptor	Serotonin	Ergot alkaloids	*Claviceps*
		Psilocin, psilocybine	*Psilocybe* and other fungi
		N,N-Dimethyltryptamine	Several plants, and toads
		Bufotenine	*Virola, Anadenanthera*
		β-Carboline alkaloids	*Banisteriopsis, Peganum*
		Mescaline	*Lophophora* and other cacti
4. Dopamine receptor	Dopamine	Ergot alkaloids	*Claviceps*
		Bulbocapnine	*Corydalis*
5. GABA receptors	GABA	Bicuculline	*Dicentra cucullaria* and other *Corydalis* species
		Corlumine	
		Nuciferine	
		Muscimol	*Amanita*
		β-Carboline alkaloids	*Peganum, Banisteriopsis*
		Argiotoxin 636	*Argiope* and other arthropods
6. Adenosine receptor	Adenosine	Caffeine	*Coffea, Camellia, Ilex, Paullinia*
		Theophylline, theobromine	*Theobroma*
7. Glycine receptor	Glycine	Brucine	*Strychnos*
		Strychnine	*Strychnos*
		Laudanosine	
		Dendrobine	
8. Opioid receptor	Endorphins	Morphine	*Papaver somniferum*

[a]More details in Wink (1993a).

Table II
Alkaloids as Inhibitors of Neurotransmitter-Degrading Enzymes

Enzyme	Natural substrate	Alkaloid	Occurrence
Acetylcholine esterase	Acetylcholine	Physostigmine (eserine)	*Physostigma venenosum*
		Berberine	Several Papaveraceae
		Coptisine	Several Papaveraceae
		Coralyne	
		Galanthamine	Several Amaryllidaceae
		Chaconine	*Solanum*
		Demissine	*Solanum*
		Solarmargine	*Solanum*
		Solanine	*Solanum*
		Solanidine	*Solanum*
		Huperzine A	*Huperzia serrata*
Monoamine oxidase (MAO)	NA, dopamine, serotonin, histamine	Harmaline	*Peganum*
		Harmine	*Peganum*
		Tetrahydro-β-carboline	*Peganum*
		Salsolinol	
		Ephedrine	*Ephedra*
Catechol-*O*-methyltransferase	NA, adrenaline, dopamine	Coralyne	
		Tetrahydroisoquinoline	

[a]More details in Wink (1993).

Table III
Alkaloids as Inhibitors of Neurotransmitter Uptake (Transport into Vesicles)[a]

Transporter	Alkaloid	Occurrence
Noradrenaline	Reserpine	*Rauwolfia*
	Ephedrine	*Ephedra*
Biogenic amines	Tetrahydro-β-carboline	*Peganum*
	Salsolinol	*Salsola*
	Stepholidine	
	Tetrahydroisoquinoline	
	Tetrahydropalmatine	
Dopamine	Cocaine	*Erythroxylum*

[a]More details in Wink (1993a).

Table IV
Alkaloids as Modulators of Na^+, K^+, and Ca^{2+} Channels[a]

Alkaloid	Occurrence (genera)	Action
Na^+ and K^+ channels		
Aconitine[b]	*Aconitum*	Activation
Sparteine[b]	*Cytisus, Lupinus, Genista*	Inhibition
Quinine	*Cinchona*	Inhibition
Quinidine[b]	*Cinchona*	Inhibition
Ajmaline[b]	*Rauwolfia*	Inhibition
Harmaline	*Peganum*	Inhibition
Protoveratrine A, B[b]	*Veratrum*	Activation
Veratridine[b]	*Veratrum*	Activation
Batrachotoxin[b]	Frogs (Dendrobatidae)	Activation
Saxitoxin[b]	*Protogonyaulax* (algae)	Inhibition
Tetrodotoxin[b]	Algae/fish	Inhibition
Ca^{2+} channels		
Ryanodine	*Ryania speciosa*	Inhibition
Bastadin 5	*Ianthella basta*	Inhibition

[a]More details in Wink (1993a).
[b]Na^+ channel.

signal can be important targets further down the pathway. These enzymes include (Fig. 3):

- Adenylyl cyclase (making cAMP)
- Phosphodiesterase (inactivating cAMP)
- Phospholipase (releasing arachidonic acid or inositol phosphates)
- Several protein kinases, such as protein kinase C (which is activated by phorbol esters and the alkaloid chelerythrine) or tyrosine kinase (activating other regulatory proteins or ion channels)

Table V lists some alkaloids that interfere with these targets.

2.3. Cytoskeleton

Many cellular activities, such as motility, endo- and exocytosis, and cell division, are mediated through elements of the cytoskeleton, including microfilaments and microtubules (for an overview see Alberts *et al.,* 1994). A number of alkaloids identified in plants and fungi can interfere with them (among others: colchicine, *Vinca* alkaloids, maytansine, maytansinine, and taxol). Any alkaloid that impairs the function of microtubules or microfilaments is likely to be toxic, and from the point of view of defense, a well-working and well-shaped molecule.

2.3.1. MICROTUBULES

Microtubules, which are important for cellular movements, vesicle transport in neurons, or the separation of chromosomes during cell division, are composed of tubulin

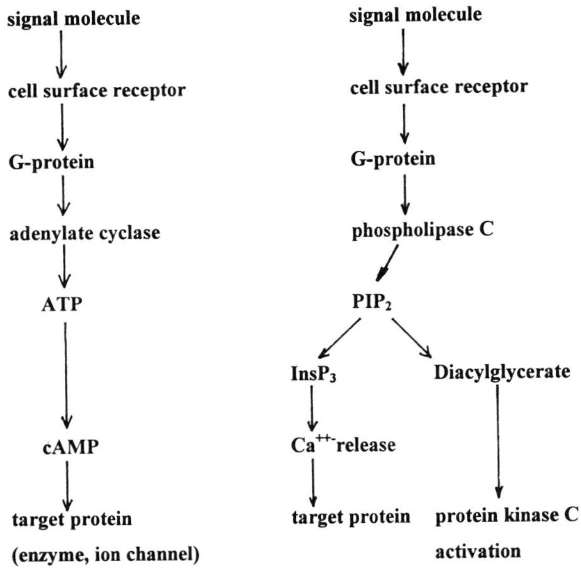

Figure 3. Signal pathway in animal cells. (Left) cAMP pathway; (right) phosphoinositol/Ca^{2+} pathway. PIP_2, phosphoinositol 4,5-bisphosphate; $InsP_3$, inositol 1,4,5-trisphosphate.

subunits. Movements and some transport processes (e.g., that of vesicles) are mediated through either the rapid assembly or disassembly of microtubules.

2.3.2. COLCHICINE

Colchicine, the major alkaloid of *Colchicum autumnale* (Liliaceae), binds tightly to tubulin (1 : 1 ratio) and thus inhibits the assembly of microtubules. As a consequence the

Table V
Alkaloids That Modulate Enzymes Involved in Signal Transduction[a]

Enzyme	Function	Alkaloid	Occurrence
Adenylyl cyclase	cAMP formation	Anonaine	
		β-Carboline-1-propionic acid	
		Isoboldine	*Peumus*
		Tetrahydroberberine	
Phosphodiesterase	cAMP inactivation	Papaverine	*Papaver*
		Caffeine, theobromine	*Coffea, Camellia, Theobroma*
		Theophylline	*Ilex paraguarensis, Paullinia*
		1-Ethyl-β-carboline	
Protein kinases	Protein phosphorylation		
		Anisomycin	*Streptomyces griseolus*
		Chelerythrine	*Chelidonium majus*
		Lyngbyatoxin A	Marine seaweeds
		Telocidin	*Streptomyces blastmyceticum*

[a]More details in Wink (1993a).

mitotic spindle of dividing cells disappears rapidly after colchicine treatment and chromatids are no longer separated. Whereas animal cells die under these conditions, plant cells become polyploid, a trait often used in plant breeding, because polyploidy leads to bigger plants.

Because of its antimitotic activity, colchicine has been tested as an anticancer drug, but has been abandoned because of its general toxicity; however, a derivative, colcemide, is less toxic and can be employed in the treatment of certain cancers. Cellular motility is impaired by colchicine. This property is exploited in the treatment of acute gout, in order to prevent the migration of macrophages to the joints (see Chapter 18). Colchicine is indeed a very toxic alkaloid which is easily resorbed because of its lipophilicity; therefore, it is not surprising that *Colchicum* plants are not attacked by herbivores to any substantial degree.

2.3.3. DIMERIC INDOLE ALKALOIDS

Another group of alkaloids with antimitotic properties are the dimeric monoterpene-indole alkaloids, such as vinblastine and vincristine, which have been isolated from *Catharanthus roseus* (Apocynaceae). These alkaloids also bind to tubulin and induce the formation of paracrystalline protein aggregates leading to microtubule depolymerization. The inhibition of cell division is similar to that described for colchicine. Both alkaloids are rather toxic but are nevertheless important antimitotic drugs for the treatment of some leukaemias and carcinomas (see Chapter 18).

2.3.4. TAXOL

From several *Taxus* species, such as *T. baccata* and *T. brevifolia* (Taxaceae), the alkaloid taxol has been isolated which also affects the architecture of microtubules, but in contrast to the compounds mentioned previously, it stabilizes them. The polymerization of tubulin is enhanced by taxol and becomes independent of GTP and microtubule-associated proteins (MAPs). The diameter of taxol-induced microtubules is 22 nm (in contrast to 24 nm for "normal" microtubules) and consists of 12 instead of 13 protofilaments. Taxol remains bound to tubulin in a ratio of 1:1. As a consequence, taxol-induced microtubules are very stable which arrests dividing cells in mitosis (overview by Reynolds, 1993). Taxol is a new antimitotic drug used in the treatment of ovarian and breast cancer (see Chapter 18).

2.3.5. MICROFILAMENTS

Cell stability, phagocytosis, cell–cell interactions, and cell movements are also controlled by actin filaments, which are rapidly assembled or disassembled from actin monomers. Cytochalasin B, an alkaloid produced by a number of molds, binds to the plus end of a growing actin filament, preventing the addition of actin monomers there. Latrunculin B from *Latrunculia magnifica* (a marine organism) is 10- to 100-fold more potent than cytochalasins in the inhibition of microfilament organization. Another nitrogenous compound, phalloidin, produced by the fatally poisonous toadstool *Amanita phalloides,* stabilizes actin filaments and inhibits their depolymerization.

Table VI
Alkaloids Interacting with DNA/RNA and Related Enzymes[a]

Target	Function	Alkaloid	Occurrence
DNA	Photoaddition	Dictamnine	*Dictamnus*
		Harman	*Peganum*
		Harmine	*Peganum*
	Alkylation	Pyrrolizidine alkaloids	Several Asteraceae, Boraginaceae
		Aristolochic acid	*Aristolochia*
		Cycasin	*Cycas*
	Intercalation	ellipticine	
		9-Methoxyellipticine	
		Quinine	*Cinchona*
		Skimmianine	
		Avicine	
		Berberine	*Berberis, Mahonia, Thalictrum, Chelidonium*
		Chelerythrine	*Chelidonium*
		Coptisine	Several Papaveraceae
		Coralyne	
		Fagaronine	
		Nitidine	
		Sanguinarine	Several Papaveraceae
		Olivacine	
DNA polymerase	Inhibition	Avicine	
		Coralyne	
		Fagaronine	
		Nitidine	
		Hippeastrine	
		Lycorine	Several Amaryllidaceae
DNA topoisomerase I	Inhibition	Camptothecine	*Camptotheca acuminata*
Reverse transcriptase	Inhibition	Berberine	Several Berberidaceae, Papaveraceae
		Chelidonine	*Chelidonium*
		Coralyne	
RNA polymerase	Inhibition	Vincristine, vinblastine	*Catharanthus roseus*
Transcription	Inhibition	Colchicine	*Colchicum, Gloriosa*
		Amanitin	*Amanita*

[a]More details in Wink (1993a).

2.4. DNA/RNA

The genetic information of most organisms is encrypted in DNA (some viruses have RNA in their genome). As DNA encodes all RNAs, proteins, and enzymes that are important for metabolism and development of an organism, DNA is a highly vulnerable target. It is not surprising that a number of secondary metabolites have been selected during evolution which interact with DNA or DNA-processing enzymes. Some alkaloids are known to bind or to intercalate with DNA (Table VI and Fig. 4). Many of these molecules are planar, hydrophobic molecules which fit between the planar stacks of AT and GC base pairs. Other alkaloids act on the level of DNA polymerases and RNA polymerases (Table VI), thus impairing the process of replication and transcription.

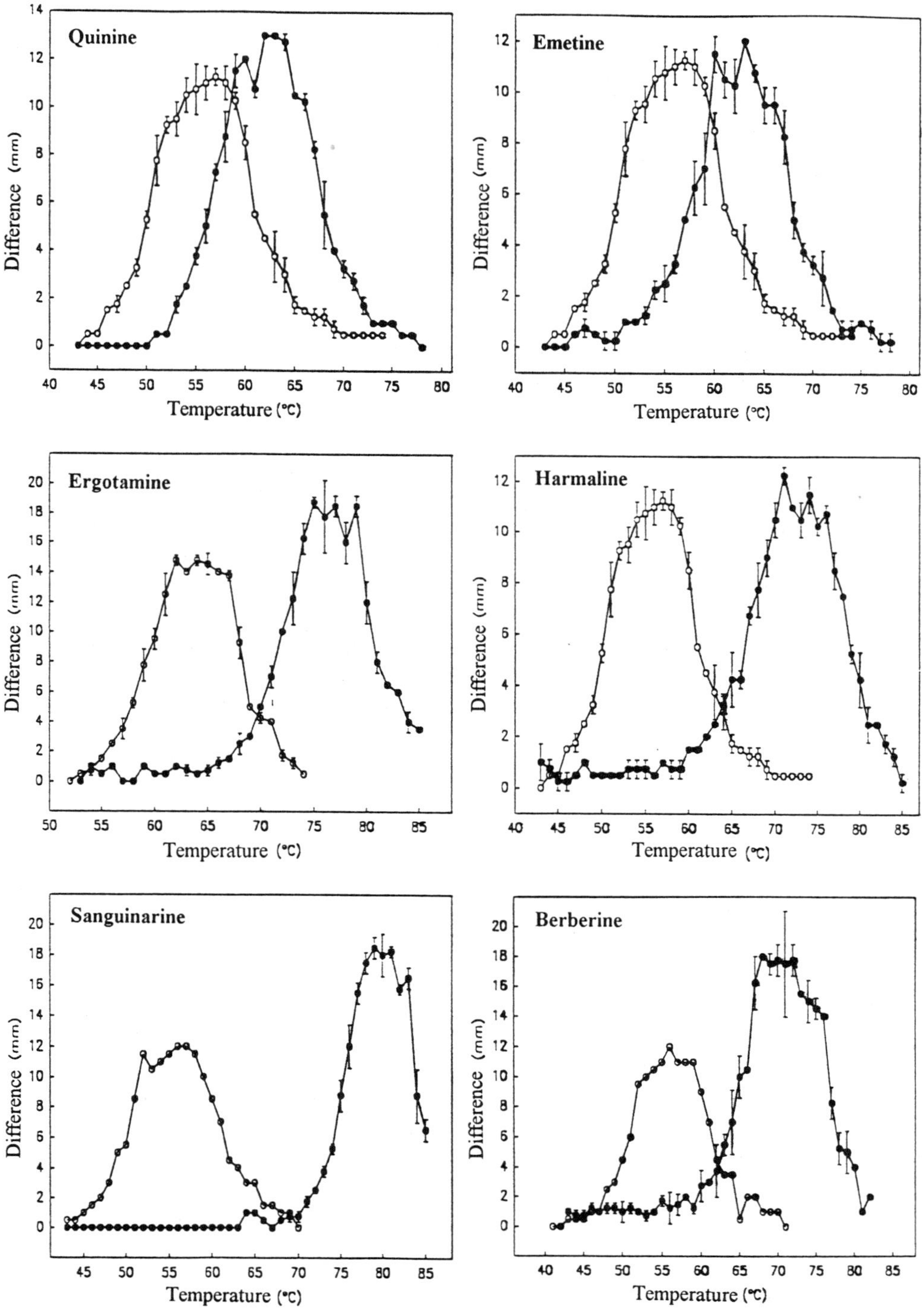

Figure 4. Intercalation of alkaloids with DNA (after Latz-Brüning and Wink, unpublished). DNA was measured at 260 nm in a spectrophotometer with (●) or without (○) alkaloids. The temperature of the solution was increased from 40 to 90°C at 1°C/min. Melting profiles were shifted to higher temperatures when alkaloids were intercalated with DNA.

The effects of DNA-binding or intercalating compounds can be mutations, which may result in malformations of newborn animals or in the initiation of cancer. In the following a few mutagenic alkaloids will be considered.

When anabasine, coniine, or anagyrine is administered to pregnant cows or sheep, a large proportion of the offspring develop malformations of the legs, the so-called "crooked calf disease" (reviewed in Wink, 1993a,b). Some alkaloids of the monocot *Veratrum,* such as jervine and cyclopamine, cause the formation of a large central eye, the cyclopean eye, which was probably known to the ancient Greeks and thus led to the mythical figure of the cyclops (see Chapter 4)

Other alkaloids are known as carcinogens, such as aristolochic acid from *Aristolochia* and pyrrolizidine alkaloids (PA) which are produced by approximately 3% of the higher plants, especially within the families of Asteraceae and Boraginaceae. Aristolochic acid has a nitro group which can be transformed into reactive intermediates in the intestine. If resorbed, these metabolites can alkylate DNA. PA are not carcinogenic in their native form, but become so when they are "detoxified" in the liver. As can be seen in Fig. 5, PA are usually present in the plant as their *N*-oxides, which are polar compounds that cannot pass biomembranes by simple diffusion. In the intestine, PA *N*-oxides are reduced by gut bacteria. The free base is readily taken up by the gut cells and transported to the liver. There, the PA are transformed into alkylating compounds, which covalently bind to DNA. As a result, mutations and cancer can be initiated. The PA story is very intriguing, as it shows how ingenious nature was in the "arms race": The herbivores invented detoxifying enzymes and the plants the compounds that are activated by this process. A herbivore feeding on PA-containing plants will eventually die, usually without reproducing properly. Only those individuals that carefully avoid the respective bitter-tasting plants maintain their fitness and will thus survive. The protection resulting from PA can easily be seen on meadows, where *Senecio* and other PA-containing plants are usually not taken by cows and sheep, at least, as long as other food is available.

2.5. Protein Biosynthesis

Protein biosynthesis is essential for all cells and thus provides another important target. Indeed, a number of alkaloids have been detected (although only a few have been studied in this context) that inhibit protein biosynthesis *in vitro,* emetine from *Cephaelis ipecacuanha* (Rubiaceae) is the most potent. Other alkaloids with the same ability include harringtonine, homoharringtonine, cryptopleurine, tubulosine, hemanthamine, lycorine, narciclasine, pretazettine, pseudolycorine, tylocrepine, and tylopherine.

Quinolizidine alkaloids, such as sparteine, lupanine, and cytisine, are relatively weak inhibitors at this target (they strongly affect ACh receptors and Na^+ and K^+ channels; see above). The stages that are inhibited are the loading of aminoacyl-tRNA with amino acids and the elongation step. The inhibitory activity was visible in heterologous systems, but protein biosynthesis in the producing plants (here lupins) was not affected. A number of antibiotics (from *Streptomyces* and other bacteria or fungi) are known that inhibit protein biosynthesis at specific steps, such as (1) initiation, (2) peptidyltransferase, or (3) elongation (Table VII). Depending on their affinity for prokaryotic or eukaryotic ribosomes, some of the antibiotics selectively inhibit microbial systems. As mitochondria also contain

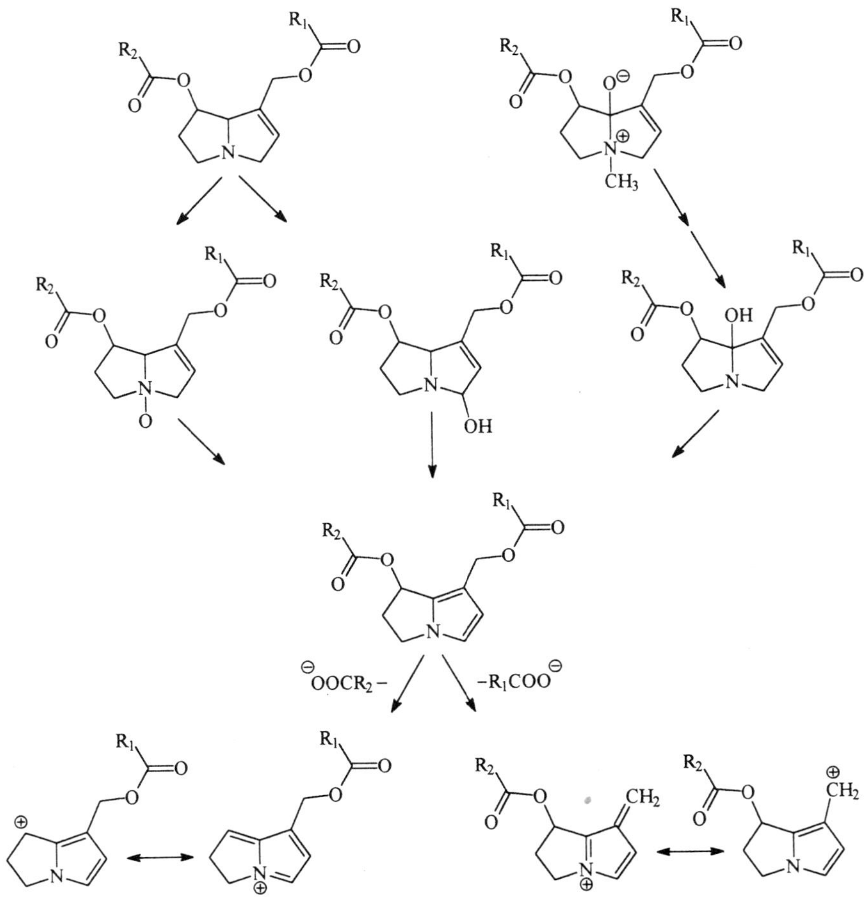

Figure 5. Conversion of pyrrolizidine alkaloids to alkylating agents in the liver.

Table VII
Inhibition of Protein Biosynthesis by Some Microbial Alkaloids ("Antibiotics")[a]

Site of inhibition of protein biosynthesis	System[b]	Compound
Blocking aa tRNA acceptor site	P/E	Tetracycline
	P	Lincomycin
Transition from initiation to chain elongation	P	Streptomycin
	P	Neomycin
	P	Kanamycin
	P	Gentamycin
Peptidyltransferase	P	Chloramphenicol
	E	Anisomycin
Translocation	P	Erythromycin
	P	Spiramycin
	E	Cycloheximide

[a]More details in Wink (1993a).
[b]P, procaryote; E, eucaryote.

Table VIII
Alkaloids That Modulate Enzyme Activity[a]

Alkaloid	Enzyme	Activity
Brucine	Lactate dehydrogenase	Inhibition
Strychnine	Lactate dehydrogenase	Inhibition
Ellipticine	Cytochrome oxidase	Inhibition
Berberine	Several enzymes	Inhibition
Canadine	Aldose reductase	Inhibition
Chelerythrine	Several enzymes	Inhibition
Castanospermine	Several hydrolases	Inhibition
Deoxynorjirimycin	Several hydrolases	Inhibition
Swainsonine	Several hydrolases	Inhibition
Ochratoxin	Glucose transport	Inhibition
Folimycin	Vacuolar H^+-ATPase	Inhibition
Calyculin A	Phosphatase (PP-1)	Inhibition

[a]More details in Wink (1993a).

ribosomes of prokaryotic origins, side effects can occur. Some of these compounds contain a nitrogen and could also be classified as alkaloids.

2.6. Electron Chains

The respiratory chain and ATP synthesis in mitochondria or photophosphorylation in chloroplasts demand the controlled flux of electrons. These targets seem to be attacked by sanguinarine, ellipticine, gramine, alpinigenine, capsaicine, and a few other alkaloids. But this activity may have been overlooked because, as has been mentioned before, only a few alkaloids have been checked in depth.

2.7. Modulation of Enzyme Activity through Alkaloids

A multitude of enzymes exist in animal cells and several alkaloids have been reported that interfere with at least one of them. A small selection of interactions is illustrated in Table VIII (see also Tables II and V).

2.8. Alkaloids Affecting More than One Target

In general, the interactions of a particular alkaloid with a molecular target (as described above) suggest a high degree of specificity. A closer look, however, shows that many alkaloids interfere with more than one target. The phenomenon will be explained for two groups of alkaloids:

2.8.1. ERGOT ALKALOIDS

Ergot alkaloids, such as ergotamine, ergometrine, or ergoclavine, are produced by fungi of the genus *Claviceps* which lives in close contact with many grasses (family

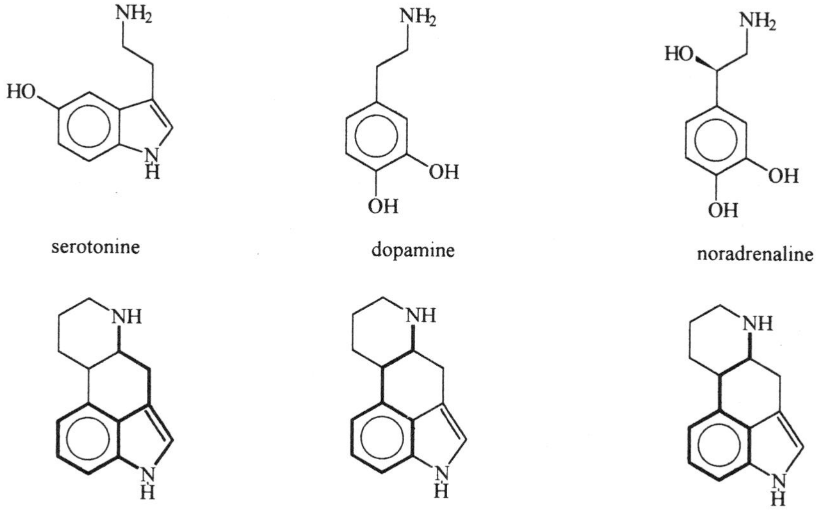

Figure 6. Structure–function relationships of ergot alkaloids with the neurotransmitters noradrenaline, dopamine, and serotonin.

Poaceae) such as the cereal *Hordeum vulgare.* These alkaloids can modulate several receptors of neurotransmitters, such as dopamine, serotonin, and noradrenaline. As a consequence, the pharmacological action of ergot alkaloids is rather broad, ranging from vasoconstriction and uterine contraction to hallucinations. We can explain these activities of alkaloids through structural similarities with the different neurotransmitters (Fig. 6). As explained in Chapter 11, it has been suggested that the interactions between *Claviceps* and its host plant are of a symbiotic nature, i.e., infected plants exploit the chemistry of the fungus for their own protection against herbivores (otherwise it would be difficult to explain why a fungal metabolite should interfere with targets that are only present in animals).

2.8.2. QUINOLIZIDINE ALKALOIDS

Quinolizidine alkaloids (QA), such as lupanine, sparteine, or cytisine, are produced by many members of the Leguminosae. QA are bitter for many animals (and plants producing them are therefore avoided as food). If ingested, QA exhibit a broad level of toxicity: They interact with ACh receptors as agonists. QA, like many other alkaloids, occur as complex mixtures in plants. We have shown recently (Schmeller *et al.,* 1994) that some QA preferentially bind to the nicotinic AChR, whereas others reveal a stronger binding to the muscarinic AChR (Table IX). Some QA exhibit a prominent cross-reactivity. Additionally, QA such as lupanine and sparteine inhibit Na^+ and K^+ channels, thus blocking the signal transduction in nerve cells at a second critical point. As mentioned above, QA slightly interfere with protein biosynthesis. A few QA, such as anagyrine, cytisine, and the bipiperidine alkaloid ammodendrine (which co-occurs with QA in many plants), are mutagenic and lead to malformations (see above). These results suggest that QA are indeed defense chemicals with a broad range of targets which might be affected simultaneously.

Table IX
Binding of Quinolizidine Alkaloids to Nicotinic and Muscarinic AChR[a]

	Alkaloid	n-AChR	m-AChR
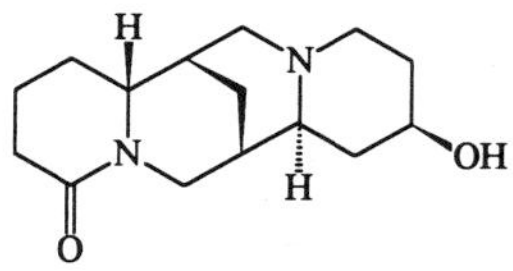	13-Hydroxylupanine	467.2	139.7
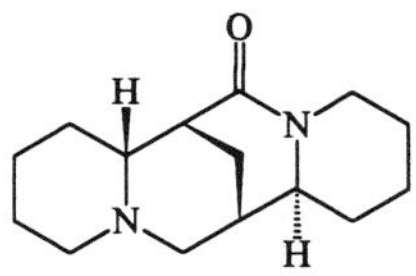	17-Oxosparteine	155.0	117.9
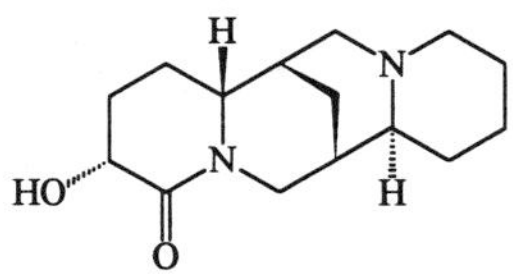	3-Hydroxylupanine	192.4	74.1
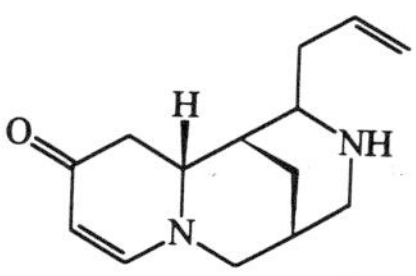	Albine	237.7	32.9
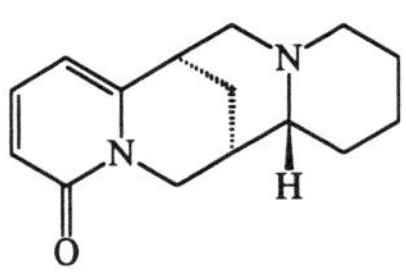	Anagyrine	2095.8	132.1
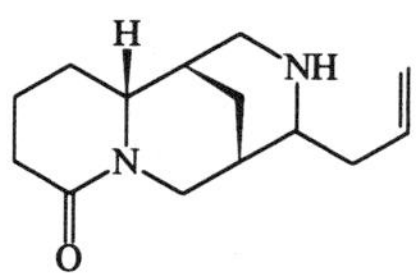	Angustifoline	>500	25.3
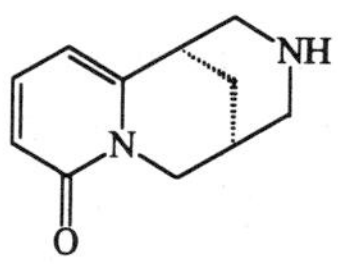	Cytisine	0.137	398.2
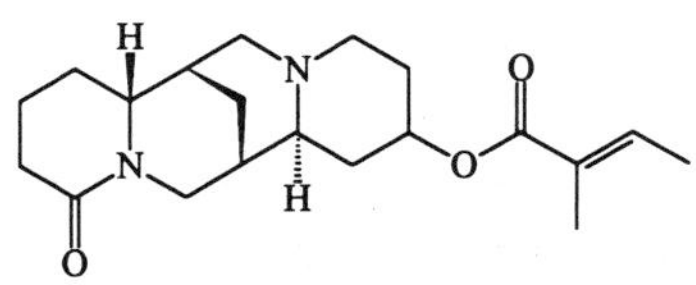	13-Tigloyloxylupanine	99.8	11.1

Table IX
(*Continued*)

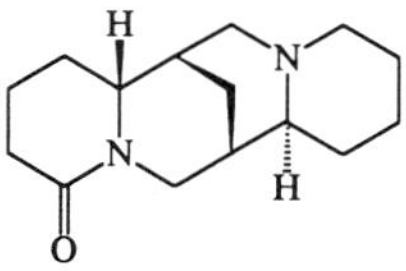	Lupanine	5.3	118.0
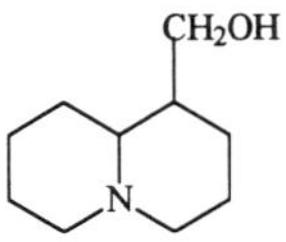	Lupinine	>500	189.9
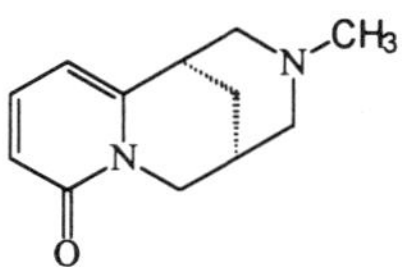	*N*-Methlcytisine	0.051	416.7
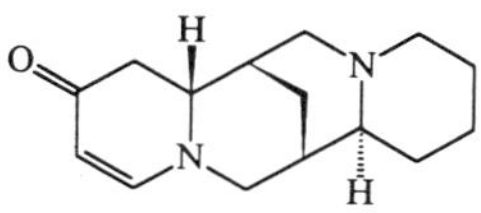	Multiflorine	>500	49.4
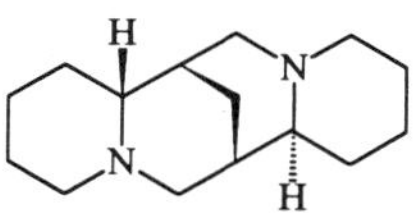	Sparteine	330.8	21.3
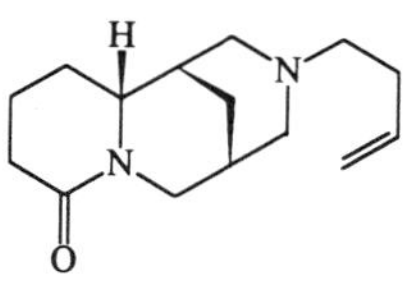	Tetrahydrorhomibifoline	347.6	128.8

[a]IC_{50} (in μM) values indicate the concentration of a particular alkaloid that displaces 50% of the specifically bound radiolabeled ligand. After Schmeller *et al.* (1994).

If we accept the hypothesis that alkaloids were developed as chemical defense compounds through a process of "evolutionary molecular modeling," the "cross-reactivity" described makes sense: Any compound that can interfere with more than one target or with more than one group of adverse organisms is likely to be more toxic and thus has a better survival value in general than a more selective allelochemical. In addition, herbivores will try to develop tolerance or resistance against the dietary toxins. For example, in the monarch butterfly (*Danaus plexippus*) which sequesters cardiac glycosides, the corresponding Na^+/K^+-ATPase has become insensitive to cardiac glycosides through a point mutation in the ouabain binding site (Holzinger *et al.,* 1992). If more than one target is affected by a defense chemical, the chances of a herbivore developing such point mutations concomitantly are much smaller than in single-target situations. In conclusion, we can say that nature has obviously tried "to catch as many flies with one clap as possible" in the selection of alkaloids during evolution.

3. TARGETS AT THE ORGAN LEVEL

Whereas the activities mentioned previously were more or less directed to molecular targets present in or on cells, we can also see some activities that are oriented against organ systems or complete organisms, although ultimately, they have molecular targets, too. In some cases, only the toxicity of an alkaloid has been reported (Table X) evidencing substantial interactions, but the exact mode of action has not yet been elucidated or is rather complex, involving several targets and organs.

3.1. Central Nervous System and Neuromuscular Junction

A remarkable number of alkaloids interfere with the metabolism and activity of neurotransmitters in the brain and nerve cells, a fact known to man for some thousands of years (see Chapter 4). The cellular interactions have been discussed in Section 2 above. A disturbance of the metabolism and binding of neurotransmitters and related signal pathways impairs learning and memory and sensory faculties (smell, vision, or hearing) or produces euphoric or hallucinogenic effects.

An animal that is no longer able to control its movements and senses properly has only a small chance of survival in nature, because it will have accidents (falling from trees or rocks or into water) or be killed by predators. Thus, euphoric and hallucinogenic compounds, which are present in a number of plants but also in fungi and the skin of toads, can be regarded as potent defense compounds. *Homo sapiens* has used and still uses these drugs for their hallucinogenic properties, but here, also, it is evident that long-term use reduces survival and fitness dramatically (see Chapter 4). Muscle activity (e.g., skeletal, heart) is controlled by ACh and NA. It is plausible that any inhibition or overstimulation of neurotransmitter-regulated ion channels will severely influence muscule activity and thus the mobility or organ function (heart, lungs, gut) of an animal. In the case of inhibition, muscles will relax, and in the case of overstimulation, muscles will be tense or in tetanus, leading to a general paralysis (which is the effect of many of the more toxic alkaloids; Table X and Chapter 4).

Alkaloids that activate (so-called parasympathomimetics) or inhibit (parasympatholytics) neuromuscular action are shown in Table I. These compounds are usually considered to be strong poisons (Table X) and it is obvious that they serve as chemical defense compounds against herbivores, for a paralyzed or anesthetized animal is an easy prey for predators. If higher doses of these alkaloids are ingested, the animal will die as a direct result of the alkaloid (see LD_{50} values in Table X). Skeletal muscles, muscle-containing organs, such as lungs, heart, and gut, and the circulatory system and the nervous system are certainly very potent targets for antiherbivore compounds although other organs have also been selected as targets during evolution.

Table X
LD_{50} Values of Some Alkaloids[a]

Alkaloid	Test system	LD_{50}[b] mg/kg
Alkaloids derived from tryptophan		
Brucine	Rat	p.o. 1
Cinchonidine	Rat	i.p. 206
Cinchonine	Rat	i.p. 152
Ellipticine	Mouse	i.v. 1.2
Ergocryptine	Rabbit	i.v. 1.1
Ergometrine	Mouse	i.v. 0.15
Ergotamine	Mouse	i.v. 62
Harman	Mouse	i.p. 50
Harmine	Mouse	i.v. 38
Physostigmine	Mouse	p.o. 4.5
Psilocybin	Mouse	i.v. 285
Quinidine	Rat	i.v. 30; p.o. 263
Quinine	*Agelaius*	p.o. 100
Reserpine	*Agelaius*	p.o. 100
Strychnine	*Agelaius*	p.o. 6
	Rat	i.v. 0.9
Vinblastine	Mouse	i.v. 9.5
Vincamine	Mouse	i.v. 75
Vincristine	Mouse	i.p. 5.2
Alkaloids derived from phenylalanine/tyrosine		
Aristolochic acid	Mouse	i.v. 38–70; p.o. 56–106
Berberine	Mouse	i.p. 23
Bulbocapnine	Mouse	p.o. 413
Canadine	Mouse	p.o. 940
Chelerythrine	Mouse	s.c. 95
Chelidonine	Mouse	i.v. 35
Codeine	Mouse	s.c. 300
Colchicine	Mouse	i.v. 4.1
	Man	p.o. 0.1–0.3
	Agelaius	p.o. 32
Emetine	Mouse	s.c. 32
Galanthamine	Mouse	i.v. 8; p.o. 18.7
Morphine	Mouse	i.v. 226–318
Papaverine	Mouse	i.v. 27.5; s.c. 150
Protopine	Mouse	i.p. 36–102
Sanguinarine	Mouse	s.c. 102; i.v. 16
Thebaine	Mouse	i.p. 20
Tubocurarine	Mouse	p.o. 33.2
Steroid alkaloid		
Batrachotoxin	Mouse	s.c. 0.002
Jervine	Mouse	i.v. 9.3
Protoveratrine	Rabbit	i.p. <0.1
Samandarine	Mouse	i.p. < 3.4
Solanine	Mouse	i.p. 42
Veratridine	Mouse	i.p. 1.4
Tropane alkaloids		
Atropine	Rat	p.o. 750
Cocaine	Rat	i.v. 17.5

(*continued*)

Table X
(Continued)

Alkaloid	Test system	LD_{50}[b] mg/kg
Pyrrolizidine alkaloids		
Echimidine	Rat	i.p. 200
Heliotrine	Rat	i.p. 300
acobine	Rat	i.p. 138
Monocrotaline	Rat	i.p. 175; p.o. 71
Senecionine	Rat	i.p. 85
Seneciphylline	Rat	i.p. 77
Quinolizidine alkaloids		
Cytisine	Mouse	i.v. 1.7
13-Hydroxylupanine	Mouse	i.p. 172
Lupanine	Mouse	i.p. 80
N-Methylcytisine	Mouse	i.v. 21; i.p. 51
Sparteine	Mouse	i.p. 55–67; p.o. 350–510
Miscellaneous alkaloids		
Aconitine	Mouse	i.v. 0.17; p.o. 1
α-Amanitin	Mouse	i.p. 0.1
Arecoline	Mouse	s.c. 100
Caffeine	Mouse	p.o. 127–137
Coniine	*Agelaius*	p.o. 56
Cycloheximide	Mouse	i.v. 150
Delphinine	Rabbit	i.p. 1.5–3.0
Maytansine	Rat	s.c. 0.48
Muscimol	Rat	p.o. 45
Nicotine	*Agelaius*	p.o. 17.8
	Mouse	i.v. 0.3; p.o. 230
Tetrodotoxin	Mouse	i.p. 0.01; s.c. 0.008

[a]More details in Wink (1993a).
[b]i.p., intraperitoneal; i.v., intravenous; p.o., oral; s.c., subcutaneous.

3.2. Inhibition of the Digestive Process

Food uptake can be reduced by pungent or bitter taste in the first instance, as was mentioned in Chapter 11. The next step can be the induction of vomiting, which is a common reaction to the ingestion of a number of alkaloids.

Causing diarrhea, or the opposite, constipation, would be another activity that negatively influences the digestive system. Many intoxications with alkaloid containing plants have diarrhea as one of the symptoms (see Chapter 4).

Another way to interfere would be the inhibition of digestive enzymes or of transport proteins for amino acids, sugars, or lipids. A recently discovered group of alkaloids are the polyhydroxy alkaloids, such as swainsonine or castanospermine, which inhibit hydrolytic enzymes, such as glucosidase, galactosidase, trehalase (trehalose is a sugar found in many insects and fungi which is hydrolyzed by trehalase), and mannosidase selectively (Table VIII).

3.3. Modulation of Liver and Kidney Function

Nutrients and xenobiotics (such as secondary metabolites) are transported to the liver after resorption in the intestine. In the liver the metabolism of carbohydrates, amino acids, and lipids, and the subsequent synthesis of proteins and glycogen takes place. The liver is also the main site for the detoxification of xenobiotics. Lipophilic compounds, which are easily resorbed from the diet, are often hydroxylated and then conjugated with a polar, hydrophilic molecule, such as glucuronic acid, sulfate, or an amino acid. These conjugates, which are more water soluble, are exported via the blood to the kidney, where they are transported into the urine for elimination. Other compounds are eliminated via the bile ducts into the intestine.

Both organ systems are affected by a variety of secondary metabolites: The pyrrolizidine alkaloids have been discussed earlier. They are activated during the detoxification process and are converted into potent carcinogens, causing liver cancer. Many other metabolic inhibitors, discussed previously, are also liver toxins.

Many alkaloids and other allelochemicals are known for their diuretic activity. For an animal, increased diuresis would also mean an increased elimination of water and essential ions. As Na^+ ions are already limited in plant food (an antiherbivore device?), long-term exposure to diuresis-inducing compounds would reduce the fitness of a herbivore substantially.

3.4. Disturbance of Reproduction

Quite a number of allelochemicals are known to influence the reproductive system of animals, which will ultimately reduce their numbers (and fitness as a species). Antihormonal effects could be achieved by mimicking the structure of sexual hormones. These effects are not known for alkaloids yet, but for other natural products: Estrogenic properties have been reported for coumarins which dimerize to dicoumarols, or isoflavones. The insect molting hormones, α- and β-ecdysone, are mimicked by many plant sterols (ecdysone itself is one of these) from the fern *Polypodium vulgare* and several *Ajuga* species or azadirachtin from the neem tree. Juvenile hormone is mimicked by a number of terpenes present in some Coniferae. Spermatogenesis is reduced by gossypol from cottonseed oil.

The next target is the gestation process itself. As outlined above, a number of alkaloids are mutagenic (see Section 2.4) and lead to malformation of the offspring or directly to the death of the embryo.

The last step would be a premature abortion of the embryo. This dramatic activity has been reported for a number of allelochemicals, including many mono- and sesquiterpenes and alkaloids. Some alkaloids achieve this by the induction of uterine contraction, as do the ergot and lupin alkaloids.

These antireproductive effects are certainly widely distributed but remain often unnoticed under natural conditions. However, they are nevertheless defense strategies with long-term consequences.

3.5. Blood and Circulatory System

All animals need to transport nutrients, hormones, ions, signal compounds, O_2 and CO_2 between the different organs of the body. This is achieved in higher animals through blood in the circulatory system. Inhibitors of its motor, the heart, were discussed earlier. But the synthesis of red blood cells is also vulnerable and can be inhibited by antimitotic alkaloids, such as vinblastine or colchicine (see Section 2.3). Some allelochemicals have hemolytic properties, such as saponins and steroidal alkaloids. If resorbed, these compounds complex membrane sterols and make the cells leaky (see Section 2.1).

3.6. Allergenic Effects

A number of secondary metabolites influence the immune system of animals, such as coumarins, furanocoumarins, hypercin, helenalin, and others. Common to these compounds is a strong allergenic effect on those parts of the skin or mucosa that have come into contact with the compounds. Activation or repression of the immune response are certainly targets that were selected during evolution as an antiherbivore strategy. A function of alkaloids in this context is hardly known.

4. MECHANISMS OF ALLELOCHEMICAL ACTIVITIES IN ANTIVIRAL, ANTIMICROBIAL, AND ALLELOPATHIC INTERACTIONS

We have circumstantial evidence that some alkaloids protect the producing plant against viruses, bacteria (see Chapter 17), fungi, and other plants. Relative to alkaloid–animal interactions, these modes of action have been studied less well or hardly at all.

A number of antimicrobial alkaloids such as sanguinarine, quinine, or berberine (Table VI) intercalate with viral and microbial DNA or bind to it. These compounds may thus inhibit processes such as DNA replication and RNA transcription which are vital for the microorganisms. Protein biosynthesis in ribosomes is another vulnerable target, attacked by emetine and several antibiotics (Table VII). The stability of biomembranes can be disturbed by steroidal alkaloids and tetrandine (as described in Section 2.1). Other targets may be electron chains or just metabolically important enzymes. Antibiotics of microbial origin (many of which could be classified as alkaloids from the chemical point of view) have similar targets, although some of them interfere with specific bacterial targets such as the biosynthesis and assembly of the bacterial cell wall.

Herbicidal properties or germination inhibition which can be observed in plant–plant interactions, can also proceed via the above-mentioned mechanisms (Wink and Latz-Brüning, 1995; Waller, 1987; Chapter 14). But interactions with growth hormones and their metabolism must also be considered.

5. CONCLUSIONS

This selection of alkaloidal activities, which is far from complete, clearly shows that many alkaloids inhibit or overstimulate central processes at the cellular and organ level. In this compilation only a limited number of structures have been discussed. In many instances, plants contain mixtures of related alkaloids, which only differ for particular substitution patterns. Very often these derivatives have properties similar to the better-known alkaloids listed in Tables I–X; therefore, with some caution their activity can be guessed. These allelochemical properties are requisite for a chemical defense compound in an ecological context but also constitute the base for their exploitation in medicine or agriculture (Wink, 1993a,b).

ACKNOWLEDGMENT

T. Schmeller kindly helped prepare some of the figures.

REFERENCES

Major Reviews

Alberts, B., Bray, D., Lewis, J., Raff, M., Roberts, K., and Watson, J. D., 1993, *Molecular Biology of the Cell,* 3rd ed., Garland, New York.

Habermehl, G., 1983, *Gifttiere und ihre Waffen,* Springer, Berlin.

Harborne, J. B., 1993, *Introduction to Ecological Biochemistry,* 4th ed., Academic Press, San Diego.

Luckner, M., 1990, *Secondary Metabolism in Microorganisms, Plants and Animals,* Springer, Berlin.

Mann, J., 1992, *Murder, Magic and Medicine,* Oxford University Press, London.

Mothes, K., Schütte, H. R., and Luckner, M., 1985, *Biochemistry of Alkaloids,* VCH, Weinheim.

Mutschler, E., 1981, *Arzneimittelwirkungen,* WVG, Stuttgart.

Rimpler, H., 1990, *Pharmazeutische Biologie II: Biogene Arzeneistoffe,* Thieme, Stuttgart.

Robinson, T. A., 1981, *The Biochemistry of Alkaloids,* Springer, Berlin.

Rosenthal, G. A., and Berenbaum, M. R., 1991, *Herbivores: Their Interactions with Secondary Plant Metabolites,* Vol. 1, Academic Press, San Diego.

Rosenthal, G. A., and Berenbaum, M. R., 1992, *Herbivores: Their Interactions with Secondary Plant Metabolites,* Vol. 2, Academic Press, San Diego.

Roth, L., Daunderer, M., and Kormann, K., 1994, *Giftpflanzen und Pflanzengifte,* 4th ed., Ecomed, Landsberg.

Teuscher, E., and Lindequist, U., 1994, *Biogene Gifte,* Fischer, Stuttgart.

Wagner, H., 1993, *Pharmazeutische Biologie. 2. Drogen und ihre Inhaltsstoffe,* Fischer, Stuttgart.

Waller, G., 1987, *Allelochemicals: Roles in Agriculture and Forestry,* ACS Symp. Ser. 330.

Wink, M., 1992a, Die chemische Verteidigung der Pflanzen und die Anpassungen der Pflanzenfresser, in: *"Lupinen 1991—Forschung, Anbau und Verwertung* (M. Wink, ed.), University of Heidelberg Press, Heidelberg, pp. 130–156.

Wink, M., 1992b, The role of quinolizidine alkaloids in plant insect interactions, in: *Insect–Plant Interactions,* Vol. IV (E. A. Bernays, ed.), CRC Press, Boca Raton, pp. 133–169.

Wink, M., 1993a, Allelochemical properties or the raison d'être of alkaloids, in: *The Alkaloids,* Vol. 43 (G. A. Cordell, ed.), Academic Press, San Diego, pp. 1–118.

Wink, M., 1993b, Production and application of phytochemicals from an agricultural perspective, in: *Phytochemistry and Agriculture,* Vol. 34, Proc. Phytochem. Soc. Eur. (T. A. van Beek and H. Breteler, eds.), Oxford University Press, London, pp. 171–213.

Key References

Holzinger, F., Frick, C., and Wink, M., 1992, Molecular base for the insensitivity of the monarch (*Danaus plexippus*) to cardiac glycosides, *FEBS Lett.* **314:**477–480.

Kebabian, J. W., and Neumeyer, J. L., 1994, *The RBI Handbook of Receptor Classification*, RBI, Natick.

Lodish, H., Baltimore, D., Berk, A., Zipursky, S. L., Matsudaira, P., and Darnell, J., 1995, *Molculear Cell Biology*. 3rd Ed., Scientific American Books, Inc., New York.

Reynolds, J. E. F. (ed.), 1993, Martindale—The Extra Pharmacopoeia, The Pharmaceutical Press, London.

Schmeller, T., Sauerwein, M., Sporer, F., Müller, W. E., and Wink, M., 1994, Binding of quinolizidine alkaloids to nicotinic and muscarinic receptors, *J. Nat. Prod.* **57:**1316–1319.

Schmeller, T., Sporer, F., Sauerwein, M., and Wink, M., 1995, Binding of tropane alkaloids to nicotinic and muscarinic receptors, *Pharmazie* **50:**493–495.

Wink, M., and Latz-Brüning, B., 1995, Allelopathic properties of alkaloids and other natural products: Possible modes of action, in: *Allelopathy: Organisms, Processes and Applications* (Inderjit, Dakshini, K. M. M., and Einhellig, F. A., eds.), ACS Symp. Ser. 582, pp. 117–126.

Wink, M., and Twardowski,T., 1992, Allelochemical properties of alkaloids. Effects on plants, bacteria and protein biosynthesis, in: *Allelopathy: Basic and Applied Aspects* (S. J. H. Rizvi and V. Rizvi, eds.), Chapman & Hall, London, pp. 129–150.

Chapter 13

Plant Parasites

F. R. Stermitz

1. INTRODUCTION

Alkaloids can be transferred from alkaloid-containing host plants to a variety of normally alkaloid-free parasitic plants. Sporadic and relatively incomplete reports of such alkaloid assimilation by parasitic plants appeared over the course of several decades before 1980. Access to these reports can be obtained from a review (Stermitz, 1990) and in typical papers of the 1980s (Wink *et al.*, 1981; Czygan *et al.*, 1988; Cordero *et al.*, 1989). The focus of this review will be on the more detailed recent studies where some attempt has been made to put data into a broader biological or ecological context, rather than simply document occurrence of the phenomenon.

The phytochemistry of secondary metabolite assimilation by parasitic plants can impact on medicinal chemistry (Okuda *et al.*, 1987; Gabius *et al.*, 1994), herbal products (Stermitz *et al.*, 1992), and plant–herbivore interactions (Stermitz *et al.*, 1989; Wink and Witte, 1993). There are over 3000 species of flowering parasitic plants and knowledge of their basic physiology and biochemistry is still somewhat limited although it is increasingly being studied (Kuijt, 1969; Atsatt, 1983; Stewart and Press, 1990). A number of parasitic plants cause substantial economic losses to agriculture and therefore there is a practical value to understanding them in more detail. They can also be model systems with which to explore more fundamental processes in plant biology and the ecology of plant–plant and plant–insect interactions (Atsatt, 1983; Stewart and Press, 1990).

Briefly, parasitic plants can be divided into obligate parasites (holoparasites) which cannot live without a host plant (they lack chlorophyll) and facultative parasites (hemiparasites) which can function at least minimally without a host. In nature, the latter most often (and perhaps ubiquitously) do adopt the parasitic habit (Heckard, 1962). Both of these major groups contain species that are exclusively root parasites and others that are only branch or stem parasites. Thus, mistletoes (*Viscum* and *Phoradendron* of the

F. R. Stermitz • Department of Chemistry, Colorado State University, Fort Collins, Colorado 80523.

Alkaloids: Biochemistry, Ecology, and Medicinal Applications, edited by Roberts and Wink. Plenum Press, New York, 1998.

Loranthaceae) are stem or branch hemiparasites, dodder (*Cuscuta* of the Convolvulaceae) is a stem holoparasite, broomrape (*Orobanche* of the Scrophulariaceae or Orobanchaceae) is a root holoparasite, and a number of figworts (*Striga, Castilleja, Pedicularis* of the Scrophulariaceae) are root hemiparasites. *Cuscuta* and especially *Striga* are genera that cause large economic losses by their parasitism on food and forage crops. Mistletoes cause losses in timber crops. Because of their economic importance, these taxa have been the most studied in terms of biology and some biochemistry.

If one goes beyond the mere documentation of alkaloid uptake by parasitic plants, a series of interesting questions that may be investigated can be asked. Are the parasites specific or selective in their use of host plants and does this or can this relate to the presence or absence of alkaloids in the host? How does alkaloid transfer relate to host and parasite anatomy? Is there any discrimination in the parasite for various alkaloid structures or classes? Does the parasite store the alkaloids in any special way or are they metabolized? Are the assimilated alkaloids of any benefit to the parasite nutritionally or as defensive substances against herbivory or disease? One would, as in any scientific area, like to establish some overall generalities or unifying concepts regarding alkaloid uptake by parasitic plants in the hope that the field would not be limited to descriptive studies of individual cases. This review will attempt to assess the current status of such concepts, while keeping in mind that, as someone has suggested, all biology can be explained in four words: "some do, some don't."

2. SPECIFICITY

Two related questions are involved: (1) Do parasitic plants have preferred hosts and, if so, are alkaloid-containing host plants among those preferred? (2) Is there any specificity for uptake among alkaloids of differing structures?

2.1. Host Plant Specificity

An early view regarding one group of parasites was that "the lack of host-specificity in nearly all parasitic Scrophulariaceae should not be forgotten" (Kuijt, 1969), but there have been sporadic reports of specialization in host utilization by this group of root parasites as well as by *Cuscuta* and mistletoe stem parasites. The key question is whether or not the parasite uses a certain host or hosts out of proportion to availability. This may be very difficult to assess. Three studies, two on a stem parasite and another on a root parasite, can serve as examples of attempts to answer the question of specialization. In one study, 88% of *Cuscuta costaricensis* use was on two hosts only, although the parasite was found on 6 of 10 potential hosts (Kelly *et al.*, 1988). The parasite had greatest vigor on these two hosts, and the relative infestation by the parasite was greater than the relative availability of the hosts. In a second study (Czygan *et al.*, 1988), two *Cuscuta* species (*C. reflexa* and *C. platyloba*) were allowed to parasitize 21 plant species from 20 different genera, 11 of which contained alkaloids and 10 of which were alkaloid-free. The *C. reflexa* failed to parasitize five alkaloid-containing hosts and six alkaloid-free hosts, while the *C. platyloba* failed to parasitize one alkaloid-containing host and six alkaloid-free hosts. There seemed to be little correlation with the alkaloid type of the host; for example, both species parasitized *Berberis vulgaris* but rejected *Mahonia aquifolia* (two phytochemically and botanically related genera) and *C. reflexa* parasitized *Datura arborea* but not *D. stramonium.*

Dwarf mistletoe parasites (*Arceuthobium*) are markedly host specific for certain conifer species in the western United States or have a very restricted principal host, but are occasionally found parasitizing individual conifers over a wide species range (Hawksworth, 1978; Atsatt, 1983).

In contrast to the case of stem parasites, the quantitation of root parasite host range in nature is hard to evaluate as the haustorial attachment is underground. Excavation to discover these attachments is difficult, especially in multispecies communities. This was, however, explored for *Rhinanthus minor* (Scrophulariaceae) in a sand dune grassland where excavation was simplified (Gibson and Watkinson, 1989). It was discovered that the natural host range of *R. minor* encompassed 50 host plant species from 18 families, but that the parasite was highly selective and showed both positive and negative correlations with host plant species availability. Root parasites can and do parasitize more than one host at a time and this also complicates analysis of their host range in nature. A general summary of these and other studies would suggest that there is a high degree of capability for multispecies use by both root and stem parasites, but also a considerable degree of specialization in actual utilization in the field.

One can then ask whether or not this degree of specialization in nature can be attributed to physical, nutritional, or secondary metabolite causes or some combination of these that increases fitness for the parasite on selected hosts. Chemical factors from the host have certainly been shown to be important in the breaking of seed dormancy and in initial root development in the parasite, especially in *Striga* (Lynn and Chang, 1990; Smith *et al.,* 1990), but alkaloids are not involved. Two poorly compatible legume hosts for the root parasite *Alectra vogelii* effectively blocked penetration of the parasite roots into the vascular tissue as compared to compatible legume hosts (Visser *et al.,* 1990). This was suggested to reflect the inability of the penetrating parasite cells to digest the host cell walls. There has been no demonstration that alkaloids *per se* can have an effect on blocking or aiding the physical attachment of parasite to host. We have observed and shown that *Castilleja integra* (Scrophulariaceae) utilizes quinolizidine or pyrrolizidine genera such as *Lupinus, Senecio,* and *Liatris* (Stermitz and Harris, 1987; Arslanian *et al.,* 1990; Mead *et al.,* 1992), but it failed to prosper in a pot with the California poppy, *Eschscholtzia californica,* a benzophenanthridine-containing taxon (Stermitz *et al.,* 1992). *Castilleja* species do grow and flower without a host (Heckard, 1962), but *C. integra* with *E. californica* remained for 6 months in an abnormally small, barely surviving rosette stage. One of the most striking associations between wildflower species in the desert southwest of the United States is between the bright yellow-orange *E. californica* and the deep red-purple *Castilleja exserta* (formerly *Orthocarpus purpurascens*). Although the parasitic *Castilleja* grows closely intermingled with the *E. californica,* it contains no benzophenanthridine alkaloids. The same *Castilleja* does parasitize *Liatris* and *Lupinus* species and assimilates their pyrrolizidine and quinolizidine alkaloids (Boros *et al.,* 1991; Mead *et al.,* 1992).

There are thus some intriguing aspects of host range and specificity for parasites, including alkaloid presence and absence, but as yet no general statement to cover more than a minimum of individual cases can apparently be made.

2.2. Specificity of Alkaloid Uptake: Root Parasites

One of the paradigms emerging from our studies is that the alkaloids that appear in the parasite are only those that are in the root, not in aboveground parts of the host. In an early study (Stermitz and Harris, 1987), the published TLC photograph showed a striking quantitative difference between the alkaloids of *Castilleja linariifolia* parasitizing *Lupinus bakeri* and the alkaloids of the lupine. The comparison was, however, between whole plant alkaloids of *Castilleja* and whole plant alkaloids of *Lupinus* and the quantitative difference disappeared when whole plant alkaloids of the *Castilleja* were compared with root alkaloids of the *Lupinus*. In none of our studies of alkaloid transmission to root parasites have we so far found any major qualitative differences between host root alkaloids and parasite alkaloids. Detailed quantitative measurements have not been made, but semiquantitative TLC analyses have also failed to show differences.

Castilleja integra often grows in the same location as *Oxytropis sericea,* which contains the biologically potent indolizidine alkaloid swainsonine. Field testing failed to show any swainsonine in *C. integra.* In the greenhouse, the *Castilleja* readily parasitized and prospered on *O. sericea,* but still contained no swainsonine. A review of the literature revealed no description of root analysis for alkaloids of the *Oxytropis.* Such an analysis revealed that *O. sericea* roots were indeed devoid of alkaloids, although swainsonine was present in aboveground parts (Stermitz *et al.,* 1992).

The iridoid glycoside content of *C. integra* parasitizing *Penstemon teucrioides* was found to correspond exactly to that of the *Penstemon* roots and not the aboveground plant (Stermitz *et al.,* 1993). Cinnamyl esters of iridoids were found in the above ground parts of the *Penstemon,* but not in the roots nor in the *C. integra* parasite.

In summary, there is a lack of evidence from any of our studies that parasites discriminate among root alkaloids of the host, nor do they assimilate alkaloids present only in aboveground parts of the host. This is apparent for quinolizidine, pyrrolizidine, piperidine (Schneider and Stermitz, 1990), and norditerpenoid (Marko, unpublished results) alkaloids. In none of these cases has a detailed quantitative study been carried out, but semiquantitative comparisons via TLC suggest that leaf alkaloid total content of the parasites may often be generally similar to that of the host plant. Stems of the parasite seem to contain much less alkaloid than do leaves, which indicates storage of the assimilated alkaloids in the parasite leaves. Alkaloids have also been detected in flowers and seeds of the parasite in some cases. This indicates that the parasites have an efficient mechanism for alkaloid transport across cell barriers.

The root holoparasite *Orobanche rapum-genistae* on *Sarothamnus scoparius* was found (Wink *et al.,* 1981) to contain a much lower total quinolizidine alkaloid content than the *S. coparius* aboveground parts, but several times more alkaloid than the roots. 13-Hydroxylupanine was the major alkaloid (43% of the total) of the *S. coparius* roots, but represented only 4% of the total in the *O. rapum-genistae* bulbs. The latter, however, contained considerably more sparteine (**1**) than the host roots (74 versus 27%, respectively) and, in addition, had 8% dehydrosparteine, an alkaloid missing from the host.

2.3. Specificity of Alkaloid Uptake: Stem Parasites

Considerable work has been done in this area, particularly with regard to *Cuscuta* assimilation of quinolizidine alkaloids from several different hosts. In some cases, there is

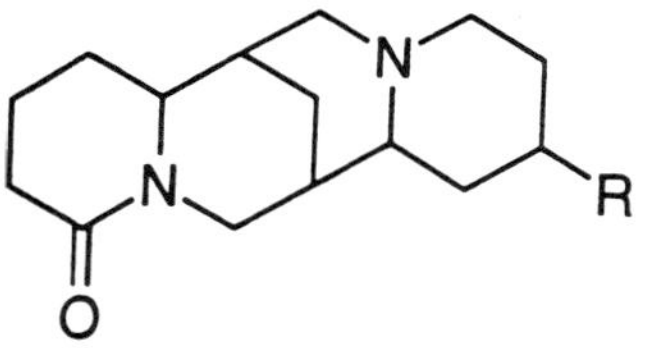

2 R = OH
4 R = H

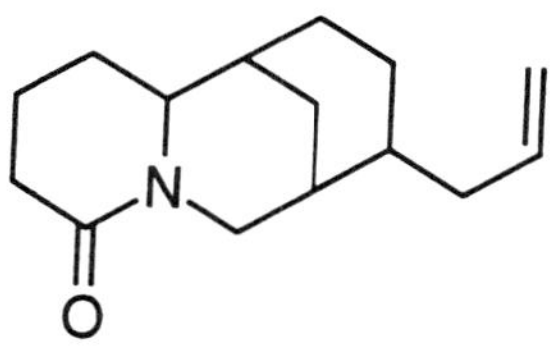

3

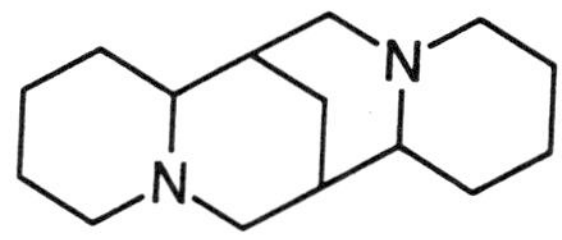

1

virtually no qualitative or even quantitative difference in alkaloid patterns of the host and the parasite while in other cases both such differences are observed.

In a preliminary report of *C. platyloba* on *Lupinus albus* (Baümel *et al.*, 1991), the alkaloid patterns were virtually the same qualitatively and quantitatively with one striking exception. The ester 13-*O*-*trans*-cinnamoyllupanine was present in the host in major amounts, but was barely detectable in the *Cuscuta.* There was a corresponding increase in 13-hydroxylupanine (**2**) in the *Cuscuta,* suggesting that hydrolysis had occurred in the parasite. An extensive later study (Baümel *et al.*, 1993a) was made of *Cuscuta reflexa* on *Lupinus angustifolius.* The alkaloid patterns were quite similar, but not identical for the haustorial region of the *Cuscuta* and the stem of the lupine, and there were some differences in total alkaloid content as well as changes with distance from the haustorium in the parasite. Two esters of 13-hydroxylupanine, which were present near the haustorium, were not found in the apex of the *Cuscuta,* where they were replaced by the alcohol. Angustifoline (**3**) was a very minor alkaloid in the lupine (fifth in concentration), but it was the major alkaloid of the *Cuscuta* haustorial region. The alkaloid content decreased from 11 mg g^{-1} dry weight (dw) at the haustorium to 2.2 mg g^{-1} dw in a middle section to 0.2 mg g^{-1} in the apex, while the host stem contained 2.1 mg g^{-1} dw. The alkaloid pattern of phloem exudate of the lupine was very similar to that of the whole stem and less similar to the xylem sap, but duplicate experiments showed considerable quantitative alkaloid variations in both cases. These experiments as well as some replication variations (both qualitative and quantitative) in at least two other reports (Wink and Witte, 1993; Baümel *et al.*, 1994) indicate that care should be taken in generalizing from a single experiment. This would be doubly true with regard to comparisons that involve the minor alkaloids.

Additional alkaloid assimilations by *C. reflexa* and *C. platyloba* from a variety of quinolizidine-containing hosts have been studied in a comparative manner (Bäumel *et al.*, 1994). Alkaloid patterns in *C. reflexa* and *C. platyloba* parasitizing *Cytisus praecox* were essentially the same, indicating that alkaloid uptake and transport were independent of the *Cuscuta* species. In addition, the only qualitative differences between the host stem and parasite were in minor or trace alkaloids and few large quantitative differences were noted. One exception was that the concentration of lupanine (**4**) represented only 5% of the total alkaloid content in the host, while it was 12% in the haustorial region of the parasite, 30% in growing ends, and 72% in flowers. A corresponding decrease in the content of sparteine was seen. *C. reflexa* was grown for 100 days on *Chamaecytisus hirsutus* and the alkaloid content and pattern of the entire parasite that grew during that time was compared to that of the host stem. Considering the variation that might occur in minor alkaloids in a repetition of this experiment, both the total alkaloid content and the qualitative and quantitative patterns for the nine quinolizidine alkaloids were essentially identical for host and parasite. Haustorial quinolizidine alkaloids of *Cuscuta platyloba* on *Spartium jun-*

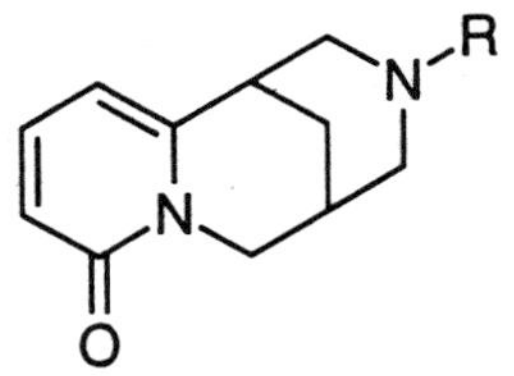

5 R = H
6 R = Me

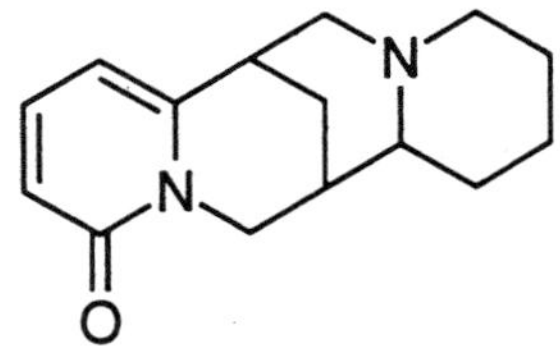

8

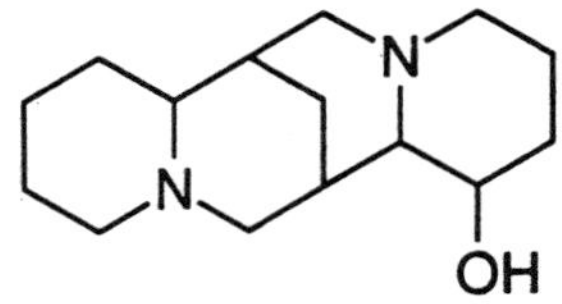

7

ceum were mostly similar except that cytisine (**5**) was increased in the parasite. Flowers and stem sections of the parasite had decreased cytisine and increased *N*-methylcytisine (**6**) relative to the haustorial region. Both of these alkaloids were also enriched in *C. reflexa* parasitizing *Petteria ramentacea.*

More major differences were reported for *Cuscuta palaestina* parasitizing *Genista acanthoclada* (Wink and Witte, 1993). Sparteine and retamine (**7**) were the major alkaloids of the host, but only traces were found in the parasite. On the other hand, lupanine, anagyrine (**8**), and *N*-methylcytisine were major alkaloids of the parasite while only traces were detected in the host. Alkaloids were analyzed on combined aerial parts of the *Genista* host, rather than just stems.

Retamine was, however, usually the major alkaloid of both *Retama* (*Lygos*) *sphaerocarpa* and its mistletoe stem parasite *Viscum cruciatum* (Cordero *et al.*, 1989, 1993) in a study that also included seasonal variations of alkaloid contents. There were a number of differences observed. *N*-Methylcytisine was a major alkaloid of the parasite in November, February, and May and a minor alkaloid in August, but it was never detected in the host. Anagyrine and cytisine were also detected at all times in the parasite, but only in May in the host. On the other hand, major amounts of genisteine (isosparteine) were found in the host in November and February, but were never detected in the parasite.

Another mistletoe, *Arceuthobium microcarpum*, contains *cis*-pinidinol (**9**), the major alkaloid of wood of the host tree *Picea engelmannii* (Engelmann spruce), but no epidihydropinidine (**10**), the major alkaloid of the needles (Schneider *et al.*, 1991).

One cause of differing alkaloid patterns in the parasite and in the host might be attributed to whether or not the parasite taps xylem or phloem (or both) in the host. An early view (Kuijt, 1969, p. 189) was that in both holoparasitic and hemiparasitic forms the emphasis is not on phloem continuity with the host but rather on xylem continuity, that "virtually all workers agree on one thing: there is no true phloem bridge" (p. 179), and that "there is nothing in the haustorium which may be called phloem" (p. 184). It was,

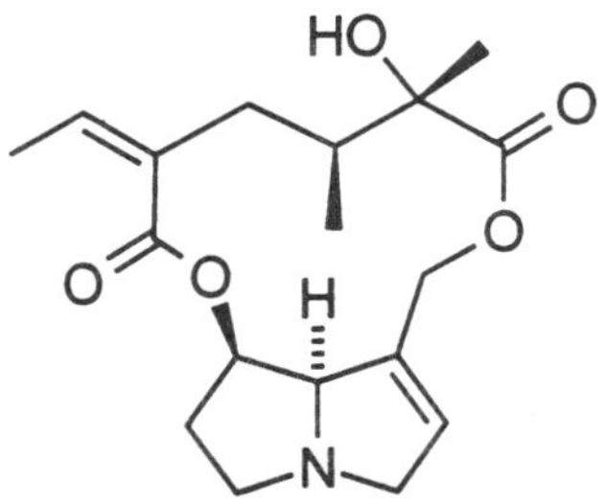

11

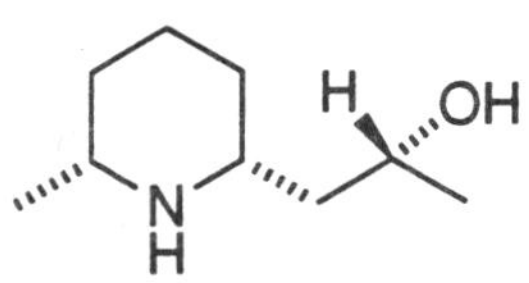

9

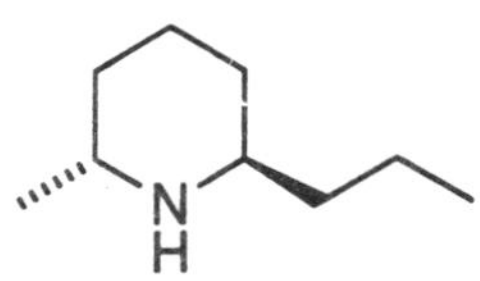

10

however, also stated (Kuijt, 1969, p. 196) that "beyond the implications of anatomy nothing was known about the tissues active in the transport of organic compounds." Electron micrographs have shown that the contacts between *Cuscuta* and *Lupinus* xylems are nearly open and separated only by a faint pit membrane (Dörr, 1972, and personal communication). No direct connection of the phloem elements of *Cuscuta reflexa* to *Lupinus angustifolius* could be found microscopically, but there was abundant such evidence for a xylem bridge (Baümel *et al.,* 1993). Yet quinolizidine alkaloid transfer has been suggested to be through phloem (Wink and Witte, 1993) or mainly through phloem with some xylem contribution (Baümel *et al.,* 1993a).

Cuscuta species parasitized two *Nicotiana* species, but did very poorly and only traces of nicotine were found in the parasite (Czygan *et al.,* 1988; Bäumel *et al.,* 1992). No nicotine was found in another study of *Cuscuta* on *Nicotiana* (Walzel, 1952). As nicotine is known to be transported through the xylem of *Nicotiana* (Dawson, 1942), this is consistent with phloem transport being the major route for alkaloid transfer in stem parasites. Mistletoes are considered to use a xylem connection to the host. The transfer of amino acids and other solutes from xylem of host *Acacia* and *Casuarina* plants to parasitic *Amyema* mistletoes was studied (Pate *et al.,* 1991). Data indicated that certain host xylem solutes were transferred directly to the parasite xylem, while others were either not absorbed or metabolized prior to transfer. As mentioned above, the major branch alkaloid (*cis*-pinidinol) of *Picea engelmannii* was found in the parasitic *Arceuthobium* (Scheider *et al.,* 1991). On the other hand, the same alkaloid was the major component of *Picea pungens* branch bark and also was found in aphids feeding on the branches (Todd and Stermitz, unpublished results). Aphids are known to be only phloem feeders.

To summarize, the anatomical evidence suggests no phloem contact but abundant xylem contact between parasite and host, while alkaloid transfer suggests the opposite. Specialized parenchyma cells do form a bridge between tracheal elements of xylem and phloem in plants and it is possible that these are involved in some way in the transfer process. Transfer may also be through the apoplast (cell wall continuum) (Dörr, 1975; Atsatt, 1983). It has also been suggested that the notion of designating parasites as xylem and phloem feeders is counterproductive as all parasites utilize products from both vascular systems of the host (Atsatt, 1983).

An additional factor to consider relates to the fact or possibility of active or passive transport of alkaloids within the parasite. Quinolizidine alkaloids are transported in *Lupinus* by carrier-mediated transport (Mende and Wink, 1987; Wink and Mende, 1987) rather than by simple diffusion. The presence of such a mechanism could account for some of the variations observed in alkaloid patterns between host and parasite in the case of stem parasites. As mentioned above, root parasites so far have not been found to exhibit any major differences in alkaloid patterns relative to their host roots and this holds for both quinolizidine and pyrrolizidine alkaloids. Transport of these alkaloids by diffusion might better explain the situation than postulation of highly efficient nonspecific carrier mechanisms which do not discriminate amoung alkaloid classes. This is certainly an open area of research.

3. ECOLOGICAL ASPECTS OF ALKALOID TRANSFER

Parallelisms in ecological and evolutionary patterns between herbivorous insects on the one hand and parasitic seed plants on the other have been proposed (Atsatt, 1977). It has also been suggested that parasites adopted their life mode as a mechanism for overcoming disease (viral and/or bacterial) (Atsatt, 1983).

Because of a series of investigations on quinolizidine alkaloids as herbivore deterrents and toxic allelochemicals, it has been suggested that the assimilation of such alkaloids also provides parasites with a defensive substance (Wink and Witte, 1993, and references therein). On the other hand, there has been no proof that the alkaloids actually function in such a manner for a parasite in a field situation. The parasite/host system should, however, provide an excellent way to probe the reality of antiherbivore chemical defenses as one can find both alkaloid-free and alkaloid-containing plants at the same field sites. Assessment of relative herbivore use between individual plants, which may be identical except for the alkaloid content, should provide an important opportunity for evaluating alkaloid deterrent effectiveness.

So far this method has only been rarely applied. Oviposition rates and larval survival were measured for the butterfly *Euphydryas editha* utilizing *Pedicularis semibarbata,* which contained α-isolupanine as a result of parasitizing *Lupinus falcatus* or lacked this quinolizidine when it parasitized alkaloid-free hosts (Stermitz *et al.,* 1989). No correlation was found in either oviposition rates or larval survival in early instars. This could have been related to the fact that α-isolupanine is not known to be a particularly toxic alkaloid. *Castilleja integra* contains senecionine (**11**) and other pyrrolizidine alkaloids from parasitizing *Liatris punctata* (Mead *et al.,* 1992). Both alkaloid-containing and alkaloid-free *Castilleja* were utilized in the field by ovipositing *Thessalia leanira* butterflies and larvae were surviving on both. The relative amounts of utilization of the two types of *Castilleja* were not measured. There was a possible trend for poorer larval growth on the alkaloid-containing *Castilleja* in a laboratory feeding experiment, however (Mead, 1992). Senecionine might be expected to be a more toxic alkaloid than α-isolupanine. Recently, *C. sulphurea* has been found to contain a series of norditerpene alkaloids from parasitizing *Delphinium occidentale* (Marko and Stermitz, unpublished data). A number of highly toxic diterpenoid alkaloids are known to be insecticides. Laboratory feeding studies have shown that larvae of *Euphydryas anicia* (a specialist) and *Trichoplusia ni* (a generalist) are indeed deleteriously affected by the presence of assimilated norditerpene alkaloids in the *Castilleja* (Marko, unpublished results). At our field site, both alkaloid-containing and alkaloid-free *Castilleja* contain aphid infestations, but the relative intensities of these infestations have not been studied.

A number of the parasitic Scrophulariaceae are used or touted as herbal medicines in the western United States. *Pedicularis,* for example, has been stated to be a useful sedative for children and a tranquilizer for adults although "large quantities may cause a befuddled lethargy and some interference with muscle control, particularly in the legs" (Moore, 1979). It was also stated that the potency of various species is variable. *Pedicularis* is a common root parasite and species have been found to contain senecionine (from *Senecio triangularis*), anagyrine and/or *N*-methylcytisine (from *Lupinus* species), and *cis*-pinidinol (from Engelmann spruce) (Stermitz *et al.,* 1992). Thus, both the toxicity and variability of potency could be a result of alkaloid uptake into the herb via parasitism.

There are other cases where genesis of alkaloids in a parasite has apparently not been considered. Thus, a novel alkaloid structurally related to tryptophan was reported from the crude Chinese drug Cuscutae semen (Yahara *et al.,* 1994). No mention was made of the parasitic nature of *Cuscuta chinensis,* the plant source of the drug; one might suspect that the alkaloid was assimilated from a host and is not inherent to the *Cuscuta.*

Recently, *in vitro* culture of shoot apices of *Cuscuta reflexa* was successful (Bäumel *et al.,* 1993b). In a nitrogen-deficient medium, death of cells occurred within 4 weeks. Addition of exogenous alkaloids to the nitrogen-deficient medium resulted in normal cell growth, thus indicating metabolism of and a nutritional value for assimilated alkaloids.

REFERENCES

General References

Atsatt, P. R., 1983, Host–parasite interactions in higher plants, in: *Encyclopedia of Plant Physiology,* Vol. 12C (O. L. Lange, C. B. Nobel, C. B. Osmond, and H. Ziegler, eds.), Springer-Verlag, Berlin, pp. 519–536.

Bäumel, P., Witte, L., Abou-Mandour, A., Proksch, P., and Czygan, F.-C., 1993, Alkaloid uptake and metabolism by *Cuscuta reflexa* grown *in vitro, Planta Med.* **59:**A643.

Dörr, I., 1975, Development of transfer cells in higher parasitic plants, in: *Phloem Transport* (S. Aronoff, J. Dainty, P. R. Gorham, L. M. Srivastava, and C. A. Swanson, eds.), Plenum Press, New York, pp. 177–186.

Kuijt, J., 1969, *The Biology of Parasitic Flowering Plants,* University of California Press, Berkeley.

Lynn, D. G., and Chang, M., 1990, Phenolic signals in cohabitation; implications for plant development, *Annu. Rev. Plant Physiol. Plant Mol. Biol.* **41:**497–526.

Moore, M., 1979, *Medicinal Plants of the Mountain West,* Museum of New Mexico Press, Santa Fe, pp. 34–36.

Stermitz, F. R., Schneider, M. J., Schell, L. D., and McGregor, S., 1992, Toxic pyrrolizidine, quinolizidine, and piperidine alkaloids of root parasitic plants used in folk medicine, in: *Poisonous Plants, Proceedings of the 3rd International Symposium* (L. F. James and R. F. Keeler, eds.), Iowa State University Press, pp. 204–207.

Stewart, G. R., and Press, M. C., 1990, The physiology and biochemistry of parasitic angiosperms, *Annu. Rev. Plant Physiol. Plant Mol. Biol.* **41:**127–151.

Specific References

Arslanian, R. L., Harris, G. H., and Stermitz, F. R., 1990, New quinolizidine alkaloids from *Lupinus argenteus* and its hosted root parasite *Castilleja sulphurea:* Stereochemistry and conformation of some naturally-occurring cyclic carbinolamides, *J. Org. Chem.* **55:**1204–1210.

Atsatt, P. R., 1977, The insect herbivore as a predictive model in parasitic seed plant biology, *Am. Nat.* **111:**579–586.

Bäumel, P., Lurz-Gresser, G., Veen, G., Witte, L., Proksch, P., and Czygan, F.-C., 1991, Uptake of host-plant alkaloids by parasitic *Cuscuta* species, *Planta Med. Suppl.* **57:**A95–A96.

Bäumel, P., Witte, L., Proksch, P., and Czygan, F.-C., 1992, Uptake and metabolism of host plant alkaloids by parasitizing *Cuscuta* species, *Planta Med. Suppl.* **58:**A671.

Bäumel, P., Jeschke, W. D., Witte, L., Czygan, F.-C., and Proksch, P., 1993a, Uptake and transport of quinolizidine alkaloids in *Cuscuta reflexa* parasitizing on *Lupinus angustifolius, Z. Naturforsch.* **48c:**436–443.

Bäumel, P., Witte, L., Czygan, F.-C., and Proksch, P., 1994, Transfer of quinolizidine alkaloids from various host plants of the Fabaceae to parasitizing *Cuscuta* species, *Biochem. Syst. Ecol.* **22:**647–656.

Boros, C. A., Marshall, D. R., Caterino, C. R., and Stermitz, F. R., 1991, Iridoid and phenylpropanoid glycosides from further *Orthocarpus* species. Alkaloid content as a consequence of parasitism on *Lupinus, J. Nat. Prod.* **54:**506–513.

Cordero, C. M., Ayuso, M. J., Richomme, P., and Bruneton, J., 1989, Quinolizidine alkaloids from *Viscum cruciatum,* hemiparasitic shrub of *Lygos sphaerocarpa, Planta Med.* **55:**196.

Cordero, C. M., Serrano, A. M. G., and Gonzalez, M. J. A., 1993, Transfer of bipiperidyl and quinolizidine alkaloids to *Viscum cruciatum* Sieber (Loranthaceae) hemiparasitic on *Retama sphaerocarpa* Boissier (Leguminosae), *J. Chem. Ecol.* **19:**2389–2393.

Czygan, F.-C., Wessinger, B., and Warmuth, K., 1988, *Cuscuta* und ihre Fähigkeit zur aufnahme und speicherung von Alkaloiden der Wirtspflanzen, *Biochem. Physiol. Pflanz.* **183:**495–501.

Dawson, R. F., 1942, Accumulation of nicotine in reciprocal grafts of tomato and tobacco, *Am. J. Bot.* **29:**66–71.

Dörr, I., 1972, Der Anschluss der *Cuscuta*-Hyphen an de Siebröhren ihrer Wirstspflanzen, *Protoplasma* **75:**167–184.

Gabius, H.-J., Gabius, S., Joshi, S. S., Koch, B., Schroeder, M., Manzke, W. M., and Westerhausen, M., 1994, From ill-defined extracts to the immunomodulatory lectin: Will there be a reason for oncological application of mistletoe? *Planta Med.* **60:**2–7.

Gibson, C. C., and Watkinson, A. R., 1989, The host range and selectivity of a parasitic plant: *Rhinanthus minor* L., *Oecologia* **78:**401–406.

Hawksworth, F. G., 1978, Biological factors of dwarf mistletoe in relation to control, in: *Proceedings of the Symposium on Dwarf Mistletoe Control through Forest Management* (R. F. Scharpf and J. R. Parmeter, eds.), University of California Press, Berkeley.

Heckard, L., 1962, Root parasitism in *Castilleja, Bot. Gaz.* September, pp. 21–29.

Kelly, C. K., Venable, D. L., and Zimmerer, K., 1988, Host specialization in *Cuscuta costaricensis*: An assessment of host use relative to host availability, *Oikos* **53:**315–320.

Mead, E. W., 1992, Chemical investigation of a multispecies plant–insect biosystem, Ph.D. dissertation, Colorado State University.

Mead, E. W., Looker, M., Gardner, D. R., and Stermitz, F. R., 1992, Pyrrolizidine alkaloids of *Liatris punctata* (Asteraceae) and its root parasite *Castilleja integra* (Scrophulariaceae), *Phytochemistry* **31:**3255–3257.

Mende, P., and Wink, M., 1987, Uptake of the quinolizidine alkaloid lupanine by protoplasts and isolated vacuoles of suspension-cultured *Lupinus polyphyllus* cells. Diffusion or carrier-mediated transport? *J. Plant Physiol.* **129:**229–242.

Okuda, T., Yoshida, T., Chen, X.-M., Xie, J.-X., and Fukushima, M., 1987, Corianin from *Coriaria japonica* and sesquiterpene lactones from *Loranthus parasiticus.* Used for treatment of schizophrenia, *Chem. Pharm. Bull.* **35:**182–187.

Pate, J. S., True, K. C., and Rasins, E., 1991, Xylem transport and storage of amino acids by S.W. Australian mistletoes and their hosts, *J. Exp. Bot.* **42:**441–451.

Schneider, M. J., and Stermitz, F. R., 1990, Uptake of host plant alkaloids by root parasitic *Pedicularis* (Scrophulariaceae) species, *Phytochemistry* **29:**1811–1814.

Schneider, M. J., Montali, J. A., Hazen, D., and Stanton, C. E., 1991, Alkaloids of *Picea, J. Nat. Prod.* **54:**905–909.

Smith, C. E., Dudley, M. W., and Lynn, D. G., 1990, Vegetative/parasitic transition: Control and plasticity in *Striga* development, *Plant Physiol.* **93:**208–215.

Stermitz, F. R., 1990, Discovery of new alkaloids by analysis of parasitic Scrophulariaceae, *Rev. Latinoam. Quím.* **21:**83–85.

Stermitz, F. R., and Harris, G. H., 1987, Transfer of pyrrolizidine and quinolizidine alkaloids to *Castilleja* (Scrophulariaceae) hemiparasites from composite and legume host plants, *J. Chem. Ecol.* **13:**1917–1925.

Stermitz, F. R., Belofsky, G. N., Ng, D., and Singer, M. C., 1989, Quinolizidine alkaloids obtained by *Pedicularis semibarbata* (Scrophulariaceae) from *Lupinus fulcratus* (Leguminosae) fail to influence the specialist herbivore *Euphydryas editha* (Lepidoptera), *J. Chem. Ecol.* **15:**2521–2530.

Stermitz, F. R., Foderaro, T., and Li, Y.-X., 1993, Iridoid glycoside uptake by *Castilleja integra* via root parasitism on *Penstemon teucrioides, Phytochemistry* **32:**1151–1153.

Visser, J. H., Dörr, I., and Kollmann, R., 1990, Compatibility of *Alectra vogelii* with different leguminous host species, *J. Plant Physiol.* **135:**737–745.

Walzel, G., 1952, *Cuscuta* auf *Nicotiana* nikotin-frei, *Phyton* **4:**121–123.

Wink, M., and Mende, P., 1987, Uptake of lupanine by alkaloid-storing epidermal cells of *Lupinus polyphyllus, Planta Med.* **53:**465–469.

Wink, M., and Witte, L., 1993, Quinolizidine alkaloids in *Genista acanthoclada* and its holoparasite, *Cuscuta palaestina, J. Chem. Ecol.* **19:**441–448.

Wink, M., Witte, L., and Hartmann, T., 1981, Quinolizidine alkaloid composition of plants and of photomixotrophic cell suspension cultures of *Sarothamnus scoparius* and *Orobanche rapum-genistae, Planta Med.* **43:**342–352.

Yahara, S., Domoto, H., Sugimura, C., Nohara, T., Niiho, Y., Nakajima, Y., and Ito, H., 1994, An alkaloid and two lignans from *Cuscuta chinensis, Phytochemistry* **37:**1755–1757.

Chapter 14

Allelopathy in Plants

J. V. Lovett and A. H. C. Hoult

1. INTRODUCTION

As Whittaker (1970) succinctly put it, "Man's use of alkaloids for flavour, mild stimulation, medicinal effect, or pleasurable self-destruction should not obscure a common theme: they are probably, although not necessarily in all cases, repellents and toxins, evolutionary expressions of quiet antagonism of a plant to its enemies." One aspect of this quiet antagonism is allelopathy, which is associated with secondary metabolites of plants, including the alkaloids. Alkaloids may, equally, be perceived as chemical messengers between the plants that produce them and an array of organisms (Levitt and Lovett, 1985).

Molisch (1937), defining allelopathy, included "any biochemical interaction, whether positive or negative, among plants of all levels of complexity, including micro-organisms." The term means "mutual harm" but, while relatively uncommon, positive interactions may be defined.

Positive allelopathic interactions are of two types. The first is exemplified by the interaction between the witchweeds, *Striga* spp., which are obligate parasites during their early growth, and their host plants. Seeds of these species will only germinate in the presence of a suitable host plant root. Ethylene, which may be produced from root exudates of a host plant, can substitute for the root itself and stimulate germination to natural levels in *Striga lutea* Lour (Egley and Dale, 1970). Similarly, the roots of *Sorghum bicolor* L., the specific host of *Striga asiatica* L., exude sorgoleones which stimulate germination of the parasite (Netzley *et al.,* 1988). In other situations, sorgoleones have negative effects on plant growth (Einhellig *et al.,* 1993). The second positive interaction between secondary metabolites of one plant and other plant species is a common phenomenon. Frequently, a compound that is toxic or inhibitory at high concentrations is stimulatory at low concentrations. Wink and Twardowski (1992) found this to be the case for alkaloids as well as other groups of allelochemicals and the phenomenon is exemplified by

J. V. Lovett • Grains Research and Development Corporation, Kingston, ACT 2604, Australia. *A. H. C. Hoult* • Department of Agronomy and Soil Science, University of New England, Armidale, NSW 2351, Australia.

Alkaloids: Biochemistry, Ecology, and Medicinal Applications, edited by Roberts and Wink. Plenum Press, New York, 1998.

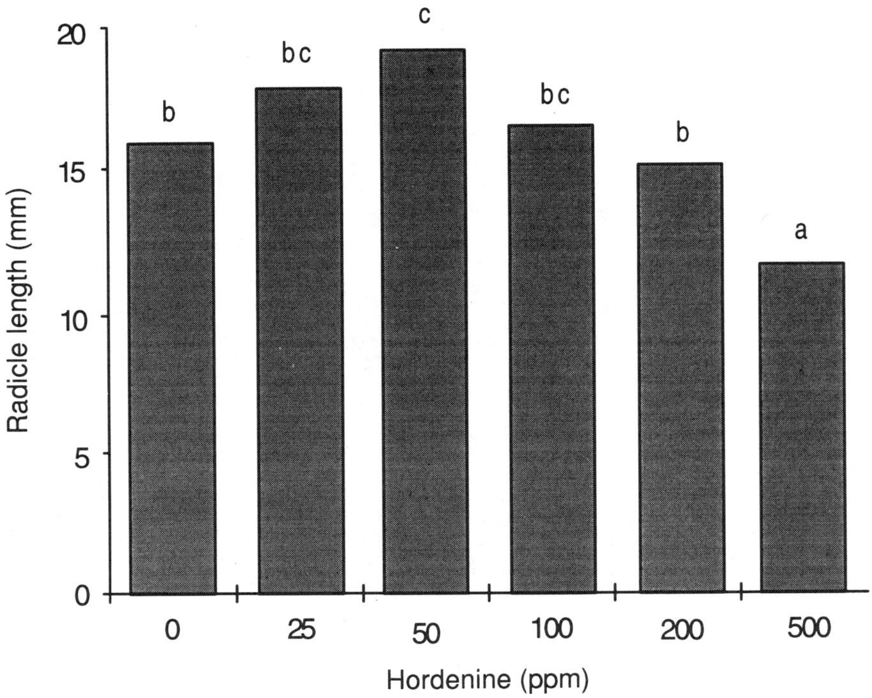

Figure 1. The effect of hordenine at different concentrations on radicle growth in *Amaranthus powellii*. Treatments bearing the same letter do not differ from each other at the 5% level.

the effect of hordenine* on radicle growth in *Amaranthus powellii* grown in bioassay (Fig. 1) (see also Lovett *et al.,* 1989).

Negative interactions are well documented and have been reviewed on numerous occasions in the past (e.g., see Rice, 1984; Waller, 1989). Since Molisch's time the use of the term has altered somewhat, the positive interactions tend to be overlooked, and the negative interactions tend to include all "self-defense" whether it be plant against plant or plant against other organisms (Lovett, 1991). Such interactions with organisms other than plants are dealt with in detail elsewhere in this book and will only be touched on here. Generally, in this chapter, the term *allelopathy* will be used in the sense of Molisch.

Allelopathy may be looked on as distinct from competition because it involves interactions between plants that result from addition to, rather than subtraction from, the environment by the plant. It may, however, contribute to the competitive ability of a plant should the interaction be disadvantageous to a neighboring plant. This is not always the case. An allelochemical may inhibit the growth of one species but not another and, as noted above, an allelochemical may be stimulatory at low concentrations but inhibitory at higher concentrations. It is not yet clear whether all plants possess allelopathic activity, although we interpret the mounting evidence for allelochemical production by plants to mean that this may well be the case.

*Strictly speaking, hordenine is not an alkaloid as it is not a heterocyclic compound; it is more properly classed as a phenol or amine.

The compounds that a plant may add to the environment and that may contribute to allelopathy are many and varied. Major groups include phenolic acids, glucosinolates, terpenes, and flavonoids. However, alkaloids form the largest group of these compounds (Swain, 1977); 20–30% of plants produce alkaloids (Wink, 1993).

Their wide distribution notwithstanding, less is known of the allelopathic activity of the alkaloids than of, for example, phenolic acids. This may be because alkaloids have been associated with medicinal uses for centuries and other possible roles have not been examined. Bode (1939) linked the alkaloid absinthin with allelopathic activity; however, for the most part, biologically active alkaloids have not been so implicated (Muller and Chou, 1971); in fact, relatively few have been tested. In the past many extensive works on allelochemicals have failed to mention alkaloids at all (e.g., see Bonner, 1950; Borner, 1960; Garb, 1961), while Haas and Hill (1928) stated that alkaloids are toxic to animals but not to plants themselves. Mothes (1955) remarks that "totally unexplored is the question of the role alkaloids play in the soil into which they are leached from the leaves or exuded from the roots. The possibility exists that they represent an ecologically important factor."

Robinson (1974) discussed the role of alkaloids in plants and considered that they had biological activity against a wide range of organisms from the plant and animal kingdoms. He cited a number of biochemical activities reported for alkaloids and pointed out that, although most of the research involved had been conducted on organisms other than plants, the processes themselves were all relevant to plants, thus raising the possibility of wide-ranging effects of alkaloids on plants. Levitt and Lovett (1985) and Lovett (1989) discuss the case of acetylcholine, an intriguing instance of such activity. Acetylcholine is a neurotransmitter in animals but is also found in plants where it contributes to the breakdown of starch. Levitt *et al.* (1984) suggested that the similarity of structure between the alkaloid scopolamine and acetylcholine may explain why scopolamine inhibits starch hydrolysis in root tips of *Helianthus annuus.* There may be analogies with the mode in which scopolamine acts in animals (Roshchina, 1987, 1988).

Wink (1987) provides evidence that the quinolizidine alkaloids, characteristic of the genus *Lupinus,* possess biochemical activity against a range of organisms from the plant and animal kingdoms. Such are the difficulties, however, of cross-disciplinary research that Biller *et al.* (1994) reported high concentrations of alkaloids to be present in the roots and flowers of *Chromolaena odorata* and discussed their insecticidal properties in the flower head, but gave no consideration to the function of the large quantities of alkaloids found in the root. The question remains to be asked in relation to many alkaloids: do they deter attack by animal pests or are they also involved in plant–plant interactions?

2. ALLELOPATHIC ACTIVITY OF THE ALKALOIDS

2.1. Against Microorganisms

Conner (1937) found that solanine in *Solanum tuberosum* is toxic to conidiospores of the fungus *Cladosporium fulvum* Link., although Manske (1950) was unable to attribute the high resistance of *Solanum racemigerum* L. (nightshade) to attack by *Phytophthora* species to its relatively high solanine content. Wippich and Wink (1985) showed that gramine and lupine alkaloids inhibited the germination of conidia of *Erisyphe graminis* f.sp. *hordei.* Grodzinski (1992) reported that barley root exudates, extracts of barley residues, and barley soils inhibited germination and germ tube growth of *Fusarium oxysporum* f. *vasinfectum,* but did not identify the active principle(s). Barley contains gramine and hordenine, which inhibit the growth of the fungal pathogen of barley, *Drechslera teres* (Lovett and Hoult, 1993). Wink (1993) listed some 100 alkaloids that show toxicity toward fungi and yeasts. It is worth noting that a substantial number of these entries refer to medically (*Candida*) and commercially (yeast) important species and very few to named plant pathogens.

Rice (1964) found that fresh extracts of *Solanum rostratum* Dun. (buffalo burr), containing various alkaloids, had inhibitory action against symbiotic nitrogen-fixing bacteria (*Rhizobium* species), with a slightly less marked inhibition of the free-living nitrogen-fixing bacterium *Azotobacter.* The inhibition of symbiotic nitrogen-fixing bacteria by this weed could aid in its competition with plants relying on nodulation for their nitrogen supply. Tyski *et al.* (1988), having tested lupin alkaloids against a number of bacteria, including one strain of *Bacillus thuringiensis* isolated from soil, concluded that these alkaloids may have an allelopathic function against bacteria. Sepulveda and Corcuera (1990) identified gramine as being inhibitory to the growth of the bacterium *Pseudomonas syringae,* and Krischik *et al.* (1991) found nicotine to be toxic to five species of *Pseudomonas,* all plant pathogens. This evidence supports the role for alkaloids in protecting plants against infection by pathogens. Wink (1993) listed 183 alkaloids with antibacterial properties but indicated that many of these may have been investigated for possible pharmaceutical use rather than to elucidate their ecological role(s).

2.2. Against Higher Plants

The families Solanaceae and Apocynaceae are known for their alkaloid production, 60–70% of their members producing alkaloids (Wink, 1993). Evanari (1949) reported that the alkaloids narcotine, scopolamine, and atropine, all of which occur in the Solanaceae, were weak inhibitors of germination.

In many examples of growth retardation by solanaceous plants, the chemicals responsible have not been identified, hence it is not known whether inhibition is related principally to alkaloids or to members of another chemical group. Loehwing (1937) noted that plants of *S. tuberosum* were often injurious to adjacent or succeeding crops and attributed this to toxic root exudates, but did not identify the factor responsible. The germination of *Medicago sativa* L. (lucerne) was inhibited by *S. tuberosum* extract (Neilson *et al.,* 1960); the germination of *Triticum aestivum* L. (wheat) was inhibited by seeds of *S. melongena* L. (eggplant) and *N. tabacum,* while a mixture of 200 g of dried plant material from *Lycopersicon esculentum* in 100 ml of water inhibited the germination of *T. aestivum* by 42%, and significantly reduced radicle and coleoptile growth (McCalla and Haskins, 1964). Neither group of workers identified the chemicals responsible for the inhibitory effects. Gressel and Holm (1964) found that aqueous extacts of ground seeds of *Datura stramonium* (thornapple) inhibited the germination of many crop species, as do intact *D. stramonium* seeds (Levitt and Lovett, 1984). Work by Hussain *et al.* (1979) showed that *D. innoxia* significantly inhibited germination and growth of test species by root exudates, aqueous extracts from various aboveground parts, and substances volatilizing from the shoots, and that soil collected under and around *D. innoxia* was also inhibitory to test species. Recently, in a preliminary study, we found that extracts of the aboveground parts of *Nicandra physaloides,* which is known to produce the alkaloid hygrine (Leete, 1985), inhibited germination of radish seeds (unpublished data).

Routley and Sullivan (1960) found that hyoscyamine and atropine were non-toxic to *Trifolium repens* L. (white clover) at 1000 μl liter^{-1}, although the nitrogen contained in these alkaloids was apparently unavailable for plant uptake. Both nicotine and trigonelline were toxic to *T. repens* at 500 μl liter^{-1}, while 200 to 500 μl liter^{-1} allowed normal growth in the absence of inorganic nitrogen. Scopolamine inhibited the growth of lettuce (*Lactuca sativa*) seedling roots at 0.1% but stimulated the growth of cress (*Lepidium sativum*) roots at the same concentration (Wink and Twardowski, 1992).

Much attention has been paid to the indole alkaloids, mainly in relation to their pharmaceutical value. These alkaloids are commonly found in the Apocynaceae (Wink,1993) but are also found in the Poaceae. Shehab (1982) reported that vinblastine and yohimbine, indole alkaloids from *Catharanthus roseus* (Apocynaceae), both used in medicine, caused temporary abnormalities in cell division in broad bean (*Vicia faba*).

Much of the recent work in our laboratory has been concerned with the identification and quantification of the indole alkaloid gramine in barley (Liu and Lovett, 1990; Hoult and Lovett, 1993). Overland (1966) found gramine to possess biological activity against a number of weeds, however, as gramine is found, in quantity, only in barley leaves (Schneider and Wightman, 1974), it is not necessarily placed to have maximum allelopathic activity against other plant species. Its activity against fungi and bacteria has already been discussed in this chapter. Gramine is found in small quantities in root exudates of barley and was shown to have some synergistic effect with hordenine in inhibiting root growth in white mustard (*Sinapis alba* L.) in bioassay (Liu and Lovett, 1993b). Results using gramine as a drench against Wimmera ryegrass (*Lolium rigidum* L.), shepherd's purse (*Capsella bursa-pastoris* L.), and white mustard were equivocal. In white mustard initial inhibition was followed, in time, by stimulation. In Wimmera ryegrass and shepherd's purse, stimulation of growth occurred with increasing dosage. Overland found gramine to be less effective against shepherd's purse than chickweed (*Stellaria media* L.).

Allelopathic potential has been claimed for L-tryptophan identified from oat (*Avena sativa* L.) shoot extract and which showed phytotoxic activity against a number of species in bioassay (Kato-Noguchi *et al.,* 1994). This amino acid is found in small quantities in proteins and has not been considered a secondary metabolite. Allelopathic activity in such a compound opens many avenues for conjecture.

Of other alkaloids reported to have allelopathic activity, among the earliest citings are cocaine from *Erythroxylum coca* Lamk (coca), strychnine from *Strychnos nux-vomica* L. (nux-vomica), and physostigmine from *Physostigma venenosum* Balf. (Calabar bean) (Evanari, 1949). The quinolizidine alkaloids, found in lupins (*Lupinus* spp.) and other genera of the Fabaceae, have also been shown to have allelopathic activity against plants. They inhibited the germination of lettuce (*Lactuca sativa* L.) and lawn grass mixture (Wink, 1983) and affected the germination of various other species (Muzquiz and de la Cuadra, 1988). The alkaloids caffeine, theobromine, and theophylline from *Coffea arabica* L. (coffee) were listed as allelopathic by Chou and Waller (1980) and were found to be autotoxic in germinating tea (*Camellia sinensis* L.) seedlings (Suzuki and Waller, 1987). Smyth (1992) found that of the three, caffeine, in particular, was inhibitory to the growth of rice seedlings. Ohigashi *et al.* (1989) reported that 3-hydroxyuridine inhibited root and hypocotyl growth of cucumber and radish.

3. THE SIGNIFICANCE OF ALKALOID-MEDIATED ALLELOPATHY IN ECOSYSTEMS

3.1. Criteria for Establishing Allelopathic Activity in Ecosystems

Waller *et al.* (1993) comment on the interest shown in the purine alkaloids, principally caffeine, but also theobromine and theophylline, and ascribe it to the popularity of the beverages containing them, for little is known of their ecological or physiological significance. If one excludes the human race from all ecosystems, this is certainly true.

It is, nevertheless, relatively easy to find examples of ecological roles played by alkaloids. These consist mainly of toxic effects on insects and other animals. Conversely, it is very hard to find examples where one plant affects the survival of another by allelopathic means. For example, Wardle (1987a) reviews allelopathy in a grassland/pasture ecosystem and argues that the evidence is inconclusive; not one alkaloid is mentioned in this article.

Wardle (1987a) also cites five criteria listed by Muller that should be met if allelopathy is to be established as a factor operating in an ecosystem. These are:

1. the plant under investigation must produce a toxin;
2. the plant must be capable of releasing the toxin;
3. means of concentrating the toxin must exist, e.g., on clay colloids;
4. other plants must be susceptible to the toxin; and
5. other factors that could be influencing any observed interference pattern must be eliminated.

One of the few studies to meet all of these criteria is the work with *Datura stramonium* discussed below.

3.2. Soil Mediation

Allelochemicals may be liberated from the aboveground parts of plants in a number of ways (Tukey, 1969); most will, eventually, find their way into the soil. As alkaloids are nonvolatile (Haas and Hill, 1928), the probable method for their release is by leaching of the aboveground parts by dew, fog, and rain. Allelochemicals may also be introduced into the soil by exudation or leaching from plant roots, living or dead, and by leaching from plant remains in general.

Woods (1960) emphasizes the probable ephemeral nature of free toxins in the soil and remarks that these, often complex, organic substances are undoubtedly subject to rapid deactivation by soil bacteria and fungi. Provided production is sustained, however, these chemicals could still play an important role in interplant relationships. Levitt and

Lovett (1984) showed that the inhibitory effect of *Datura stramonium* (thornapple) seeds or seed washings on *Helianthus annuus* seedlings continued in a lateritic podsolic soil for 20 weeks under controlled conditions and for 8 months in a black earth soil under field conditions. The alkaloids scopolamine and hyoscyamine were identified in leachates of *D. stramonium* seeds (Lovett *et al.*, 1981) and both were isolated from the soil at a field site infested with *D. stramonium.*

Although it has been claimed (Borner, 1960) that, under natural conditions, chemicals are probably not liberated from seeds, fruits, or other parts in quantities sufficient to affect other plants, evidence to the contrary was found by Muller *et al.* (1964), who determined that toxic substances liberated from plants were deposited on the soil by the action of dew. The depositions, although not quantified, were sufficiently concentrated to inhibit nearby species.

Wink (1983) reported that alkaloids (mainly lupanine and 13-tigloyloxylupanine) were excreted by *Lupinus albus* seedlings into a "hydroculture" system for 39 days after imbibition. Liu and Lovett (1990) reported that barley roots released the alkaloids gramine and hordenine into a bioassay system at concentrations that were inhibitory to the growth of white mustard, *Sinapis alba* L. (Liu and Lovett, 1993b). When barley was included in each system, the growth of white mustard was reduced whether it was grown in a hydroponic system or a stairstep apparatus (Liu and Lovett, 1993a).

3.3. Other Interactions

In aquatic ecosystems the means of concentrating an allelochemical are minimal if not absent, however, Elakovich and Wooten (1991) cited evidence for allelopathy in such ecosystems and showed that extracts of both roots and shoots of *Nuphar lutea* L., known to produce alkaloids, inhibited the growth of duckweed (*Lemna minor*). Wink and Twardowski (1992) confirm that some alkaloids negatively affect the growth of duckweed, in this case *Lemna gibba.* Welch *et al.* (1990) and Gibson *et al.* (1990) showed that microbial decomposition of barley straw created compounds that inhibited the growth of algae and noted that blue-green algae, which are of particular concern as they release toxins into waterways, may also be susceptible to allelochemicals produced from straw. Although the presence of gramine in aboveground parts of barley is well established, there was no evidence for the involvement of alkaloids in this work, only that allelopathy may be a factor in aquatic ecosystems.

In terrestrial ecosystems alkaloids make plants bitter-tasting (Lamp and Collet, 1976). This factor may be of importance in the interference between alkaloid-producing species and more desirable, palatable plants in natural grasslands and pasture situations. These observations are confirmed by Harborne (1982) who, further, suggests that grazing behavior may be stimulated at low concentrations of an alkaloid while it is inhibited as the concentration increases. Wardle (1987b) speculated that the production of the alkaloid jacobine by ragwort (*Senecio jacobaea* L.) may be allelopathic in nature because, although toxic to stock, defoliation has little effect on its survival.

Castanospermine inhibits glycogen hydrolysis in animals (Reichmann *et al.*, 1987) (Fig. 2), just as scopolamine inhibits starch hydrolysis in plants (Levitt and Lovett, 1985).

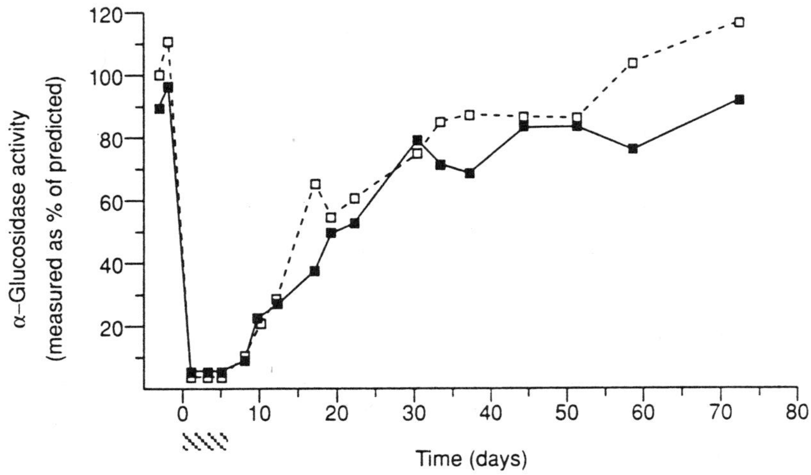

Figure 2. Changes in α-glucosidase activity (measured as percentage of predicted) in two animals dosed with *Castanospermum australe* seeds (0.6 g kg^{-1} body weight) daily for 6 days. Shaded area indicates days animals were dosed. After Reichmann *et al.* (1987); by permission of the *Australian Veterinary Journal.*

While similar in effect on contrasting organisms, these phenomena do not depend on the inhibition of an identical reaction.

4. CONCLUSION

There is an increasing body of evidence to suggest that alkaloids are capable of causing negative effects on the growth of many life forms be they microorganism, plant, or animal both invertebrate and vertebrate. Extensive evidence to support this contention is tabulated by Wink (1993). This multiplicity and variety of susceptible target suggests mechanisms of action for the alkaloids that are either common to all types of organism (see Robinson, 1974) or, perhaps, depend on parallel modes of action. Wink and Latz-Bruning (1995) referred to many basic plant processes that might be targeted by allelochemicals, e.g., membrane stability, protein synthesis, and electron transport chains, and developed bioassays, at least two of which used material of animal origin, to elucidate the mode of action of 14 alkaloids that inhibited radicle growth in cress. Many of these alkaloids were found to be inhibitory in more than one assay, some were active in none. Much more of this type of evidence is necessary; for example, we still do not know if an alkaloid that causes membrane leakage in erythrocytes will do so in plants. Far more necessary, however, is evidence that conclusively links the production of alkaloids with ecologically significant allelopathic activity in the field.

ACKNOWLEDGMENTS

We wish to thank the Grains Research and Development Corporation of Australia for funding one of the authors (A.H.C.H.) and our students and colleagues, Ms. N. Payne and Mr. A. Davidson, for their contributions to this chapter.

REFERENCES

General Reviews

Bonner, J., 1950, The role of toxic substances in the interaction of higher plants, *Bot. Rev.* **16:**51–65.

Borner, H., 1960, Liberation of organic substances from higher plants and their role in the soilsickness problem, *Bot. Rev.* **26:**393–424.

Evanari, M., 1949, Germination inhibitors, *Bot. Rev.* **15:**153–194.

Garb, S., 1961, Differential growth inhibitors produced by plants, *Bot. Rev.* **27:**422–443.

Grodzinski, A. M., 1992, Allelopathic effects of cruciferous plants in crop rotation, in: *Allelopathy: Basic and Applied Aspects* (S. J. H. Rizvi and V. Rizvi, eds.), Chapman & Hall, London, pp. 77–85.

Haas, P., and Hill, T. G., 1928, *An Introduction to the Chemistry of Plant Products,* Longmans, Green, New York.

Harborne, J. B., 1982, *Introduction to Ecological Biochemistry,* 2nd ed., Academic Press, New York.

Lamp, C., and Collet, F., 1976, *A Field Guide to Weeds in Australia,* Inkata Press, Melbourne.

Liu, De Li, and Lovett, J. V., 1990, Allelopathy in barley: Potential for biological suppression of weeds, in: *Alternatives to the Chemical Control of Weeds* (C. J. Bassett, L. J. Whitehouse, and J. A. Zabiewicz, eds.), Proceedings of an International Conference, Rotorua, New Zealand, July 1989, Ministry of Forestry, FRI Bulletin 155, pp. 85–92.

Loehwing, F., 1937, Root interactions of plants, *Bot. Rev.* **3:**195–239.

Lovett, J. V., 1982, Allelopathy and self-defence in plants, *Aust. Weeds* **2:**33–36.

Lovett, J. V., 1989, Allelopathy research in Australia: An update, in: *Phytochemical Ecology: Allelochemicals, Mycotoxins and Insect Pheromones and Allomones,* Institute of Botany, Academia Sinica Monograph Series No. 9 (C. H. Chou and G. R. Waller, eds.), Taipei, pp. 49–67.

Lovett, J. V. 1991, Changing perceptions of allelopathy and biological control, *Biol. Agric. Hortic.* **8:**89–100.

Lovett, J. V., and Hoult, A. H. C., 1993, Biological activity of barley secondary metabolites, in: Proceedings of the 7th Australian Agronomy Conference, Adelaide, S. Australia, pp. 158–161.

Lovett, J. V., Ryuntu, M. Y., and Liu, D. L., 1989, Allelopathy, chemical communication, and plant defense, *J. Chem. Ecol.* **15:**1193–1202.

Manske, R. H. F., 1950, Sources of alkaloids and their isolation, in: *The Alkaloids, Chemistry and Physiology,* Vol. 1 (R. H. F. Manske and H. R. Holmes, eds.), Academic Press, New York, p. 525.

Molisch, H., 1937, *Der Einfluss einer Pflanze auf die andere-Allelopathie,* Fischer, Jena.

Mothes, K., 1955, Physiology of alkaloids, *Annu. Rev. Plant Physiol.* **6:**393–492.

Muller, C. H., and Chou, C. H., 1971, Phytotoxins: An ecological phase of phytochemistry, in: *Phytochemical Ecology* (J. B. Harborne, ed.), Academic Press, New York, pp. 201–216.

Muller, C. H., Muller, W. H., and Haines, B. L., 1964, Volatile growth inhibitors produced by aromatic shrubs, *Science* **143:**471–473.

Rice, E. L., 1984, *Allelopathy,* 2nd ed., Academic Press, New York.

Robinson, T., 1974, Metabolism and function of alkaloids, *Science* **184:**430–435.

Suzuki, T., and Waller, G. R., 1987, Purine alkaloids in tea seeds during germination, in: *Allelochemicals: Role in Agriculture and Forestry* (G. R. Waller, ed.), ACS Symp. Ser. 330, pp. 289–294.

Swain, T., 1977, Secondary compounds as protective agents, *Annu. Rev. Plant Physiol.* **28:**479–501.

Tukey, H. B., 1969, Implications of allelopathy in agricultural plant science, *Bot. Rev.* **35:**1–16.

Waller, G. R., 1989, Allelochemical action of some natural products, in: *Phytochemical Ecology: Allelochemicals, Mycotoxins and Insect Pheromones and Allomones,* Institute of Botany, Academia Sinica Monograph Series No. 9 (C. H. Chou and G. R. Waller, eds.), Taipei, pp. 129–153.

Whittaker, R. H., 1970, The biochemical ecology of higher plants, in: *Chemical Ecology* (E. Sondheimer and J. B. Simeone, eds.), Academic Press, New York, pp. 43–70.

Wink, M., 1983, Inhibition of seed germination by quinolizidine alkaloids, *Planta* **158:**365–368.

Wink, M., 1987, Chemical ecology of quinolizidine alkaloids, in: *Allelochemicals: Role in Agriculture and Forestry* (G. R. Waller, ed.), ACS Symp. Ser. 330, pp. 524–533.

Wink, M., 1993, Allelochemical properties or the raison d'être of alkaloids, in: *The Alkaloids,* Vol. 43 (G. Cordell, ed.), Academic Press, San Diego, pp. 1–118.

Wink, M., and Latz-Bruning, B., 1995, Allelopathic properties of alkaloids and other natural products: Possible

modes of action, in: *Insights into Allelopathy* (Inderjit, K. M. M. Dakshini, and F. A. Einhellig, eds.), *American Chemical Society Symp. Ser.* **582,** 117–126.
Wink, M., and Twardowski, T., 1992, Allelochemical properties of alkaloids. Effects on plants, bacteria and protein biosynthesis, in: *Allelopathy: Basic and Applied Aspects* (S. J. H. Rizvi and V. Rizvi, eds.), Chapman & Hall, London, pp. 129–150.

Specific References

Biller, A., Boppre, M., Witte, L., and Hartmann, T., 1994, Pyrrolizidine alkaloids in *Chromolaena odorata.* Chemical and chemoecological aspects, *Phytochemistry* **35:**615–619.
Bode, N. R., 1939, Uber die Blattausscheidung des Wermuts und ihre Wirkung auf andere Pflanzen, *Planta* **30:**566–589.
Chou, C. H., and Waller, G. R., 1980, Possible allelopathic constituents of *Coffea arabica, J. Chem. Ecol.* **6:**643–654.
Conner, H. W., 1937, Effect of light on solanine synthesis in the potato tuber, *Plant Physiol.* **12:**79–98.
Egley, G. H., and Dale, J. E., 1970, Ethylene, 2-chloro-ethylphosphonic acid and witchweed germination, *Weed Sci.* **18:**586–589.
Einhellig, F. A., Rasmussen, J. A., Hejl, A. M., and Souza, I. F., 1993, Effects of root exudate sorgoleone on photosynthesis, *J. Chem. Ecol.* **19:**369–375.
Elakovich, S. D., and Wooten, J. W., 1991, Allelopathic potential of *Nuphar lutea* (L.) Sibth. and SM. (Nymphaceae), *J. Chem. Ecol.* **17:**707–714.
Gibson, M. T., Welch, I. M., Barrett, P. R. F., and Ridge, I., 1990, Barley straw as an inhibitor of algal growth. II. Laboratory studies, *J. Appl. Phycol.* **2:**241–248.
Gressel, J. B., and Holm, L. G., 1964, Chemical inhibition of crop germinaton by weed seeds and the nature of inhibition by *Abutilon theophrasti, Weed Res.* **4:**44–53.
Hoult, A. H. C., and Lovett, J. V., 1993, Biologically active secondary metabolites of barley. III. A method for identification and quantification of hordenine and gramine in barley by high-performance liquid chromatography, *J. Chem. Ecol.* **19:**2245–2254.
Hussain, F., Mubarak, B., Imtiaz-ul-Haq, and Naqvi, H. H., 1979, Allelopathic effects of *Datura innoxia, Pak. J. Bot.* **11:**141–154.
Kato-Noguchi, H., Kosemura, S., Yamamura, S., Mizutani, J., and Koji, H., 1994, Allelopathy of oats. I. Assessment of allelopathic potential of extract of oat shoots and identification of an allelochemical, *J. Chem. Ecol.* **20:**309–314.
Krischik, V. A., Goth, R. W., and Barbosa, P., 1991, Generalized plant defense: Effects on multiple species, *Oecologia* **85:**562–571.
Leete, E., 1985, Biosynthesis of hygrine from [5-^{14} C]ornithine via a symmetrical intermediate in *Nicandra physaloides, Phytochemistry* **24:**953–955.
Levitt, J., and Lovett, J. V., 1984, Activity of allelochemicals of *Datura stramonium* L. (thornapple) in contrasting soil types, *Plant Soil* **79:**181–189.
Levitt, J., and Lovett, J. V., 1985, Alkaloids, antagonism and allelopathy, *Biol. Agric. Hortic.* **2:**289–301.
Levitt, J., Lovett, J. V., and Garlick, P. R., 1984, *Datura stramonium* allelochemicals: Longevity in soil and ultrastructural effects on root tip cells of *Helianthus annuus* L., *New Phytol.* **97:**213–218.
Liu, De Li, and Lovett, J. V., 1993a, Biologically active secondary metabolites of barley. I. Developing techniques and asessing allelopathy in barley, *J. Chem. Ecol.* **19:**2217–2230.
Liu, De Li, and Lovett, J. V., 1993b, Biologically active secondary metabolites of barley. II. Phytotoxicity of barley allelochemicals, *J. Chem. Ecol.* **19:**2231–2244.
Lovett, J. V., Levitt, J., Duffield, A. M., and Smith, N. G., 1981, Allelopathic potential of *Datura stramonium* L. (thorn-apple), *Weed Res.* **21:**165–170.
McCalla, T. M., and Haskins, F. A., 1964, Phytotoxic substances from soil microorganisms and crop residues, *Bacteriol. Rev.* **28:**181–207.
Muzquiz, M., and de la Cuadra, C., 1988, personal communication.
Neilson, K. F., Cuddy, T., and Woods, W., 1960, The influence of the extract of some crops and soil residues on germination and growth, *Can. J. Plant Sci.* **40:**188–197.

Netzley, D. H., Riopel, J. L., Ejeta, G., and Butler, L. G., 1988, Germination stimulants of witchweed (*Striga asiatica*) from hydrophobic root exudate of sorghum (*Sorghum bicolor*), *Weed Sci.* **36:**441–446.
Ohigashi, H., Kaji, M., Sakaki, M., and Koshimizu, K., 1989, 3-Hydroxyuridine, an allelopathic factor of an African tree, *Baillonella toxisperma, Phytochemistry* **28:**1365–1368.
Overland, L., 1966, The role of allelopathic substances in the "smother crop", barley, *Am. J. Bot.* **53:**423–432.
Reichmann, K. G., Twist, J. O., McKenzie, R. A., and Rowan, K. J., 1987, Inhibition of bovine α-glucosidase by *Castanospermum australe* and its effects on the biochemical identification of heterozygotes for generalised glycogenesis II (Pompe's disease) in cattle, *Aust. Vet. J.* **9:**274–276.
Rice, E. L., 1964, Inhibition of nitrogen fixing and nitrifying bacteria by seed plants, *Ecology* **45:**824–832.
Roshchina, V. V., 1987, Action of acetylcholine and antagonists on reactions of photosynthetic membranes, *Photosynthetica* **21:**296–300.
Roshchina, V. V., 1988, Characterization of pea chloroplast cholinesterase: Effect of inhibitors of animal enzymes, *Photosynthetica* **22:**2–26.
Routley, D. G., and Sullivan, J. T., 1960, Toxic and nutritional effects of organic compounds on Ladino clover seedlings, *Agron. J.* **52:**317–319.
Schneider, E. A., and Wightman, F., 1974, Amino acid metabolism in plants. V. Changes in basic indole compounds and the development of tryptophan decarboxylase activity in barley (*Hordeum vulgare*) during germination and seedling growth, *Can. J. Biochem.* **52:**698–705.
Sepulveda, B. A., and Corcuera, L. J., 1990, Effect of gramine on the susceptibility of barley leaves to *Pseudomonas syringae, Phytochemistry* **29:**465–468.
Shehab, A. S., 1982, Effect of vinblastine and yohimbine alkaloids on meiosis of *Vicia faba, Egypt. J. Bot.* **25**(1–3):205–209.
Smyth, D. A., 1992, Effect of methylxanthine treatment on rice seedling growth, *J. Plant Growth Regul.* **11:**125–128.
Tyski, S., Markiewicz, M., Gulewicz, K., and Twardowski, T., 1988, The effect of lupin alkaloids and ethanol extracts from seeds of *Lupinus angustifolius* on selected bacterial strains, *J. Plant Physiol.* **133:**240–242.
Waller, G. R., Ashihara, H., and Suzuki, T., 1993, Updated review of purine alkaloid metabolism in *Coffea* and *Camellia* plants, ASIC, 15th Colloque, Montpellier, pp. 141–154.
Wardle, D. A., 1987a, Allelopathy in the New Zealand grassland/pasture ecosystem, *N. Z. J. Exp. Agric.* **15:**243–255.
Wardle, D. A., 1987b, The ecology of ragwort (*Senecio jacobaea* L.)—A review, *N. Z. J. Ecol.* **10:**67–76.
Welch, I. M., Barrett, P. R. F., Gibson, M. T., and Ridge, I.,1990, Barley straw as an inhibitor of algal growth. I. Studies in the Chesterfield canal, *J. Appl. Phycol.* **2:**231–239.
Wippich, C., and Wink, M., 1985, Biological properties of alkaloids. Influence of quinolizidine alkaloids and gramine on the germination and development of powdery mildew, *Erysiphe graminis* f. sp. *hordei, Experientia* **41:**1477–1479.
Woods, F. W., 1960, Biological antagonisms due to phytotoxic root exudates, *Bot. Rev.* **26:**546–569.

Chapter 15

Alkaloids in Animals

J. C. Braekman, D. Daloze, and J. M. Pasteels

1. INTRODUCTION

For a long time, the ability to produce alkaloids was thought to be restricted to the plant kingdom. It is now clear that at least some animals are able to biosynthesize these compounds, even if the remarkable variety of structures found in plants has no parallel in the animal kingdom. Generally, alkaloids are defined as being basic, nitrogen-containing natural compounds. However some compounds, such as colchicine, which contain a nonbasic nitrogen atom, are nevertheless considered as alkaloids. In this chapter, we will use the term *alkaloid* in a broad sense, but we will not consider nitrogen-containing derivatives such as amino acids, purine and pyrimidine bases, pigments, common polyamines (e.g., spermine, putrescine), nitro compounds, or cyanogenic glycosides. Similarly, biogenic amines which are widespread in mammals (see Brossi, 1993; Collins, 1983) and alkaloids sequestered or derived from plants by insect herbivores and their natural enemies which are covered elsewhere (see Wink, Chapter 11, this volume) will not be included in this chapter. Even so, it will not be possible to present an exhaustive list of species producing alkaloids, nor an exhaustive list of the alkaloids identified in animals. All major taxa in which alkaloids were described will be considered, however, as well as all major classes of alkaloids in terrestrial animals. Because of the huge number of alkaloids from marine animals, this overview is restricted to a few examples illustrating the diversity of compounds isolated from this source. When possible, review papers in which more detailed information can be found will be quoted. After having surveyed alkaloids in the different taxa, a brief account of our present knowledge of their biosynthesis will be given.

Alkaloids in animals act generally as defensive compounds and/or as chemical signals (pheromones). Evidence for these functions will be incorporated in the survey. For more comprehensive reviews on the ecological significance of alkaloids, the reader is referred to the recent paper of Brown and Trigo (1995) and to the contribution of Proksch and Ebel in this volume.

J. C. Braekman and D. Daloze • Laboratory of Bio-Organic Chemistry, Free University of Brussels, 50-1050 Brussels, Belgium. *J. M. Pasteels* • Laboratory of Animal and Cellular Biology, Free University of Brussels, 50-1050 Brussels, Belgium.

Alkaloids: Biochemistry, Ecology, and Medicinal Applications, edited by Roberts and Wink. Plenum Press, New York, 1998.

2. ARTHROPODS

2.1. Insecta

In the insects, most of the alkaloids are produced in two orders, Coleoptera (beetles) and Hymenoptera (mostly ants).

Because of the diversity and the large number of alkaloids isolated from beetles and ants, only a general overview will be given here. For a more comprehensive treatment, the reader is referred to the reviews of King and Meinwald (1996) and Daloze *et al.* (1995), Braekman and Daloze (1990), Numata and Ibuka (1987), Schmidt (1986), Attygalle and Morgan (1984), Jones and Blum (1983), Blum and Hermann (1978), Ayer and Browne (1977), and Tursch *et al.* (1976).

2.1.1. COLEOPTERA

2.1.1a. Coccinellidae

Many ladybird beetles are brightly colored (aposematic) and, when molested, emit hemolymph droplets at their joints. This mechanism, known as reflex bleeding, constitutes an efficient protection against predators, resulting from the presence in the discharged blood of bitter-tasting alkaloids (King and Meinwald, 1996; Daloze *et al.,* 1995; Tursch *et al.,* 1975; Pasteels *et al.,* 1973). Coccinellid beetles synthesize many types of defensive alkaloids, including acyclic amines, pyrrolidines, piperidines, 9-azabicyclo[3.3.1]nonanes (homotropanes), 2-methylperhydro-9*b*-azaphenalenes and azamacrolides. The most characteristic alkaloids found in these beetles, such as precoccinelline (**1**) and its *N*-oxide coccinelline (**2**), from *Coccinella septempunctata,* hippodamine (**3**) and its *N*-oxide convergine (**4**), from *Hippodamia convergens,* myrrhine (**5**), from *Myrrha octodecimguttata,* and propyleine (**6**), from *Propylaea quatuordecimpunctata,* are based on the 2-methylperhydro-9*b*-azaphenalene ring system (Fig. 1) (Tursch *et al.,* 1975, 1976). This skeleton is also found in the complex hexacyclic alkaloid exochomine (**7**) from *Exochomus quadripustulatus* (Timmermans *et al.,* 1992) and the heptacyclic chilocorine A (**8**) from *Chilocorus cacti* (McCormick *et al.,* 1994).

Two 9-azabicyclo[3.3.1]nonane derivatives, adaline (**9**) and euphococcinine (**10**) (Fig. 2), have been isolated from the two-spotted ladybird *Adalia bipunctata* (Tursch *et al.,* 1976) and from the Mexican bean beetle, *Epilachna varivestis* (Attygalle *et al.,* 1993a), respectively. The latter is a rich source of alkaloids, producing, besides euphococcinine an array of pyrrolidine and piperidine alkaloids (e.g., **11** to **13**) (Fig. 2) (Attygalle *et al.,* 1993a; Proksch *et al.,* 1993). In a study of the ontogenetic variations of defensive alkaloids in this phytophagous beetle, the eggs, larvae, and pupae were found to contain only the pyrrolidine alkaloids. Imagoes, for the first 2 days after adult emergence, resembled pupae with regard to the alkaloids present. In older adults, however, the pyrrolidine alkaloids (e.g., **12** and **13**) were progressively replaced by piperidine (e.g., **11**) and homotropane (euphococcinine) alkaloids (Proksch *et al.,* 1993). In another study of *E. varivestis,* azamacrolides such as epilachnene (**14**) have been evidenced as components of defensive droplets from glandular hairs of the pupae (Attygalle *et al.,* 1993b). Such qualitative

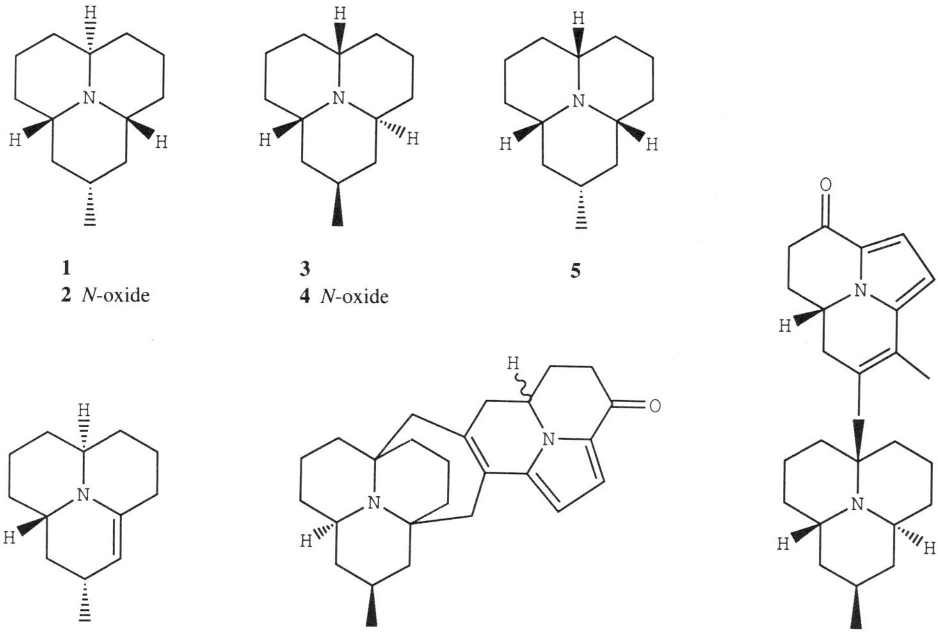

1
2 *N*-oxide
3
4 *N*-oxide
5
6
8
7

Figure 1. Coccinellidae alkaloids based on the 2-methylperhydro-9*b*-azaphenalene ring system.

evolution of the alkaloids during development was not observed in predatory ladybirds, e.g., *Coccinella septempunctata* and *Exochomus quadripustulatus*: eggs, larvae, and pupae contain the same alkaloids as the adults (Daloze *et al.*, 1995; Pasteels *et al.*, 1973).

Harmonine (**15**), a long-chain primary diamine, was found in several species of ladybirds (Braconnier *et al.*, 1985) (Fig. 2). Compounds **1–6, 9, 10,** and **15** have been synthesized by different research groups (e.g., see Yue *et al.*, 1994; Enders and Bartzen, 1991; Mueller *et al.*, 1984; Ayer and Browne, 1977).

From a survey of 30 species and varieties of ladybirds, it was concluded that the presence of alkaloids is correlated with the existence of aposematic colors and not with being carnivorous or phytophagous. Bitter-tasting alkaloids certainly play a major role in the defense strategy of coccinellid beetles against arthropod or vertebrate predators. Coccinelline (**2**) and convergine (**4**) were found to effectively deter ant workers from drinking, the threshold concentrations being 5×10^{-4} and 2×10^{-4} M, respectively (Pasteels *et al.*, 1973). Similar deterrence activity toward ants was demonstrated for euphococcinine which also protects *Epilachna varivestis* from jumping spiders (Eisner *et al.*, 1986), whereas the azamacrolide secretions of pupal hairs of this ladybird are also strongly deterrent to ants (Attygale *et al.*, 1993b).

Five species of alkaloid-containing ladybird beetles were systematically rejected by European quails, whereas three species devoid of alkaloids were readily accepted (Pasteels *et al.*, 1973). *Coccinella septempunctata* proved to be toxic to nestling blue tits, causing severe liver damage (Marples *et al.*, 1989). Toxicity of ladybirds to bird nestlings

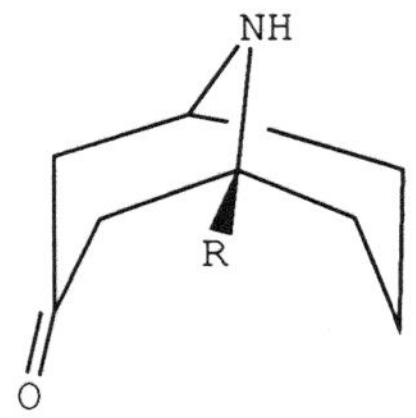

9 R = C_5H_{11}

10 R = CH_3

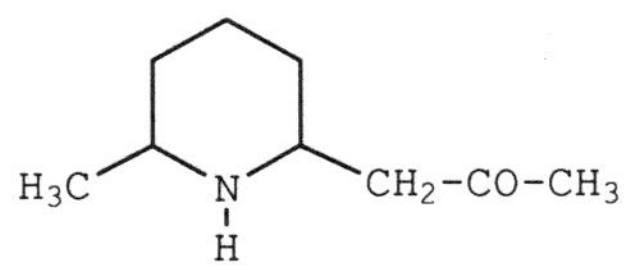

11

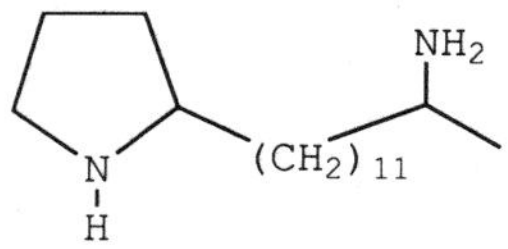

12

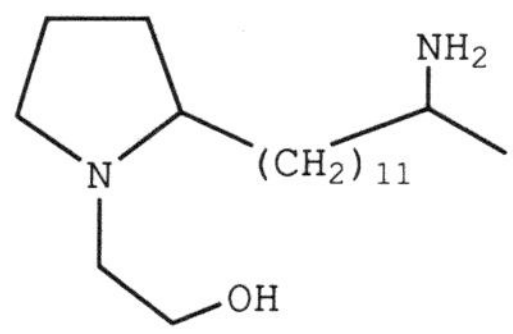

13

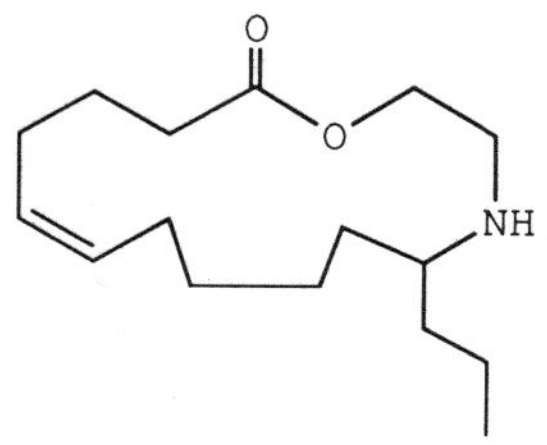

14

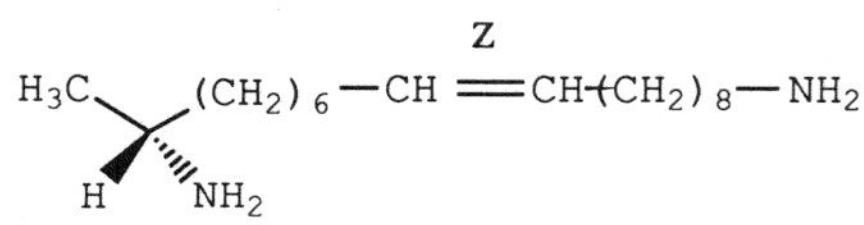

15

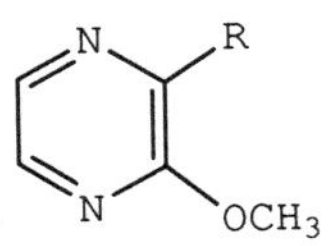

16 a R = i-Pr
b R = *s*-Bu
c R = i-Bu

Figure 2. Miscellaneous Coccinellidae alkaloids.

seems quite variable, however, monomorphic species being more toxic than polymorphic species suggesting Müllerian and Batesian mimetic relationships among species, monomorphic species being models for polymorphic species (Marples, 1993).

2-Methoxy-3-alkylpyrazines (**16**) are responsible of the typical odor of coccinellids when they reflex bleed and act as warning signals for predators together with aposematic coloration (Moore *et al.*, 1990) (Fig. 2).

2.1.1b. Staphylinidae

The hemolymph of staphylinid beetles of the genus *Paederus* contains a powerful cytotoxin, pederin (**17**) (Fig. 3) (Furusaki *et al.*, 1968). It causes severe dermotoxic symptoms when applied to the skin of warm-blooded animals and is toxic on ingestion

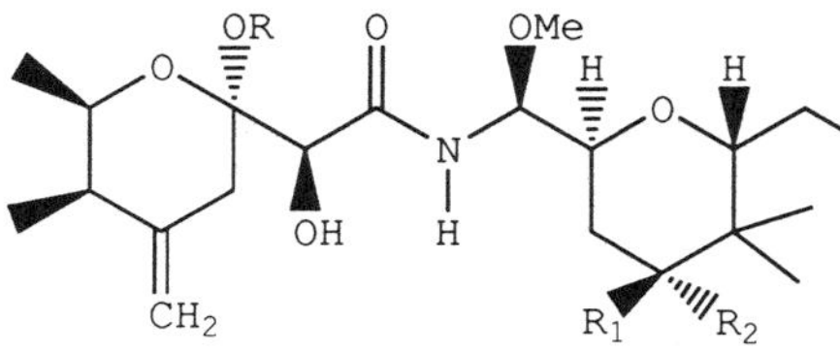

17 R = CH_3; R_1 = H; R_2 = OH
18 R = H; R_1 = H; R_2 = OH
19 R = CH_3; R_1 = R_2 = O

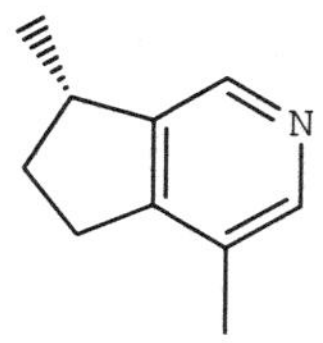

20

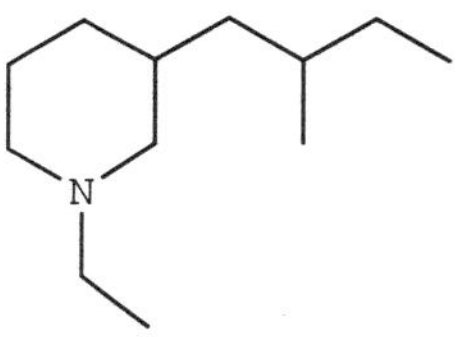

21

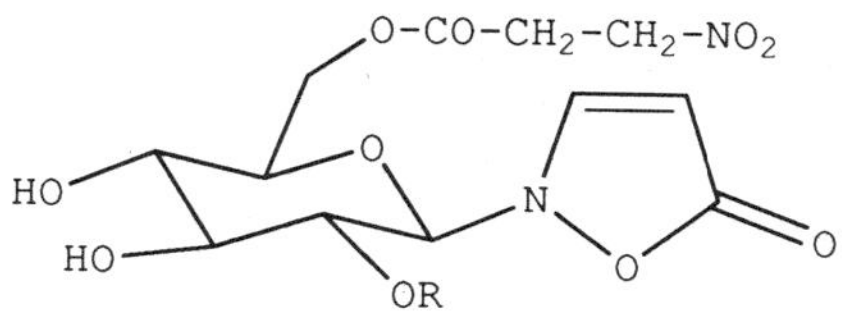

22 R = H
23 R = CO-CH_2-CH_2-NO_2

Figure 3. Alkaloids of Staphylinidae and Chrysomelidae.

(Capelli *et al.,* 1982). Two closely related compounds, pseudopederin (**18**) and pederone (**19**), have also been isolated from *Paederus fuscipes* (Cardani *et al.,* 1973a). It is intriguing that several metabolites closely related to pederin (e.g., mycalamide A) have recently been isolated from marine sponges (Perry *et al.,* 1988). The total synthesis of (+)-pederin has been achieved (Matsuda *et al.,* 1988).

The monoterpene alkaloid actinidine (**20**) (Fig. 3) is present in the defensive secretion of several species of staphylinid beetles, Staphylininae and Xantholininae (Huth and Dettner, 1990), accompanied by iridodial from which it is presumably derived. It is usually a trace or minor constituent, except in the Philonthina in which it is the major constituent of the secretion (Dettner, 1983).

Stenusine (**21**) (Fig. 3) is emitted from the pygidial glands of an aquatic staphylinid beetle, *Stenus comma.* This compound has a high spreading power and can efficiently propel the beetle on water surfaces, thus providing it with an elegant escape mechanism (Schildknecht *et al.,* 1976). Enantioselective synthesis has shown that natural stenusine is a mixture of the four possible stereoisomers (Enders *et al.,* 1993).

2.1.1c. Chrysomelidae

Four out of the nineteen subfamilies of Chrysomelidae (leaf beetles) possess pronotal and elytral glands that produce and release defensive compounds (Pasteels *et al.,* 1988, 1994). Several species belonging to the subtribe Chrysomelina secrete *N*-glucosides of Δ^3-isoxazolin-5-one such as **22** and **23** bearing one or more 3-nitropropanoic acid residues on the sugar moiety (Fig. 3) (Pasteels *et al.,* 1982). These compounds, which are also present in the eggs of the beetles, are strongly repellent for ants. Emission of the defensive secretion by the adults is accompanied by release of free 3-nitropropanoic acid, probably through an enzyme-mediated reaction.

Some larvae of leaf beetles also possess defensive glands characterized by the autogenous production of volatile iridoid monoterpenes, accompanied in some cases by a trace of actinidine (**20,** Fig. 3) (Veith *et al.,* 1994).

2.1.1d. Other Beetles

Few alkaloids have been isolated from beetles in groups other than those discussed above. A quinoline derivative, **24** (Fig. 4), is present in the prothoracic defensive glands of the water beetle *Ilybius fenestratus* (Dytiscidae). It is a powerful antiseptic, which could be used by the insect to prevent penetration of microorganisms (Schildknecht, 1970). The ladybird alkaloids precoccinelline (**1**), hippodamine (**3**), and propyleine (**6**) (Fig. 1) have been found in a soldier beetle (Cantharidae) (Moore and Brown, 1978). 3-Phenylpropanamide (**25**) and 1-methyl-2-quinolone (**26**) (Fig. 4) are the major bitter principles in the hemolymph of *Metriorrhynchus rhipidius* (Lycidae) (Moore and Brown, 1981). Bitter glucosides [e.g., buprestin A (**27**), Fig. 4] have been found in *Stigmodera maculata* (Buprestidae) (Brown *et al.,* 1985).

Various aposematic beetles, e.g., Lycidae, Cantharidae, Endomychidae, Oedemeridae, Meloidae (list in Moore *et al.,* 1990), release 2-methoxy-3-alkylpyrazines (**16,** Fig. 2) acting as warning odor as in the ladybirds and other aposematic insects.

2.1.2. HYMENOPTERA

2.1.2a. Formicidae

Ants are highly evolved social insects often organized into large colonies. Members of the ant colony coordinate their activities by means of a complex communication

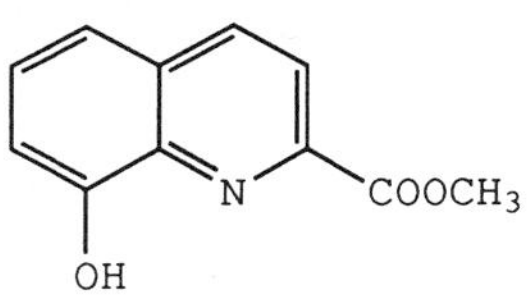

24

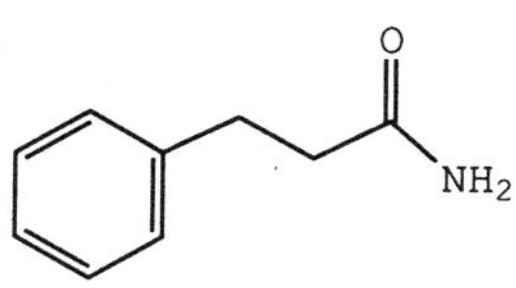

25

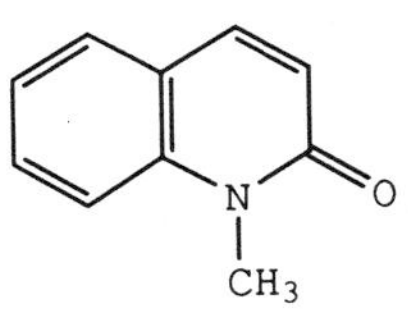

26

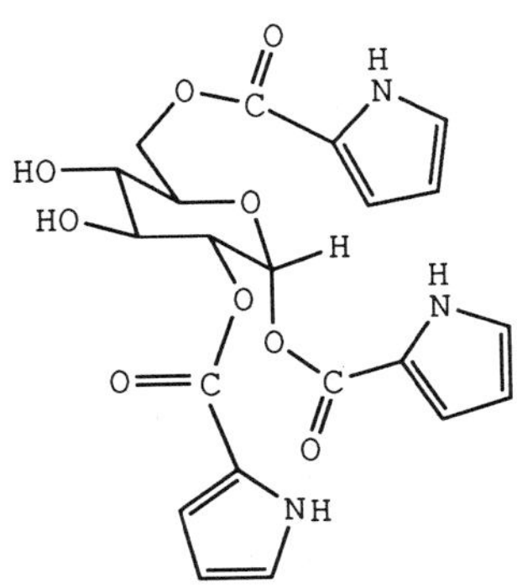

27

Figure 4. Alkaloids from other beetles.

system, mostly based on chemical signals. These are produced in specialized glands and are used by ants in relationship with their nestmates, members of other colonies and species of ants, and other organisms. Besides their complex intraspecific communication system, the vast majority of ant species use chemical secretions for defensive and offensive purposes. Among the different exocrine glands, the poison gland which is attached to the sting plays a major role in defense mechanisms. In many subfamilies, e.g., Myrmeciinae, Pseudomyrmecinae, Ponerinae, Dorylinae, and Myrmicinae, the constituents of the poison gland are usually proteinaceous (Schmidt, 1986; Blum and Hermann, 1978). However, in some groups, these protein venoms have been superseded by low-molecular-weight organic compounds including alkaloids acting as toxins or pheromones. Besides the poison gland the mandibular and anal (pygidial) glands are also the source of alkaloidic pheromones and defensive compounds in some ants. In this survey, we have followed the ant classification of Hölldobler and Wilson (1990).

Myrmicinae

Solenopsidini

Solenopsis. The large *Solenopsis* genus-group consists of numerous species of fire ants in the subgenus *Solenopsis* plus the much smaller thief ants in the subgenus *Diplorhoptrum,* plus others (e.g., *Euophthalma*).

The venom of fire ants is characterized by a predominance of *cis*- and *trans*-2-methyl-6-alkylpiperidines, for which the trivial name *solenopsins* has been coined. The structures of these compounds are remarkably similar, with only the relative configuration at carbons 2 and 6 of the piperidine ring and the length and unsaturation of the side chain varying (Fig. 5). The alkyl side chain contains from 7 to 17 carbon atoms; *trans*- and *cis*-C_{11} (solenopsin A, **28a,** and isosolenopsin A, **28b**), *trans*- and *cis*-C_{13} (solenopsin B, **29a,** and isosolenopsin B, **29b**), and *trans*-C_{15} (solenopsin C, **30a**) are the most frequently encountered. Most of the unsaturated derivatives have a *Z* double bond nine carbons from the end of the side chain (**31a, 31b, 32a, 32b,** Fig. 5). One piperideine derivative (**33**) is present in small quantities in the venom of *S. xyloni.* The absolute configuration of the solenopsins was recently determined to be always 2*R*,6*R* for the *trans* and 2*R*,6*S* for the *cis* alkaloids (Leclercq *et al.,* 1994), a finding that could have biosynthetic implications. The blends of alkaloids in the fire ant venoms are species specific. The venoms of the species that induce the greatest pain in stung humans, *S. invicta* and *S. richteri,* are both dominated by *trans* alkaloids (e.g., **29a** and **30a**), thus suggesting a correlation between the relative configuration of the solenopsins and their potency.

The venoms of *Solenopsis* exhibit pronounced necrotic, hemolytic, antibiotic, insecticidal, and toxic properties (references in Attygalle and Morgan, 1984, and Escoubas and Blum, 1990). It is used to disperse foreign ant competitors and presumably as an antiseptic in the brood chamber, being dispersed as an aerosol by "gaster flagging" during which the gaster is raised and vibrated, the sting extruded (Obin and Vander Meer, 1985). The queen of *S. invicta* secretes in its poison gland *cis*-2-methyl-6-undecylpiperidine which is deposited on its eggs during egg-laying, acting as potent fungicidal and antibacterial agents (Vander Meer and Morel, 1994).

Solenopsis (*Diplorhoptrum* and *Euophthalma*). Thief ants in these two subgenera are small and their sting is ineffective to humans. Their name derives from their ability to enter into the brood chamber of other ant species, where they repel the host workers by releasing their venom, and then steal the host's brood (Blum *et al.,* 1980). Their venoms, together with those of *Monomorium* spp., are the most complex of any alkaloid-producing ants (Fig. 6), *trans*-2,5-dialkylpyrrolidines (e.g., **34** to **37,** Fig. 6) with saturated side

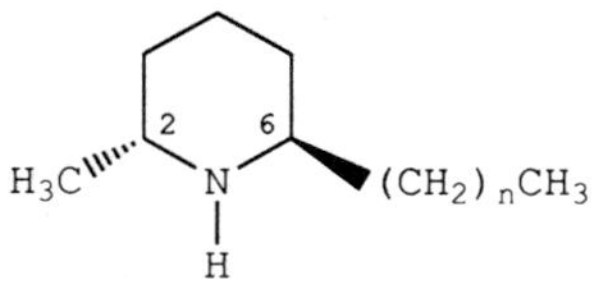

28a $n = 10$
29a $n = 12$
30a $n = 14$

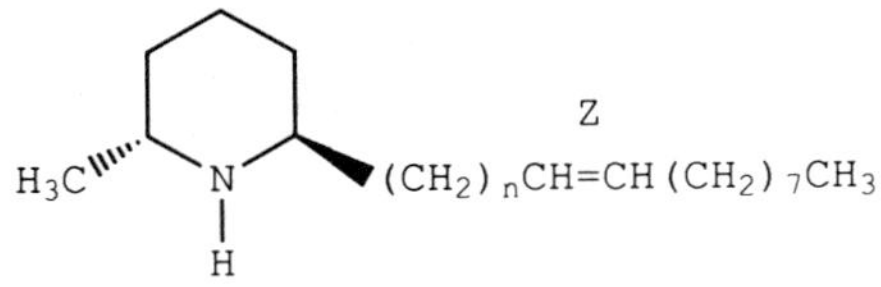

31a $n = 3$
32a $n = 5$

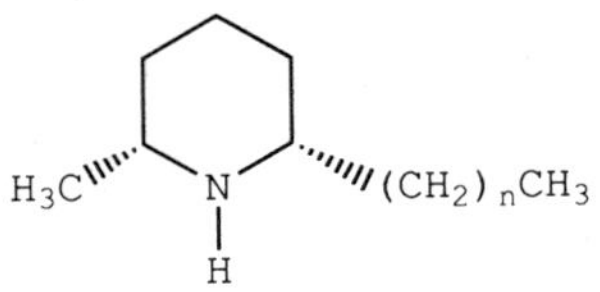

28b $n = 10$
29b $n = 12$
30b $n = 14$

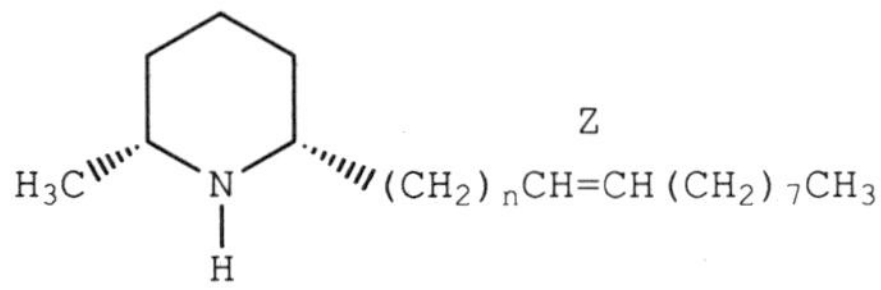

31b $n = 3$
32b $n = 5$

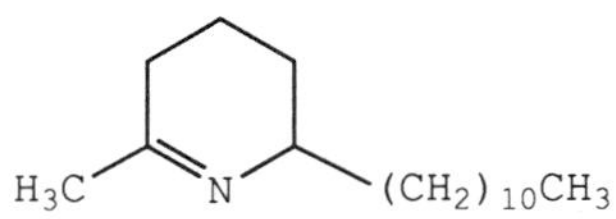

33

Figure 5. Piperidine alkaloids from *Solenopsis* (*Solenopsis*) ants.

chains of two to seven carbon atoms being the major constituents. Several 1-pyrrolines (e.g., **38** to **41**) have also been found as well as a few *N*-methylated analogues. In addition to producing these five-membered ring alkaloids, thief ants also synthesize piperidines similar to those found in fire ants (Fig. 5), *N*-methylpiperidines, 2-(4-penten-1-yl)-1-piperideine (**42**), and two (5*Z*,9*Z*)-3-alkyl-5-methylindolizidines (**43** and **44**) (Jones *et al.*, 1984). In *Solenopsis* (*Diplorhoptrum*) sp. near *tennesseensis*, (5Z,8*E*)-3-heptyl-5-methylpyrrolizidine (**45**) has been evidenced (Jones *et al.*, 1980).

Monomorium and *Megalomyrmex*. As is the case with thief ants, the venoms of *Monomorium* species are characterized by the presence of *trans*-2,5-dialkylpyrrolidines (e.g., **46–48, 50**), *N*-methyl-*trans*-2,5-dialkylpyrrolidines (e.g., **49, 51**), and 2,5-dialkylpyrrolines (e.g., **52–57,** Fig. 7). However, the *Monomorium* compounds usually differ from those of the thief ants by having generally longer side chains and by the presence, in many cases, of an unsaturation in the terminal position. Besides 2,5-dialkylpyrrolidines, the venoms of several *Monomorium* spp. also contain indolizidine and pyrrolizidine alkaloids, e.g., (+)-(3*R*,5*S*,9*S*)-3-butyl-5-methylindolizidine (**58**) (monomorine I), the related **59,** and the pyrrolizidines **60** and **61** (Jones *et al.,* 1988). The composition of the venoms is again species specific.

Investigations of the related genus *Megalomyrmex* showed that its alkaloidal content is closely related to that of *Monomorium* (Jones *et al.,* 1991).

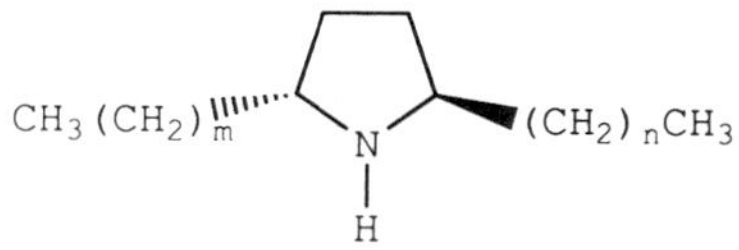

34 $m = 4, n = 3$
35 $m = 5, n = 4$
36 $m = 6, n = 1$
37 $m = 6, n = 3$

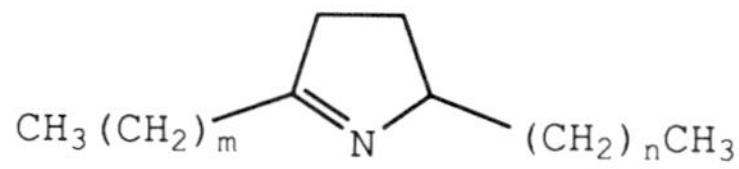

38 $m = 4, n = 1$
39 $m = 1, n = 4$
40 $m = 6, n = 1$
41 $m = 1, n = 6$

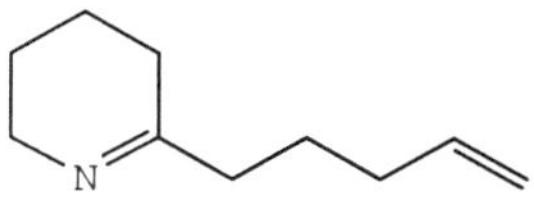

42

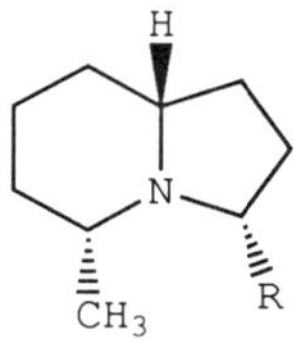

43 R = C_2H_5
44 R = n-C_6H_{11}

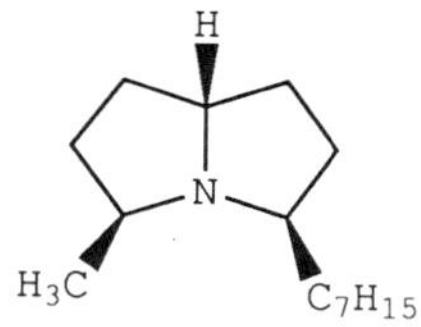

45

Figure 6. Alkaloids from *Solenopsis* (*Euophthalma* and *Diplorhoptrum*) ants.

Monomorium alkaloids are deposited by the ants on their prey (e.g., termites) and demonstrated an insecticidal activity higher than that of nicotine. However, the ants themselves proved to be remarkably immune to their alkaloids (Escoubas *et al.,* 1987).

The remarkable biological activities of these alkaloids and the minute amounts available from natural sources prompted many groups to develop new and efficient methods for their synthesis (for reviews, see Braekman and Daloze, 1990; Numata and Ibuka, 1987).

Leptothoracini

The poison gland of the females of the social parasitic ants *Harpagoxenus sublaevis* and *Doronomyrmex goesswaldi* and host species *Leptothorax accrorum* and *L. muscorum* contains *N*-alkylated 3-methylpyrrolidines in nanogram or picogram quantities. At least in the parasitic species, they act as sex pheromones (Reder *et al.,* 1995).

Myrmicini, Attini, and Tetramoriini

Trail pheromones in ants are produced by various exocrine glands and include quite diverse compounds (references in Hölldobler and Wilson, 1990). In members of these three myrmicine subtribes, the trail pheromone is alkaloidic and secreted by the poison gland (references in Schmidt, 1986; Attygalle and Morgan, 1984). For example, the pheromone of *Atta texana* has been identified as 4-methylpyrrole-2-carboxylate (**62,** Fig. 8). In other *Atta* species, the latter compound is accompanied by 3-ethyl-2,5-dimethylpyrazine (**63a**) or 3-ethyl-2,6-dimethylpyrazine (**64**). Not all of these alkaloids function as trail pheromone even when present in the poison gland, and the alkaloid acting as pheromone depends on the species. In eight species of *Myrmica* and some *Atta* species, it is **63a** that is the pheromone. In *Manica rubida,* methylpyrazine, 2,5-dimethylpyrazine (**63b**), and trimethylpyrazine are also present, but only **63a** acts as trail pheromone (Attygalle *et al.,* 1986). In *Tetramorium caespitum,* **63a** is used synergistically with **63b**, whereas in *T. meridionale* the trail pheromone is a synergistic mixture of **63a,** methylpyrazine, trimethylpyrazine and indole (Jackson *et al.,* 1990). These alkaloids are present

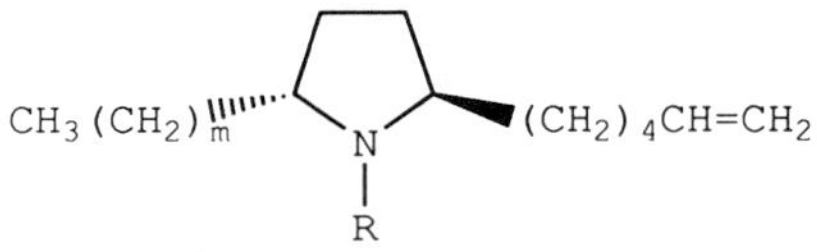

46 $m = 4$, R = H
47 $m = 6$, R = H
48 $m = 8$, R = H
49 $m = 8$, R = CH_3

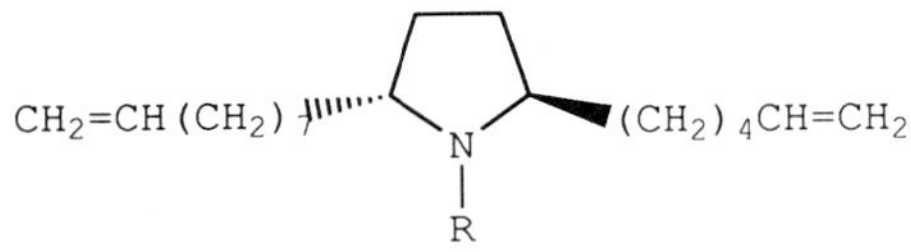

50 R = H
51 R = CH_3

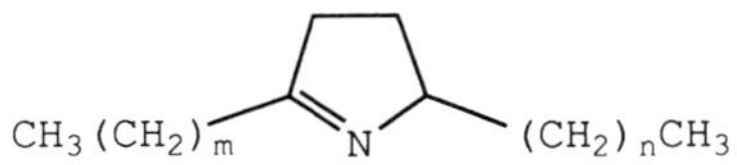

52 $m = 4$, $n = 3$
53 $m = 6$, $n = 1$
54 $m = 1$, $n = 6$
55 $m = 6$, $n = 3$

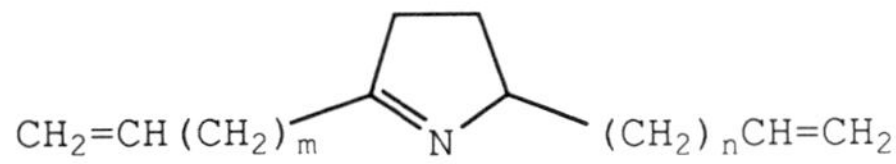

56 $m = 4$, $n = 7$
57 $m = 7$, $n = 4$

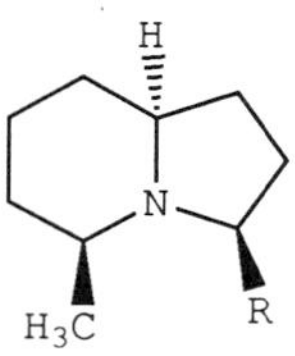

58 R = C_4H_9
59 R = C_2H_5

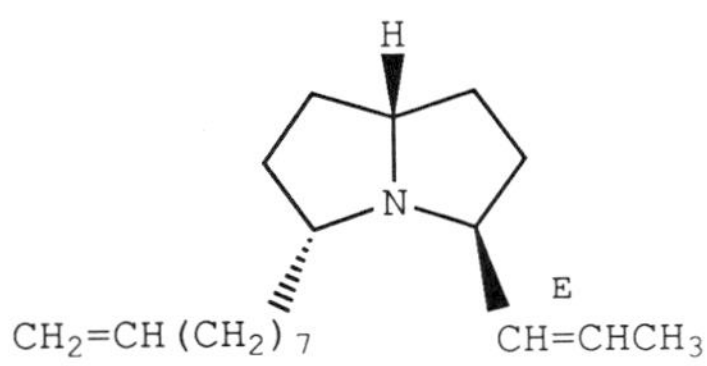

60

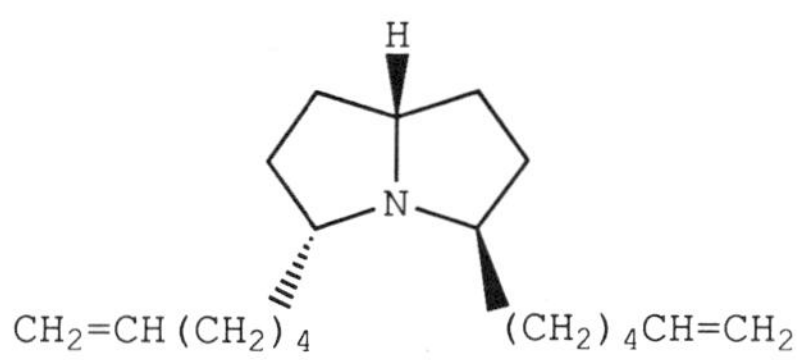

61

Figure 7. Pyrrolidine, indolizidine, and pyrrolizidine alkaloids from *Monomorium* and *Megalomyrmex* ants.

in nanogram or picogram quantities, but are remarkably active compounds. In *Atta texana,* the threshold response to **62** was evaluated as 0.08 pg/cm, meaning that one-third of 1 mg would be sufficient to draw an active trail around the world (Tumlinson *et al.,* 1971). Pheidolini (references in Schmidt, 1986) and *Wasmannia*

The venoms of two *Aphaenogaster* species contain the tobacco alkaloid anabaseine (**65**), which acts as an attractant for *Aphaenogaster fulva.* In *Messor capensis,* **65** is accompanied by the related compound anabasine (**66**) (Brand and Mpuru, 1993). Skatole (**67**) has been found in *Pheidole fallax.* The exact function of these alkaloids remains to be investigated. Trialkylpyrazines and 3-isopentyl-2,5-dimethylpyrazine (**74,** Fig. 10) have been reported respectively from the mandibular glands of *Aphaenogaster rudis* and *Wasmannia auropunctata* (references in Jones and Blum, 1983). In the latter species, **74** acts as repellent and alarm pheromone.

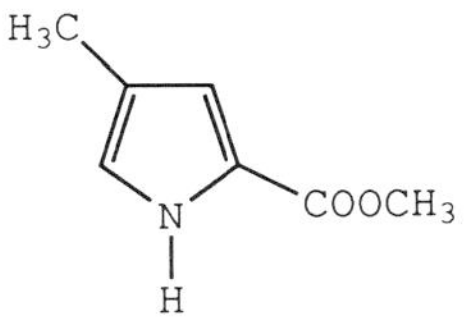

62

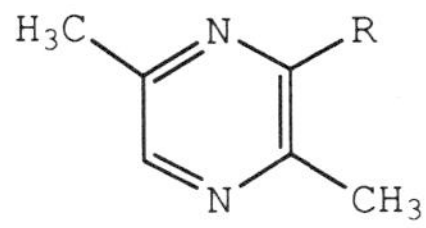

63a R = C_2H_5
63b R = H

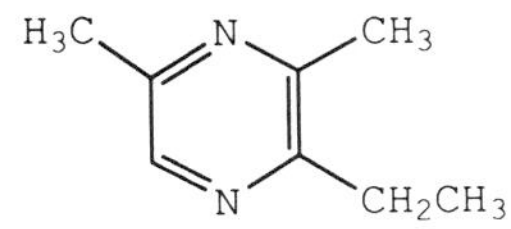

64

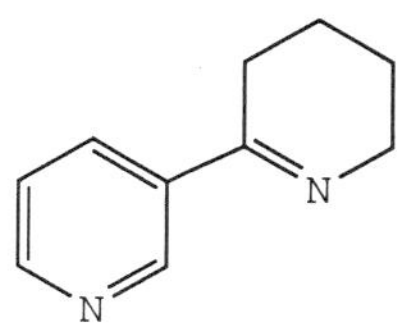

65

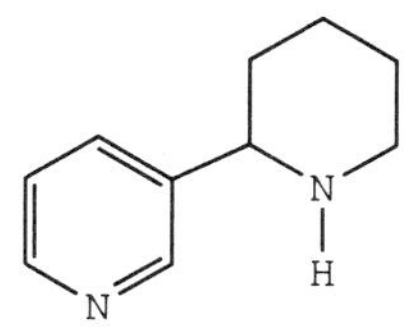

66

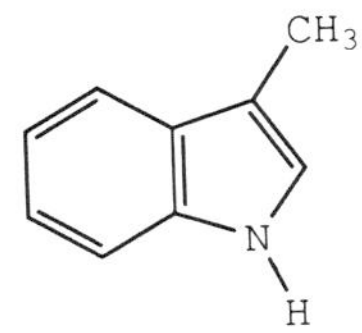

67

Figure 8. Miscellaneous alkaloids from ants.

Pseudomyrmecinae

Tetraponera sp.: The New Guinean *Tetraponera* sp. has a modified sting that is not suitable as an injection device, but well suited to deposit its venom on an enemy. The venom contains eight related tricyclic alkaloids, the tetraponerines (T-1 to T-8), which are the more structurally complex alkaloids found in ants so far. The structure of (+)-T-8 (**68,** Fig. 9), the major alkaloid of the venom, was determined by an X-ray diffraction analysis (Braekman *et al.,* 1987). Structures were subsequently proposed for five other tetraponerines, T-3 to T-7, by comparison of their spectroscopic properties with those of T-8 (Merlin *et al.,* 1988). Recent synthetic work led to a revision of the structures of T-3, T-5, T-6, and T-7. The structures and absolute configuration of (+)-T-8 (**68**), (+)-T-4 (**69**), (+)-T-7 (**70**), (+)-T-3 (**71**), (+)-T-6 (**72**), and (+)-T-5 (**73**) are depicted in Fig. 9 (Macours *et al.,* 1995; De Vijver *et al.,* 1995). The venom of *Tetraponera* sp. is highly efficient in deterring aggressive sympatric ants; T-8 shows toxic properties against other insects when applied topically (Braekman *et al.,* 1987).

Dolichoderinae, Formicinae, and Ponerinae

A number of alkylpyrazines were isolated from the mandibular glands of ants belonging to these subfamilies, where they seem to act as alarm pheromones and/or defensive substances (references in Attygalle and Morgan, 1984, and Jones and Blum, 1983). In ponerine ants of the genus *Mesoponera,* cephalic extracts contained mostly 3-isopentyl-2,5-dimethylpyrazine (**74,** Fig. 10), whereas thoracic and gaster extracts yielded a series of secondary dialkylamines and amides, of which *N*-isopentylnonylamine (**75,** Fig. 10) was the major component (Fales *et al.,* 1984).

Males of *Camponotus sansabaneus,* but not workers or females, secrete **74** in their mandibular glands, possibly acting as sex pheromone during swarming (Blum, 1981).

Secondary amines characterized by terminal unsaturations in the two alkyl chains (e.g., **76** to **78,** Fig. 10) were found in the venom of a dolichoderine ant, *Technomyrmex albipes* (Brophy *et al.,* 1993). Actinidine (**20,** Fig. 3) was also found as a constituent of the defensive secretions of some dolichoderine ants, produced in their anal (pygidial) gland rich in various iridoid monoterpenes (references in Attygalle and Morgan, 1984).

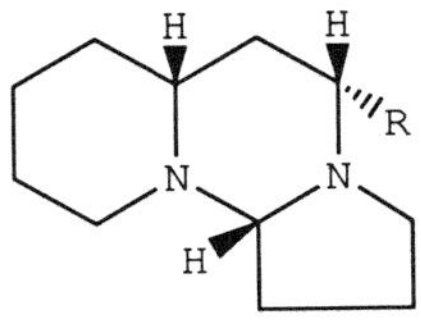

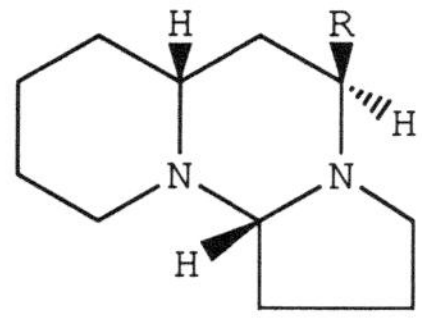

68 R = *n*-C_5H_{11}
69 R = *n*-C_3H_7

70 R = *n*-C_5H_{11}
71 R = *n*-C_3H_7

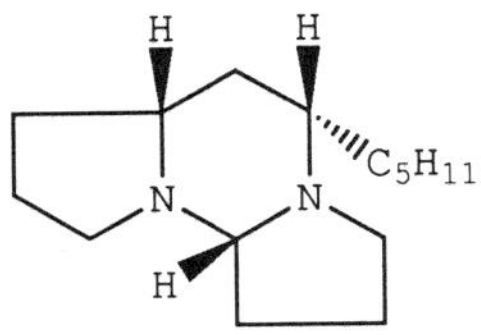

72

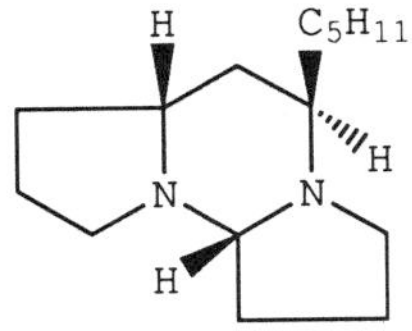

73

Figure 9. Alkaloids from *Tetraponera* sp. ants.

Myrmeciinae, Nothomyrmeciinae, and Ecitoninae

Indole in nanogram quantities was recently reported as a minor constituent of the pygidial glands of *Myrmecia nigriceps* and *Nothomyrmecia macrops* (Jackson *et al.*, 1990).

The army ant *Neivarmyrmex nigrescens* secretes skatole (**67**) presumably in its mandibular gland. Skatole possibly acts as a defensive compound against microorganisms and predators (Brown *et al.*, 1979).

2.1.2b. Other Hymenoptera (Bees and Wasps) (references in Numata and Ibuka, 1987)

The venoms of wasps and bees are mostly proteinaceous, but they usually also contain small biogenic amines such as histamine, dopamine, norepinephrine, 5-hydroxytryptamine, and acetylcholine (Schmidt, 1986), which are outside the scope of this chapter. An interesting low-molecular-weight polyamine, philanthotoxin-433 (**79,** Fig. 11), which is a potent blocker of the glutamate and nicotinic acetylcholine receptors, has been isolated from the venom of a wasp, *Philanthus triangulum* (Kalivretenos and Na-

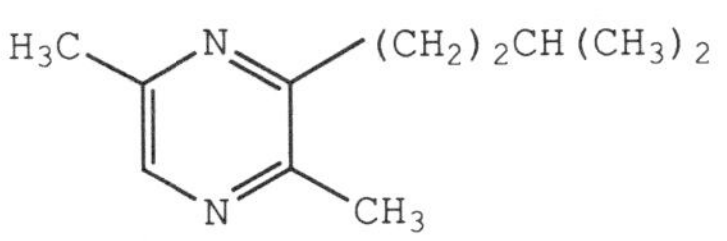

74

$CH_2{=}CH(CH_2)_8{-}NH{-}(CH_2)_9CH{=}CH_2$

77

$(CH_3)_2CH(CH_2)_2{-}NH{-}C_9H_{19}$

75

$CH_2{=}CH(CH_2)_6{-}NH{-}(CH_2)_7CH{=}CH_2$

76

$CH_2{=}CH(CH_2)_8{-}NH{-}(CH_2)_6CH{=}CH_2$

78

Figure 10. A pyrazine and secondary amines from *Camponotus* and *Technomyrmex* ants.

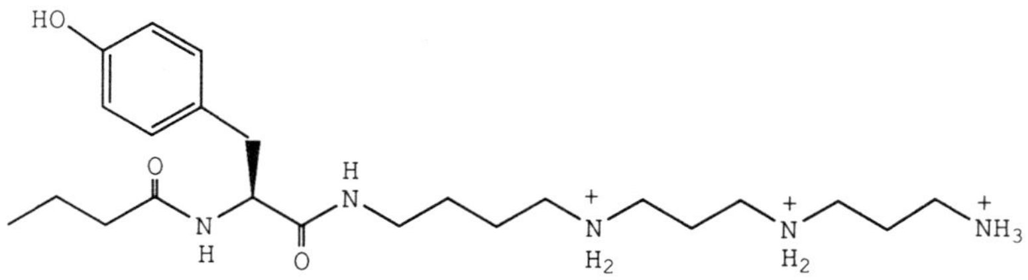

79

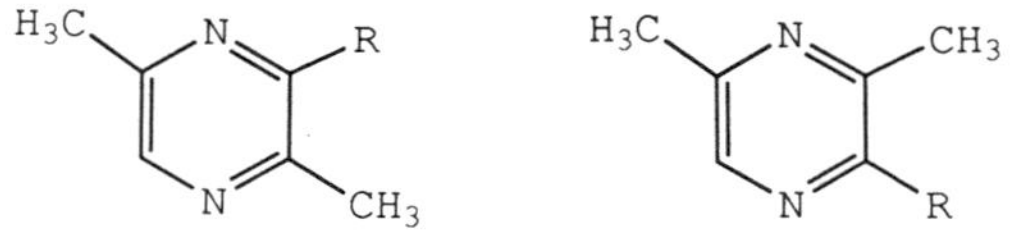

80 R = *n*-Pr
81 R = *n*-Bu
82 R = i-Bu

83 R = *n*-Pr
84 R = *n*-Bu
85 R = i-Pe

Figure 11. Alkaloids from wasps.

kanishi, 1993). Interestingly, polyamine **79** is closely related to some toxins found in the venoms of spiders (e.g., argiopine, see later).

Many wasps species (Tiphiidae, Eumanidae, Sphecidae *sensu lato,* see list in Numata and Ibuka, 1987) contain various alkylpyrazines (e.g., **74** and **80–85**), which could play a role in some species as aggregating pheromone during nocturnal roosting.

2.1.3. OTHER INSECTS (REFERENCES IN NUMATA AND IBUKA, 1987)

Actinidine (**20**) also found in some beetles (see above) is the major volatile constituent of the defensive secretion of a stick insect, *Megacrania tsudai* (Phasmida) from Taiwan (Ho and Chow, 1993). Mature males of the locust *Schistocerca gregaria* (Orthoptera, Acrididae) secrete as a volatile Δ^1-pyrroline responsible for the maturation-accelerating effect observed in the males of their species.

Skatole (**67,** Fig. 8) is secreted as defensive compound in the prothoracic gland of the lacewing, *Chrysopa oculata* (Chrysopidae, Neuroptera), and in the sternal glands of the caddisfly, *Pycropsyche scabripennis* (Linnephilidae), in admixture with indole.

Δ^1-Pyrroline and pyrazines are constituents of the sex pheromone secreted by the rectal gland of male fruit flies (Trypetidae).

Finally, 2-methoxy-3-alkylpyrazines (e.g., **16**) function as warning odors in many toxic insects, e.g., Lepidoptera (Papilionidae, Nymphalidae, including monarch butterflies and *Helioconius* spp., Zygaenidae, and Aretiidae) as well as Orthoptera (*Poekilocerus bufonius*) (list in Moore *et al.,* 1990).

2.2. Arachnida

2.2.1. ARANEA (SPIDERS)

Until now, about 30,000 species of spiders have been identified. All are carnivorous and have a pair of fangs to which, in most cases, a venom gland is associated. The venom

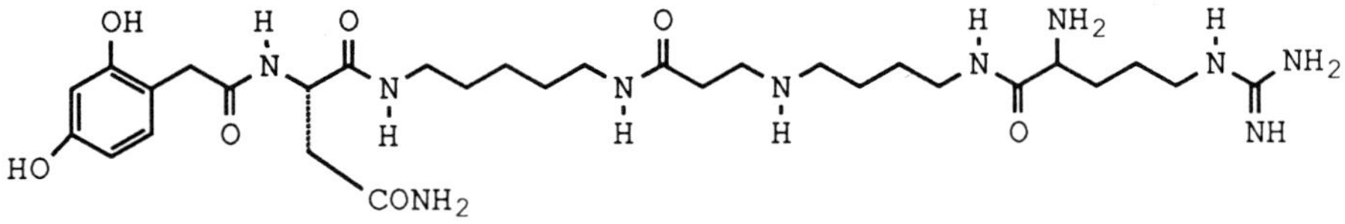

86

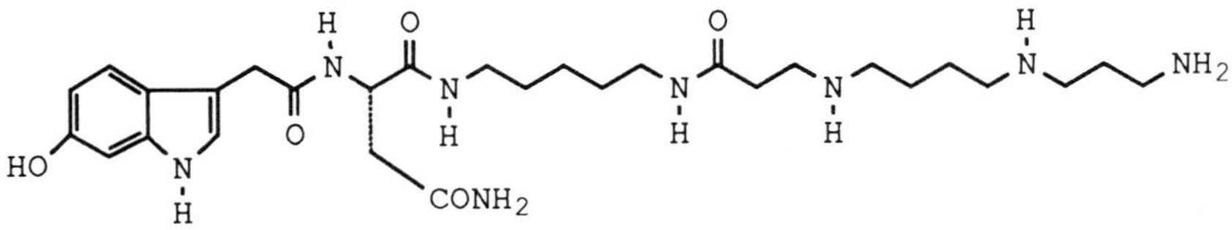

87

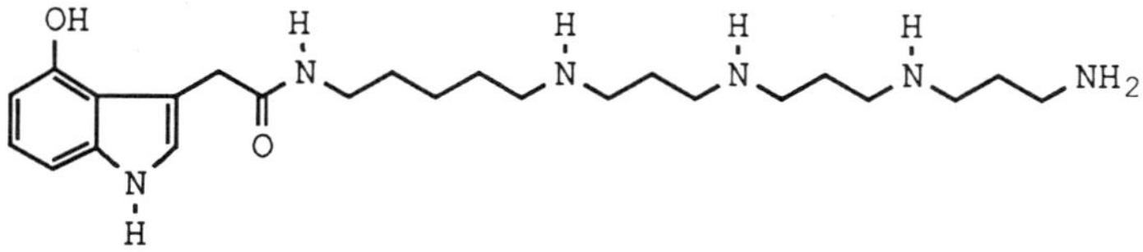

88

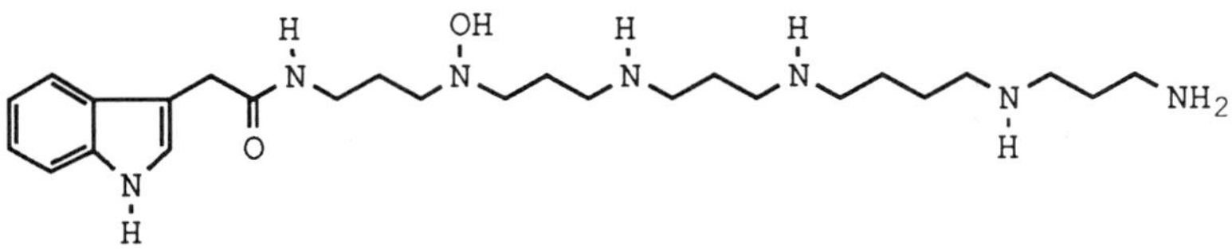

89

Figure 12. Polyamine alkaloids from spiders.

is used by the spider for hunting prey but also for defense. Generalities about the distribution, biology, pharmacology, and mode of action of spider venoms have been reviewed several times (Tu, 1984; Habermehl, 1981; Bettini, 1978).

The spider venoms are now receiving considerable attention because they function as neurotoxins and exhibit remarkable specificities and binding affinities for particular neural receptors or ion channels. McCormick and Meinwald (1993) have reported in a recent review on spider venom that "venom components from over 85 spiders representing 15 different families have been investigated, although fewer than a dozen of these species have been studied in depth." The crude venoms are always complex mixtures. They contain proteins or polypeptides responsible in general for the neurotoxicity of the venom, accompanied by small molecules such as α-amino acids and biogenic amines (e.g., histamine, serotonin).

In the venom of a few spiders, a new class of neuroactive acylpolyamines has been discovered. Their structures are characterized by a polyamine chain acylated by an aromatic acid. To date about 50 of these alkaloids have been fully characterized. An exhaustive list of these compounds and of their origin is given in McCormick and Meinwald (1993). They differ as to the nature of the aromatic acid (e.g., 3-indoleacetic acid, *p*-hydroxybenzoic acid) and of the polyamine chain. As far as the polyamine chain is concerned, the spider acylpolyamines can be divided into two main groups, those containing one or two amino acids between the acyl end and the polyamine chain [e.g., NSTX-3 (**86**) (Teshima *et al.*, 1991) and NPTX-1 (**87**) (Toki *et al.*, 1988)] and those without amino acid moieties [e.g., CNS 2103 (**88**) (McCormick *et al.*, 1993) and AGEL 489 (**89**) (Jasys *et al.*, 1990) (Fig. 12)].

Most of the spider acylpolyamines function by blocking glutamate-sensitive calcium channels. To better understand these activities and the way these channels are working, larger amounts of individual alkaloids were needed. This has encouraged many synthetic organic chemists to perform their synthesis. A compilation of all of the successful syntheses published before 1993 is presented in the review of McCormick and Meinwald (1993).

2.2.2. OPILIONES (HARVESTMEN)

The defensive secretion of *Sclerobunus robustus* contains, in admixture with terpenic compounds, *N,N*-dimethyl-β-phenylethylamine as a major constituent and nicotine as a minor one (Ekpa *et al.*, 1984).

2.3. Myriapoda–Diplopoda

The millipedes (Diplopoda) are a very ancient group of arthropods dating back to Devonian times. They possess a very effective defense mechanism consisting of secretions that are excreted from exocrine glands located in the segments, when they are disturbed. Chemically they consist of low molecular weight benzene and quinone derivatives frequently mixed with proteinaceous material that hardens quickly on exposure to air and sticks to attacking animals.

Of particular interest in the context of this review are the species *Glomeris marginata* and *Polyzonium rosalbum* whose defensive secretions possess characteristic nitrogen-containing compounds. The secretion of the European millipede *G. marginata* contains two unique quinazolinones (Schildknecht *et al.*, 1967; Schildknecht and Wenneis, 1967; Meinwald *et al.*, 1966): glomerin (**90**) and homoglomerin (**91**). Both alkaloids have been synthesized (Schildknecht and Wenneis, 1967). The effectiveness of the defensive secretion of *G. marginata* in repelling predators is well established. Ingestion of *G. marginata* is reported to cause delayed general symptoms and even death in mice as well as paralysis in spiders (Eisner, 1970).

The secretion of *P. rosalbum* has been shown to contain as a major constituent the spirocyclic alkaloid (+)-polyzonimine (**92**) (Smolanoff *et al.*, 1975). The structure of this unusual imine, first deduced from its spectral properties, was further confirmed by synthesis. It acts as a topical irritant to insects. Polyzonimine is accompanied in the secretion by a minor, less volatile derivative: (+)-nitropolyzonamine (**93,** Fig. 13) (Meinwald *et al.*, 1975). In addition to the same spirocyclic nucleus, the latter has another unusual feature not often found among animal metabolites, an aliphatic nitro group. Its structure and relative stereochemistry were based on an X-ray diffraction analysis of its perchlorate salt. The absolute configuration was revised by Miller and McPhail (1978) and shown to be as

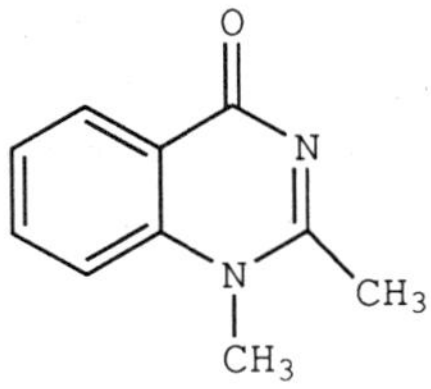

90

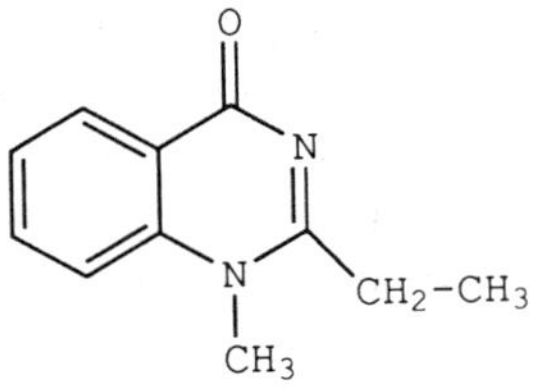

91

92

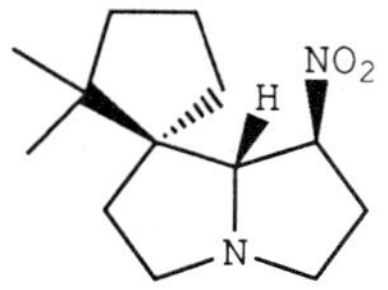

93

Figure 13. Alkaloids from Myriapoda–Diplopoda.

reported in **93.** Treatment of synthetic (+)-polyzonimine with 1-nitro-3-iodopropane gives a crystalline salt that cyclized readily to (+)-nitropolyzonamine (Miller and McPhail, 1978).

3. VERTEBRATES

With only few exceptions, all alkaloids observed in Vertebrata are secreted by salamanders, frogs, and toads.

3.1. Amphibia

Many amphibians have been found to produce noxious and sometimes poisonous substances in their skin secretions. It is commonly held that these secretions are used for protection against microbial infections and natural predators. A large variety of bioactive compounds of great chemical diversity including biogenic amines, peptides, proteins, steroids, and alkaloids have been isolated from the skin secretions of amphibians (Tu, 1984; Habermehl, 1971). Alkaloids have been found in salamanders of the genus *Salamandra* and in frogs of the family Dendrobatidae (*Phyllobates, Dendrobates, Epipedobates,* and *Minyobates*). Dendrobatid alkaloids have also been isolated from frogs of the genera *Mantella* (Mantellidae) and *Pseudophryne* (Myobatrachidae) as well as in toads of the genus *Melanophryniscus* (Bufonidae).

3.1.1. *SALAMANDRA*

Salamanders have been known to be toxic since ancient times. Most of our knowledge on the chemical constituents of the secretion of the parotid glands of the salamanders

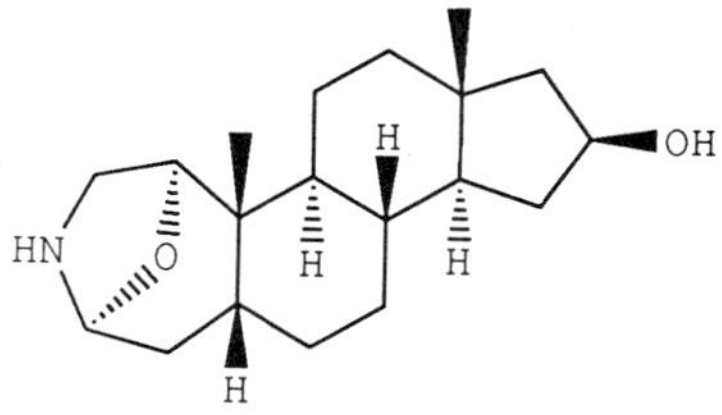

94

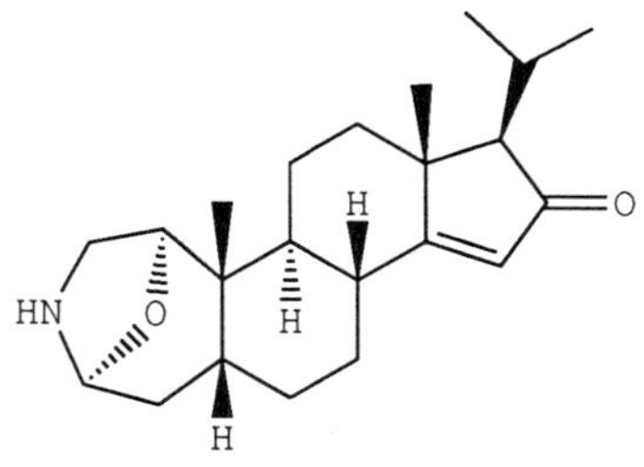

95

Figure 14. Alkaloids from *Salamandra*.

follows from the study of the European black and yellow spotted fire salamanders: *Salamandra maculosa taeniata* from western Europe and *S. maculosa maculosa* from southeastern Europe. The salamander toxins turned out to be steroidal alkaloids linked to the androstane skeleton, all distinguished by a ring A expanded to a seven-membered azaheterocycle *cis*-fused to ring B. The only other amphibian species known to contain samandarine-type alkaloids is the alpine salamander (*Salamandra atra*). Nine different alkaloids have been isolated [e.g., samandarine (**94**) and samandenone (**95**)], and their isolation and structure elucidation have been reviewed several times (e.g., Daly *et al.*, 1993; Witkop and Gossinger, 1983; Habermehl, 1971; Habermehl, 1967). The structure and absolute configuration of some of the *Salamandra* alkaloids have been confirmed by partial syntheses. Samandarine affects the CNS. It possesses remarkable hypertensive and local anesthetic properties. It is one-third as toxic as strychnine (Habermehl, 1971).

3.1.2. DENDROBATIDAE

Dendrobatid frogs have long been known to be venomous and their skin secretions are used by some Indian tribes of Colombia as dart poisons. The composition of the secretions have been very thoroughly studied in recent years and an astonishing array of alkaloids have been discovered. Occurrences, structure elucidations, syntheses, and biological activities of the dendrobatid alkaloids have already been reviewed in detail several times (Daly *et al.*, 1987, 1993; Daly and Spande, 1986; Witkop and Gossinger, 1983; Daly, 1982; Witkop, 1971). We will here only briefly report some general information about the main classes of dendrobatid alkaloids.

On the basis of their structures the dendrobatid alkaloids have been divided into five main classes: (1) the batrachotoxins, (2) the histrionicotoxins, (3) the simple indolizidine derivatives, (4) the pumiliotoxin-A-type alkaloids, and (5) the decahydroquinoline derivatives.

The batrachotoxins are highly toxic steroidal alkaloids which were the first class of dendrobatid alkaloids to be characterized. Six batrachotoxins have been isolated and the structures of five of these have been defined. Their structures are complex and based on the pregna-7,16-diene skeleton (**96** to **100**). The batrachotoxins have been detected from amphibians only in the skin of five species of *Phyllobates*.

The LD_{50} of batrachotoxin (**97**) on subcutaneous injection in mice is about 40 ng. The nature of the ester function is of critical importance to toxicity. The latter is the result

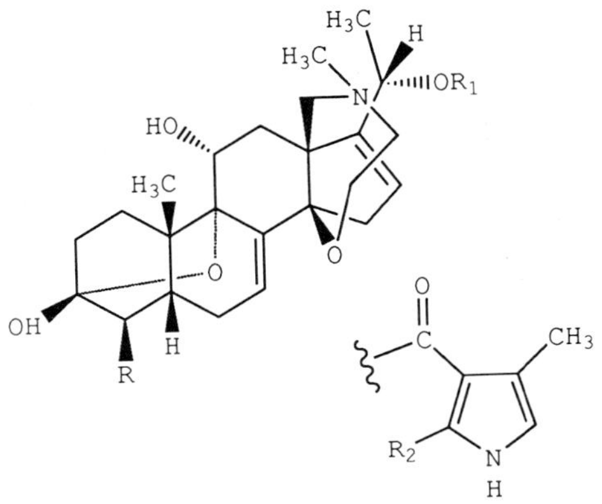

	R	R_1
96	H	H
97	OH	A
98	H	B
99	OH	A
100	OH	B

A $R_2 = CH_3$
B $R_2 = C_2H_5$

Figure 15. Batrachotoxin alkaloids from dendrobatid frogs.

of a specific high-affinity interaction of batrachotoxin with a site on voltage-dependent sodium channels of nerve and muscle, that keeps the sodium channels open. Because of this property, batrachotoxin has been particularly useful for the study of the function of these channels (Daly and Spande, 1986).

The histrionicotoxins represent a unique structural class of alkaloids characterized by the 1-azaspiro[5.5]undecan-8-ol system (**101**). They differ in the length and the number of unsaturations of the alkyl side chains in positions 2 and 7. They are restricted to various species of *Dendrobates, Phyllobates* and *Epipedobates*. About 16 histrionicotoxins have been detected, e.g., histrionicotoxin (**102**) and isotetrahydrohistrionicotoxin (**103**). The latter possesses an allenic unsaturation which is without precedent in animal alkaloids. Histrionicotoxins have relatively low toxicities.

The amphibian alkaloids also include a great variety of indolizidines which were originally placed in a single group, the gephyrotoxins (Daly, 1982). This group has now been split into two groups, the simple indolizidines and the pumiliotoxin-A type of alkaloids (Daly *et al.,* 1993).

The simple indolizidine class comprises 3,5- and 5,8-dialkylated indolizidines, e.g., **104** and **105**. They differ in the length and the relative configuration of the side chains as well as in the number of unsaturations.

The pumiliotoxins, the allopumiliotoxins, and the homopumiliotoxins compose the pumiliotoxin-A class of frog alkaloids. They are characterized by the substructures **106, 107,** and **108,** respectively. The allopumiliotoxins are pumiliotoxins with a hydroxyl group in the 7 position of the indolizidine ring and the homopumiliotoxins have a quinolizidine ring instead of the indolizidine ring. Pumiliotoxin-A (**106a**) and pumiliotoxin-B (**106b**) are relatively toxic compounds causing death in mice at subcutaneous doses as low as 20 and 50 mg, respectively.

The last main class of amphibian alkaloids is composed of a range of 2,5-dialkyl-decahydroquinolines having low toxicities. Both *cis-* and *trans-*fused derivatives have been isolated. Compounds **109** and **110** are two representatives of this class of frog alkaloids.

Next to these alkaloids most of which are unknown elsewhere in nature, a further large number of nitrogenous derivatives occur as minor or trace alkaloids in the skin extract of some frogs. Among these, 2,5-dialkylpyrrolidines, 2,6-dialkylpiperidines, and

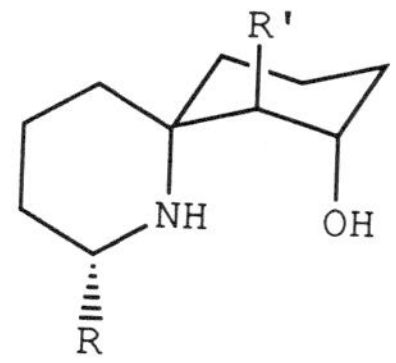

	R	R′
101	H	H
102	UCH$_2$CHuCHCuCH	UCHuCHCuCH
103	UCH$_2$CH$_2$CHuCuCH$_2$	UCHuCHCHuCH$_2$

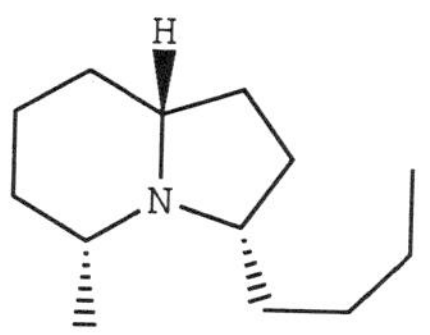

104

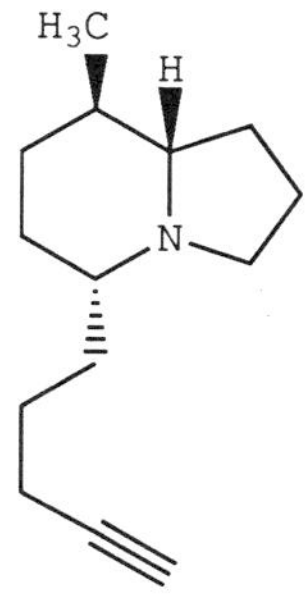

105

Figure 16. Histrionicotoxins and indolizidine alkaloids from frogs.

coccinelline alkaloids are well known as insect defensive compounds. A comprehensive list of the amphibian alkaloids can be found in Daly *et al.* (1993).

3.2. Other Vertebrates

A batrachotoxin, homobatrachotoxin (**98**), has been discovered recently in three passerine birds of New Guinea (*Pitohui* species). Concentration of alkaloid is highest in the skin and feathers, and this compound most probably acts as a chemical defense. The local New Guineans call the birds "rubbish birds" which should not be eaten unless skinned and specially prepared (Dumbacher *et al.*, 1992).

Finally, 2,5-dimethylpyrazine (**63b**) is a primer pheromone secreted by grouped female mice (*Mus domesticus*) in their urine. It inhibits sexual maturation in juvenile female and male mice at a concentration of 2000 ppm v/v (Jemiolo and Novotny, 1994).

4. MARINE ANIMALS

Among marine animals, alkaloids have been found mainly in tunicates, bryozoans, and sponges, the last phylum outclassing all others in terms of number and variety of structures. A few alkaloids have also been reported from other phyla (e.g., acorn worms, mollusks, and echinoderms), but their occurrence is rather restricted.

Most of the early research in marine chemistry was focused on compounds of low to medium polarity, such as terpenes. The last decade, however, has seen a phenomenal

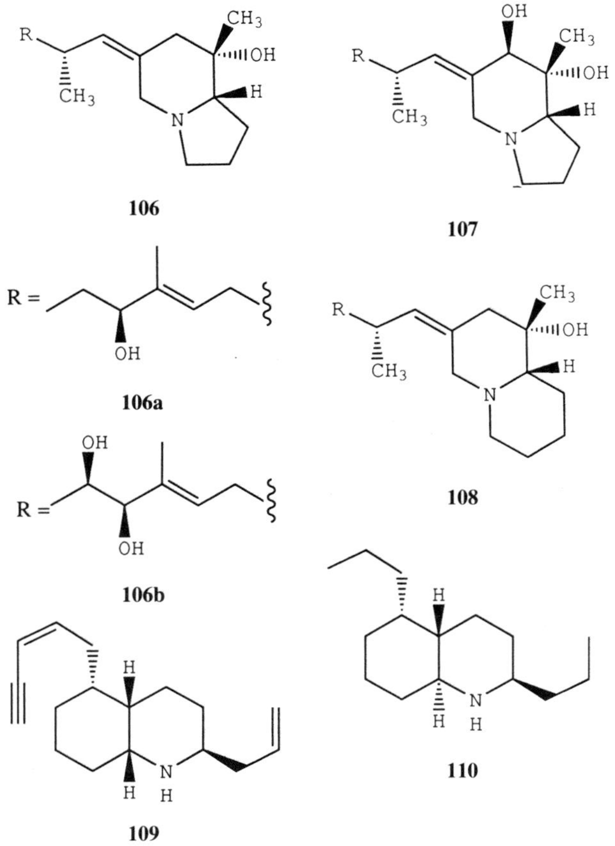

Figure 17. Pumiliotoxin-A and decahydroquinoline alkaloids from frogs.

development in the field of the more polar marine alkaloids. This development has been prompted by the novelty of many of the structures discovered and by the search for pharmacologically active compounds, at a time where new discoveries stemming from terrestrial sources were diminishing. Whereas many of the alkaloid skeletons found in marine animals have no terrestrial equivalent, the great wealth and variety of plant indole alkaloids has no counterpart in marine animals, most of the marine indoles described so far being structurally much simpler. Another difference between marine and terrestrial alkaloids is the frequent occurrence of halogen atoms (mostly chlorine and bromine) in the former.

It is important to mention that, in many cases, the precise origin of marine alkaloids is not known. Indeed, many marine animals, and this is particularly true for sponges, live in symbiosis with blue-green algae (Cyanophyceae), bacteria, or fungi, which are suspected to be the true producers of some of the compounds isolated from the animal–microorganism association. This microbial origin has been proved in a few cases, and these will not be considered here (Garson, 1993). However, at the present time we cannot be sure that all of the compounds that will be presented are really produced by the animals themselves.

Because of the large number of marine invertebrate alkaloids already described, it will only be possible here to highlight some of the major types of alkaloids that have been found to date. For more comprehensive reports, the interested reader is referred to the ongoing series of reviews on marine natural products published by Faulkner (1986–1995). Other reviews of general interest by Davidson (1993), Molinski (1993), Alvarez *et al.* (1991a,b), Christophersen (1983, 1985a,b), Arai and Kubo (1983), and Chevolot (1981) should also be consulted.

4.1. Bryozoa (Ectoprocta or Moss Animals)

Bryozoans are rather poorly known colonial animals, although they are widely distributed in both freshwater and marine environments. The secondary metabolite chemistry of these animals has been reviewed by Christophersen (1983, 1985a,b). The species that has been most investigated is undoubtedly *Flustra foliacea,* which produces complex mixtures of bromoindole alkaloids based on the physostigmine skeleton, such as flustramines A and B (**111** and **112**) (Fig. 18). As these compounds have structural features already found in mold metabolites, their precise origin is still controversial. Other characteristic alkaloids from these animals include the chartellines (e.g., **113**) from *Chartella papyracea* and a series of β-carbolines (e.g., **114**) (Fig. 18).

4.2. Tunicates (Ascidians)

These animals have been the subject of numerous chemical studies prompted by the frequent occurrence of biologically active peptides and alkaloids. In tunicates, the two

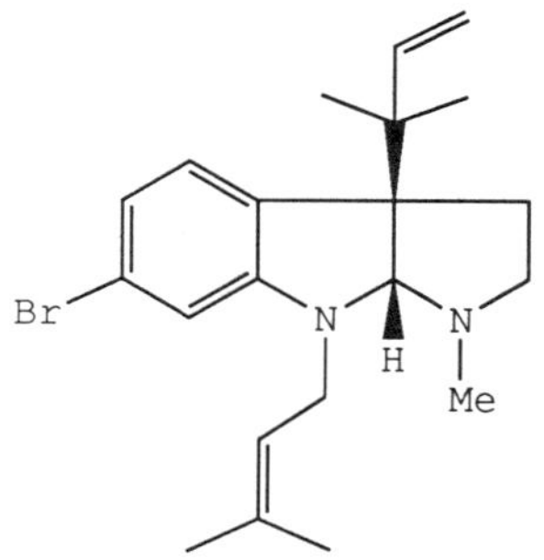

111

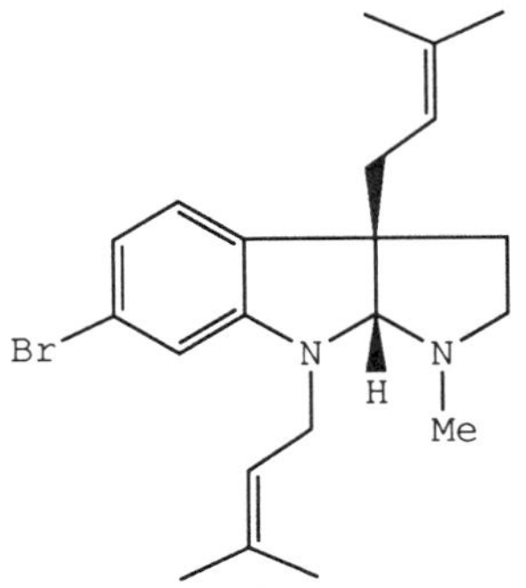

112

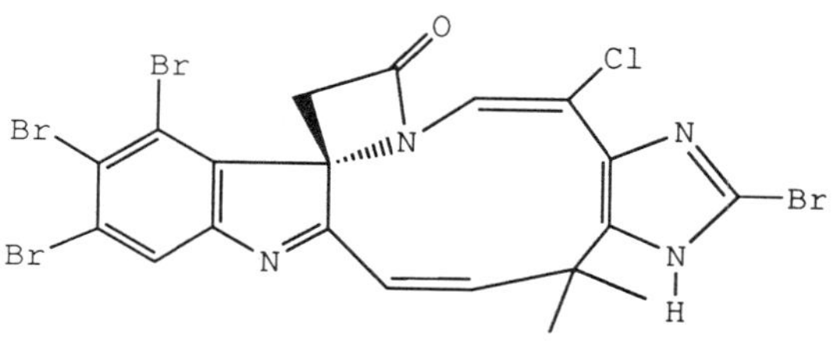

113

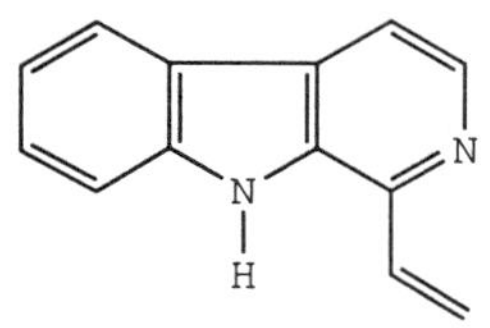

114

Figure 18. Alkaloids from Bryozoa.

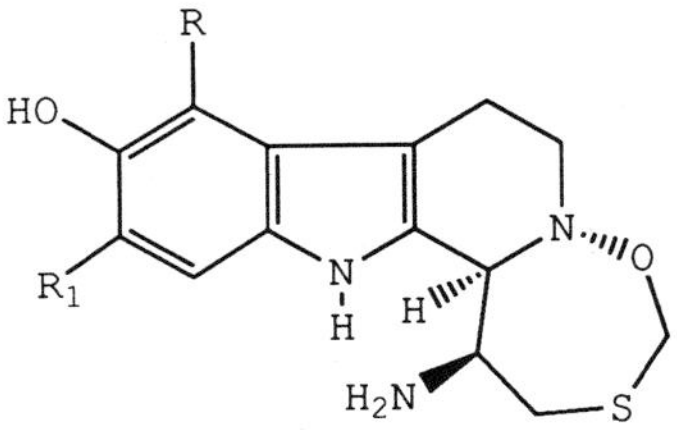

115 R = H; R_1 = Br
116 R = Br; R_1 = H

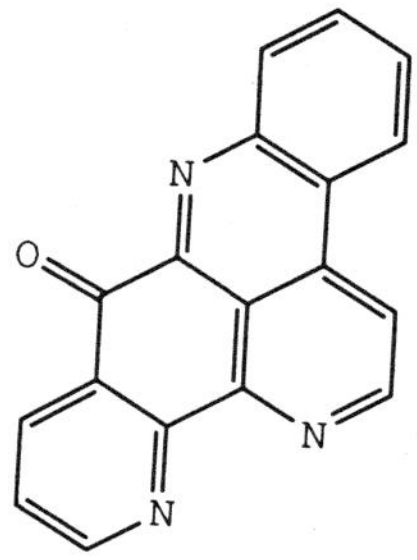

117

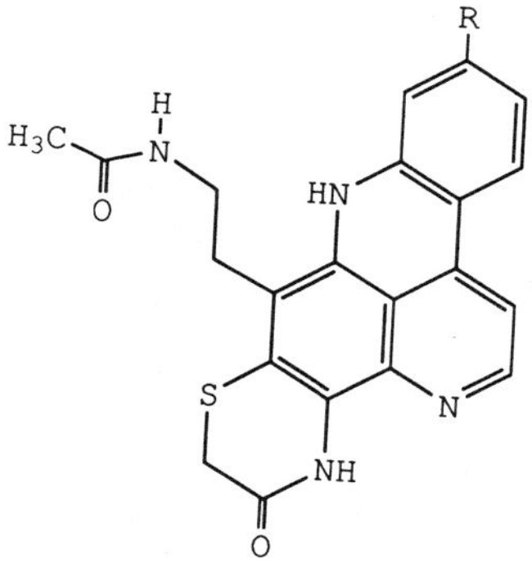

118 R = Br
119 R = H

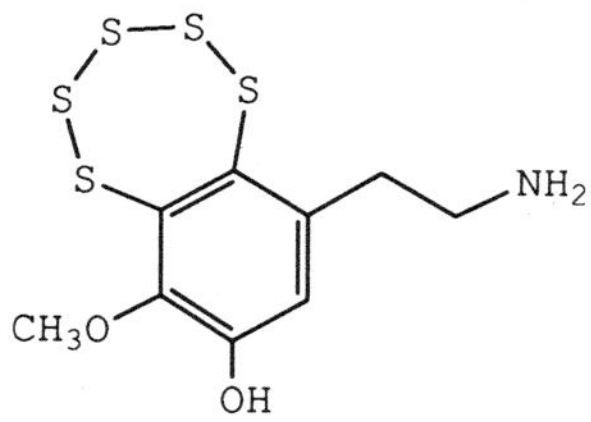

120

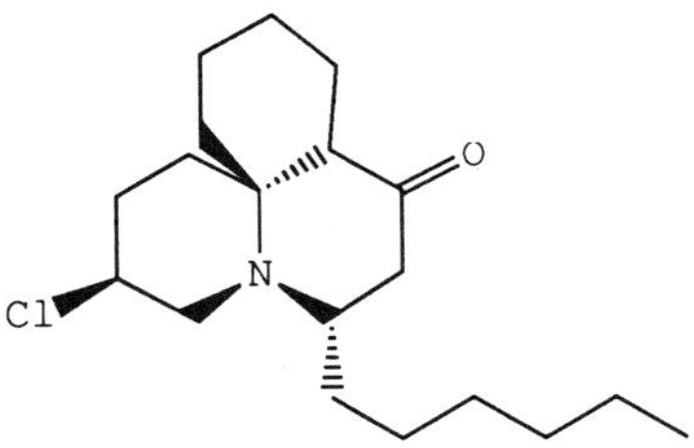

122

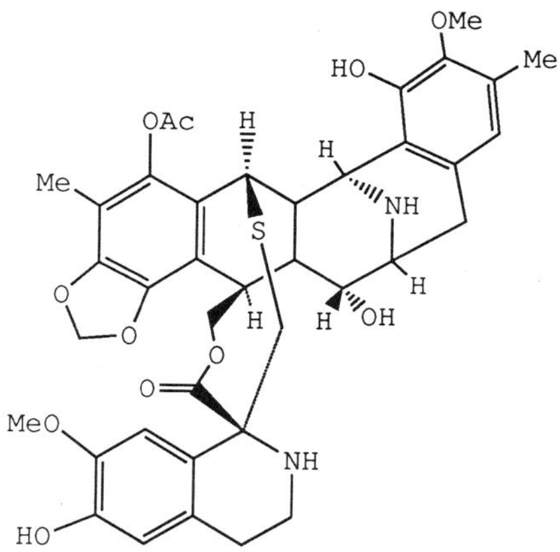

121

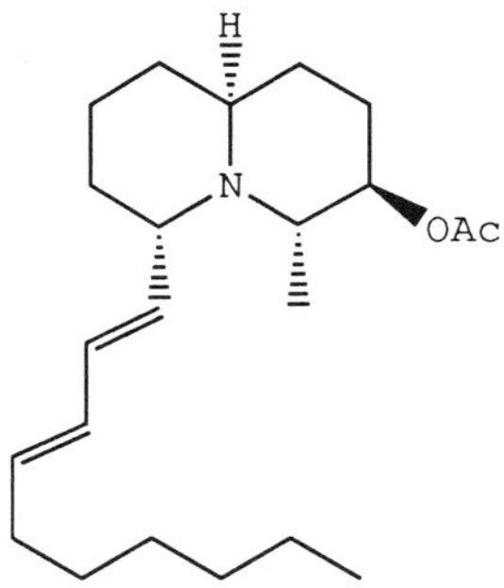

123

Figure 19. Alkaloids from tunicates.

most frequently encountered alkaloid groups are β-carbolines and the so-called pyridoacridines, which seem to be exclusive to marine invertebrates and endowed with numerous interesting biological properties (Molinski, 1993). The first group is exemplified by the eudistomins, having various degrees of complexity, among which eudistomins C and E (**115** and **116**) (Fig. 19), characterized by a condensed oxathiazepine

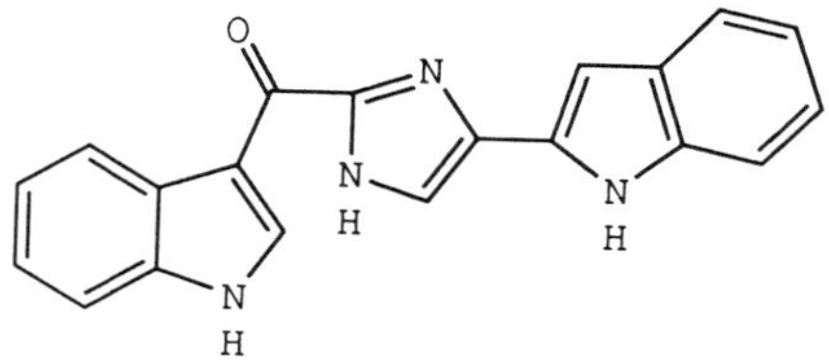

124

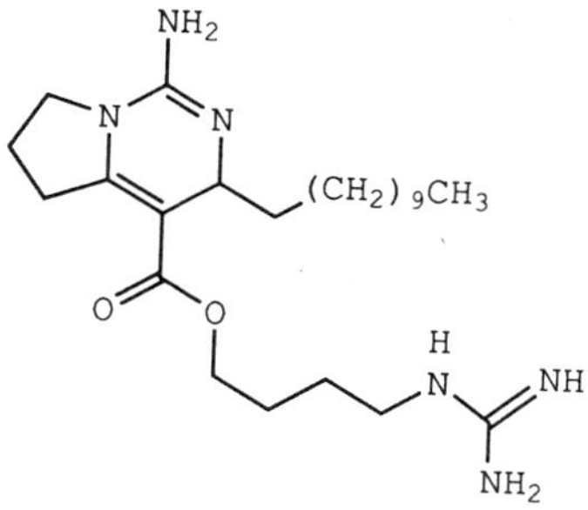

125

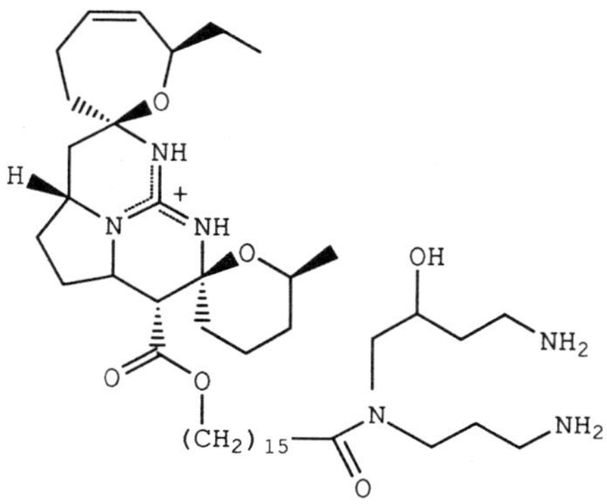

126

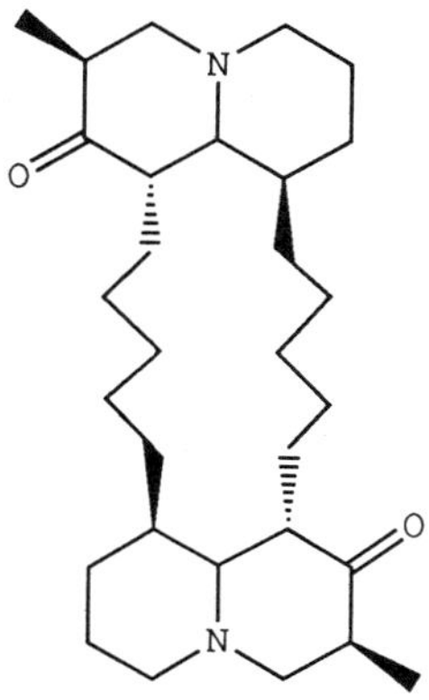

127

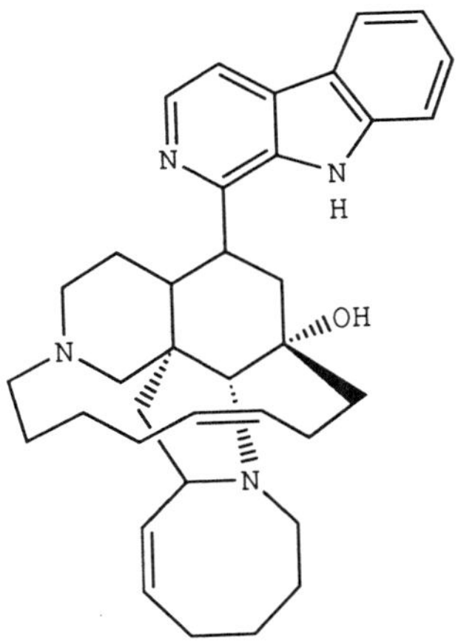

128

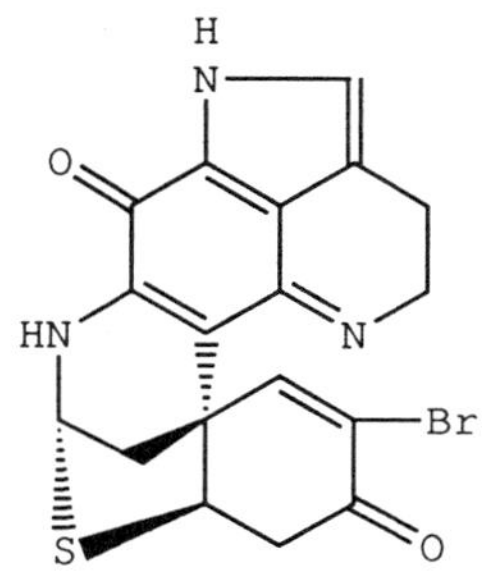

129

Figure 20. Alkaloids from sponges.

ring, are good representatives. About 30 pyridoacridine alkaloids have been isolated so far from tunicates, among them ascididemnin (**117**) and shermilamines A and B (**118** and **119**), the two latter again containing a sulfur atom. The presence of sulfur in tunicate alkaloids is not infrequent, as witnessed by compounds like the lissoclinotoxins (e.g., **120**). Other alkaloid types also exist in tunicates, e.g., tetrahydroisoquinoline alkaloids, like the complex, sulfur-containing, **121,** and quinolizidine alkaloids, e.g., cylindricine B (**122**), and clavepictine A (**123**) (Fig. 19).

4.3. Porifera

The phylum Porifera, comprising the most primitive multicellular animals, contains some 5000 species, most of them marine. They are surely the most thoroughly investigated marine invertebrates from a chemical point of view. This is related not only to the high level of incidence of pharmacologically active species, but also to the number, variety, and structural complexity of secondary metabolites produced by these animals, which precludes a detailed treatment of the subject. For an exhaustive survey of the field, see Faulkner (1986–1995). Thus, we shall limit ourselves to a few representatives of the major groups of sponge alkaloids, many of which have been isolated in our laboratory. Thus, the topsentins (e.g., **124**) are representative of the indole, the crambescins and crambescidins (e.g., **125** and **126**) of the guanidine, the petrosins (e.g., **127**) of the quinolizidine, the manzamines (e.g., **128**) of the macrocyclic and discorhabdins (e.g., **129**) (1986), of the pyridoacridine group of alkaloids (Fig. 20).

Finally, the symbiotic associations of sponges mentioned above pose a number of challenging ecological and biosynthetic problems, which are only beginning to be tackled.

5. BIOSYNTHESIS

Few works have been devoted to the biosynthesis of alkaloids from animals. This is undoubtedly related to the difficulties associated with the administration of labeled precursors. Moreover, the low level of incorporation requires the use of radioisotopes rather than stable isotopes. Nevertheless, some biosynthetic studies on animal alkaloids have been conducted and are briefly reported here.

Despite their structural diversity, the different alkaloid skeletons found to date in Coccinellidae can be seen to consist of a chain of carbon atoms of varying length, joined at one or more positions to a nitrogen atom. This points to a polyketide origin for all of these alkaloids, although other pathways cannot be excluded. Circumstantial evidence indicates that these compounds are most likely biosynthesized *de novo* by the beetles. This has been proved in *Coccinella septempunctata* by feeding experiments with [1-^{14}C]- and [2-^{14}C]acetate. Isolation of labeled coccinelline (**2**) (Fig. 1) and selective degradation to labeled acetic acid led to the conclusion that the 2-methylperhydro-9*b*-azaphenalene skeleton probably originates from seven acetate units (Tursch *et al.*, 1975). In a recent study using deuterated precursors (Attygalle *et al.*, 1994), the azamacrocycle epilachnene (**14**) (Fig. 2) was shown to derive from oleic acid and serine. Oleic acid loses four carbon atoms from its carboxyl end whereas serine is decarboxylated to ethanolamine. Other studies are needed, however, before a clear picture of the biosynthetic pathways in coccinellids will emerge. In staphylinid beetles, feeding of *P. fuscipes* with [1-^{14}C]- and [2-^{14}C]acetate, followed by isolation of radioactive pederin and selective degradation, suggests that the toxin is probably biosynthesized through a polyketide pathway (Cardani *et al.*, 1973b). In the case of the Δ^3-isoxazolin-5-one glucoside **22** (Fig. 3) of adult chrysomelid beetles, it has been shown that both the heterocycle and the 3-nitropropanoic acid moiety are biosynthesized *de novo* by the beetles from L-aspartic acid (Randoux *et al.*, 1991).

Considering the number and variety of ant alkaloids isolated thus far, it is intriguing that only one paper has been published dealing with their biosynthesis. Administration of sodium [1-^{14}C]- and [2-^{14}C]acetate, L-[U-^{14}C]glutamic acid, γ-amino[U-^{14}C]butyric acid, L-[U-^{14}C]ornithine hydrochloride, and [1,4-^{14}C]putrescine dihydrochloride to *Tetraponera* ants, followed by chemical degradation of the labeled tetraponerine 8 (**68**) (Fig. 9), led to the conclusion that this compound has a mixed biogenetic origin: the pyrrolidine ring derives from glutamic acid via L-ornithine and putrescine, whereas the remaining 12-carbon chain is derived from six acetate units (Renson *et al.,* 1994).

In the millipedes *Glomeris marginata,* glomerin (**90**) and homoglomerin (**91**) (Fig. 13) are biosynthesized *de novo,* by using anthranilic acid as a precursor (Schildknecht and Wenneis, 1967).

In salamanders, steroidal alkaloids are biosynthesized by the amphibian, and in agreement with their steroidal nature, [4-^{14}C]cholesterol was incorporated. The nitrogen atom is derived from glutamine (Habermehl and Haaf, 1968). By contrast, Daly *et al.* (1992, 1994a) have observed that captive-raised dendrobatid frogs, unlike wild specimens, do not contain skin alkaloids. Moreover, captive frogs fed fruit flies dusted with alkaloid-containing powder readily accumulated alkaloids into their skin (Daly *et al.,* 1994b). Similarly, frogs raised on leaf-litter arthropods also had substantial amounts of several alkaloids. These experiments demonstrate the occurrence of a very effective uptake system and suggest that some if not all such dendrobatid alkaloids may have a dietary origin. They suggest also that leaf-litter arthropods may have a key contribution to the alkaloid profile of these frogs.

The few reports on marine alkaloid biosynthesis are covered in the recent review of Garson (1993) and will not be considered here.

REFERENCES

General Reviews

Alvarez, M., Salas, M., and Joule, J. A., 1991a, Marine, nitrogen-containing heterocyclic natural products. Structures and syntheses of compounds containing quinoline and/or isoquinoline units, *Heterocycles* **32:**759–795.

Alvarez, M., Salas, M., and Joule, J. A., 1991b, Marine, nitrogen-containing heterocyclic natural products. Structures and syntheses of compounds containing indole units, *Heterocycles* **32:**1391–1429.

Arai, T., and Kubo, A., 1983, Isoquinolinequinones from Actinomycetes and sponges, in: *The Alkaloids,* Vol. 21 (A. Brossi, ed.), Academic Press, San Diego, pp. 56–100.

Attygalle, A. B., and Morgan, E. D., 1984, Chemicals from the glands of ants, *Chem. Soc. Rev.* **13:**245–278.

Bettini, S., (ed.), 1978, *Arthropod Venoms,* Springer-Verlag, Berlin, Chapters 6–11.

Blum, M. S., 1981, Sex pheromones in social insects: Chemotaxonomic potential, in: *Biosystematics of Social Insects* (P. E. House and J.-L. Clément, eds.), Academic Press, San Diego, pp. 163–174.

Blum, M. S., and Hermann, H. R., 1978, Venoms and venom apparatuses of the Formicidae: Myrmeciinae, Ponerinae, Dorylinae, Pseudomyrmecinae, Myrmicinae and Formicinae, in: *Arthropod Venoms* (S. Bettini, ed.), Springer-Verlag, Berlin, pp. 801–869.

Braekman, J. C., and Daloze, D., 1990, Chemical defense in ants, in: *Studies in Natural Products Chemistry,* Vol. 6, Part D (Atta ur Rahman, ed.), Elsevier, Amsterdam, pp. 421–466.

Brossi, A., 1993, Mammalian alkaloids II, in: *The Alkaloids,* Vol. 43 (G. A. Cordell, ed.), Academic Press, San Diego, pp. 119–183.

Brown, K. S., and Trigo, J. R., 1995, The ecological activity of alkaloids, in: *The Alkaloids,* Vol. 47 (G. A. Cordell, ed.), Academic Press, San Diego, pp. 227–354.

Chevolot, L., 1981, Guanidine derivatives, in: *Marine Natural Products, Chemical and Biological Perspectives,* Vol. 4 (P. J. Scheuer, ed.), Academic Press, New York, pp. 54–91.

Christophersen, C., 1983, Marine indoles, in: *Marine Natural Products, Chemical and Biological Perspectives,* Vol. 5 (P. J. Scheuer, ed.), Academic Press, New York, pp. 259–285.

Christophersen, C., 1985a, Bioactive metabolites from marine bryozoans. A review, *Acta Chem. Scand.* **39:**517–529.

Christophersen, C., 1985b, Marine alkaloids, in: *The Alkaloids,* Vol. 24 (A. Brossi, ed.), Academic Press, San Diego, pp. 25–112.

Collins, M. A., 1983, Mammalian alkaloids I, in: *The Alkaloids,* Vol. 21 (A. Brossi, ed.), Academic Press, San Diego, pp. 329–358.

Daloze, D., Braekman, J. C., and Pasteels, J. M., 1995, Ladybird defence alkaloids: structural, chemotaxonomic and biosynthetic aspects, *Chemoecology* **5/6:**173–183.

Daly, J. W., 1982, Alkaloids of neotropical poison frogs (Dendrobatidae), in: *Progress in the Chemistry of Organic Natural Products,* Vol. 41 (W. Hertz, H. Grisebach, and G. W. Kirby, eds.), Springer-Verlag, Berlin, pp. 205–340.

Daly, J. W., and Spande, T. F., 1986, Amphibian alkaloids: chemistry, Pharmacology, and biology, in: *Alkaloids: Chemical and Biological Perspectives,* Vol. 4 (S. W. Pelletier, ed.), Wiley, New York, pp. 1–274.

Daly, J. W., Garraffo, H. M., and Spande, T. F., 1993, Amphibian alkaloids, in: *The Alkaloids,* Vol. 43 (G. A. Cordell, ed.), Academic Press, New York, pp. 185–288.

Davidson, B. S., 1993, Ascidians: Producers of amino acid derived metabolites, *Chem. Rev.* **93:**1771–1791.

Eisner, T., 1970, Chemical defense against predation in arthropods, in: *Chemical Ecology* (E. Sondheimer and J. B. Simeone, eds.), Academic Press, San Diego, pp. 157–217.

Faulkner, D. J., 1986–1995, Marine natural products, *Nat. Prod. Rep.* **3:**1–34; **4:**539–576; **5:**613–663; **7:**269–310; **8:**97–148; **9:**323–364; **10:**497–539; **11:**355–394; **12:**223–270.

Garson, M. J., 1993, The biosynthesis of marine natural products, *Chem. Rev.* **93:**1699–1734.

Habermehl, G. G., 1967, The steroid alkaloids: The *Salamandra* group, in: *The Alkaloids,* Vol. 9 (R. H. F. Manske and H. L. Holmes, eds.), Academic Press, San Diego, pp. 427–439.

Habermehl, G., 1971, Toxicology, pharmacology, chemistry and biochemistry of salamander venom, in: *Venomous Animals and Their Venoms,* Vol. 2 (W. Bucherl and E. E. Buckley, eds.), Academic Press, San Diego, pp. 569–584.

Habermehl, G. G., 1981, *Venomous Animals and Their Toxins,* Springer-Verlag, Berlin.

Hölldobler, B., and Wilson, E. O., 1990, *The Ants,* Springer-Verlag, Berlin.

Jones, T. H., and Blum, M. S., 1983, Arthropod alkaloids: Distribution, functions, and chemistry, in: *Alkaloids: Chemical and Biological Perspectives,* Vol. 1 (S. W. Pelletier, ed.), Wiley, New York, pp. 33–84.

King, A. G., and Meinwald, J., 1996, Review of the defensive chemistry of coccinellids, *Chem. Rev.* **96:**1105–1122.

McCormick, K. D., and Meinwald, J., 1993, Neurotoxic acylpolyamines from spider venoms, *J. Chem. Ecol.* **19:**2411–2451.

Molinski, T. F., 1993, Marine pyridoacridine alkaloids: Structure, synthesis and biological activity, *Chem. Rev.* **93:**1825–1834.

Numata, A., and Ibuka, T., 1987, Alkaloids of ants and other insects, in: *The Alkaloids,* Vol. 31 (A. Brossi, ed.), Academic Press, San Diego, pp. 193–315.

Pasteels, J. M., Braekman, J. C., and Daloze, D., 1988, Chemical defense in the Chrysomelidae, in: *The Biology of Chrysomelidae* (P. Jolivet, E. Petitpierre, and T. H. Hsiao, eds.), Kluwer, Dordrecht, pp. 233–252.

Pasteels, J. M., Braekman, J. C., and Daloze, D., 1994, Chemical defense in adult leaf beetles updated, in: *Novel Aspects of the Biology of Chrysomelidae* (P. H. Jolivet, M. L. Cox, and E. Petitpierre, eds.), Kluwer, Dordrecht, pp. 289–301.

Schmidt, J. O., 1986, Chemistry, pharmacology, and ecology of ant venoms, in: *Venoms of the Hymenoptera: Biochemical, Pharmacological, and Behavioural Aspects* (T. Piek, ed.), Academic Press, San Diego, pp. 425–508.

Tu, T., (ed.), 1984, *Handbook of Natural Toxins,* Vol. 2, Dekker, New York, Chapters 12–14.

Tursch, B., Braekman, J. C., and Daloze, D., 1976, Arthopod alkaloids, *Experientia* **32:**401–407.

Witkop, B., and Gossinger, E., 1983, Amphibian alkaloids, in: *The Alkaloids,* Vol. 21 (A. Brossi, ed.), Academic Press, San Diego, pp. 139–253.

Specialist References

Attygalle, A. B., Cammaerts, M.-C., Cammaerts, R., Morgan, E. D., and Ollett, D. G., 1986, Chemical and ethological studies of the trail pheromone of the ant *Manica rubida* (Hymenoptera: Formicidae), *Phys. Entomol.* **11:**125–132.

Attygalle, A. B., Xu, S. C., McCormick, K. D., Meinwald, J., Blankespoor, C. L., and Eisner, T., 1993a, Alkaloids of the Mexican bean beetle, *Epilachna varivestis* (Coccinellidae), *Tetrahedron* **49:**9333–9342.

Attygalle, A. B., McCormick, K. D., Blankespoor, C. L., Eisner, T., and Meinwald, J., 1993b, Azamacrolides: A family of alkaloids from the pupal defensive secretion of a ladybird beetle (*Epilachna varivestis*), *Proc. Natl. Acad. Sci. USA* **90:**5204–5208.

Attygalle, A. B., Blankespoor, C. L., Eisner, T., and Meinwald, J., 1994, Biosynthesis of a defensive insect alkaloid: Epilachnene from oleic acid and serine, *Proc. Natl. Acad. Sci. USA* **91:**12790–12793.

Ayer, W., and Browne, L. M., 1977, The ladybug alkaloids including synthesis and biosynthesis, *Heterocycles* **7:**685–707.

Blum, M. S., Jones, T. H., Hölldobler, B., Fales, H. M., and Jaouni, T., 1980, Alkaloidal venom mace: Offensive use by a thief ant, *Naturwissenschaften* **67:**144–145.

Braconnier, M. F., Braekman, J. C., Daloze, D., and Pasteels, J. M., 1985, (Z)-1,17-diaminooctadec-9-ene, a novel aliphatic diamine from Coccinellidae, *Experientia* **41:**519–520.

Braekman, J. C., Daloze, D., Pasteels, J. M., Vanhecke, P., Declercq, J. P., Sinnwell, V., and Francke, W., 1987, Tetraponerine-8, an alkaloidal contact poison in a Neoguinean pseudomyrmecine ant, *Tetraponera* sp., Z. Naturforsch. **42c:**627–630.

Brand, J. M., and Mpuru, S. P., 1993, Dufour's gland and poison gland chemistry of the myrmicine ant, *Messor capensis* (Mayr), *J. Chem. Ecol.* **19:**1315–1321.

Brophy, J. J., Clezy, P. S., Leung, C. W. F., and Robertson, P. L., 1993, Secondary amines isolated from venom gland of dolichoderine ant, *Technomyrmex albipes, J. Chem. Ecol.* **19:**2183–2192.

Brown, C. A., Watkins, J. F., II, and Eldridge, D. W., 1979, Repression of bacteria and fungi by the army ant secretion: Skatole, *J. Kansas Entomol. Soc.* **52:**119–122.

Brown, W. V., Jones, A. J., Lacey, M. J., and Moore, B. P., 1985, The chemistry of buprestins A and B. Bitter principles of jewel beetles (Coleoptera: Buprestidae), *Aust. J. Chem.* **38:**197–206.

Capelli, E., Attolini, C., Mazza, G., Pavan, M., and De Carli, L.,1982, Studies on the genotoxic effects of pederin, *Publ. Istit. Entomol. Univ. Pavia* 1982 (**21**)**:**1–16.

Cardani, C., Ghiringhelli, D., Mondelli, R., and Selva, A., 1973a, On the methanolysis products of pederin, *Gazz. Chim. Ital.* **103:**247–253.

Cardani, C., Fuganti, C., Ghiringhelli, D., Grasselli, P., Pavan, M., and Valcurone, M. D., 1973b, The biosynthesis of pederin, *Tetrahedron Lett.* **14:**2815–2818.

Daly, J. W., Myers, C. W., and Whittaker, N., 1987, Further classification of skin alkaloids from neotropical poison frogs (Dendrobatidae), with a general survey of toxic/ noxious substances in the Amphibia, *Toxicon* **25:**1023–1095.

Daly, J. W., Secunda, S. I., Garraffo, H. M., Spande, T. F., Wisnieski, A., Nishihira, C., and Cover, J. F., Jr., 1992, Variability in alkaloid profiles in neotropical poison frogs (Dendrobatidae): Genetic versus environmental determinants, *Toxicon* **30:**887–898.

Daly, J. W., Garraffo, H. M., Spande, T. F., Jaramillo, C., and Rand, A. S., 1994a, Dietary source for skin alkaloids of poison frogs (Dendrobatidae)? *J. Chem. Ecol.* **20:**943–955.

Daly, J. W., Secunda, S. I., Garraffo, H. M., Spande, T. F., Wisnieski, A., and Cover, J. F., Jr., 1994b, An uptake system for dietary alkaloids in poison frogs (Dendrobatidae), *Toxicon* **32:**657–663.

Dettner, K., 1983, Vergleichende Untersuchungen zur Wehrchemie und Drüsenmorphologie von Kurzflüglern aus dem Subtribus Philonthina (Coleoptera, Staphylinidae), *Z. Naturforsch.* **38c:**319–328.

De Vijver, C., Macours, P., Braekman, J. C., Daloze, D., and Pasteels, J. M., 1995, Short syntheses of (±)-tetraponerines 5 and 6. The structures of tetraponerines 1 and 2, and a revision of the structures of tetraponerines 5 and 6, *Tetrahedron* **51:**10913–10922.

Dumbacher, J. A., Beeler, B., Garraffo, H. M., Spande, T. F., and Daly, J. W., 1992, Homobatrachotoxin in the genus *Pitohui*: Chemical defense in birds, *Science* **258:**799–801.

Eisner, T., Goetz, M., Aneshansley, D., Ferstandig-Arnold, G., and Meinwald, J., 1986, Defensive alkaloids in blood of Mexican bean beetle (*Epilachua varivestis*) *Experientia,* **42:** 204–207.

Ekpa, O., Wheeler, J. W., Cokendelpher, J. C., and Duffield, R. M., 1984, *N,N*-dimethyl-β-phenylethylamine and bornyl esters from the harvestman *Sclerobunus robustus* (Arachnida: Opiliones), *Tetrahedron Lett.* **25:**1315–1318.

Enders, D., and Bartzen, D., 1991, Enantioselective total synthesis of harmonine, a defence alkaloid of ladybugs (Coleoptera: Coccinellidae), *Liebigs Ann.* 569–574.

Enders, D., Tiebes, J., De Kimpe, N., Keppens, M., Stevens, C., Smagghe, G., and Betz, O., 1993, Enantioselective synthesis and determination of the configuration of stenusine, the spreading agent of the beetle *Stenus comma, J. Org. Chem.* **58:**4881–4884.

Escoubas, P., and Blum, M. S., 1990, The biological activities of ant-derived alkaloids, in: *Applied Myrmecology: A World Perspective* (R. K. Vander Meer, K. Jaffe, and A. Cedeno, eds.), Westview Press, Boulder, pp. 482–489.

Escoubas, P., Le Maire, M., Clément, J.-L., Lhommet, G., and Celerier, J. P., 1987, European *Monomorium* alkaloids: Toxicity against preys and emitters, in: *Chemistry and Biology of Social Insects* (J. Eder and H. Rembold, eds.), Verlag J. Peperny, Munich, pp. 419–420.

Fales, H. M., Blum, M. S., Bian, Z., Jones, T. H., and Warwick Don, A., 1984, Volatile compounds from ponerine ants in the genus *Mesoponera, J. Chem. Ecol.* **10:**651–665.

Furusaki, A., Watanabe, T., Matsumoto, T., and Yanagiya, M., 1968, The crystal and molecular structure of pederin di-*p*-bromobenzoate, *Tetrahedron Lett.* **9:**6301–6304.

Habermehl, G. G., and Haaf, A., 1968, Samandarine and related alkaloids. XVIII. Cholesterol as precursor in the biosynthesis of salamander alkaloids, *Chem. Ber.* **101:**198–200.

Ho, H. Y., and Chow, Y. S., 1993, Chemical identification of defensive secretion of stick insect, *Megacrania tsudai* Shiraki, *J. Chem. Ecol.* **19:**39–46.

Huth, A., and Dettner, K., 1990, Defense chemicals from abdominal glands of 13 rove beetles species of subtribe Staphylina (Coleoptera: Staphylinidae, Staphylinae), *J. Chem. Ecol.* **16:**2691–2711.

Jackson, B. D., Morgan, E. D., and Billen, J. P. J., 1990, Contents of the pygidial gland of the ant *Myrmecia nigriceps, Naturwissenschaften* **77:**187–188.

Jasys, V. J., Kelbaugh, P. R., Nason, D. M., Phillips, D., Rosnack, K. J., Saccomano, N. A., Stroh, J. G., and Volkmann, R. A., 1990, Isolation, structure elucidation and synthesis of novel hydroxylamine-containing polyamines from the venom of the *Agelenopsis aperta* spider, *J. Am. Chem. Soc.* **112:**6696–6704.

Jemiolo, B., and Novotny, M., 1994, Inhibition of sexual maturation in juvenile female and male mice by a chemosignal of female origin, *Physiol. Behav.* **55:**519–522.

Jones, T. H., Blum, M. S., Fales, H. M., and Thompson, C. R., 1980, (5Z,8E)-3-heptyl-5-methylpyrrolizidine from a thief ant, *J. Org. Chem.* **45:**4778–4780.

Jones, T. H., Highet, R. J., Blum, M. S., and Fales, H. M., 1984, (5Z,9Z)-3-alkyl-5-methylindolizidines from *Solenopsis* (*Diplorhoptrum*) species, *J. Chem. Ecol.* **10:**1233–1249.

Jones, T. H., Stahly, S. M., Warwick Don, A., and Blum, M. S., 1988, Chemotaxonomic implications of the venom chemistry of some *Monomorium "antarcticum"* populations, *J. Chem. Ecol.* **14:**2197–2212.

Jones, T. H., Blum, M. S., Fales, H. M., Brandao, C. R. F., and Lattke, J., 1991, Chemistry of venom alkaloids in the ant genus *Megalomyrmex, J. Chem. Ecol.* **17:**1897–1908.

Kalivretenos, A. G., and Nakanishi, K., 1993, Synthesis of philanthotoxin analogs with a branched polyamine moiety, *J. Org. Chem.* **58:**6596–6608.

Leclercq, S., Thirionet, I., Broeders, F., Daloze, D., Vander Meer, R. K., and Braekman, J. C., 1994, Absolute configuration of the solenopsins, venom alkaloids of the fire ants, *Tetrahedron* **50:**8465–8478.

McCormick, K. D., Kobayashi, K., Goldin, S. M., Reddy, N. L., and Meinwald, J., 1993, Characterization and synthesis of a new calcium antagonist from the venom of a fishing spider, *Tetrahedron* **49:**11155–11168.

McCormick, K. D., Attygalle, A. B., Xu, S. C., Svatos, A., Meinwald, J., Houck, M. A., Blankespoor, C. L., and Eisner, T., 1994, Chilocorine: Heptacyclic alkaloid from a coccinellid beetle, *Tetrahedron* **50:**2365–2372.

Macours, P., Braekman, J. C., and Daloze, D., 1995, Concise asymmetric syntheses of (+)- and (-)-tetraponerine-8, (+)- and (−)-tetraponerine-7, and their ethyl homologues. A correction of the structures of tetraponerine-3 and -7, *Tetrahedron* **51:**1415–1428.

Marples, N. M., 1993, Toxicity assays of ladybirds using natural predators, *Chemoecology* **4:**33–38.

Marples, N. M., Brakefield, P. M., and Cowie, R. J., 1989, Differences between the 7-spot and 2-spot ladybird beetles (Coccinellidae) in their toxic effects on a bird predator, *Ecol. Entomol.* **14:**79–84.

Matsuda, F., Tomiyoshi, N., Yanagiya, M., and Matsumoto, T., 1988, Stereocontrolled total synthesis of (+)-pederine, *Tetrahedron* **44:**7063–7080.

Meinwald, J., Smolanoff, J., McPhail, A. T., Miller, R. W., Eisner, T., and Hicks, K., 1975, Nitropolyzonamine: A spirocyclic nitro compound from the defensive glands of a millipede (*Polyzonium rosalbum*), *Tetrahedron Lett.* 2367–2370.

Meinwald, Y. C., Meinwald, J., and Eisner, T., 1966, 1,2-Dialkyl-4(3H)-quinazolinones in the defensive secretion of a millipede (*Glomeris marginata*), *Science* **154:**390–391.

Merlin, P., Braekman, J. C., Daloze, D., and Pasteels, J. M., 1988, Tetraponerines, toxic alkaloids in the venom of the Neoguinean pseudomyrmecine ant, *Tetraponera* sp., *J. Chem. Ecol.* **14:**517–527.

Miller, R. W., and McPhail, A. T., 1978, Structure and revised absolute stereochemistry for nitropolyzonamine, 2′,2′-dimethyl-6-nitrospiro{1-azabicyclo[3.3.0]octane-4,1′-cyclopentane}, a novel spirocyclic nitro compound from the defensive glands of the millipede *Polyzonium rosalbum*: X-ray analysis of nitropolyzonammonium perchlorate, *J. Chem. Res.* (S) 76.

Moore, B. P., and Brown, W. V., 1978, Precoccinelline and related alkaloids in the Australian soldier beetle, *Chauliognathus pulchellus* (Coleoptera: Cantharidae), *Insect Biochem.* **8:**393–395.

Moore, B. P., and Brown, W. V., 1981, Identification of warning odour components, bitter principles and antifeedants in an aposematic beetle: *Metriorrhynchus rhipidius* (Coleoptera: Lycidae), *Insect Biochem.* **11:**493–499.

Moore, B. P., Brown, W. V., and Rothschild, M., 1990, Methylalkylpyrazines in aposematic insects, their hostplants and mimics, *Chemoecology* **1:**43–51.

Mueller, R. H., Thompson, M. E., and Di Pardo, R. M., 1984, Stereo- and regioselective total synthesis of the hydropyrido[2,1,6-*de*]quinolizine ladybug defensive alkaloids, *J. Org. Chem.* **49:**2217–2231.

Obin, M. S., and Vander Meer, R. K., 1985, Gaster flagging by fire ants (*Solenopsis* spp.): Functional significance of venom dispersal behavior, *J. Chem. Ecol.* **11:**1757–1768.

Pasteels, J. M., Deroe, C., Tursch, B., Braekman, J. C., Daloze, D., and Hootele, C., 1973, Distribution et activités des alcaloïdes défensifs des Coccinellidae, *J. Insect Physiol.* **19:**1771–1784.

Pasteels, J. M., Braekman, J. C., Daloze, D., and Ottinger, R., 1982, Chemical defence in chrysomelid larvae and adults, *Tetrahedron* **38:**1891–1897.

Perry, N. B., Blunt, J. W., Munro, M. H. G., and Pannell, L. K., 1988, Mycalamide A, an antiviral compound from a New Zealand sponge of the genus *Mycale, J. Am. Chem. Soc.* **110:**4851–4853.

Proksch, P., Witte, L., Wray, V., and Hartmann, T., 1993, Ontogenic variation of defensive alkaloids in the Mexican bean beetle, *Epilachna varivestis* (Coleoptera: Coccinellidae), *Entomol. Gener.* **18:**1–7.

Randoux, T., Braekman, J. C., Daloze, D., and Pasteels, J. M., 1991, *De novo* biosynthesis of Δ^3-isoxazolin-5-one and 3-nitropropanoic acid derivatives in *Chrysomela tremulae, Naturwissenschaften* **78:**313–314.

Reder, E., Veith, H. J., and Buschinger, A., 1995, Neuartige Alkaloide aus dem Giftdrüsensekret sozialparasitischer Ameisen (Myrmicinae: Leptothoracini), *Helv. Chim. Acta* **78:**73–79.

Renson, B., Merlin, P., Daloze, D., Braekman, J. C., Roisin, Y., and Pasteels, J. M., 1994, Biosynthesis of tetraponerine 8, a defence alkaloid of the ant *Tetraponer*a sp., *Can. J. Chem.* **72:**105–109.

Schildknecht, H.,1970, The defensive chemistry of land and water beetles, *Angew. Chem. Int. Ed. Engl.* **9:**1–9.

Schildknecht, H., and Wenneis, W. F., 1967, Anthranilsaure als Precursor der Arthropodenalkaloide Glomerin und Homoglomerin, *Tetrahedron Lett.* 1815–1818.

Schildknecht, H., Maschwitz, U., and Wenneis, W. F., 1967, Neue Stoffe aus dem Wehrsekret der Diplopodengattung *Glomeris, Naturwissenschaften* **54:**196–197.

Schildknecht, H., Berger, D., Krauss, D., Connert, J., Gelhaus, J., and Essenbreis, H., 1976, Defense chemistry of *Stenus comma* (Coleoptera: Staphylinidae). LXI, *J. Chem. Ecol.* **2:**1–11.

Smolanoff, J., Kluge, A. F., Meinwald, J., McPhail, A., Miller, R. W., Hicks, K., and Eisner, T., 1975, Polyzonimine: A novel terpenoid insect repellent produced by a millipede, *Science* **188:**734–736.

Teshima, T., Matsumoto, T., Wakamiya, T., Shiba, T., Aramaki, Y., Nakajima, T., and Kawai, N., 1991, Total synthesis of NSTX-3, spider toxin of *Nephila maculata, Tetrahedron* **47:**3305–3312.

Timmermans, M., Braekman, J. C., Daloze, D., Pasteels, J. M., Merlin, J., and Declercq, J. P., 1992, Exochomine, a dimeric ladybird alkaloid, isolated from *Exochomus quadripustulatus* (Coleoptera: Coccinellidae), *Tetrahedron Lett.* **33:**1281–1284.

Toki, T., Yasuhara, T., Aramaki, Y., Osawa, K., Miwa, A., Kawai, N., and Nakajima, T., 1988, Isolation and chemical characterization of a series of new spider toxin (nephilatoxins) in the venom of Joro spider, *Nephila clavata, Biomed. Res.* **9:**421–428.

Tumlinson, J. H., Silverstein, R. M., Moser, J. C., Brownlee, R. G., and Ruth, J. M., 1971, Identification of the trail pheromone of a leaf-cutting ant, *Atta texana, Nature* **234:**348–349.

Tursch, B., Daloze, D., Braekman, J. C., Hootele, C., and Pasteels, J. M., 1975, The structure of myrrhine and the biosynthesis of coccinelline, *Tetrahedron* **31:**1541–1543.

Vander Meer, R. K., and Morel, L., 1994, Ant queens deposit pheromones and antimicrobial agents on eggs, *Naturwissenschaften* **82:**93–95.

Veith, M., Lorenz, M., Boland, W., Simon, H., and Dettner, K., 1994, Biosynthesis of iridoid monoterpenes in insects: Defensive secretions from larvae of leaf beetles (Coleoptera: Chrysomelidae), *Tetrahedron* **50:**6859–6874.

Witkop, B., 1971, New directions in the chemistry of natural products: The organic chemist as a pathfinder for biochemistry and medicine, *Experientia* **27:**1121–1138.

Yue, C., Nicolay, J. F., Royer, J., and Husson, H. P., 1994, A new stereoselective synthesis of ladybug defence alkaloid precoccinelline, *Tetrahedron* **50:**3139–3148.

Chapter 16

Ecological Significance of Alkaloids from Marine Invertebrates

Peter Proksch and Rainer Ebel

1. INTRODUCTION

In the marine environment invertebrates constitute the richest source of natural products (Faulkner, 1993), in sharp contrast to the terrestrial habitat where plants yield the largest number of natural products (Luckner, 1990). Several thousand different natural products from marine invertebrates have been identified in the last 25 years (Faulkner, 1993) [the discovery of prostaglandins in the Caribbean gorgonian *Plexaura homomalla* by Weinheimer and Spraggins (1969) is usually considered as a starting point of marine natural product chemistry]. The frequent occurrence (and in many cases high yields) of natural products in sessile or slow-moving invertebrates such as sponges, bryozoans, or holothurians that often live exposed to potential predators but lack effective morphological defense mechanisms is considered to reflect the evolution of chemically mediated defense mechanisms that protect these soft-bodied organisms from environmental dangers such as predation (Paul, 1992; Proksch, 1994).

The occurrence of bioactive natural products in marine invertebrates often correlates with the risk of predation which corroborates their proposed defensive functions (Bakus, 1981; Bakus and Green, 1974; Green, 1977). In field studies conducted at the Great Barrier Reef (Australia), 73% of all exposed coral reef invertebrates (e.g., sponges, soft corals, ascidians, or holothurians) were found to be toxic to fish whereas 75% of the unexposed, cryptic species that are less likely to be discovered by predators were nontoxic (Bakus, 1981). Toxicity of marine invertebrates was further found to increase in correlation with the total number of fish species present in the respective habitat. For example, screening experiments that were conducted along a north–south gradient from the state of Washington (United States) to Vera Cruz (Mexico) indicated a significant increase of the fish toxicity of marine sponges. Whereas only 9% of all analyzed sponges from the coast

Peter Proksch and Rainer Ebel • Julius von Sachs Institute for Biological Sciences, Department of Pharmaceutical Biology, University of Würzburg, D-97082 Würzburg, Germany.

Alkaloids: Biochemistry, Ecology, and Medicinal Applications, edited by Roberts and Wink. Plenum Press, New York, 1998.

of Washington were ichthyotoxic, the number of toxic species increased to 75% at Vera Cruz (Green, 1977). The high percentage of ichthyotoxic sponges in the tropics (e.g., at Vera Cruz) compared to colder waters is in congruence with the number of fish species occurring in the respective habitats. Along the Pacific coast of Canada, for example, 227 species of fish were recorded (Clemens and Wilby, 1946). In the Gulf of Mexico the count increased to about 440 (Starck, 1968) whereas at Queensland (Australia) more than 1000 species of nearshore and deepwater fish were recorded (Scott, 1962). The high incidence of toxic invertebrates (e.g., sponges) in tropical waters may have been the result of the high species diversity of marine fish which (because of pronounced competition for food resources) will select for optimized defense strategies in prey organisms (Green, 1977). Chemical defense has apparently been selected as one efficient strategy to cope with the intense grazing pressure found for example on tropical coral reefs (Sale, 1991).

The chemical nature of natural products from marine invertebrates that have been shown to be deterrent or toxic to fish and other potential predators is diverse and ranges from various terpenoids and polyketides to nitrogenous compounds (Paul, 1992; Proksch, 1994) which according to more recent definitions (Pelletier, 1983) may be considered as "alkaloids."

Unlike the terrestrial environment where higher plants are the most important producers of alkaloids (Southon and Buckingham, 1989), relatively few alkaloids have been isolated from marine plants (Fenical, 1986). The vast majority of known marine alkaloids are found in the numerous invertebrates that inhabit the marine environment (Fenical, 1986; Faulkner, 1993). The adaptive significance of alkaloids for survival and competition of marine invertebrates as well as the possible origin of these compounds will be discussed in the present review.

2. CHEMICAL DEFENSE AGAINST FOULING AND SPATIAL COMPETITION

Marine surfaces are usually covered by fouling organisms. Although the complex interactions underlying biofouling are not understood in all details, three major steps seem to be involved: (1) "conditioning" of a surface by a glycoproteinaceous layer, (2) formation of the so-called "primary film" by bacteria, diatoms, and protozoa, and (3) colonization by macrophytes and epibiotic invertebrates (Bakus *et al.*, 1986; Davis *et al.*, 1989; Melton and Bodnar, 1988). Fouling is a continuous and massive event in the marine environment that is observed on artificial (e.g., ship hulls) as well as natural substrates (e.g., marine algae or invertebrates). Fouling of marine organisms may be highly disadvantageous especially for the numerous invertebrate filter feeders as obstruction of pores by epibionts may result in reduced feeding capacity. Alternatively, filter-feeding epibiotic invertebrates may even compete with their hosts for food. Although numerous marine invertebrates such as sponges or bryozoans provide excellent potential settlement sites for fouling organisms, their surfaces are often remarkably clean suggesting the existence of effective mechanisms for the suppression of epibionts. According to Davis *et al.* (1989), two major defense strategies that both involve the production of deterrent or toxic metabolites may be realized in nonfouled marine organisms: (1) epibionts may be deterred from settling at early stages by defensive chemicals that are released from the potential host into

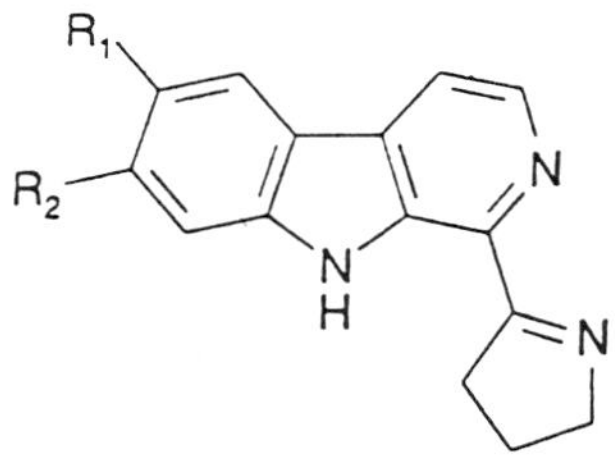

1 $R_1 = H, R_2 = Br$
2 $R_1 = Br, R_2 = H$

the surrounding water column; (2) at later stages settling epibionts may be deterred or even poisoned on encounter with chemicals present on the surface of the host. In the following, several recent examples that primarily highlight the second proposed defense strategy will be presented.

Numerous marine invertebrates including sponges, bryozoans, or ascidians contain a wealth of antibiotically active natural products (Krebs, 1986; Faulkner, 1993) that have frequently been suggested to be involved in the suppression of epibiotic bacteria. This hypothesis is corroborated by observations of Thompson *et al.* (1985) who reported that marine sponges with clean surfaces usually exhibit greater antimicrobial properties than those with fouled surfaces. However, experiments on antibiotically active natural products from marine invertebrates conducted to date were usually confined to the laboratory using easily accessible terrestrial bacteria. Results derived from these studies cannot necessarily be extrapolated to the marine environment (Bakus *et al.,* 1990).

Recently, however, ecologically more relevant studies were conducted on antifouling properties of alkaloids from several marine invertebrates including the ascidian *Eudistoma olivaceum* that accumulates a series of β-carbolines called eudistomins. The eudistomins are known for their pronounced antiviral and cytotoxic activities (Schmitz *et al.,* 1993; Rinehart *et al.,* 1993). Eudistomins G (**1**) and H (**2**) were found to be responsible for the epibiont-free surface of *E. olivaceum* (Davis, 1991). Concentrations of eudistomins G and H of less than one-fifth compared to those found in the ascidian inhibited the settlement of larvae of five invertebrate species including the bryozoans *Bugula neritina* and *B. stolonifera* as well as the barnacle *Balanus amphritite* in the laboratory as well as in field trials.

Alkaloids were also reported as antifouling constituents present in several bryozoan species from Australia (Walls *et al.,* 1993). From four species of bryozoans studied *Amathia wilsoni* and *Orthoscuticella ventricosa* had the most active antibacterial extracts and exhibited the lowest levels of fouling in the field. Amathamide alkaloids (**3**) (present in *A. wilsoni*) or β-carboline alkaloids (**4**) (present in *O. ventricosa*) were suggested as causative antibiotic agents present in the extracts of *A. wilsoni* and *O. ventricosa* although no bioassays were performed with the isolated natural products (Walls *et al.,* 1993). Interestingly, the content and distribution of amathamide alkaloids within single colonies of *A. wilsoni* were found to vary depending on the location in the colony (Walls *et al.,* 1991). The outermost, exposed tips of the colonies had the highest concentrations of alkaloids (up to 9% of dry weight) whereas the basal parts were devoid of alkaloids. This preferential accumulation of alkaloids in the exposed and most vulnerable parts of the colonies supports the suggested defensive function of these metabolites.

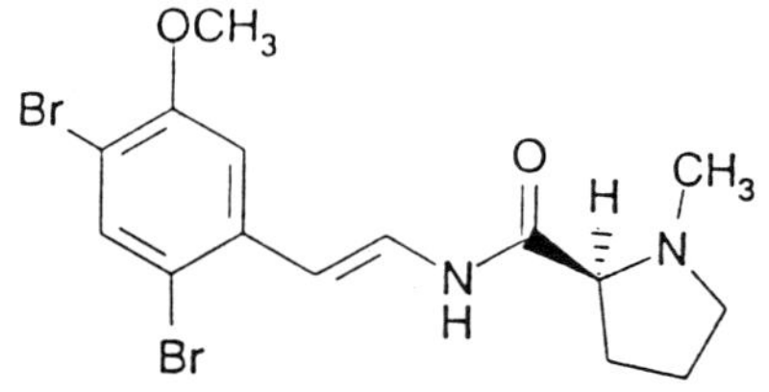

3

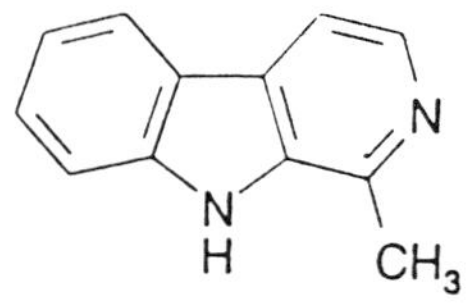

4

For an optimized defense against fouling organisms, however, a concentration of the respective natural products on the surface or a continuous release of the compounds into the seawater should be expected (Paul, 1992). Only a few studies on marine natural products that have been suggested to be involved in chemical defense against antifouling organisms have addressed the localization of the assumed defensive compounds. The sponge *Aplysina fistularis* (syn. *Verongia fistularis*) has been shown to exude the antimicrobially active alkaloids aerothionin (**5**) and homoaerothionin (**6**) into seawater at rates of 8.9×10^{-3} to 7.7×10^{-4} μg min^{-1} g^{-1} (Walker *et al.*, 1985). Following mechanical injury of the sponge the exudation was increased by a factor of 10–100 compared to controls. Thompson *et al.* (1985) had shown earlier that low concentrations of aerothionin and homoaerothionin can deter potential fouling organisms from settling or overgrowth of *A. fistularis.* A similar injury-induced exudation of biologically active natural products

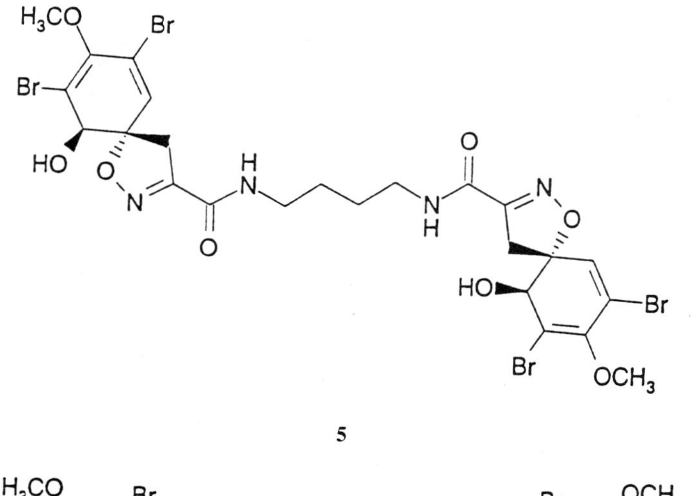

5

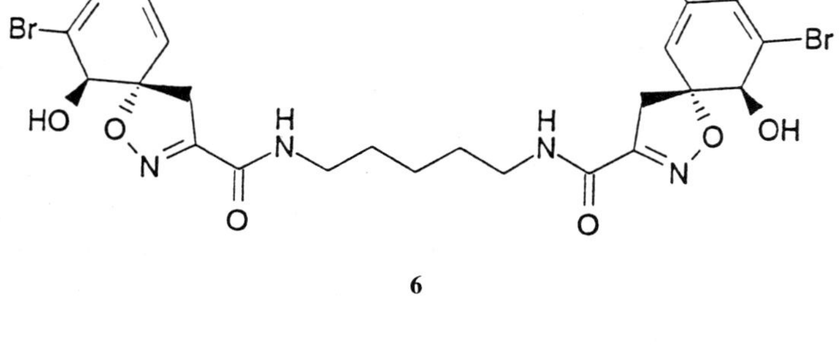

6

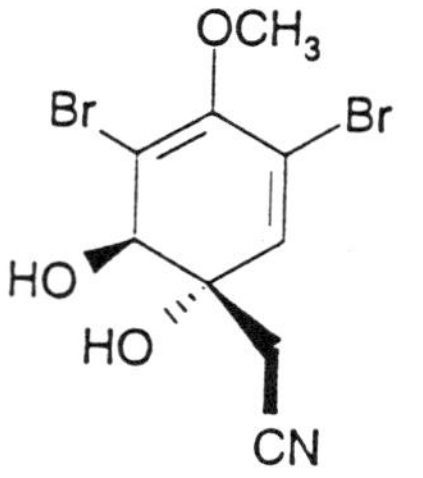

7

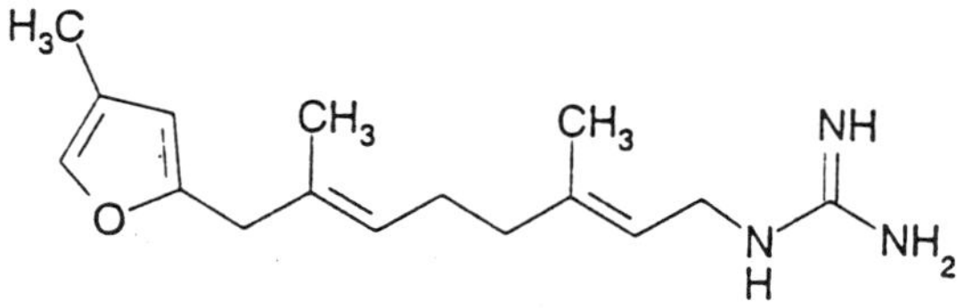

8

was reported for the related sponge species *Verongia aerophoba* (Kreuter *et al.*, 1992). Cubes of sponges (100 g) kept in aquaria released a total of 13 mg of aeroplysinin-1 (**7**) into seawater during a 10-day observation period. As shown recently, aeroplysinin-1 strongly suppresses the growth of several marine bacteria as well as microalgae suggesting that aeroplysinin-1 might be involved in the suppression of epibionts.

There is considerable circumstantial evidence for an involvement of natural products in spatial competition between marine invertebrates (Jackson and Buss, 1975; Porter and Targett, 1988). Spatial competition is especially pronounced on tropical coral reefs that are characterized by an exceedingly high species diversity and a density of population that is unmatched in terrestrial ecosystems with the possible exception of tropical rain forests (Mebs, 1989; Sale, 1991). Homogenates, especially of sponges, have often been reported to inhibit the growth of other sessile invertebrates, such as corals, indicating the presence of allelopathy in the marine environment (Jackson and Buss, 1975). This suggestion is corroborated by the occurrence of distinct bare zones around sponges that are observed in the natural habitat (Porter and Targett, 1988). However, natural products responsible for the allelopathic activity of marine invertebrates have rarely been characterized.

A notable exception are burrowing sponges of the genus *Siphonodictyon. S. coralliphagum* burrows into the heads of living corals. Overgrowth of *S. coralliphagum* by corals is prevented by exudation of a mucus that apparently acts as a carrier for several toxic metabolites including siphonodictidin (**8**) (Sullivan *et al.*, 1981). Siphonodictidin inhibits the photosynthesis of zooxanthellae in the hard coral *Acropora formosa* at concentrations > 10 ppm resulting in suppressed coral growth (Sullivan *et al.*, 1983). The related sponge species *S. mucosa* grows in dead corals. The mucus of *S. mucosa* lacks siphonodictidin, which supports the suggested allelopathic role of this metabolite (Sullivan *et al.*, 1983).

3. CHEMICAL DEFENSE AGAINST FISH

Chemical defense against fish through accumulation of toxic or deterrent alkaloids has frequently been reported for marine invertebrates (Braekman and Daloze, 1986; Mebs, 1989; Pawlik, 1993). A striking example of the significance of marine alkaloids for chemical defense against fish is provided by the red-colored sponge *Latrunculia magnifica* from the Red Sea. Even though *L. magnifica* grows exposed, it is apparently avoided by fish whereas other sponges from the same habitat that are cryptic are readily consumed by fish when artificially exposed (Neeman *et al.*, 1975). When specimens of *L. magnifica* are squeezed, the reddish exudate causes fish to flee immediately. When squeezed into aquar-

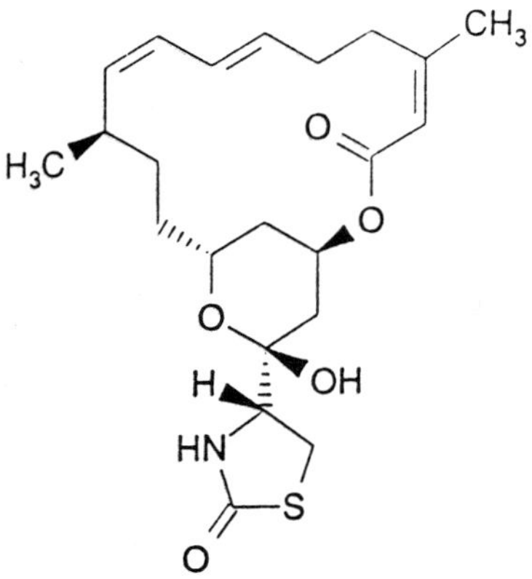

9

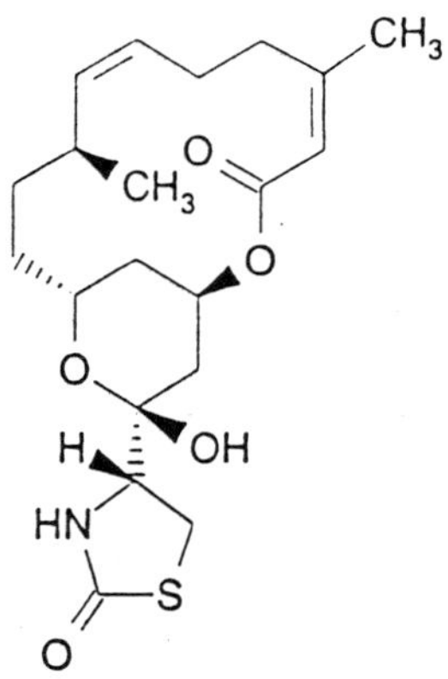

10

ia, fish (*Gambusia affinis*) died within minutes (Neeman *et al.,* 1975). Chemical analysis revealed the toxic macrocyclic alkaloids latrunculin-A (**9**) and -B (**10**) that are natural inhibitors of acetylcholinesterase (Kashman *et al.,* 1980; Groweiss *et al.,* 1983).

Although the latrunculins are effective in deterring predatory fish, the vividly colored sponge-feeding nudibranch *Glossodoris quadricolor* is not at all deterred. Nudibranchs are exclusively marine sluglike invertebrates in the molluskan subclass Opisthobranchia that have lost the protective shell characteristic of other gastropods (Karuso, 1987). *G. quadricolor* was found not only to feed unharmed on *L. magnifica* but to incorporate latrunculins into its mucus (Mebs, 1985). It is tempting to assume that the aposematically colored nudibranch derives protection from its own predators through sequestration of host-derived toxins even though no experiments have been conducted so far to prove this hypothesis.

Sequestration of host-derived toxins for defensive purposes is by no means limited to *G. quadricolor* but has also frequently been described for nudibranchs that prey on other marine invertebrates (Karuso, 1987). One of the earliest examples was provided by the sponge-feeding nudibranch *Phyllidia varicosa.* Specimens of *P. varicosa* exude a mucus that is toxic to crustaceans as well as to fish (Johannes, 1963). Attempts to isolate and characterize the toxins failed because of the small number of nudibranchs available. By chance, however, a specimen of *P. varicosa* from Hawaii was observed feeding on a sponge of the genus *Ciocalypta* from which the toxin 9-isocyanopupukeanane (**11**) as well as an isomer were eventually isolated and identified (Burreson *et al.,* 1975). Apparently, the nudibranch had sequestered the toxin from its food source.

Sponges of the genus *Halichondria* were traced as dietary source of macrolide

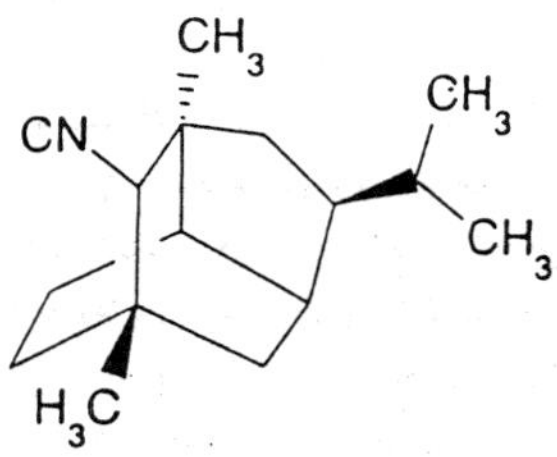

11

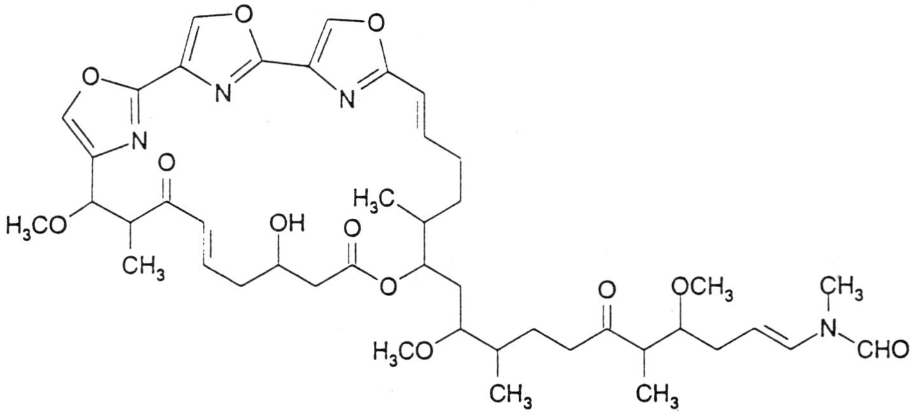

12

oxazole alkaloids isolated from *Hexabranchus sanguineus* (Kernan *et al.*, 1988). *H. sanguineus* or "Spanish dancer" is a large and brightly colored nudibranch found on Indo-Pacific coral reefs. Even though morphologically defenseless, the nudibranchs are rejected in feeding experiments by sympatric predators including the Indo-Pacific reef fish *Thalassoma lunare* and the reef hermit crab *Dardanus megistos* (Pawlik *et al.*, 1988). The unpalatability of the nudibranchs was traced to macrocyclic alkaloids including halichondramide (**12**) sequestered from *Halichondria* species (Pawlik, 1993). The alkaloids deterred feeding of potential predators at concentrations of 0.01–0.02% dry weight, which is one order of magnitude below the concentrations that are present in the sponge or in the nudibranch (Pawlik, 1993). The alkaloids are concentrated in the dorsal mantle (thereby strengthening the proposed defensive function) which is most exposed to potential predators but are also passed on to the likewise conspicuous egg ribbons (Pawlik *et al.*, 1988).

In a recent study by Lindquist *et al.* (1992), fish-deterrent alkaloids were reported from larvae of ascidians. Although ascidians release large, conspicuous larvae during daylight when exposure to fish predation is high, field observations indicate that many conspicuous ascidian larvae are distasteful to potential predators (Lindquist *et al.*, 1992). A vivid example is provided by the aposematically colored larvae of the ascidian *Ecteinascidia turbinata.* The pinfish *Lagodon rhomboides* after tasting larvae of *E. turbinata* was found to avoid palatable larvae of *Clavelina oblonga* when the latter were dyed orange to resemble larvae of *E. turbinata* (Young and Bingham, 1987). Whereas the deterrent compounds present in larvae of *E. turbinata* were not elucidated, the unpalatability of the ascidian *Trididemnum solidum* was shown to be related to alkaloids of the didemnin type that are also present in the adults (Lindquist *et al.*, 1992). Didemnin B (**13**) and nordidemnin B (**14**) inhibited feeding of reef fishes in field assays at concentrations below those found in the ascidians (Lindquist *et al.*, 1992).

4. INDUCED CHEMICAL DEFENSE OF THE SPONGE *VERONGIA AEROPHOBA*

The chemical defense mechanisms elucidated for marine organisms so far are constitutive. They rely on preformed toxic or deterrent natural products that are either continu-

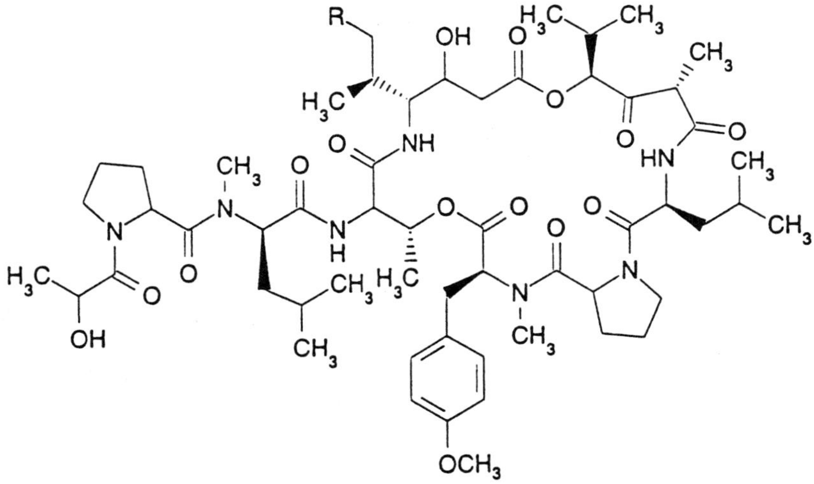

13 R = CH_3
14 R = H

ously exuded as shown, for example, for the sponge *Aplysina fistularis* (Walker *et al.*, 1985) or are liberated when tissues are injured. In the terrestrial environment most chemical defense mechanisms rely also on constitutively formed natural products that deter potential herbivores (in the case of plants) or predators (in the case of animals) (Harborne, 1993). However, induction of chemical defense either by *de novo* synthesis of defense metabolites (e.g., phytoalexins) (Bailey and Mansfield, 1982) or by enzymatically catalyzed activation of preformed compounds (e.g., liberation of HCN from cyanogenic glycosides) (Jones, 1988) as a direct response to environmental stress factors (such as herbivory) is also well established for plants as well as animals in the terrestrial environment (Harborne, 1993). In the marine environment the sponge *Verongia aerophoba* (syn. *Aplysina aerophoba*) is one of the few examples for an induced defense mechanism which relies on enzymatic conversion of inactive storage compounds into toxic defense metabolites (Proksch, 1994).

The conspicuously yellow-colored sponge *V. aerophoba* which is frequently found in the Mediterranean as well as around the Canary islands is exceptionally rich in brominated isoxazoline alkaloids (Teeyapant *et al.*, 1993a). Healthy specimens of *V. aerophoba* contain isofistularin-3 (**15**) as well as aerophobin-2 (**16**) as major brominated alkaloids that may account for 10% of the dry weight of the sponge (Teeyapant *et al.*, 1993a). Following disruption of the cellular organization a rapid, enzymatically catalyzed bioconversion of the storage compounds **15** and **16** is observed leading to the lower molecular weight compound aeroplysinin-1 (**7**), which under mild alkaline conditions (e.g., in seawater) spontaneously gives rise to the dienone (**17**) (Teeyapant and Proksch, 1993; Ebel *et al.*, unpublished data).

The described reaction leading to the formation of aeroplysinin-1 is observed both *in vivo* (Kreuter *et al.*, 1992) as well as *in vitro* using a cell free extract of *V. aerophoba* (Teeyapant, 1994). Cubes of *V. aerophoba* (fresh weight approximately 100 g) that were cut from specimens collected in the Adriatic sea and subsequently kept in aquaria were found to release **7** (13 mg during a 10 day period) (Kreuter *et al.*, 1992) whereas intact

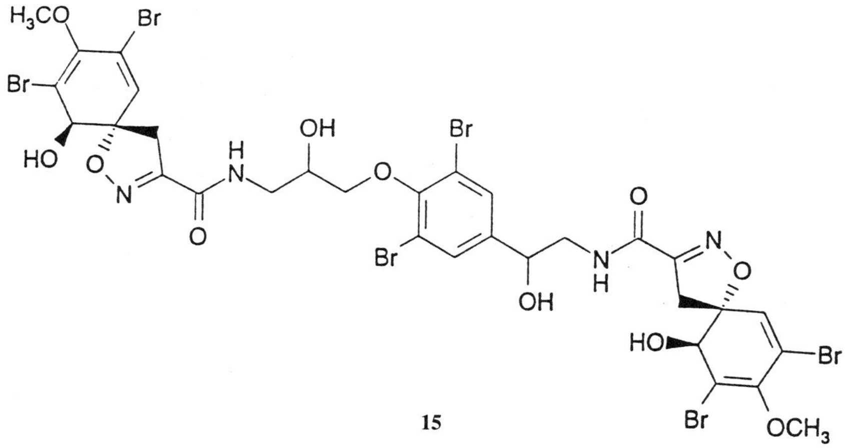

15

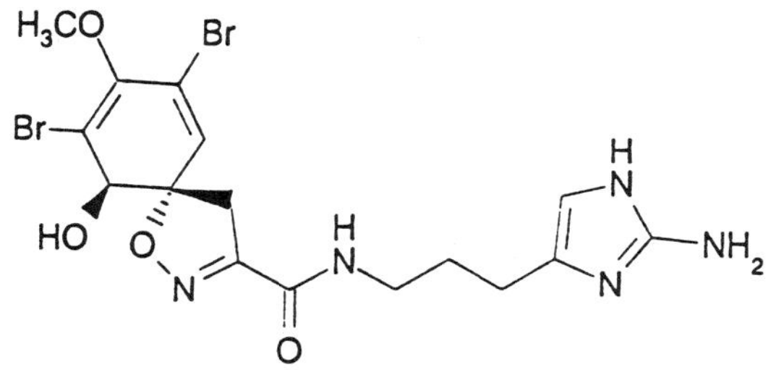

16

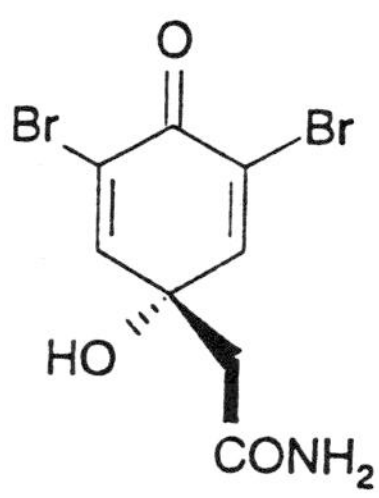

17

specimens from the Adriatic that were immediately frozen following collection contain only **15** and **16** as brominated alkaloids (Teeyapant, 1994). Preliminary *in vitro* experiments using a cell-free extract of *V. aerophoba* suggested that the enzyme responsible for the bioconversion of **15** or **16** into 1 is membrane associated ($M_r > 600{,}000$), has a pH optimum between 5.5 and 6.0 and a temperature optimum between 55 and 60°C (Teeyapant, 1994).

The described bioconversion of **16** or **15** to **7** and **17** was shown to result in a marked increase of the biological activities of the products compared to their substrates. For example, **7** as well as **17** were antibiotically active against a broad spectrum of marine (e.g., *Vibrio anguillarum* or *Serratia plymuthica*) as well as terrestrial gram-positive and gram-negative bacteria (e.g., *Bacillus subtilis, Staphylococcus aureus*) whereas **16** and **15** were inactive (Teeyapant *et al.,* 1993b; Weiss *et al.,* unpublished results). The minimum inhibitory concentrations for bacteria were in the range of 68–136 μM for **7** and **17** depending on the bacteria used for the bioassays. The toxicity of **7** and **17** extended to several species of marine microalgae (e.g., *Coscinodiscus wailesii* or *Prorocentrum minimum*) that together with bacteria are responsible for the formation of the "primary film" during fouling of a marine surface (Davis *et al.,* 1989). Whereas the EC_{50} (concentrations that reduce cell division by 50% compared to controls) of **7** and **17** were between 5.9 and 22 μM (depending on the species of microalgae used for the experiments), **16** (**15** was not tested because of its poor solubility) was inactive (Weiss *et al.,* unpublished results). Further experiments on the metabolites of *V. aerophoba* were conducted using the polyphagous marine snail *Littorina littorea* as test organism. When **7** or **17** were added to aquaria containing specimens of *L. littorea* the snails retreated into their shells whereas addition of **16** at similar concentrations (0.10 mM) caused no retraction of the soft bodies of *L. littorea* (Weiss *et al.*, unpublished results).

The data presented suggest an effective chemical defense mechanism of the sponge *V. aerophoba* that operates through an enzymatically triggered activation of preformed metabolites as a consequence of disruption of the cellular organization similar to the formation of HCN from cyanogenic glycosides (Jones, 1988).

5. MARINE ALKALOIDS AS WATERBORNE SIGNALS IN INTER- AND INTRASPECIFIC COMMUNICATION

Marine alkaloids are not only involved in chemical defense but also are important as alarm pheromones in intraspecific communication. The nudibranchs *Tambje abdere* and *T. eliora* both feed on the bryozoan *Sessibugula translucens* and sequester the alkaloids tambjamines A–D (**18–21**) from their prey (Carté and Faulkner, 1986). Even though exudation of tambjamines by the bryozoan was not examined the alkaloids are apparently used by the nudibranchs to detect their bryozoan prey, as *T. eliora* was shown to be attracted toward seawater containing *S. translucens* or a 10^{-10} M mixture (1 : 1) of tambjamines A and B (Carté and Faulkner, 1986). The carnivorous nudibranch *Roboastra tigris* preys on both *Tambje* species and is able to follow their slime trails using contact chemoreception. When the slime trail is broken, *R. tigris* is able to reverse its direction. At least the slime trail of *T. abdere* was shown to contain tambjamines in low concentrations (Carté and Faulkner, 1986). Thus, the tambjamines may also be responsible for prey location by *R. tigris*. As the tambjamines are easily hydrolyzed yielding less active aldehydes, *R. tigris* is probably able to differentiate "old trails" from "new trails" (Carté and Faulkner, 1986).

When attacked by its predator, *T. abdere* secretes a mucus rich in tambjamines which may cause *R. tigris* to break off the attack. The substantial quantities of tambjamines released by *T. abdere* (approximately 3 mg/individual) during defense against *R. tigris* may trigger escape behavior at least in *T. eliora* that contains smaller amounts of tambjamines in its mucus (compared to *T. abdere*) insufficient for deterring *R. tigris* (Carté and

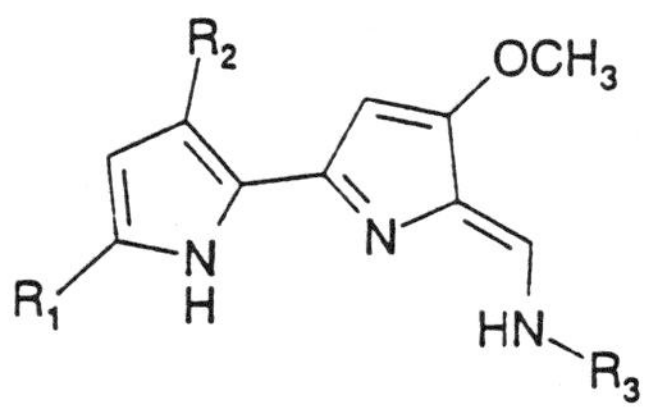

18 R_1 = H; R_2 = H; R_3 = H
19 R_1 = Br; R_2 = H; R_3 = H
20 R_1 = H; R_2 = H; R_3 = i-Bu
21 R_1 = H; R_2 = Br; R_3 = i-Bu

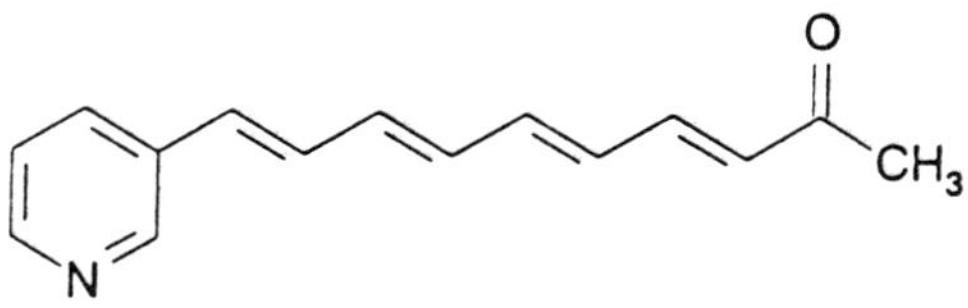

22

Faulkner, 1986). Specimens of *T. eliora* are repelled by tambjamine concentrations > 10^{-8} M whereas they are attracted by tambjamine concentrations of 10^{-10} M (*vide supra*) (Carté and Faulkner, 1986). Thus, whereas small concentrations of tambjamines may attract *T. eliora* to its food source, larger concentrations (two orders of magnitude higher) may be recognized as alarm pheromones and may trigger escape behavior.

Alarm pheromones have also been reported from the marine opisthobranch *Navanax inermis* (Sleeper *et al.*, 1980). When molested, *N. inermis* secretes a mixture of related compounds called navenones including the alkaloid navenone A (**22**) from a special gland (called "yellow gland") into its slime trail. These compounds induce an avoidance response in trail-following *Navanax* specimens at concentrations of 10^{-5} M (Fenical *et al.*, 1979; Sleeper *et al.*, 1980). The navenone containing slime trail of *N. inermis* was shown to persist up to several hours depending on currents and bottom conditions (Sleeper *et al.*, 1980). When released into the slime trail, the navenones were restored within the opisthobranchs within 3–7 days dependent on vitality and feeding of the animals.

6. ORIGIN OF ALKALOIDS FROM MARINE INVERTEBRATES

Sponges, ascidians, and bryozoans which are among the major sources of marine alkaloids (Faulkner, 1993) are filter feeders which feed on microscopic food particles (e.g., bacteria, microalgae, detritus). Furthermore, these filter feeders usually live in close association with symbiotic microorganisms that may occasionally constitute more than 50% of the biomass of the respective invertebrate, as reported for the sponge *V. aerophoba* (Vacelet, 1975). Even though frequently assumed (Garson, 1994), it is still unclear in most cases if symbiotic or filtered microorganisms contribute to the numerous alkaloids isolated from marine invertebrates. Evidence for an involvement of microorganisms in the biosynthesis of natural products isolated from marine invertebrates is indirect in most cases. For example, the frequent occurrence of alkaloids in marine invertebrates as compared to marine algae has been suggested to result from nitrogen-fixing cyanobacteria that often live as endosymbionts in the respective invertebrates (Paul, 1992). Nitrogen-fixing endosymbionts of invertebrates can be expected to concentrate nitrogen which in turn may be utilized for the biosynthesis of alkaloids. In nitrogen-poor environments such as the tropics (Paul, 1992), an increased nitrogen supply may substantially influence the rate of alkaloid biosynthesis. Further arguments in favor of a microbial origin of natural products from invertebrates focus on chemical similarities between compounds isolated from bacteria and those isolated from marine invertebrates. For example, the alkaloids renierone (**23**) and renierol (**24**) isolated from the sponge *Xestospongia caycedoi* were suggested to be microbial metabolites based on structural similarities with mimosamycin (**25**) which is

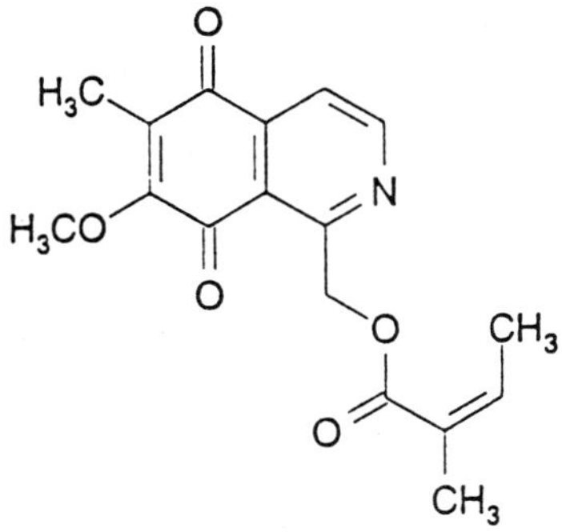

23

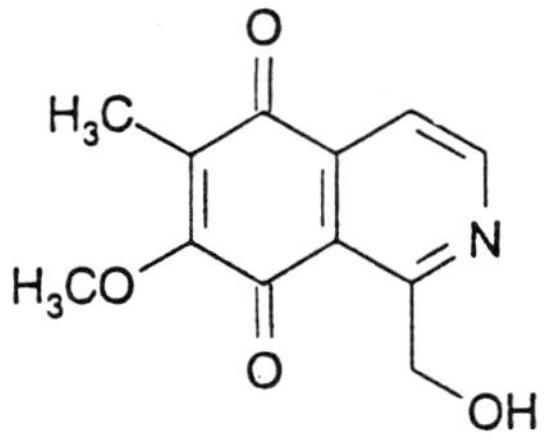

24

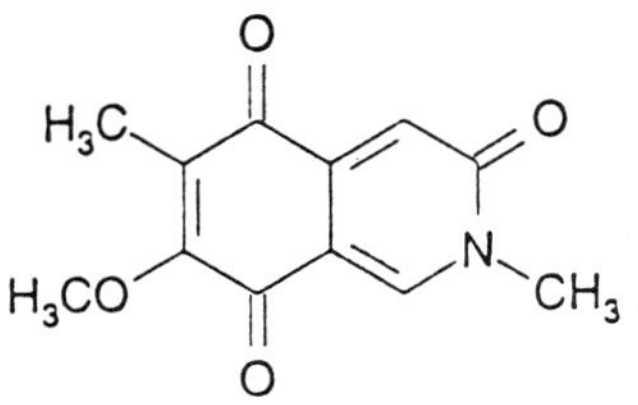

25

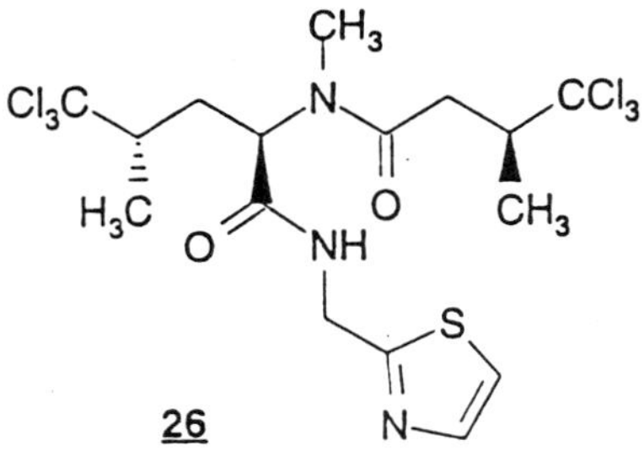

26

produced by bacteria of the genus *Streptomyces* (McKee and Ireland, 1987). Direct proof for symbionts as the real producers of natural products originally ascribed to invertebrates, however, is rare. Recently, however, using flow cytometric separation of the symbiotic cyanobacterium *Oscillatoria spongeliae* from cells of the sponge *Dysidea herbacea,* it could be demonstrated that only the symbiont contains the polychlorinated alkaloids including 13-demethylisodysidenin (**26**) that are present among the "sponge" metabolites (Unson and Faulkner, 1993). However, even the presence of a natural product in a symbiont is no unequivocal proof for the symbiont as producer of the compound because transport of natural products within invertebrates cannot be excluded. Thus, axenic culture of symbionts seems still inevitable for the elucidation of the biosynthetic capacity of symbiotic microorganisms. The first successful attempts at culturing isolated endosymbionts from marine invertebrates have been described (Kobayashi and Ishibashi, 1993). It is to be expected that this rapidly growing field of marine biotechnology will help to clarify the real origin of natural products from marine sources.

7. CHEMICAL DEFENSE IN THE MARINE VERSUS THE TERRESTRIAL ENVIRONMENT

Marine chemical ecology, even though still in its infancy when compared to the long tradition of terrestrial chemical ecology (e.g., Zukal, 1895; Stahl, 1904), has firmly established the significance of natural products in inter- as well as intraspecific interactions. As in the terrestrial environment, alkaloids, which usually possess pronounced physiological activities, play an important role as chemical mediators at least for the numerous invertebrate organisms that accumulate these natural products. Chemical defense in the marine environment, which is most pronounced in exposed species that are subjected to an intense grazing pressure (Bakus, 1981; Bakus and Green, 1974; Green, 1977), seems to fit the "plant apparency model" originally developed for higher terrestrial plants (Feeny, 1976; Rhoades and Cates, 1976) which suggests a tight correlation of chemical defense and the plant's risk of being discovered by herbivores. The "optimal defense theory" (McKey, 1979; Rhoades, 1979) that was also developed for higher terrestrial plants suggests that chemical defenses that are metabolically costly are especially pronounced in the most valuable plant parts (usually those with a high nitrogen content). Seeds, for example, fall in this category as they guarantee the next generation and hence survival of the species. Consequently many seeds including those of alkaloid-containing members of the Fabaceae (Szentesi and Wink, 1991) are extremely toxic because of the presence of quinolizidine alkaloids. The same holds true for chemically defended eggs of herbivorous insects such as *Epilachna varivestis* (Proksch *et al.,* 1993) and apparently also for eggs and larvae of marine invertebrates (Pawlick *et al.,* 1988; Lindquist *et al.,* 1992) even though further studies are still needed for marine systems.

Sesquestration of toxic metabolites from dietary sources by specialized predators and employment of the sequestered compounds for chemical defense is likewise known from marine as well as terrestrial systems. In the marine environment, nudibranchs such as the aposematically colored *Glossodoris quadricolor* (Mebs, 1985) provide a vivid example for this counteradaption of specialists to their prey organisms. In the terrestrial environment numerous aposematically colored herbivorous insects such as the arctiid moth *Tyria jacobaeae* which sequesters pyrrolizidine alkaloids from *Senecio* species (Ehmke *et al.,* 1990) provide a well known paradigm for the transfer and utilization of toxic natural products through the food chain.

In conclusion: the specific physicochemical properties of the marine environment have selected for certain chemical features of natural products (e.g., frequent halogenation like that observed for alkaloids **5–7** from *Verongia aerophoba*) that are unprecedented or rare in the terrestrial environment. The strategies of chemical defense, however, appear to be very similar on land and in the sea.

ACKNOWLEDGMENT

The senior author wishes to thank the Deutsche Forschungsgemeinschaft for financial support of his research on the inducible defense system of the marine sponge *Verongia aerophoba.*

REFERENCES

General Reviews

Bailey, J. A., and Mansfield, J. W., (eds.), 1982, *Phytoalexins,* Blackie, Glasgow.

Bakus, G. J., 1981, Chemical defense mechanisms on the Great Barrier Reef, Australia, *Science* **211:**497–499.

Bakus, G. J., and Green, G., 1974, Toxicity in sponges and holothurians: A geographic pattern, *Science* **185:**951–953.

Bakus, G. J., Targett, N. M., and Schulte, B., 1986, Chemical ecology of marine organisms: An overview, *J. Chem. Ecol.* **12:**951–987.

Bakus, G. J., Schulte, B., Wright, M., Green, G., and Gomez, P., 1990, Antibiosis and antifouling in marine sponges: Laboratory versus field studies, in: *New Perspective in Sponge Biology* (K. Rützler, ed.), Smithsonian Institution Press, Washington, DC, pp. 102–108.

Davis, A. R., Targett, N. M., McConnell, O. J., and Young, C. M., 1989, Epibiosis of marine algae and benthic invertebrates: Natural products chemistry and other mechanisms inhibiting settlement and overgrowth, in: *Bioorganic Marine Chemistry,* Vol. 3 (P. J. Scheuer, ed.), Springer-Verlag, Berlin, pp. 85–114.

Faulkner, D. J., 1993, Marine natural products, *Nat. Prod. Rep.* **9:**323–539. [Preceding reviews by same author]

Fenical, W., 1986, Marine alkaloids and related compounds, in: *Alkaloids: Chemical and Biological Perspectives, 4* (S. W. Pelletier, ed.), Wiley, New York, pp. 275–330.

Garson, M., 1994, The biosynthesis of sponge secondary metabolites: Why it is important. in: *Sponges in Time and Space* (R. W. M. van Soest, T. M. G. van Kempen, and J. C. Braekman, eds.), Balkema, Rotterdam, pp. 427–440.

Harborne, J. B., 1993, *Introduction to Ecological Biochemistry,* 4th ed., Academic Press, San Diego.

Jones, D. A., 1988, Cyanogenesis in animal–plant interactions, in: *Cyanide Compounds in Biology* (D. Evered and S. Harnett, eds.), Wiley, New York, pp. 151–170.

Karuso, P., 1987, Chemical ecology of the nudibranchs, in: *Bioorganic Marine Chemistry,* Vol. 1, (P. J. Scheuer, ed.), Springer-Verlag, Berlin, pp. 32–60.

Luckner, M., 1990, *Secondary Metabolism in Microorganisms, Plants, and Animals,* Springer-Verlag, Berlin.

McKey, D., 1979, The distribution of secondary compounds within plants, in: *Herbivores: Their Interactions with Secondary Plant Metabolites* (G. A. Rosenthal and D. H. Janzen, eds.), Academic Press, New York, pp. 56–133.

Paul, V. J., 1992, Chemical defenses of benthic marine invertebrates, in: *Ecological Roles of Marine Natural Products* (V. J. Paul, ed.), Cornell University Press (Comstock), Ithaca, NY, pp. 164–188.

Pelletier, S. W., 1983, The nature and definition of an alkaloid, in: *Alkaloids: Chemical and Biological Perspectives, 1* (S. W. Pelletier, ed.), Wiley, New York, pp. 1–32.

Rhoades, D. F., 1979, Evolution of plant chemical defense against herbivores, in: *Herbivores: Their Interactions with Secondary Plant Metabolites* (G. A. Rosenthal and D. H. Janzen, eds.), Academic Press, New York, pp. 3–54.

Rhoades, D. F., and Cates, R. G., 1976, Toward a general theory of plant antiherbivore chemistry, *Recent Adv. Phytochem.* **10:**168–213.

Rinehart, K. R., Shield, L. S., and Cohen-Parsons, M., 1993, Antiviral substances, in: *Marine Biotechnology* (D. H. Attaway and O. R. Zaborsky, eds.), Plenum Press, New York, pp. 309–342.

Sale, P. F., (ed.), 1991, *The Ecology of Fishes on Coral Reefs,* Academic Press, San Diego.

Schmitz, F. J., Bowden, B. F., and Toth, S. I., 1993, Antitumor and cytotoxic compounds from marine organisms, in: *Marine Biotechnology* (D. H. Attaway and O. R. Zaborsky, eds.), Plenum Press, New York, pp. 197–308.

Scott, T. D., 1962, *The Marine and Fresh Water Fishes of South Australi*a, W. L. Hames, Adelaide.

Southon, I. W., and Buckingham, J., (eds.), 1989, *Dictionary of Alkaloids,* Chapman & Hall, London.

Stahl, E., 1904, Die Schutzmittel der Flechten gegen Tierfraβ, in: *Festschrift zum 70. Geburtstag von Ernst Haeckel,* Fischer Verlag, Jena, pp. 357–375.

Specific References

Braekman, J. C., and Daloze, D., 1986, Chemical defence in sponges, *Pure Appl. Chem.* **58:**357–364.

Burreson, B. J., Clardy, J., Finer, J., and Scheuer, P. J., 1975, 9-Isocyanopupukeanane, a marine invertebrate allomone with a new sesquiterpene skeleton, *J. Am. Chem. Soc.* **97:**4763–4764.

Carté, B., and Faulkner, D. J., 1986, Role of secondary metabolites in feeding associations between a predatory nudibranch, two grazing nudibranchs, and a bryozoan, *J. Chem. Ecol.* **12:**795–804.

Clemens, W. A., and Wilby, G. V., 1946, The fishes of the Pacific coast of Canada, *Bull. Fish. Res. Bd. Can.* **68:**1–368.

Davis, A. R., 1991, Alkaloids and ascidian chemical defense: Evidence for the ecological role of natural products from *Eudistoma olivaceum, Mar. Biol.* **111:**375–379.

Ehmke, A., Witte, L., Biller, A., and Hartmann, T., 1990, Sequestration, N-oxidation and transformation of plant pyrrolizidine alkaloids by the arctiid moth *Tyria jacobaeae* L., *Z. Naturforsch.* **45c:**1185–1192.

Feeny, P., 1976, Plant apparency and chemical defense, *Recent Adv. Phytochem.* **10:**1–40.

Fenical, W., Sleeper, H. L., Paul, V. J., Stallard, M. O., and Sun, H. H., 1979, Defensive chemistry of *Navanax* and related opisthobranch molluscs, *Pure Appl. Chem.* **51:**1865–1874.

Green, G., 1977, Ecology of toxicity in marine sponges, *Mar. Biol.* **40:**207–215.

Groweiss, A., Shmueli, U., and Kashman, Y., 1983, Marine toxins of *Latrunculia magnifica, J. Org. Chem.* **48:**3512–3516.

Jackson, J. B. C., and Buss, L., 1975, Allelopathy and spatial competition among coral reef invertebrates, *Proc. Natl. Acad. Sci. USA* **72:**5160–5163.

Johannes, R. E., 1963, A poison-secreting nudibranch (Mollusca: Opisthobranchia), *Veliger* **5:**104–105.

Kashman, Y., Groweiss, A., and Shmueli, U., 1980, Latrunculin, a new 2-thiazolidinone macrolide from the marine sponge *Latrunculia magnifica, Tetrahedron Lett.* **21:**3629–3632.

Kernan, M. R., Molinski, T. F., and Faulkner, D. J., 1988, Macrocyclic antifungal metabolites from the Spanish dancer nudibranch *Hexabranchus sanguineus* and sponges of the genus *Halichondria, J. Org. Chem.* **53:**5014–5020.

Kobayashi, J., and Ishibashi, M., 1993, Bioactive metabolites of symbiotic marine microorganisms, *Chem. Rev.* **93:**1753–1769.

Krebs, H. C., 1986, Recent developments in the field of marine natural products with emphasis on biologically active compounds, *Prog. Chem. Org. Nat. Prod.* **49:**151–363.

Kreuter, M. H., Robitzki, A., Chang, S., Steffen, R., Michaelis, M., Kljajic, Z., Bachmann, M., Schröder, H. C., and Müller, W. E. G., 1992, Production of the cytostatic agent aeroplysinin by the sponge *Verongia aerophoba* in *in vitro* culture, *Comp. Biochem. Physiol.* **101C:**183–187.

Lindquist, N., Hay, M. E., and Fenical, W., 1992, Defense of ascidians and their conspicuous larvae: Adult vs. larval chemical defenses, *Ecol. Monogr.* **62:**547–568.

McKee, T. C., and Ireland, C. M., 1987, Cytotoxic and antimicrobial alkaloids from the Fijian sponge *Xestospongia caycedoi, J. Nat. Prod.* **50:**754–756.

Mebs, D., 1985, Chemical defense of a dorid nudibranch, *Glossodoris quadricolor,* from the Red Sea, *J. Chem. Ecol.* **11:**713–716.

Mebs, D., 1989, *Gifte im Riff,* Wissenschaftliche Verlagsgesellschaft, Stuttgart.

Melton, T., and Bodnar, J. W., 1988, Molecular biology of marine microorganisms: Biotechnological approaches to naval problems, *Nav. Res. Rev.* **40:**24–39.

Neeman, I., Fishelson, L., and Kashman, Y., 1975, Isolation of a new toxin from the sponge *Latrunculia magnifica* in the Gulf of 'Aqaba (Red Sea), *Mar. Biol.* **30:**293–296.

Pawlik, J. R., 1993, Marine invertebrate chemical defenses, *Chem. Rev.* **93:**1911–1922.

Pawlik, J. R., Kernan, M. R., Molinski, T. F., Harper, M. K., and Faulkner, D. J., 1988, Defensive chemicals of the Spanish dancer nudibranch *Hexabranchus sanguineus* and its egg ribbons: Macrolides derived from a sponge diet, *J. Exp. Mar. Biol. Ecol.* **119:**99–109.

Porter, J. W., and Targett, N. M., 1988, Allelochemical interactions between sponges and corals, *Biol. Bull.* **175:**230–239.

Proksch, P., 1994, Defensive roles for secondary metabolites from marine sponges and sponge-feeding nudibranchs, *Toxicon* **32:**639–655.

Sleeper, H. L., Paul, V. J., and Fenical, W., 1980, Alarm pheromones from the marine opisthobranch *Navanax inermis, J. Chem. Ecol.* **6:**57–70.

Starck, W. A., 1968, A list of fishes of Alligator Reef, Florida with comments on the nature of the Florida reef fish fauna, *Undersea Biol.* **1:**1–40.

Sullivan, B., Djura, P., McIntyre, D. E., and Faulkner, D. J., 1981, Antimicrobial constituents of the sponge *Siphonodictyon coralliphagum, Tetrahedron* **37:**979–982.

Sullivan, B., Faulkner, D. J., and Webb, L., 1983, Siphonodictidine, a metabolite of the burrowing sponge *Siphonodictyon* sp. that inhibits coral growth, *Science* **221:**1175–1176.

Szentesi, A., and Wink, M., 1991, Fate of quinolizidine alkaloids through three trophic levels: *Laburnum anagyroides* (Leguminosae) and associated organisms, *J. Chem. Ecol.* **17:**1557–1574.

Teeyapant, R., 1994, Brominated secondary metabolites of the marine sponge *Verongia aerophoba* Schmidt and the sponge feeding gastropod *Tylodina perversa* Gmelin: Identification, biological activities and biotransformation, Ph.D. thesis, Würzburg.

Teeyapant, R., and Proksch, P., 1993, Biotransformation of brominated compounds in the marine sponge *Verongia aerophoba*: Evidence for an induced chemical defense?, *Naturwissenschaften* **80:**369–370.

Teeyapant, R., Kreis, P., Wray, V., Witte, L., and Proksch, P., 1993a, Brominated secondary compounds from the marine sponge *Verongia aerophoba* and the sponge feeding gastropod *Tylodina perversa, Z. Naturforsch.* **48c:**640–644.

Teeyapant, R., Woerdenbag, H. J., Kreis, P., Hacker, J., Wray, V., Witte, L., and Proksch, P., 1993b, Antibiotic and cytotoxic activity of brominated compounds from the marine sponge *Verongia aerophoba, Z. Naturforsch.* **48c:**939–945.

Thompson, J. E., Walker, R. P., and Faulkner, D. J., 1985, Screening and bioassays for biologically-active substances from forty marine sponge species from San Diego, California, USA, *Mar. Biol.* **88:**11–21.

Unson, M. D., and Faulkner, D. J., 1993, Cyanobacterial symbiont biosynthesis of chlorinated metabolites from *Dysidea herbacea* (Porifera), *Experientia* **49:**349–353.

Vacelet, J., 1975, E'tude en microscopie electronique de l'association entre bacteries et spongiaires du genre *Verongia* (Dictyoceratida), *J. Micr. Biol. Cell.* **23:**271–288.

Walker, R. P., Thompson, J. E., and Faulkner, D. J., 1985, Exudation of biologically-active metabolites in the sponge *Aplysina fistularis.* II. Chemical evidence, *Mar. Biol.* **88:**27–32.

Walls, J. T., Blackman, A. J., and Ritz, D. A., 1991, Distribution of amathamide alkaloids within single colonies of the bryozoan *Amathia wilsoni, J. Chem. Ecol.* **17:**1871–1881.

Walls, J. T., Ritz, D. A., and Blackman, A. J., 1993, Fouling, surface bacteria and antibacterial agents for bryozoan species found in Tasmania, Australia, *J. Exp. Mar. Biol. Ecol.* **169:**1–13.

Weinheimer, A. J., and Spraggins, R. L., 1969, The occurrence of two new prostaglandin derivatives (15-epi-PGA_2 and its acetate, methyl ester) in the gorgonian *Plexaura homomalla.* Chemistry of coelenterates. XV, *Tetrahedron Lett.* **15:**5185–5188.

Young, C. M., and Bingham, B. L., 1987, Chemical defense and aposematic coloration in larvae of the ascidian *Ecteinascidia turbinata, Mar. Biol.* **96:**539–544.

Zukal, H., 1895, Morphologische und biologische Untersuchungen über die Flechten, *Ber. K. Böhm. Ges. Wiss. Math.-Nat. Kl.* **104:**1303–1395.

Part IV

Alkaloids in Medicine

Chapter 17

Antimicrobially Active Alkaloids

R. Verpoorte

1. INTRODUCTION

Plants produce a broad variety of natural products. The data bank NAPRALERT listed about 88,000 natural products in 1988, of which about 16,000 were alkaloids. These so-called secondary metabolites most likely play a role in the interaction of the plant with its environment, e.g., to defend the plant against microorganisms or various predators. In the former case the plant secondary metabolites concerned are expected to have an antimicrobial activity, i.e., the phytoalexins. Phytoalexins are low-molecular-weight compounds, which are synthesized and accumulated in the plant after microbial infection (Paxton, 1981). Among the compounds recognized as phytoalexins are some alkaloids (reviewed by Kuc, 1992; Whitehead and Threlfall, 1992); however, the number of alkaloids that have been shown to possess antimicrobial activity is in fact much larger. There has long been an interest in screening plants and the derived natural products for biological activity, with the aim of producing new drugs. Since ancient times man has been using plants to treat all kinds of diseases. As infectious diseases are easy to diagnose, and the effects of an antibiotic can be observed clearly, it is likely that many traditional medicines have useful biological activity. Screening such plants has indeed shown that the incidence of antimicrobial activity found is much higher than for a random screening (e.g., Dornberger and Lich, 1982; Elmi *et al.*, 1986; Le Grand *et al.*, 1988; Mitscher *et al.*, 1987; Verpoorte *et al.*, 1982b, 1983a). Studies of antimicrobially active plants have resulted in identifying many alkaloids that have such activity. Whether these compounds also play a role in the defense against micro-organisms in the plant is not known in most cases, unfortunately.

Here I will only consider alkaloids from higher plants as marine organisms have been reviewed elsewhere (Chapter 16). Antiamoebic, antiviral, antimalarial, and antitumor activities will not be dealt with here, but it must be stressed that plants are a valuable and interesting source for such compounds as well.

R. Verpoorte • Division of Pharmacognosy, Leiden/Amsterdam Center for Drug Research, University of Leiden, 2300 RA Leiden, The Netherlands.

Alkaloids: Biochemistry, Ecology, and Medicinal Applications, edited by Roberts and Wink. Plenum Press, New York, 1998.

Presently, only two alkaloids are used for their antimicrobial activity: berberine because of its antidiarrhetic properties, and sanguinarine because of its anticaries properties. Some bisbenzylisoquinoline alkaloids have been used in the past, and there is renewed interest in them.

Numerous reviews have been published on the antimicrobial activity of plants and natural products (Burkholder and Sharma, 1969; Cavallito, 1951; Dornberger and Lich, 1982; Hiller, 1964; Irving, 1949; Korzybski *et al.*, 1967; Mitscher *et al.*, 1972a, 1975, 1987; Nickell, 1959; Skinner, 1965; Stoessl, 1970). The antimicrobial activity of alkaloids has also been extensively reviewed (Clark and Hufford, 1992; Simeon *et al.*, 1989; Verpoorte, 1986, 1987; Wink, 1993).

2. INDOLE ALKALOIDS

2.1. Terpenoid Indole Alkaloids

A number of terpenoid indole alkaloids have been reported to have antimicrobial activity (Table I). These alkaloids, found mainly in plants belonging to the Apocynaceae, Loganiaceae, and Rubiaceae, are found in all parts of the plants. Each plant and each type of tissue has its own particular types of alkaloids. Verpoorte *et al.* (1983a) and Van Beek *et al.* (1984a) screened a large number of plants belonging to the Apocynaceae and Loganiaceae for antimicrobial activity. Many extracts showed strong activity, some from plants well known for containing alkaloids.

Some *Aspidosperma* species, in particular, revealed strong activity. A series secamine-type of alkaloids (Fig. 1) were shown to be among the active compounds in these plants (Verpoorte *et al.*, 1982a, 1983b). Activity was only found against some gram-positive bacteria with MIC values of about 0.1 mg ml^{-1}. The quasidimeric alkaloid ochrolifuanine A (Fig. 2) was also found to be active. Further studies on the structure–activity relationship of 34 different quasidimeric alkaloids (Fig. 2), differing in stereochemistry and substitution pattern, were made by Caron *et al.* (1988). The alkaloids showed similar activities, and minor changes were observed influenced by the C/D ring junction, aromatic substitution, and the oxidation of the C-ring. This type of alkaloid has also been shown to exhibit antiprotozoal activity (Keene *et al.*, 1987; Phillipson *et al.*, 1993; Wright *et al.*, 1991) and cytotoxic activity (Keene *et al.*, 1987; Seguin *et al.*, 1985) as well as insect-antifeedant activity (Aerts *et al.*, 1992). In these respects they resemble emetine, a biosynthetically related quasidimeric isoquinoline alkaloid (Keene *et al.*, 1987).

Antimicrobially dimeric alkaloids were also isolated from several *Tabernaemontana* species (Achenbach *et al.*, 1980; Achenbach, 1986; Hernandez, 1979; Munoz *et al.*, 1994; Perera *et al.*, 1985; Van Beek *et al.*, 1984b, 1985a,b). The voacamine type (Fig. 3), consisting of an iboga- and a vobasine-type moiety, showed particularly strong antimicrobial activity. Such alkaloids also have antileishmanial activity (Munoz *et al.*, 1994). Vobparicine, a dimer consisting of a vobasinyl and an apparicine moiety, also has antimicrobial activity. Interestingly, even the monomers related to the building blocks of these dimers, have antimicrobial activity, although at a lower level than the dimers. Among both the dimers and the iboga-type monomers quite a few 3-hydroxy derivatives are found. Such a substitution leads to an increase of activity for the monomers. A carbomethoxy

Table I
Terpenoid Indole Alkaloids Reported to Have Antimicrobial Activity

	Activity[a]					
	+	−	y	f	MIC (μg/ml)[b]	References
Secamine type (Fig. 1)						
tetrahydrosecamine	+				110	Verpoorte *et al.* (1982a)
and derivatives						
16-decarbomethoxy	+				70	Verpoorte *et al.* (1982a)
didecarbomethoxy	+				100	Verpoorte *et al.* (1982a)
16-OH	+					Verpoorte *et al.* (1982a)
16-OH-decarbomethoxy	+					Verpoorte *et al.*, 1982a
Aspidospermine type						
aspidospermine	+	−			*1000*	Hernandez (1979), Hernandez *et al.* (1977), Mitscher *et al.* (1972a)
vindoline	+	−		f		Hernandez (1979)
vindolinine	+	−				Hernandez (1979)
Yohimbine–heteroyohimbine type						
ajmalicine		−		f		Hernandez (1979)
isoraunescine	+				*1000*	Mitscher *et al.* (1972a)
tetrahydroalstonine	+	−				Hernandez (1979)
reserpine		−		f		Hernandez (1979)
dihydrocorynantheol	+					Verpoorte *et al.* (1982a)
usambarensine	+					Caron *et al.* (1988)
and derivatives						
dihydro- (3α)	+				64	Caron *et al.* (1988)
tetrahydro- (3α,17α)	+					Caron *et al.* (1988)
tetrahydro- (3α,17β)	+				32	Caron *et al.* (1988)
tetrahydro- (3β,17?)	+				16	Caron *et al.* (1988)
tetrahydro- (3β,17?)	+					Caron *et al.* (1988)
tetrahydrousambarensine derivatives (Fig. 2)						
10′-OH- (3α,17α)	+					Caron *et al.* (1988)
10′-OH- (3α,17β)	+					Caron *et al.* (1988)
10,10′-diOMe- (3α,17α)	+					Caron *et al.* (1988)
10,10′-diOMe,NMe- (3α,17α)	+					Caron *et al.* (1988)
10′-OH,10-OMe,NMe-(3α,17α)	+					Caron *et al.* (1988)
10,10′-diOH,NMe- (3α,17α)	+					Caron *et al.* (1988)
10-OH,10′-OMe,NMe-(3α,17α)	+					Caron *et al.* (1988)
ochrolifuanine A	+				32	Verpoorte *et al.* (1983b), Caron *et al.* (1988)
ochrolifuanine E	+					Caron *et al.* (1988)
ochrolifuanine F	+	−			32	Caron *et al.* (1988)
18,19dihydroochrolifuanine F	+	−				Caron *et al.* (1988)
cinchophylline (3α,17α)	+					Caron *et al.* (1988)
cinchophylline (3α,17β)	+					Caron *et al.* (1988)
cinchophylline (3β,17α)	+				16	Caron *et al.* (1988)
cinchophylline (3β,17β)	+				32	Caron *et al.* (1988)
and derivatives						
17,4′-didehydro (3α)	+					Caron *et al.* (1988)
18,19-dihydro (3α,17α)	+					Caron *et al.* (1988)
18,19-dihydro (3α,17β)	+					Caron *et al.* (1988)

(continued)

Table I
(Continued)

	Activity[a]					
	+	−	y	f	MIC (μg/ml)[b]	References
18,19-dihydro (3β,17α)	+					Caron *et al.* (1988)
18,19-dihydro (3β,17β)	+					Caron *et al.* (1988)
19-OH,18,19-diH (3α,17α)	+					Caron *et al.* (1988)
19-OH,18,19-diH (3α,17β)	+				32	Caron *et al.* (1988)
19-OH,18,19-diH (3β,17α)	+					Caron *et al.* (1988)
19-OH,18,19-diH (3β,17β)	+					Caron *et al.* (1988)
diploceline	+				500	Coune (1980)
Iboga-type (Fig. 3)						
catharanthine	+	−		f		Hernandez (1979)
3-OH-coronaridine		−			0.01–100*	Achenbach *et al.* (1980) Achenbach (1986)
3-OH-ibogamine		−				Achenbach *et al.* (1980) Achenbach (1986)
ibogamine	+	−	y		*1000*	Mitscher *et al.* (1972a)
iboxygaine	+				*1000*	Mitscher *et al.* (1972a)
isovoacangine	+				*1000*	Mitscher *et al.* (1972a)
tabernanthine	+				*1000*	Mitscher *et al.* (1972a)
3-OH-conopharyngine	+				60–140	Van Beek *et al.* (1985a)
3-OH-isovoacangine	+	−			50–500	Van Beek *et al.* (1985a)
3-OH-19*R*-heyneanine	+					Perera *et al.* (1985)
ibogaine	+					Van Beek *et al.* (1985b)
conoduramine	+	−			15–400	Van Beek *et al.* (1985a), Munoz *et al.* (1994)
3-OH-conoduramine	+	−			8–170	Van Beek *et al.* (1985a)
conodurine	+	−			4–400	Van Beek *et al.* (1985a), Munoz *et al.* (1994)
3-OH-conodurine	+	−			14–750	Van Beek *et al.* (1985a)
voacamine	+	−			20–400	Van Beek *et al.* (1984b)
3-OH-voacamine	+	−			20–400	Van Beek *et al.* (1985a)
3′-OH-tabernamine	+					Perera *et al.* (1985)
3′-OH-NdeMe-tabernamine	+					Perera *et al.* (1985)
3′-OH-NdeMe-ervahanine B	+					Perera *et al.* (1985)
vobparicine	+	−			50–100	Van Beek *et al.* (1985a)
vobparicine *N*-oxide	+					Van Beek *et al.* (1985a)
Sarpagan-type						
affinisine	+	−	y		*1000*	Mitscher *et al.* (1972a)
perivine	+	−				Hernandez (1979)
tabernaemontanine		−				Hernandez (1979)
Toxiferine type (Fig. 4)						
bisnordihydrotoxiferine	+	−	y		270–3000	Verpoorte *et al.* (1978)
-di-*N*-oxide	+					Verpoorte *et al.* (1978)
caracurine V	+	−	y		210–1400	Verpoorte *et al.* (1978)
caracurine V di-*N*-oxide	+					Verpoorte *et al.* (1978)
Miscellaneous terpenoid indoles						
11-methoxytubotaiwine	+				230	Verpoorte *et al.* (1983b)
9-OH-ellipticine (Fig. 5)	+	−			1–250	Michel *et al.* (1975a,b)

Table I
(*Continued*)

	Activity[a] +	−	y	f	MIC (μg/ml)[b]	References
apparicine	+	−		f		Hernandez (1979), Hernandez *et al.* (1977)
stemmadenine	+	−	y		*1.2–37.5*	Mariee *et al.* (1988)
borreverine	+	−			*1.5–400*	Maynart *et al.* (1980)
strictosidine + glucosidase	+			f	*500*	Luijendijk (1995)
Canthinone derivatives (Fig. 7)						
canthin-6-one	+	−			*12.5–100*	Mitscher *et al.* (1972b) Odebiyi and Sofowora (1979)
5-OH-4-OMe-canthinone	+					Yang *et al.* (1979)
4,5-diOMe-canthinone	+					Yang *et al.* (1979)
Clavine-type alkaloids						
agroclavine	+	−			*200*	Eich *et al.* (1985)
festuclavine	+	−			*200–500*	Eich *et al.* (1985)
Various indoles						
1-carboxymethoxy-β-carboline	+					Yang *et al.* (1979)
harmaline	+	−	y	f		Ahmad *et al.* (1992), Ross *et al.* (1980)
harmalol	+	−		f		Ahmad *et al.* (1992), Ross *et al.* (1980)
harmine	+	−	y	f		Ahmad *et al.* (1992), Al-Shamma *et al.* (1981)
harmol	+	−	y	f		Ahmad *et al.* (1992), Al-Shamma *et al.* (1981)
brevicolline	+	−	y			Towers and Abramowski (1983)
3-dimethylallylindole	+	−	y	f	*0.78–25*	Adeoye *et al.* (1986)
glycozolidol	+	−				Bhattacharyya *et al.* (1985)
yuehchukene	+				20–25*	Waterman (1990)
cryptolepine (Fig. 6)	+	−	y		7.8–500	Cimanga *et al.* (1991)
	+	−			*1.5–100*	Paulo *et al.* (1994)
cryptoheptine	+	−			*6.2–100*	Paulo *et al.* (1994)
hydroxycryptoleptine	+	−			*100*	Paulo *et al.* (1994)
quindoline	+	−			*100*	Paulo *et al.* (1994)
cryptoquindoline	+	−			*100*	Paulo *et al.* (1994)
murrayanine						Das *et al.* (1965)
quadrigemine B	+	−			*125*	Mahmud *et al.* (1993)

[a]+, gram-positive bacteria; −, gram-negative bacteria; y, yeasts; f, fungi.
[b]Values in italic were determined using an agar dilution method; the others using agar diffusion. Asterisk indicates that no details of the method were given.

group at C-16 results in less activity than for corresponding alkaloids with a hydrogen at this position.

Borreverine is another type of dimeric alkaloid having bactericidal activity against both gram-positive and gram-negative bacteria (Maynart *et al.*, 1980).

Stemmadenine was isolated as an antimicrobially active constituent of *Rhazya stricta*

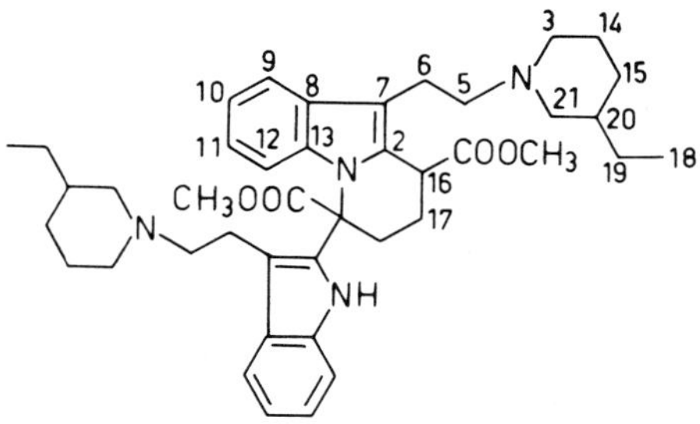

Figure 1. Tetrahydrosecamine.

leaves (Mariee *et al.*, 1988), having MIC values similar to benzalkonium and chlorhexidine for some gram-positive and gram-negative bacteria and a yeast.

Catharanthus roseus has been one of the best studied plants because it contains the antitumor alkaloids vinblastine and vincristine. Among the other alkaloids present in this plant, several have been found to be antimicrobially active (Hernandez, 1979), the most active being vindoline and apparicine. Recently it was found that strictosidine in combination with strictosidine glucosidase has activity against some fungi and gram-positive bacteria (Luijendijk, 1995). As this alkaloid is found to be stored at high levels in the vacuoles of cells in young tissues, which also contain high levels of the specific glucosidase, the role of this alkaloid as a phytoanticipin in plant defense was postulated.

Several antimicrobially active alkaloids have been isolated from plants of the genus *Strychnos* (Loganiaceae). Dimeric tertiary toxiferine-type alkaloids (Fig. 4) were identified as the active compounds in some African chewing sticks (Verpoorte *et al.*, 1978). These alkaloids have activity against both gram-positive and gram-negative bacteria, and in particular against some *Streptococcus* species connected with caries. *N*-oxides of these

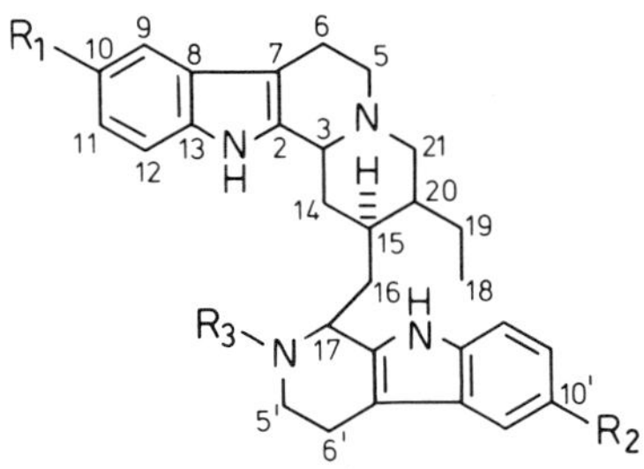

Figure 2. Usambarensine (γ19,20, $R_1 = R_2 = R_3$ = H, 17,4′,5′,6′-tetradehydro) and derivatives: dihydro-(3α, 19 = 20*E*), tetrahydro-(3α,17α, 19 = 20*E*), tetrahydro-(3α,17β, 19 = 20*E*), tetrahydro-(3β,17?, 19 = 20*E*), tetrahydro-(3β,17?, 19 = 20*E*). Tetrahydrousambarensine derivatives: 10′-OH-(3α,17α, 19 = 20*E*), 10,10′-diOMe-(3α,17α, 19 = 20*Z*), 10,10′-diOMe,NMe-(3α,17α, 19 = 20*Z*), 10′-OH,10-OMe,NMe-(3α,17α, 19 = 20*Z*), 10-OH,10′-OMe,NMe-(3α,17α, 19 = 20*Z*). Ochrolifuanines (γ18,19, 20βH, $R_1 = R_2 = R_3$ = H): ochrolifuanine A (3α,17β), ochrolifuanine E (3β,?), ochrolifuanine F (3β,?), 18,19dihydroochrolifuanine F (3β,?). Cinchophyllines (γ18,19, 20βH, $R_1 = R_2$ = OMe, R_3 = H): cinchophylline (3α,17α), cinchophylline (3α,17β), cinchophylline (3β,17α), cinchophylline (3β,17β); and derivatives: 17,4′-didehydro (3α), 18,19-dihydro (3α,17α), 18,19-dihydro (3α,17β), 18,19-dihydro (3β,17α), 18,19-dihydro (3β,17β), 19-OH,18,19-diH (3α,17α), 19-OH,18,19-diH (3α,17β), 19-OH,18,19-diH (3β,17α), 19-OH,18,19-diH (3β,17βA).

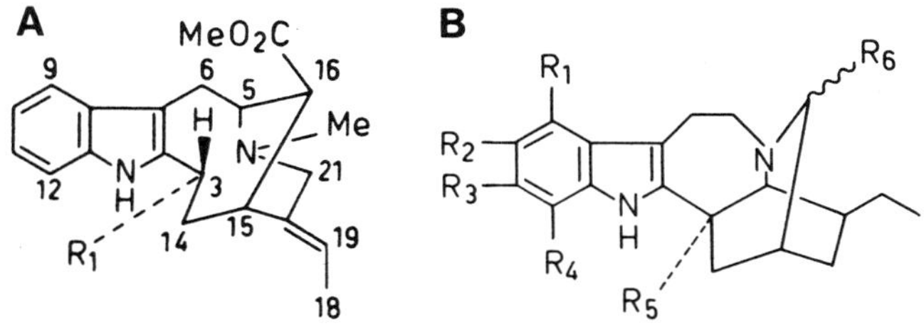

Figure 3. Voacamine-type alkaloids.

	R_1	R_2	R_3	R_4	R_5	R_6
3-Hydroxyconopharyngine	H	OMe	OMe	H	COOMe	OH
3-Hydroxyisovoacangine	H	H	OMe	H	COOMe	OH
Conoduramine	H	Vob	OMe	H	COOMe	H
3-Hydroxyconoduramine	H	Vob	OMe	H	COOMe	OH
Conodurine	H	H	OMe	Vob	COOMe	H
3-Hydroxyconodurine	H	H	OMe	Vob	COOMe	OH
Voacamine	H	OMe	Vob	H	COOMe	H

Vob = vobasinyl (Fig. 3B)

alkaloids, also found in the plant, were less active. Thomas and co-workers reported an antidiarrheal activity for bisnordihydrotoxiferine (Melo *et al.,* 1988; Thomas *et al.,* 1992). The activity was thought to be related to an antagonistic effect on the stimulant activity of endogenous compounds on the gastrointestinal smooth muscle. Coune and co-workers (Coune, 1980; Gasquet *et al.,* 1992) reported a weak activity against gram-positive bacteria for the quaternary alkaloid diploceline.

9-Hydroxyellipticine (Fig. 5), known as an antitumor drug (for a review see Bourgois *et al.,* 1991), has antimicrobial activity against some gram-positive and gram-negative bacteria. The bactericidal concentrations were close to those required for bacteriostatic activity (Alazard *et al.,* 1976; Michel *et al.,* 1975a,b). The cytotoxic activity is probably related to the DNA-intercalating properties of this type of alkaloid.

In a broad screening of natural products, Mitscher *et al.* (1972a) found antimicrobial activity for several indole alkaloids (Table I).

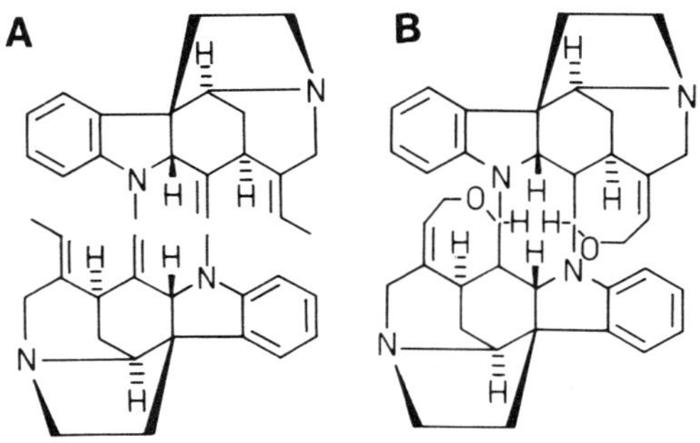

Figure 4. (A) Bisnordihydrotoxiferine; (B) caracurine V.

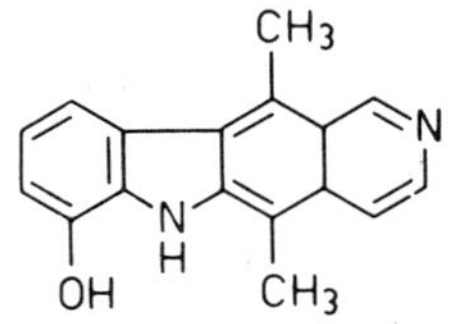

Figure 5. 9-Hydroxyellipticine.

2.2. Other Indole Alkaloids

Harmane-type alkaloids have been reported to exhibit antimicrobial activity, although the results are contradictory. Ross *et al.* (1980) found harmaline and harmalol to be active and harmine to be inactive, whereas Al-Shamma *et al.* (1981) reported the opposite. Ahmad *et al.* (1992) tested these alkaloids against a series of microorganisms, including gram-positive and gram-negative bacteria, fungi (e.g., dermatophytes) and yeasts. Harmine was the most active (MIC value: 75–500 $\mu g\ ml^{-1}$), harmaline and harmol also showed activity, whereas harmalol and the tetrahydro derivatives of harmine and harmol only showed very weak activities.

The harmane-type alkaloids are known to be phototoxic to bacteria and insects (Larson *et al.,* 1988; McKenna and Towers, 1981; Towers and Abramowski, 1983). Similarly, brevicolline, an *N*-methylpyrrolidine substituted harmane derivative, is phototoxic for microorganisms (Towers and Abramowski, 1983). Larson *et al.* (1988) studied the mechanism of the phototoxicity on *E. coli.* Harmalol was found not to be active in this testsystem, whereas harmaline, harmine, harmane, and norharmane were active. Although oxygen was required for the phototoxicity, no clear correlation was found between photoproduction of singlet oxygen or hydrogen peroxide by the alkaloids in the cells and phototoxicity. Lipophilicity, as measured by the chromatographic behavior of the alkaloids, gave a better correlation with the phototoxicity. Transport through cell membranes to the target site in the cells, where the formation of reactive oxygen species or oxygenated derivatives occurs, thus seems to play an important role. Harmaline and harmalol are the most susceptible to photodegradation, which may explain their lower toxicity; they might be decomposed before they reach the target site in the cells.

3-Dimethylallylindole was shown to exhibit strong antifungal activity against both plant and human pathogenic fungi, and was also active against gram-positive and gram-negative bacteria (Adeoye *et al.,* 1986).

Cryptolepine (Fig. 6) isolated from a *Cryptolepis* species (Periplocaceae) was report-

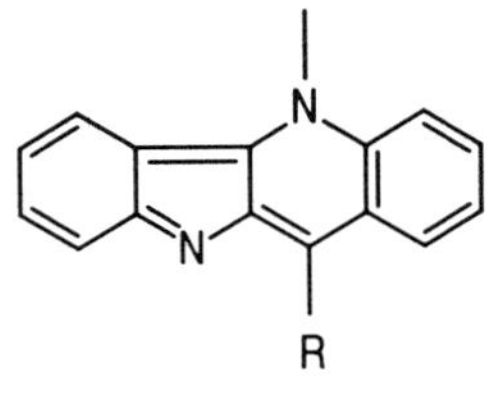

Figure 6. Cryptolepine (R = H; hydroxycryptolepine, R = OH).

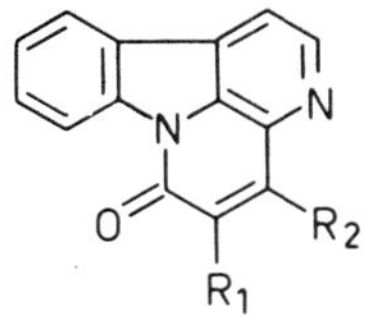

Figure 7. Canthinones.

	R_1	R_2
Canthin-6-one	H	H
5-Hydroxy-4-methoxy-canthin-6-one	OH	OMe
4,5-Dimethoxycanthin-6-one	OMe	OMe

ed to be active against a series of bacteria and a yeast, and the MIC value was close to the concentration at which a bactericidal effect was observed (Cimanga *et al.,* 1991). The related compound cryptoheptine had similar activities, whereas several others isolated from the same plant only had activity at much higher levels (Paulo *et al.,* 1994).

Strong antimicrobial activity was found for canthinone and derivatives (Fig. 7) (Mitscher *et al.,* 1972b; Yang *et al.,* 1979), and the activity of these alkaloids is enhanced by light (Towers and Abramowski, 1983). Waterman (1990) isolated an antimicrobially active dimeric prenylindole alkaloid, yuehchukene, from a *Murraya* species (Rutaceae). Murrayanine, a carbazole alkaloid having an indole nucleus, was found to exhibit antimicrobial activity (Das *et al.,* 1965). Bhattacharyya *et al.* (1985) reported on a new carbazol alkaloid having weak antibacterial activity against gram-positive bacteria.

Clavine alkaloids, which are produced by some fungi as well as some plants of the Convolvulaceae, and a series of semisynthetic derivatives were shown to have activity against various human pathogenic bacteria and *Candida albicans* (Eich *et al.,* 1985).

The polypyrroloindolinic alkaloid quadrigemine B isolated from *Psychotria* species has both cytotoxic and bactericidal activities (Mahmud *et al.,* 1993).

Although quite a few indole alkaloids have been reported to have antimicrobial activity, very little is known about the mechanisms behind the reported effects. In some cases the compounds have also been shown to have cytotoxic activity and some have been studied extensively as potential antitumor drugs. 9-Hydroxyellipticine, in fact, is used as such. Given these results, further studies on the active alkaloids seem of great interest. Considering the role of the indole alkaloids in the plants, little can be said. The fact that most of the terpenoid indole alkaloids are always present in the plant means that they are probably not acting as phytoalexins. Studies on their production in cell cultures also support this conclusion (Meijer *et al.,* 1993).

3. ISOQUINOLINE ALKALOIDS

The isoquinoline alkaloid group is second in size to the indole alkaloids. Of the various classes of isoquinoline alkaloids, several include quite a few compounds with strong antimicrobial activity (Table II). Several authors have studied a broad range of isoquinoline alkaloids (Abbasoglu *et al.,* 1991; Simeon *et al.,* 1990; Tsai *et al.,* 1989; Villar *et al.,* 1986, 1987). Abbasoglu *et al.* (1991) tested a series of (tetrahydro)protober-

Table II
Isoquinoline Alkaloids Reported to Have Antimicrobial Activity

	Activity[a]					
	+	−	y	f	MIC (μg/ml)[b]	References
Bisbenzylisoquinoline type (Fig. 8)						
N-desmethylthalistyline	+				*100–1000*	Wu *et al.* (1977a)
thalistyline	+				*50*	Wu *et al.* (1977a)
thalistylinemethodiiodide	+				*100–1000*	Wu *et al.* (1977a)
O-methyldauricine	+				250–1000†	Kuroda *et al.* (1976)
pennsylvanine	+				*1000*	Wu *et al.* (1977b)
thalmelatine	+				*100*	Wu *et al.* (1977b)
thalicarpine	+		y		*100–1000*	Wu *et al.* (1977b)
thalrugosidine	+	−			*100–200*	Mitscher *et al.* (1971a,b, 1972a)
thalidasine	+	−			*100–200*	Mitscher *et al.* (1971a,b, 1972a)
O-methylthalmetine	+				*100*	Wu *et al.* (1977b)
thalicberine	+	−			250–1000†	Kuroda *et al.* (1976)
thalrugosaminine	+				*50–100*	Wu *et al.* (1976, 1977b)
oxyacanthine	+	−			62–100†	Kuroda *et al.* (1976)
			y		*1000*	Mitscher *et al.* (1972a)
	+				*20*	Bersch and Dopp (1955)
cepharanthine	+	−			8–1000†	Kuroda *et al.* (1976)
	+				*5–10*	Bersch and Dopp (1955)
berbamine	+	−			125–1000†	Kuroda *et al.* (1976)
	+				*50–100*	Bersch and Dopp (1955)
	+				*250–1500*	Lahiri and Sen (1958)
obamegine	+	−			*50–200*	Mitscher *et al.* (1971a,b, 1972a)
N-desmethylthalidezine	+		y		*50–1000*	Wu *et al.* (1977a)
thalidezine	+	−	y		*100*	Wu *et al.* (1977a)
isotetrandrine	+	−			62–1000*	Kuroda *et al.* (1976)
	+	−			*100–200*	Mitscher *et al.* (1972a)
hernandizine (thalicsimine)	+		y		*25–100*	Wu *et al.* (1977a)
tetrandrine	+	−			15–1000†	Kuroda *et al.* (1976)
thalrugosine	+	−			*100–2000*	Mitscher *et al.* (1971a,b, 1972a)
isotrilobine	+	−			8–500†	Kuroda *et al.* (1976)
l-curine	+				*1000*	Mitscher *et al.* (1972a)
funiferine	+				*100*	Mitscher *et al.* (1972a)
dimethylgrisabine	+				62.5*	Sedmera *et al.* (1990)
thalmirabine	+				*100*	Wu *et al.* (1980)
thalostine	+				*100*	Wu *et al.* (1980)
O-methylthalibrine	+		y		*100–500*	Wu *et al.* (1980)
dehatrine	+			f	*300–1000*	Tsai *et al.* (1989)
Aporphine alkaloids (Figs. 9 and 10)						
anolobine	+	−	y		*3–200*	Villar *et al.* (1984b, 1987), Simeon *et al.* (1990)

Table II
(Continued)

	Activity[a]				MIC (μg/ml)[b]	References
	+	−	y	f		
anonaine	+	−	y	f	*3–125*	Villar *et al.* (1984a,b, 1987), Simeon *et al.* (1990), Paulo *et al.* (1992), Tsai *et al.* (1989)
nornantenine	+		y		*3–100*	Villar *et al.* (1984a, 1987)
xylopine	+		y		*25–100*	Tsai *et al.* (1989), Villar *et al.* (1986, 1987), Simeon *et al.* (1990)
N-methylxylopine methoiodide	+	−			*50–300*	Tsai *et al.* (1989)
isopiline	+		y		*25–100*	Villar *et al.* (1987)
O-methylisopiline	+		y		*25–100*	Villar *et al.* (1986, 1987)
puterine	+		y		*25–50*	Villar *et al.* (1986)
norushinsunine	+		y		*6–100*	Simeon *et al.* (1990)
anhydroushinsunine			y	f	*125–1000*	Tsai *et al.* (1989)
anhydroushinsunine methoiodide			y	f	*62.5–1000*	Tsai *et al.* (1989)
asimilobine	+		y		*12–100*	Simeon *et al.* (1990)
N-methylasimilobine	+				*25*	Simeon *et al.* (1990)
dehydroglaucine	+		y		*25–50*	Hufford *et al.* (1975)
thaliglucinone	+	−	y		*25–200*	Wu *et al.* (1976), Gharbo *et al.* (1973)
bulbocapnine	+	−	y		*1000*	Mitscher *et al.* (1972a), Abbasoglu *et al.* (1991)
O-methylbulbocapnine			y		*500–1000*	Tsai *et al.* (1989)
O-methylbulbocapnine methoiodide	+				*300*	Tsai *et al.* (1989)
nuciferine	+				*50–1000*	Mitscher *et al.* (1972a), Villar *et al.* (1986), Simeon *et al.* (1990)
hydroxynornuciferine	+				*25–50*	Villar *et al.* (1986)
belemine	+				*50*	Villar *et al.* (1986)
corydine	+				*25–50*	Villar *et al.* (1987)
glaucine methoiodide	+				*4300*	Hufford *et al.* (1975), Tsai *et al.* (1989)
thalphenine	+		y		*1000*	Wu *et al.* (1976, 1977b)
isoboldine	+	−		f	500	Abbasoglu *et al.* (1991), Paulo *et al.* (1992)
laurelliptine				f	500	Paulo *et al.* (1992)
actinodaphnine	+	−	y	f	*50–1000*	Tsai *et al.* (1989)
N-methylactinodaphnine	+	−	y	f	*50–1000*	Tsai *et al.* (1989)
laurotetanine	+	−	y		*100–1000*	Tsai *et al.* (1989)
N-methyllaurotetanine	+	−	y		*100–1000*	Tsai *et al.* (1989)
dicentrine methoiodide	+				*50*	Tsai *et al.* (1989)
roemerine methoiodide	+				*50–100*	Tsai *et al.* (1989)
magnoflorine			y		*250*	Tsai *et al.* (1989)

(continued)

Table II
(Continued)

	Activity[a]				MIC (μg/ml)[b]	References
	+	−	y	f		
liriodenine	+	−	y	f	*0.4–100*	Hufford *et al.* (1975, 1980), Villar *et al.* (1984a, 1987), Simeon *et al.* (1990), Pabuccuoglu *et al.* (1991), Clark *et al.* (1987)
liriodendronine			y		*25*	Pabuccuoglu *et al.* (1991)
2-*O*-methylliriodendronine			y		*50*	Pabuccuoglu *et al.* (1991)
liriodenine methoiodide	+		y		*0.4–6*	Hufford *et al.* (1975), Pabuccuoglu *et al.* (1991), Clark *et al.* (1987)
lysicamine	+		y	f	*12–26*	Hufford *et al.* (1980), Pabuccuoglu *et al.* (1991), Villar *et al.* (1987), Simeon *et al.* (1990)
lysicamine methoiodide			y		*0.8–6.2*	Pabuccuoglu *et al.* (1991)
cassameridine	+		y	f	*25–50*	Hufford *et al.* (1980)
oxonantenine	+		y	f	*6–25*	Hufford *et al.* (1980)
oxoglaucinemethoiodide	+	−	y		*1.3–25*	Hufford *et al.* (1975, 1980), Clark *et al.* (1987)
O-methylmoschatoline	+					Villar *et al.* (1987)
lanuginosine	+			f	*25–100*	Villar *et al.* (1986), Simeon *et al.* (1990), Ferdous *et al.* (1992)
oxoputerine	+		y		*3–25*	Villar *et al.* (1986)
subsessiline	+		y		*3–50*	Villar *et al.* (1986)
glaziovine	+				*12–50*	Simeon *et al.* (1990)
oxostephanine	+	−			50–150	Ferdous *et al.* (1992)
Benzophenanthridine alkaloids (Fig. 11)						
chelidonine	+	−	y	f	*1000*	Mitscher *et al.* (1972a), Zbierska and Kowalewski (1979a), Wolters (1969)
chelidonine *N*-oxide	+	−	y	f	*1000*	Zbierska and Kowalewski (1980)
chelidoninemethoiodide	+	−	y	f	*1000*	Zbierska and Kowalewski (1979a,b)

Table II
(Continued)

	Activity[a]					
	+	−	y	f	MIC (μg/ml)[b]	References
chelerythrine	+	−	y	f	*6.3–100*	Mitscher *et al.* (1972a, 1978), Lenfeld *et al.* (1981), Hejtmankova *et al.* (1984), Abbasoglu *et al.* (1991), Stermitz *et al.* (1975)
sanguinarine	+	−	y	f	*0.5–100*	Mitscher *et al.* (1978), Vallejos and Roveri (1972), Bersch and Dopp (1955), Wolters (1969), Hejtmankova *et al.* (1984), Abbasoglu *et al.* (1991), Stermitz *et al.* (1975)
sanguinarine pseudomethanolate	+	−	y		*6.3–100*	Mitscher *et al.* (1978)
sanguinarine pseudoethanolate	+	−	y		*6.3–100*	Mitscher *et al.* (1978)
sanguinarine pseudocyanide	+	−	y		*6.3–100*	Mitscher *et al.* (1978)
nitidine				f	15	Vallejos and Roveri (1972)
fagaronine	+	−			*112–225*	Stermitz *et al.* (1975)
Protoberberine alkaloids (Fig. 12)						
berberine	+	−	y		*1000*	Mitscher *et al.* (1972a)
	+	−	y	f	*3.1–100*	Amin *et al.* (1969)
	+	−	y	f	0.31–100	Okunade *et al.* (1994)
				f	25‡	Wolters (1969), Mahajan *et al.* (1982)
palmatine	+	−	y		*1000*	Mitscher *et al.* (1972a), Villar *et al.* (1986), Abbasoglu *et al.* (1991)
				f	250‡	Horn and Steffen (1968a,b)
tetrahydropalmatine	+				*50*	Simeon *et al.* (1990)
jatrorrhizine				f	250‡	Horn and Steffen (1968a,b)
columbamine	+		y		*100*	Wu *et al.* (1976), Andronescu *et al.* (1973)
berberubine	+				*100*	Gharbo *et al.* (1973)
dihydroberberine	+	−	y		*1000*	Mitscher *et al.* (1972a)
scoulerine	+	−				Abbasoglu *et al.* (1991)
corydaline	+	−				Abbasoglu *et al.* (1991)
stylopine	+	−				Abbasoglu *et al.* (1991)
ophiocarpine	+	−				Abbasoglu *et al.* (1991)
canadine	+	−				Abbasoglu *et al.* (1991)

(continued)

Table II
(Continued)

	Activity[a]					
	+	−	y	f	MIC (μg/ml)[b]	References
prothalipine	+				*1000*	Wu *et al.* (1976)
protopine	+	−			*100*	Bersch and Dopp (1955), Abbasoglu *et al.* (1991)
cryptopine	+	−				Abbasoglu *et al.* (1991)
β-allocryptopine	+	−				Abbasoglu *et al.* (1991)
Miscellaneous isoquinoline alkaloids						
hydrastine	+	−	y		*1000*	Mitscher *et al.* (1972a), Abbasoglu *et al.* (1991)
hydrastinine	+	−				Abbasoglu *et al.* (1991)
corydaldine	+	−				Abbasoglu *et al.* (1991)
bicuculline	+	−				Abbasoglu *et al.* (1991)
adlumidine	+	−				Abbasoglu *et al.* (1991)
lycorine				f		Miyakado *et al.* (1975)
diacetyllycorine				f		Miyakado *et al.* (1975)
dihydrolycorine				f		Miyakado *et al.* (1975)
2-acetyllycorine				f		Miyakado *et al.* (1975)
crinamine	+				10	Adesanya *et al.* (1992)
diacetylcrinamine	+				10	Adesanya *et al.* (1992)
diacetylhamayne	+				10	Adesanya *et al.* (1992)
3-methoxysampangine (Fig. 13)			y	f	*0.2–3.1*	Liu *et al.* (1990), Peterson *et al.* (1992)
eupolauridine			y		*1.56*	Hufford *et al.* (1987), Liu *et al.* (1990)

[a]+, gram-positive bacteria; −, gram-negative bacteria; y, yeasts; f, fungi.
[b]Values in italic were determined using an agar dilution method, the others using agar diffusion. *No details of the method were given. †Method according to the Japanese Society of Chemotherapy. ‡Channel test (Wolters, 1963).

berines, protopines, aporphines, benzophenanthridines, and phthalideisoquinolines and concluded that alkaloids containing methoxy and methylenedioxy groups were more active than those that did not.

3.1. Bisbenzylisoquinoline Alkaloids

One of the few alkaloids that has been used as an antibiotic is cepharanthine (Fig. 8), which was employed in Japan as a prophylactic drug against tuberculosis during World War II. It was also used against leprosy (Bersch and Dopp, 1955; Hiller, 1964, and references cited therein). Although a tuberculostatic activity was confirmed by Bersch and Dopp (1955), the *in vitro* active concentration was much higher than the prophylactic levels used. Several *Thalictrum* alkaloids were reported to have activity against gram-positive bacteria, in particular *Mycobacterium smegmatis*. Some were also active against *Candida* sp. (Mitscher *et al.*, 1971a,b; Wu *et al.*, 1976, 1977a,b). A structure–activity

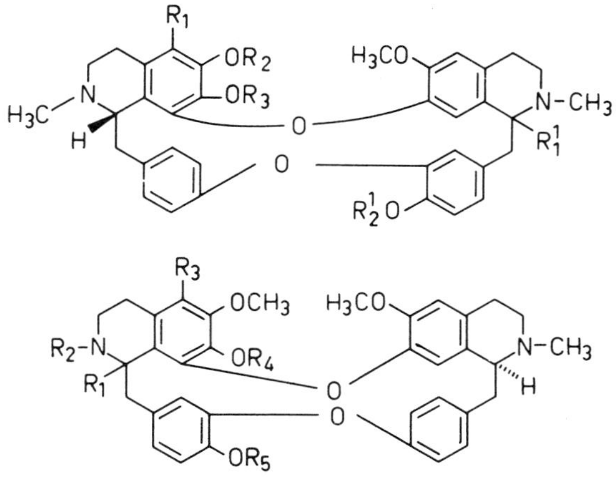

Figure 8. Bisbenzylisoquinoline alkaloids.

A

	R_1	R_2	R_3	R_1^1	R_2^1
Thalrugosamine	OMe	Me	Me	αH	Me
Oxyacanthine	H	Me	Me	βH	H
Cepharanthine	H	$-CH_2-$		βH	Me

B

	R_1	R_2	R_3	R_4	R_5
Berbamine	αH	Me	H	Me	H
Isotetrandrine	αH	Me	H	Me	Me
Tetrandrine	βH	Me	H	Me	Me

relationship study for antitumor and antimicrobial activity was made with 23 bisbenzylisoquinoline alkaloids. While several alkaloids showed strong activity, particularly against gram-positive bacteria (Kuroda *et al.*, 1976), the antitumor and antimicrobial activity was not always coupled. The results differed to some extent from the conclusions of Kupchan and Altland (1973) considering antitumor effects.

Several other studies reported antibacterial and antifungal activity for various bisbenzylisoquinoline alkaloids (Sedmera *et al.*, 1990; Tsai *et al.*, 1989; Wu *et al.*, 1980).

3.2. Aporphine Alkaloids

Several alkaloids having an oxoaporphine skeleton are strongly antimicrobially active (Table II). Liriodenine (Fig. 9) and dehydroglaucine are active against a broad spectrum of organisms, with MIC values similar to those of known antibiotics such as streptomycin. Their methiodides have similar but weaker activities, except against yeasts, where they are more active (Hufford *et al.*, 1975). Lysicamine, which has two *O*-methyl groups instead of the methylenedioxy group in liriodenine, was inactive against *Candida albicans*. Conversion into the quaternary nitrogen derivative resulted in an active compound, though not as active as the liriodenine metho compound. The corresponding

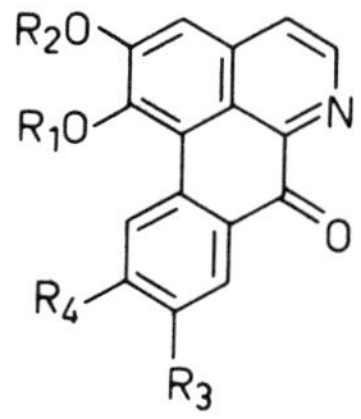

Figure 9. Aporphine alkaloids.

	R_1	R_2	R_3	R_4
Liriodenine	$-CH_2-$		H	H
Lysicamine	OMe	OMe	H	H

compounds having one or two hydroxy substituents instead of the *O*-methyl group(s) did not show any activity, even after quaternization (Pabuccuoglu *et al.,* 1991).

Another study showed that the MIC value of liriodenine against *Trichophyton* species is similar to that of griseofulvin (2 μg ml^{-1}). Of a series of oxoaporphines, liriodenine was the most active. Activity was also found against several plant pathogenic microorganisms (Hufford *et al.,* 1980). The activity against *C. albicans* was further explored in *in vivo* tests using a mouse model of disseminated candidiasis (Clark *et al.,*1987). Liriodenine methiodide, with an *in vitro* MIC value of 0.78 μg ml^{-1}, was shown to be more active than liriodenine and oxoglaucine methiodide (MIC values about 3.12 μg ml^{-1}). In a narrow dose range (ca. 0.1–1 mg kg^{-1}), liriodenine methiodide was found to be effective after parenteral administration in the *in vivo* model. The less toxic liriodenine had activity in the range of 0.5–1 mg kg^{-1}; this compound can also be administered orally and a multidose regimen is most effective. Interestingly, doses above the mentioned ranges had lower efficacy.

In a study of a series of aporphines, Villar and co-workers (Villar *et al.,* 1984a,b, 1986, 1987; Simeon *et al.,* 1990) found that noraporphines also have strong antimicrobial activity. Anonaine and nornantenine (Fig. 10) were even more active than liriodenine against *C. albicans* and some gram-negative bacteria. Introduction of a methyl group, yielding a tertiary alkaloid, resulted in loss of activity. The alkaloids containing a phenolic

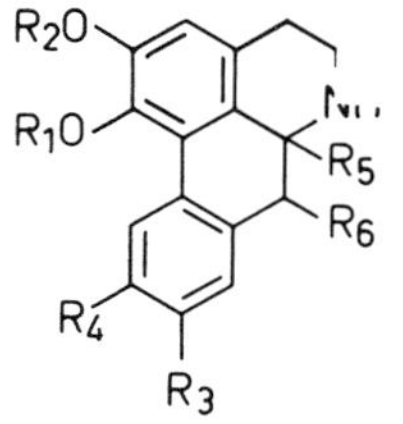

Figure 10. Aporphine alkaloids.

	R_1	R_2	R_3	R_4	R_5	R_6
Anonaine	$-CH_2-$		H	H	βH	H
Nornantenine	Me	Me	$-OCH_2O-$		αH	H

hydroxy substituent were less active (Simeon *et al.*, 1990; Villar et al., 1986). The basic structure, a noraporhine skeleton with a 1,2-methylenedioxy substitution, as in the oxoaporphine series, seemed to be the most active.

Tsai *et al.* (1989) found anonaine to be one of the most active compounds among a series of isoquinoline alkaloids, including 26 aporphines. Its quaternary dimethyl derivative had similar activities.

It has been shown that some aporphine alkaloids inhibit respiration in yeasts, and this effect is counteracted by Ca^{2+} (Vallejos and Roveri, 1972).

In considering the reported activities for the various aporphines, it is obvious that the activity increases going from quaternary to tertiary to secondary amine function. In the highly conjugated oxoaporphines, the introduction of a quaternary function results in a decrease of activity.

3.3. Benzophenanthridine Alkaloids

The biological activities of benzophenanthridine (Fig. 11) and phenanthrene alkaloids were reviewed by Simeon *et al.* (1989). The antimicrobial activity of chelidonine and sanguinarine (Fig. 11) has long been known, particularly the tuberculostatic properties of these alkaloids (Bersch and Dopp, 1955). Sanguinarine was more active, although the tuberculostatic concentration was difficult to determine because of the instability of this alkaloid. These two alkaloids (Wolters, 1969) and chelerythrine (Hejtmankova *et al.*, 1984) have antifungal activity, and the inhibitory effect is antagonized by ergosterol. The antimicrobial activity of sanguinarine, chelerythrine, and their tertiary pseudoalcoholate derivatives against some gram-positive, and gram-negative bacteria and *Candida* species was reported by Mitscher *et al.* (1978). The pseudoalcoholates were more active than the corresponding quaternary alkaloids, probably because they are able to pass through cell membranes and act as prodrugs, which in the cells are converted into the active quaternary alkaloids. This hypothesis is supported by the fact that real tertiary analogues, like dihydro and oxo analogues, which cannot be converted easily into the quaternary compounds, have much lower activity. The quaternary alkaloids nitidine, chelerythrine, and sanguinarine are strong inhibitors of respiration in yeasts; mitochondria are thought to be the site of action and this effect is counteracted by Ca^{2+} (Vallejos and Roveri, 1972). The effects of

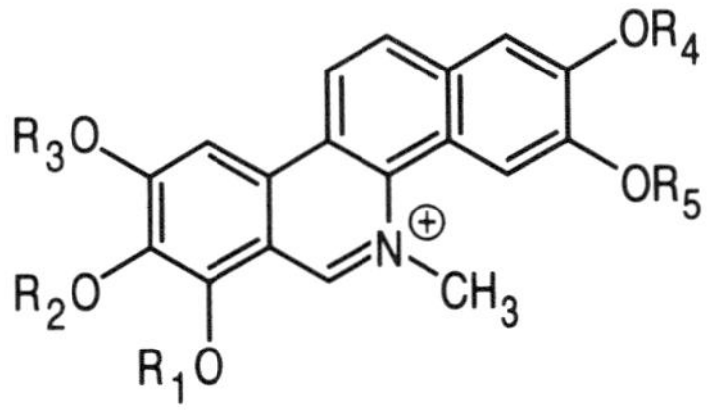

Figure 11. Benzophenanthridine alkaloids.

	R_1	R_2	R_3	R_4	R_5
Chelerythrine	OMe	OMe	H	$-OCH_2O-$	
Sanguinarine	$-OCH_2O-$		H	$-OCH_2O-$	
Nitidine	H	OMe	OMe	$-OCH_2O-$	
Fagaronine	H	OMe	OMe	OH	OMe

sanguinarine were reported to be caused by intercalation of the alkaloid with DNA (Maiti *et al.*, 1982).

Chelidonine and its *N*-methyl and *N*-oxide derivatives were found to have a broad antimicrobial activity, although at rather high MIC values (Zbierska and Kowalewski, 1979a,b, 1980). Neal (1989) reported that chelidonine is an inhibitor of microsomal monooxygenases.

Fagaronine, a potential new antitumor drug (for a review see Barret and Sauvaire, 1992; Phillips and Castle, 1981) has bactericidal activity as well (Pezzuto *et al.*, 1983). Stermitz *et al.* (1975) reported on the synthesis of a series of antitumor benzophenanthridine alkaloids, including fagaronine, and found these to exhibit antimicrobial activity as well. The most active were sanguinarine and chelerythrine; however, the antitumor activity of these alkaloids was much less than other compounds having a different substitution pattern in the A-ring (e.g., fagaronine and nitidine).

Smekal *et al.* (1986) studied the binding of some benzophenanthridine alkaloids to DNA. They proposed that sanguinarine, chelerythrine, fagaronine, and nitidine may intercalate with DNA. According to these authors, the antibacterial activity of these alkaloids decreases in the order mentioned by the ratio 100:10:5:1.

Sanguinarine is applied as an antimicrobial agent in dentifrice and oral rinses because of its antiplaque activity (Southard *et al.*, 1984); however, Harper *et al.* (1990) could not find significant differences in dental yeast flora after 6 months' use of sanguinarine containing dental products.

3.4. Protoberberine Alkaloids

Berberine (Fig. 12) is the major compound in a number of medicinal plants employed since ancient times against diseases such as diarrhea, dysentery, cholera, and eye infections. In a number of studies the activity of this alkaloid has been confirmed. An antitrachoma action for berberine was proved by Sabir *et al.* (1976). Das Gupta and Dikshit (1929) and Gupta and Kahali (1944) showed an antileishmania activity. Amoebicidal activity was also confirmed (Dutta *et al.*, 1968; Subbaiah and Amin, 1967). Berberine has several interesting pharmacological activities, and is not very toxic (Chopra *et al.*, 1932). The efficacy of berberine for the treatment of cholera was shown in several studies (Ahkter *et al.*, 1977; Dutta and Panse, 1962; Dutta *et al.*, 1972; Lahiri and Dutta, 1967; Mekawi, 1968; Modak *et al.*, 1970a,b; Nair *et al.*, 1969). For maximum effect the alkaloid should be administered in an early stage of infection. Its mode of action was obviously different from that of chloramphenicol (Lahiri and Dutta, 1967). The effect of the alkaloid

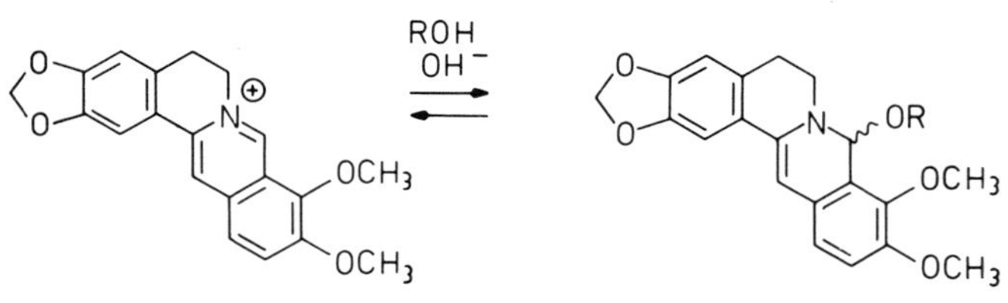

Figure 12. Berberine.

on *Vibrio cholerae* is bacteriostatic according to Nair *et al.* (1969), but Amin *et al.* (1969) reported that it is more bactericidal than bacteriostatic in nature. Moreover, berberine inhibits the reaction of the cholera toxin on the cells and inactivates the toxin (Akhter *et al.*, 1977; Dutta *et al.*, 1972; Modak *et al.*, 1970a,b; Nair *et al.*, 1969). Several studies concerned the activity of berberine on diarrhea (Homma *et al.*, 1961; Purohit and Rao, 1969; Sharda, 1970). It was successfully applied in the treatment of gastroenteritis in children (Sharda, 1970). Intestinal staphylococci, some antibiotic-resistant *Shigella* strains, and pathogenic *E. coli* were found to be sensitive to berberine (Homma *et al.*, 1961).

Other related protoberberine alkaloids have also shown antimicrobial activity (Abbasoglu *et al.*, 1991).

The antimicrobial activity of berberine against a broad range of organisms is extensively documented (see Table II). With MIC values of 3.1–100 $\mu g\ ml^{-1}$ it comes close to broad-spectrum antibiotics like tetracycline. The activity increases with the pH, a pH increase from 7 to 8 resulting in a two- to fivefold decrease of the MIC. This probably occurs via a mechanism similar to that reported for the benzophenanthridine alkaloids (see above): the formation of the tertiary 8-hydroxy compound at elevated pH, which is able to pass membranes and thus to act as a prodrug.

In an extensive study of the effect of berberine on a large number of *Staphylococcus aureus* strains, a bactericidal activity was found, though at higher MIC values than known antibiotics. There was no cross-resistance other than with streptomycin (Kowalewski *et al.*, 1972).

Berberine has tuberculostatic properties (Amin *et al.*, 1969; Bersch and Dopp, 1955; Jezowa *et al.*, 1966). The related alkaloid protopine has a similar activity (Bersch and Dopp, 1955). Berberine is also active against plant pathogens (Amin *et al.*, 1969; Gilliver, 1946).

Horn and Steffen (1968a,b) and Wolters (1969) reported antifungal activity for berberine; the related alkaloids jatrorrhizine and palmatine had similar activities, but at higher concentrations. Berberine is also active against some human pathogenic fungi (Mahajan *et al.*, 1982). Okunade *et al.* (1994) found *in vitro* activity against some pathogenic yeasts such as *Candida albicans* and *Cryptococcus neoformans* and the bacterium *Mycobacterium intracellulare.* However, in an *in vivo* mouse model of disseminated cryptococcosis, no activity could be detected probably because the quaternary alkaloid is unable to pass the blood–brain barrier. The isohomoprotoberberine alkaloid puntarenine had similar activities, and further some activity against dermatophytic fungi.

Several other related protoberberines have been reported to exhibit antimicrobial activity (see Table II).

According to Kuwano and Yamauchi (1960a,b), the bacteriostatic activity of berberine could result from a competition for pyridoxal phosphate with the apoenzyme(s), e.g., tyrosine decarboxylase was inhibited. It is likely that berberine intercalates with DNA (Krey and Hahn, 1969). Smekal and Kubova (1982) measured and characterized the binding of several protoberberines to DNA. Although, in an extensive review, Hahn and Ciak (1975) concluded that the antibiotic activity of berberine is not related to the effects of berberine on DNA, Maiti and Chaudhuri (1981) concluded that the exact mode and mechanism of antimicrobial activity of berberine is not clear, but they found a highly specific interaction of berberine with DNA.

The mechanism of the antibacterial effect of berberine remains unknown and it remains an interesting topic for further study. With its broad, though modest, antimicrobial activity, but relatively low toxicity, it indeed deserves further exploration.

3.5. Protopines

Abbasoglu *et al.* (1991) tested some protopine alkaloids and found them to have activities similar to the benzophenanthridine alkaloids.

3.6. Miscellaneous Isoquinoline Alkaloids

In Table II several antimicrobially active isoquinoline alkaloids of different structural types are summarized. Phthalideisoquinolines were reported to have antimicrobial activity (Abbasoglu *et al.,* 1991).

Several Amaryllidaceae alkaloids of the lycorine type and some derivatives have been reported to possess antimicrobial activity against a plant pathogenic *Phytophthora* species (Miyakado *et al.,* 1975). Adesanya *et al.* (1992) reported antibacterial activity for some lycorine- and crinamine-type alkaloids, crinamine itself being most active. No antifungal activity was noted against the organisms tested.

The copyrines are an interesting group of alkaloids (Fig. 13). They were isolated, together with liriodenine, from an annonaceous plant (Liu *et al.,* 1990). 3-Methoxysampangine showed strong effects against some yeasts and fungi at MIC values similar to that of amphotericin B. Eupolauridine, another related alkaloid isolated from the same plant, showed strong inhibition of *Candida albicans* (Hufford *et al.,* 1987; Liu *et al.,* 1990). In a subsequent study a series of sampangine derivatives was synthesized and tested for antifungal and antimycobacterial activity (Peterson *et al.,* 1992). The introduction of a substituent at C-3 increases activity, whereas substitution at C-4 leads to reduced activity against various AIDS-related fungal and mycobacterial opportunistic infection pathogens.

4. STEROIDAL ALKALOIDS

Antibiotics having a steroidal skeleton have been reviewed by Hüttenrauch (1962). In several Solanaceae plants, such as tomato and potato, glycoalkaloids are found that exhibit antifungal activities (Table III). Tomatine (Fig. 14) has even been used as a topical

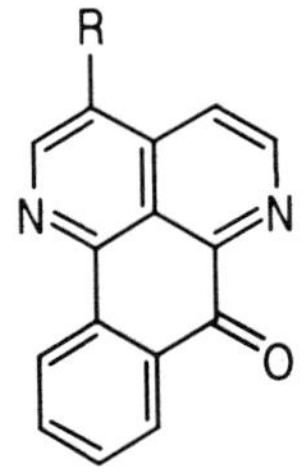

Figure 13. Sampangine (R = H; 3-methoxysampangine, R = OCH_3).

Table III
Steroidal Alkaloids Reported to Have Antimicrobial Activity (Fig. 14)

	Activity[a]					
	+	−	y	f	MIC (μg/ml)[b]	References
soladulcidine tetraosid				f	10–50†	Wolters (1964, 1965a,b, 1966)
soladulcidine				f	15†	Wolters (1964, 1966)
solasodine				f	15†	Wolters (1964, 1966)
tomatine				f	2–40†	Wolters (1964, 1966)
			y	f	*100–250*	Fontaine *et al.* (1948), Ma and Fontaine (1948)
tomatidine				f	2–22†	Wolters (1964, 1966)
δ^5-tomatidenol				f	1–40†	Wolters (1964, 1966)
solasonine				f	40†	Wolters (1964, 1966)
solamargine				f	40†	Wolters (1964, 1966)
			y		*1000*	Mitscher *et al.* (1972a)
solanidine				f	5†	Wolters (1964, 1966, 1970)
	+	−	y		*1000*	Mitscher *et al.* (1972a)
solanine				f	40†	Wolters (1966)
			y	f	*250–1000*	Fontaine *et al.* (1948)
				f	*80–100 μM*	Fewell and Roddick (1993)
α-chaconine				f	*60–100 μM*	Fewell and Roddick (1993)
solanocapsine				f		Wolters (1969)
	+				*3*	Boll *et al.* (1955)
	+		y		*100*	Mitscher *et al.* (1976)
solacasine	+		y		*2.5–12.5*	Mitscher *et al.* (1976)
solacongestidine			y	f	*0.4–1.5*	Kusano *et al.* (1987b)
solafloridine			y	f	*6.2–25*	Kusano *et al.* (1987b)
verazine			y	f	*3.1–12.5*	Kusano *et al.* (1987b)
demissidine				f	5†	Wolters (1966, 1969)
conessine				f	30–250†	Wolters (1966, 1969)
	+	−	y		*100–1000*	Mitscher *et al.* (1972a)
protoveratrine A				f	1000†	Wolters (1970)
protoveratrine B				f	1000†	Wolters (1970)
cevadine				f	1000†	Wolters (1970)
veratridine				f	1000†	Wolters (1970)
jervine				f	0.8–100†	Wolters (1970)
			y	f	9–120*	Han and Woo (1973a,b)
pseudojervine				f		Wolters (1970)
			y	f	18.5–54*	Han and Woo (1973a,b)
rubijervine				f	150†	Wolters (1970)
isorubijervine				f	150†	Wolters (1970)
			y	f	72–200*	Han and Woo (1973a,b)
veratramine			y	f	72–200*	Han and Woo (1973a,b)
veratrobasine			y	f	72–200*	Han and Woo (1973a,b)

[a] +, gram-positive bacteria; −, gram-negative bacteria; y, yeasts; f, fungi.
[b] Values in italic were determined using an agar dilution method, the others using agar diffusion. *No details of the method were given. †Channel test (Wolters, 1963).

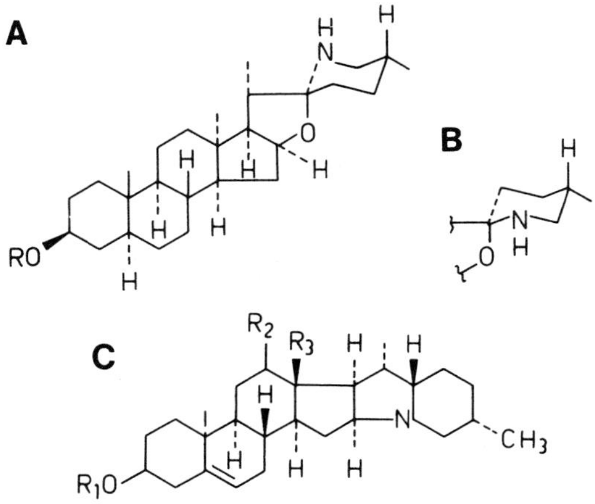

Figure 14. Steroidal alkaloids.

A

	R
Soladulcidine	H
Solasodine	$H_{\Delta 5}$
Solasonine	rhamnose-galactose-glucose$_{\Delta 5}$
Solamargine	rhamnose-rhamnose-glucose$_{\Delta 5}$

B

Tomatine	H
Tomatidine	galactose-(glucose)$_2$-xylose

C

	R_1	R_2	R_3
α-Solanine	-galactose-rhamnose glucose	H	Me
α-Chaconine	-glucose-rhamnose rhamnose	H	Me

antimycoticum (Allen and Kuc, 1968; Fontaine *et al.,* 1948; Wolters, 1964, 1965a, 1966). Boll *et al.* (1955) reported a selective activity against tubercle bacteria for solanocapsine. The structural requirements for antifungal activity of the alkaloids from *Solanum dulcamara* were studied in detail by Wolters (1964, 1965a,b, 1966, 1969, 1970). These alkaloids have activities similar to some common commercial fungicides (Wolters, 1969); the alkaloids with a Δ^5 double bond (solasodine type, Fig. 14) showed the broadest spectrum of antifungal activity, but the tomatine-type alkaloids were active at lower concentration. The corresponding glucosides were active at similar or even lower concentrations (Wolters, 1964, 1966), and glucosides having four sugars bonded to the 3-OH were more active than those with three sugars. The combination of aglucones and glucosides was found to be more active than each of the single components. Related saponins, lacking the amine function, are active though at higher levels (Wolters, 1965b). The mode of antifungal action of the steroidal alkaloids is postulated to be similar as for polyene antibiotics, through complexing with cell wall steroids (Wolters, 1966). The addition of ergosterol to the steroidal alkaloids reduced the activity (Wolters, 1969).

Kusano *et al.* (1987b) reported on the antifungal properties of some *Solanum* alkaloids. Solacongestidine, verazine, and solafloridine were much more active than, e.g., solasodine and tomatidine. These alkaloids were found to inhibit the ergosterol and cholesterol biosynthesis from lanesterol (Kusano *et al.*, 1987a). It has been reported that the fungitoxic activity of α-tomatine and some of its hydrolysis products was closely correlated with the ability to complex with sterols, and not with the surfactant activity (Arneson and Durbin, 1968).

Sinden *et al.* (1973) suggested that the glycoalkaloids play a role in restricting the lesions in young leaves after infection with *Alternaria solani.* The effects of α-solanine, α-chaconine, tomatine (Fig. 14), and related *Solanum* alkaloids on synthetic membranes, organelles, and cells have been studied extensively (Fewell and Roddick, 1993; Roddick, 1978; Roddick and Drysdale, 1984; Roddick and Rijnenberg, 1986, 1987; Roddick *et al.*, 1988, 1990; Steel and Drysdale, 1988). Tomatine disrupts membranes containing 3β-hydroxy sterols causing electrolyte leakage, e.g., in fungi, mammalian red blood cells, protozoa, and plant organelles. In tomato and potato cells, membranes consist mainly of substituted sterols, which might explain why these plants can contain high levels of such membrane disrupting compounds (Roddick and Drysdale, 1984; Steel and Drysdale, 1988).

α-Solanine and α-chaconine have been studied for their membrane disrupting properties (Roddick *et al.*, 1988; Roddick and Rijnenberg, 1986, 1987). α-Chaconine disrupts phosphatidyl/cholesterol liposomes, whereas α-solanine had very little activity; however, a mixture of these alkaloids had a clearly stronger effect. There is no synergistic effect with tomatine. Similar results were found with beet cells, erythrocytes, and fungal protoplasts (Roddick *et al.*, 1988). In studies on other *Solanum* alkaloids from other plant species, solamargine was found to be more active in disrupting artificial vesicle membranes than solasonine, but together they show a synergistic effect (Roddick *et al.*, 1990). All of these alkaloids bind 3β-hydroxy sterols. They also lyse protoplasts of *Penicillium* and erythrocytes, with a similar pattern of structure–activity relationship and synergism as in the *in vitro* tests. From this the conclusion was drawn that the sugar moiety plays an important role in the activity. In solamargine and chaconine the sugar is chacotriose, whereas in solasonine and solanine it is solatriose. The synergism of α-solanine and α-chaconine has also been proven for the antifungal activity of these alkaloids (Fewell and Roddick, 1993). The synergistic effect is over a wide range of ratios (10–20% of either compound) and is much less affected by a decrease in pH than are each of the individual compounds.

The finding of synergistic effects of compounds co-occurring in a plant is of great general interest. Activities of combinations of compounds have been studied in only a few cases, other than in crude extracts. Such studies have not been performed for any of the other antimicrobial active alkaloids reported here.

Several of the alkaloids found in *Veratrum* have antimicrobial activity (Han and Woo, 1973a,b; Wolters, 1966, 1969, 1970). Jervine is particularly active against Basidiomycetes, whereas its glucoside, pseudojervine, is about 10-fold less active (Wolters, 1970). In the test system of Han and Woo (1973a,b), the glucoside showed higher activity, probably because of its better solubility. The hydroxysolanidine alkaloids rubijervine and isorubijer-

Table IV
Miscellaneous Types of Alkaloids Reported to Have Antimicrobial Activity

	Activity[a]				MIC (μg/ml)[b]	References
	+	−	y	f		
Quinoline alkaloids						
dihydrocinchonine	+		y			Mitscher *et al.* (1972a)
pteleatinium	+	−	y		*100–1000*	Mitscher *et al.* (1971c, 1975)
dehydropteleatinium	+	−	y		*50–100*	Mitscher *et al.* (1975)
veprisinium	+					Ayafor *et al.* (1982)
dictamnine	+			f	10–100[†]	Wolters and Eilert (1981)
γ-fagarine	+			f	20–100[†]	Wolters and Eilert (1981)
skimmianine				f	>100[†]	Wolters and Eilert (1981)
Acridone alkaloids (Fig. 15)						
rutacridone epoxide	+	−	y	f	0.25–5[†]	Wolters and Eilert (1981)
hydroxyrutacridone epoxide	+	−	y	f	0.1–10[†]	Wolters and Eilert (1981)
Terpenoid alkaloids						
chaksine	+				*100*	Gupta and Chopra (1953)
taxol				f	*0.4–5.9*	Young *et al.* (1992)
Piperidine alkaloids						
wisanine	+				*128*	Linke *et al.* (1978)
dihydrowisanine	+				*128*	Linke *et al.* (1982)
dihydrookolasine	+				*128*	Linke *et al.* (1982)
juliflorine (Fig. 16)	+	−	y	f	*1.5–50*	Khursheed *et al.* (1986), Aqeel *et al.* (1986)
julifloricine	+	−	y		*0.5–100*	Aqeel *et al.* (1989)
euphococcinine		−			1000*	Tawara *et al.* (1993)
Quinolizidine alkaloids						
sparteine	+	−		f	*20–2000*	Wink (1984), Wippich and Wink (1985)
lupanine	+	−		f		Wink (1984), Wippich and Wink (1985)
13-tigloyl-oxylupanine	+	−		f		Wink (1984)
angistifoline	+	−			*5000*	Tyski *et al.* (1988)
13-hydroxylupanine	+	−			*5000*	Tyski *et al.* (1988)
Pyrrolizidine alkaloids						
supinine			y			Marquina *et al.* (1989)
heliotrine			y			Marquina *et al.* (1989)
lasiocarpine	+		y	f		Marquina *et al.* (1989)
		−				Jain and Sharma (1987)
lasiocarpine *N*-oxide		−		f		Jain and Sharma (1987)
9-angeloylretronecine	+					Marquina *et al.* (1989)
europine		−	y	f		Jain and Sharma (1987)
Miscellaneous						
cryptopleurine (Fig. 17)			y		*0.1*	Al-Shamma *et al.* (1982), Foldeak (1971)
antofine	+			f		Baumgartner *et al.* (1990)
ficuseptine	+			f		Baumgartner *et al.* (1990)
julandine				f	*12.5*	Al-Shamma *et al.* (1982)

Table IV
(*Continued*)

	Activity[a] +	−	y	f	MIC (μg/ml)[b]	References
3,4-dimethoxy-ω-(piperid-2-yl)-acetophenone				f	*3.1*	Al-Shamma *et al.* (1982)
6,6′-dihydroxythiobinupharidine				f	*0.1*	Cullen *et al.* (1973)
tryptanthrin	+			f	*3–6*	Honda and Tabata (1979)
onychine			y		*3.1*	Hufford *et al.* (1987)
scutiane A–E	+			f	200*	Tscheche and Ammermann (1974)
sinharine				f		Greger *et al.* (1992, 1993)
N-methylsinharine				f		Greger *et al.* (1992, 1993)
illukumbin B				f		Greger *et al.* (1992, 1993)
methylillukumbin B				f		Greger *et al.* (1992, 1993)
methylillukumbin A				f		Greger *et al.* (1992, 1993)

[a]+, gram-positive bacteria; −, gram-negative bacteria; y, yeasts; f, fungi.
[b]Values in italic were determined using an agar dilution method, the others using agar diffusion. *No details of the method were given. †TLC disk test.

vine were particularly active against Ascomycetes and Fungi imperfecti. Esterified alkaloids were less active. The alkaloids did not show antibacterial activity (Wolters, 1970).

5. QUINOLINE ALKALOIDS

Pteleatinium and its dehydro derivative were identified as the active principles in *Ptelea trifoliata* (Mitscher *et al.*, 1971c, 1975) (Table IV). These quaternary furoquinoline alkaloids have very strict structural requirements for activity, e.g., substitution with a methyl group at the 8-OH results in loss of activity. For veprisinium weak activity against *Staphylococcus* species has been reported (Ayafor *et al.*, 1982). In *Ruta graveolens* some furoquinolines were found to exhibit antimicrobial activity (Wolters and Eilert, 1981). Dictamine, with the lowest degree of substitution, was most active. With increasing substitution a decrease in activity was found. Towers and Abramowski (1983) reported that the activity of these alkaloids is tremendously enhanced by light.

6. ACRIDONE ALKALOIDS

The major antimicrobial components in *Ruta graveolens* cell cultures are acridone alkaloids (Fig. 15).The epoxides of rutacridone and hydroxyrutacridone, in particular, are highly active against a wide spectrum of bacteria and fungi (Table IV). These alkaloids were more active than the synthetic antibiotic ethacridine lactate (Wolters and Eilert, 1981). The activity is linked with the substituents in the acridone skeleton.

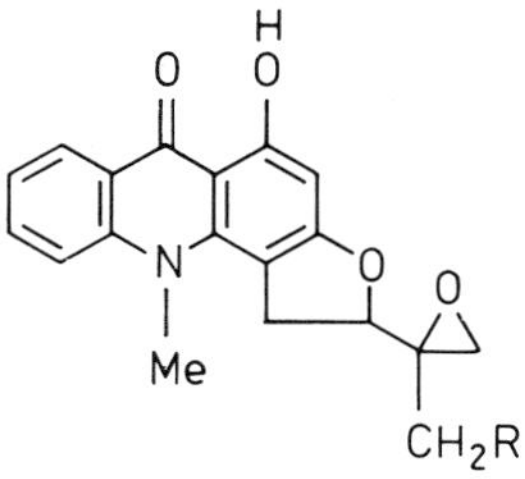

Figure 15. Rutacridone epoxide (R = H; hydroxyrutacridone epoxide, R = OH).

7. TERPENOID ALKALOIDS

The monoterpenoid alkaloid chaksine has weak antimicrobial activity (Gupta and Chopra, 1953). Taxol, well known as a new antitumor drug, also possesses antifungal activity (Young *et al.,* 1992). Oomycetes were particularly sensitive to taxol (EC_{50} 0.4–5.9 μM), compared to other types of fungi ($EC_{50} > 50$ μM). This activity is similar to the toxicity for mammalian cells. Inhibition of mitosis, resulting from interaction with microtubules, seemed to be involved in the antifungal activity, the same effect as in mammalian cells. A series of taxol analogues (baccatin III derivatives) was also studied, but all were clearly less active than taxol. The taxol side chain is thus crucial for activity.

8. PIPERIDINE ALKALOIDS

Piper alkaloids, such as wisanine, dihydrookolasine, and dihydrowisanine, have been shown to possess antimicrobial activity (Linke *et al.,* 1978, 1982) (Table IV). The piperidine alkaloid juliflorine (Fig. 16) was reported to have a strong effect against dermatophytic fungi (MIC 1.5 μg ml^{-1}, versus griseofulvin 0.1–0.5 μg $ml^{-}M^{1}$) (Ahmad *et al.,* 1989; Khursheed *et al.,* 1986). It also has activity against gram-positive bacteria at levels similar to those of known antibiotics; gram-negative bacteria are less sensitive (Aqeel *et al.,* 1986). The compound is nonmutagenic and also has activity against *Candida albicans.* The isomer julifloricine is active against gram-positive bacteria and *Candida,* but not against gram-negative bacteria or dermatophytes (Aqeel *et al.,* 1989).

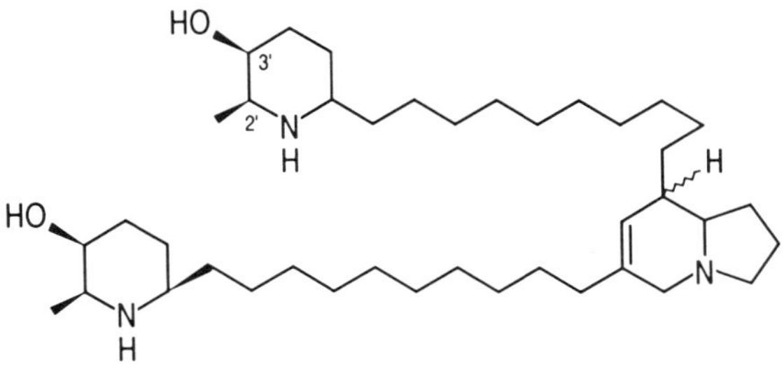

Figure 16. Juliflorine.

Euphococcinine, an alkaloid occurring in pine and spruce trees, was shown to have weak antibacterial activity (Tawara *et al.,* 1993).

9. QUINOLIZIDINE ALKALOIDS

The alkaloids from lupin were reported to exhibit antibacterial activities (Tyski *et al.,* 1988; Wink 1984) (Table IV). This activity is bacteriostatic rather than bactericidal. Also, some phytopathogenic fungi were inhibited in their growth at sparteine concentrations in the range of 1 to 20 mM. The quinolizidine alkaloids were also active against powdery mildew (*Erysiphe graminis*) at levels of 1–5 mM, i.e., at levels lower than those found in the plant (Wippich and Wink, 1985).

10. PYRROLIZIDINE ALKALOIDS

Besides their hepatotoxic properties, the *Heliotropium* alkaloids possess antimicrobial activity (Jain and Sharma, 1987; Marquina *et al.,* 1989) (Table IV).

11. MISCELLANEOUS

Cryptopleurine (Fig. 17) and the related alkaloids julandine and 3,4-dimethoxy-ω--(piperid-2-yl)acetophenone are highly active against *Candida* species (Al-Shamma *et al.,* 1982). Cryptopleurine and derivatives are active against fungi (Foldeak, 1971), cytotoxic (Hoffmann *et al.,* 1978), and have antiviral activity (Krmpotic *et al.,* 1972). Their effect on the protein-synthesizing system in some microorganisms was studied by Haslam *et al.* (1968). In a series of 27 synthetic analogues, activities were found to be lower than reported for cryptopleurine when the quinolizidine ring was opened (Hegyes *et al.,* 1985). Antofine, a phenanthroindolizidine alkaloid a structurally very similar to cryptopleurine, but isolated from a different source, has antimicrobial activity. Ficuseptine, a novel alkaloid from the same plant, has a similar activity (Baumgartner *et al.,* 1990).

A sulfur-containing alkaloid from *Nuphar luteum,* 6,6′-dihydroxythiobinupharidine, was reported to be active against various human pathogenic fungi (Cullen *et al.,* 1973).

From a folk remedy for athlete's foot (leaves of *Strobilanthes cusia*), the antifungal compound tryptanthrin was isolated (Honda and Tabata, 1979). It showed strong activity

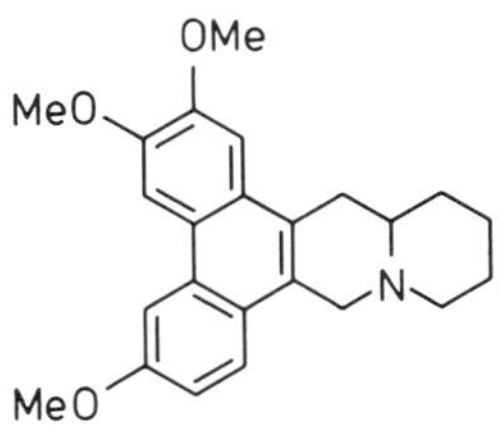

Figure 17. Cryptopleurine.

against dermatophytes, such as *Trichophyton, Microsporum,* and *Epidermophyton* species. This compound is also found in a Chinese medicinal plant, *Isatis indigotica* (Li *et al.,* 1983).

A series of carbazole alkaloids was reported to be antimicrobially active (Das *et al.,* 1965; Chakraborty *et al.,* 1975), 3-methyl-6-hydroxycarbazole being the most active. Some synthetic tetrahydrocarbazole derivatives were also found to have antifungal and antibacterial activity (Chowdhury *et al.,* 1978).

The alkaloid onychine has anticandida activity similar to eupolaridine (Hufford *et al.,* 1987).

Cyclopeptide alkaloids isolated from the bark of *Scutia buxifolia* were shown to have antimicrobial activity (Tschesche and Ammermann, 1974).

Although not alkaloids in the strict sense of the word, amides as N-containing compounds can be mentioned as antimicrobial agents. Particularly the hydroxamic acids found as defense compounds in Gramineae are worth mentioning. For an extensive review, the reader is referred to Niemeyer (1988). DIBOA and DIMBOA are respectively the main compounds in rye or wheat and maize. These compounds are derived from chorismate through anthranilate. Besides being involved in the resistance against various pests, DIMBOA also plays a role in the plant's resistance against fungi and bacteria. Greger and co-workers (Greger *et al.,* 1992, 1993) isolated some sulfur cinnamides with antifungal activity.

12. CONCLUSION

This review has considered about 300 alkaloids with antimicrobial activity. As only a small number of all known alkaloids have been studied for antimicrobial activity, one may conclude that within this class of compounds quite a large percentage has such activity. Unfortunately little is known about the mode of action for most compounds, in many cases this might be a general cytotoxic activity which excludes them as antibiotic drugs. Further studies on the mechanism of action might be useful for development of new antibiotics, as plants in general have an efficient defense against microorganisms, in which the secondary metabolites play a key role (phytoalexins). Antifungal activity is quite common among plant compounds, a type of activity that is of particular interest for drug development, as only a few antifungal antibiotics are known. In this connection one has to bear in mind that the plant defense systems are multifactorial, and the antimicrobial activity of a single pure compound is not necessarily representative of the efficiency of the compound in combination with other factors, such as light, cell-wall-weakening enzymes, and membrane-disturbing compounds. Synergistic effects between closely related compounds may play an important role in activity, whereas the occurrence of mixtures of related compounds may be of importance for avoiding rapid development of resistance by microorganisms. There may be interesting lessons from nature for future development of antibiotics. Apparently the present practice of using a single pure compound leads, in evolutionary terms, to the rapid development of resistance by microorganisms.

For understanding plants' defense systems as well as for drug development, further studies in this field will be of great interest. From the present review it is clear that standardization of methods for determining antimicrobial activity is important. The results described here are in many cases difficult to compare because of the totally different methodologies applied. A broad screening, similar to that undertaken by the NCI for antitumor and anti-HIV compounds, should be of great interest for developing totally new classes of antibiotics. The use of cell cultures in combination with elicitation with fungal elicitors is an approach that could yield interesting leads for new antifungal compounds.

REFERENCES

General Reviews

Abbasoglu, U., Sener, B., Gunay, Y., and Temizer, H., 1991, Antimicrobial activity of some isoquinoline alkaloids, *Arch. Chem.* **324:**379–380.

Al-Shamma, A., Drake, S., Flynn, D. L., Mitscher, L. A., Park, Y. H., Rao, G. S. R., Simpson, A., Swayze, J. K., Veysoglu, T., and Wu, S. T. S., 1981, Antimicrobial agents from higher plants. Antimicrobial agents from *Peganum harmala* seeds, *J. Nat. Prod.* **44:**745–747.

Amin, A. H., Subbaiah, T. V., and Abbasi, A. M., 1969, Berberine sulfate: Antimicrobial activity, bioassay and mode of action, *Can. J. Microbiol.* **15:**1067–1076.

Burkholder, P. R., and Sharma, G. M., 1969, Antimicrobial agents from the sea, *Lloydia* **32:**466–483.

Cavallito, G. J., 1951, Antibiotics from plants, in: *Medicinal Chemistry,* Vol. 1. (C. M. Suter, ed), Wiley, New York, pp. 222–240.

Chopra, D. N., Dikshit, B. B., and Chowhan, J. S., 1932, Pharmacological action of berberine, *Indian J. Med. Res.* **19:**1193–1203.

Chowdhury, D. N., Basak, S. K., and Das, B. P., 1978, Studies on the insecticidal and antimicrobial properties of some carbazole alkaloids, *Curr. Sci.* **47:**490–491.

Clark, A. M., and Hufford, C. D., 1992, Antifungal alkaloids, in: *The Alkaloids,* Vol. 42 (G. A. Cordell ed.), Academic Press, San Diego, pp. 117–150.

Dornberger, K., and Lich, H., 1982, Screening nach antimikrobiell sowie potentiell cancerostatisch wirksamen Pflanzeninhaltstoffen, *Pharmazie* **37:**215–221.

Elmi, A. S., Baerheim Svendsen, A., Scheffer, J. J. C., and Verpoorte, R., 1986, Screening of some Somalian medicinal plants for antimicrobial activity, *J. Ethnopharmacol.* **17:**283–288.

Hahn, F. E., and Ciak, J., 1975, Berberine, *Antibiotics* **3:**577–584.

Hejtmankova, N., Walterova, D., Preininger, V., and Simanek, V., 1984, Antifungal activity of quaternary benzophenanthridine alkaloids from *Chelidonium majus, Fitoterapia* **55:**291–294.

Hernandez, N. M. R., 1979, Assessment of the antimicrobial activity of indole alkaloids, *Rev. Cubana Med. Trop.* **31:**199–204.

Hiller, K., 1964, Antimikrobielle Stoffe in Bluetenpflanzen. Eine Uebersicht, *Pharmazie* **19:**167–188.

Hüttenrauch, R., 1962, Antibiotica mitt steroidstruktur, *Pharmazie* **17:**590–595.

Jain, S. C., and Sharma, R., 1987, Antimicrobial activity of pyrrolizidine alkaloids from *Heliotropium ellipticum, Chem. Pharm. Bull.* **35:**3487–3489.

Korzybski, T., Kowszyk-Gindifer, Z., and Kurylowicz, W., 1967, Antibiotics from higher plants, in: *Antibiotics,* Vol. 2, Pergamon Press, New York, pp. 1438–1451.

Kuc, J., 1992, Antifungal compounds from plants, in: *Phytochemical Resources for Medicine and Agriculture* (H. N. Nigg and D. S. Seigler, eds.), Plenum Press, New York, pp. 159–184.

Kusano, G., Takahashi, A., Sugiyama, K., and Nozoe, S., 1987b, Antifungal properties of Solanum alkaloids, *Chem. Pharm. Bull.* **35:**4862–4867.

Lahiri, S. C., and Dutta, N. K., 1967, Berberine and chloramphenicol in the treatment of cholera and severe diarrhoea, *J. Indian Med. Assoc.* **48:**1–10.

Le Grand, A., Wondergem, P. A., Verpoorte, R., and Pousset, J. L., 1988, Antiinfectious phytotherapies of the tree-savannah of Senegal. II. Antimicrobial activity of 33 species, *J. Ethnopharmacol.* **22:**25–31.

Lenfeld, J., Kroutil, M., Marsalek, E., Slavik, J., Preininger, V., and Simanek, V., 1981, Antiinflammatory activity of quaternary benzophenanthridine alkaloids from *Chelidonium majus, Planta Med.* **43:**161–165.

Mitscher, L. A., Leu, R. P., Bathala, M. S., Wu, W. N., Beal, J. L., and White, R., 1972a, Antimicrobial agents from higher plants. I. Introduction, rationale and methodology, *Lloydia* **35:**157–176.

Mitscher, L. A., Clark, G. W., and Beal, J. L., 1975, Antimicrobial agents from higher plants: The quaternary alkaloids from *Ptelea trifoliata, Lloydia* **38:**109–116.
Mitscher, L. A., Drake, S., Gollapudi, S. R., and Okwute, S. K., 1987, A modern look at folkloric use of anti-infective agents, *J. Nat. Prod.* **50:**1025–1040.
Nickell, L. G., 1959, Antimicrobial activity of vascular plants, *Econ. Bot.* **13:**281–318.
Paxton, J. D., 1981, Phytoalexins—A working redefinition, *Phytopathol. Z.* **101:**106–109.
Perera, P., Sandberg, F., Van Beek, T. A., and Verpoorte, R., 1985, Alkaloids in the stem and root bark of *Tabernaemontana dichotoma, Phytochemistry* **24:**2097–2104.
Phillipson, J. D., Wright, C. W., Kirby, G. C., and Warhurst, D. C., 1993, Tropical plants as sources of antiprotozoal agents, in: *Phytochemical Potential of Tropical Plants* (K. R. Downum, ed.), Plenum Press, New York, pp. 1–40.
Stoessl, A., 1970, Antifungal compounds produced by higher plants, in: *Recent Advances in Phytochemistry,* Vol. 3 (C. Steelink and V. C. Runeckles, eds.), Plenum Press, New York, pp. 143–181.
Verpoorte, R., 1986, Antimicrobially active alkaloids in: *Proceedings XXI-eme Rencontre Internationale de Chemie Therapeutique*(J. Y. Laronze and J. Levy, eds.), University of Reims, Reims, France, pp. 195–209.
Verpoorte, R., Tsin a Tsoi, A., van Doorne, H., and Baerheim Svendsen, A., 1982b, Medicinal plants of Suriname I: Antimicrobial activity of some medicinal plants, *J. Ethnopharmacol.* **5:**221–226.
Verpoorte, R., van Beek, T. A., Thomassen, P. H. A. M., Aandewiel, J., and Baerheim Svendsen, A., 1983a, Screening of antimicrobial activity of some plants belonging to the Apocynaceae and Loganiaceae, *J. Ethnopharmacol.* **8:**287–302.
Villar, A., Rios, J. L., Recio, M. C., Cortes, D., and Cave, A., 1986, Antimicrobial activity of benzylisoquinoline alkaloids. II. Relation between chemical composition and antimicrobial activity, *Planta Med.* **52:**556–557.
Villar, A., Mares, M., Rios, J. L., Canton, E., and Gobernado, M., 1987, Antimicrobial activity of benzylisoquinoline alkaloids, *Pharmazie* **42:**248–250.
Waterman, P. G., 1990, Searching for bioactive compounds: Various strategies, *J. Nat. Prod.* **53:**13–22.
Whitehead, I. M., and Threlfall, D. R., 1992, Production of phytoalexins by plant tissue cultures, *J. Biotechnol.* **26:**63–81.
Wink, M., 1984, Chemical defense of Leguminosae. Are quinolizidine alkaloids part of the antimicrobial defense system of lupins? *Z. Naturforsch.* **39C:**548–552.
Wink, M., 1993, Allelochemical properties or the raison d'être of alkaloids, in: *The Alkaloids,* Vol. 43 (G. A. Cordell, ed.), Academic Press, San Diego, pp. 1–118.
Wright, C. W., Bray, D. H., O'Neill, M. J., Warhurst, D. C., Phillipson, J. D., Quetin-Leclercq, J., and Angenot, L., 1991, Antiamoebic and antiplasmodial activities of alkaloids isolated from *Strychnos usambarensis, Planta Med.* **57:**337–340.
Young, D. H., Michelotti, E. L., Swindell, C. S., and Krauss, N. E., 1992, Antifungal properties of taxol and various analogues, *Experientia* **48:**882–885.
Zbierska, J., and Kowalewski, Z., 1979a, Anticancer and antibiotic properties of chelidonine methyliodide, *Herba Pol.* **25:**209–217.

Key References

Achenbach, H., 1986, Investigations on West African medicinal plants, *Pure Appl. Chem.* **58:**653–662.
Achenbach, H., Raffelsberger, B., and Brillinger, G., 1980, Constituents of West-African medicinal plants. Part 4. 19-Hydroxycoronaridine and 19-hydroxyibogamine, two antibiotic alkaloids of the ibogamine type, *Phytochemistry* **19:**2185–2188.
Adeoye, A. O., Oguntimein, B. O., Clark, A. M., and Hufford, C. D., 1986, 3-Dimethylallylindole: An antibacterial and antifungal metabolite from *Monodora tenuifolia, J. Nat. Prod.* **49:**534–537.
Adesanya, S. A., Olugbade, T. A., Odebiyi, O. O., and Aladesanmi, J. A., 1992, Antibacterial alkaloids in *Crinum jagus, Int. J. Pharmacog.* **30:**303–307.
Aerts, R. J., Stoker, A., Beishuizen, M., Jaarsma, I., Van de Heuvel, M., Van der Meijden, E., and Verpoorte, R., 1992, Detrimental effects of *Cinchona* leaf alkaloids on larvae of the polyphagous insect *Spodoptera exigua, J. Chem. Ecol.* **18:**1955–1964.
Ahmad, A., Khursheed, A. K., Sahiba, Q., and Viqaruddin, A., 1989, Antifungal activity of some hydrosoluble *Prosopis juliflora* alkaloids, *Fitoterapia* **60:**86–89.

Ahmad, A., Khan, K. A., Sultana, S., Siddiqui, B. S., Begum, S., Faizi, S., and Siddiqui, S., 1992, Study of the in vitro antimicrobial activity of harmine, harmaline and their derivatives, *J. Ethnopharmacol.* **35:**289–294.

Akhter, M. H., Sabir, M., and Bhide, N. K., 1977, Antiinflammatory effect of berberine in rats injected locally with cholera toxin, *Indian J. Med. Res.* **65:**133–140.

Alazard, R., Boquet, L., and Paoletti, C., 1976, Effects of 9-hydroxyellipticine on growth and macromolecular synthesis in *Escherichia coli, FEBS Lett.* **62:**278–282.

Allen, E. H., and Kuc, J., 1968, α-Solanine and α-chaconine as fungitoxic compounds in extracts of Irish potatoes tubers, *Phytopathology* **58:**776–781.

Al-Shamma, A., Guagliardi, L. E., Mitscher, L. A., and Swayze, J. K., 1982, Antimicrobial alkaloids from *Boehmeria cylindrica, Phytochemistry* **21:**485–487.

Andronescu, E., Petcu, P., Goina, T., and Radu, A., 1973, Antibiotic activity of the extract and the alkaloid isolate from *Berberis vulgaris, Clujul Med.* 627–631.

Aqeel, A. A., Kursheed, A. K., Viqaruddin, A., and Sabiha, Q., 1986, Antibacterial activity of juliflorine isolated from *Prosopis juliflora, Planta Med.* 285–288.

Aqeel, A. Kursheed, A. K., Viqaruddin, A., and Sahiba, Q., 1989, Antimicrobial activity of julifloricine isolated from *Prosopis juliflora, Arzneim. Forsch.* **39:**652.

Arneson, P. A., and Durbin, R. D., 1968, Studies on the mode of action of tomatine as a fungitoxic agent, *Plant Physiol.* **43:**683–686.

Ayafor, J. F., Sondengam, B. L., and Ngadjui, B. T., 1982, Veprisinium salt, a novel antibacterial quaternary alkaloid from *Vepris louisii, Planta Med.* **44:**139–142.

Barret, Y., and Sauvaire, Y., 1992, Fagaronine, a novel antileukaemic alkaloid, *Phytother. Res.* **6:**59–63.

Baumgartner, B., Erdelmeier, C. A. J., Wright, A. D., Ralli, T., and Sticher, O., 1990, An antimicrobial alkaloid from *Ficus septica, Phytochemistry* **29:**3327–3330.

Bersch, H., and Dopp, W., 1955, Pruefung einiger Alkaloide in vitro auf Tuberkulostase, *Arzneim.-Forsch.* **5:**77–78.

Bhattacharyya, P., Chakrabartty, P. K., and Chowdhury, B. K., 1985, Glycozolidol, an antibacterial carbazole alkaloid from *Glycosmis pentaphylla, Phytochemistry* **24:**882–883.

Boll, P. M., Lillevik, H. A., Gottshall, R. Y., and Lucas, E. H., 1955, Antibacterial substances in seed plants against tubercle bacilli. III. Solanocapsine, the antibacterial alkaloid of *Solanum pseudocapsicum, Antibiot. Ann.* 255–259.

Bourgois, J., Bourgois, M., and Formisyn, P., 1991, Promises and prospects of an indolic derivative: The 9-hydroxy-ellipticine, *Trends Heterocycl. Chem.* **2:**205–232.

Caron, C., Hoizey, M. J., Le Men-Olivier, L., Massiot, G., Zeches, M., Choisy, C., Le Mayre, E., and Verpoorte, R., 1988, Antimicrobial and antifungal activity of quasidimeric and related alkaloids, *Planta Med.* **54:**409–412.

Chakraborty, D. P., Das, K., Das, B. P., and Chowdhury, B. K., 1975, Antibiotics from higher plants. Part IV. On the antibiotic properties of some carbazole alkaloids, *Trans. Bose Res. Inst. Calcutta* **38:**1–4.

Cimanga, K., Pieters, L., Claeys, M., Vanden Berghe, D., and Vlietinck, A. J., 1991, Biological activities of cryptolepine. An alkaloid from *Cryptolepis sanguinolenta, Planta Med.* **57:**A98–A99.

Clark, A. M., Watson, E. S., Ashfaq, M. K., and Hufford, C. D., 1987, In-vivo efficacy of antifungal oxo-aporphines alkaloids in experimental disseminated candidiasis, *Pharm. Res.* **4:**495–498.

Coune, C., 1980, Contribution a l'etude de *Strychnos gossweileri* du Zaire. Isolement et determination de la structure des nouveaux alcaloides. These de Doctorat, Liege.

Cullen, W. P., LaLonde, R. T., Wang, C. J., and Wong, C. F., 1973, Isolation and in-vitro antifungal activity of 6,6′-dihydroxythionupharidine, *J. Pharm. Sci.* **62:**826–827.

Das, K. C., Chakraborty, D. P., and Bose, P. K., 1965, Antifungal activity of some constituents of *Murraya koenigii* Spreng, *Experientia* **21:**340.

Das Gupta, B. M., and Dikshit, B. B., 1929. Berberine in the treatment of oriental sore, *Int. Med. Gaz.* **64:**67.

Dutta, N. K., and Panse, M. V., 1962, Usefulness of berberine, an alkaloid from *Berberis aristata,* in the treatment of cholera, *Indian J. Med. Res.* **50:**732–736.

Dutta, N., Iyer, K., and Natatrajan, S., 1968, Anti-amoebic value of berberine and kurchi alkaloids, *Indian J. Med. Assoc.* **50:**349–352.

Dutta, N. K., Marker, P. H., and Rao, N. R., 1972, Berberine in toxin-induced experimental cholera, *Br. J. Pharmacol.* **44:**153–159.

Eich, E., Eichberg, D., Schwarz, G., Clas, F., and Loos, M., 1985, Antimicrobial activity of clavines, *Arzneim. Forsch.* **35:**1760–1762.

Ferdous, A. J., Islam, M. O., Hasan, C. M., and Islam, S. N., 1992, In vitro antimicrobial activity of lanuginosine and oxostephanine, *Fitoterapia* **63:**549–550.

Fewell, A. M., and Roddick, J. G., 1993, Interactive antifungal activity of the glycoalkaloids α-solanine and α-chaconine, *Phytochemistry* **33:**323–328.

Foldeak, S., 1971, Synthesis of analogues of cryptopleurine, *Tetrahedron* **27:**3465–3467.

Fontaine, T. D., Irving, G. W., Ma, R., Poole, J. B., and Doolittle, S. P., 1948, Isolation and characterization of crystalline tomatine, an antibiotic agent from the tomatoe plant, *Arch. Biochem. Biophys.* **18:**467–475.

Gasquet, M., Quetin-Leclercq, J., Timon-David, P., Balansard, G., and Angenot, L., 1992, Antiparasitic properties of diploceline, a quaternary alkaloid from *Strychnos gossweileri, Planta Med.* **58:**276–277.

Gharbo, S. A., Beal, J. L., Doskotch, R. W., and Mitscher, L. A., 1973, Alkaloids of *Thalictrum.* XIV. Isolation of alkaloids having antibacterial activity from *Thalictrum polyganum, Lloydia* **36:**349–351.

Gilliver, K., 1946, The inhibitory action of antibiotics on plant pathogenic bacteria and fungi, *Ann. Bot.* **10:**271.

Greger, H., Hofer, O., Kählig, H., and Wurz, G., 1992, Sulfur containing cinnamides with antifungal activity from *Glycosmis cyanocarpa, Tetrahedron* **48:**1209–1218.

Greger, H., Zechner, G., Hofer, O., Hadacek, F., and Wurz, G., 1993, Sulphur containing amides from *Glycosmis* species with different antifungal activity, *Phytochemistry* **34:**175–179.

Gupta, K. C., and Chopra, I. C., 1953, A short note on antibacterial properties of chaksine: An alkaloid from *Cassia absus* L, *Indian J. Med. Res.* **41:**459–460.

Gupta, K. C., and Kahali, B. S., 1944, General pharmacology of umbellatine, a new alkaloid isolated from *Berberis umbellata* Wall and *Berberis insignis* Hook f. and its use in oriental sore, *Ind. J. Med. Res.* **32:**53–56.

Han, Y. B., and Woo, W. S., 1973a, Antimicrobial principles of *Veratrum grandiflorum, Yakhak Hoeji* **17:**13–16.

Han, Y. B., and Woo, W. S., 1973b, Antimicrobial activity of *Veratrum* alkaloids, *Yakhak Hoeji* **17:**137–140.

Harper, D. S., Mueller, L. J., Fine, J. B., Gordon, J., and Laster, L. L., 1990, Effect of 6 months use of a dentifrice and oral rinse containing sanguinaria extract and zinc chloride upon the microflora of the dental plaque and oral soft tissues, *J. Periodontol.* **61:**359–363.

Haslam, J. M., Davey, P. J., Linnane, A. W., and Atkinson, M. R., 1968, Differentiation in-vitro by phenanthrene alkaloids of yeast mitochondrial protein synthesis from ribosomal systems of both yeast and bacteria, *Biochem. Biophys. Res. Commun.* **33:**368–373.

Hegyes, P., Foldeak, S., and Feher, L., 1985, Investigation of phenanthroquinolizidine analogues, *Arzneim. Forsch.* **35**(Suppl. 2):1758–1759.

Hernandez, N. M. R., Perez, C. D., and Perez, O. D., 1977, Antimicrobial activity of *Vallesia antilana* wood. I. Alkaloids, *Rev. Cubana Farmac.* **11:**249–255.

Hoffmann, J. J., Luzbetak, D. J., Torrance, S. J., and Cole, J. R., 1978, Cryptopleurine, cytotoxic agent from *Boehmeria caudata* (Urticaceae) and *Cryptocarya laevigata* (Lauraceae), *Phytochemistry* **17:**1448.

Homma, N., Kono, M., Kadohira, H., Yoshihara, S., and Masuda, S., 1961, Der Einfluss des Berberin-chlorids auf die Darmflora des Sauglings, *Arzneim. Forsch.* **11:**450–454.

Honda, G., and Tabata, M., 1979, Isolation of fungal principle tryptanthrin, from *Strobilanthus cusia* O. Kuntze, *Planta Med.* 36, 85–86.

Horn, I., and Steffen, K., 1968a, Antifungale Wirkungen der Inhaltstoffe von Radix Colombo, *Pharm. Z.* **113:**85–86.

Horn, I., and Steffen, K., 1968b, Antifungale Wirkungen der Inhaltstoffe von Radix Colombo, *Pharm. Z.* **113:**945–950.

Hufford, C. D., Funderburk, M. J., Morgan, J. M., and Robertson, L. W., 1975, Two antimicrobial alkaloids from heartwood of *Liriodendron tulipifera, J. Pharm. Sci.* **64:**789–792.

Hufford, C. D., Sharma, A. S., and Oguntimein, B. O., 1980, Antibacterial and antifungal activity of liriodenine and related oxoaporphine alkaloids, *J. Pharm. Sci.* **69:**1180–1182.

Hufford, C. D., Liu, S., Clark, A. M., and Oguntimein, B. O., 1987, Anticandidal activity of eupolauridine and onychine, alkaloids from *Cleistopholis patens, J. Nat. Prod.* **50:**961–964.

Irving, G. W., Jr., 1949, Antibiotics from higher plants. in: Antibiotics (G. W. Irving, Jr. and M. T. Herrick, eds.) Chemical Publishing Company, New York, pp. 157–203.

Jezowa, L., Kaczmarek, F., and Rafinski, T., 1966, The synergistic action of berberine-HCl and tuberculostatic preparation upon the growth of *Mycobacterium tuberculosis* in-vitro, *Herba Pol.* **12:**266–272.

Keene, A. T., Phillipson, J. D., Warhurst, D. C., Koch, M., and Seguin, E., 1987, In-vitro amoebicidal testing of natural products; Part 2. Alkaloids related to emetine, *Planta Med.* **49:**201–206.

Khursheed, A. K., Arshad, H. F., Viqaruddin, A., Sahiba, Q., Sheikh, A. R., and Tahir, S. H., 1986, In-vitro studies of antidermatophytic activity of juliflorine and its screening as carcinogen in Salmonella/microsome test system, *Arzneim. Forsch.* **36:**17–19.

Kowalewski, Z., Kedzia, W., and Mirska, I., 1972, Effect of berberine sulfate on staphylococci, *Arch. Immunol. Ther. Exp.* **20:**352–360.

Krey, A. K., and Hahn, F. E., 1969, Berberine: Complex with DNA, *Science* **166:**755–757.

Krmpotic, E., Farnsworth, N. R., and Messmer, W. M., 1972, Cryptopleurine, an active antiviral alkaloid from *Boehmeria cylindrica* (L.) Sw. (Urticaceae), *J. Pharm. Sci.* **61:**1508–1509.

Kupchan, S. M., and Altland, H. W., 1973, Structural requirements for tumor-inhibitory activity among bis-benzylisoquinoline alkaloids and related synthetic compounds, *J. Med. Chem.* **15:**913–917.

Kuroda, H., Nakazawa, S., Katagiri, K., Shiratori, O., Kozuka, M., Fujitani, K., and Tomita, M., 1976, Antitumor effects of bisbenzylisoquinoline alkaloids, *Chem. Pharm. Bull.* **24:**2413–2420.

Kusano, G., Takahashi, A., Nozoe, S., Sonoda, Y., and Sato, Y., 1987a, Solanum alkaloids as inhibitors of enzymatic conversion of dihydrolanosterol into cholesterol, *Chem. Pharm. Bull.* **35:**4321–4323.

Kuwano, S., and Yamauchi, K., 1960a, Effect of berberine on tyrosine decarboxylase activity of *Streptococcus faecalis, Chem. Pharm. Bull.* **8:**491–496.

Kuwano, S., and Yamauchi, K., 1960b, Competition of berberine with pyridoxal phosphate in the tryptophanase system of *E. coli, Chem. Pharm. Bull.* **8:**497–503.

Lahiri, S. C., and Sen, K. N., 1958, Antibacterial properties of berbamine, *Ann. Biochem. Exp. Med. (Calcutta)* **18:**95–96.

Larson, R. A., Marley, K. A., Tuveson, R. W., and Berenbaum, M. R., 1988, β-Carboline alkaloids: Mechanisms of phototoxicity to bacteria and insects, *Photochem. Photobiol.* **48:**665–674.

Li, Q., Jin, J., Chong, M., and Song, Z., 1983, Studies on the antifungal constituent of Qing Dai (*Isatis indigotica*), *Zhongcaoyao* **14:**440. Through *Chem. Abstr.* 100(1984)12501k.

Linke, S., Kurz, J., and Zeiler, H. J., 1978, Wisanin und andere Amide der 5-(2-methoxy-4,5-methylendioxy-phenyl)-2,4-pentadiensaüre, *Tetrahedron* **34:**1979–1983.

Linke, S., Kurz, J., and Zeiler, H. J., 1982, Synthesis, spectroscopic examination and testing for antibacterial activity of some pepper alkaloids. Olefination reactions with phosphorylacetamides, *Liebigs Ann. Chem.* 1142–1149.

Liu, S., Oguntimein, B. O., Hufford, C. D., and Clark, A. M., 1990, 3-Methoxysampangine, a novel antifungal copyrine alkaloid from *Cleistopholis patens, Antimicrob. Agents Chemother.* **34:**529–533.

Luijendijk, T. J. C., 1995. Strictosidine glucosidase in indole alkaloids biosynthesis, Ph.D. thesis, Leiden University.

Ma, R., and Fontaine, T. D., 1948, In-vitro antibiotic activity of crystalline tomatine towards *Candida albicans.* Antagonistic effect of rutin and quercitin. *Arch. Biochem. Biophys.* **16:**399–402.

McKenna, D. J., and Towers, G. H. N., 1981, Ultraviolet mediated cytotoxic activity of β-carboline alkaloids, *Phytochemistry* **20:**1001–1004.

Mahajan, V. M., Sharma, A., and Rattan, A., 1982, Antimycotic activity of berberine sulfate: An alkaloid from an Indian medicinal herb, *Sabouraudia* **20:**79–81.

Mahmud, Z., Musa, M., Ismail, N., and Lajis, N. H., 1993, Cytotoxic and bactericidal activities of *Psychotria rostrata, Int. J. Pharmacog.* **31:**142–146.

Maiti, M., and Chaudhuri, K., 1981, Interaction of berberine chloride with naturally occurring deoxyribonucleic acids, *Indian J. Biochem. Biophys.* **18:**245–250.

Maiti, M., Nandi, R., and Chaudhuri, K., 1982, Sanguinarine: A monofunctional intercalating alkaloid, *FEBS Lett.* **142:**280–284.

Mariee, N. K., Khalil, A. A., Nasser, A. A., Al-Hitti, M. M., and Ali, W. M., 1988, Isolation of the antimicrobial alkaloid stemmadenine from Iraqi *Rhazya stricta, J. Nat. Prod.* **51:**186–187.

Marquina, G., Laguna, A., Franco, P., Fernandez, L., Perez, R., and Valiente, O., 1989, Antimicrobial activity of pyrrolizidine alkaloids from *Heliotropium bursiferum* Wr ex Grisebach, *Pharmazie* **44:**870–871.

Maynart, G., Pousset, J. L., Mboup, S., and Denis, F., 1980, Antibacterial effect of borreverine, an alkaloid isolated from *Borreria verticillata* (Rubiaceae), *C. R. Soc. Biol. (Paris)* **174:**925–928.

Meijer, A. H., Verpoorte, R., and Hoge, J. H. C., 1993, Regulation of enzymes and genes in terpenoid indole alkaloid biosynthesis in *Catharanthus roseus.* Third Special Issue "Cellular and Molecular Biology in Plant Cell Cultures," *J. Plant Res.* Tokyo (A. Komamine, H. Fukuda, Y. Komeda, U. Sankawa, and K. Syono, eds.), pp. 145–164.

Mekawi, M., 1968, Effect of berberine alkaloid on *Vibrio cholerae* and its endotoxin, *J. Egypt. Med. Assoc.* **49:**554–559.

Melo, M. de F. F., Thomas, G., and Mukherjee, R., 1988, Antidiarrhoeal activity of bisnordihydrotoxiferine isolated from the root bark of *Strychnos trinervis* (Vell.) Mart. (Loganiaceae), *J. Pharm. Pharmacol.* **40:**79–82.

Michel, M. M. G., Lemoine, A., Dat-Xuong, N., and Olle, J., 1975a, Hydroxy-9-ellipticine: Activite bacteriostatique et bactericide in vitro, *C. R. Acad. Sci. Paris Ser. D* **280:**1493–1496.

Michel, M. M. G., Lemoine, A., Dat-Xuong, N., and Olle, J., 1975b, Study on the antibacterial effects of 9-hydroxy-ellipticine, *Rev. Inst. Pasteur Lyon* **8:**53–68.

Mitscher, L. A., Wu, W. N., Doskotch, R. W., and Beal, J. L., 1971a, Antibiotics from higher plants. *Thalictrum rugosum*: New bisbenzylisoquinoline alkaloids active against *Mycobacterium smegmatis, Chem. Commun.* **11:**589–590.

Mitscher, L. A., Wu, W. N., Doskotch, R. W., and Beal, J. L., 1971b, Antibiotisch wirksame Alkaloide aus *Thalictrum rugosum, Dtsch. Apoth. Ztg.* **111:**1704.

Mitscher, L. A., Bathala, M. S., and Beal, J. L., 1971c, Antibiotics from higher plants: Pteleatinium chloride a new quaternary quinoline alkaloid from *Ptelea trifoliata* with antitubercular and antiyeast activity, *J. Chem. Soc.* **D17:**1040.

Mitscher, L. A., Showalter, H. D. H., Shipchandler, M. T., Leu, R. P., and Beal, J. L., 1972b, Antimicrobial agents from higher plants. IV. *Zanthoxylum elephantiasis.* Isolation and identification of canthin-6-one, *Lloydia* **35:**177–180.

Mitscher, L. A., Juvarkar, J. V., and Beal, J. L., 1976, Solacassine, a new steroidal alkaloid from *Solanum pseudocapsicum* possessing antimicrobial activity, *Experientia* **76:**415–416.

Mitscher, L. A., Park, Y. H., Clark, C. D., Clark, G. W., Hammesfahr, P. D., Wu, W. N., and Beal, J. L., 1978, Antimicrobial agents from higher plants. An investigation of *Hunnemannia fumariaefolia.* Pseudoalcoholates of sanguinarine and chelerythrine, *Lloydia* **41:**145–150.

Miyakado, M., Kato, T., Ohno, N., and Koshimuzi, K., 1975, Alkaloids of *Urginea altissima* and their antimicrobial activity against *Phytophthora capsici, Phytochemistry* **14:**2717.

Modak, S., Modak, M. J., and Venkataraman, A., 1970a, Mechanism of action of berberine on *Vibrio cholerae* and *Vibrio cholerae* biotype Eltor, *Indian J. Med. Res.* **58:**1510–1521.

Modak, S., Modak, M. J., and Venkataraman, A., 1970b, Effect of berberine on the fatty acid composition of *Vibrio cholerae* and *Vibrio cholerae* biotype Eltor, *Indian J. Med. Res.* **58:**1523–1527.

Munoz, V., Moretti, C., Sauvain, M., Caron, C., Porzel, A., Massiot, G., Richard, B., and Le Men-Olivier, L., 1994, Isolation of bis-indole alkaloids with antileishmanial and antibacterial activities from *Peschiera van heurkii* (Syn. *Tabernaemontana van heurkii*), *Planta Med.* **60:**455–459.

Nair, S., Modak, M. J., and Venkataraman, A., 1969, Mechanism of action of berberine, part one and two, *Indian J. Pharm.* **13:**173–178.

Neal, J. J., 1989, Methylenedioxyphenyl-containing alkaloids and autosynergism, *Phytochemistry* **28:**451–453.

Niemeyer, H. M., 1988, Hydroxamic acids (4-hydroxy-1,4-benzoxazin-3-ones), defence chemicals in the Gramineae, *Phytochemistry* **27:**3349–3358.

Odebiyi, O. O., and Sofowora, E. A., 1979, Antimicrobial alkaloids from a Nigerian chewing stick (*Fagara zanthoxyloides*), *Planta Med.* **36:**204–207.

Okunade, A. L., Hufford, C. D., Richardson, M. D., Peterson, J. R., and Clark, A. M., 1994, Antimicrobial properties of alkaloids from *Xanthorhiza simplicissima, J. Pharm. Sci.* **83:**404–406.

Pabuccuoglu, V., Rozwadowska, M. R., Brossi, A., Clark, A., Hufford, C. D., George, C., and Flippen-Anderson, J. L., 1991, Oxoaporphine alkaloids: Conversion of lysicamine into liriodenine and its 2-O-methyl ether and antifungal activity, *Arch. Pharm.* **324:**29–33.

Paulo, A., Duarte, A., and Gomes, E. T., 1994, In vitro antibacterial screening of *Cryptolepis sanguinolenta* alkaloids, *J. Ethnopharmacol.* **44:**127–130.

Paulo, M. de Q., Barbosa-Filho, J. M., Lima, E. O., Maia, R. F., Barbosa, R. de C. B. B. C., and Kaplan, M. A. C., 1992, Antimicrobial activity of benzylisoquinoline alkaloids from *Annona salzmanii, J. Ethnopharmacol.* **36:**39–41.

Peterson, J. R., Zjawiony, J. K., Liu, S., Hufford, C. D., Clark, A. M., and Rogers, R. D., 1992, Copyrine alkaloids: Synthesis, spectroscopic characterization, and antimycotic/antimycobacterial activity of A- and B-ring functionalized sampangines, *J. Med. Chem.* **35:**4069–4077.

Pezzuto, J. M., Antosiak, S. K., Messmer, W. M., Slaytor, M. B., and Honig, G. R., 1983, Interaction of the antileukemic alkaloid, 2-hydroxy-3.8.0-trimethoxy-5-methylbenzo(c)phenanthridine (fagaronine) with the nucleic acids, *Chem. Biol. Interact.* **43:**323–339.

Phillips, S. D., and Castle, R. N., 1981, A review of the chemistry of the antitumor benzo(c)phenanthridine alkaloids nitidine and fagaronine and of the related antitumor alkaloid coralyne, *J. Heterocycl. Chem.* **18:**223–232.

Purohit, R. K., and Rao, K. S., 1969, Neomycin compound with berberine. Trial in infantile diarrhoea, *Antiseptic* **66:**855–863.

Roddick, J. G., 1978, Effect of α-tomatine on the integrity and biochemical activities of isolated plant organelles, *J. Exp. Bot.* **29:**1371–1381.

Roddick, J. G., 1987, Antifungal activity of plant steroids, in: *Lipid Interactions among Plants and Their Pests and Pathogens* (G. Fuller and W. D. Nes, eds.), ACS Symp. Ser. 325, American Chemical Society, Washington, DC, pp. 201–222.

Roddick, J. G., and Drysdale, R. B., 1984, Destabilization of liposome membranes by the steroidal glycoalkaloid α-tomatine, *Phytochemistry* **23:**543–547.

Roddick, J. G., and Rijnenberg, A. L., 1986, Effect of steroidal glycoalkaloids of the potato on the permeability of liposome membranes, *Physiol. Plant* **68:**436–440.

Roddick, J. G., and Rijnenberg, A. L., 1987, Synergistic interaction between the potato glycoalkaloids α-solanine and α-chaconine in relation to lysis of phospholipid/sterol liposomes, *Phytochemistry* **26:**1325–1328.

Roddick, J. G., Rijnenberg, A. L., and Osman, S. F., 1988, Synergistic interaction between potato glycoalkaloids α-solanine and α-chaconine in relation to destabilization of cell membranes: Ecological implications, *J. Chem. Ecol.* **14:**889–902.

Roddick, J. G., Rijnenberg, A. L., and Weissenberg, M., 1990, Membrane-disrupting properties of the steroidal glycoalkaloids solasonine and solamargine, *Phytochemistry* **29:**1513–1518.

Ross, S. A., Megalla, S. E., Bishay, D. W., and Awad, A. H., 1980, Studies for determining antibiotic substances in some Egyptian plants. Part II. Antimicrobial alkaloids from the seeds of *Peganum harmala* L, *Fitoterapia* **6:**309–312.

Sabir, M., Mahajan, V. M., Mohapatra, L. N., and Bhide, N. K., 1976, Experimental study of the antitrachoma action of berberine, *Indian J. Med. Res.* **64:**1160–1167.

Sedmera, P., Nguyen, T. N., Valka, I., Cave, A., Cortes, D., and Simanek, V., 1990, A new bisbenzylisoquinoline alkaloid from *Phaeanthus vietnamensis* and its antibacterial activity, *Heterocycles* **30:**206–209.

Seguin, E., Chau, T. M., Koch, M., Andre, S., Farjaudon, N., Pareyre, C., Tempete, C., Robert-Gero, M., Bourut, C., Chenu, E., and Maral, R., 1985, Ochrolifuanines, alcaloides analogues de l'emetine et de la tubulosine: hemisynthèse et recherche de propriétés cytotoxiques et antitumorales, *Ann. Pharm. Fr.* **43:**301–310.

Sharda, D. C., 1970, Berberine in the treatment of diarrhoea of infancy and childhood, *Indian J. Med. Res.* **54:**22–24.

Simeon, S., Rios, J. L., and Villar, A., 1989, Pharmacological activities of benzophenanthridine and phenanthrene alkaloids, *Pharmazie* **44:**593–597.

Simeon, S., Rios, J. L., and Villar, A., 1990, Antimicrobial activity of *Annona cherimolia* stem bark alkaloids,*Pharmazie* **45:**442–443.

Sinden, S. L., Goth, R. W., and O'Brien, M. J., 1973, Effect of potato alkaloids on the growth of *Alternaria solani* and their possible roles as resistance factors in potatoes, *Phytopathology* **63:**303–307.

Skinner, F. A., 1965, Antibiotics, in: *Modern Methods of Plant Analysis,* Vol. 3 (K. Paech and M. V. Tracey, eds.), Springer, Berlin, pp. 626–725.

Smekal, E., and Kubova, N., 1982, Interactions and DNA binding parameters of selected alkaloids of protoberberine group, *Stud. Biophys.* **92:**73–81.

Smekal, E., Kubova, N., Kleinwächter, V., and Cushman, M., 1986, DNA binding studies: Selected benzophenanthridine alkaloids as binding probes, *Stud. Biophys.* **114:**257–262.

Southard, G. L., Boulware, R. T., Walborn, D. R., Groznik, W. J., Thorne, E. E., and Yankell, S. L., 1984, Sanguinarine, a new antiplaque agent: Retention and plaque specificity, *J. Am. Dent. Assoc.* **108:**328–341.

Steel, C. C., and Drysdale, R. B., 1988, Electrolyte leakage from plant and fungal tissues and disruption of liposome membranes by α-tomatine, *Phytochemistry* **27:**1025–1030.

Stermitz, F. R., Gillespie, J. P., Amoros, L. G., Romero, R., and Stermitz, T. A., 1975, Synthesis and biological activity of some antitumor benzophenanthridinium salts, *J. Med. Chem.* **18:**708–713.

Subbaiah, T. V., and Amin, A. H., 1967, Effect of berberine sulfate on *Entamoeba histolytica, Nature* **215:**527–528.

Tawara, J. N., Blokhin, A., Foderaro, T. A., and Stermitz, F. R., 1993, Toxic piperidine alkaloids from pine (*Pinus*) and spruce (*Picea*) trees. New structures and a biosynthetic hypothesis, *J. Org. Chem.* **58:**4813–4818.

Thomas, G., Melo Diniz, M. de F. F., and Mukherjee, R., 1992, Further studies on the antidiarrhoeal activity of bisnordihydrotoxiferine, a tertiary indole alkaloid in rodents, *Phytother. Res.* **6:**84–88.

Towers, G. H. N., and Abramowski, Z., 1983, UV-mediated genotoxicity of furanoquinoline and of certain tryptophan-derived alkaloids, *J. Nat. Prod.* **46:**576–581.

Tsai, I. L., Liou, Y. F., and Lu, S. T., 1989, Screening of isoquinoline alkaloids and their derivatives for antibacterial and antifungal activities, *Kaohsing J. Med. Sci.* **5:**132–145.

Tschesche, A., and Ammermann, E., 1974, Scutianin-C, -D und E, drei weitere Cyclopeptidalkaloide aus *Scutia buxifolia* Reiss, *Chem. Ber.* **107:**2274–2283.

Tyski, S., Markiewicz, M., Gulewicz, K., and Twardowski, T., 1988, The effect of lupin alkaloids and ethanol extracts from seeds of *Lupinus angustifolius* on selected bacterial strains, *J. Plant Physiol.* **133:**240–242.

Vallejos, R. H., and Roveri, O. A., 1972, Alkaloid inhibition of yeast respiration. Prevention by Ca^{2+}, *Biochem. Pharmacol.* **21:**1160–1167.

Van Beek, T. A., Deelder, A. M., Verpoorte, R., and Baerheim Svendsen, A., 1984a, Antimicrobial, antiviral and antiamoebial activity of *Tabernaemontana* species, *Planta Med.* **50:**180–185.

Van Beek, T. A., Kuijlaars, F. L. C., Thomassen, P. H. A. M., Verpoorte, R., and Baerheim Svendsen, A., 1984b, Antimicrobial active alkaloids from *Tabernaemontana pachysiphon, Phytochemistry* **23:**1771–1778.

Van Beek, T. A., de Smidt, C., and Verpoorte, R., 1985a, Phytochemical investigation of *Tabernaemontana crassa, J. Ethnopharmacol.* **14:**315–318.

Van Beek, T. A., Verpoorte, R., Baerheim Svendsen, A., and Fokkens, R., 1985b, Antimicrobially active alkaloids from *Tabernaemontana chippii, J. Nat. Prod.* **48:**400–423.

Verpoorte, R., 1987, Antimicrobially active alkaloids, in: *Handbuch der Antiseptik,* Vol. II, Part 3 (A. Kramer, W. Weuffen, A. P. Krasilnikov, D. Groeschel, E. Bulke, and D. Rehn, eds.), VEB Verlag Volk und Gesundheit, Berlin, pp. 385–422.

Verpoorte, R., Kodde, E. W., van Doorne, H., and Baerheim Svendsen, A., 1978, Antimicrobial effect of the alkaloids from *Strychnos afzelii, Planta Med.* **33:**237–242.

Verpoorte, R., Ruigrok, C. L. M., and Baerheim Svendsen, A., 1982a, Medicinal plants of Suriname. II. Antimicrobial active alkaloids from *Aspidosperma marcgravianum, Planta Med.* **46:**149–152.

Verpoorte, R., Kos-Kuick, E., Tsin a Tsoi, A., Ruigrok, C. L. M., de Jong, G., and Baerheim Svendsen, A., 1983b, Medicinal plants of Suriname. III. Antimicrobially active alkaloids from *Aspidosperma marcgravianum, Planta Med.* **48:**283–289.

Villar, A., Rios, J. L., and Mares, M., 1984a, *Annona cherimolia* alkaloids: Isolation, identification and antimicrobial activity, *Farm. Tijdschr. Belg.* **61:**299.

Villar, A., Mares, M., Rios, J. L., and Gobernado, M, 1984b, Antimicrobial activity of benzylisoquinoline alkaloids, *Farm. Tijdschr. Belg.* **61:**300.

Wippich, C., and Wink, M., 1985, Biological properties of alkaloids. Influence of quinolizidine alkaloids and gramine on the germination and development of powdery mildew, *Erysiphe graminis,* f.sp. *hordei, Experientia* **41:**1477–1479.

Wolters, B., 1963, Biologische Wertbestimmung von Antimykotika mit dem Kanaltest, *Planta Med.* **11:**183–190.

Wolters, B., 1964, Beziehungen zwischen Struktur und antibiotischer Wirkung bie einigen Steroidalkaloiden, *Arch. Pharm. (Weinheim Ger.)* **297:**748–754.

Wolters, B., 1965a, Die antibiotische Wirkung von *Solanum dulcamara* L., *Pharm. Ztg.* **110:**1718–1720.

Wolters, B., 1965b, Der Anteil der Steroid-saponinen an der antibiotischen Wirkung von *Solanum dulcamara, Planta Med.* **13:**189–193.

Wolters, B., 1966, Zur antimikrobiellen Wirksamkeit pflanzlicher Steroide und Triterpene, *Planta Med.* **14:**392–401.

Wolters, B., 1969, Zur Verwendung vorbeschichteter Folien bei der Dunnschicht-chromatografischen Untersuchung pflanzlicher Fungistatica, *Planta Med.* **17:**42–50.

Wolters, B., 1970, Antimicrobial activity of Veratrum alkaloids, *Planta Med.* **19:**189–193.

Wolters, B., and Eilert, U., 1981, Antimicrobial substances in cell cultures of *Ruta graveolens, Planta Med.* **43:**166–174.

Wu, W. N., Beal, J. L., Clark, G. W., and Doskotch, R. W., 1976, Antimicrobial agents from higher plants. Additional alkaloids and antimicrobial agents from *Thalictrum rugosum, Lloydia* **39:**65–75.

Wu, W. N., Beal, J. L., Leu, R. P., and Doskotch, R. W., 1977a, Isolation and characterization of alkaloids from the roots of *Thalictrum podocarpum, Lloydia* **40:**384–394.

Wu, W. N., Beal, J. L., and Doskotch, R. W., 1977b, Alkaloids of Thalictrum. XXII. Isolation of alkaloids with hypotensive and antimicrobial activity from *Thalictrum revoltum, Lloydia* **40:**508–514.

Wu, W. N., Liao, W. T., Mahmoud, Z. F., Beal, J. L., and Doskotch, R. W., 1980, Alkaloids of Thalictrum. XXXIV. Three new alkaloids, thalmirabine, thalistine and O-methylthalibrine, and others from roots of *Thalictrum minus* race B, *J. Nat. Prod.* **43:**472–481.

Yang, J. S., Luo, S. R., Chen, X. L., and Li, Y. X., 1979, Chemical composition of the alkaloids of Ku-mu (*Picrasma quassioides* (D. Don) Benn, *Acta Pharm. Sin.* **14:**167–177.

Zbierska, J., and Kowalewski, Z., 1979b, Anticancer and antibiotic properties of N-methylchelidonine methylsulfate, *Herba Pol.* **25:**311–316.

Zbierska, J., and Kowalewski, Z., 1980, Anticancer and antibiotic properties of chelidonine N-oxide, *Herba Pol.* **26:**61–66.

Chapter 18

Utilization of Alkaloids in Modern Medicine

T. Schmeller and M. Wink

1. INTRODUCTION

As outlined in Part III, alkaloids have evolved as a chemical defense against herbivores, microorganisms, and viruses or against other plants (Chapters 11, 14, 15, and 16). Assuming an evolutionary molecular modeling, the structures of many alkaloids have been shaped through natural selection so that they can interfere with a wide variety of molecular or physiological targets of animals and microorganisms (Chapter 12).

Homo sapiens encountered these active molecules, first in their vegetal food items and later as a means to hunt, kill or murder other animals or his contemporaries (see examples given in Chapters 2 and 3). In addition we have to assume that early humans used alkaloidal and other plants as painkillers, stimulants, or hallucinogens. The treatment of disease or illness with medicinal plants has been rather well documented for the last 3000 to 5000 years (see Chapter 2).

Finding useful properties in plants certainly proceeded by trial and error but does not seem to be restricted to humans: It has been documented by E. Rodrigues (1985, 1993) that chimpanzees in the wild selectively pick plants rich in antimicrobial thiophenes when suffering from severe diarrhea. Many birds of prey decorate their nests with oak leaves rich in tannins. Because the birds defecate into their nests, the antimicrobial tannins might help to control the development of bacteria and fungi.

The medicinal use of alkaloids and other natural products could be regarded as an exploitation of properties that originally had been selected and developed in an ecological or evolutionary context (Wink, 1993). The right plants and concentrations became excellent means for the therapeutic treatment of illness and disease. In this chapter we describe those alkaloids that are used in modern medicine (relevant handbooks include Reynolds, 1993; Budavari, 1989; Roth *et al.*, 1994; Teuscher and Lindequist, 1994; Gilman, 1991;

T. Schmeller and M. Wink • Institute for Pharmaceutical Biology, University of Heidelberg, D-69120 Heidelberg, Germany.

Alkaloids: Biochemistry, Ecology, and Medicinal Applications, edited by Roberts and Wink. Plenum Press, New York, 1998.

Harborne and Baxter, 1993). Alkaloids are discussed in alphabetical order with a short summary of their occurrence in plants and their therapeutic and traditional use in medicine (including important trade names). The biochemical/molecular mechanisms of action are given whereever known. Some of these have been discussed in more detail in Chapters 2 and 12.

2. ALKALOIDS USED IN MODERN MEDICINE

2.1. Aconitine

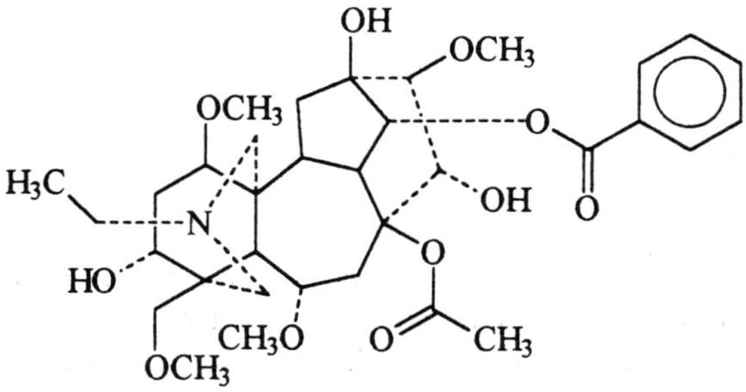

aconitine

Source: Aconitum napellus (Ranunculaceae); tubers contain about 0.3–3% and the leaves 0.1–1% alkaloids of which aconitine is the major component.
Pharmaceuticals (examples): Aconitysat™, Bronpax™, Etermol™, Pectovox™, Pulmo-Xeedol™, Sirop Famel™, Vocadys™.
Therapeutic use: Liniments have been used in the treatment of rheumatism, neuralgia, and sciatica.
Mechanism of action: Aconitine increases the permeability of excitable membranes for Na^+ ions and prolongs the Na^+ influx during the action potential. As a consequence, sensible nerve endings and motor endplates are first activated but later blocked.

2.2. Ajmaline

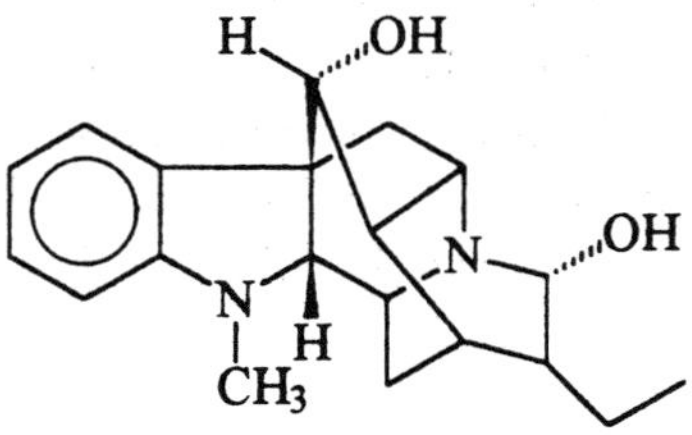

ajmaline

Source: Industrially, ajmaline is isolated from *R. vomitoria* or *Catharanthus roseus.*
Pharmaceuticals (examples) : Aritmina™, Gilurytmal™, Rauwopur™, Ritmos™.

Therapeutic use: Ajmaline is a class Ia antiarrhythmic agent and used in the treatment of supraventricular and ventricular arrhythmias. It might also be useful for patients with myasthenia gravis.
Mechanism: Ajmaline prolongs the refraction phase of the heart through a blockade of Na^+ channels. In addition, the action potential is prolonged and an increase of the depolarization threshold occurs. It has negative inotropic properties accompanied by a small decrease of the heart rate.

2.3. Atropine

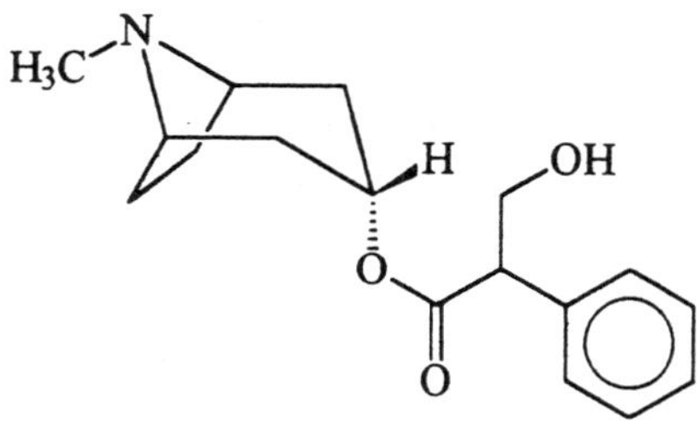

atropine

Source: Mainly *Atropa belladonna* (Solanaceae); its roots contain about 0.45–0.85% alkaloids. Alkaloid content of the leaves varies between 0.2 and 1.5%. The major component is L-hyoscyamine, which racemizes to DL-hyoscyamine (= atropine) during storage or alkaline isolation. Industrially atropine is prepared from leaves of Australian *Duboisia* species (e.g., *D. myoporoides* and *D. leichhardtii;* Solanaceae) or from seeds of *Hyoscyamus muticus* and *H. niger.*
Pharmaceuticals (examples): The following preparations contain mostly atropine sulfate: Abdominol™, Atropinol™, Atropisol™, Dilaudid-Atropin™, Espasmo™, Protecor™, Ru-Tuss™, Spasmosol™, Spersatropine™, Tonaton™.
Therapeutic use: (1) As an antispasmodic in the treatment of gastrointestinal, urinary, and biliary colics; (2) in preparation of general anesthesia, to reduce bronchial and salivary secretion and to avoid bronchospasms; (3) in combination with morphine, to suppress its effect on the vomitory center and the gastrointestine; (4) against parkinsonism (displaced by benzhexol); (5) as an antidote against phosphoric acid ester and mushroom poisoning; (6) against myasthenia gravis; (7) in the management of arrhythmias; (8) in the local treatment of muscular rheumatism, sciatica, and neuralgia; (9) in ophthalmology as a mydriatic and cycloplegic drug.
Use in traditional medicine: The extracts from *A. belladonna* were used as cosmetics, because of their dilatatory effect on the pupils. In the treatment of bronchial asthma, the dried leaves of *Datura stramonium* (Solanaceae) were burned and the smoke inhaled (see also Chapter 2).
Mechanism: Atropine blocks muscarinic cholinergic receptors competitively, and therefore acts as a parasympatholytic at parasympathetically innervated organs. Atropine also has central effects, because as a tertiary amine it is able to cross the blood–brain barrier.

2.4. Berberine

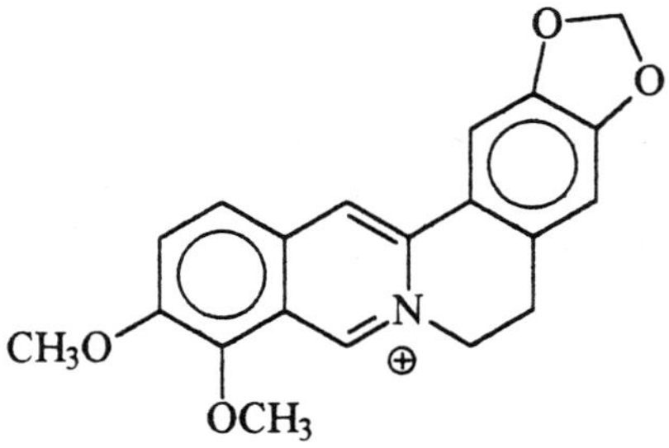

berberine

Source: Berberis vulgaris (Berberidaceae); the bark contains up to 8% alkaloids, with berberine as the major component (about 5%). Berberine also occurs in several *Hydrastis* species (Papaveraceae) and *Chelidonium majus* (Papaveraceae).
Pharmaceuticals (examples): Kollyr™, Murine™, Pastilles Jessel™, Sedacollyre™.
Therapeutic use: In the treatment of infected eyes and other eye irritations; it might also be useful in the treatment of AIDS, because of its inhibitory properties on HIV-1 reverse transcriptase. It also has a favorable effect on toxic hepatitis induced by intoxication.
Use in traditional medicine: The extract of the root has been used as an antiamoebic in the treatment of cholelithiasis and liver disorders. It is also used as a bitter in alcoholic drinks and as flavor in food.
Mechanism: Berberine exhibits antimicrobial activity, which might be related to its effects on several molecular targets: DNA intercalation, inhibition of esterases, inhibition of DNA and RNA polymerases, inhibition of cellular respiration. Some affected organisms are: *Shigella dysenteriae, Neisseria gonorrhoeae, Phymatotrichum omnivorum, Candida albicans,* several species of *Escherichia* and *Aerobacter, Bacillus subtilis, Streptococcus pneumoniae, Pseudomonas aeruginosa,* and *Staphylococcus aureus.* Berberine also increases biliary secretion.

2.5. Boldine

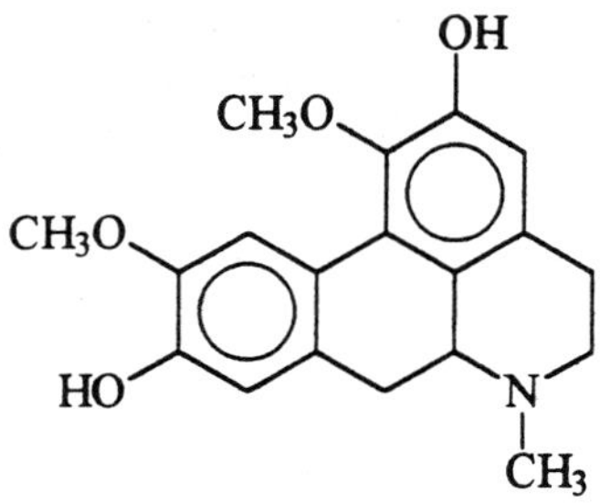

boldine

Source: Peumus boldo (Monimiaceae); leaves contain 0.4–0.5% alkaloids. Boldine is the major component (20–25%) of the alkaloidal fraction.
Pharmaceuticals (examples): Boldina Houde™, Boldoflorine™, Boldosal™, Digedryl™, Ibsesal™, Menabil Complex™, Oxyboldine™, and Sambil™.

Therapeutic use: Treatment of cholelithiasis, stomachic disorders, vomiting, constipation, and dyspepsia.
Use in traditional medicine: In Chile boldine is used as an anthelmintic drug; it has diuretic properties and stimulates liver metabolism.
Mechanism: Boldine exhibits morphinelike antinociceptive properties. Besides opioid receptors, alpha receptors might also be involved in this action (Zetler, 1988).

2.6. Caffeine

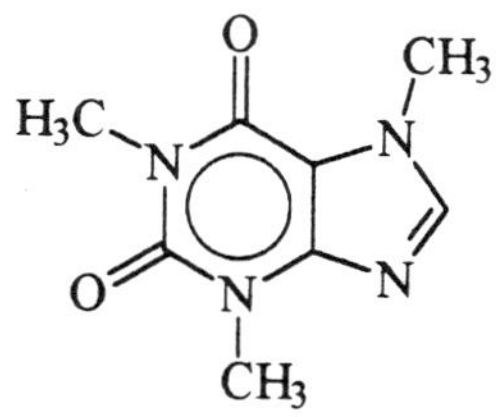

caffeine

Source: Coffea arabica (Rubiaceae), *Paullinia cupana* (Sapindaceae), *Cola nitida* and *C. acuminata* (Sterculiaceae). The caffeine content of the above-mentioned plants varies between 0.6 and 5.0%. Caffeine is also synthetically available.
Pharmaceuticals (examples): (A small selection of more than 300 preparations) Agevis™, Analgen™, Antigrippalin™, Coffein-Enervit™, Doppel-Spalt N™, Kontragripp™, Panax™, Percoffedrinol™, Thomapyrine™, Vomex A™.
Therapeutic use: Caffeine is added to analgesics to increase their activity. It is also used in the treatment of neonatal apnea and atopic dermatitis.
Use in traditional medicine: As a stimulant and in the treatment of headache.
Mechanism: Caffeine inhibits phosphodiesterase and thus influences calcium-mediated signaling through an increase of cAMP levels. In addition, caffeine exhibits competitive antagonistic effects on central adenosine receptors. Its diuretic action is based on its vasodilatory activity on renal blood vessels. The use against headache is related to a constriction of brain vessels and decrease of liquor pressure.

2.7. Cathine

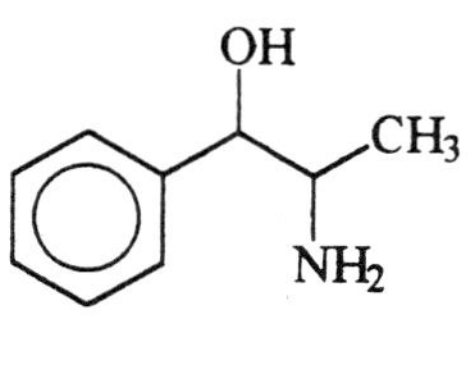

cathine

Source: Catha edulis (Celastraceae); the leaves contain 1% alkaloids. Cathine [(+)-norpseudoephedrine] is the major component. It is also a minor component of several *Ephedra* species.

Pharmaceuticals (examples): Amorphan™, Antiadipositum X 112™, Appetrol™, Eetless™, Recatol™, Vitamin-Schlanktropfen™.
Therapeutic use: As an anorectic drug.
Use in traditional medicine: In Africa and the Middle East the drug is chewed for its stimulant effects.
Mechanism: Cathine is central stimulant, acting as an indirect sympathomimetic. It increases the release of serotonin, noradrenaline, and dopamine and decreases their reuptake. An inhibitory activity on monoamine oxidase (MAO) has also been discussed.

2.8. Cocaine

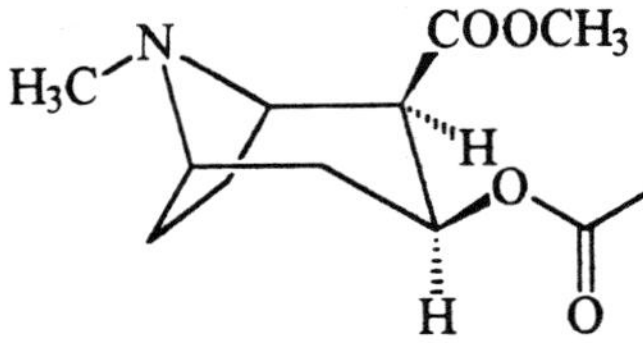

cocaine

Source: Erythroxylon coca (Erythroxylaceae); leaves contain 0.7–2.5% alkaloids. In general, cocaine is the major component.
Pharmaceuticals (examples): There are none readily available. They are prepared as described in several pharmacopoeias (e.g., USP, BPC).
Therapeutic use: As a local surface anesthetic, in ophthalmology for corneal anesthesiology, and for treatment of epistaxis.
Use in traditional medicine: Cocaine is abused for its stimulant effects on the central nervous system. It causes euphoria and mood elevation. For this purpose, either leaves are chewed or pure cocaine is sniffed or injected.
Mechanism: Cocaine blocks the uptake of catecholamines (especially of dopamine) at adrenergic nerve endings and potentiates the action of catecholamines.

2.9. Codeine

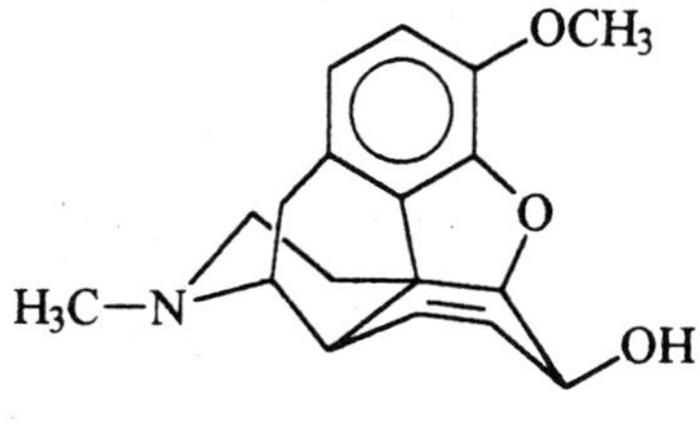

codeine

Source: Codeine is a component of opium, which is obtained from the dried or partly dried latex by incision of the unripe capsules of *Papaver somniferum* (Papaveraceae). Opium contains 0.2–3.0% codeine. Because of its low content in the plant, codeine is

obtained through methylation of morphine. It is also preparable from thebaine, which is the major component of roots, aerial parts (1%) and the capsules (3.5%) of *Papaver bracteatum* (Papaveraceae).
Pharmaceuticals (examples): (A small selection of more than 200 preparations: Antituss™, Bisolvon-Gribletten™, Bronchicum Tropfen mit Codein™, Codicaps™, Codipront™, Contrapect™, Contrapect Infant N™, Dolodens™, Dolorol Forte™, Spasmo-Cibalgin comp.™, Tussipax™.
Therapeutic use: As antitussive in the treatment of cough, and as an analgesic and mild sedative.
Mechanism: Codeine binds to opiate receptors, diminishes bronchial secretion, and acts as a suppressant on the cough center of the medulla oblongata.

2.10. Colchicine

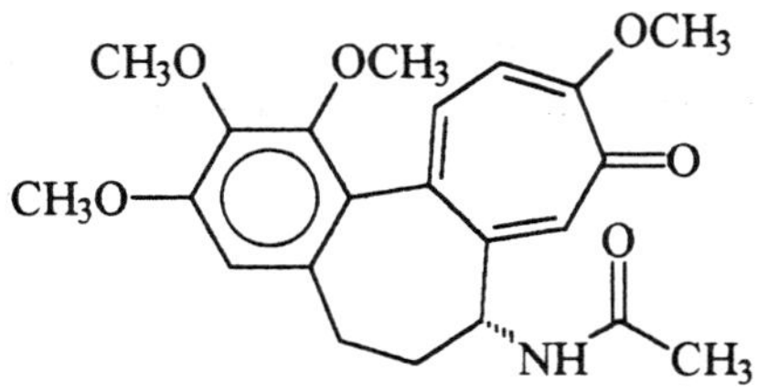

colchicine

Source: Colchicum autumnale (Liliaceae); seeds contain 0.2–1.2% and the tubers 0.1–0.6% colchicine.
Pharmaceuticals (examples): ColBenemid™, Colchimax™, Colgout™, Verban™.
Therapeutic use: Colchicine is used in the treatment of acute gout and the prophylaxis of recurrent gout. It can also be used for the treatment of amyloidosis.
Mechanism: Colchicine inhibits the polymerization of tubulin and thus arrests mitosis in the metaphase (Capraro and Brossi, 1984). But colchicine has too many side effects to be employed for the treatment of cancer. In acute gout colchicine decreases the mobility of phagocytes through inhibition of microtubule formation, important for cell movements. Phagocytes are attracted to sites of inflammation and normally get ruptured after phagocytosis of urate crystals, which causes in turn a decrease of pH followed by further precipitation of urate. This will lead to a stronger inflammation and will attract even more leukocytes. This vicious circle can be disrupted by colchicine.

2.11. Emetine

Source: Cephaelis acuminata (Rubiaceae); roots contain 1.8–3.5% alkaloids. Emetine content varies between 35 and 80%. It is also prepared from cephaeline through methylation.
Pharmaceuticals (examples): Cophylac™, Ipecac™, Rectopyrine™.
Therapeutic use: As a second choice in the treatment of severe intestinal amoebiasis and

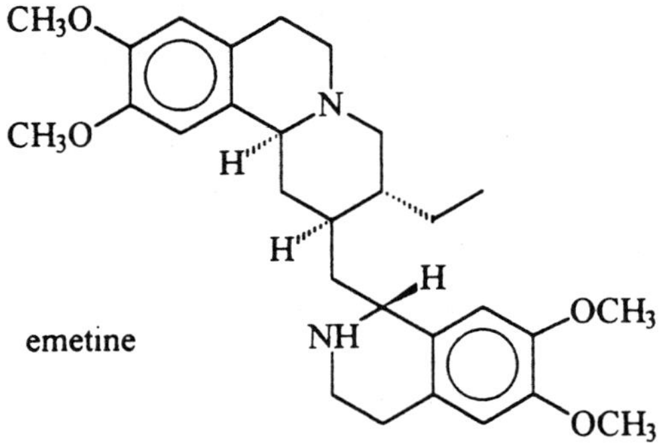

emetine

hepatic amoebiasis when nitroimidazoles are not effective or contraindicated. It is also used in low doses as an expectorant drug with secretolytic and secretomotoric properties.
Use in traditional medicine: As an emetic drug.
Mechanism: Emetine acts indirectly on the bronchia through activation of nervus vagus. Because emetine is a strong inhibitor of protein biosynthesis, it has a strong toxic potential (Fujii and Ohba, 1983).

2.12. Ephedrine

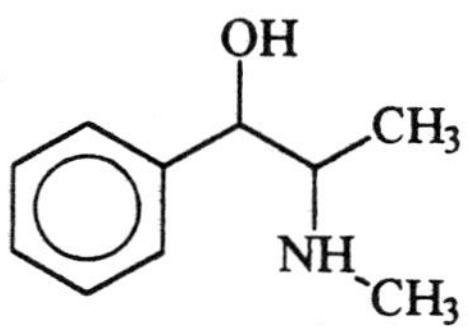

ephedrine

Source: Ephedra sinica and *E. shunnungiania* (Ephedraceae) contain about 0.2–2.0% alkaloids with L-ephedrine as the major component.
Pharmaceuticals (examples): Amidoyna™, Bronchicum™, Bronchisan™, Coughcod™, Dorex™, Endrine™, Ephepect™, Grippon™, Nasenöl-ratiopharm™, Peripherin™, Rectinol™, Solamin™, Vicks Medinite™.
Therapeutic use: As a nasal decongestant for the relief of cold symptoms. Ephedrine may also be employed in the treatment of edema in insulin-dependent diabetics. As a bronchodilator it is used in the treatment of bronchial asthma.
Use in traditional medicine: The herb for the treatment of asthma.
Mechanism: Ephedrine has alpha- and beta-adrenergic activity. In therapeutic doses it increases blood pressure by raising cardiac output and by causing peripheral vasoconstriction. Its central actions include stimulant activity on the respiratory center. It reduces motility and intestinal tone and uterus tone, causes bronchodilation, relaxes the bladder wall and the detrusor muscle while contracting the sphincter muscle.

2.13. Ergometrine

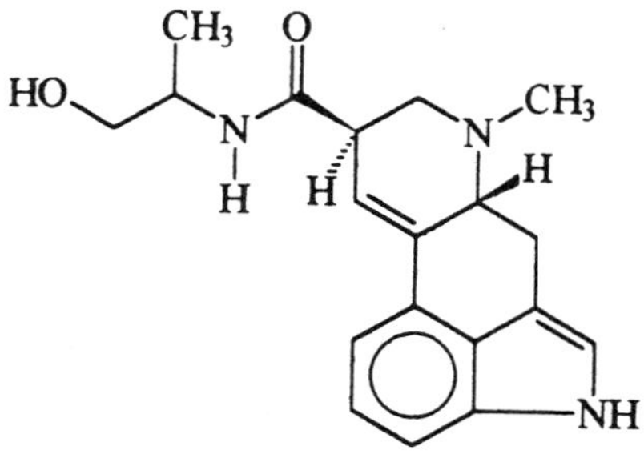

ergometrine

Source: Ergometrine is prepared semisynthetically from amides of *Claviceps paspali* (Clavicipitaceae). The ascomycete can be cultivated on nutritive solutions for alkaloid production.
Pharmaceuticals (examples): Ergometron™, Ergotrate Maleat™, Syntometrine™.
Therapeutic use: Ergometrine is used in the prevention and treatment of postpartum or postabortal hemorrhage.
Use in traditional medicine: Ergometrine and other ergot alkaloids are smoked for their hallucinogenic properties (see Chapter 2).
Mechanism: Ergometrine acts directly on the uterus muscle, causing rhythmic contraction, because of its high affinity to uterine $alpha_2$ receptors.

2.14. Ergotamine

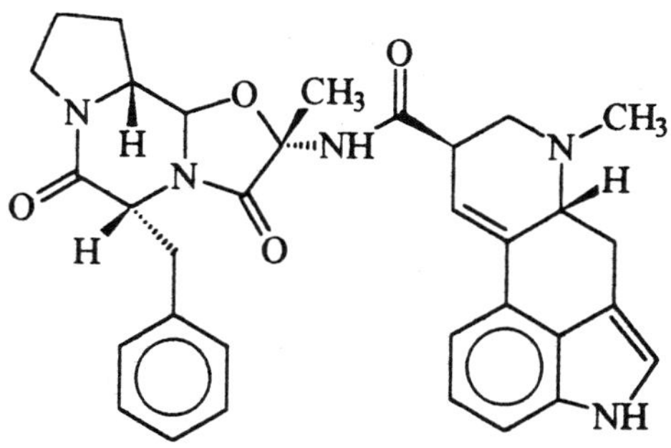

ergotamine

Source: Claviceps purpurea (Clavicipitaceae), the alkaloid content in cultivated species is about 1%.
Pharmaceuticals (examples): Ergo-Kranit mono™, Ergo sanol SL™, Ergostat™, Lingraine™, Migral™, Migräne-Kranit spezial™, Regium™, Virdex™.
Therapeutic use: Ergotamine is used in the treatment of migraine.
Use in traditional medicine: See ergometrine.
Mechanism: Ergotamine is a strong vasoconstrictor through blockade of alpha receptors.

It exhibits a complex effect on the cardiovascular system, which consists of a combination of central inhibitory activity on the vasomotoric center, a peripheral vasoconstriction through partial agonism at serotonergic, alpha-adrenergic, and tryptaminergic receptors and an antagonism at alpha receptors.

2.15. Eserine (= Physostigmine)

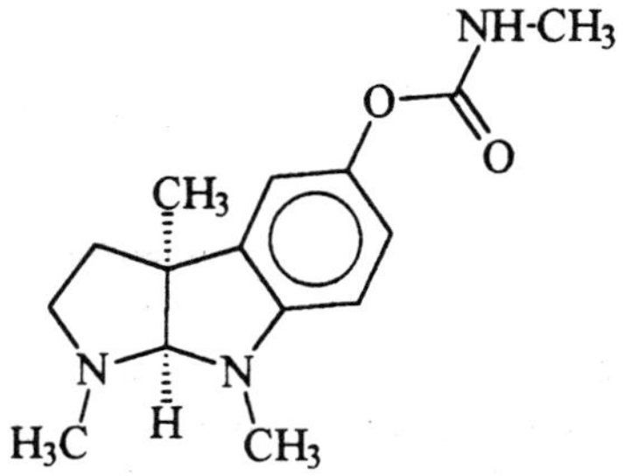

eserine

Source: Physostigma venenosum (Leguminosae); seeds contain 0.3–0.5% alkaloids with eserine as the major component (about 0.15%).
Pharmaceuticals (examples): Anticholium™, Antilirium™, Isopto Pilomin™, Miopos-POS stark™, Miosica Doble™, Pilo-Eserin™, Pilo/Eserin in der Opthiole™, Piloserin™.
Therapeutic use: Eserine is used in ophthalmology to reduce intraocular pressure in the treatment of glaucoma and as a miotic. It is also given as an antidote in acute poisoning with antimuscarinic agents such as atropine.
Use in traditional medicine: In some African tribes the beans are used for ordeals.
Mechanism: Eserine acts through blockade of acetylcholine esterase.

2.16. Galanthamine

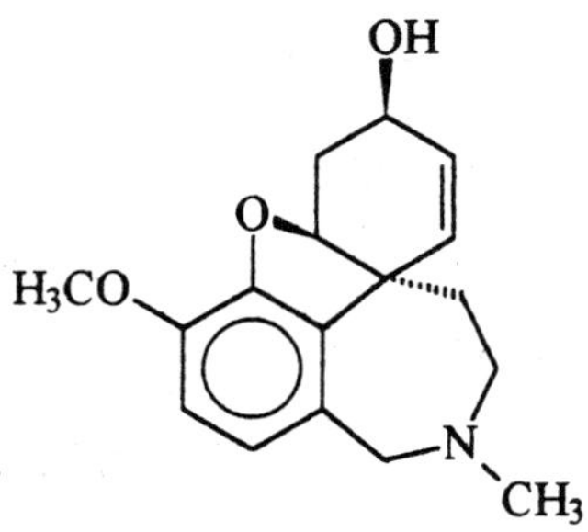

galanthamine

Source: Galanthus woronowii and *G. nivalis* (Amaryllidaceae). Alkaloid contents of bulbs are about 0.1%.
Pharmaceuticals (examples): Nivalina™.
Therapeutic use: It is used to curtail the muscle relaxant activity of nondepolarizing drugs

such as gallamine and tubocurarine. Recently, its use in the treatment of Alzheimer's disease has been proposed.
Mechanism: Galanthamine blocks acetylcholine esterase activity.

2.17. Hydrastine

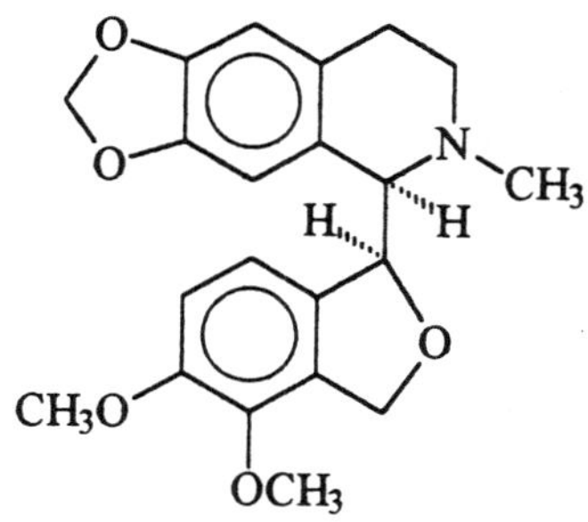

hydrastine

Source: Dried rhizome and roots of cultivated *Hydrastis canadensis* (Berberidaceae), indigenous to Canada and the eastern United States, containing 1.5–4.0% hydrastine.
Pharmaceuticals (examples): Gine Sedans™, Kollyr™.
Therapeutic use: Treatment of gastrointestinal disorders.
Mechanism: Hydrastine acts as a sympatholytic and thus causes a slowdown of heart action. It is paralyzing to the CNS and causes limp paralysis of all muscles.

2.18. Hyoscine (= Scopolamine)

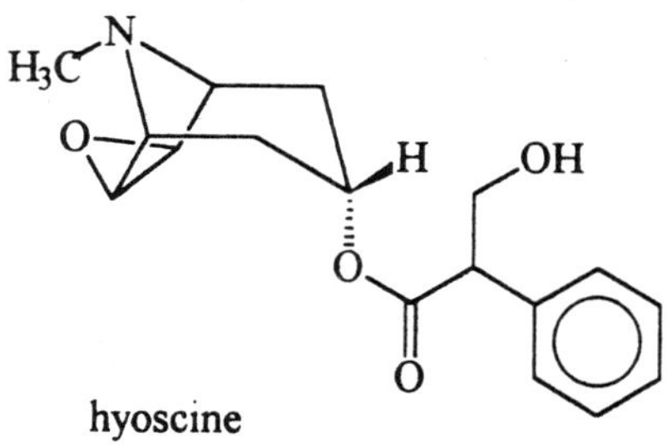

hyoscine

Source: Duboisia myoporoides and *D. leichhardtii* (Solanaceae), but also *Datura* and *Hyoscyamus.*
Pharmaceuticals (examples): Buscopan™, Hyospasmol™, Lotanal™, Oportunin™, Scop™, Scopoderm T™, Spasmofen™, Transcop™, Transderm™.
Therapeutic use: Hyoscine is used in transdermal plasters in the prophylactic treatment of acute motion sickness. As a premedicant, hyoscine is injected subcutaneously or intramuscularly, usually in combination with papaveretum, up to one hour before induction of general anesthesia.
Mechanism: Hyoscine blocks muscarinic cholinergic receptors competitively. It is more powerful than atropine. The central effects differ from those of atropine. In the brain hyoscine depresses the motor areas of cerebral cortex.

2.19. Hyoscyamine

Source: See atropine.
Pharmaceuticals (examples): bella sanol™, Bellatard™, Cystospaz™, Donnatab™, Neurovegetalin™, Solamin™, Tropax™, Urised™, Wigraine™.
Therapeutic use and mechanism: See atropine.

2.20. Lobeline

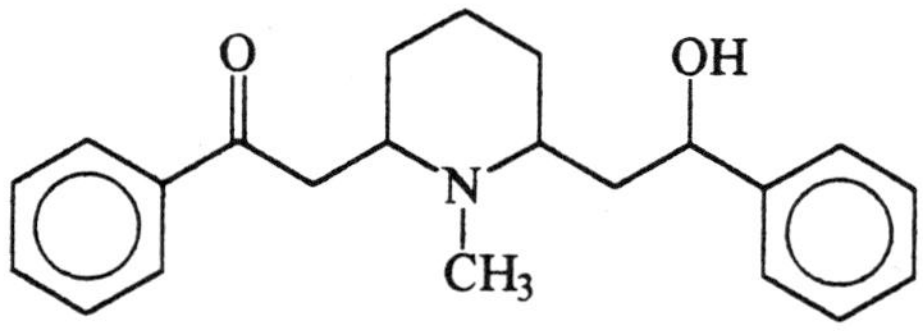

lobeline

Source: Lobelia inflata (Lobeliaceae); aerial parts contain 0.2–0.6% alkaloids.
Pharmaceuticals (examples): Cig-Ridettes™, Citotal™, Lobatox™, Refrane™, Smokeless™, Stopsmoke™.
Therapeutic use: Lobeline is used to discourage smoking. It also has been used in preparations against bronchial asthma, chronic bronchitis, emphysema, cough, nervous and vascular disorders, and insomnia. An injection of lobeline hydrochloride is used to resuscitate newborns.
Use in traditional medicine: Indians smoked it, like tobacco, as cigarettes.
Mechanism: It has similar central and peripheral activities as nicotine, because it affects nicotinic acetylcholine receptors. In therapeutic doses it activates respiration through an increase of excitability of chemoreceptors to the oxygen pressure of the blood.

2.21. Morphine

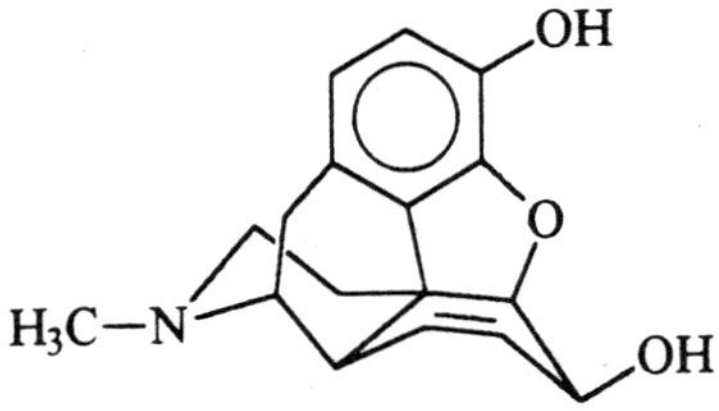

morphine

Source: Papaver somniferum (Papaveraceae); morphine is a component of opium, which is obtained from the dried or partly dried latex after incision of the unripe capsules. The incision is done 8–10 days after the corolla leaves have fallen.
Pharmaceuticals (examples): Collis Brown's™, Diastat™, Diocalm™, Duromorph™, Enterosan™, Morphalgin™, Nepenthe™, Oramorph™, Roxanol™, Spasmofen™.

Therapeutic use: Morphine is used in the control of moderate and severe pain, e.g., in cancer patients. It is also used in the treatment of diarrhea.
Use in traditional medicine: Morphine has been used for its euphorizing properties. For this purpose it is smoked in cigarettes and water pipes. It is also a long-known analgesic.
Mechanism: It has agonistic activity at μ-opiate receptors and also, to a lesser extent, at κ-receptors. Morphine also acts directly on smooth muscles as, for example, in the intestine.

2.22. Narceine

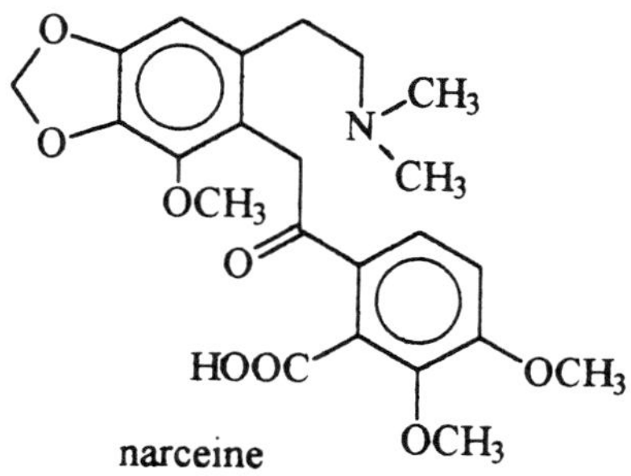

narceine

Source: Papaver somniferum (Papaveraceae). It is a minor component of opium (0.1–0.7%). For preparation of opium see morphine.
Pharmaceuticals (examples): Paneraj™.
Therapeutic use: Treatment of cough.
Mechanism: As in codeine (see above).

2.23. Nicotine

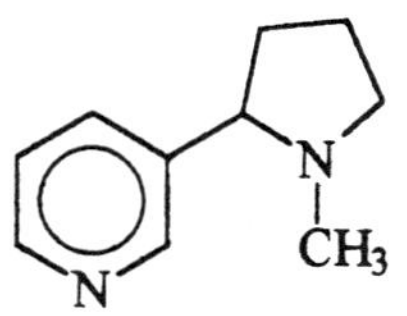

nicotine

Source: Nicotiana tabacum and *N. rustica* (Solanaceae); leaves contain 0.5–9% nicotine, depending on species, developmental stages, and environmental conditions.
Pharmaceuticals (examples): Nicabate™, Nicoderm™, Nicorette™, Nicotinell™, Nicotrol™, nikofrenon™, ProStep™, Resolution™, Stubit™.
Therapeutic use: It is used in gums and transdermal plasters as aids to giving up smoking.
Traditional use: Tobacco is smoked for its stimulatory effects. Nicotine has also been used as a natural insecticide.
Mechanism: Nicotine exhibits agonistic activity at nicotinic acetylcholine receptors. In low doses it stimulates ganglia through depolarization of the postsynaptic membranes of vegetative ganglia; in higher doses it blocks them through permanent depolarization.

2.24. Noscapine (= Narcotine)

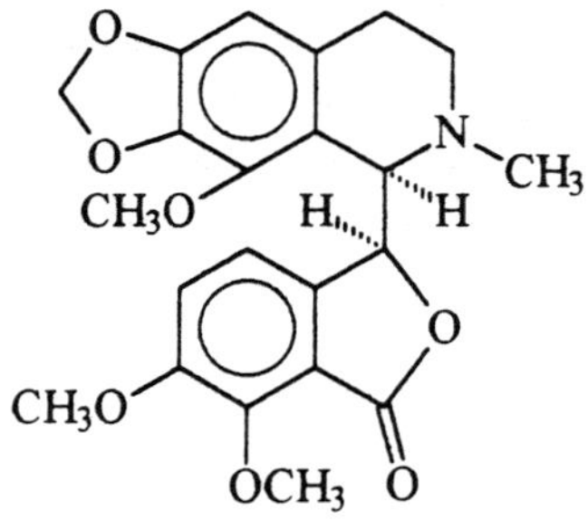

noscapine

Source: Papaver somniferum (Papaveraceae): opium (see morphine) contains about 2–10% noscapine.
Pharmaceuticals (examples): Bequitusin™, Capval™, Degoran™, Difmetus™, Finipect™, Lyobex retard™, Nipaxon™, Nitepax™, Noscalin™, Rea Tos™, Rectolmin Bronquial™, Ribelfan™, Tossamine™, Tussisedal™, Tussoretard™.
Therapeutic use: It is used as a central cough suppressant.
Mechanism: Noscapine affects the cough center directly, increases respiration, and acts as a weak bronchodilator.

2.25. Papaverine

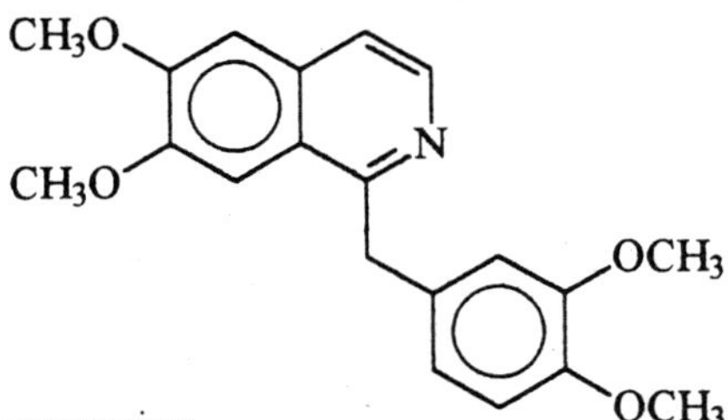

papaverine

Source: Papaver somniferum (Papaveraceae). Papaverine is a component of opium (about 1%) and also of *Chelidonium majus* (Papaveraceae); the aerial parts contain about 0.1–1.0% alkaloids.
Pharmaceuticals (examples): Acticarbine™, Opdensit™, Optenyl™, Pameion™, Panergon™, Pavabid™, Pavasule™, Paveron™, Riddobron™, Vasocalm™.
Therapeutic use: It is used as a vasodilator, because of its smooth muscle relaxant properties. Papaverine is also used in the treatment of impotence by intracavernosal injection. It has been given to patients with cerebral, peripheral, and vascular disorders. In the treatment of gastrointestinal disorders it is employed as an antispasmodic.
Mechanism: Papaverine is a direct relaxant on all smooth muscles. It inhibits the activity of phosphodiesterase and thus causes an increase in cAMP level. In turn, Ca^{2+} concentrations are reduced, which are an important trigger in muscle contraction.

2.26. **Physostigmine** (see Eserine)

2.27. Pilocarpine

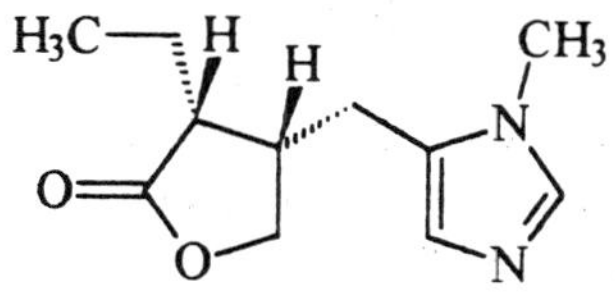

pilocarpine

Source: Pilocarpus jaborandi, P. pennatifolius, P. racemosus, and *P. microphyllus* (Rutaceae). *P. microphyllus* has the highest alkaloid content (0.7–0.8% alkaloids). Pilocarpine is the major component.
Pharmaceuticals (examples): (A small selection of about 60 preparations) Frikton™, Ocusert™, Piladren™, Pilocarpol™, Pilo/Eserin in der Ophtiole™, Pilomann-Öl™, Resorpil™, Spersacarpin™, Thiloadren™, Vistacarpin™.
Therapeutic use: It is used as a miotic in the treatment of open-angle glaucoma. The miotic action is also exploited to antagonize the effects of mydriatics (e.g., phenylephrine) on the eye, but it is ineffective against mydriasis from antimuscarinic agents, such as atropine. In the treatment of leprosy it is used to induce sweat secretion. For this purpose the nitrate salt is injected into the skin of patients with leprotic skin lesions.
Mechanism: Pilocarpine has the antimuscarinic actions of acetylcholine. This leads to contraction of the musculus sphincter pupillae and the ciliary muscle and thus to a pupil contraction. In addition, the drain of chamber water is enhanced, resulting in a decrease of intraocular pressure.

2.28. Quinidine

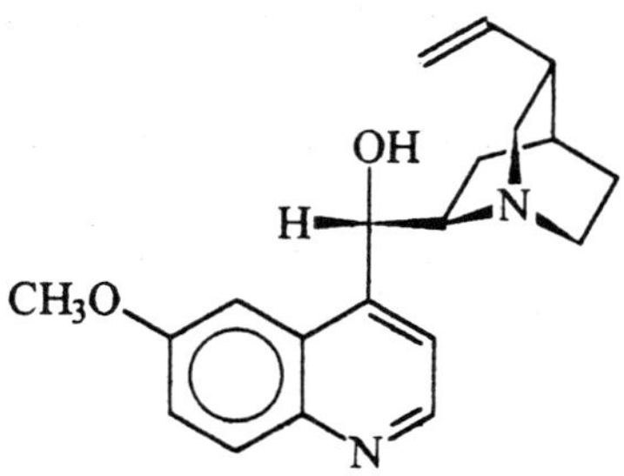

quinidine

Source: Cinchona succirubra (Rubiaceae); bark should not contain less than 6.5% alkaloids. 30–60% of the alkaloidal fraction are quinine-type alkaloids.
Pharmaceuticals (examples): Biquin Duriles™, Cardioquin™, Duraquin™, Quinalan™, Quinidex™, Rhythmochin I™.

Therapeutic use: Quinidine is a class Ia antiarrhythmic, which is used in the treatment of ventricular and supraventricular arrhythmias, malaria (as an alternative to quinine), and in the prevention of nocturnal cramps.
Use in traditional medicine: As an antimalarial bark extract.
Mechanism: Antimuscarinic and alpha-adrenergic blocking reagent. The schizonticide action is based on the inhibition of nucleic acid synthesis through DNA intercalation and of carbohydrate metabolism. Quinidine binds to sarcoplasmic reticulum vesicles and diminishes binding and uptake of Ca^{2+}. It also binds to Na^+/K^+-ATPase and inhibits it partially (Besch and Watanabe, 1977).

2.29. Quinine

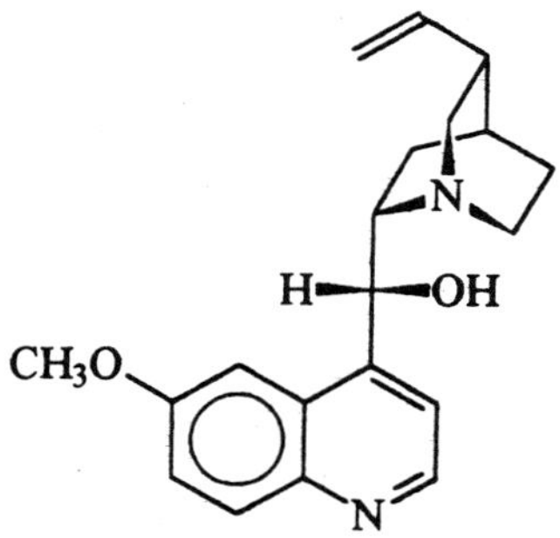

quinine

Source: Cinchona succirubra (Rubiaceae); bark contains not less than 6.5% alkaloids. 30–60% of the alkaloidal fraction are quinine-type alkaloids.
Pharmaceuticals (examples): Adaquin™, Biquinate™, Quinoctal™, Witte Kruis™, Zynedo-B™.
Therapeutic use: Treatment of Resochine-resistant malaria, babesiosis, and myotonic disorders.
Traditional use: Quinine is used for treatment of night cramps, fever, and as a bitter in tonic water.
Mechanism: Schizonticide (for exact mechanism see quinidine) against *Plasmodium falciparum, P. ovale, P. vivax, and P. malariae.* Quinine also blocks alpha-adrenergic receptors to the same degree as quinidine.

2.30. Raubasine (= Ajmalicine)

Source: Rauwolfia serpentina (Apocynaceae); roots contain about 0.8–2% alkaloids. Raubasine is also isolated from *R. vomitoria.*
Pharmaceuticals (examples): Card-Lamuran™, Circolene™, Cristanyl™, Duxil™, Duxor™, Hydrosarpon™, Iskedyl™, Isosarpan™, Isquebral™, Lamuran™, Melanex™, Saltucin Co™, Salvalion™, Sarpan™.

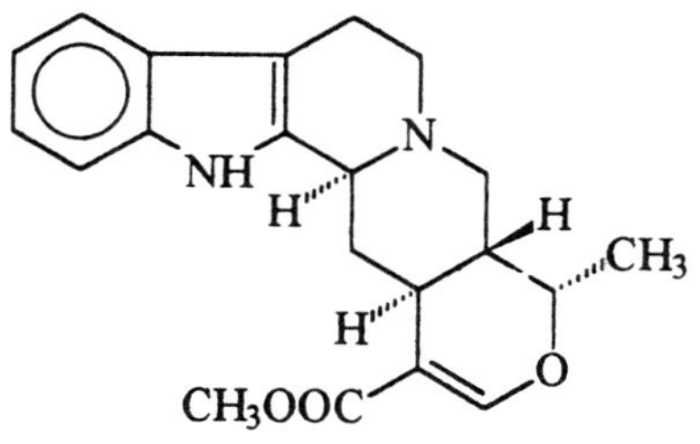

raubasine

Therapeutic use: Treatment of peripheral and cerebral vascular disorders.
Mechanism: Raubasine acts as a selective $alpha_1$ sympatholytic drug. It depletes peripheral noradrenaline stores, resulting in a decrease of peripheral resistance and blood pressure. It also causes depletion of catecholamine and serotonin stores in the brain, heart, and many other organs.

2.31. Rescinnamine

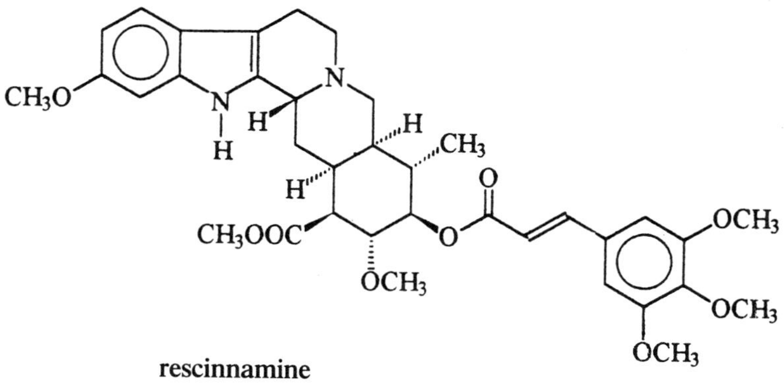

rescinnamine

Source: Rauwolfia serpentina (Apocynaceae); roots contain about 0.8–2% alkaloids with rescinnamine as a major component (0.15%).
Pharmaceuticals (examples): Detensitral™, Diuraupur™, Diu Rauwiplus™, Modenol™, Moderil™, Rauwopur™, Regulaserp™, Saltucin Co™.
Therapeutic use: Treatment of hypertension; properties and applications are similar to those of reserpine (see below).
Mechanism: See reserpine.

2.32. Reserpine

Source: Rauwolfia serpentina (Apocynaceae); roots contain about 0.15% alkaloids of the reserpine group.
Pharmaceuticals (examples): Abicol™, Adelphan-Esidrix™, Briserin™, Diutensen-R™, Pressimed™, Sandril™, Serpasil™, Serpasol™, Terbolan™.

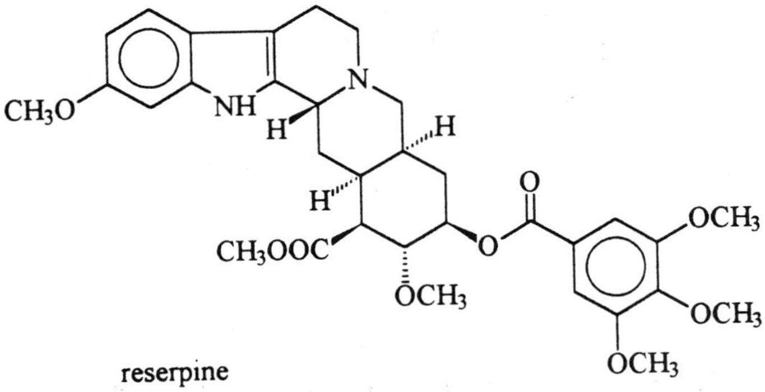

reserpine

Therapeutic use: Reserpine is applied in cases of mild to moderate hypertension and has also been used in the treatment of chronic psychoses.
Use in traditional medicine: In the Indian literature whole roots are employed in the treatment of psychoses and hypertension.
Mechanism: Reserpine acts as an alpha sympatholytic. It depletes peripheral noradrenaline stores, thus decreasing peripheral resistance and blood pressure. This action is explained through inhibition of a Mg^{2+}-dependent ATPase at the vesicle membrane, which pumps protons into the storage vesicles. Thus, the concentration of H^+ decreases and basic substances as noradrenaline and dopamine cannot be protonated intravesicularly. The remaining noradrenaline follows the concentration gradient and diffuses into the cytoplasm where it is destroyed by intraneuronal monoamine oxidase, so that little or no neurotransmitter remains. It also causes depletion of catecholamine and serotonin stores in the brain, heart, and many other organs, in a similar way.

2.33. Sanguinarine

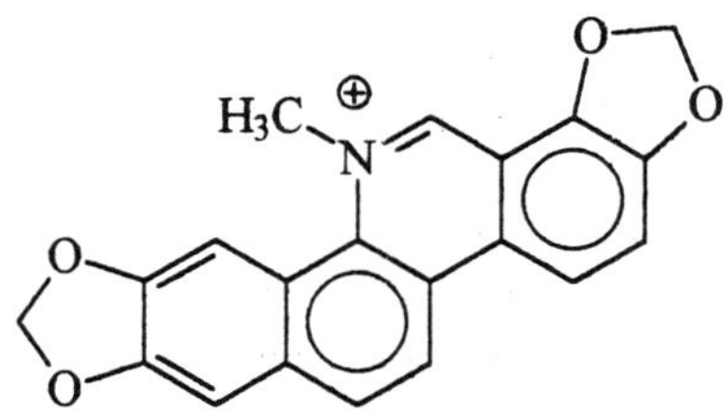

sanguinarine

Source: Sanguinaria canadensis (Papaveraceae); rhizomes contain 4–7%, roots about 1.8% alkaloids.
Pharmaceuticals (examples): Sanguinarine is used in toothpastes and mouthwashes.
Therapeutic use: As an antiplaque agent and also as an expectorant.
Mechanism: Sanguinarine modulates a number of molecular targets such as: intercalation to DNA, complexation of SH groups in enzymes, inhibition of esterases, Na^+/K^+ ATPase, alanine aminotransferase, human sputum elastase and others. Some of the affected

organisms are: several *Escherichia* and *Aerobacter* species, *Bacillus anthracis, Trichomonas vaginalis, Trypanosoma lewisii,* several *Vibrio* species. It is also active against several oral microbial isolates, such as several *Actinomyces, Bacteroides, Eubacterium, Streptococcus,* and *Propionibacterium* species, *Leptotrichia buccalis, Campylobacter concisus,* and others (Simeon *et al.*, 1989). Furthermore it has adrenolytic, sympatholytic, anti-inflammatory, cytotoxic, antifungal, and local anesthetic effects. Antibacterial and antiviral activities appear plausible.

2.34. Scopolamine (see Hyoscine)

2.35. Sparteine

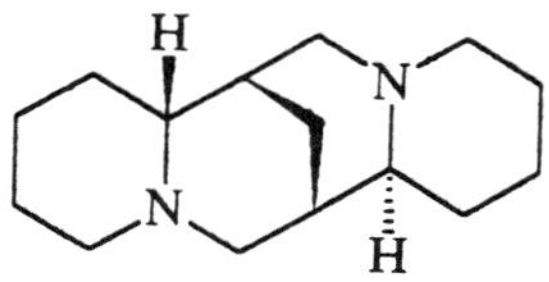

sparteine

Source: Cytisus scoparius (Leguminosae); aerial parts contain about 0.8–2% alkaloids, with sparteine as the major component.
Pharmaceuticals (examples): Anxoral™, Ariven™, Cardiopax™, cordi sanol forte™, Depasan™, Diffucord™, Digi-Pulsnorma™, Gelsadon™, Hypolind™, Hypotonin™, Jatamansin™, Kanovenol™, Morfi™, Normotin™, Palpipax™, Perivar™, Pulsnorma™, RR-plus™, Sedol™, Tachynerg™.
Therapeutic use: In the treatment of cardiac arrhythmias and to induce uterine contractions. Because approximately 10% of the patients cannot metabolize sparteine, the latter application has become obsolete.
Use in traditional medicine: The herb is used as a diuretic. This action might be related to the positive inotropic action of sparteine, but also to the diuretic flavonoids present.
Mechanism: Sparteine inhibits Na^+ channels and thus Na^+ transport through the membranes. Sparteine activates muscarinergic acetylcholine receptors (Schmeller *et al.*, 1994). As a result, conductivity of the cardiac muscle is modulated. Small doses stimulate and large doses paralyze the autonomic ganglia.

2.36. Strychnine

Source: Strychnos nux-vomica (Loganiaceae); seeds contain 1.5–5% alkaloids.
Pharmaceuticals (examples): Dysurgal™, Neuroftal™, Pastilles Jessel™, Pasuma™, Retinovix™, Rubistenol™, Senirakt™, Sulfa-Dysurgal™.
Therapeutic use: Strychnine is used in patients with eye disorders and optic nerve atrophy. It has been tried in the treatment of nonketotic hyperglycinemia.
Traditional use: As a poison against several animals, such as moles and mice.

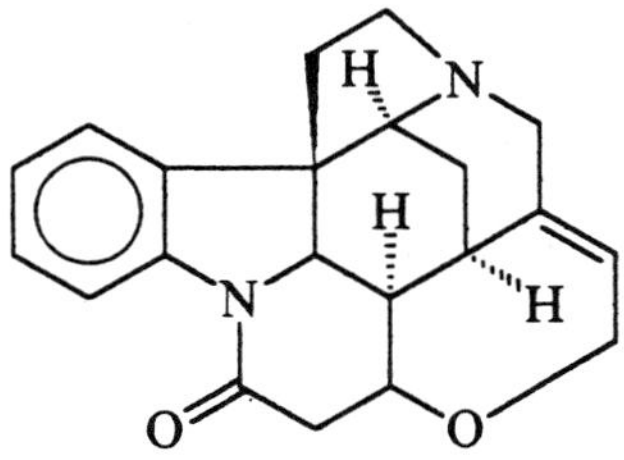

strychnine

Mechanism: Strychnine binds to glycine receptors and competes with glycine, which is an inhibitory neurotransmitter to motoneurons and interneurons in the spinal cord. In therapeutic doses strychnine increases the muscle tonus and stimulates respiration and circulation, which leads to an increase of activity in physically weakened patients.

2.37. Taxol

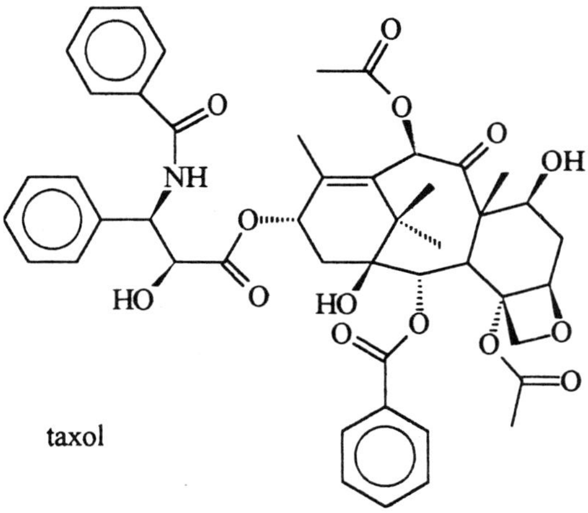

taxol

Source: Taxus brevifolia (Taxaceae); bark contains several diterpene alkaloids, including taxol.
Pharmaceuticals (examples): Taxol™.
Therapeutic use: Taxol was recently introduced into medicine for the treatment of mamma and ovary carcinoma and several other malignancies.
Mechanism: Taxol inhibits the disassembly of microtubules into tubulins and promotes the assembly of microtubules. Thus, it disrupts normal cell division in the G_2 and the M phase of the cell cycle.

2.38. Theobromine

Source: Theobroma cacao (Sterculiaceae); testae contain about 1.5% theobromine. Pharmaceuticals (examples): Atrofed™, Circovegetalin™, Dynamol™, Eupond™, Menstraleve™, Nephronorm™, Opticardon™, Seominal™, Spasdialt™, Urodonal™.

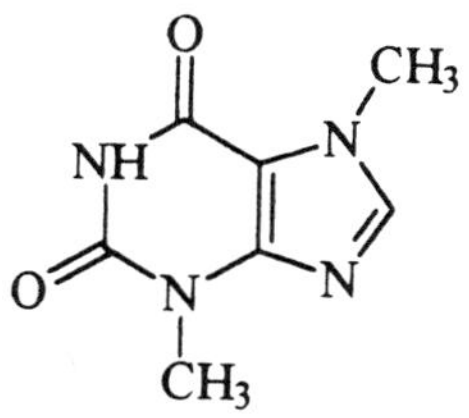

theobromine

Therapeutic use: Treatment of asthma and in combination with other drugs for several purposes. Theobromine was used as a diuretic and in the treatment of angina pectoris.
Traditional use: Important ingredient of chocolate and cacao drinks.
Mechanism: Similar to caffeine as an inhibitor of phosphodiesterase and as a ligand for the adenosine receptor. Theobromine relaxes various smooth muscles, notably the bronchial muscle, stimulates cardiac muscle, stimulates the CNS, and increases blood flow in the kidney. Because it displays only low activities, it has almost been displaced by other drugs.

2.39. Theophylline

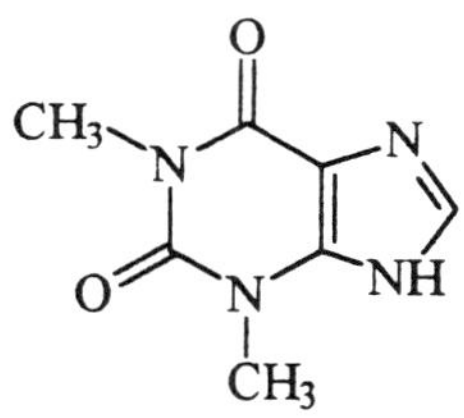

theophylline

Source: Because the yield from natural sources is very low, theophylline is prepared synthetically.
Pharmaceuticals (examples): (A small selection of about 180 preparations) Adenovasin™, Aerobin™, Asthma-Bisolvon™, Bronchoretard™, Euphyllin™, Perspiran™, Pulmeno™, Solphyllin™, Theochron™.
Therapeutic use: It is used for relief of bronchospasms in asthma, bronchitis, and emphysema. In acute asthma theophylline is administered intravenously. It is also used to prevent chronic bronchospasms.
Traditional use: Minor component of green and black tea (*Camellia sinensis*).
Mechanism: See caffeine and theobromine. The smooth muscle relaxant activity is the highest among the purines.

2.40. Tubocurarine

Source: Chondodendron tomentosum (Menispermaceae); the drug contains 2–7% of (+)-tubocurarine.
Pharmaceuticals (examples): Jexin™, Tubarine™.

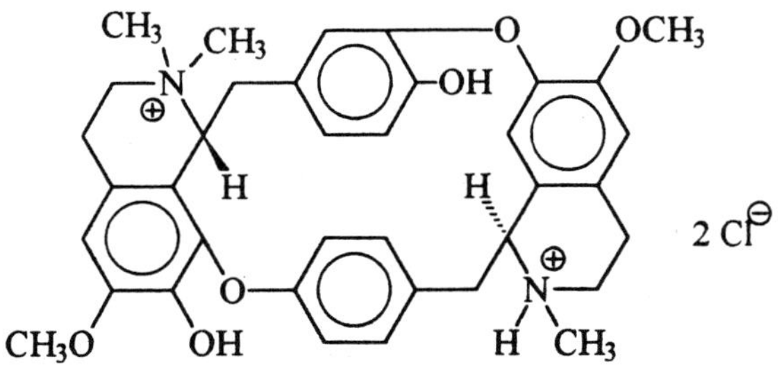

(+)-tubocurarine chloride

Therapeutic use: In surgical procedures to produce muscle relaxation. Tubocurarine is also used intravenously or intramuscularly to control muscle spasms and convulsions of tetanus.
Traditional use: As an arrow poison (see Chapter 3).
Mechanism: Tubocurarine is a competitive inhibitor at nicotinic acetylcholine receptors on the motor endplate. Thus, it blocks the transmission of the nerval impulse to the muscle fiber.

2.41. Vinblastine

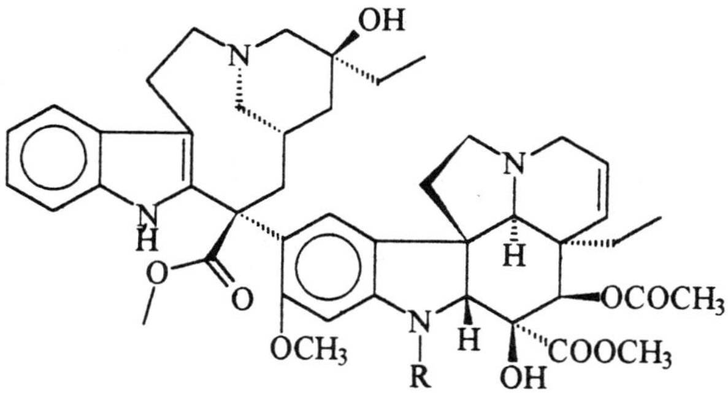

vinblastine, R= CH_3
vincristine, R= CHO

Source: Catharanthus roseus (Apocynaceae). In aerial parts with low content of about 0.001–0.005%.
Pharmaceuticals (examples): Periblastine™, Velban™, Velbe™, Velsar™.
Therapeutic use: Treatment of Hodgkin's disease and other lymphoma, testicular cancer, and a variety of solid neoplasms. It is also used in patients with neuroblastoma, choriocarcinoma, and Kaposi's sarcoma. Vinblastine has been used in several autoimmune blood disorders.
Mechanism: Vinblastine binds to microtubules of the spindle apparatus, thus arresting mitosis in the metaphase. It also affects amino acid metabolism and DNA synthesis, and has some immunosuppressant properties.

2.42. Vincamine

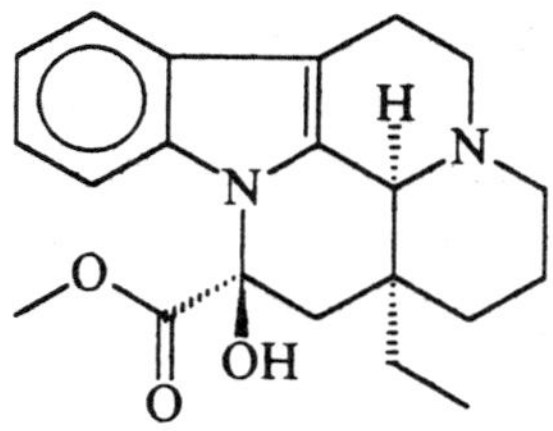

vincamine

Source: Vinca minor (Apocynaceae); aerial parts contain 0.2–0.7% alkaloids with vincamine as a major component. Vincamine is also prepared synthetically.
Pharmaceuticals (examples): (Selection of about 50 preparations) Aethroma™, Angiopac™, Arteriovinca™, Cerebramina™, Dilarterial™, Pervin™, Vasonett™, Vincabrain™, Vincimax™.
Therapeutic use: As a vasodilator in the therapy of cerebral disorders.
Mechanism: Vincamine increases cerebral circulation and utilization of oxygen and glucose.

2.43. Vincristine

Source: See vinblastine.
Pharmaceuticals (examples): Norcristine™, Oncovin™, Pericristine™, Vincasar PSF™, Vincrisul™.
Therapeutic use: See vinblastine. It is also used in the treatment of Burkitt's lymphoma, Wilms' tumor, myeloma, rhabdomyosarcoma, and in patients with tumors of the brain and lung.
Mechanism: See vinblastine.

2.44. Yohimbine

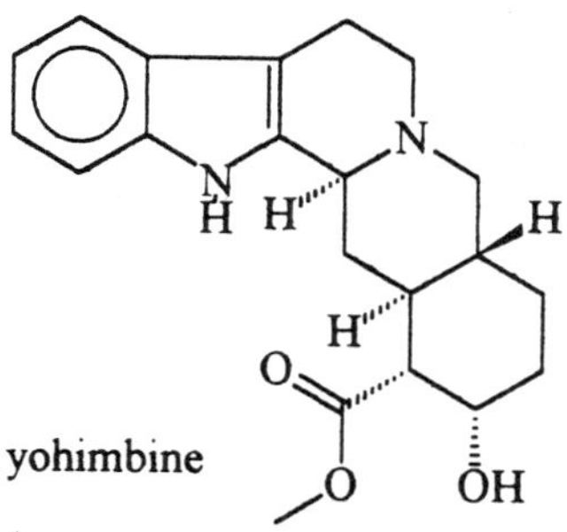

yohimbine

Source: Rauwolfia vomitoria (Apocynaceae).
Pharmaceuticals (examples): Aphrodyne™, Dayto Himbin™, Ichtho-Himbin™, Pasuma™, Pluriviron™, Prowess™, Yocon™, Yohimex™.

Therapeutic use: For impotent male patients and as an aphrodisiac. It is also used in the treatment of urinary incontinence.
Use in traditional medicine: As in therapy.
Mechanism: Yohimbine blocks presynaptic alpha$_2$ adrenoceptors, and thus it increases the noradrenaline release at sympathetic nerve endings. It therefore increases heart rate and blood pressure and produces orthostatic hypotension. The antidiuretic effect can be explained through an increase of vasopressin release from the neurohypophysis.

3. CONCLUSIONS

As shown in this chapter, many alkaloids are used in modern medicine and the exploitation of alkaloids and alkaloidal plants is economically interesting and important. It has been estimated that approximately 20% of all plants produce alkaloids. As only 10 to 15% of all plants have been studied phytochemically and even less pharmacologically, it is safe to assume that we probably know and utilize only a small proportion of naturally occurring alkaloids and their properties. From a medical and economic standpoint it would be interesting to systematically evaluate the composition of natural products in plants (and also in some animals; see Chapters 15 and 16), not only those from temperate climates but also those from tropical rain forests (which still hold many unknown species). It seems to be important that new alkaloids be screened using modern pharmacological assays to fully understand their targets and their potential action and application. Furthermore, as the underlying biology and biochemistry are usually very interesting and challenging, these new discoveries will open a wide range of basic research, including biotechnological approaches that might be useful for producing the alkaloids in question on an industrial scale.

REFERENCES

Major Reviews

Budavari, S., (ed.), 1989, *The Merck Index,* 11th ed. Merck, Rahway, NJ.

Gilman, A. G., 1991, *Goodman and Gilman's: The Pharmacological Basis of Therapeutics,* 8th ed., Pergamon Press, New York.

Harborne, J. B., and Baxter, H., 1993, *Phytochemical Dictionary: A Handbook of Bioactive Compounds from Plants,* Taylor & Francis, London.

Reynolds, J. E. F., (ed.), 1993, *Martindale—The Extra Pharmacopoeia,* The Pharmaceutical Press, London.

Roth, L., Daunderer, M., and Kormann, K., 1994, *Giftpflanzen Pflanzengifte,* Ecomed, Landsberg/Lech.

Teuscher, E., and Lindequist, U., 1994, *Biogene Gifte,* Fischer Verlag, Stuttgart.

Wink, M., 1993, Allelochemical properties or the raison d'être of alkaloids, in: *The Alkaloids,* Vol. 43 (G. A. Cordell, ed.), Academic Press, San Diego, pp. 1–118.

Key References

Besch, H., Jr., Watanabe, A. M., 1977, Binding and effect of tritiated quinidine on cardiac subcellular enzyme systems: Sarcoplasmic reticulum vesicles, mitochondria and Na^+,K^+-adenosine triphosphatase. *J. Pharmacol. Exp. Ther.* **202,** 354–364.

Capraro, H.-G., and Brossi, A., 1984, Troplonic colchicum alkaloids, in: *The Alkaloids,* Vol. 23 (A. Brossi, ed.), Academic Press, San Diego, pp. 1–70.

Fujii, T., and Ohaba, M., 1983, Ipecac alkaloids and β-carboline congeners, in: *The Alkaloids,* Vol. 22 (A. Brossi, ed.), Academic Press, San Diego, pp. 1–50.

Rodriguez, E., Aregullin, M., Nishida, T., Uehara, S., Wrangham, R. W., Abramowski, Z., Finlayson, A., and Towers, G. H. N., 1985, Thiarubrin A, a bioactive constituent of *Aspilia* (Asteraceae) consumed by wild chimpanzees. *Experientia* **41:**419–420.

Rodriguez, E., and Wrangham, R. W. 1993, Zoopharmacognosy: The use of medicinal plants by animals: in *Phytochemical Potentials of Tropical Plants* (K. R. Downum, J. T. Romeo, and H. A. Stafford, eds.), Plenum Press, New York, pp. 89–105.

Schmeller, T., Sauerwein, M., Sporer, F., Wink, M., and Müller, W. E., 1994, Binding of quinolizidine alkaloids to nicotinic and muscarinic acetylcholine receptors, *J. Nat. Prod.* **57:**1316–1319.

Simeon, S., Rios, J. L., and Villar, A., 1989, Pharmacological activities of benzophenanthridine and phenanthrene alkaloids, *Pharmazie* **44:**593–597.

Zetler, G., 1988, Neuroleptic-like, anticonvulsant, and antinociceptive effects of aporphine alkaloids: Bulbocapnine, corytuberine, boldine, glaucine, *Arch. Int. Pharmacodyn. Ther.* **296:**255–281.

Subject Index Including Orders, Families, and Common Names

Index of Alkaloids, Amines, and Other Molecules

Organism Index

A REMINDER
Materials are due
on the last date
stamped below